CELL TECHNOLOGY FOR CELL PRODUCTS

# Cell Technology for Cell Products

## Proceedings of the 19th ESACT Meeting, Harrogate, UK, June 5-8, 2005

*Edited by*

Rodney Smith
*CTM BioTech Ltd, Babraham Cambridge, UK*

A C.I.P. Catalogue record for this book is available from the Library of Congress.

ISBN-10 1-4020-5475-0 (HB)
ISBN-13 978-1-4020-5475-4 (HB)
ISBN-10 1-4020-5476-9 (e-book)
ISBN-13 978-1-4020-5476-1 (e-book)

---

Published by Springer,
P.O. Box 17, 3300 AA Dordrecht, The Netherlands.

*www.springer.com*

*Printed on acid-free paper*

# Contents

# 19th ESACT Meeting Committees

*Organising Committee*

| | |
|---|---|
| Rodney Smith (Chair) | CTM BioTech Ltd., UK |
| Fiona Godsman (Trade Fair/Sponsorship) | BioReliance, UK |
| John Bonham-Carter (Trade Fair/Sponsorship) | Magellan Instruments Ltd., UK |
| Tracey Sambrook (Social Arrangements) | CTM BioTech Ltd., UK |
| Glenda Bland (Conference Manager) | Global Meeting Planning |

*Scientific Committee*

| | |
|---|---|
| Dr. Rod Smith (Chair) | CTM BioTech Ltd., UK |
| Prof. Mohammed Al-Rubeai | University of Dublin, Ireland |
| Prof. Dr. Martin Fussenegger | ETH Zürich - Institute of Biotechnology, Switzerland |
| Prof. Dr. Ing. Jürgen Lehmann | Universität Bielefeld, Germany |
| Prof. James M. Piret | University of British Columbia, Canada |
| Prof. Glyn Stacey | National Institute for Biological Standards and Control, UK |
| Dr. David Venables | BioReliance, UK |

*Poster Award Committee*

| | |
|---|---|
| Nathalie Chatzissavidou | Biovitrum, Sweden |
| Paula Marques Alves | ITQB/IBET, Portugal |

# ESACT Executive Committee

| | |
|---|---|
| Otto-Wilhelm Merten, Chairman | Généthon III, France |
| Alain Bernard, Secretary | Serono Biotech Center, Switzerland |
| Martin Fussenegger, Treasurer | ETH- Hönggerberg , Zurich, Switzerland |
| Rodney Smith, Meeting Chair | CTM BioTech, UK |
| Francesc Gòdia | Universitat Autonoma de Barcelona, Spain |
| Hansjörg Hauser | GBF, Germany |
| Stefanos Grammatikos | Boehringer Ingelheim Pharma, Germany |
| Florian Wurm | EPFL, Switzerland |
| Christophe Losberger, Web Master | Serono Pharmaceutical, Research Institute, Switzerland |
| Steve Oh, Newsletter Chief Editor | Bioprocessing Technology Institute, Singapore |
| Merlin Goldman, Newsletter Co-Editor | Wyeth Biopharma, Ireland |
| Bryan Griffiths | ESACT Secretariat, UK |

# Sponsors

ESACT and the Organising Committee wish to thank the following companies for their generous support.

Aber Instruments

AstraZeneca

Aventis Pasteur SA

ChemoMetic A/S

Covance

CTM BioTech Ltd.

DSM Biologics

GE Healthcare

Genetech

Genzyme Flanders NV

Invitrogen

JRH Europe

Magellan Instruments

Merck & Co., Inc., USA

Merck Research Laboratories

Rütten Engineering AG/Ltd.

Sartorius BBI Systems GmbH

Serono

Sigma Aldrich

Wave BioTech

# Companies in Trade Exhibition

Aber Instruments Ltd.

Applikon. BV/AppliSens

Areta International s.r.l.

Bayer HealthCare AG

Bayer Technology Services GmbH

BDD Diagnostic Systems Europe

Beckman Coulter

Bioengineering AG

BUA/BioProcess UK

BioProcess Engineering Services Ltd.

BioProcess International

BioProcessors Corporation

Bioquell

BioXplore

Blue Membranes GmbH

Broadley Technologies Ltd

Cambrex Bio Science

Celliance

Cellon SA

Celonic GmbH

Genetix Ltd.

Greiner Bio-One Ltd.

Guava Technologies

Helena Biosciences

Henogen

HyClone/Perbio Science

Infors UK Ltd./Hi Tec-Zang

Innovatis AG

Invitrogen

Irvine Scientific

J M Separations BV

JRH Europe

Kerry Bioscience

Lonza Biologics plc

Magellan Instruments

Moregate TCS Ltd

New Brunswick Scientific (UK) Ltd

NewLab BioQuality AG

NUNC A/S

PAA Laboratories GmbH

Charter Medical Ltd.

Chemicon Europe

ChemoMetec A/S

Cobra BioManufacturing

Consolidated Polymer Technologies (C-Flex)

Corning Life Sciences

Covance

CTM BioTech Ltd.

DASGIP AG

Delta Biotechnology Ltd.

Eden Biodesign

European Collection of Animal Cell Cultures (ECACC)

Evotec Technologies

GE Healthcare

Genetic Engineering News

Pall Life Sciences

Panacea

Polyplus Transfection

Quest Biomedical

Rütten Engineering

Sartorius BBI Systems GmbH

Sigma Aldridch

SLS

Stedim S.A.

Tecan Deutschland GmbH

The Automation Partnership

The Williamsburg BioProcessing

Wave Biotech AG

Wave Europe Pvt. Ltd.

Xenova Biomanufacturing

YSI (UK) Limited

# Acknowledgements

In the planning, organising and running the meeting a great number of people have been involved. The outcome was a successful meeting and these proceedings of the science presented. The 19$^{th}$ ESACT Meeting built on the success of the previous ones and lays the foundation for the 20$^{th}$ and beyond.

In making the meeting so successful I would like to acknowledge and thanks the following people for their effort hard work and most of all friendship over the period. These people provided immense support to me during the events and I shall never forget that.

My personal thanks must go to **Fiona Godsman** for staying on with the Trade Organisation and whose invaluable comments were appreciated and hopefully listened too. **John Bonham-Carter** (Trade Organisation) who was so keen in Tylosand to be involved and relished the chance of developing his (sometimes wild) ideas on the Harrogate meeting. **Glenda Bland** whose efforts in keeping us on track and focussed was extremely well received. **Tracey Sambrook** for her advice support and constant council on how the meeting was destined to be a success.

These people had the vision and energy to ensure that no part of the Harrogate meeting was forgotten or left out. I owe them the greatest of thanks for being part of the team.

I would also like to thanks the other people who helped in laying down the science for the community to ensure the programme was balanced and thought provoking. The main scientific committee comprising James Piret, Glynn Stacey, Martin Fussenegger, Mohammed Al-Rubeai and Jürgen Lehmann. We should not forget the organisers of the now ever present Poster award and my thanks go to Nathalie Chatissavidou and Paula Alves for the excellent organisation and selection of some truly exciting posters.

The biggest acknowledgment and thanks must go to the delegates and all those scientists that wanted to participate in some way to the meeting. Without your contributions the meeting would not be the same.

A very big Thank you to you all.

**Rodney Smith**

# List of Participants

**Prof. Adriano Aguzzi**, Institute of Neuropathology, University Hospital of Zurich, Schmelzbergstrasse 12, Zurich CH-8091, SWITZERLAND, Telephone: +4112552107, Email: adriano@pathol.unizh.ch

**Mr. Yoshihiro Aiba**, Laboratory of Cellular Regulation Technology, Department of Genetic Resources Technology, Kyushu University, Fukuoka, JAPAN, Telephone: +84 92 642 3046, Email: aiba@grt.kyushu-u.ac.jp

**Mr. Peter Aizawa**, Octapharma AB, Nordenflychtvägen 55 Stockholm 12137, SWEDEN, Telephone: +46 8 56643436, Email: peter.aizawa@octapharma.se

**Mr. Franz Alig**, Sartorius BBI Systems GmbH, Schwarzenberger Weg 73 - 79, Melsungen D-34212, GERMANY

**Mr. Tony Allman**, Infors UK Ltd., The Courtyard Business Centre, Dovers Farm, Lonesome Lane, Reigate Surrey RH2 7QT, UK

**Prof. Mohammed Al-Rubeai**, Department of Chemical & Biochemical Engineering, Falculty of Engineering & Architecture, University College Dublin, Belfield, Dublin 4, Ireland, Email: m.al-rubeai@ucd.ie

**Dr. Paula Alves**, ITQB/IBET, Apartado 12, OEIRAS 2781-901, PORTUGAL, Telephone: + 351 21 4469421, Email: marques@itqb.unl.pt

**Dr. Madawi Alwazzan**, Flowgen Bioscience, Wilford Industrial Estate, Ruddington Lane, Nottingham NG11 7EP, UK, Telephone: +44 115 977 5494, Email: malwazzan@flowgen.co.uk

**Dr. Hagit Amitai**, Compugen Ltd., 72 Pinchas Rosen Street, Tel-Aviv 69512, ISRAEL, Telephone: +972 3 7658513, Email: hagita@compugen.co.il

**Dr. Anette Amstrup Pedersen**, Novo Nordisk A/S, Novo Alle, Bagsvared 2880, DENMARK, Telephone: +45 44 42 00 08, Email: aamp@novonordisk.com

**Dr. Dana Andersen**, Director, Late Stage Cell Culture, Genentech, Inc., 1 DNA Way - MS 32, South San Francisco, CA 94080, USA, Telephone: +1 650 225 4443, Email: andersen@gene.com

**Mr. Jorn Sondrup Andersen**, Rocket Management, Ganloseparken 16, Stenlose DK-3660, DENMARK, Telephone: +45 22 16 67 26, Email: sondrup@rocketmail.com

**Mr. William Anderson**, Synthecon, Inc., 8042 El Rio, Houston, TX 77054, USA, Telephone: +1 713 741 2582, Email: abill@synthecon.com

**Dr. Carlo Andretta**, Biospectra AG, Zuercherstrasse 137, Schlieren 8952, SWITZERLAND, Telephone: +41 44 7556700, Email: andretta@biospectra.ch

**Mrs. Emilie Ansel**, GlaxoSmithKline Biologicals S. A., rue de Institut 89, Bixensart 1330, BELGIUM, Telephone: +32 2 656 9890, Email: emilie.ansel@gskbio.com

**Ms. Fabienne Anton**, Insitut für Technische Chemie - Universitat Hannover, Callinstrasse 3, Hannover 30167, GERMANY, Email: anton@iftc.uni-hannover.de

**Mr. Didier Argentin**, Henogen, Rue des Professeurs Jeener et Brachet 12, Charleroi B-6041, BELGIUM

**Mr. Paul Armstrong**, Celliance, 5655 Spalding Drive, Norcross, GA 3009, USA

**Mr. William Arts**, New Brunswick Scientific B.V., Kerkenbos 1101, Nijmegen 6546 BC, THE NETHERLANDS, Telephone: +31 24 371 7600, Email: w.arts@nbsbv.nl

**Dr. Soheil Asgari**, Blue Membranes GmbH, Rheingaustrasse 190-196, Wiesbaden D-65203, GERMANY

**Ms. Kelly Astley**, 41 Wealdstone Drive, Lower Gornal, Dudley, West Midlands DY3 2QR, UK, Telephone: +441384 821 487, Email: kla266@bham.ac.uk

**Mr. Alex Baker**, Maverick DCM, Suite 12-16 Macquarie Chambers, 69-79 Macquarie Street, PO Box 2997, Dubbo 2831, AUSTRALIA, Telephone: +61 2 6885 1200, Email: ahanslow@maverickdcm.com

**Mr. Osman Bal**, Git Verlag, Roessler Str. 90, Darmstadt 64283, GERMANY

**Mr. John Baldwin**, Rütten Engineering Ltd., Industriestrasse 9, Stäfa 8712, SWITZERLAND, Telephone: +41 1 928 29 30, Email: info@rutten.com

**Mr. Jean-Philippe Barbe**, Sanofi Pasteur, 1541 av. M. Mérieux, Marcy l'Etoile 69280, FRANCE, Telephone: +33 4 3737 3294, Email: jean-philippe.barbe@sanofipasteur.com

**Dr. Jacquiline Barnes-Locke**, Helena Biosciences, UK

**Dr. Barney Barnett**, Biogen Idec, 5200 Research Place, San Diego 92122, USA, Telephone: +1 858 401 8650, Email: barney.barnett@biogenidec.com

**Mr. Timothy Barrett**, University College London, UK, Email: timothy.barrett@ucl.ac.uk

**Ms. Natalie Baseley**, European Collection of Cell cultures (ECACC), Health Protection Agency, Porton Down, Salisbury SP4 0JG, UK

**Miss. Vera Basto**, ECBio-R&D in Biotechnology, SA, Ap 98, Lab 4.11, Ed ITQB, Oeiras 2781-901, PORTUGAL, Telephone: +351214469466, Email: vera.basto@ecbio.com

**Dr. Thierry Battle**, Serono Pharmaceutical Research, 14 Chemin Des Aulx, Plan-les-Ouates 1228, SWITZERLAND, Telephone: +41 22 706 9644, Email: thierry.battle@serono.com

**Dr. Fabrizio Baumann**, SLS, Schönbühlstr. 8, Chur 7000, SWITZERLAND, Telephone: +41 81 286 9541, Email: fabrizio.baumann@integra-biosciences.com

**Mr. Andreas Bauwens**, University of Bielefeld, Institute of Cell Culture Technology, Roggenkamp 20, Bielefeld 33605, GERMANY, Telephone: +49 521 2702996, Email: andreas.bauwens@web.de

**Mrs. Amandine Beaucourt**, Stedim S.A., Z. I. Des Paluds, Avenue De Jouques, Aubagne F-13781, FRANCE

**Mr. Eric Becker**, Boehringer Ingelheim Pharma GmbH & Co. KG, Birkendorger Strasse 65, Biberach an der Riss 88397, GERMANY, Telephone: +49 7351 54 96 590, Email: eric.becker@bc.boehringer-ingelheim.com

**Mr. Alexander Beetz**, JRH Europe, Smeaton Road, West Portway, Andover Hampshire, UK, Telephone: + 44 1264 333311

**Dr. Mark Beggs**, The Automation Partnership, York Way, Royston Hertfordshire SG8 5WY, UK

**Dr. Pascale Belguise**, Polyplus Transfection, BioParc, Boulevard Sebastien Brand, PB 90018, Illkirch 67401, FRANCE, Telephone: +33 3 90 40 61 80, Email: pbelguise@polyplus-transfection.com

**Ms. Jacki Bennett**, Invitrogen, 3 Fountain Drive, Inchinnan Business Park, Paisley PA4 9RF, UK, Telephone: +44 141 814 5814, Email: jacki.bennet@invitrogen.com

**Mr. Dan Benson**, Kerry Bioscience, Huizerstraatweg 28, Naarden 1411 GP, THE NETHERLANDS

**Mr. Paul Bentley**, BioProcess Engineering Services Ltd., 186 Carver Drive, Kent Science Park, Sittingbourne Kent, UK,

**Dr. Guy Berg**, NewLab BioQuality AG, Max-Planck-Strasse 15A, Erkrath D-40699, GERMANY

**Prof. Alain Bernard**, Serono Biotech Center, Route de Fenil Z1B, Corsier-sur-Vevey CH-1804, SWITZERLAND, Telephone: +41 21 923 2357, Email: alain.bernard@serono.com

**Mr. Olivier Berteau**, New Brunswick Scientific (UK) Ltd, 3 Rue dex Deux-Boules, Paris 75001, FRANCE, Telephone: +33 40 25 22 46, Email: o.berteau@nbsnv-sa.be

**Mr. Martin Bertschinger**, EPFL, SV IGBB LBTC, J2-496 (Building CH), Station 6, Lausanne 1015, SWITZERLAND, Telephone: +41 21 693 61 37, Email: sylvia.fabris@epfl.ch

**Mrs. Iris Besancon**, Bio-Springer, 103 Jean Jaures, Maisons Alfort 94704, FRANCE, Email: iris.besancon@springer.fr

**Prof. Michael Betenbaugh**, Johns Hopkins University, Department of Chemical and Biomolecular Engineering, 3400 N. Charles Street, Baltimore, MD 21218, USA, Telephone: +1 410 516 5461, Email: beten@jhu.edu

**Dr. John Birch**, Lonza Biologics, 228 Bath Road, Slough SL1 4DX, UK, Telephone: +44 1753 777000, Email: john.birch@lonza.com

**Mr. David Birch**, Covance, Langhurstwood Road, Horsham Sussex, RH12 4QD, UK, Telephone: +44 1423 500011, Email: david.birch@covance.com

**Mr Robert Bird**, Chemicon Europe, The Science Centre, Eagle Close, Chandlers Ford, Southampton SO53 4NF, UK

**Mr. Henri Blachere**, New Brunswick Scientific Co. Inc., PO Box 4005, 44 Talmadge Road, Edison, NJ 08818-4005, USA, Telephone: +1 732 287 1200, Email: eppstein@nbsc.com

**Dr. Simone Blayer**, Univalid Bioprocessing, 17 Archimedesweg, Leiden 2333CN, THE NETHERLANDS, Email: s.blayer@crucell.com

**Mrs. Rachel Blogg**, NUNC A/S, Kamstrupvej 90, Postbox 280, Roskilde DK-4000, DENMARK

**Dr. Gerald Blueml**, GE Healthcare, Bjorkgatan 30, Uppsala S-75182, SWEDEN

**Mrs. Margaret Blundy**, Cambridge Antibody Technology, Milstein Building, Granta Park, Great Abington, Cambridge CB1 6GH, UK, Telephone: +44 1223 898428, Email: susan.messenger@cambridgeantibody.com

**Mr. Andreas Bock**, MPI-Institute, Sandtorstr. 1, Magdeburg 39106, GERMANY, Telephone: +493916110206, Email: bock@mpi-magdeburg.mpg.de

**Dr. Berthold Bödeker**, Bayer HealthCare, PH-OP-BT / Elberfeld, Geb. 46, Raum 403, Friedrich-Ebert-Strasse 217, Wuppertal 42096, GERMANY, Telephone: +49 202 36 7531, Email: berthold.boedeker@bayerhealthcare.com

**Ms. Elisabeth Boehlen**, Bioengineering AG, Sagentainstrasse 7, Wald 8636, SWITZERLAND

**Dr. Christoph Boehme**, Norbitec GmbH, Pinnauallee 4, Uetersen 25436, GERMANY, Telephone: +49 4122 712 427, Email: christoph.boehme@norbitec.de

**Mr. Sthen Boisen**, ChemoMetec A/S, Gydevang 43, Alleroed DK-3450, DENMARK, Telephone: +45 48 13 10 20, Email: Info@chemometec.dk

**Mr. Olivier Boizet**, GE Healthcare, Bjorkgatan 30, Uppsala, S-75182, SWEDEN

**Dr. Mariela Bollati-Fogolin**, German Research Centre for Biotechnology (GBF), Mascheroder WEG 1, Braunschweig 38124, GERMANY, Telephone: +49 531 6181 919, Email: mrb@gbf.de

**Mr. Fred Bollin**, Serono Pharmaceutical Research, 14 Chemin Des Aulx, Plan-les-Ouates 1228, SWITZERLAND, Telephone: +41 22 706 9637, Email: frederic.bollin@serono.com

**Mr. Gert Bolt**, Novo Nordisk A/S, Mammalian Cell Technology, Bagsvaerd 2880, DENMARK, Telephone: +45 444 36654, Email: bolt@novonordisk.com

**Mr. John Bonham-Carter**, Magellan Instruments Ltd, Rockfort, Tower Road, Hindhead Surrey, GU26 6SL, UK, Telephone: +44 79 7383 5224, Email: john.bonham-carter@magellaninstruments.com

**Prof. Nicole Borth**, Universität für Bodenkultur, Dept. Biotechnology, Muthgasse 18, Vienna A-1190, AUSTRIA, Telephone: +43 1 36006 6232, Email: nicole.borth@boku.ac.at

**Dr. Michael Borys**, Gala Design / Cardinal Health, 8137 Forsythia Street, Middleton, WI 53562, USA, Telephone: +1 608 821 6245, Email: michael.borys@cardinal.com

**Ms. Marianne Boulakas**, Nature Publishing Group, 4 Crinan Street, London, N1 9XW, UK, Telephone: +44 20 7843 4969, Email: m.boulakas@nature.com

**Mr. Jean Francois Bouquet**, Sanofi Pasteur, 1541 Avenue Marcel Merieux, March L'Etoile 69280, FRANCE, Telephone: +33 4 37 37 07 41, Email: jean-françois.bouquet@sanofipasteur.com

**Mr. Thierry Bovy**, Cambrex Bio Science, Parc Industriel de Petit Rechain, Verviers B-4800, BELGIUM

**Mr. Aston Bowles**, Helena Biosciences, UK, Email: ab@helena-biosciences.com

**Mr. Harry Brack**, PAA Laboratories GmbH, Unterm Bornrain 2, Colbe D-35091, GERMANY, Telephone: +49 6421 17539 0

**Prof. Allan Bradley**, The Wellcome Trust Sanger Institute, Wellcome Trust Genome Campus, Hinxton, Cambridge CB10 1SA, UK, Telephone: +44 1223 494 884, Email: skh@sanger.ac.uk

**Dr. Tony Bradshaw**, BIA, 14/15 Belgrave Square, London SW1X 8PS, UK, Telephone: +44 20 7565 7190, Email: tbradshaw@bioindustry.org

**Dr. Joye Bramble**, Merck Research Laboratories, P.O. Box 4, West Point, PA, 19486, USA, Telephone: +1 215 652 2282, Email: joye_bramble@merck.com

**Mr. Thomas Braschler**, EPFL, Ave de Coar 12, Lausanne 1007, SWITZERLAND, Telephone: +41 21 617 03 30, Email: thomas.braschler@epfl.ch

**Mr. Jean-Michel Brasnu**, Irvine Scientific, 2511 Daimler Street, Santa Ana, CA, 92705-5588, USA

**Mr. Malcolm Brattle**, CTM BioTech, Babraham Research Campus, Babraham, Cambridge CB2 4AT, UK, Telephone: +44 1223 49 60 70, Email: malcolm.brattle@ctmbiotech.co.uk

**Dr. Rene Brecht**, ProBioGen AG, Goethestr. 54, Berlin 13086, GERMANY, Telephone: +49 30 9240060, Email: rene.brecht@probiogen.de

**Mr. Michel Brik Ternbach**, Centocor B.V., Einsteinweg 92, Leiden 2333CD, THE NETHERLANDS, Telephone: +31 71 5242817, Email: mbrikter@cntnl.jnj.com

**Dr. Helmut Brod**, Bayer Technology Services GmbH, Building K 9, Leverkusen D-51368, GERMANY

**Mrs. Julie Brogestam**, Biovitrum, Strandbergsgatan 49, Stockholm 11276, SWEDEN, Telephone: +4686973027, Email: julie.brogestam@biovitrum.com

**Dr. William Brondyk**, Genzyme Corporation, One Mountain Road, Framingham, MA 01701-9322, USA, Telephone: +1 508 271 2847, Email: william.brondyk@genzyme.com

**Dr. Peter Brown**, Biotechnology Solutions, 200 Woodcrest Dr, Orinda 94563, USA, Telephone: +1 925 254 1601, Email: P_Brown@comcast.net

**Dr. Michael Brown**, Invotrogen, 228 Bath Road, Slough SL1 4DX, UK, Telephone: +44 1753 777000, Email: mike.brown@lonza.com

**Mr. Peter Brown**, Cicero Limited, 8 Exchange Quay, Manchester M5 3EQ, UK, Telephone: +44 161 876 5522, Email: prb@cicero-marketing.com

**Dr. Sharon Brownlow**, Sartorius - Biotechnology Division, Longmead Business Centre, Blenheim Road, Epsom Surrey KT19 9QQ, UK, Email: sharon.brownlow@sartorius.com

**Mr. Daniel Bruecher**, Infors UK Ltd., The Courtyard Business Centre, Dovers Farm, Lonesome Lane, Reigate Surrey RH2 7QT, UK

**Dr. Barry Buckland**, Merck & Co., Inc, PO Box 4, WP17-101, 770 Sumneytown Pike, West Point,19486, USA, Telephone: +1 215 652 3612, Email: barry_buckland@merck.com

**Miss. Angela Buckler**, University of Birmingham, Department of Chemical Entineering, Edgbaston, Birmingham B15, UK, Telephone: +44 121 414 3887, Email: AXB303@bham.ac.uk

**Dr. Heino Buentemeyer**, University of Bielefeld, Institute of Cell Culture Technology, Bielefeld 33594, GERMANY, Email: hb@zellkult.techfak.uni-bielefeld.de

**Ms. Joanne Bull**, Lonza Biologics, 101 International Drive, Portsmouth, NH 03801, USA, Telephone: +1 603 610 4754, Email: joanne.bull@lonza.com

**Ms. Sandy Bulloch**, Invitrogen, 3 Fountain Drive, Inchinnan Business Park, Paisley PA4 9RF, UK, Telephone: +44 141 814 5814, Email: sandy.bulloch@invitrogen.com

**Ms. Shani Bullock**, Irvine Scientific, 2511 Daimler Street, Santa Ana, CA 92705-5588, USA

**Mr. Colin Burdick**, BioProcess Engineering Services Ltd., 186 Carver Drive, Kent Science Park, Sittingbourne Kent, UK, Telephone: +44 1795 411 531, Email: cburdick@bioprocess-eng.co.uk

**Mr. Thomas Burdick**, BioProcess Engineering Services Ltd., 186 Carver Drive, Kent Science Park, Sittingbourne Kent, UK

**Dr. Christa Burger**, Merck KGaA, Frankfurter Str.250, TRB A32/410, Darmstadt 64271, GERMANY, Telephone: +49 6151 726032, Email: christa.burger@merck.de

**Dr. Julian Burke**, Genetix Ltd., Queensway, New Milton Hampshire BH25 5NN, UK

**Mr. Robert Burns**, Scottish Agricultural Science Agency, 82 Craigs Road, East Craigs, Edinburgh EH12 8NJ, UK, Telephone: +44 131 244 8911, Email: Robert.Burns@sasa.gsi.gov.uk

**Mr. Bob Burrier**, Gibco Invitrogen Corporation, 3175 Staley Road, Grand Island, NY 14221, USA, Telephone: +1 716 774 3131

**Dr. Claudia Buser**, Genzyme Corp, PO Box 9322, 45 New York Ave., Framingham, MA 01701, USA, Telephone: +1 508 271 3691, Email: claudia.buser@genzyme.com

**Dr. Mike Butler**, University of Manitoba, 418 Buller Bldg, Winnipeg R3T 2N2, CANADA, Telephone: +1 204 474 9372, Email: butler@cc.umanitoba.ca

**Mr. Claude Cabau**, Irvine Scientific, 2511 Daimler Street, Santa Ana, CA 92705-5588, USA

**Dr. David Calabrese**, Lab of Molecular Biotechnology, Station 6, FSB-ISP, EPFL, Lausanne 1015, SWITZERLAND, Email: david.calabrese@unil.ch

**Mr. Eric Calvosa**, Sanofi Pasteur, 1541 Avenue Marcel Merieux, Marcy L' Etoile 69280, FRANCE, Telephone: +33 4 37 37 32 14, Email: eric.calvosa@sonofipasteur.com

**Ms. Susan Cameron**, Invitrogen, 3 Fountain Drive, Inchinnan Business Park, Paisley PA4 9RF, UK, Telephone: +44 141 814 5814, Email: susan.cameron@invitrogen.com

**Mr. Joseph Camire**, HyClone Labs, 925 W. 1800 S., Logan 84321, USA, Telephone: +1 4357928000, Email: joseph.camire@perbio.com

**Mr. Matt Caple**, JRH Europe, Smeaton Road, West Portway, Andover Hampshire, UK, Telephone: + 44 1264 333311

**Mr. Matt Caple**, SAFC, ZI Chesnes BP 701, St Quentin Fallavier 38297, FRANCE

**Prof. Manuel Carrondo**, ITQB/IBET, Apartado 12, Oeiras 2781-901, PORTUGAL, Telephone: +351 21 44 69 363, Email: mjtc@itqb.unl.pt

**Keith Carson**, The Williamsburg BioProcessing Foundation, PO Box 1229, Virginia Beach, VA 23451, USA, Telephone: +1 757 423 8823, Email: kcarson@wilbio.com

**Mr. Curt Carste**, Charter Medical Ltd., 14 rue du Rhone, 4th floor, Geneva 1204, SWITZERLAND, Telephone: +41 22 819 1836

**Mrs. Jane Carter**, Amgen, 1201 Amgen Court West, Seattle, WA 98107, USA, Telephone: +1 206 265 7685, Email: carterj@amgen.com

**Dr. John Carvell**, Aber Instruments Ltd., 5, Science Park, Aberystwyth SY23 3AH, UK, Telephone: +44 1970 636 300, Email: jc@aber-instruments.co.uk

**Mr. Scott Carver**, Regeneron Pharmaceuticals, 81 Columbia Turnpike, Rensselear, NY 12144, USA, Telephone: +1 518 488 6187, Email: james.cowman@regeneron.com

**Dr. Leda dos Reis Castilho**, COPPE/UFERJ, Programa de Engenharia, Quimica Caixa, Postal 68502, Rio de Janeiro 21941 972, BRAZIL, Telephone: +21 2562 8336, Email: leda@peq.coppe.ufrj.br

**Mr. Adolfo Castillo-Vitlloch**, Center of Molecular Immunology, Calle 216 y 15, Atabey, Playa, Havana 11600, CUBA, Telephone: +53 7 2716867, Email: adolfo@ict.cim.sld.cu

**Ms. Susana Alexandra Castro**, Instituto Superior Técnico – Lisboa, Rua Ilha dos Amores, Lote 4.39.01 - 1o A, Moscavide 1990-371, PORTUGAL, Telephone: +351965476382, Email: susanacastro721@hotmail.com

**Dr. Luigi Cavenaghi**, Areta International s.r.l., Via Roberto Lepetit, 34, Gerenzano I-21040, ITALY, Telephone: +39 02 96489264, Email: sbaldi@aretaint.com

**Ms. Wendy Chaderjian**, Genentech, Inc., 1 DNA Way, Mail Stop 32, South San Francisco, CA 94080, USA, Telephone: +1 650 225 5518, Email: wendy@gene.com

**Dr. Gilliane Chadeuf**, Laboratoire de Thérapie Génique, INSERM U649, Bâtiment Jean Monnet-Hôtel Dieu, 30 Bd Jean Monnet, Cedex 01 Nantes 44035, FRANCE, Telephone: +33 40 08 74 89, Email: gilliane.chadeuf@univ-nantes.fr

**Mr. Martin Chan**, Pfizer Ltd., Ramsgate Road, Sandwich Kent CT13 9NJ, UK, Email: martin.chan@pfizer.com

**Dr. Roop Chandwani**, Aggio Partners, UK, Telephone: +44 7786 554177, Email: rc@aggiopartners.com

**Dr. Dale Charlton**, Flowgen Bioscience, Wilford Industrial Estate, Ruddington Lane, Nottingham NG11 7EP, UK, Telephone: +44 115 982 1111, Email: dcharlton@flowgen.co.uk

**Dr. Michel Chartrain**, Merck Research Laboratories, RY-80Y-105, PO Box 2000, Rahway, NJ 07065, USA, Telephone: +1 732 594 4945, Email: chartrain@merck.com

**Ms. Nathalie Chatzissavidou**, Biovitrum AB, Lindhagensgatan 133, Stockholm SE-11276, SWEDEN, Telephone: +46 8 6973038, Email: nathalie.chatzissavidou@biovitrum.com

**Mr. Jean-Francois Chaubard**, Sanofi Pasteur, 1541 Avenue Marcel Merieux, Marcy L'Etoile 69280, FRANCE, Telephone: +33 4 37 37 07 29, Email: jean-françois.chaubard@sanofipasteur.com

**Mr. Gautam Chaudhary**, HyClone/Perbio Science, Industrielaan 27, Industriezone III, Erembodegem-Aalst B-9320, BELGIUM

**Mr. Paul Chen**, Biogen Idec, 14 Cambridge Center, Cambridge, MA 02142, USA, Telephone: +1 617 679 2415, Email: paul.chen@biogenidec.com

**Mr. Sebastien Chenuet**, EPFL, SV IGBB LBTC, J2-496 (Building CH), Station 6, Lausanne 1015, SWITZERLAND, Telephone: +41 21 693 55 42, Email: sylvia.fabris@epfl.ch

**Mrs. Yuen-Ting Chim**, GlaxoSmithKline, South Eden Park Road, Beckenham, Kent BR3 3BS, UK, Telephone: +44 208 639 6309, Email: yuen-ting.t.chim@gsk.com

**Dr. Myung-Sam Cho**, Celltrion, Inc., 1001-5, Dongchun-dong, Yeonsu-gu, Incheon 406-130, KOREA, Telephone: +82 32 850 5215, Email: mscho@celltrion.com

**Mr. John Chon**, Genzyme Flanders NV, Cipalstraat 8, Geel 2440, BELGIUM, Telephone: +32 1 456 4829, Email: ann.verheyden@genzyme.com

**Ms. Lilly Chong**, Informa, Telephone House, 69-77 Paul Street, London EC2A 4LQ, UK, Telephone: +44 20 797 5011, Email: lily.chong@informa.com

**Dr. Veronique Chotteau**, Biovitrum Biopharmaceuticals Process Development, Lindhagensgatan 133, Stockholm SE-11276, SWEDEN, Telephone: +46 8 69 72 758, Email: veronique.chotteau@biovitrum.com

**Mr. Richard Christison**, Astrazeneca Pharmaceuticals, Structural Chemistry Laboratory, Pepparedsleden 1, 431 83 Mölndal, SWEDEN, Telephone: +46 31 7065542, Email: richard.christison@astrazeneca.com

**Dr. Charles Christy, Covance**, Langhurstwood Road, Horsham Sussex RH12 4QD, UK, Telephone: +44 1423 500011, Email: charles.christy@covance.com

**Dr. Bok-Hwan Chun**, Celltrion, 1001-5, Dongchun-dong, Yeonsu-gu, Incheon 406-130, KOREA, Telephone: +82 32 850 5161, Email: bhchun@celltrion.com

**Mr. Martin Clarkson**, Maxygen, Agern Alle 1, Hørsholm 2970, DENMARK, Telephone: +45 7020 5550, Email: mpc@maxygen.dk

**Mr. Paul Clee**, University of Birmingham, Birmingham, UK, Email: PXC235@bham.ac.uk

**Mrs. Catherine Cleuziat**, Merial, 254 Rue Marcel Merieux, Lyon 69007, FRANCE, Telephone: +33 4 72 72 34 01, Email: Catherine.cleuziat@merial.com

**Dr. Jose Coco Martin**, Animal Sciences Group Wur, Products Division, PO Box 65, Lelystad 8200AB, THE NETHERLANDS, Telephone: + 31 320 238 249, Email: jose.cocomartin@wur.nl

**Dr. Trevor Collingwood**, Sangamo BioSciences, 501 Canal Blvd, Ste. A100, Richmond 94804, USA, Telephone: +1 510 970 6000 x261, Email: tcollingwood@sangamo.com

**Mr. James Cooper**, European Collection of Cell cultures (ECACC), Health Protection Agency, Porton Down, Salisbury SP4 0JG, UK, Telephone: +44 1980 612 707, Email: jim.cooper@hpa.dg.uk

**Miss. Ana Coroadinha**, ITQB/IBET, Apartado 12, Oeiras 2781-901, PORTUGAL, Telephone: +351 21 44 69 424, Email: avalente@itqb.unl.pt

**Mr. Mesbah Creitz**, Cellon SA, 29 AM Bechler, Bereldange, Luxembourg 7213, LUXEMBOURG

**Ms. Jennifer Cresswell**, Sigma-Aldrich Biotechnology, 2909 Laclede Ave., Saint Louis, MO 63103, USA, Telephone: +1 314 289 8496 x1379, Email: jcresswell@sial.com

**Mr. Jason Cronwall**, C-Flex cpt, 4451 110th Ave, North, Clearwater, FL 33762, USA

**Mr. Tori Crook**, Cambridge Antibody Technology, Milstein Building, Granta Park, Abington CB1 6H, UK, Telephone: +44 1223 898737, Email: collette.king@cambridgeantibody.com

**Mr. John Crowley**, Wyeth Medica Ireland, Grange Castle Business Park, Clondalkin, Dublin 22, IRELAND, Telephone: +353 1 469 7952, Email: crowleyj@wyeth.com

**Ms. Alexandra Croxford**, University of Manchester, The Michael Smith Bldg., Oxford Road, Manchester M13 9PT, UK, Telephone: +44 161 275 5101, Email: alexandra.croxford@postgrad.manchester.ac.uk

**Dr. Pedro Cruz**, ITQB/IBET, Apartado 12, Oeiras 2781-901, PORTUGAL, Telephone: +351 21 44 69 417, Email: pcruz@itqb.unl.pt

**Dr. Hans-Joachim Daniel**, Boehringer Ingelheim GmbH & Co. KG, Birkendorger Str. 65, Biberach 88397, GERMANY, Telephone: +49 7351 54 7377, Email: alexandra.neff@bc.boehringer-ingelheim.com

**Mr. Terri Danisevich**, Charter Medical Ltd., 14 rue du Rhone, 4th floor, Geneva 1204, SWITZERLAND, Telephone: +41 22 819 1836

**Dr. Carol David**, Teva Pharmaceuticals, PO Box 4079, Ness Ziona 4140, ISRAEL, Telephone: +972 8 938 7002, Email: carol.david@teva.co.il

**Mr. Ian Davies**, Applikon. BV/AppliSens, De Brauwweg 13, Schiedam 3125 AE, THE NETHERLANDS

**Dr. John Davis,** Bio Products Laboratory, Dagger Lane, Elstree, Hertfordshire WD6 3BX, UK, Telephone: +44 20 8258 2323, Email: John.Davis@bpl.co.uk

**Dr. Pieter de Geus**, DSM Biologics, Zuiderweg 72/2, Groningen 9700AL, THE NETHERLANDS, Telephone: +31 50 5222258, Email: pieter.geus-de@dsm.com

**Ms. Esther de Graaf-Groenendijk**, Crucell, Archimedesweg 4, PO box 2048, 2301CA, Leiden 2333CN, THE NETHERLANDS, Telephone: +31 71 5248864, Email: e.degraaf@crucell.com

**Dr. Maria Joao De Jesus**, ExcellGene SA, Rte de l'Ile au Bois 1A, Monthey 1870, SWITZERLAND, Telephone: +41 24 471 96 60, Email: maria.dejesus@excellgene.com

**Dr. Greg Dean**, Cambridge Antibody Technology, Milstein Building, Granta Park, Abington, Cambridge Cambs, CB1 6GH, UK, Telephone: +44 1223 898757, Email: colette.king@cambridgeantibody.com

**Mr. Zach Deeds**, SAFC, ZI Chesnes BP 701, St Quentin Fallavier 38297, FRANCE

**Ms. Katrien Deknopper**, Cambrex Bio Science, Parc Industriel de Petit Rechain, Verviers B-4800, BELGIUM

**Mrs. Neslihan DelaCruz**, Genentech Inc., 1 DNA Way, MS# 75B, South San Francisco, CA 94080-4918, USA, Telephone: +1 650 225 2029, Email: delacruz.neslihan@gene.com

**Dr. Christine DeMaria**, Genzyme Corporation, One Mountain Road, Framingham, MA 01701, USA, Telephone: +1 508 270 2254, Email: christine.demaria@genzyme.com

**Mr. Sam Denby**, UCL, ACBE, UCL Department of Biochemical Engineering, Torrington Place, London WC1E 7JE, UK, Telephone: +44 207 679 4410, Email: samuel.denby@ucl.ac.uk

**Dr. Richard Dennetts**, Eden Biodesign, Unit D5, Stanlaw Abbey Business Centre, Dover Drive, Ellsemere Port Cheshire CH65 9BF, UK, Telephone: +44 151 356 5632, Email: richard.dennett@edenbiodesign.com

**Mrs. Veronique Deparis**, Laboratoires Serono SA, ZiB, Fenil-sur-Corsier 1809, SWITZERLAND, Email: veronique.deparis@serono.com

**Dr. Madiha Derouazi**, EPFL, SV IGBB LBTC, J2-496 (Building CH), Station 6, Lausanne 1015, SWITZERLAND, Telephone: +41 21 693 61 45, Email: sylvia.fabris@epfl.ch

**Dr. Rohini Deshpande**, Amgen Inc., One Amgen Center Drive, Thousand Oaks, CA 91320, USA, Telephone: +1 805 447 7583, Email: rohinid@amgen.com

**Ms. Maree Devine**, NUNC A/S, Kamstrupvej 90, Postbox 280, Roskilde DK-4000, DENMARK

**Mr. Miguel Diez Ibanez**, Laboratoires Serono S.A., Zone Industrielle de l'Ouriettaz, Aubonne CH-1170, SWITZERLAND, Telephone: +41 21 821 70 00, Email: miguel.diez@serono.com

**Mr. Matt Dillingham,** Celliance Ltd., 4, Fleming Rd., Kirkton Campus, Livingston, West Lothian EH54 7BN, UK, Email: mdillingham@celliancecorp.com

**Ms. Lisa Dixon**, Cobra Biomanufacturing Plc, Stephenson Building, The Science Park, Keele Staffordshire ST5 5SP, UK, Telephone: +44 1782 877 298, Email: lisa.dixon@cobrabio.com

**Ms. Miranda Docherty**, Bioprocess International Magazine, Oakhouse Business Centre, 39-41 The Parade, Claygate KT10 0PD, UK, Telephone: +44 1372 471 906, Email: mirandabpi@yahoo.co.uk

**Miss. Emily Dorckel**, CYTHERIS, 21 rue aristide briand, Vanves 92170, FRANCE, Telephone: +3333158883800, Email: avsauvezie@cytheris.com

**Dr. Rosemary Drake**, The Automation Partnership, York Way, Royston Hertfordshire, SG8 5WY, UK

**Mr. Jurgen Drewelies**, Cambrex Bio Science, Parc Industriel de Petit Rechain, Verviers B-4800, BELGIUM

**Mr. Patrick Dumas**, GlaxoSmithKline Biologicals S. A., rue de Institut 89, Bixensart 1330, BELGIUM, Telephone: + 32 2 656 9890, Email: patrick.dumas@gskbio.com

**Dr. Reinhard Dunker**, MERCK KGaA, Glob. Technol. / TRB, Q27 236, Frankfurter Strasse 250, DARMSTADT D-64271, GERMANY, Telephone: +49 6151 72 7449, Email: reinhard.dunker@merck.de

**Mr. James Dunster**, Moregate TCS Ltd, Botolph Claydon, Buckingham MK18 2LR, UK

**Dr. Roland Durner**, Tecan Deutschland GmbH, Theodor-Storm-Strasse 17, Crailsheim D-74564, GERMANY, Email: roland.durner@tecan.com

**Dr. Charlotte Dyring**, Pharmexa A/S, Kogle Alle 6, Horsholm 2970, DENMARK, Email: cd@pharmexa.com

**Mr. Richard Eagling**, Stedim S.A., Z. I. Des Paluds, Avenue De Jouques, Aubagne F-13781, FRANCE

**Dr. Horst Eberhardt**, Roche Diagnostics GmbH, Penzberg, Nonnenwald 2, Penzberg 82377, GERMANY, Telephone: +49 8856 60 2648, Email: horst.eberhardt@roche.com

**Dr. Chandrika Ediriwickrema**, Diosynth Biotechnology, 3000, Weston Parkway, Cary, NC 27513, USA, Telephone: +1 919 388 5666, Email: Chandrika Ediriwickrema@diosynth-rtp.com

**Ms. Jenny Eggenschwiler**, Hochschule Wädenswil, Einsiedlerstrasse 29b, Wädenswil 8820, SWITZERLAND, Telephone: +41 1 789 9675, Email: j.eggenschwiler@hsw.ch

**Dr. Iman Elmahdi**, Pall Life Sciences, Walton Road, Portsmouth PO6 1TD, UK, Telephone: +44 2392 303 438, Email: iman.elmahdi@europe.pall.com

**Dr. Barbara Enenkel**, Boehringer Ingelheim Pharma GmbH & Co. KG, Birkendorfer Strasse 65, Biberach an der Riss 88397, GERMANY, Telephone: +49 7351 54 96117, Email: barbara.enenkel@bc.boehringer-ingelheim.com

**Dr. Alfred Engel**, Roche Diagnostics GmbH, Nonnenwald 2, Penzberg 82372, GERMANY, Telephone: +49 8856 603509, Email: alfred.engel@roche.com

**Ms. Annette England**, JRH Europe, Smeaton Road, West Portway, Andover Hampshire, UK, Telephone: + 44 1264 333311

**Mr. Lee Eppstein**, New Brunswick Scientific Co. Inc., PO Box 4005, 44 Talmadge Road, Edison, NJ 08818-4005, USA, Telephone: +1 732 287 1200, Email: eppstein@nbsc.com

**Dr. David Epstein**, Centocor, Inc., 145 King of Prussia Road, R-1-1, Radnor, PA 19087, USA, Email: depstei4@cntus.jnj.com

**Miss. Ulrika Eriksson**, Biotechnology, Royal Institute of Technology, Roslagstullsbacken 21, Stockholm 10691, SWEDEN, Telephone: +46855378306, Email: ulrika@biotech.kth.se

**Dr. Böhm Ernst**, Baxter AG, Vienna 1220, AUSTRIA, Email: ernst_boehm@baxter.com

**Mr. Holger Essig**, Blue Membranes GmbH, Rheingaustrasse 190-196, Wiesbaden D-65203, GERMANY, Telephone: +49 611 962 8723, Email: he@blue-membranes.com

**Mr. Geoffrey Esteban**, FOGALE Nanotech, Park Kennedy - Bat A3, 285, rue Gilles Roberval, CS 32028, NIMES Cedex 2, FRANCE, Telephone: +33 466 62 71 73, Email: g.esteban@fogale.fr

**Karen Fallen**, Lonza Biologics, 101 International Drive, Portsmouth, NH 03801, USA, Telephone: +1 603 610 4754, Email: karen.fallen@lonza.com

**Dr. Elizabeth Fashola-Stone**, European Collection of Cell cultures (ECACC), Health Protection Agency, Porton Down, Salisbury SP4 0JG, UK

**Mr. Ralf Fehrenbach**, Cell Culture / USP, Rentschler, Biotechnologie GmbH, Erwin-Rentschler-Strasse 21, Laupheim 88471, GERMANY, Telephone: +49 7392/701 879, Email: karin.schoch-hofmann@rentschler.de

**Ms. Delia Fernandez**, Gibco Invitrogen Corporation, 3175 Staley Road, Grand Island, NY 14221, USA, Telephone: +1 716 774 3131, Email: delia.fernandez@invitrogen.com

**Ms. Pat Ferns**, Irvine Scientific, 2511 Daimler Street, Santa Ana, CA 92705-5588, USA

**Mr. Bruno Ferreira**, Animal Cell Technology Laboratory, ITQB/IBET, Apartado 12, Oeiras 2780-901, PORTUGAL, Telephone: +351 21 446 94 24, Email: tbpsf@itqb.unl.pt

**Mr. Guilherme Ferreira**, Centre for Molecular and Structural Biomedicine, University of Algarve, Campus de Gambelas, Faro 8000 Faro, PORTUGAL, Telephone: +351 289 800 900, Email: gferrei@ualg.pt

**Mr. Tiago Ferreira**, ITQB/IBET, Apartado 12, Oeiras 2781-901, PORTUGAL, Telephone: +351 214469424, Email: tbpsf@itqb.unl.pt

**Mr. Gary Finka**, GlaxoSmithKline, South Eden Park Road, Beckenham, Kent BR3 3BS, UK, Telephone: +44 20 8639 6372, Email: gary.b.finka@gsk.com

**Mr. David Fiorentini**, Biological Industries Ltd., Kibbutz Belt, Haemek 25115, ISRAEL, Telephone: +972 4 996 0595

**Mr. Daniel Fleischanderl**, Polymun/ Institute of Applied Microbiology, University of Natural Resources and Applied Life Sciences, Muthgasse 18 ( House B), Vienna 1190, AUSTRIA, Email: daniel.fleischanderl@blackbox.net

**Mr. Tom Fletcher**, Irvine Scientific, 2511 Daimler Street, Santa Ana, CA 92705-5588, USA

**Dr. Marcel Flikweert**, Centocor B.V., Einsteinweg 92, P.O.Box 251, 2300 AG, THE NETHERLANDS, Telephone: +31 71 5242719, Email: mflikwee@cntnl.jnj.com

**Mr. Ingo Focken**, Aventis Pharma Deutschland GmbH, Lead Identification Technologies, Gebaeude H811, Industriepark Hoechst, Frankfurt 65926, GERMANY, Telephone: +49 69 305 24594, Email: andrea.goldapp@sanofi-aventis.com

**Mr. Andrew Ford**, Greiner Bio-One Ltd., Brunel Way, Stroudwater Business Park, Stonehouse Gloustershire GL10 3SX, UK, Email: andy.ford@gbo.com

**Mr. Pietro Forgione**, J M Separations BV, Boschoijk 780, PO Box 6286, Eindhoven 5600 HG, THE NETHERLANDS

**Alex Forrest-Hay**, Beckman Coulter UK Ltd., Kingsmead Business Park, London Road, High Wycombe HP11 1JU, UK, Telephone: +44 1494 429 196

**Mr. Patrick Fox**, Genetic Engineering News, 36 Orchard Road, Hook Norton Oxfordshire OX15 5LX, UK

**Ms. Sophia Fox**, Genetic Engineering News, 36 Orchard Road, Hook Norton Oxfordshire OX15 5LX, UK

**Dr. Elisabeth Fraune**, Sartorius BBI Systems GmbH, Schwarzenberger Weg 73 - 79, Melsungen D-34212, GERMANY

**Mrs. Tracy Freeman**, Covance Laboratories Ltd., Otley Road, Harrogate HC8 1PY, UK, Telephone: +44 1423 500 011, Email: tracy.freeman@covance.com

**Prof. Ruth Freitag**, University of Bayreuth, Chair for Process Biotechnology, Universitätsstr. 30, Bayreuth 95447, GERMANY, Telephone: +49 921 557371, Email: bioprozesstechnik@uni-bayreuth.de

**Ms. Kathie Fritchman**, BDD - Diagnostic Systems Europe, 11 rue Aristide Berges, Le Pont de Claix 38800, FRANCE, Telephone: +33 4 76 68 32 60, Email: marjorie_herbin@europe.bd.com

**Mr. Thomas Fröberg**, AstraZeneca, B841, Södertälje 15185, SWEDEN, Email: thomas.froberg@astrazeneca.com

**Mr. Steve Froud**, Lonza Biologics, 101 International Drive, Portsmouth, NH 03801, USA, Telephone: +1 603 610 4754, Email: steve.froud@lonza.com

**Mr. Richard Fry**, Cellon SA, 29 AM Bechler, Bereldange, Luxembourg 7213, LUXEMBOURG, Telephone: +352 263 3731, Email: cellon@gms.lu

**Mr. Erwin Fuerst**, Berna Biotech Ltd., Rehhagstr. 79, Thorwhauf, Bern 3174, SWITZERLAND, Telephone: +41 31 888 5123, Email: erwin.fuerst@bernabiotech.com

**Dr. Antje Fuhrmann**, Greiner Bio-One Ltd., Brunel Way, Stroudwater Business Park, Stonehouse, Gloustershire GL10 3SX, UK, Email: Antje.fuhrmann@gbo.com

**Mr. James Furey**, PendoTech, 2193 Commonwealth Avenue, #192, Boston, MA 02135, USA, Telephone: +1 617 817 0350, Email: request@pendotech.com

**Prof. Martin Fussenegger**, ETH Zürich, ETH-Hönggerberg, HPT, Zürich CH-8093, SWITZERLAND, Telephone: +41 01 633 34 48, Email: fussenegger@biotech.biol.ethz.ch

**Dr. Zbigniew Gadek**, Center for Holistic Medicine and Naturopathy Co.Ltd, Talweg 14, Schmallenberg-Nordenau 57392, GERMANY, Telephone: +49 2975 9622190, Email: ZGNGmbH@t-online.de

**Mr. Douglas Galbraith**, University of Queensland, Chemical Engineering, Building 74, Brisbane Queensland 4072, AUSTRALIA, Telephone: +61 7 3365 3568, Email: douglasg@cheque.uq.edu.au

**Mr. Daniel Galbraith**, Covance, Langhurstwood Road, Horsham Sussex RH12 4QD, UK, Telephone: +44 1423 500011, Email: daniel.galbraith@covance.com

**Dr. Gilad Gallili**, Abic Biological Laboratories Teva, P.O.B 489, Beit Shemesh 99100, ISRAEL, Telephone: +972 54 888 5938, Email: gilad.gallili@teva.co.il

**Mr. Jordi Gálves Sánchez**, Universitat Autònoma de Barcelona, C/Noguera Pallaresa, Num. 9, Sabadell, Barcelona 08202, SPAIN, Telephone: +34 935812694, Email: jordi.galvez@uab.es

**Mr. Paul Ganglberger**, Igeneon AG Immunotherapy of Cancer, Krebs-immuntherapie Forschungs und, Entwicklungs AG, Brunner Strasse 69/ Obj. 3, Vienna A-1230, AUSTRIA, Telephone: +43 699 18904 256, Email: p.ganglberger@gmx.at

**Dr. Subinay Ganguly**, Centocor, Inc., Pharmaceutical Development, 145 King of Prussia Road, Mail Stop: R-1-2, Radnor 19087, USA, Telephone: +1 610 240 4090, Email: sgangul2@CNTUS.jnj.com

**Mrs. Catherine Gardner**, Moregate TCS Ltd, Botolph Claydon, Buckingham MK18 2LR, UK

**Mr. Alun Garner**, GE Healthcare, 7 Capesthorne Close, Northwich CW9 8FA, UK, Telephone: +44 1606 333 574, Email: alun.garner@ge.com

**Ms. Annika Garnes**, Infors UK Ltd., The Courtyard Business Centre, Dovers Farm, Lonesome Lane, Reigate Surrey, RH2 7QT, UK

**Mr. Patrice Garnier**, BioProcess Engineering Services Ltd., 186 Carver Drive, Kent Science Park, Sittingbourne, Kent, UK,

**Prof. Alain Garnier**, Université Laval, Pavillon Pouliot, Québec G1K 7P4, CANADA, Telephone: +1 418 656 3106, Email: alain.garnier@gch.ulaval.ca

**Ms. Montserrat Garrell**, Biokit, Can Male S/N, Lliçà D'Amunt 08186, SPAIN, Telephone: +34938609000, Email: mgarrell@biokit.com

**Mr. Dominik Gaser**, Lek Pharmaceuticals did., Verovskova 57, Ljubljana SL-1526, SLOVENIA, Telephone: +386 1 7217 635, Email: dominik.gaser@lek.si

**Dr. Martin Gawlitzek**, Genentech, Inc., 1 DNA Way, South San Francisco, CA 94080, USA, Telephone: +1 650 225 8869, Email: martin@gene.com

**Mrs. Cécile Geny**, Généthon, 1bis, rue de l'internationale, BP 60, Evry F-91002, FRANCE, Email: fiamma@genethon.fr

**Dr. Yvonne Genzel**, Max Planck Institute for Dynamics of Complex Technical Systems, Sandtorstraße 1, Magdeburg 39106, GERMANY, Telephone: +49 391 6110257, Email: genzel@mpi-magdeburg.mpg.de

**Mr. Ron Geven**, BDD - Diagnostic Systems Europe, 11 rue Aristide Berges, Le Pont de Claix 38800, FRANCE, Telephone: +33 4 76 68 32 60, Email: marjorie_herbin@europe.bd.com

**Mr. Christoph Giese**, ProBioGen AG, Goethestr. 54, Berlin D-13086, GERMANY, Telephone: +49 3092400624, Email: christoph.giese@probiogen.de

**Mr. Simeon Gill**, MHRA, Room 12-203, Market Towers, 1 Nine Elms Lane, London SW8 5NQ, UK, Telephone: +44 20 7084 2659, Email: simeon.gill@mhra.gsi.gov.uk

**Mr. Arnaud Glacet**, LFB, 59 Rue de trevise, BP 2006, Lille Cedex 59011, FRANCE, Telephone: +33 3 20494473, Email: glacet@lfb.fr

**Dr. Bernd Glauner**, Schaerfe System GmbH, Kraemerstrasse 22, Reutlingen 72764, GERMANY

**Mr. Loic Glez**, Serono Pharmaceutical Research, 14 Chemin Des Aulx, Plan-les-Ouates 1228, SWITZERLAND, Telephone: +41 22 706 9646, Email: loic.glez@serono.com

**Prof. Francesc Gòdia**, Universitat Autònoma de Barcelona, Dept d'Enginyeria Quimica, Vitubrio 8, Edifici C, Bellaterra (Barcelona) E-08093, SPAIN, Telephone: +34 93 5812692, Email: francesc.godia@uab.es

**Dr. Fiona Godsman**, BioReliance Ltd, Todd Campus, West of Scotland Science Park, Glasgow G20 0XA, UK, Telephone: +44 141 946 9999, Email: fgodsman@bioreliance.com

**Mrs. Astrid Goedde**, ORPEGEN Pharma, Czerny-Ring 22, Heidelberg 69115, GERMANY, Telephone: +49 6221 9105 60, Email: agoedde@web.de

**Dr. Anuj Goel**, Biocon Limited, 20th KM, Hosur Road, Electronic City P.O., Bangalore 560100, INDIA, Telephone: +918028082371, Email: anuj.goel@biocon.com

**Prof. Jean-Louis Goergen**, CNRS-INPL, ENSAIA, 2 avenue de la foret de Haye, Vandoeuvre 54505, FRANCE, Telephone: +33 383 59 58 44, Email: jean-louis.goergen@ensaia.inpl-nancy.fr

**Mr. Tony Goguillon**, SAFC, ZI Chesnes BP 701, St Quentin Fallavier 38297, FRANCE

**Dr. Charles Goochee**, Merck & Company, Inc., P.O. Box 4, Mail Code WP17-201, 770 Sumneytown Pike, West Point, PA 19486, USA, Telephone: +1 215 652 7159, Email: charles_goochee@merck.com

**Dr. Margaret Goodall**, University of Birmingham. U.K., Division of Immunity and Infection, The Medical School, Vincent Drive, Birmingham B15 2TT, UK, Telephone: +44 121 414 6849, Email: d.m.goodall@bham.ac.uk

**Mr. Roel Gordijn**, Cambrex Bio Science, Parc Industriel de Petit Rechain, Verviers B-4800, BELGIUM, Telephone: +32 87 321 611, Email: roel.gordinjn@cambrex.com

**Dr. Volker Gorenflo**, Sanofi Pasteur, 1755 Steeles Avenue West, Toronto, M2R3T4, CANADA, Email: volker.gorenflo@aventis.com

**Dr. Uwe Gottschalk**, Sartorius AG -- Biotechnology, Weeder Landstr. 94-108, Goettingen 37075, GERMANY, Telephone: +49 551 308 2016, Email: uwe.gottschalk@sartorius.com

**Mr. Tony Gougullion**, JRH Europe, Smeaton Road, West Portway, Andover Hampshire, UK, Telephone: + 44 1264 333311

**Mr. Gerard Gourdon**, Applikon. BV/AppliSenc, De Brauwweg 13, Schiedam 3125AE, THE NETHERLANDS

**Dr. Beatrice Goxe**, Euroscreen, Rue Adrienne Bolland, 47, Gosselies B-6041, BELGIUM, Telephone: +32 1 71 348 554

**Prof. Stefanos Grammatikos**, Boehringer Ingelheim Pharma GmbH, Birkendorfer Straße 65, Biberach D-88397, GERMANY, Email: stefanos.grammatikos@bc.boehringer-ingelheim.com

**Mr. Lothar Grannemann**, Innovatis AG, Meisenstrasse 96, Bielefeld D-33607, GERMANY

**Mr. Martin Graw**, Genzyme Flanders NV, Cipalstraat 8, Geel 2440, BELGIUM, Telephone: +32 1 456 4829, Email: ann.verheyden@genzyme.com

**Dr. James Greaves**, Delta Biotechnology Ltd., Castle Court, 59 Castle Boulevard, Nottingham NG7 1FD, UK

**Mr. Jonathan Green**, Cambridge Antibody Technology, Milstein Building, Granta Park, Cambridge CB1 6GH, UK, Telephone: +44 1223 471471, Email: jon.green@cambridgeantibody.com

**Dr. Anne Gregoire**, CYTHERIS, 21 rue aristide Briand, Vanves 92170, FRANCE, Telephone: +33158883800, Email: avsauvezie@cytheris.com

**Mr. Bryan Griffiths**, ESACT Office, PO Box 1723, 5 Bourne Gardens, Salisbury Wilts SP4 0NU, UK, Telephone: +44 19806 10405, Email: griff@evemail.net

**Mr. Philippe Grimm**, BioReliance Ltd, Todd Campus, West of Scotland Science Park, Glasgow G20 0XA, UK, Telephone: +44 141 946 9999

**Dr. Silke Grueneberg**, Sanofi-Aventis, Industriepark Hoechst, H812, Frankfurt 65926, GERMANY, Telephone: +49 69 305 80133, Email: silke.grueneberg@sanofi-aventis.com

**Dr. Emmanuel Guedon**, CNRS-INPL, ENSAIA, 2 avenue de la foret de Haye, Vandoeuvre 54505, FRANCE, Telephone: +33 383 59 58 42, Email: emmanuel.guedon@ensaia.inpl-nancy.fr

**Dr. Fabienne Guéhenneux**, VIVALIS, SITE BIO OUEST, RUE DU MOULIN DE LA ROUSSELIERE, SAINT-HERBLAIN 44805, FRANCE, Telephone: +33 2 28 07 37 10, Email: fabienneguehenneux@vivalis.com

**Dr. Frank Gundermann**, Innovatis AG, Meisenstrasse 96, Bielefeld D-33607, GERMANY

**Nr, Laurent Gutzwiller**, Infors UK Ltd., The Courtyard Business Centre, Dovers Farm, Lonesome Lane, Reigate Surrey, RH2 7QT, UK

**Ms. Mary Jane Guy**, Wave BioTech LLC, 999 Frontier Rd., Bridgewater, NJ 08807, USA, Telephone: 1 908 707 9210, Email: info@wavebiotech.com

**Mr. Iggy Gyepi-Garbrah**, Pall Life Sciences, Walton Road, Portsmouth PO6 1TD, UK, Telephone: +44 2392 303 438, Email: iggy.gyepi-garbrah@europe.pall.com

**Ms. Nadia Haag,** Rütten Engineering, Industriestrasse 9, Stäfa 8712, SWITZERLAND, Telephone: +41 1 928 29 30, Email: info@rutten.com

**Dr. Jürgen Haas**, Boehringer Ingelheim Pharma GmbH & Co. KG, Birkendorfer Strasse 65, Biberach an der Riss 88397, GERMANY, Telephone: +49 7351 54 4476, Email: juergen.haas@bc.boehringer-ingelheim.com

**Dr. David Hacker**, EPFL, SV IGBB LBTC, J2-496 (Building CH), Station 6, Lausanne 1015, SWITZERLAND, Telephone: +41 21 693 61 42, Email: sylvia.fabris@epfl.ch

**Mr. Paul Haffenden**, TerraCell International S.A., Chateau de Vaumarcus, Vaumarcus CH-2028, SWITZERLAND, Telephone: +1 905 859 4991, Email: paulhaff@terracell.ca

**Prof. Lena Häggström**, Biotechnology, The Royal Institute of Technology, School of Biotechnology, AlbaNova University Centre, Stockholm S-106 91, SWEDEN, Email: lenah@biotech.kth.se

**Dr. Adrian Haines**, ML Laboratories PLC, Med IC4, Keele University & Business Park, Keele ST5 5NL, UK, Telephone: +44 0845 0060920, Email: adrian.haines@MLresearch.co.uk

**Mr. Christian Hakemeyer**, University of Bielefeld, Institute for Cell Culture Technology, Universitätsstr. 25, Bielefeld 33615, GERMANY, Telephone: +490521 1066320, Email: cha@zellkult.techfak.uni-bielefeld.de

**Mr. Paul Hallet**, BioReliance Ltd, Todd Campus, West of Scotland Science Park, Glasgow G20 0XA, UK, Telephone: +44 141 946 9999

**Mr. Takeki Hamasaki**, Department of Genetic Resources Technology, Kyushu University, 6-10-1 Hakozaki, Higashi-ku, Fukuoka 812-0063, JAPAN, Telephone: +81 92 642 3046, Email: takeki@grt.kyushu-u.ac.jp

**Mr. Toby Hamblin**, Charter Medical Ltd., 14 rue du Rhone, 4th floor, Geneva 1204, SWITZERLAND, Telephone: +41 22 819 1836, Email: thamblin@lydall.com

**Dr. Bruce Hamilton**, GlaxoSmithKline, South Eden Park Road, Beckenham BR3 3BS, UK, Telephone: +44 2086396102, Email: bjh44674@gsk.com

**Dr. Petra Hanke**, Roche Diagnostics GmbH, Nonnenwald 2, Penzberg 82372, GERMANY, Telephone: +49 8856 602907, Email: petra.hanke@roche.com

**Dr. Karen Hansen**, Novo Nordisk A/S, Department of Cell Biology, Hagedornsvej 1, Gentofte DK 2820, DENMARK, Telephone: +45 4444 8888, Email: kha@novonordisk.com

**Ms. Jeanette Hartshorn**, JRH Europe, Smeaton Road, West Portway, Andover Hampshire, UK, Telephone: + 44 1264 333311

**Dr. Lorenz Hartwieg**, Corning Life Sciences, Koolhovenlean 12, Schiphol-Rijk 119-N3, THE NETHERLANDS, Telephone: +44 1494 714 949, Email: allenjm@corning.com

**Prof. Hansjörg Hauser**, German Research Centre for Biotechnology, Mascheroder Weg 1, Braunschweig D-38124, GERMANY, Telephone: +49 531 6181250, Email: hha@gbf.de

**Mr. Nicolas Havelange**, Henogen, Rue des Professeurs Jeener et Brachet 12, Charleroi B-6041, BELGIUM, Telephone: +32 71 37 89 01, Email: nicolas.havelange@henogen.com

**Mr. Mischa Hawrylenko**, Infors UK Ltd., The Courtyard Business Centre, Dovers Farm, Lonesome Lane, Reigate Surrey RH2 7QT, UK

**Mr. Steven Hawrylik**, Pfizer PGRD, MS4062, Eastern Point Road, Groton 06340, USA, Telephone: +1 860 441 5103, Email: steven.j.hawrylik@pfizer.com

**Ms. Megumi Hayashi**, University of Fukui, 3-9-1 Bunkyo, Fukui 910-8507, JAPAN, Telephone: +81 776 27 8645, Email: megumi01370468@hotmail.com

**Mr. David Haynes**, Regeneron Pharmaceuticals, 81 Columbia Turnpike, Rensselear, NY 12144, USA, Telephone: +1 518 488 6187, Email: james.cowman@regeneron.com

**Mr. Andrew Hayward**, Broadley Technologies Ltd, Chain Hill Lodge, Wrest Park, Silsoe Bedford MK43 4HS, UK, Telephone: +44 1525 862 518, Email: ahayward@broadleyjames.com

**Dr. Jochen Heberle**, Boehringer Ingelheim Pharma GmbH & Co. KG, Purchasing Department, Birkendorfer Str. 65, Biberach an der Riss 88397, GERMANY

**Ms. Mary Heenan**, Wyeth Medica Ireland, The Wyeth BioPharma Campus at Grange Castle, Grange Castle International Business Park, Clondalkin, Dublin 22, IRELAND, Telephone: +353 1 4696814, Email: heenanm@wyeth.com

**Ms. Anne Charlotte Hegelund**, Novo Nordisk A/S, Novo Nordisk Park, F9.1.14, Maaloev DK-2760, DENMARK, Telephone: +45 44 42 75 48, Email: ache@novonordisk.com

**Mr. Delf Heger**, Helena Biosciences, UK

**Mr. Daniel Heid**, Wave Biotech AG, Ringstrasse 24, Tagelswangen 8317, SWITZERLAND

**Dr. Rüdiger Heidemann**, Bayer HealthCare, Biological Products, 800 Dwight Way, Berkeley, CA 94710, USA, Telephone: +1 510 705 5617, Email: rudiger.heidemann.b@bayer.com

**Dr. Holger Heine**, Novartis Pharma AG, NIBR / DT / BMP / ACE, WSJ-506.1.12B, Basel CH-4002, SWITZERLAND, Telephone: +41 61 324 30 34, Email: holger.heine@novartis.com

**Dr. Robin Heller-Harrison**, Wyeth BioPharma, 1 Burtt Road, Andover, MA 01810, USA, Telephone: +1 978 247 1406, Email: rhharrison@wyeth.com

**Mr. Herbert Hermann**, PAA Laboratories GmbH, Unterm Bornrain, Coelbe 35091, GERMANY

**Dr. Andreas Herrmann**, Celonic GmbH, Karl-Heinz-Beckurts-Str. 13, Julich 522428, GERMANY

**Dr. Friedemann Hesse**, Austrian Center of Biopharmaceutical Technology, Muthgasse 18, Vienna A-1190, AUSTRIA, Telephone: +43 1 36006 6806, Email: fhesse@edv2.boku.ac.at

**Mr. Mark Hessey**, GlaxoSmithKline, Medicines Research Centre, Gunnels Wood Road, Stevenage, Herts SG1 2NY, UK, Telephone: +44 1438 763502, Email: mark.p.hessey@gsk.com

**Dr. Gregory Hiller**, Wyeth BioPharma, 1 Burtt Road, Andover, MA 01810, USA, Telephone: +1 978 247 3173, Email: ghiller@wyeth.com

**Mr. Peter Hinterleitner**, Igeneon AG, Brunner Strasse 69/3, Vienna 1230, AUSTRIA, Telephone: +43 699 18904212, Email: peter.hinterleitner@igeneon.com

**Prof. Nobutaka Hirokawa**, The University of Toyko, Graduate School of Medicine, Dept of Cell Biology and Anatomy, Hongo, 7-3-1, Bunkyo-ku, Tokyo 113-003, JAPAN, Telephone: +81 3 54841 3326, Email: hirokawa@m.u-tokyo.ac.jp

**Mr. Win Ho**, Helena Biosciences, UK

**Ms. Keri Hodgkinson**, Covance, Langhurstwood Road, Horsham Sussex, RH12 4QD, UK, Telephone: +44 1423 500011, Email: k.hodgkinson@covance.com

**Dr. Simon Hoerstrup**, University Hospital, Dept of Surgical Research and, Clinic for Cardiovascular Surgery, Raemistrasse 100, Zurich CH-8091, SWITZERLAND, Email: simon_philipp.hoerstrup@usz.ch

**Ms. Stacy Holdread**, BD, 54 Loveton Circle, Baltimore, MD 21152, USA, Telephone: +1 410 316 3628, Email: Stacy_Holdread@bd.com

**Dr. James Hope**, BioProcessors Corporation, 12-A Cabot Road, Woburn, MA, USA

**Mr. Werner Höra**, University of Bielefeld, Institute for Cell Culture Technology, Universitaetsstr 25, Bielefeld 33675, GERMANY, Telephone: +49 521 1066338, Email: who@zellkult.techfak.uni-bielefeld.de, werner.hoera@gmx.de

**Mr. Anthony Hornby**, BDD - Diagnostic Systems Europe, 11 rue Aristide Berges, Le Pont de Claix 38800, FRANCE, Telephone: +33 4 76 68 32 60, Email: Tony_HORNBY@europe.bd.com

**Mr. Ralf Hornscheidt**, HyClone/Perbio Science, Industrielaan 27, Industriezone III, Erembodegem-Aalst B-9320, BELGIUM

**Prof. Jeffrey Hubbell**, Laboratory for Regenerative Medicine and Pharmacobiology, AAB039, Integrative Biosciences Institute, Station 15, Swiss Federal Institute of Technology (EPFL), Lausanne CH-1015, SWITZERLAND, Email: jeffrey.hubbell@epfl.ch

**Ms. Joanna Hudson**, Stedim S.A., Z. I. Des Paluds, Avenue De Jouques, Aubagne F-13781, FRANCE

**Mr. Richard Hughes**, Celliance, 5655 Spalding Drive, Norcross, GA 30092, USA

**Mr. Stefan Hummel**, Evotec Technologies, Schnackenburgallee 114, Hamburg 22525, GERMANY

**Mr. Boris Hundt**, Otto-von-Guericke-University, Chair of Bioprocess Engineering, Universitätsplatz 2, Magdeburg 39106, GERMANY, Telephone: +49 391 6110 213, Email: boris.hundt@vst.uni-magdeburg.de

**Mr. George Hutchinson**, Genetix Ltd., Queensway, New Milton, Hampshire BH25 5NN, UK

**Mr. Hans Huttinga**, Kerry Bioscience, Huizerstraatweg 28, Naarden 1411 GP, THE NETHERLANDS

**Mrs. Kuniko Ikura**, Kyoto Institute of Technology, Sakyo-ku Matsugasaki, Kyoto 606-8585, JAPAN, Telephone: +81 75 724 7535, Email: ikura@kit.ac.jp

**Prof. Koji Ikura**, Kyoto Institute of Technology, Sakyo-ku Matsugasaki, Kyoto 606-8585, JAPAN, Telephone: +81 75 724 7535, Email: ikura@kit.ac.jp

**Mr. Atsutoshi Ina**, Coastal Bioenvironment Center, Saga University, 152-1 Shonan-cho, Karatsu 847-0021, JAPAN, Telephone: +81 955 77 4484, Email: 04972002@edu.cc.saga-u.ac.jp

**Dr. Yuichi Inoue**, Kitakyushu National College of Technology 5-20-1, Shii, Kokuraminami-ku, Kitakyushu 802-0985, JAPAN, Email: inoue@kct.ac.jp

**Mr. Bojan Isailovic**, Bioprocessing Centre, Centre of Formulation Engineering, University of Birmingham, Edgbaston Park Road, Birmingham B15 2TT, UK, Email: BXI897@bham.ac.uk

**Dr. Adiba Ishaque**, Bayer Healthcare, 800 Dwight Way, Berkeley, CA 94710, USA, Telephone: +1 510 705 4614, Email: adiba.ishaque.b@bayer.com

**Mr. John Islas**, Broadley Technologies Ltd, Chain Hill Lodge, Wrest Park, Silsoe, Bedford MK43 4HS, UK

**Dr. Volker Jäger**, G B F, Mascheroder Weg 1, Braunschweig D-38124, GERMANY, Telephone: +49 531 6181102, Email: vja@gbf.de

**Prof. David James**, University of Queensland, Chemical Engineering, Building 74, Brisbane Queensland 4072, AUSTRALIA, Telephone: +61 7 3365 4638, Email: davidj@cheque.uq.edu.au

**Radh Jani**, Sartorius - Biotechnology Division, Longmead Business Centre, Blenheim Road, Epsom Surrey KT19 9QQ, UK

**Dr. Rachel Jarman-Smith**, Bioquell UK Limited, 34 Walworth Road, Andover, Hants SP10 5AA, UK

**Dr. Sushma Jassal**, Cobra Biomanufacturing Plc, Stephenson Building, The Science Park, Keele, Staffordshire ST5 5SP, UK, Telephone: +44 1782 714 181, Email: sushma.jassal@cobrabio.com

**Mrs. Nanni Jelinek**, Molecular and Cell Biology, Janderstr. 3, Mannheim 68199, GERMANY, Telephone: +49 621 87556 23, Email: christina.blach@biogenerix.com

**Dr. Valérie Jérôme**, University of Bayreuth, Chair for Process Biotechnology, Universitätsstr. 30, Bayreuth 95447, GERMANY, Telephone: +49 921 557372, Email: valerie.jerome@uni-bayreuth.de

**Dr. Yun Jiang**, Biovitrum AB, Lindhagensgatan 133, Stockholm 112 76, SWEDEN, Telephone: +46 8 6972647, Email: yun.jiang@biovitrum.com

**Mr. Miguel Jimenez**, Genetix Ltd., Queensway, New Milton, Hampshire BH25 5NN, UK, Telephone: +44 1425 624 629, Email: miguel.jimenez@genetix.com

**Mr. Alexander Jockwer**, Research Centre Juelich GmbH, Biotechnology Institute 2, Leo-Brandt-Strasse, Juelich D-52425, GERMANY, Telephone: +49 2461 61 3948, Email: a.jockwer@fz-juelich.de

**Dr. Klaus Joeris**, Bayer Healthcare LLC, 800 Dwight Way, B28A, Berkeley, CA 94701, USA, Email: klaus.joeris.b@bayer.com

**Mr. Laust Bruun Johnsen**, Novo Nordisk, novo alle, bygn 6B3.99, Bagsvaerd 2880, DENMARK, Telephone: +4544422836, Email: lbjh@novonordisk.com

**Dr. Mike Johnston, Xenova**, 310 Cambridge Science Park, Cambridge Cambs CB4 0WG, UK, Telephone: +44 1223 436 564, Email: mike_johnston@xenova.co.uk

**Mr. Simon Jones**, BioProcess International, One Research Drive, Suite 400A, Westborough, MA MA-1581, USA, Telephone: +1 508 614 1273, Email: sshaffer@bioprocessintl.com

**Mr. Nigel Jones-Blackett**, Irvine Scientific, 2511 Daimler Street, Santa Ana, CA, 92705-5588, USA

**Mr. Ole Zander Jorgensen**, ChemoMetec A/S, Gydevang 43, Alleroed DK-3450, DENMARK

**Mr. Reg Joseph**, Invitrogen Corporation, 3175 Staley Road, Grand Island, NY 14221, USA, Telephone: +1 716 774 3131

**Dr. Christine Jung**, Roche Diagnostics GmbH, Nonnenwald 2, Penzberg 82377, GERMANY, Telephone: +49 8856 60 4533, Email: diana.strakeljahn@roche.com

**Mr. Wolfgang Kahlert**, Sartorius BBI Systems GmbH, Schwarzenberger Weg 73 - 79, Melsungen D-34212, GERMANY

**Ms. Kozue Kaito**, University of Fukui, 3-9-1, Bunkyo, Fukui 910-8507, JAPAN, Telephone: +81 776 27 8645, Email: k-kaito@acbio2.acbio.fukui-u.ac.jp

**Mr. Erik Kakes**, Applikon. BV/AppliSens, De Brauwweg 13, Schiedam 3125 AE, THE NETHERLANDS

**Ms. Rebecca Kale**, Pall Life Sciences, Europa House, Havant Street, Portsmouth PO1 3PD, UK, Telephone: +44 2392 302 309

**Dr. Hela Kallel**, Institut Pasteur de Tunis, 13, place Pasteur. BP 74, Tunis 1002, TUNISIA, Telephone: +21671848903, Email: hela.kallel@pasteur.rns.tn

**Dr. Stephan Kalwy**, Lonza Biologics, 228 Bath Road, Slough SL1 4DX, UK, Telephone: +44 1753 777000, Email: stephan.kalwy@lonza.com

**Prof. Yuto Kamei**, Coastal Bioenvironment Center, Saga University, 152-1 Shonan-cho, Karatsu Saga 847-0021, JAPAN, Email: kameiy@cc.saga-u.ac.jp

**Mr. Takanori Kanayama**, University of Fukui, 3-9-1, Bunkyo, Fukui 910-8507, JAPAN, Telephone: +81 776 27 8645, Email: kanayama@acbio2.acbio.fukui-u.ac.jp

**Mr. Palo Karlsson**, Infors UK Ltd., The Courtyard Business Centre, Dovers Farm, Lonesome Lane, Reigate, Surrey RH2 7QT, UK

**Dr. Cornelia Kasper**, Institut für Technische Chemie Uni Hannover, Callinstr. 3, Hannover 30167, GERMANY, Telephone: +49 511 7622967, Email: kasper@iftc.uni-hannover.de

**Mr. Joerg Kauling**, Bayer Technology Services GmbH, Gebaeude K9, Leverkusen D-51368, GERMANY

**Mr. Jarlath Keating**, HyClone/Perbio Science, Erembodegem-Aalst B-9320, BELGIUM

**Mrs. Alison Keen**, Cambridge Antibody Technology, Milstein Building, Granta Park, Great Abington, Cambridge CB1 6GH, UK, Telephone: +44 1223 898428, Email: susan.messenger@cambridgeantibody.com

**Dr. Klaus Kellings**, NewLab BioQuality AG, Max-Planck-Strasse 15A, Erkrath D-40699, GERMANY

**Dr. Paul Kemp**, Intercytex, Innovation House, Oaks Business Park, Crewe Road, Manchester M23 9QR, UK, Telephone: +44 161 904 4500

**Dr. Ralph Kempken**, Boehringer Ingelheim Pharma GmbH & Co. KG, Birkendorfer Strasse 65, Biberach an der Riss 88397, GERMANY, Telephone: +49 7351 54 4818, Email: ralph.kempken@bc.boehringer-ingelheim.com

**Dr. Gary Kennerley**, Greiner Bio-One Ltd., Brunel Way, Stroudwater Business Park, Stonehouse, Gloustershire GL10 3SX, UK, Email: gary.kennerley@gbo.com

**Jim Kenworthy**, Sartorius - Biotechnology Division, Longmead Business Centre, Blenheim Road, Epsom Surrey KT19 9QQ, UK

**Ms. Nicole Kessler**, Cambrex Bio Science, Parc Industriel de Petit Rechain, Verviers B-4800, BELGIUM

**Mr. Ken Ketley**, JRH Europe, Smeaton Road, West Portway, Andover, Hampshire, UK, Telephone: + 44 1264 333311

**Dr. Jamshad Khan**, Cobra Biomanufacturing Plc, Stephenson Building, The Science Park, Keele Staffordshire ST5 5SP, UK, Telephone: +44 1782 877 298, Email: lisa.dixon@cobrabio.com

**Mr. Gary Khoo**, Animal Cell Technology Group, University of Birmingham, Edgbaston Park Road, Birmingham B15 2TT, UK, Email: SHK378@bham.ac.uk

**Mr. Sung Hyun Kim**, Department of Biological Sciences, Korea Advanced Institute of Science and Technology, Department of Biological Sciences, 373-1 Kusong-Don, Yusong-Gu, Daejeon 305-701, KOREA, Telephone: +82 42 869 2658, Email: holyblack@kaist.ac.kr

**Mr. Makoto Kitano**, Nichirei, JAPAN, Email: kitanom@nichirei.co.jp

**Mr. Christensen Klaus**, F. Hoffmann La Roche, PRBD-E Blg. 70/010, Grenzacherstrasse, Basel 4070, SWITZERLAND, Telephone: +41 61 6884356, Email: klaus.christensen@roche.com

**Mr. Philippe Klein**, LFB, 59 Rue de trevise, BP 2006, Lille Cedex 59011, FRANCE, Telephone: +33 3 20493432, Email: klein@lfb.fr

**Mr. Ralph Klein**, BioReliance Ltd, Todd Campus, West of Scotland Science Park, Glasgow G20 0XA, UK, Telephone: +44 141 946 9999

**Mrs. Anne Klein-Vehne**, Evotec Technologies GmbH, Schnackenburgallee 114, Hamburg 22525, GERMANY, Telephone: +49 40 56081 338, Email: andreas.niewoehner@evotec-technologies.com

**Dr. Christof Knocke**, DASGIP AG, Rudolf-Schulten-Strasse 5, Juelich 52428, GERMANY, Telephone: +49 2461 980 0, Email: c.knocke@dasgip.de

**Mrs. Isabelle Knott**, GlaxoSmithKline Biologicals S. A., rue de Institut 89, Bixensart 1330, BELGIUM, Telephone: + 32 2 656 9890, Email: isabelle.knott@gskbio.com

**Mr. Kevin Kolell**, JRH Europe, Smeaton Road, West Portway, Andover, Hampshire, UK, Telephone: + 44 1264 333311

**Dr. Manfred Koller**, Cyntellect, Inc., 6199 Cornerstone Court, Ste 111, San Diego, CA 92121, USA, Telephone: +1 858 550 1770, Email: mhu@cyntellect.com

**Ms. Bhargavi Kondragunta**, Human Genome Sciences, Inc., 14200 Shady Grove Road, Rockville, MD 20850, USA, Telephone: +1 240 314 4400 ext 1627, Email: bhargavi_Kondragunta@hgsi.com

**Dr. Rashmi Korke**, Biogen Idec, US, Email: rashmi.korke@biogenidec.com

**Dr. Rob Kotin**, Laboratory of Biochemical Genetics, Rockville, MD, USA, Email: kotin@nhlbi.nih.gov

**Dr. Ekaterini Kotsopoulou**, Senior Scientist / Cell Line Development, GSK, 126 Crosslet Vale, London SE10 8DL, UK, Telephone: +44 208 639 6590, Email: nina.a.kotsopoulou@gsk.com

**Mr. Georg Kox**, DASGIP AG, Rudolf-Schulten-Strasse 5, Juelich 52428, GERMANY

**Mr. Beat Kramer**, Institute of Bio- and Chemical Engineering, ETH Zurich, HCI F114, ETH Zurich, Hoenggerberg Campus, Zurich 8093, SWITZERLAND, Email: beat.kramer@gmx.ch

**Ms. Daniella Kranjac**, Wave Europe Pvt. Ltd., IRELAND, Email: info@wavebiotech.com

**Dr. Claus Kristensen**, Novo Nordisk, Novo Alle 6B2.107, Bagsvaerd 2880, DENMARK, Telephone: +4544423572, Email: clak@novonordisk.com

**Mr. Edwin Krowinkel**, Applikon. BV/AppliSens, De Brauwweg 13, Schiedam 3125 AE, THE NETHERLANDS

**Mr. Bernd Krueger**, JRH Europe, Smeaton Road, West Portway, Andover, Hampshire, UK, Telephone: + 44 1264 333311

**Dr. Maya Kuchenbecker**, Molecular Biology, Rentschler, Biotechnologie GmbH, Erwin-Rentschler-Strasse 21, Laupheim 88471, GERMANY, Telephone: +49 7392 701 879, Email: karin.schoch-hofmann@rentschler.de

**Mr. Luc Kupers**, Genzyme Flanders NV, Cipalstraat 8, Geel 2440, BELGIUM, Telephone: +32 1 456 4829, Email: ann.verheyden@genzyme.com

**Dr. Simona La Seta Catamancio**, Molmed S.p.A., via Olgettina, 58, Milan 20100, ITALY, Telephone: +39 02 21277223, Email: simona.laseta@molmed.com

**Mr. Shiraz Ladiwali**, HyClone/Perbio Science, Industrielaan 27, Industriezone III, Erembodegem-Aalst B-9320, BELGIUM, Telephone: +32 53 85 72 27, Email: Sofie.Wijmeersch@perbio.com

**Miss. Michelle LaFond**, Regeneron Pharmaceuticals, 777 Old Saw Mill River Rd., Tarrytown, NY 10591, USA, Telephone: +1 914 345 7387, Email: michelle.lafond@regeneron.com

**Dr. Haley Laken**, Wyeth BioPharma, 1 Burtt Road, Andover, MA 01810, USA, Telephone: +1 978 247 2864, Email: hlaken@wyeth.com

**Ms. Jutta Lamlé**, University of Bielefeld, Institute of Cell Culture Technology, Universitätsstrasse 25, Technische Fakultät EO, Bielefeld 33605, GERMANY, Telephone: +49 521 601 6338, Email: jla@zellkult.techfak.uni-bielefeld.de

**Dr. Karlheinz Landauer**, Igeneon AG, Brunner Strasse 69/3, Vienna 1230, AUSTRIA, Telephone: +43 699 18904242, Email: karlheinz.landauer@igeneon.com

**Mrs. Silke Langhammer**, ProBioGen AG, Goethestr. 54, Berlin 13086, GERMANY, Telephone: +49 30 9240060, Email: silke.langhammer@probiogen.de

**Ms. Christine Lattenmayer**, Austrian Center of Biopharmaceutical Technology (ACBT), Muthgasse 18, Vienna 1190, AUSTRIA, Email: h9640202@edv2.boku.ac.at

**Dr. Shawn Lawrence**, Regeneron Pharmaceuticals, 777 Old Saw Mill River Rd., Tarrytown, NY 10591, USA, Telephone: +1 914 345 7716, Email: shawn.lawrence@regeneron.com

**Dr. Jemma Lawson**, GroPep, PO Box 10065 BC, Adelaide 5001, AUSTRALIA, Telephone: +61 8 83547787, Email: jemma.lawson@gropep.com.au

**Dr. Arye Lazar**, Scientist, Israel Inst Biol Res, PO Box 19, Ness Ziona 70410, ISRAEL, Telephone: +972 8 9381485, Email: larye@iibr.gov.il

**Ms. Gwenaëlle Le Chapellier**, Stedim, Z. I. Des Paluds, Av. De Jouques, Aubagne F-13781, FRANCE, Telephone: +33 4 42 84 56 31, Email: c-vedova@stedim.fr

**Mr. Richard Leach**, Sigma Aldrich, ZI Chesnes BP 701, St. Quentin Fallavier F-38297, FRANCE

**Mr. Matthew LeClair**, Baxter BioScience, USA, Email: matthew_leclair@baxter.com

**Dr. Brian Lee**, Amgen Inc., One Amgen Center Drive, B18S-1A, Thousand Oaks, CA 91320, USA, Telephone: +1 805 313 4083, Email: brian.lee@amgen.com

**Mr. Yih Yean Lee**, Bioprocessing Technology Institute, 220 Biopolis Way, #06-01 Centros, Singapore 138668, SINGAPORE, Telephone: +65 64788891, Email: lee_yih_yean@bti.a-star.edu.sg

**Prof. Gyun Min Lee**, KAIST, 373-1 Kusong-dong, Yusong-gu, Daejon 305-701, KOREA, Email: gyunminlee@kaist.ac.kr

**Dr. Gene Lee**, Wyeth BioPharma, One Burtt Road, Andover, MA 01820, USA, Telephone: +1 978 247 1963, Email: gwlee@wyeth.com

**Mr. Jong Min Lee**, Samsung Fine Chemicals Co., LTD., Rm 6117, College of Medicine, SKKU, 300, Chunchun-Dong, Jangan-Gu, Suwon, Kyunggi-Do, REPUBLIC OF KOREA, Telephone: +82 31 299 6454, Email: jongmin85.lee@samsung.com

**Mr. Pascal Lefebvre**, Cambrex Bio Science, Parc Industriel de Petit Rechain, Verviers B-4800, BELGIUM

**Prof. Jürgen Lehmann**, Zellkulturtechnik, University of Bielefeld, Technische Fakultät, Postfach 10 01 31, Bielefeld D-33501, GERMANY, Telephone: +49 521 106 6319, Email: jl@zellkult.techfak.uni-bielefeld.de

**Mr. Michael Lehnerer**, Octagene gmbH, Am Klopferspitz 19, Mastinstied 82-152, GERMANY, Telephone: +49 82 700 769 86, Email: lehnerer@octagene.com

**Mrs. Maria João Leite Costa**, CEBQ - Instituto Superior Técnico, Av. Rovisco Pais, Lisboa 1049-001, PORTUGAL, Telephone: +351 218 419 065, Email: mjcosta@ff.ul.pt

**Dr. Mark Leonard**, Wyeth BioPharma, 1 Burtt Rd, Andover, MA 01810, USA, Telephone: +1 978 247 2131, Email: mleonard@wyeth.com

**Ms. Christine Lettenbauer**, Wave Biotech AG, Ringstrasse 24, Tagelswangen 8317, SWITZERLAND

**Ms. Stacy Leugars**, JRH Europe, Smeaton Road, West Portway, Andover, Hampshire, UK, Telephone: +44 1264 333311

**Ms. Patricia Leung-Tack**, Project Manager - Virology platform - Manufacturing Technology, sanofi pasteur, 1541, av. M. Mérieux, Marcy l'Etoile ( Lyon ) 69280, FRANCE, Telephone: +33 4 3737 0918, Email: patricia.leung-tack@sanofipasteur.com

**Ms. Fangyu Li**, Sartorius BBI Systems GmbH, Schwarzenberger Weg 73 - 79, Melsungen D-34212, GERMANY

**Mrs. Elisabeth Lindner**, Octapharma AB, Stockholm S-112 75, SWEDEN, Telephone: +46 8 566 43128, Email: elisabeth.lindner@octapharma.se

**Mrs. Catherine Ljung**, Biovitrum AB, Lindhagensgatan 133, Stockholm 112 76, SWEDEN, Telephone: +46 8 6972913, Email: catherine.ljung@biovitrum.com

**Mr. Klaus Löchner**, SAFC, ZI Chesnes BP 701, St Quentin Fallavier 38297, FRANCE

**Mr. Christophe Losberger**, Serono, 14, chemin des Aulx, Plan-les-Ouates 1228, SWITZERLAND, Telephone: +41 22 7069637, Email: christophe.losberger@serono.com

**Dr. Holger Luebben**, Chiron Vaccines, Emil von Behring Str. 76, Marburg 35041, GERMANY, Telephone: +49 6421 39 5807, Email: holger._uebben@chiron.com

**Dr. Regine Luemen**, Norbitec GmbH, Pinnauallee 4, Uetersen 25436, GERMANY, Telephone: +49 4122 712 426, Email: regine.luemen@norbitec.de

**Dr. Dirk Luetkemeyer**, Bibitec GmbH, Pinnauallee 4, Uetersen 25436, GERMANY, Telephone: +49 4122 712 340, Email: dirk.luetkemeyer@norbitec.de

**Mr. Alfred Luitjens**, Crucell Holland BV, PO BOX 2048, Archimedesweg 4, Leiden 8242CG, THE NETHERLANDS, Telephone: +31 71 5248831, Email: a.luitjens@crucell.com

**Mr. Björn Lundgren**, GE Healthcare, Bjorkgatan 30, Uppsala S-75182, SWEDEN, Telephone: +46 18 612 05 05, Email: bjorn.lundgren@amershambiosciences.com

**Dr. Mats Lundgren**, BioPR&D, AstraZeneca, b841, Södertälje S-15185, SWEDEN, Telephone: +46 8 55253977, Email: Mats. X.Lundgren@astrazeneca.com

**Mr. Shun Luo**, JRH Europe, Smeaton Road, West Portway, Andover, Hampshire, UK, Telephone: +44 1264 333311

**Ms. Belinda Luscombe**, Tissue Therapies Limited, School of Life Sciences, Queensland University of Technology, Level 5, Q Block, 2 George St, Brisbane 4000, AUSTRALIA, Telephone: +617 3864 4071, Email: b.luscombe@tissuetherapies.com

**Mr. Ian Lyall**, Invitrogen, 3 Fountain Drive, Inchinnan Business Park, Paisley PA4 9RF, UK, Telephone: +44 141 814 5814, Email: ian.lyall@invitrogen.com

**Ms. Caroline MacDonald**, Glasgow Caledonian University, City Campus, Cowcaddens Road, Glasgow G4 0BA, UK, Telephone: +44 141 331 3624, Email: caroline.macDonald@gcal.ac.uk

**Mr. Lars Macke**, Department of Molecular Biotechnology, GBF, Mascheroder Weg 1, Braunschweig 38124, GERMANY, Telephone: +49 531 6181293, Email: Lars.Macke@gbf.de

**Dr. Chris Mann**, Genetix Ltd., Queensway, New Milton Hampshire BH25 5NN, UK

**Mr. Rolf Manser**, Bioengineering AG, Sagentainstrasse 7, Wald 8636, SWITZERLAND

**Dr. Annie Marc**, Laboratoire des Sciences du Génie Chimique, CNRS-ENSAIA, 2, av. de la Forêt de Haye, Vandoeuvre-lès-Nancy 54505, FRANCE, Telephone: +33 383595785, Email: Annie.Marc@ensic.inpl-nancy.fr

**Miss. Isabel Marcelino**, ITQB/IBET, Apartado 12, Oeiras 2781-901, PORTUGAL, Telephone: +351 21 4469424, Email: isabelm@itqb.unl.pt

**Ms. Corinne Marchand**, Cambrex Bio Science, Parc Industriel de Petit Rechain, Verviers B-4800, BELGIUM

**Dr. Lara Marchetti**, Greiner Bio-One Ltd., Brunel Way, Stroudwater Business Park, Stonehouse Gloustershire GL10 3SX, UK, Email: lara.marchetti@gbo.com

**Ms. Martine Marigliano**, TRANSGENE, 11 rue de molsheim, Strasbourg 67000, FRANCE, Telephone: +33 3 88 27 91 13, Email: marigliano@transgene.fr

**Ms. Maura Mariko**, Department of Genetic Resources Technology, Kyushu University, 6-10-1 Hakozaki, Higashi-ku, Fukuoka 812-8581, JAPAN, Telephone: +81 92 642 3046, Email: mariko@grt.kyushu-u.ac.jp

**Ms. Julia Markusen**, Merck & Co., Inc, RY80Y-120, PO Box 2000, Rahway, NJ 07065, USA, Telephone: +1 732 594 6021, Email: julia_markusen@merck.com

**Dr. Carl Martin**, Covance, Langhurstwood Road, Horsham Sussex, RH12 4QD, UK, Telephone: +44 1423 500011, Email: carl.martin@covance.com

**Dr. Kristina Martinelle**, Biopharmaceuticals, Octapharma AB, Nordenflychtsvägen 55, Stockholm SE-112 75, SWEDEN, Telephone: +46 8 566 43412, Email: kristina.martinelle@octapharma.se

**Mr. Alfredo Martinez**, Genzyme Flanders NV, Cipalstraat 8, Geel 2440, BELGIUM, Telephone: +32 1 456 4829, Email: ann.verheyden@genzyme.com

**Mr. Ralph Martinke**, Innovatis AG, Meisenstrasse 96, Bielefeld D-33607, GERMANY

**Dr. Uwe Marx**, ProBioGen AG, Goethestr. 54, Berlin 13086, GERMANY, Telephone: +49 30 9240060, Email: uwe.marx@probiogen.de

**Mr. Bernard Massie**, Institut de recherche en biotechnologie, CNRC, 6100, Royalmount ave, Montreal H4P 2R2, CANADA, Telephone: +1 514 496 6131, Email: bernard.massie@cnrc-nrc.gc.ca

**Miss. Ana Matias**, Animal Cell Technology Laboratory, ITQB/IBET, Apartado 12, Oeiras 2780-901, PORTUGAL, Telephone: +351 21 446 9728, Email: amatias@itqb.unl.pt

**Mr. Shinei Matsumoto**, Department of Genetic Resources Technology, Kyushu University, Fukuoka, JAPAN, Telephone: +84 92 642 3046, Email: shin-ei@grt.kyushu-u.ac.jp

**Dr. Hiroshi Matsuoka**, Teikyo University of Science & Technology, 2525 Uenohara, Yamanashi-ken 409-0193, JAPAN, Telephone: +81 554634411, Email: matsuoka@ntu.ac.jp

**Dr. Patricia McAlernon**, Innovatis AG, Meisenstrasse 96, Bielefeld D-33607, GERMANY

**Dr. Martin McCall**, Acyte Biotech, Dept Biotechnology University, of South Wales, Kensington NSW, Sydney 2052, AUSTRALIA, Telephone: +61 2 93853869, Email: m.mccall@unsw.edu.au

**Mr. Kevin McCormack**, Guava Technologies, Guava House, Drope Rd., St. Georges Super Ely, Cardiff CF5 6EP, USA, Telephone: +44 1446 760 112, Email: kmccormack@guavatechnologies.com

**Dr. Joanne McCudden**, Lark Technologies, Langhurstwood Road, Horsham Sussex RH12 4QD, UK, Telephone: +44 1423 500011, Email: j.mccudden@genaissance.com

**Ms. Kristina McEwan**, Moregate TCS Ltd, Botolph Claydon, Buckingham MK18 2LR, UK

**Ms. Sandy McNorton**, JRH Europe, Smeaton Road, West Portway, Andover Hampshire, UK, Telephone: + 44 1264 333311, Email: Sandy.McNorton@jrhbio.com

**Dr. David Mead**, Delta Biotechnology Ltd., Castle Court, 59 Castle Boulevard, Nottingham NG7 1FD, UK

**Ms. Clare Medlow**, JRH Europe, Smeaton Road, West Portway, Andover, Hampshire, UK, Telephone: + 44 1264 333311

**Prof. Ricardo Medronho**, Federal University of Rio de Janeiro, Escola de Química, CT, Bloco E, Rio de Janeiro 21949-900, BRAZIL, Telephone: +55 21 25627635, Email: medronho@eq.ufrj.br

**Dr. Petra Meissner**, Micromet AG, Staffelseestr.2, Munich D-81477, GERMANY, Telephone: +49 89 89 52 77 110, Email: petra.meissner@micromet.de

**Ms. Maria Candida Maia Melado**, Cell Culture Engineering Laboratory, Chemical Engineering Program - COPPE, Federal University of Rio de Janeiro, Bloco G, sala G-115, Rio de Janeiro 21945-970, BRAZIL, Telephone: +55 21 2562 7163, Email: maria@peq.coppe.ufrj.br

**Mr. Jacques Meler**, Stedim S.A., Z. I. Des Paluds, Avenue De Jouques, Aubagne F-13781, FRANCE

**Mr. Michael Mellor-Clark**, BioReliance, Innovation Park, Hillfoots Road, Stirling FK9 4NF, UK, Telephone: +44 1786 451 318, Email: mmellor-clark@bioreliance.com

**Mr. Jim Mercer**, Wyetyh BioPharma, 1 Burtt, Andover, MA, US, Email: jmercer@wyeth.com

**Dr. Otto-Wilhelm Merten**, Genethon, 1, rue de l'Internationale, BP 60, Evry F-91000, FRANCE, Telephone: +33169472590, Email: omerten@genethon.fr

**Dr. Ferruccio Messi**, Cell Culture Technologies, via al Chioso 12, Gravesano 6929, SWITZERLAND, Telephone: +41 91 6045322, Email: info@cellculture.com

**Dr. Jim Mills**, Xenova, 310 Cambridge Science Park, Cambridge Cambs CB4 0WG, UK

**Ms. Jean-Maurice Mimran**, JRH Europe, Smeaton Road, West Portway, Andover,Hampshire, UK, Telephone: + 44 1264 333311

**Dr. Stephen Minger**, King's College London, Stem Cell Biology Laboratory, Centre for Neuroscience Research, Guy's Campus/Hodgkin Building, London SE1 1UL, UK, Telephone: +44 207 848 6169, Email: stephen.minger@kcl.ac.uk

**Dr. Torsten Minuth**, Bayer HealthCare, PH-OP-BT / Elberfeld, Geb. 46, Raum 403, Friedrich-Ebert-Strasse 217, Wuppertal 42096, GERMANY, Telephone: +49 202 36 7531, Email: torsten.minuth@bayerhealthcare.com

**Ms. Regine Mistou Mistou**, Invitrogen, 3 Fountain Drive, Inchinnan Business Park, Paisley PA4 9RF, UK, Telephone: +44 141 814 5814, Email: regine.mistou@invitrogen.com

**Ms. Miriam Monge**, Stedim S.A., Z. I. Des Paluds, Avenue De Jouques, Aubagne F-13781, FRANCE, Telephone: +33 4 42 84 56 31, Email: m-monge@stedim.fr

**Mr. Bryan Monroe**, Invitrogen Corporation, 3175 Staley Road, Grand Island, NY 14221, USA, Telephone: +1 716 774 3131

**Ms. Renae Moomjian**, C-Flex cpt, 4451 110th Ave, North, Clearwater, FL, 33762, USA

**Ms. Angela Moraes**, State University of Campinas, Faculdade de Engenharia Quimica Universidade, Estadual de Campinas Cidade Universitaria Zeferino Vaz, Caixa Postal 6066 CEP 13083-970, Campinas, BRAZIL, Telephone: +55 19 3788 3920, Email: ammoraes@feq.unicamp.br

**Dr. Ana Maria Moro**, Instituto Butantan, Av. Vital Brazil 1500, São Paulo 05503-900, BRAZIL, Telephone: +55 011 3726 7222, Email: anamoro@butantan.gov.br

**Dr. Arvia Morris**, Amgen, 1201 Amgen Court West, Seattle, WA, 98119-3105, USA, Telephone: +1 206 265 7679, Email: morrisae@amgen.com

**Ms. Sarah Michelle Mortellaro**, GroPep Ltd, 28 Dalgleish St Thebarton, Adelaide 5031, AUSTRALIA, Telephone: +61 8 8354 7784, Email: sarah.mortellaro@gropep.com.au

**Dr. Jon Mowles**, CTM BioTech, Babraham Research Campus, Babraham Hall, Babraham, Cambridge CB2 4AT, UK, Telephone: +44 1223 496 070, Email: jon.mowles@ctmbiotech.co.uk

**John Moys**, Sartorius - Biotechnology Division, Longmead Business Centre, Blenheim Road, Epsom Surrey KT19 9QQ, UK, Email: john.moys@sartorius.com

**Dr. Dethardt Mueller**, Austrian Center Of Biopharmaceutical Technologies, c/o Institute of Applied Microbiology, University of Natural Resources and Applied Life Sciences, Muthgasse 18, Vienna A- 1190, AUSTRIA, Telephone: +43 1 36006 6202, Email: dethardt.mueller@boku.ac.at

**Mr. Steffen Mueller**, BioProcess Engineering Services Ltd., 186 Carver Drive, Kent Science Park, Sittingbourne, Kent, UK

**Prof. Masanobu Munekata**, Graduate School of Eng.,Hokkaido Univ., N13, W8, Kita-ku, Sapporo 060-8628, JAPAN, Telephone: +81 11 706 7815, Email: munekata@eng.hokudai.ac.jp

**Mr. Stephen Munt**, Beckman Coulter UK Ltd., Kingsmead Business Park, London Road, High Wycombe HP11 1JU, UK, Telephone: +44 1494 429 196

**Dr. Daniele Murith**, Laboratoires Serono S.A., Zone Industrielle de l'Ouriettaz, Aubonne CH-1170, SWITZERLAND, Telephone: +41 21 821 70 00, Email: daniele.murith@serono.com

**Mr. Raymond Murray**, Corning Life Sciences, Koolhovenlean 12, Schiphol-Rijk 119-N3, THE NETHERLANDS, Telephone: +44 1494 714 949, Email: allenjm@corning.com

**Mr. Nick Musgrove**, Infors UK Ltd., The Courtyard Business Centre, Dovers Farm, Lonesome Lane, Reigate Surrey RH2 7QT, UK, Email: n.musgrove@infors-ht.com

**Mrs. Ann Merete Mygind**, Novo Nordisk A/S, Hagedornsvej 1, HAB1.38, Gentofte 2820, DENMARK, Telephone: +45 44439434, Email: ammy@novonordisk.com

**Mr. Nobuhiro Nagai**, Division of Molecular Chemistry, Graduate School of Engineering, Hokkaido University, N13-W8, Kita-ku, Sapporo, Hokkaido 060-8628, JAPAN, Telephone: +81 11 706 7118, Email: nagai419@eng.hokudai.ac.jp

**Mr. Stefan Naschberger**, Rütten Engineering Ltd., Industriestrasse 9, Stäfa 8712, SWITZERLAND, Telephone: +41 1 928 29 30, Email: info@rutten.com

**Dr. Stephen Navran**, Synthecon, Inc., 8042 El Rio, Houston, TX 77054, USA, Telephone: +1 713 741 2582, Email: navran@synthecon.com

**Mr. Anders Nelving**, BioProcess R&D, AstraZeneca Södertälje, AstraZeneca R&D Södertälje, b841 Biotech Laboratory, Södertälje S-15185, SWEDEN, Telephone: +46 552 51979, Email: anders.nelving@astrazeneca.com

**Dr. Michael Nemecek**, Bayer Technology Services GmbH, Gebaeude K 9, Leverkusen 51368, GERMANY, Email: michael.nemecek@bayertechnology.com

**Ms. Rachel Newby**, Infors UK Ltd., The Courtyard Business Centre, Dovers Farm, Lonesome Lane, Reigate Surrey RH2 7QT, UK

**Mrs. Helen Newton**, GlaxoSmithKline, South Eden Park Road, Beckenham, KENT BR3 3BS, UK, Telephone: +44 20 8639 6454, Email: helen.l.newton@gsk.com

**Ms. Pauline Nicholson**, Pall Life Sciences, Walton Road, Portsmouth PO6 1TD, UK, Telephone: +44 2392 303 438, Email: pauline.nicholson@europe.pall.com

**Ms. Sylvia Niebrügge**, University of Bielefeld, Institute of Cell Culture Technology, Universitätsstrasse 25, Bielefeld 33615, GERMANY, Telephone: +49 521 1066320, Email: sylvia.niebruegge@gmx.de

**Ms. Karin Nitzche**, Greiner Bio-One Ltd., Brunel Way, Stroudwater Business Park, Stonehouse Gloustershire GL10 3SX, UK, Email: karin.nitzche@gbo.com

**Dr. Thomas Noll**, Research Center Julich, Institute of Biotechnology 2, Leo-Brandt-Strasse, Julich 52425, GERMANY, Telephone: +49 2461 613955, Email: th.noll@fz-juelich.de

**Dr. Maria Luisa Nolli**, Areta International s.r.l., Via Roberto Lepetit, 34, Gerenzano I-21040, ITALY, Telephone: +39 02 96489264, Email: sbaldi@aretaint.com

**Mr. Tobias Nottorf**, University of Bielefeld, Institute of Cell Culture Technology, Universitätsstrasse 25, Bielefeld 33615, GERMANY, Telephone: +49 521 1066323, Email: tno@zellkult.techfak.uni-bielefeld.de

**Ms. Lotte Nyman Göransson**, AstraZeneca Bio PR&D, B 841, Södertälje 151 85, SWEDEN, Telephone: +46 8 552 51181, Email: lotte.nyman@astrazeneca.com

**Mr. Chris Oakley**, Chemicon Europe, The Science Centre, Eagle Close, Chandlers Ford, Southampton SO53 4NF, UK

**Mr. Bernadette O'Connel**, Guava Technologies, 25801 Industrail Blvd., Hayward CA 94545, USA

**Prof. Kim O'Connor**, Tulane University, Dept. of Chemical & Biomolecular Engineering, Lindy Boggs Center Room 300, New Orleans, LA, 70118, USA, Telephone: +1 504 865 5740, Email: koc@tulane.edu

**Mr. Shane O'Dwyer**, Valcon Thechnologies Ltd., Ashley Cottage, Tagoe's Mills, Kinsale County Cork, IRELAND, Telephone: +353 87 247 0468, Email: SHANEODWYER@EIRCOM.NET

**Mr. Phil Offin**, The Automation Partnership, York Way, Royston, Hertfordshire SG8 5WY, UK

**Dr. Duk Jae Oh**, Sejong University, Dept. of Bioscience and Biotechnology, 98 Gunja-dong, Gwangjin-ku, Seoul 143-747, SOUTH KOREA, Telephone: +82 2 3408 3764, Email: djoh@sejong.ac.kr

**Mr. Steve Oh**, Bioprocessing Technology Institute, 20 Biopolis Way, No. 06-01 Centros, Singapore 138668, SINGAPORE, Telephone: +65 6478 8888

**Mr. Eduardo Ojito**, Center of Molecular Immunology, Calle 216 esq 15, Atabey, Playa, Ciudad Habana, CUBA, Telephone: +53 7 271 7933, Email: ojito@ict.cim.sld.cu

**Mr. Gary Oliff**, Covance, Langhurstwood Road, Horsham Sussex, RH12 4QD, UK, Telephone: +44 1423 500011, Email: gary.oliff@covance.com

**Dr. Adekunle Onadipe**, Pfizer Global Research & Development, Pfizer Limited, Ramsgate Road, Sandwich, Kent CT13 9NJ, UK, Telephone: +44 1304 61 61 61, Email: kunle.onadipe@pfizer.com

**Dr. Sheldon Oppenheim**, Millennium Pharmaceuticals, 40 Landsdowne Street, Cambridge 02467, USA, Telephone: +1 617 444 1623, Email: sheldon.oppenheim@mpi.com

**Mr. Laurie Overton**, GlaxoSmithKline, 5 Moore Dr., RTP, NC, 27709, USA, Telephone: +1 919 483 6209, Email: laurie.k.overton@gsk.com

**Dr. Sadettin Ozturk**, Centocor Inc, 200 Great Valley Parkway, Malvern, PA 19355, USA, Telephone: +1 610 240 8496, Email: sozturk@cntus.jnj.com

**Mr. Stefan Papadileris**, PAA Laboratories GmbH, Unterm Bornrain 2, Coelbe 35091, GERMANY

**Ms. Sandy Parten**, Irvine Scientific, 2511 Daimler Street, Santa Ana, CA, 92705-5588, USA

**Mr. Chandra Patel**, New Brunswick Scientific (UK) Ltd, 17 Alban Park, Hatfield Road, St. Albans AL4 0JJ, UK, Email: Patel@nbsuk.co.uk

**Ms. Iris Pavenstaedt**, Invitrogen, 3 Fountain Drive, Inchinnan Business Park, Paisley PA4 9RF, UK, Telephone: +44 141 814 5814, Email: iris.pavenstaedt@invitrogen.com

**Dr. Martin Peacock**, BioXplore, 50, Moxon Street, Barnet Hertfordshire, EN5 5TS, UK, Telephone: +44 2 20 8441 6778, Email: mpeacock@helgroup.com

**Dr. Dermot Pearson**, Delta Biotechnology Ltd., Castle Court, 59 Castle Boulevard, Nottingham NG7 1FD, UK, Telephone: +44 115 955 3347, Email: dermot.pearson@aventis.com

**Dr. Nels Eric Pederson**, Biogen Idec, 14 Cambridge Center B6-6, Cambridge, MA 02142, USA, Telephone: +1 617 914 4858, Email: nels.pederson@biogenidec.com

**Mr. Derek Pendlebury**, Charter Medical Ltd., 14 rue du Rhone, 4th floor, Geneva 1204, SWITZERLAND, Telephone: +41 22 819 1836

**Dr. Angelo Perani**, Ludwig Institute for Cancer Research, Biological Production Facility, AMRC Studley road, HSB Level 6, Heidelberg 3084, AUSTRALIA, Telephone: +61 394965463, Email: angelo.perani@ludwig.edu.au

**Dr. Joerg Peters**, Bayer HealthCare AG, PH-OP-BT / Elberfeld, Geb. 46, Raum 403, Friedrich-Ebert-Str. 217, Wuppertal 42096, GERMANY, Telephone: +49 202 36 7531

**Mr. Jorn Meidahl Petersen**, Novo Nordisk, Novo Allé, Bldg. 6B3.99, Bagsvaerd 2880, DENMARK, Telephone: +4544422634, Email: jmp@novonordisk.com

**Ms. Louisa Pickering**, Informa, Telephone House, 69-77 Paul Street, London EC2A 4LQ, UK, Telephone: +44 20 797 5011, Email: louisa.pickering@informa.com

**Ms. Leigh N. Pierce**, PacificGMP, 8810 Rehco Road, Suite 3, San Dieg, CA, 92121, USA, Telephone: 1 619 251 6515, Email: piercebiodev@cox.net

**Dr. Mangino Pierluigi**, Alfa Wassermann S.P.A., Via Ragazzi Del '99, N. 5, Bologna 40133, ITALY, Telephone: +39 51 64 89 592, Email: dirricerchesg@alfawassermann.it

**Mr. Warren Pilbrough**, Merck & Co., Inc., Sumneytown Pike, West Point, PA 19486-0004, USA, Telephone: +1 215 652 6066, Email: warren_pilbrough@merck.com

**Mr. Hervé Pinton**, Sanofi-Pasteur, 1541, av M Mérieux, Marcy l'Etoile ( Lyon ) 69280, FRANCE, Telephone: +33 4 3737 9377, Email: herve.pinton@sanofipasteur.com

**Prof. James Piret**, University of British Columbia, Department of Chemical and Biological Engineering, 2216 Main Mall, Vancouver British Columbia V6T 1Z4, CANADA, Telephone: +1 604 822 5835, Email: jpiret@chml.ubc.ca

**Dr. Barbara Plaimauer**, Baxter AG, Industriestr. 72, Vienna 1220, AUSTRIA, Telephone: +43 1 20100 3681, Email: plaimab@baxter.com

**Mrs. Davone Platz**, Rinat Neuroscience Corporation, 3155 Porter Drive, Palo Alto, CA 94304, USA, Telephone: 1 650 213 5317

**Mr. Christopher Plows**, YSI (UK) Limited, Lynchford House, Lynchford lane, Farnborough Hampshire GU14 6LT, UK, Telephone: +44 12542 514 711, Email: cplows@analyticaltechnologies.co.uk

**Mr. Michael Pohlscheidt**, Roche Diagnostics GmbH, Penzberg, Nonnenwald 2, Penzberg 82377, GERMANY, Telephone: +49 8856 60 7361, Email: michael.pohlscheidt@roche.com

**Dr. Andreas Popp**, MorphoSys AG, Lena-Christ-Strasse 13a, Martinsried/Planegg 82152, GERMANY, Telephone: +49 89 899 27 121, Email: gorissen@morphosys.com

**Mr. Zsolt Popse**, BioProcess Engineering Services Ltd., 186 Carver Drive, Kent Science Park, Sittingbourne, Kent, UK

**Ms. Alison Porter**, Lonza Biologics, 228 Bath Road, Slough SL1 4DX, UK, Telephone: +44 1753 777000, Email: alison.porter@lonza.com

**Dr. Ralf Pörtner**, TU Hamburg-Harburg, Biotechnologie 1, Denickestr. 15, Hamburg 21073, GERMANY, Telephone: +49 40 42878 2886, Email: poertner@tuhh.de

**Mr. Dave Potts**, YSI (UK) Limited, Lynchford House, Lynchford lane, Farnborough, Hampshire GU14 6LT, UK

**Mr. Stuart Prime**, Pfizer Ltd., Ramsgate Road, Sandwich Kent, CT13 9NJ, UK, Email: stuart.prime@pfizer.com

**Mr. John Pring**, Charter Medical Ltd., 14 rue du Rhone, 4th floor, Geneva 1204, SWITZERLAND, Telephone: +41 22 819 1836

**Mr. David Prudhoe**, Cellon SA, 29 AM Bechler, Bereldange, Luxembourg 7213, LUXEMBOURG

**Dr. Gianluca Quintini**, ALTANA Pharma AG, Byk-Gulden-Strasse 2, Konstanz 78467, GERMANY, Telephone: +49 7531 843508, Email: gianluca.quintini@altanapharma.de

**Dr. Pippa Radcliffe**, Oxford BioMedica (UK) Ltd, Medawar Centre, Robert Robinson Avenue, Oxford Science Park, Oxford OX4 4GA, UK, Telephone: +44 1865 783 000, Email: p.radcliffe@oxfordbiomedica.co.uk

**Mr. Soheil Rahmati**, Biogen Idec, One Antibody Way, Oceanside, CA 92056, USA, Telephone: +1 760 231 2153, Email: rahmati@biogenidec.com

**Dr. Colette Ranucci**, Merck Research Labs, WP17-201, 770 Sumneytown Pike, PO Box 4, West Point 19486, USA, Telephone: +1 215 652 6052, Email: colette_ranucci@merck.com

**Mr. Poul Baad Rasmussen**, Maxygen Aps, Agern Allé 1, Hoersholm 2970, DENMARK, Telephone: +45 7020 5550, Email: pbr@maxygen.dk

**Mr. Kenrik Rechmann**, Infors UK Ltd., The Courtyard Business Centre, Dovers Farm, Lonesome Lane, Reigate Surrey RH2 7QT, UK

**Miss. Alison Rees**, Xenova Ltd., 310 Cambridge Science Park, Milton Road, Cambridge, Cambridgeshire CB4 0WG, UK, Telephone: +44 1223 423413, Email: Alison_Rees@Xenova.co.uk

**Prof. Udo Reichl**, Max Planck Institute for Dynamic of Complex Technical Systems, Sandtorstraße 1, Magdeburg 39106, GERMANY, Telephone: +49 391 6110230, Email: behling@mpi-magdeburg.mpg.de

**Mr. Jon Reid**, HyClone/Perbio Science, Industrielaan 27, Industriezone III, Erembodegem-Aalst B-9320, BELGIUM

**Ms. Dorothea Reilly**, Genentech, Inc., 1 DNA Way, Mail stop 32, South San Francisco, CA 94080, USA, Email: der@gene.com

**Dr. Mark Rendall**, Lonza Biologics, 228 Bath Road, Slough SL1 4DX, UK, Telephone: +44 1753 777000, Email: mark.rendall@lonza.com

**Mr. Oscar Repping**, Intervet International B.V., Wim de Körverstraat 35, P.O. Box 31, BOXMEER 5830 AA, THE NETHERLANDS, Telephone: +31 485587600, Email: oscar.repping@intervet.com

**Mr. Anthony Richards**, HyClone/Perbio Science, Industrielaan 27, Industriezone III, Erembodegem-Aalst B-9320, BELGIUM

**Dr. Andreas Richter**, NewLab BioQuality AG, Max-Planck-Strasse 15A, Erkrath D-40699, GERMANY

**Ms. Alison Ridley**, Cambridge Antibody Technology, Milstein Building, Granta Park, Abington CB1 6H, UK, Telephone: +44 1223 898737, Email: alison.ridley@cambridgeantibody.com

**Mr. Philip Ridley-Smith**, Cobra Biomanufacturing Plc, Stephenson Building, The Science Park, Keele Staffordshire ST5 5SP, UK, Telephone: +44 1782 877 298, Email: lisa.dixon@cobrabio.com

**Mr. Joachim Ritter**, MPI for Dynamics of Complex Technical Systems, Sandtorstrasse 1, Magdeburg 39106, GERMANY, Telephone: +49 391 6110 227, Email: ritter@mpi-magdeburg.mpg.de

**Mr. Graham Roberts**, Origin Pharmaceutical Services Ltd., 20 Milton Park, Abingdon Oxfordshire OX14 4SH, UK, Telephone: +44 1235 437 400, Email: groberts@originpharm.com

**Mr. Glenn Robinson**, Sartorius BBI Systems GmbH, Schwarzenberger Weg 73 - 79, Melsungen D-34212, GERMANY

**Mr. Alan Robinson**, BIOQUELL UK Limited, 34 Walworth Road, Andover Hants SP10 5AA, UK

**Ms. Niamh Roche**, Wyeth Medica Ireland, Grange Castle Business Park, Clondalkin, Dublin 22, IRELAND, Telephone: +353 1 469 7952, Email: rochen@wyeth.com

**Mr. Robert Rodewald**, Evotec Technologies, Schnackenburgallee 114, Hamburg 22525, GERMANY, Telephone: +49 40 56081 338, Email: andres.niewoehner@evotec-technologies.com

**Dr. Seth Rodgers**, BioProcessors Corp, 12-A Cabot Road, Woburn, MA, 01801, USA

**Miss. Teresa Rodrigues**, ITQB/IBET, Av. da República (EAN), Oeiras 2781-901, PORTUGAL, Telephone: +351 214469426, Email: teresar@itqb.unl.pt

**Mr. Juli Rodriguez Bagó**, Universitat Autònoma de Barcelona, Department Enginyeria Quimica, Edifici C, Campus UAB, Cerdanyola del Vallés, Barcelona 08193, SPAIN, Telephone: +34 935 811808, Email: julio.rodriguez@uab.es

**Mr. Marcel Röll**, Wave Biotech AG, Ringstrasse 24, Tagelswangen 8317, SWITZERLAND

**Mr. Thomas Rose**, ProBioGen AG, Goethestr. 54, Berlin 13086, GERMANY, Telephone: +49 30 9240060, Email: thomas.rose@probiogen.de

**Ms. Monika Rothweiler**, Rütten Engineering, Industriestrasse 9, STAFA 8712, SWITZERLAND, Email: monika.rothweiler@rutten.com

**Mr. Simon Routledge**, Eden Biodesign, Unit D5, Stanlaw Abbey Business Centre, Dover Drive, Ellsemere Port Cheshire CH65 9BF, UK, Telephone: +44 151 356 5632, Email: simon.routledge@edenbiodesign.com

**Mr. Paul Rowan**, PAA Laboratories GmbH, Unterm Bornrain 2, Colbe D-35091, GERMANY

**Mrs. Vibeke Rowell**, NUNC A/S, Kamstrupvej 90, Postbox 280, Roskilde DK-4000, DENMARK

**Mr. Guido Rudolph**, Institute for Technical Chemistry, University of Hannover, Callinstrasse 3, Hannover 30167, GERMANY, Telephone: +49 511 762 2381, Email: rudolph@iftc.uni-hannover.de

**Mr. Kurt Russ**, Cell Culture / USP Production, Rentschler, Biotechnologie GmbH, Erwin-Rentschler-Strasse 21, Laupheim 88471, GERMANY, Telephone: +49 7392/701 879, Email: karin.schoch-hofmann@rentschler.de

**Dr. Kurt Rütten**, Rütten Engineering Ltd., Industriestrasse 9, Stäfa 8712, SWITZERLAND, Telephone: +41 1 928 29 30, Email: info@rutten.com

**Mr. Michael Rütten**, Rütten Engineering Ltd., Industriestrasse 9, Stäfa 8712, SWITZERLAND, Telephone: +41 1 928 29 30, Email: info@rutten.com

**Mr. Thomas Rysiok**, MERCK KGaA, Glob. Technol. / TRB, A17/311, Frankfurter Strasse 250, Darmstadt D-64271, GERMANY, Telephone: +49 6151 72 7898, Email: thomas.rysiok@merck.de

**Mrs. Sónia Sá Santos**, ITQB/IBET, Apartado 12, Oeiras 2781-901, PORTUGAL, Telephone: +351 21 4469434, Email: sasantos@itqb.unl.pt

**Mr. Hiroyuki Saitoh**, Kirin Brewery Co., Ltd., 100-1 Hagiwara, Takasaki 370-0013, JAPAN, Telephone: +81 27 353 7386, Email: hsaitoh@kirin.co.jp

**Dr. Patrick Salou**, Novartis Pharma S.A.S., Centre de Biotechnologie, 8, Rue de l'Industrie - DP350, Hunigue Cedex F 68333, FRANCE, Telephone: +33 389 895 224, Email: patrick.salou@novartis.com

**Ms. Tracey Sambrook**, CTM BioTech, Babraham Research Campus, Babraham Hall, Babraham, Cambridge CB2 4AT, UK, Telephone: +44 1223 496 070, Email: ta_sambrook@yahoo.co.uk

**Kalbinder Singh Sandhu**, University of Birmingham, Edgbaston Park Road, Birmingham B15 2TT, UK, Telephone: +44 121 44 3889, Email: kss226@bham.ac.uk

**Dr. Volker Sandig**, ProBioGen AG, Goethestr. 54, Berlin 13086, GERMANY, Telephone: +49 30 9240060, Email: volker.sandig@probiogen.de

**Mr. Enric Sarró Casanovas**, Universitat Autònoma de Barcelona, C/Sant Josep 56 2n 3a, Manresa, Barcelona 08240, SPAIN, Telephone: +34 935811808, Email: enric,sarro@uab.es

**Dr. Kedarnath Sastry**, Biocon Limited, 20th KM, Hosur Road, Electronic City P.O., Bangalore 560100, INDIA, Telephone: +918028082306, Email: kedarnath.sastry@biocon.com

**Mr. Mitch Scanlan**, BioProcessors Corporation, 12-A Cabot Road, Woburn, MA, USA

**Dr. Juergen Schaerfe**, Schaerfe System GmbH, Kraemerstrasse 22, Reutlingen 72764, GERMANY, Telephone: +49 7121 38 78 60, Email: mail@casy-technology.com

**Ms. Janina Schaper**, Research Centre Juelich GmbH, Biotechnology Institute 2, leo-Brandt-Strasse, Juelich D-52425, GERMANY, Telephone: +49 2461 61 2255, Email: j.schaper@fz-juelich.de

**Mrs. Dana Schiefelbein**, Institut fuer Technische Chemie, Callinstr. 3, Hannover 30167, GERMANY, Email: Dana.Schiefelbein@gmx.de

**Dr. Stefan Schlatter**, University of Queensland, Bldg. 74, Chemical Engineering, Brisbane Queensland 4072, AUSTRALIA, Telephone: +61 7 33469490, Email: s.schlatter@uq.edu.au

**Dr. Georg Schmid**, F. Hoffmann-La Roche AG, Preclinical Biotechnology Bldg. 66/112A, Basel 4070, SWITZERLAND, Telephone: +41 61 688 2986, Email: georg.schmid@roche.com

**Mr. Joerg Schmidt**, Novartis Pharma AG, WKL-681.1.42, Werk Klybeck, Basle CH-4002, SWITZERLAND, Telephone: +41 61 69 62 8 25, Email: joerg.schmidt@pharma.novartis.com

**Mr. Heinz Schmidt**, Greiner Bio-One Ltd., Brunel Way, Stroudwater Business Park, Stonehouse Gloustershire, GL10 3SX, UK, Email: heinz.schmidt@gbo.com

**Dr. Sebastian Schmidt**, Bayer Technology Services GmbH, Gebaeude K 9, Leverkusen D-51368, GERMANY

**Prof. Yves-Jacques Schneider**, Laboratoire de Biochimie cellulaire, Pl. L. Pasteur, 1, Louvain-la-Neuve B 1348, BELGIUM, Telephone: +32 10 47 27 91, Email: yjs@bioc.ucl.ac.be

**Dr. Karl-Heinz Schneider**, Bayer HealthCare, PH-OP-BT / Elberfeld, Geb. 46, Raum 403, Friedrich-Ebert-Strasse 217, Wuppertal 42096, GERMANY, Telephone: +49 202 36 7531, Email: berthold.boedeker@bayerhealthcare.com

**Dr. Erik Schneider**, Miltenyi Biotec GmbH, Friedrich-Ebert-Str 68, Bergisch Gladbach 51429, GERMANY, Telephone: +49 221 8306 764, Email: eriks@miltenyibiotec.de

**Mr. Christopher Schofield**, GlaxoSmithkline, New Frontires Science Park North, Third Avenue, Harlow CM19 5AW, UK, Email: Christopher.A.Schofield@gsk.com

**Mr. Michael Schomberg**, University of Bielefeld, Institute of Cell Culture Technology, PO Box 100131, Bielefeld 33501, GERMANY, Telephone: +49 521 1066338, Email: msc@zellkult.techfak.uni-bielefeld.de

**Mrs. Kornelia Schriebl**, Austrian Center of Biopharmaceutical Technology, Muthgasse 18, Vienna 1190, AUSTRIA, Telephone: +43 1 36006 6229, Email: h9540012@edv1.boku.ac.at

**Dr. Bernd Schröder**, Miltenyi Biotec GmbH, Robert Koch Str. 1, Teterow 17166, GERMANY, Telephone: +49 3996 158 246, Email: bernds@miltenyibiotec.de

**Mr. Carl Schrott**, SAFC, 3050 Spruce Street, St. Louis, MO 63103, USA, Telephone: +1 314 301 2054, Email: schrott@sial.com

**Mr. Roland Schucht**, German Research Centre for Biotechnology, Mascheroder Weg 1, Braunschweig 38124, GERMANY, Telephone: +49 531 6181 256, Email: roland.schucht@gbf.de

**Dr. Diana Maria Sofia Schuhbauer**, Fa Hoffmann La Roche AG, Biotechnology, Pharma Research, Bldg 66/018, Postfach, Basel 4070, SWITZERLAND, Telephone: +41 61 688 85871, Email: diana.schuhbauer@roche.com

**Mr. Josef Schulze-Horsel**, MPI Dynamics of complex technical systems, Sandtorstrasse 1, Magdeburg 39106, GERMANY, Telephone: +493916110208, Email: horsel@mpi-magdeburg.mpg.de

**Mr. Edwin Schwander**, NUNC A/S, Kamstrupvej 90, Postbox 280, Roskilde DK-4000, DENMARK

**Mr. Jörg Schwinde**, DASGIP AG, Rudolf-Schulten-Strasse 5, Juelich 52428, GERMANY

**Ms. Cheryl Scott**, Bioprocess International Magazine, Oakhouse Business Centre, 39-41 The Parade, Claygate, KT10 0PD, UK, Telephone: +44 1372 471 906, Email: cherylbpi@yahoo.co.uk

**Mr. Anders Sehested**, NUNC A/S, Kamstrupvej 90, Postbox 280, Roskilde DK-4000, DENMARK

**Ms. Victoria Sergeant**, Aggio Partners, UK, Telephone: +44 7786 554177, Email: vs@aggiopartners.com

**Miss. Irene Shackel**, Raven Biotechnologies, Inc., 1140 Veterans Blvd., South San Francisco, CA 94080, USA, Telephone: +1 650 624 2606, Email: ishackel@ravenbio.com

**Dr. Dimpalkumar Shah**, University of Birmingham, 89 Falconhurst Road, Selly Oak, Birmingham B29 6SB, UK, Telephone: +44 777 376 282, Email: dimpalkumar.shah@gmail.com

**Mr. Martyn Shaw**, Lonza Biologics, 228 Bath Road, Slough SL1 4DX, UK, Telephone: +44 1753 777000, Email: martyn.shaw@lonza.com

**Mr. Mick Shaw**, AstraZeneca, B841, Södertälje 15185, SWEDEN, Email: mick.shaw@astrazeneca.com

**Mrs. Amy Shen**, Genetech, Inc., 1 DNA Way, South San Francisco, CA, 94080, USA, Telephone: +1 650 225 6446, Email: shen.amy@gene.com

**Mr. Jerry Shevitz**, BioProcess Engineering Services Ltd., 186 Carver Drive, Kent Science Park, Sittingbourne Kent, UK

**Dr. Najam Shezad**, BioXplore, 50, Moxon Street, Barnet Hertfordshire EN5 5TS, UK

**Dr. Sanetaka Shirahata**, Department of Genetic Resources Technology, Kyushu University, 6-10-1 Hakozaki, Higashi-ku, Fukuoka 812-8581, JAPAN, Telephone: +81 92 642 3045, Email: sirahata@grt.kyushu-u.ac.jp

**Dr. Shilpa Shroff**, BioMarin Pharmacueticals, 105 Digital Drive, Novato, CA 94949, USA, Telephone: +1 415 506 6495, Email: sshroff@bmrn.com

**Miss. Carina Silva**, ITQB/IBET, Apartado 12, Oeiras 2780-901, PORTUGAL, Telephone: + 351 21 44 69 422, Email: carinas@itqb.unl.pt

**Mr. Patrick Simpson**, Innovatis AG, Meisenstrasse 96, Bielefeld D-33607, GERMANY

**Dr. Marty Sinacore**, Wyeth BioPharma, One Burtt Road, Andover, MA 01810, USA, Telephone: +1 978 247 2028, Email: msinacore@wyeth.com

**Mr. Scott Sinclair**, Thermo Electron Corporation, The Exchange, Haslucks Green Road, Shirley, Sulihull B90 2EL, UK, Telephone: +44 1684 311 335, Email:

**Dr. Mallika Singh**, Human Genome Sciences, Inc., 14200 Shady Grove Road, Rockville, MD 20850, USA, Telephone: +1 240 314 4400 ext 1627, Email: mallika_singh@hgsi.com

**Dr. Vijay Singh**, Wave BioTech LLC, 999 Frontier Rd., Bridgewater, NJ 08807, USA, Telephone: +1 908 707 9210, Email: info@wavebiotech.com

**Mr. Kalbinder Singh Sandhu**, University of Birmingham, Biochemical Engineering, Edgbaston, Birmingham B15 2TT, UK, Telephone: +44 121 414 3889, Email: kss226@bham.ac.uk

**Dr. William Sisk**, Biogen Idec, 12 Cambridge Center, B-6-620, Cambridge, MA 02142, USA, Telephone: +1 617 679 2799, Email: william.sisk@biogenidec.com

**Mr. Dick Smit**, Animal Sciences Group WUR, Products Division, PO Box 65, Lelystad 8200AB, THE NETHERLANDS, Telephone: +31 320 238 675, Email: dick.smit@wur.nl

**Dr. Rodney Smith**, CTM BioTech, Babraham Research Campus, Babraham Hall, Babraham, Cambridge CB2 4AT, UK, Telephone: +44 1223 496 070, Email: rodney.smith@ctmbiotech.co.uk

**Ms. Victoria Smith**, JRH Europe, Smeaton Road, West Portway, Andover, Hampshire, UK, Telephone: + 44 1264 333311, Email: victoria.smith@jrhbio.com

**Mr. Shawn Smith**, Gibco Invitrogen Corporation, 3175 Staley Road, Grand Island, NY 14221, USA, Telephone: +1 716 774 3131, Email:

**Mr. John Smith**, Flowgen Bioscience, Wilford Industrial Estate, Nottingham NG11 7EP, UK, Telephone: +44 7917370333, Email: jsmith@flowgen.co.uk

**Dr. André Sobczyk**, VIVALIS S.A., Chu Hotel Dieu, Place A Ricordeau, Nantes 44093, FRANCE, Telephone: +33 2 2807 3710, Email: andresobczyk@vivalis.com

**Dr. Corinna Sonderegger**, Sandoz GmbH BTD/cell culture, Plant Schaftenau, Building 520, Biochemiestrasse 3, Langkampfen A-6336, AUSTRIA, Telephone: +43 5372 6996 5049, Email: corinna.sonderegger@gx.novartis.com

**Dr. Zhi Wei Song**, Bioprocessing Technology Institute, 20 Biopolis Way, #06-01 Centros, Singapore 138668, SINGAPORE, Telephone: +6564788844, Email: song_zhiwei@bti.a-star.edu.sg

**Miss. Erika Spens**, Biotechnology, The Royal Institute of Technology, School of Biotechnology, AlbaNova University Centre, Stockholm S-106 91, SWEDEN, Email: erika@biotech.kth.se

**Mr. Robert Spokane**, YSI (UK) Limited, Lynchford House, Lynchford lane, Farnborough Hampshire GU14 6LT, UK, Telephone: +1 937 767 7241x230, Email: RSpokane@ysi.com

**Dr. Glyn Stacey**, National Institute for Biological Standards and Control, Blanche Lane, South Mimms, Potters Bar, Herts EN6 3QG, UK, Telephone: +44 1707 641000, Email: gstacey@nibsc.ac.uk

**Mr. Jerry Stadt**, C-Flex cpt, 4451 110th Ave, North, Clearwater, FL 33762, USA, Telephone: +1 727 531 4191, Email: glstadt@c-flex-cpt.com

**Dr. Olaf Stamm**, NewLab BioQuality AG, Max-Planck-Strasse 15A, Erkrath D-40699, GERMANY

**Mr. Mike Stanforth**, Infors UK Ltd., The Courtyard Business Centre, Dovers Farm, Lonesome Lane, Reigate Surrey RH2 7QT, UK, Telephone: +44 1737 223 100, Email: m.stanforth@infors-ht.com

**Mr. Scott Stansfield**, UNIVERSITY OF QUEENSLAND, CHEMICAL ENGINEERING, BLDG. 74, St Lucia Queensland 4072, AUSTRALIA, Telephone: +61 7 33653568, Email: s.stansfield@uq.edu.au

**Mr. Jan Steels**, Cellon SA, 29 AM Bechler, Bereldange, Luxembourg 7213, LUXEMBOURG

**Mr. John Sterling**, Genetic Engineering News, 2 Madison Avenue, Larchmont, NY 10538, USA

**Dr. Beate Stern**, UniTargeting Research AS, HIB, Thormohlensgate 55, Bergen NO-5008, NORWAY, Email: beate@unitargeting.com

**Mr. Josef Stettner**, SANDOZ GmbH / BTD - cell culture, Plant Schaftenau, Building 520, Biochemiestr. 10, Langkampfen 6336, AUSTRIA, Telephone: +43 5372 6996 5046, Email: josef.stettner@gx.novartis.com

**Ms. Katie Stewart**, Human Genome Sciences, Inc., 14200 Shady Grove Road, Rockville, MD 20850, USA, Telephone: +1 240 314 4400 ext 1627

**Mr. Scott Storms**, Irvine Scientific, 2511 Daimler Street, Santa Ana, CA 92705-5588, USA

**Mr. Patrick Stragier**, Cambrex Bio Science, Parc Industriel de Petit Rechain, Verviers B-4800, BELGIUM

**Dr. Lars Strandberg**, Biovitrum AB, Lindhagensgatan 133, Stockholm SE 112 76, SWEDEN, Telephone: +46 8 697 2675, Email: lars.strandberg@biovitrum.com

**Mr. Claudio Strebel**, CePower GmbH, Einsiedlerstrasse 29, Wädenswil 8820, SWITZERLAND, Telephone: +41 43 4778422, Email: c.strebel@cepower.ch

**Mr. Marc Strijbos**, HyClone/Perbio Science, Industrielaan 27, Industriezone III, Erembodegem-Aalst B-9320, BELGIUM

**Mr. Marc Stuerz**, Innovatis AG, Meisenstrasse 96, Bielefeld D-33607, GERMANY

**Ms. Kirstin Suck**, Institut für Technische Chemie, Universität Hannover, Callinstrasse 3, Hannover 30167, GERMANY, Telephone: +49 511 762 2966, Email: suck@iftc.uni-hannover.de

**Mr. Neil Sullivan**, Yorkshire Forward, Yorkshire and Humber Regional Development Agency, Victoria House, Victoria Place, Leeds LS11 5AE, UK, Telephone: +44 1226 209 952, Email: neil.sullivan@yorkshire-forward.com

**Mr. Shrikumar Suryanarayan**, Biocon Limited, 20th KM, Hosur Road, Electronic City P.O., Bangalore 560100, INDIA, Telephone: +918028082301, Email: shrikumar.suryanarayan@biocon.com

**Dr. Jelto Swaving**, DSM Biologics, Zuiderweg 72/2, Groningen 9700AL, THE NETHERLANDS, Telephone: +31 50 5222327, Email: jelto.swaving@dsm.com

**Dr. Bernie Sweeney**, UCB, 216 Bath Road, Slough SL1 4EN, UK, Telephone: +44 1753 534 655 x 2218

**Dr. Berthold Szperalski**, Roche Diagnostics GmbH, Penzberg, Nonnenwald 2, Penzberg 82377, GERMANY, Telephone: +49 8856 60 2835, Email: berthold.szperalski@roche.com

**Dr. Hisahiro Tabuchi**, Chugai Pharmaceutical Co., LTD, 5-5-1 Ukima Kita-ku, Tokyo 115-8543, JAPAN, Telephone: +81 3 3968 8602, Email: tabuchihsh@chugai-pharm.co.jp

**Mr. Naoki Takada**, University of Fukui, 3-9-1, bunkyo, Fukui 910-8507, JAPAN, Telephone: +87 776 27 8645, Email: takada_naoki1004@yahoo.co.jp

**Mr. Yoshinori Takagi**, Chugai Pharmaceutical Co., Ltd., 5-5-1 Ukima Kita-ku, Tokyo 115-8543, JAPAN, Telephone: +81 3 3968 8599, Email: takagiysn@chugai-pharm.co.jp

**Dr. Mylene Talabardon**, Laboratoires Serono SA, ZiB, Fenil-sur-Corsier 1809, SWITZERLAND, Email: mylene.talabardon@serono.com

**Dr. Ian Taylor**, Genetix Ltd., Queensway, New Milton Hampshire BH25 5NN, UK, Telephone: +44 1425 624 629, Email: ian.taylor@genetix.com

**Mr. Ian Tedder**, Lonza Biologics, 228 Bath Road, Slough SL1 4DX, UK, Telephone: +44 1753 777000, Email: hilary.metcalfe@lonza.com

**Prof. Satoshi Terada**, University of Fukui, 3-9-1, Bunkyo, Fukui 910-8507, JAPAN, Telephone: +81 776 27 8645, Email: terada@acbio2.acbio.fukui-u.ac.jp

**Miss. Linda Thompson**, HyClone/Perbio Science, Industrielaan 27, Industriezone III, Erembodegem-Aalst B-9320, BELGIUM

**Mrs. Sian Thompson**, Eden Biodesign, Unit D5, Stanlaw Abbey Business Centre, Dover Drive, Ellsemere Port Cheshire CH65 9BF, UK, Telephone: +44 151 356 5632, Email: sian.thompson@edenbiodesign.com

**Ms. Ann Thompson**, Aggio Partners, UK, Telephone: +44 7786 554177, Email: at@aggiopartners.com

**Mr. Stephen Tingley**, BioProcessors Corporation, 12-A Cabot Road, Woburn, MA, USA

**Ms. Anne Bondgaard Tolstrup**, Symphogen A/S, Elektrovej 375, Kgs Lyngby 2800, DENMARK, Telephone: +45 4526 5050, Email: abt@symphogen.com

**Mr. Tomohiro Toyosawa**, Department of Applied Chemistry and Biotechnology, University of Fukui, 3-9-1, Bunkyou, Fukui 910-8507, JAPAN, Telephone: +81 776 27 8645, Email: toyosawa@acbio2.acbio.fukui-u.ac.jp

**Mrs. Manisha Trivedi**, The Williamsburg BioProcessing Foundation, PO Box 1229, Virginia Beach, VA, 23451, USA, Telephone: +1 757 423 8823, Email: mtrived@wilbio.com

**Ms. Evelyn Trummer**, Austrian Center of Biopharmaceutical Technology (ACBT), Muthgasse 18, Vienna 1190, AUSTRIA, Telephone: +43 1 36006 6587, Email: h9740207@edv1.boku.ac.at

**Mrs. Karin Tsiobanelis**, AstraZeneca R&D Lund, Scheelevägen 8, Lund S-221 87, SWEDEN, Telephone: +46 46 336050, Email: karin.tsiobanelis@astrazeneca.com

**Ms. Hilary Turnbull**, Genetic Engineering News, 2 Madison Avenue, Larchmont, NY 10538, USA

**Mr. Peter Turner**, Cambrex Bio Science, Parc Industriel de Petit Rechain, Verviers B-4800, BELGIUM

**Mr. Angus Turner**, Apptec, 4 Merchiston Crescent, Edinburgh EH10 5AN, UK, Telephone: +44 131 2211323, Email: angus.turner@apptecis.com

**Mr. Ola Tuvesson**, AstraZeneca, B841, Södertälje 15185, SWEDEN, Email: ola.tuvesson@astrazeneca.com

**Mr. Jochen Uhlenkueken**, Innovatis AG, Meisenstrasse 96, Bielefeld D-33607, GERMANY

**Dr. Michele Underhill**, University of Kent, Department of Biosciences, Canterbury CT2 7NJ, UK, Telephone: +44 1227 764 000 ext. 3562, Email: m.f.underhill@kent.ac.uk

**Dr. Florian Unterluggauer**, Sandoz, Biochemiestrasse 10, Werk Schaftenau, Langkampfen A-6336, AUSTRIA, Telephone: +43 5372 6996 5470, Email: florian.unterluggauer@gx.novartis.com

**Dr. Todd Upton**, Corning, Inc., 2 Alfred Rd., Kennebunk, ME, 04043, USA, Telephone: +1 2007 985 5315, Email: uptonTM@corning.com

**Dr. Ulrich Valley**, Chiron Vaccines, P.O. Box 16 30, Marburg 35006, GERMANY, Telephone: + 49 6421 39 5337, Email: Ulrich_Valley@chiron.com

**Dr. Jana van de Goor**, Genentech, Inc., 1 DNA way, South San Francisco, CA 94080, USA, Telephone: +1 650 225 2018, Email: goor@gene.com

**Dr. Rene Van de Griend**, Biocult, Niels Bohrweg 11-13, Leiden 2333 LA, THE NETHERLANDS, Telephone: +31 71 521 5443, Email: biocult@sand&courses.nl

**Mr. Hans van den Berg**, Applikon. BV/AppliSens, De Brauwweg 13, Schiedam 3125 AE, THE NETHERLANDS, Telephone: +31 10 2983576, Email: hbe@applikon.com

**Mr. John van der Veeken**, J M Separations BV, Hondsruglaan 99c, PO Box 6286, Eindhoven 5600 HG, THE NETHERLANDS, Telephone: +31 40 290 1570, Email: jvanderveeken@jmseparations.com

**Mrs. Christina A.M. van der Velden**, NVI (Netherlands Vaccine Institute ), A. van Leeuwenhoeklaan 11, P.O.Box 457, Bilthoven 3720 AL, THE NETHERLANDS, Telephone: +31 30 2742360, Email: Tiny.van.der.Velden@nvi-vaccin.nl

**Mr. Jerry van Diest**, BioProcess Engineering Services Ltd., 186 Carver Drive, Kent Science Park, Sittingbourne Kent, UK

**Ms. Miranada van Iersel-Snijder**, Kerry Bioscience, Huizerstraatweg 28, Naarden 1411 GP, THE NETHERLANDS

**Dr. David Venables**, BioReliance Ltd, Todd Campus, West of Scotland Science Park, Glasgow G20 0XA, UK, Email: dvenables@bioreliance.com

**Mr. Francis Ver Hoeye**, Henogen, Rue des Professeurs Jeener et Brachet 12, Charleroi B-6041, BELGIUM

**Mr. Farlan Veraitch**, University College London, Advanced Centre for biochemical Engineering, Department of Biochemical Engineering, Torrington Place, London WC1E 7JE, UK, Telephone: +44 20 7679 3785

**Dr. Ute Vespermann**, Corning Life Sciences, Koolhovenlean 12, Schiphol-Rijk 119-N3, THE NETHERLANDS, Telephone: +44 1494 714 949, Email: allenjm@corning.com

**Mr. Simon Vincent**, JRH Europe, Smeaton Road, West Portway, Andover Hampshire, UK, Telephone: +44 1264 333311

**Mr. Tom Vink**, Genmab B.V., Yalelaan 60, Utrecht 3584CM, NETHERLANDS, Telephone: +31 30 2123313, Email: t.vink@nl.genmab.com

**Dr. Joaquim Vives**, Centre Development in Stem Cell Biology, Institute for Stem Cell Research, University of Edinburgh, King's Buildings, West Mains Road, Edinburgh EH9 3JQ, UK, Telephone: +44 131 651 7249, Email: quimv@hotmail.com

**Dr. Angelika Viviani**, Hochschule Wädenswil, Einsiedlerstrasse 29b, Wädenswil 8820, SWITZERLAND, Telephone: +41 1 789 9717, Email: a.viviani@hsw.ch

**Dr. Bénédicte Vonach**, Novartis Pharma AG, WKL-681.1.05, Basel 4002, SWITZERLAND, Telephone: +41 61 696 12 70, Email: benedicte.vonach@pharma.novartis.com

**Dr. Jürgen Vorlop**, Chiron Behring GmbH & Co KG, Emil-von-Behring-Str. 76, Marburg 35041, GERMANY, Telephone: +49 6421 39 4188, Email: juergen_vorlop@chiron.com

**Prof. Roland Wagner**, Miltenyi Biotec GmbH, Robert-Koch-Straße 1, Teterow 17166, GERMANY, Telephone: +49 3996 158 260, Email: rolandw@miltenyibiotec.de

**Dr. Richard Wales**, The Automation Partnership, York Way, Royston Hertfordshire SG8 5WY, UK

**Mr. Jeff Walker**, Helena Biosciences, 3 Steeple Close, Wigginton York O32 2FQ, UK, Telephone: +44 7841 215327, Email: jimwalker@helena-biosciences.com

**Ms. Jo Wallace**, Quest Biomedical Limited, The Exchange, Haslucks Green Road, Shirley, Sulihull B90 2EL, UK, Telephone: +44 1684 311 335, Email: jowallace@questbiomedical.com

**Mr. David Wallder**, Aggio Partners, UK, Telephone: +44 7786 554177, Email: dw@aggiopartners.com

**Ms. Tracy Walshe**, Wyeth Medica Ireland, Grange Castle Business Park, Clondalkin, Dublin 22, IRELAND, Telephone: +353 1 469 7952, Email: walshet@wyeth.com

**Ms. Johanna Walter**, Institut für Technishche Chemie, Universität Hannover, Callinstr. 3, Hannover 30167, GERMANY, Telephone: +49 511 762 2955, Email: walter@iftc.uni-hannover.de

**Dr. Shue-Yuan Wang**, Abbott Bioresearch Center, 100 Research Drive, Worcester, MA 01605, USA, Telephone: +1 508 849 2586, Email: shue-yuan.wang@abbott.com

**Mr. Tim Ward**, The Automation Partnership, York Way, Royston Hertfordshire, SG8 5WY, UK

**Dr. Vishal Warke**, Himedia Laboratories PVT. Ltd, 23 Vadhani Industrial Estate, LBS MARG, Ghatkopar (W), Mumbai 400086, INDIA, Telephone: +91 22 2500 1607, Email: vwarke@himedialabs.com

**Mr. Mark Watson**, Raumedic UK, Hermann-Staudinger-Strasse 2, Helmbrechts, GERMANY, Email: mark.watson@raumedic.com

**Mr. Matthew Webb**, Tecan Deutschland GmbH, Theodor-Storm-Strasse 17, Crailsheim D-74564, GERMANY, Email: matthew.webb@tecan.com

**Mr. Urs Weber**, Laboratoires Serono SA, ZiB, Fenil-sur-Corsier 1809, SWITZERLAND, Email: urs.weber@serono.com

**Mr. Visti Wedege**, NUNC A/S, PO Box 280, Kamstrupvej 90, Roskilde DK-4000, DENMARK

**Mr. Dietmar Weilguny**, Symphogen A/S, Elektrovej, building 375, Kgs. Lyngby 2800, DENMARK, Telephone: +45 4526 5050, Email: dw@symphogen.com

**Mr. Nicole Susann Werner**, Novartis Institutes for Biomedical Research, DT/Biomolecules Production, WSJ-506.304, Basel CH-4002, SWITZERLAND, Telephone: +4161 324 3806, Email: Nicole_Susann.Werner@pharma.novartis.com

**Dr. Finn Christoph Wiberg**, Symphogen A/S, DTU, Elektrovej buildung 375, Lyngby DK-2800, DENMARK, Telephone: +45 4526 5078, Email: fcw@symphogen.com

**Dr. Philipp Wiedemann**, Boehringer-Ingelheim Pharma GmbH & Co KG, Birkendorferstr. 65, Biberach 88397, GERMANY, Telephone: +49 7351 54 4035, Email: margret.raichle@bc.boehringer-ingelheim.com

**Ms. Sofie Wijmeersch**, HyClone/Perbio Science, Industrielaan 27, Industriezone III, Erembodegem-Aalst B-9320, BELGIUM, Telephone: +32 53 85 72 27, Email: Sofie.Wijmeersch@perbio.com

**Mr. Bernd Ulrich Wilhelm**, Sartorius BBI Systems GmbH, Schwarzenberger Weg 73 - 79, Melsungen D-34212, GERMANY

**Mr. John Williams**, Aber Instruments Ltd., 5, Science Park, Aberystwyth SY23 3AH, UK

**Ms. Wellae Williams-Dalson**, University College London, Department of Biochemical Engineering, Torrington Place, London WC1E 7JE, UK, Telephone: +44 20 7679 3794, Email: w.williams-dalson@ucl.ac.uk

**Dr. Mark Wilson**, Xenova, 310 Cambridge Science Park, Cambridge Cambs, CB4 0WG, UK

**Dr. Giles Wilson**, Novo Nordisk, Hagedornsvej 1, Gentofte, Løbenhavn G2820, DENMARK, Telephone: +45 44438299, Email: gcw@novonordisk.com

**Mr. Jim Wilson**, Celliance Ltd., 4, Fleming Road, Kirkton Campus, Livingston, West Lothian EH54 7BN, UK

**Dr. Karsten Winkler**, ProBioGen AG, Goethestr. 54, Berlin 13086, GERMANY, Telephone: +49 30 9240060, Email: karsten.winkler@probiogen.de

**Mr. Jacques Winter**, Infors UK Ltd., The Courtyard Business Centre, Dovers Farm, Lonesome Lane, Reigate Surrey RH2 7QT, UK

**Ms. Christelle Wittische**, Genentech, Inc., 1 DNA Way, South San Francisco, CA 94080-4990, USA, Email: crystal@gene.com

**Dr. Harald Wizemann**, Roche Diagnostics GmbH, Nonnenwald 2, Penzberg 82377, GERMANY, Telephone: +49 8856 60 4533, Email: diana.strakeljahn@roche.com

**Prof. Wilfried Woehrer**, Baxter Vaccine AG, Biomedical Research Center, Uferstrasse 15, Orth A-2304, AUSTRIA, Telephone: +43 1 20100 4649, Email: wilfried_woehrer@baxter.com

**Mr. Frank Wolpers**, HyClone/Perbio Science, Industrielaan 27, Industriezone III, Erembodegem-Aalst B-9320, BELGIUM

**Dr. Kathy Wong**, Bioprocessing Technology Institute, 20 Biopolis Way, #06-01 Centros, Singapore 138668, SINGAPORE, Telephone: +65 64788833, Email: kathy_wong@bti.a-star.edu.sg

**Mr. Chee Furng Wong**, Bioprocessing Technology Institute, 20 Biopolis Way, #06-01 Centros, Singapore, Singapore 138688, SINGAPORE, Telephone: +65 6478 8891, Email: danny_wong@bti.a-star.edu.sg

**Mr. Tom Woods**, Greiner Bio-One Ltd., Brunel Way, Stroudwater Business Park, Stonehouse Gloustershire GL10 3SX, UK, Email: tom.woods@gbo.com

**Dr. Jason Wright**, Rinat Neuroscience Corporation, 3155 Porter Drive, Palo Alto, CA 94304, USA, Telephone: +1 650 213 5317, Email: wright@rinatneuro.com

**Mr. Andrew Wu**, University College London, Dept of Biochemical Engineering, Torrington Place, London WC1E 7JE, UK, Email: a.wu@ucl.ac.uk

**Mr. Karsten Wunschel**, Kerry Bioscience, Huizerstraatweg 28, Naarden 1411 GP, THE NETHERLANDS

**Prof. Florian Wurm**, Swiss Federal Institute of Technology, Laboratory of Cellular Biotechnology, EPFL SV-IGBB-LBTC, CH J2-496 (Building CH), Lausanne CH-1015, SWITZERLAND, Telephone: +41 21 693 61 41, Email: florian.wurm@epfl.ch

**Mr. John Wynne**, ML Laboratories PLC, MED IC4, Science Park, Keele ST5 5NL, UK, Telephone: +448450060923, Email: john.wynne@mlresearch.co.uk

**Mr. Koichi Yamamoto**, KIRIN BREWERY Co., Ltd., 100-1 Hagiwara, Takasaki, Gunma 370-0013, JAPAN, Telephone: +81 27 353 7388, Email: ykoichi@kirin.co.jp

**Mr. Tatsuya Yamashita**, University of Fukui, 3-9-1 Bunkyo, Fukui 910-8507, JAPAN, Telephone: +81 776 27 8645, Email: tatsuya19820703@yahoo.co.jp

**Dr. Catherine Anne Yandell**, GroPep, PO Box 10065 BC, Adelaide 5001, AUSTRALIA, Telephone: +61 8 83547765, Email: catherine.yandell@gropep.com.au

**Dr. Xiaoming (Jerry) Yang**, Amgen Inc., one amgen center drive, 18S-1-A, Thousand Oaks, CA 91360, USA, Telephone: +1 805 447 4076, Email: xyang@amgen.com

**Dr. Jeffrey Yant**, Amgen, Inc., One Amgen Center Drive, MS 185-1-A, Thousand Oaks, CA 91320-1799, USA, Telephone: +1 805 447 2690, Email: jyant@amgen.com

**Prof. Miranda Yap**, Bioprocessing Technology Institute, 20 Biopolis Way, #06-01 Centros 138668, SINGAPORE, Telephone: +65 64788888, Email: miranda_yap@bti.a-star.edu.sg

**Mrs. Stefanie Zahn**, Celonic GmbH, Karl-Heinz-Beckurts-Str. 13, Julich 522428, GERMANY

**Dr. Andrey Zarur**, BioProcessors Corporation, 12-A Cabot Road, Woburn, MA, USA, SA

**Dr. Wenlin Zeng**, Abgenix Inc, 6701 Kaiser Dr., Fremont, CA 94555, USA, Telephone: +1 510 284 6466, Email: wenlin.zeng@abgenix.com

**Ms. Nadia Zghoul**, GBF, German Research Center for Biotechnology, Cell Culture Technology Department, Mascheroder Weg 1, Braunschweig 38124, GERMANY, Telephone: +49 531 6181 181, Email: nzg@gbf.de

**Dr. Chun Zhang**, Bayer Biological Products, 800 Dwight Way, Berkeley, CA, 94701, USA, Telephone: +1 510 705 4416, Email: chun.zhang.b@bayer.com

**Dr. Marie Zhu**, Xencor Inc., 111 W Lemon Street, Monrovia, CA 91006, USA, Telephone: +1 626 737 8156, Email: mzhu@xencor.com

**Mr. Andres Zimmer**, Git Verlag, Rossler str. 90, Dormstadt 64283, GERMANY

# CHAPTER I: PERSPECTIVES OF THE BIOTECH INDUSTRY

# The UK bioprocessing sector

Tony Bradshaw[1]

[1] *Affiliation Director, bioProcessUK, BioIndustry Association, 14/15 Belgrave Square, London SW1X 8PS*

**Abstract:** Biopharmaceuticals are the fastest growing sector of the pharmaceutical industry, currently accounting for ten per cent of total pharmaceutical sales, and over one third of drugs in development. They are frequently breakthrough products, where no effective treatment was available before. Biologic medicines are significantly more complex than chemical drugs, costing more to develop and manufacture Thus bioprocessing (devising the process and manufacturing the goods) is important both in its own right, as a high value manufacturing sector, and as a component of the overall biosciences sector.

**Key words:** biopharmaceuticals, bioprocessing, manufacture, process development

## 1. INTRODUCTION

According to the most recent UK government figures the UK bioscience sector currently comprises of over 480 companies, employing 26,000 people, and generating annual revenues of GBP 4 billion. The bioprocessing sub sector is a significant part of the overall bioscience and healthcare sector, consisting of 1,100 companies employing 100,000 staff, with annual revenues of £11 billion.

The range of goods produced in the UK ranges from blood products to recombinant proteins (from mammalian and microbial production systems), cell therapies, tissue replacements, antibodies, fusion proteins, conjugates



*R. Smith (ed.), Cell Technology for Cell Products*, 3–7.

(for example, antibody-targeted radiotherapy and chemotherapy) and enzymes.

The UK is also at the forefront of upcoming developments in manufacturing gene therapy products, DNA vaccines, transgenic animal production platforms, biogenerics, bionanotechnology, tissue engineering and stem cell research.

Two significant infrastructure projects, each five years in gestation, are due to come on stream within the next year. Once they are both up and running, the resources provided by the National Biomanufacturing Centre in Liverpool and Biocampus in Edinburgh will strengthen the UK's bioprocess development and manufacturing capabilities.

All this activity is underpinned by a world class academic base, which is not only generating products and technology platforms, but also developing new bioprocesses and providing ways to improve existing manufacturing techniques. The multidisciplinary expertise on offer covers all aspects of bioprocessing from lab-scale development through to large-scale manufacturing.

## 2. CREATING A NATIONAL BIOPROCESSING COMMUNITY

Despite this strength and depth, the UK is not generally viewed as a leader in bioprocessing. There are several reasons for this misperception, but I believe the key factor is that the component parts of the sector are geographically dispersed. There are pockets of strength throughout the regions in the UK but a single voice is needed.

In the past couple of years the trade body, the BioIndustry Association (BIA), with backing from the Government, has been working to build a national bioprocessing community.

The move to unify the UK bioprocessing sector gathered pace following the investigation into the overall bioscience sector carried out by the Bioscience Innovation and Growth Team (BIGT), led by the BIA, the Department of Trade and Industry and the Department of Health. The BIGT's report, Bioscience 2015, reflected the views of more than 70 individuals working in biosciences and the government has since provided funding for several of the initiatives it proposed.

One such initiative was my appointment as Bioprocessing Industry Development Director under the auspices of the BIA, with a brief to coordinate strategic initiatives, build links within the sector and shape the knowledge infrastructure.

In January 2005 the BIA was awarded further government funding of GBP three million to establish a National Bioprocessing Knowledge

Transfer Network called bioProcessUK. Its key objective is to develop a robust and productive national network that underpins partnerships between academia and industry. I lead that initiative.

## 3. CREATING A POWERHOUSE IN BIOPROCESSING

In the past few years multinational companies including Eli Lilly, MedImmune and Avecia have made significant investments in their UK bioprocessing plants. To ensure that this trend continues, the UK strategy is to become the number one player in the skills, expertise and intellectual property needed to move compounds from research, into manufacturing and the marketplace.

## 4. SPURRING INNOVATION IN BIOPROCESSING

While few may care to admit it, bioprocessing is technically immature and there is potential for huge improvements in existing processes to drive down costs and improve productivity.

Brendan Fish, who is responsible for process development at Cambridge Antibody Technology Group, provides a specific example – that of adopting disposable containers. "We conduct regular reviews to see if there is a business case for setting up our own antibody production plant. At the last review three years ago, the answer was no. Now we are doing the review again and one of the key considerations is the development of disposable technology that enables you to swop stainless steel tanks for single-use plastic bags. This promises to cut the capital cost of setting up a plant and the cost of running it, while increasing the number of products that can be manufactured."

## 5. FUTURE THERAPIES IMPLY EVER MORE COMPLEX BIOPROCESSES

While there is obvious room for improvements in current processes, establishing robust and safe manufacturing processes for novel treatments such as gene and cell therapies will require an even higher level of innovation in bioprocessing techniques.

The UK is recognised for its proactive stance on stem cells, and apart from creating a supportive legislative environment and funding academic research, the Government has provided GBP five million for three multi-partner bioprocessing projects that aim to develop manufacturing processes and delivery systems.

More recently, the Government announced £6 million funding for a variety of industry/academic collaborations in tissue engineering, biopharmaceutical formulation and delivery, and bioscience underpinning bioprocessing.

## 6. A SKILLED LABOUR FORCE IS THE KEY FACTOR IN BIOPROCESSING

Even the major biomanufacturers are finding it difficult to recruit experienced staff, and one of the objectives of UK strategy is to present bioprocessing as an attractive career and entice more people to work in this knowledge intensive area.

Dr John Birch, CSO at Lonza Biologics has been chairing a group of industry representatives to advise the Biotechnology and Biological Sciences Research Council (BBSRC) on its bioprocessing research priorities. The BBSRC has responded positively with a proposal for a bioprocessing research club with the Engineering and Physical Sciences Research Council (EPSRC), DTI and industry. The GBP ten million will focus on funding research partnerships between universities, research councils and industry with one output being high quality researchers in the bioprocessing field.

The club approach, with several universities carrying out research under the direction of a steering committee from several companies, has been tried successfully before in the UK. In the past though, the clubs were asked to concentrate on a specialized topic such as animal cell culture. The new club will have broader horizons than previous academic-industry clubs in the 1980s. The initiative from BBSRC and EPSRC will ease the current skills shortages which are inhibiting business development. The collaborative approach proposed by the BBSRC is vital. To train good people, you have got to do so in the context of leading edge research.

UCL's Centre for Biochemical Engineering is a role model for the rest of the world. All the students get hands on experience in UCL's bioprocessing unit, built at a cost of more than £40 million. For example, the most recent batch of trainees spent time earlier this year collecting missing bioprocess data for a DNA vaccine for pandemic flu.

## 7. PROGRESS TO DATE

The BIGT strategy is succeeding in fostering a community and advancing the UK's profile in bioprocessing. The BIGT report was published in

October 2003 and we are already seeing real results – people are doing something, and there is a collaborative atmosphere. It is particularly helpful in the way it has pulled in academics.

The Bioscience 2015 report has had impact in the UK with all stakeholders showing strong commitment to implementing the bioprocessing recommendations. With continued commitment the UK can play a significant global role in the development of innovative medicines.

# CHAPTER II: TRANSCRIPTION TO SECRETION

# A Novel AIR-induced Gene Expression System in HEK.EBNA Cells and its Applications

Nicola Susann Werner[1], Wilfried Weber[2], Martin Fussenegger[2], and Sabine Geisse[1]

[1] *Novartis Institutes for BioMedical Research, DT/BMP, CH-4002 Basel, Switzerland, email:* nicole_susann.werner@novartis.com, [2]*Institute for Chemical and Bio-Engineering, Swiss Federal Institute of Technology, ETH Hönggerberg, Wolfgang-Pauli Str. 10, CH-8093 Zürich, Switzerland*

**Abstract:** This work describes the first application of an efficient inducible gene expression system in HEK.EBNA cells. The transgene control system of choice is the novel acetaldehyde-inducible regulation (AIR) technology which has recently been shown to modulate transgene levels following exposure of cells to gaseous acetaldehyde (Weber *et al.*, 2004). The AIR technology was engineered for gas-adjustable transgene expression in HEK.EBNA cells. AlcR transactivator and reporter plasmids containing a chimeric mammalian promotor ($P_{AIR}$) were constructed for acetaldehyde-inducible expression of intracellular and secreted proteins. Several highly inducible transactivator encoding cell lines were established. Following transient transfection with reporter constructs and induction with acetaldehyde an up to 80-fold increase in transgene expression was achieved whereas leakiness of the system only accounted for 1-2% of the induction level. Thus, the AIR technology has been successfully applied to HEK.EBNA cells with tight repression and high induction levels. This system can be used for inducible expression of any desired recombinant protein.

One application of the inducible gene expression system is controlled proliferation technology. An inverse correlation between growth rate and productivity of recombinant cell lines has been previously described (Fussenegger *et al.*, 1998). Directing the metabolic capacity of HEK.EBNA cells towards production of recombinant proteins either by expression of cell cycle regulators or anti-apoptotic genes could be advantageous. This project comprises metabolic engineering of HEK.EBNA cells which upon expression of conditional cell cycle regulators uncouple the cell proliferation phase from the protein production phase. The $p27^{Kip1}$ protein has been selected as a



*R. Smith (ed.), Cell Technology for Cell Products*, 11–18.

powerful cell cycle regulator for induction of growth arrest in HEK.EBNA cells.

Data on inducible expression of recombinant proteins as well as on controlled proliferation strategies for enhancement of productivity are presented.

**Key words:** AIR-induced gene expression, HEK.EBNA cells

# 1. INTRODUCTION

Inducible gene expression technologies are widely used in research *in vitro* and *in vivo*. Gene regulation systems based on steroid hormones or antibiotics as inducing agents are commercially available and well established in cell lines such as HeLa and CHO-K1. For the successful HEK.EBNA protein production system based on episomal long-term propagation of several copies of recombinant plasmids no inducible system has been described so far. Former attempts with the commonly used Tet/ON and Tet/OFF systems were unsuccessful for application in large-scale transient expression processes in HEK.EBNA cells. The Tet/ON system applied to HEK.EBNA cells resulted in high background expression levels. In contrast, maximum expression levels were obtained with the Tet/OFF system only after 5-10 days of cultivation following removal of doxycycline. Moreover, the inducible phenotype was lost upon scale-up to large-scale serum-free suspension culture (Geisse *et al.*, unpublished). The novel AIR-adjustable regulation technology based on the chimeric mammalian promoter $P_{AIR}$, a constitutively expressed *Aspergillus nidulans*-derived transactivator protein, and acetaldehyde as inducing agent has been recently described (Felenbok *et al.*, 1988; Flipphi *et al.*, 2002; Weber *et al.*, 2004). In the presence of the inducer the AlcR transactivator protein binds to specific operator modules within $P_{AIR}$ and induces transcription whereas in the absence of acetaldehyde the gene expression remains silent (Figure 1). In this study we describe the first application of this efficient inducible gene expression system in HEK.EBNA cells by engineering of AlcR transactivator and AIR-adjustable HEK.EBNA production cell lines.

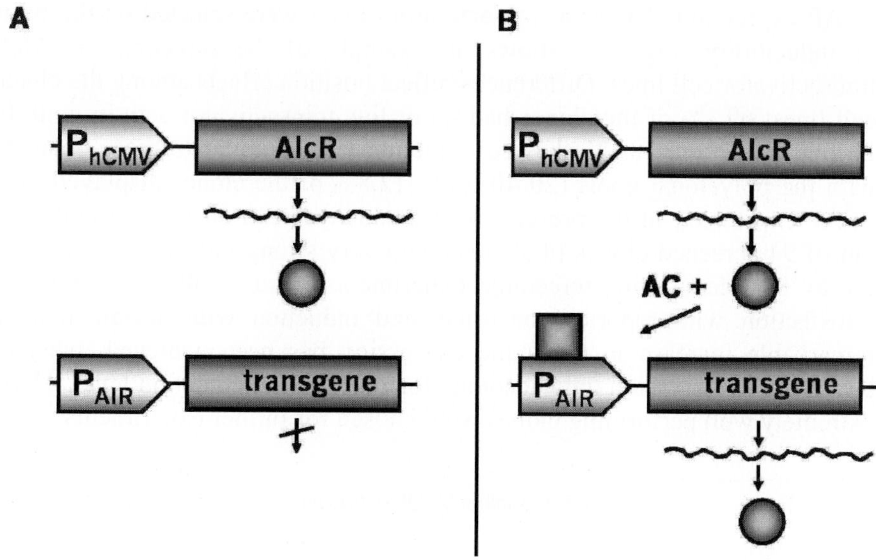


*Figure 1.* Mechanism of AIR-inducible transgene expression (ON-system). A. Absence of inducing compound acetaldehyde (AC). B. Presence of acetaldehyde. $P_{hCMV}$: human cytomegalovirus immediate-early promoter, AlcR: fungal transactivator, $P_{AIR}$: chimeric mammalian promoter engineered for inducible gene expression. Rectangular boxes represent genes, curved lines mRNA, and circles proteins. The putative conformational changes of the AlcR protein upon AC induction are reflected by modified geometric shaping.

# 2. RESULTS AND DISCUSSION

## 2.1 AIR-inducible Transgene Expression

The AIR technology was engineered for gas-adjustable transgene expression in HEK.EBNA cells. AlcR transactivator plasmids were constructed with and without $His_6$-tag and transactivator cell lines were engineered via stable transfections and subsequent cloning steps. 94 stable transactivator clones as well as 80 pools were analysed for inducibility and production rates by transient transfection of $P_{AIR}$-SEAP. This reporter plasmid features a chimeric mammalian promotor ($P_{AIR}$) for acetaldehyde-inducible expression of the human model glycoprotein secreted alkaline phosphatase (SEAP).

After transfection of reporter constructs, cells were incubated for 4 hours following addition of acetaldehyde to the cells. Subsequently, cells were cultivated in culture dishes placed in acetaldehyde-inert 2L polypropylene boxes equilibrated in a 5% $CO_2$ containing humide atmosphere. After 72h hours, cell supernatants were harvested for chemiluminescent analysis of

SEAP expression. The best transactivator clones were selected on the basis of inducibility. Figure 2 shows an example of the screening of AlcR transactivator cell lines. Differences reflect position effects among the clonal cell lines: 69.1% of the clones had no or low transactivator activity with 0-20-fold induction, 18.1% showed induction levels similar to those obtained with the polyclonal pools (20-40-fold), 12.8% of the clones displayed high SEAP expression in the presence of acetaldehyde (40-80-fold). 4 cell lines out of 94 screened clones (4.3%) showed very strong inductive capacity of up to 81-fold in the screening experiments. Thus, following transient transfection with reporter constructs and induction with acetaldehyde a remarkable increase in transgene expression has been achieved, whereas background expression only accounts for 1-2% of the induction level. These extremely well performing clones were chosen for further experiments.

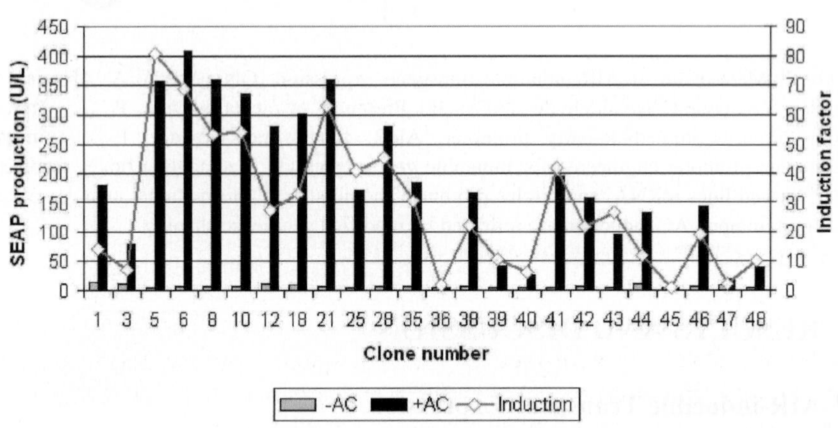


*Figure 2.* Analysis of stable transactivator clones for inducibility and production rates by transient transfection of $P_{AIR}$-SEAP. Transactivator cell lines were seeded at similar densities, transiently transfected with $P_{AIR}$-SEAP and incubated in the absence (-AC) and presence of acetaldehyde (+AC). After 72h supernatants were harvested and assayed for SEAP expression. From these data the induction factor was calculated for each clone.

The inducible system was fine-tuned with regard to parameters such as acetaldehyde concentration, cell density and incubation periods in order to achieve high induction levels. Figure 3 shows an example of three different acetaldehyde concentrations used to induce SEAP expression in HEK.EBNA cells stably expressing the fungal transactivator protein AlcR. Inducible SEAP expression is dependent on the acetaldehyde concentration, reflected by increasing levels of secreted SEAP up to an acetaldehyde concentration of 0.04µl/ml medium. Peak SEAP expression rates of nearly 120-fold compared to control cells were achieved with an inoculation cell density of $5*10^5$/ml while preserving high cell viability (>90%). Thus, AIR technology has successfully been applied to HEK.EBNA cells with tight repression and adjustable medium to high induction levels in dependence on the acetaldehyde concentration and the transactivator cell line used.

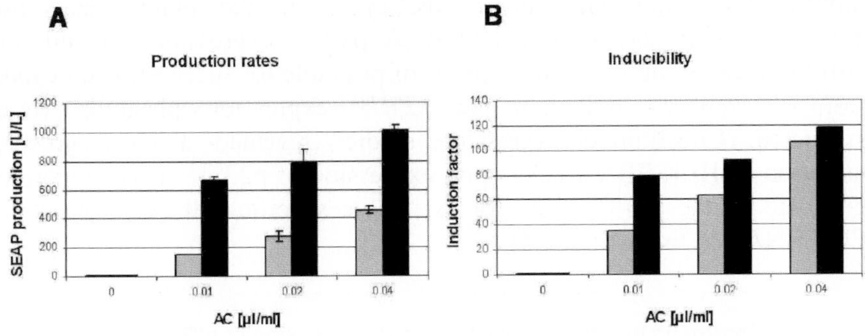


*Figure 3.* Effect of different acetaldehyde concentrations on AIR-inducible SEAP expression (A) and inducibility (B). Two transactivator cell lines (grey and black bars) were seeded at similar densities, transiently transfected with $P_{AIR}$-SEAP and incubated in the absence and presence of acetaldehyde (AC). After 72h supernatants were harvested and assayed for SEAP concentration. From these data the induction factor for each clone was calculated.

## 2.2 Controlled Proliferation Strategies

For several commonly used mammalian expression systems an inverse correlation between growth rate of cells and specific productivity has been demonstrated (Fussenegger *et al.*, 1998) and attempts have been made to uncouple the cell expansion phase from the production phase via controlled proliferation strategies. In such a biphasic production process, cells are first grown to a desired cell density and are then growth arrested to enable maximum production of the protein of choice. Recently, it has been shown that metabolic engineering of CHO and NSO cells overexpressing either anti-apoptotic genes or cell cycle regulators in a regulated fashion results in increased specific protein productivity although volumetric productivity of these cell lines is insufficient without gene amplification of the target gene (Mazur *et al.*, 1998; Meents *et al.*, 2002, Watanabe *et al.*, 2002, Ibarra *et al.*, 2003, Bi *et al.*, 2004). For the HEK.EBNA protein production system a controlled proliferation strategy has as yet not been shown, but, if successful, could significantly impact on protein yields.

We aim to obtain higher cellular productivity by arresting HEK.EBNA cells at the restriction point at the G1/S border of the cell cycle. Early G1 phase cells are characterized by an enlarged cytosol and high metabolic activity and are not committed to undergo DNA replication (Ho and Dowdy, 2002). The restriction point is controlled by the activity of several cyclin-dependent kinases being responsible for Rb phosphorylation and its inactivation in order to permit entry into S-phase (Massagué, 2004). Therefore, candidate growth suppressing genes are members of the kinase inhibitor protein (KIP) family. The $p27^{Kip1}$ protein regulates the cell cycle by inhibiting the checkpoint kinase cdk2/cyclin E and other cdk/cyclin complexes (Sherr and Roberts, 1999). A $p27^{Kip1}$ expression plasmid for constitutive expression serving as proof of principle for successful cell cycle arrest as well as AIR-inducible $p27^{Kip1}$ expression plasmids were constructed. Constitutive expression studies revealed a clear growth retardation of HEK.EBNA cells upon expression of $p27^{Kip1}$. This supported the choice of the $p27^{Kip1}$ gene for G1 phase specific cell cycle arrest in HEK.EBNA cell lines (Figure 4).

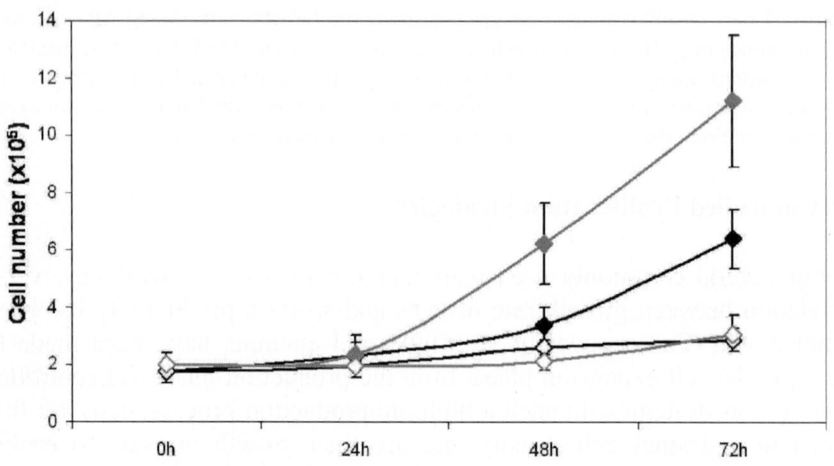


*Figure 4.* Controlled proliferation of HEK.EBNA cells by transient expression of $p27^{Kip1}$. Two HEK.EBNA cell lines (grey and black lines) were seeded at similar densities in triplicate, transiently transfected with $P_{hCMV}$-$p27^{Kip1}$ and cultivated for further 72h. Numbers of control cells (filled squares) and corresponding growth arrested cells (open squares) were determined every 24h.

First results of transient productivity assays using recombinant cell pools revealed that controlling cell proliferation via expression of $p27^{Kip1}$ is possible and can lead to an increase of productivity per cell of tagged recombinant proteins (data not shown). As the results of transient expression studies were very promising the next step of engineering of AIR-adjustable production cell lines was initiated. We generated engineered HEK.EBNA cell lines, which upon expression of the conditional cell cycle regulator $p27^{Kip1}$ uncoupled the cell proliferation phase from the protein production phase. Double/triple stable clonal HEK.EBNA cell lines were developed expressing the transactivator constitutively, $p27^{Kip1}$ inducibly and in part the reporter gene SEAP constitutively.

Stable pools on the genetic background of two different transactivator cell lines were screened for inducibility of $p27^{Kip1}$ by Western Blot analysis, specific cell cycle arrest by FACS analysis and production rates by determination of SEAP expression (data not shown). Figure 5 shows the application of the controlled proliferation technology via induction of $p27^{Kip1}$ in stably engineered AIR-adjustable production cell lines.

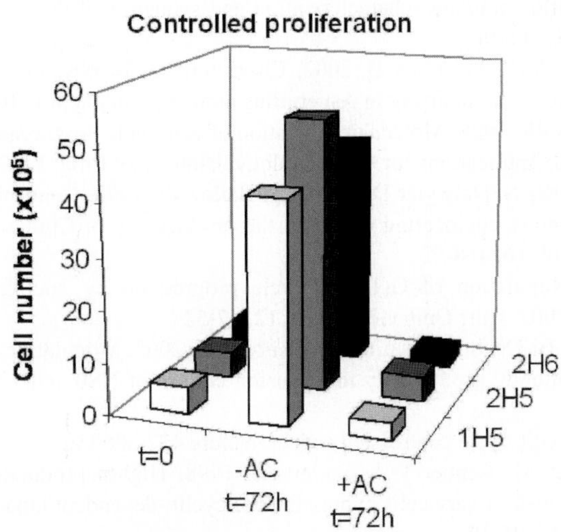


*Figure 5.* Controlled proliferation via induction of $p27^{Kip1}$ in stably engineered AIR-adjustable HEK.EBNA cell lines. Two transactivator cell lines were stably transfected with $P_{AIR}$-$p27^{Kip1}$ and $P_{hCMV}$/SEAP plasmids for inducible and constitutive expression, respectively. Cells were seeded at similar densities and incubated in the absence (-AC) and presence of acetaldehyde (+AC). After 72h cell numbers were determined.

First results indicate that the controlled proliferation technology via induction of $p27^{Kip1}$ leads to an increase of productivity per cell in stably engineered AIR-adjustable SEAP production cell lines after 72h (Werner *et al.*, manuscript in preparation).

In summary, this work describes the first application of an efficient inducible gene expression system in HEK.EBNA cells. The highly innovative AIR technology has been successfully applied to HEK.EBNA cells with tight repression and high induction levels. This system can be used for controlled proliferation studies or for inducible expression of a recombinant protein of choice. $p27^{Kip1}$ has been selected as a specific cell cycle regulator for induction of growth arrest in HEK.EBNA cell lines. First results indicate that controlled proliferation via expression/induction of $p27^{Kip1}$ leads to an increase of productivity per cell in recombinant production pools and stable engineered AIR-adjustable production cell lines.

## REFERENCES


Bi JX, Shuttleworth J, Al-Rubeai M. 2004, Uncoupling of cell growth and proliferation results in enhancement of productivity in p21CIP1-arrested CHO cells. Biotechnol Bioeng. **85,** 741-9

Felenbok B, Sequeval D, Mathieu M, Sibley S, Gwynne DI, Davies RW. 1988, The ethanol regulon in Aspergillus nidulans: characterization and sequence of the positive regulatory gene alcR. Gene **73,** 385-96

Flipphi M, Kocialkowska J, Felenbok B. 2002, Characteristics of physiological inducers of the ethanol utilization (alc) pathway in Aspergillus nidulans. Biochem J. **364,** 25-31

Fussenegger M, Bailey JE. 1998, Molecular regulation of cell-cycle progression and apoptosis in mammalian cells: implications for biotechnology. Biotechnol Prog. **14,** 807-33

Fussenegger M, Schlatter S, Datwyler D, Mazur X, Bailey JE. 1998, Controlled proliferation by multigene metabolic engineering enhances the productivity of Chinese hamster ovary cells. Nat Biotechnol. **16,** 468-72

Ho A, Dowdy SF. Regulation of G(1) cell-cycle progression by oncogenes and tumor suppressor genes. 2002, Curr Opin Genet Dev. **12,** 47-52

Ibarra N, Watanabe S, Bi JX, Shuttleworth J, Al-Rubeai M. 2003, Modulation of cell cycle for enhancement of antibody productivity in perfusion culture of NS0 cells. Biotechnol Prog **19**, 224-8

Massagué J. 2004, G1 cell-cycle control and cancer. Nature **432,** 298-306

Mazur X, Fussenegger M, Renner WA, Bailey JE. 1998, Higher productivity of growth-arrested Chinese hamster ovary cells expressing the cyclin-dependent kinase inhibitor p27. Biotechnol Prog. **14,** 705-13

Meents H, Enenkel B, Werner RG, Fussenegger M. 2002, p27Kip1-mediated controlled proliferation technology increases constitutive sICAM production in CHO-DUKX adapted for growth in suspension and serum-free media. Biotechnol Bioeng. **79,** 619-27

Sherr CJ, Roberts JM. 1999, CDK inhibitors: positive and negative regulators of G1-phase progression. Genes Dev. **13**, 1501-12

Watanabe S, Shuttleworth J, Al-Rubeai M. 2002, Regulation of cell cycle and productivity in NS0 cells by the over-expression of p21CIP1. Biotechnol Bioeng **77**, 1-7

Weber W, Rimann M, Spielmann M, Keller B, Daoud-El Baba M, Aubel D, Weber CC, Fussenegger M. 2004, Gas-inducible transgene expression in mammalian cells and mice. Nat Biotechnol. **22**, 1440-4

# Towards Stronger Gene Expression - a Promoter's Tale

*In search of improving the GS-vector expression system*

Stephan Kalwy, James Rance, Alison Norman and Robert Gay
*Lonza Biologics plc, 228 Bath Rd, Slough SL1 4DX*

**Abstract:** The creation of highly productive cell lines is the sum of many parts including choice of expression system, host cell line, selection strategies and media. The expression system itself is made up of a range of components which when assembled together must achieve gene expression in the transfected cells. Components from promoter and enhancer element, polyadenylation signal, 5'- and 3'- untranslated regions to mechanism of action of the selectable marker have been shown to have a profound effect on the level of mRNA expressed from a transfected vector.

The aim of this study was to identify and evaluate modifications to the vectors of Lonza's Glutamine Synthetase (GS) expression system which may enhance expression from these vectors. A number of alterations to the current GS-vectors were evaluated, including replacing the human cytomegalovirus major immediate early (CMV-MIE) promoter with a mouse CMV-MIE promoter; altering the order of the heavy chain and light chain transcript units within the vector; introduction of a transcription blocker sequence; evaluation of matrix-associated region elements; and coding-region optimisation. The alternative constructs were evaluated relative to the current GS vector by comparing the production of a model chimeric antibody in stably transfected CHOK1SV cells, a suspension variant of CHO-K1 cells.

Most of the alternative configurations were of no benefit or in some cases detrimental. However, significant improvements in product yields were obtained following optimisation of the heavy chain and light chain coding regions.

**Key words:** Glutamine synthetase, Chinese hamster ovary, expression, antibody, promoter, matrix associated elements, cytomegalovirus, transcription, codon optimisation, NS0, mRNA, intron, CHO-K1, CHOK1SV



*R. Smith (ed.), Cell Technology for Cell Products*, 19–28.

# 1. INTRODUCTION

Lonza's GS-expression vectors utilise a strong promoter, the human CMV-MIE promoter (hCMV) to drive the gene(s) of interest, whilst expression of the GS selectable marker is driven by a weaker promoter, SV40E.

Stringent selection in a glutamine-free medium with the addition of methionine sulphoximine (MSX), an irreversible inhibitor of GS, selects for rare integrations of the vector into transcriptionally active sites of the genome. Since the selection marker is on the same vector as the genes of interest, these are also likely to be expressed at high levels.

The current vector utilizes the hCMV promoter to drive expression of both the heavy chain (HC) and the light chain (LC), with the LC placed in front of the HC within the vector. To further improve these vectors, we investigated a number of alternative expression configurations expressing a model antibody, chimeric $IgG_4$ cB72.3.

# 2. METHODS AND MATERIALS

CHO-K1 cells were propagated in static flasks in a DMEM-based medium supplemented with 10% fetal calf serum, 2 mM L-glutamine. The cells were detached and transfected by electroporation with GFP-expressing vectors. Each transfection was plated into a number of T175 flasks and grown for 26 days in selective medium (growth medium containing 50 µM MSX). The pools of cells were subsequently detached and analysed for fluorescence intensity at 510 nm (excitation 488 nm) using a Coulter Elite flow cytometer, and recording the mean channel number values.

CHOK1SV cells, a variant of CHO-K1 that has been adapted to grow in suspension and protein-free medium, were propagated in CD-CHO (Invitrogen) medium supplemented with 6 mM L-glutamine, in shaker flasks. These cells were transfected by electroporation with antibody-coding vectors and plated into 96-well plates in a glutamine-free DMEM-based medium supplemented with 10% fetal calf serum and 50 µM MSX. The cells were grown for 3-4 weeks in selective medium until individual colonies were evident. For each of the different vectors evaluated, approximately 100 cultures arising from single colonies were transferred to 24-well plates containing selective medium, as above. The cultures were grown for two weeks, after which the concentration of antibody in the medium was determined by Protein-A HPLC.

# 3. RESULTS AND DISCUSSION

## 3.1 Mouse CMV

The mouse CMV-MIE (mCMV) promoter has previously been reported to be superior to the human CMV promoter for gene expression in mouse or hamster cells (Rotodaro *et al.,* 1996; Addison *et al.,* 1997). We therefore compared the performance of the mCMV promoter relative to the hCMV promoter within the context of Lonza's GS-vector system. The mouse and human CMV promoters comprise an upstream enhancer region, a minimal promoter region, and a downstream portion that forms the 5'-untranslated region of the transcripts. A number of vectors were generated, in which GFP expression was driven from the mCMV promoter including either the 5'-upstream region (mMCV long) or without this region (mMCV short), and both promoters were also assessed with or without the associated mCMV intron. The control vector contained the hCMV plus the associated hCMV intron (see Figure 1a).

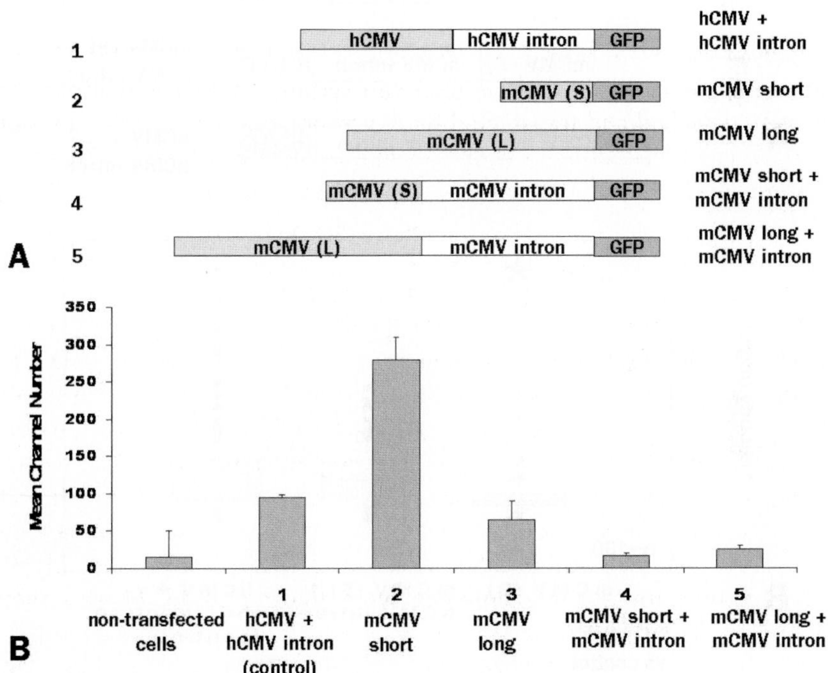


*Figure 1.* mCMV vs hCMV promoters driving expression of GFP.

The preliminary investigations using pooled transfectants of CHO-K1 cells expressing GFP suggested that the highest expression was from the short mouse promoter region; the region upstream of the short mCMV promoter was not beneficial whereas the intron was detrimental to expression. These results are summarised in Figure. 1b and are consistent with previously reported studies (eg Rotodaro *et al.,* 1996; Addison *et al.,* 1997).

In order to verify that the enhanced expression seen with the mCMV-short promoter would also be applicable to expression of an antibody, vectors were constructed in which the expression of the HC and LC genes of the model antibody were driven by the mCMV short promoter (see Figure 2a). In addition, since the mCMV intron region was detrimental to transcription, the potential benefit of the hCMV intron in conjunction with the mCMV promoter was also evaluated. The human intron is reported to be critical for the effective expression from the hCMV promoter (Chapman *et al.,* 1991) and is present as part of the current GS-vectors.

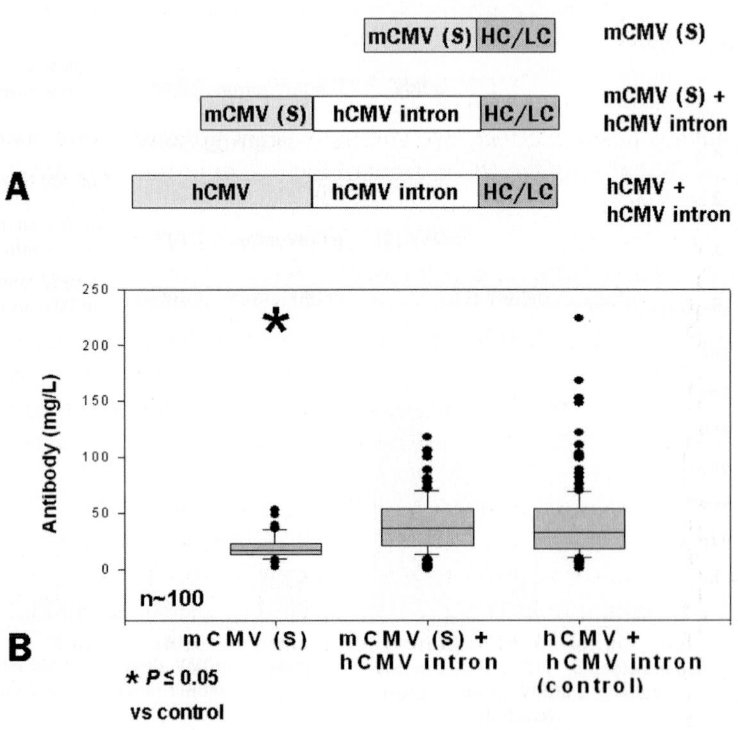


*Figure 2.* mCMV vs hCMV driving expression of antibody.

These constructs were used to generate double-gene vectors expressing the cB72.3 model antibody and used to transfect CHOK1SV cells. The productivity of individual cell lines was evaluated, as summarised in Figure 2b. Surprisingly, the mean productivity levels from the resultant cell lines containing the mCMV-short was lower than that seen with the control vector (hCMV plus human intron). However, the combination of the mCMV short promoter plus the hCMV intron resulted in a mean productivity level equivalent to that of the control vector. Thus, the mCMV short promoter is able to drive transcription of antibody genes in CHO cells but the human intron region is required for effective transcription from this promoter, as it is for the hCMV promoter. These results emphasize the importance of using the correct reporter protein(s), whose characteristics should closely mimic that the final desired product.

### 3.2 Gene position effect – gene order and transcription blocker

The current configuration of the antibody genes in the GS-vector double-gene construct places the LC gene in front of the HC gene. Since it is common to observe excess LC protein secreted from stable cell lines, relative to HC protein, it has been suggested that transcriptional read-through from the LC gene could lead to promoter occlusion of the HC gene. To investigate this position effect, two alternative configurations were generated (see Figure 3a): placing the HC in front of the LC, or introducing an artificial transcription blocker sequence (Eggermont and Proudfoot, 1993) between the LC and the HC genes. These constructs were used to transfect CHOK1SV cells, and the concentration of secreted antibody determined, as summarised in Figure 3b.

Reversing the order of the HC and LC genes within the vector had a detrimental effect, resulting in a decrease in the mean product concentration for the selected cell lines. This suggests that while the position of the genes might influence their levels of expression, excess levels of HC protein may be detrimental to the formation of stable cell lines.

The introduction of the transcription blocker sequence had no beneficial effect compared to the original construct. Therefore, the hypothesized transcriptional read-through from the LC gene into the HC gene does not appear to be a limiting factor in the secretion of whole antibody. In summary, the original double-gene vector configuration appears to be the better option.

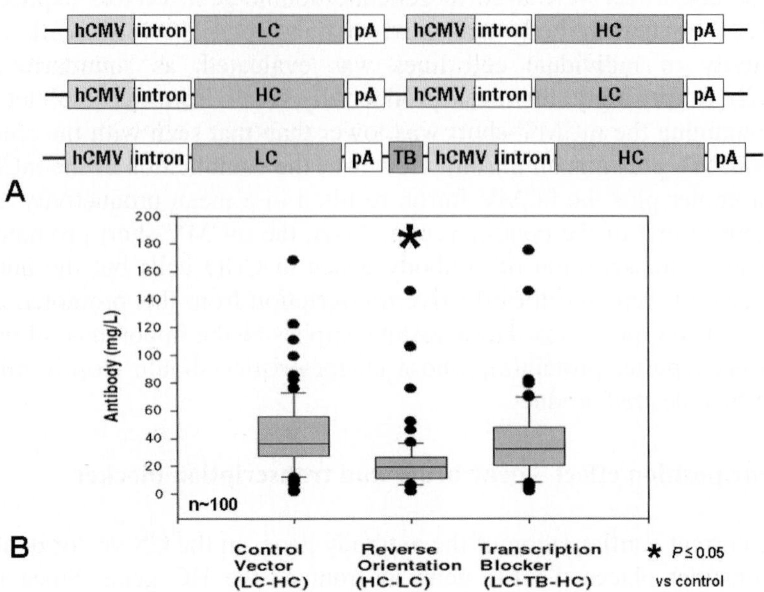


*Figure 3.* Gene position – HC, LC gene order and transcription blocker.

### 3.3 Matrix associated region elements

Matrix associated region (MAR) elements are DNA elements that bind to the nuclear scaffolding matrix found associated with many house-keeping genes and are proposed to maintain the local chromatin regions in an open state, enabling long-term transcription from the adjacent genes. When used in the context of an expression vector, they may serve to facilitate both higher and more persistent expression from the genes of interest (Zahn-Zabal *et al.*, 2001; Kim *et al.*, 2004).

We evaluated the effect of placing a 3 kb MAR element associated with the chicken lysozyme gene (Phi-Van and Strätling, 1988) at one of three different locations within the GS-vector: upstream of the LC promoter (5'), downstream of the HC (3'), between the LC and HC (middle), and a two-MAR construct with insertion at both the 5' and 3' positions (see Figure 4a). These constructs were used to generate double-gene vectors expressing the cB72.3 model antibody and used to transfect CHOK1SV cells. The concentration of secreted antibody from cultures of individual cell lines was determined, as summarised in Figure 4a.

The inclusion of MAR elements within the GS vectors was not beneficial (3' or middle positions) or in the worst case, detrimental (5' position). For the construct with MAR elements in both the 5' plus 3' positions, the effect was as detrimental as that seen with the single MAR in the 5' position.

These findings are in contrast to previous studies, which have reported significant improvements from the introduction of MAR elements (eg Zahn-Zabal *et al.,* 2001). The lack of improvement seen with MAR elements may be due in part to the stringent selection protocol used to generate these stable cells but this does not explain the different detrimental effects of the MARs in each of the three different positions within the vector.

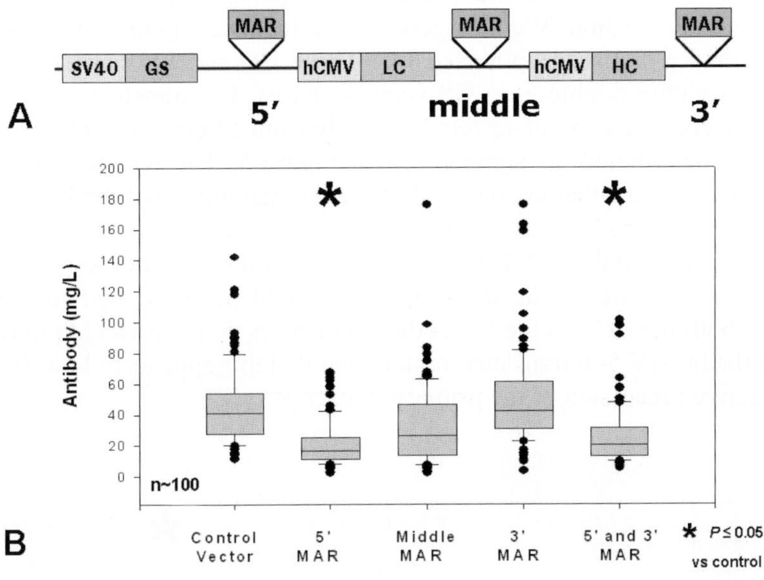


*Figure 4.* MAR elements - effect of different positions within vector.

## 3.4 Gene coding region optimisation

Recent studies in GS-NS0 cells show that the level of antibody production in stably transfected cells does not necessarily correlate with their levels of mRNA encoding the HC and LC proteins (Smales *et al.*, 2004), suggesting that there are rate-limiting events downstream of the primary transcript processing steps. We therefore considered strategies to improve the processing of the transcripts. The sequences encoding HC and LC genes may contain multiple introns, depending on their source. These multiple splicing sites may lead to erroneous splice variants, especially if the variable regions have been altered during the course of codon optimisation, humanisation or site-directed mutagenesis. The model antibody cB72.3 is encoded by genomic-derived HC constant regions, and hence has internal introns; the chimeric LC gene does not contain internal introns. We therefore evaluated the potential benefits of removing these additional introns from the HC sequence. Additionally, we evaluated the potential benefit of altering the codons of the gene-coding region to a codon usage optimised for CHO cells, but importantly, whilst also avoiding unfavourable features such as cryptic slice sites, high GC content regions and undesirable folding domains, using an algorithm developed by GeneArt GmbH.

Accordingly, a non-optimised cDNA version of the HC-coding sequence of cB72.3 was generated and evaluated compared to the original intron-containing HC version. We also generated a sequence-optimised version of this HC and a similarly optimised LC. These HC and LC sequences were expressed within double-gene GS-vectors and used to transfect CHOK1SV cells. The concentration of secreted antibody from cultures of individual cell lines was determined, as summarised in Figure 5. The results from these studies demonstrate that the removal of the internal introns of the HC did not affect the mean productivity or range of antibody levels seen with a range of cell lines. This indicated that the internal introns are not essential for effective processing of the transcripts. It should be noted that in the GS vectors, both the HC and the LC-coding sequences do have a single intron as part of the hCMV 5-untranslated region, and that this appears to be sufficient for effective processing of the primary transcript.

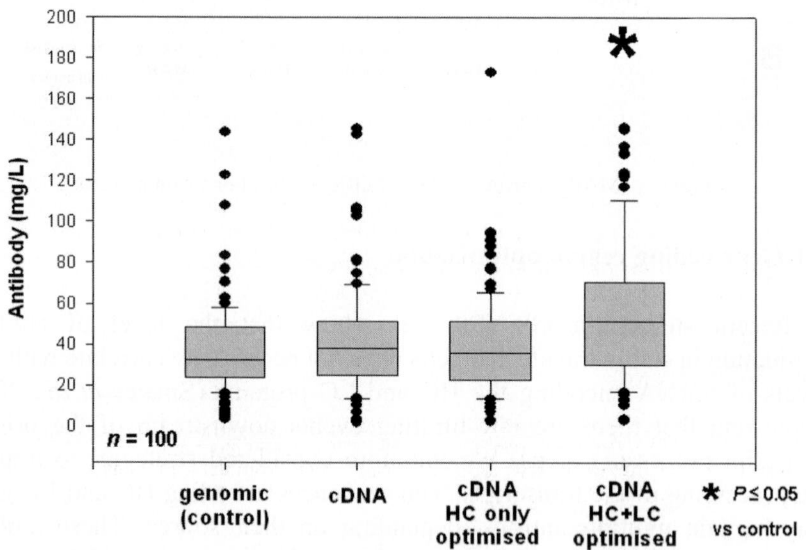


*Figure 5.* Gene optimisation - effects of different HC, LC coding sequences.

Sequence-optimisation of the HC-coding sequence alone was not beneficial, relative to the non-optimised control, but optimising both the HC and the LC-coding sequences led to a statistically significant improvement in the mean productivity of the cell lines. Since the cB72.3 antibody is already well expressed in these cells, an even greater improvement in yields might be expected from optimizing otherwise poorly expressed antibodies.

## 4. CONCLUSIONS

Of the various modifications to Lonza's GS-expression vectors, few have lead to a positive benefit; most have been either non-beneficial or in some cases have lead to an overall decrease in the mean productivity of cell lines. Only the efforts to increase efficiency of transcript processing and translation lead to an increase in product expression.

The current vector configurations therefore appear to be approaching optimal, in terms of maximal transcription efficiency. Further efforts to increase transcription levels are unlikely to result in significant increases in product secretion. Instead, it is envisaged that the next rate-limiting steps to be tackled will lie downstream in the transcript processing, or in the protein translation, secretion and modification pathways. Once these putative bottlenecks have been alleviated, it may be again useful to evaluate methods to further improve both the absolute levels of transcription as well as generating more high-expressing cell lines.

## ACKNOWLEDGEMENTS

The authors wish to acknowledge the members of the Cell Culture Process Development and Assay Development departments, Lonza Biologics, for their excellent technical support.

## REFERENCES

Addison, C.L., Hott, M., Kunsken, D. and Graham, F.L., 1997, Comparison of the human versus murine cytomegalovirus immediate early gene promoters for transgene expression by adenoviral vectors, J. Virol., **78**:1653-1661.

Chapman, B.S. Thayer, R.M., Vincent, K.A. and Haigwood, N.L., 1991, Effect of intron A from human cytomegalovirus (Towne) immediate-early gene on heterologous expression in mammalian cells, Nucl.Acid.Res., **19**:3979-3986.

Eggermont, J. and Proudfoot, N.J., 1993, Poly(A) signals and transcriptional pause sites combine to prevent interference between RNA polymerase II promoters, EMBO J., **12**:2539-2548.

Kim, J.M., Kim, J.S., Park, D.H., Kang, H.S., Yoon, J., Baek, K. and Yoon, Y., 2004, Improved recombinant gene expression in CHO cells using matrix attachment regions, J. Biotech., **107**:95-105.

Phi-Van, L. and Strätling, W.H., 1988, The matrix attachment regions of the chicken lysozyme gene co-map with the boundaries of the chromatin domain, EMBO J., **7**: 655-664.

Rotondaro, L., Mele, A. nad Rovera, G., 1996, Efficiency of different viral promoters in directing gene expression in mammalian cells: effect of 3'-untrnaslated sequences, Gene, **168**:195-198.

Smales, C.M., Dinnis, D.M., Stanfield, S.H., Alete, D., Sage, E.A., Birch, J.R., Racher, A.J., Marchall., C.T. and James, D.C., 2004, Comparative proteomic analysis of GS-NS0 murine myeloma cell lines with varying recombinant monoclonal antibody production rates, Biotech. Bioeng. **88**:474-488.

Zahn-Zabal, M., Kobr, M., Gorod, P-A., Imhof, M., Chatellard, P., de Jesus, M., Wurm, F. and Mermod, N., 2001, Development of stable cell lines for production or regulated expression using matrix attachment regions, J. Biotech., **87**:29-42.

# Improving the Expression of a Soluble Receptor:Fc fusion Protein in CHO Cells by Coexpression with the Receptor Ligand

Gene W. Lee, Jill K. Fecko, Ann Yen, Deb Donaldson, Clive Wood, Scott Tobler, Suresh Vunuum, Yin Luo and Mark Leonard
*Wyeth BioPharma, One Burtt Rd., Andover, MA 01810, U.S.A.*

**Abstract:** A novel soluble receptor:Fc fusion protein was designed as a recombinant therapeutic targeting human IL-13, a central mediator of asthma. This protein, sIL-13R, was expressed in stable CHO cell lines for the purpose of producing sufficient amounts of material to support early clinical trials. However, expression levels of sIL-13R were poor, and the protein was prone to misfolding and aggregation. Culturing the recombinant CHO cell lines at reduced temperatures significantly enhanced expression levels and lowered the levels of aggregated material. An even more dramatic improvement was observed when sIL-13R was coexpressed with its natural ligand, IL-13, in CHO cells. Protein expression was enhanced at both 37°C and 31°C, and product aggregation was significantly reduced. The sIL-13R protein secreted by the coexpressing cell line adopted a conformation that remained stable even after removal of the ligand during purification. Together, these results suggest a possible strategy for improving the expression of other difficult to express soluble proteins.

**Key words:** Coexpression, expression, aggregation, soluble receptor, fusion protein



*R. Smith (ed.), Cell Technology for Cell Products*, 29–39.

# 1. INTRODUCTION

The pro-inflammatory cytokine IL-13 has been implicated as a central mediator of allergy-induced asthma (1,2). A high affinity cell surface receptor to human IL-13, designated as IL-13Rα2, was identified and cloned (3). A soluble version of the receptor, designated as sIL-13R, was chosen as a candidate for development and evaluation as a therapeutic agent. Because the anticipated material needs to evaluate the safety and efficacy of sIL-13R in the clinic were high, a large scale manufacturing process was developed.

The soluble receptor was expressed as a fusion protein created by combining the extracellular domain of IL-13Rα2 with the Fc region of human $IgG_1$. The gene encoding sIL13R was cloned into an expression vector containing an optimized promoter and enhancer sequences to drive high level expression, and the dihydrofolate reductase (DHFR) selectable marker to allow amplification of the integrated gene to increase gene copy number. However, initial results in COS cell transient transfection assays indicated that sIL-13R was expressed poorly. In contrast, a soluble receptor containing the extracellular domain of the low affinity receptor to IL-13 (IL-13Rα1) was well expressed in the same system. These results highlight the fact that soluble receptors designed as fusion proteins are artificial constructs that do not appear in nature, and some constructs may therefore be susceptible to problems with folding, secretion and/or expression. Proteins that aid in folding nascent peptides in the lumen of the endoplasmic reticulum, such as protein disulfide isomerase (PDI) or BiP/Grp78, have had mixed success when overexpressed in cells to assist in protein folding or secretion (4,5). In order to ensure that a productive manufacturing process could be developed, it was necessary to identify methods that would allow high expression levels of sIL-13R, and that the sIL-13R produced was of acceptable product quality.

# 2. RESULTS AND DISCUSSION

*Expression of sIL13R in CHO cells is poor.* Stable CHO cell lines were developed for the expression of recombinant sIL-13R. When cultured at 37°C, the expression levels of sIL-13R in the conditioned medium were found to be surprisingly low, despite the observation that these cell lines expressed high levels of sIL-13R specific mRNA (data not shown). Moreover, the sIL-13R expressed in the conditioned medium was extremely prone to formation of high molecular weight aggregate species, as evident on size exclusion chromatography of Protein A purified sIL-13R (Figure 1).

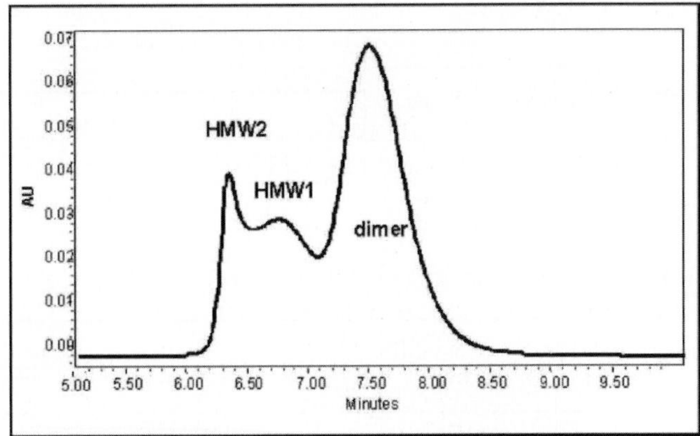


*Figure 1.* SEC-HPLC chromatogram of Protein A purified sIL-13R. Conditioned medium from CHO cells expressing sIL-13R was processed through a Protein A column, and the purified sIL-13R protein was analyzed by SEC-HPLC.

*Expression of sIL-13R is significantly improved by reducing the cell culture temperature.* A common cell culture strategy for large-scale production of recombinant proteins in mammalian cell culture is to reduce, or "shift" the cell culture temperature from 37°C to a lower culture temperature. The lower temperature is thought to induce growth arrest in the cell culture and potentially prolong the viability of cells in the production bioreactor, and may also have a modest positive effect on recombinant protein expression (6,7). Interestingly, when CHO cells expressing sIL-13R were cultured at 31°C in a fed-batch production assay, a significant increase in the cell specific productivity and titer was also observed, as compared to cells cultured at 37°C (Figure 2). These results indicate that lowering the cell culture temperature can increase the expression levels of sIL-13R in the conditioned medium.

*Temperature shift reduces formation of sIL-13R HMW species.* To better understand the mechanism(s) by which lowering the cell culture temperature can improve the expression of sIL-13R, northern blot analysis was performed to assess the steady-state levels of sIL-13R transcripts and metabolic labeling experiments were performed to assess the intracellular levels of sIL-13R protein. These results indicated that temperature shift did not induce an increase in either the transcript levels of sIL-13R, nor did it induce a significant increase in the intracellular levels of sIL-13R protein (data not shown). However, subcellular fractionation experiments did suggest that at reduced cell culture temperature, the sIL-13R protein moved more readily through the secretory pathway (data not shown). Moreover,

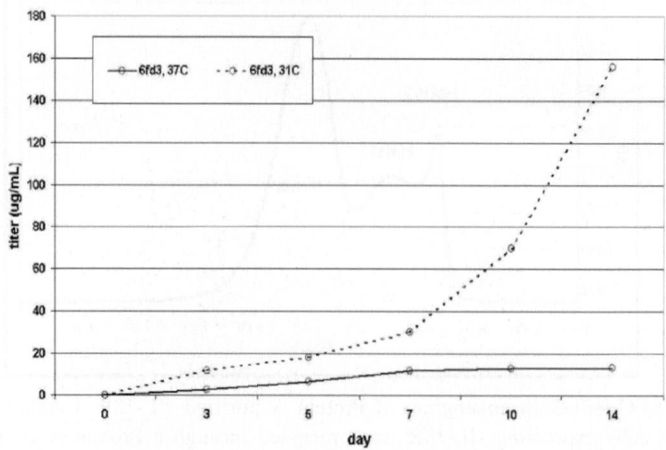


*Figure 2.* Temperature shift significantly improves sIL-13R expression. A stable CHO cell line expressing sIL-13R (6FD3) was assessed in a 14-day fed-batch production assay, with or without a temperature shift from 37°C to 31°C on day 3 (dashed or solid line, respectively). Product titer was assessed by Protein A-HPLC.

differential scanning calorimetry indicated that sIL-13R undergoes its first secondary structure transition at ~38-40°C (data not shown). Finally, SEC-HPLC analysis indicated that there is a moderate reduction in the levels of high molecular weight (HMW) aggregates of sIL-13R in the conditioned medium when cells are cultured at 31°C, as compared to when cells are cultured at 37°C (Figure 3). Together, these results suggest that sIL-13R may adopt a more stable folding configuration at lower temperatures. Similar observations have been made with other proteins, most notably the cystic fibrosis mutant CFTR delta508 (8,9). However, the levels of HMW species present even at the lower temperatures still remained above 10% in the conditioned medium. While it was possible to achieve a further reduction in HMW species (target: ≤5% HMW) through the use of large-scale size exclusion chromatography columns during purification, this approach also led to a substantial loss in the overall yield of sIL-13R protein. We therefore explored alternative methods for improving the expression of sIL-13R and reducing the levels of HMW species.

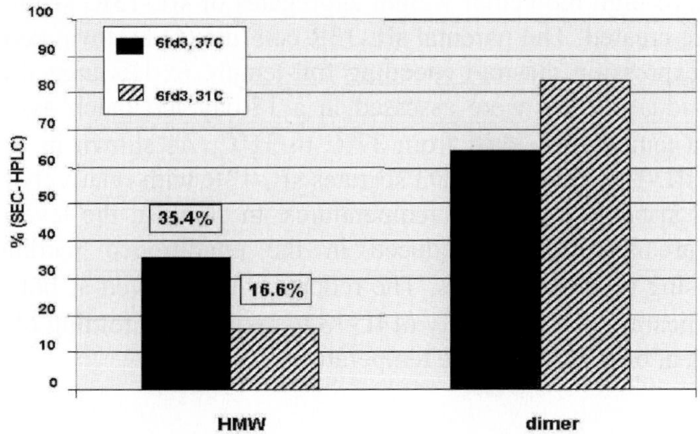


*Figure 3.* Temperature shift reduces the formation of HMW aggregates of sIL-13R in the conditioned medium. A stable CHO cell line expressing sIL-13R (6FD3) was evaluated in a 12-day fed batch production assay, with or without a temperature shift from 37°C to 31°C on day 3 (black or hatched bars, respectively). sIL-13R was purified from day 12 conditioned medium by Protein A chromatography, and analyzed by SEC-HPLC. The levels of high molecular weight species (HMW) or the desired dimer species are shown.

*Coexpression of IL-13 with sIL-13R significantly reduces formation of high molecular weight aggregates.* The reduction in the formation of high molecular weight aggregates of sIL-13R in the conditioned medium when cells are cultured at reduced temperature, together with the improved overall expression of sIL-13R, strongly suggested that the sIL-13R protein was prone to misfolding when cultured at the standard cell culture temperature of 37°C. Over-expression of molecular chaperones or foldases has been described previously as one method of improving the folding of some proteins. However, these approaches may also lead to decreased expression of recombinant protein (5).

Earlier work indicated that the expression levels of sIL-13R in COS cell transient transfection experiments could be significantly increased by coexpressing human IL-13 (data not shown). This effect was not observed if a non-specific cytokine, such as IL-6, was coexpressed with sIL-13R. Nor was this effect observed if recombinant human IL-13 was added to the cell culture exogenously. These data suggest that IL-13 may serve as an intracellular folding chaperone for sIL-13R, presumably as the proteins are translated and translocated into the lumen of the endoplasmic reticulum.

To test if coexpression of IL-13 with sIL-13R would also lead to a reduction of high molecular weight aggregates of sIL-13R, stable CHO cell lines were created. The parental sIL-13R cell line (6FD3) was re-transfected with an expression plasmid encoding full-length, 6x-His tagged human IL-13. Individual clones were assessed in a 12-day fed batch assay, with or without a temperature shift from 37°C to 31°C. As shown in Table 1, the parental sIL-13R cell line 6FD3 secretes sIL-13R with relatively high levels of HMW at both cell culture temperatures. In contrast, the levels of HMW species are significantly reduced in the conditioned medium of the coexpressing (CXP) cell lines. The reduction is observed at both 31°C and 37°C, indicating that the ability of IL-13 to assist in the folding of sIL-13R is independent of the cell culture temperature.

*Table 1.* Coexpression with IL-13 significantly reduces formation of HMW species of sIL-13R. Conditioned medium from the parental sIL-13R cell line (6FD3) or four IL-13 coexpressing cell lines derived from the parental cell line (CXP) were Protein A purified and %HMW was assessed by SEC-HPLC analysis.

| Clone | %HMW (31°C) | %HMW (37°C) |
|---|---|---|
| 6FD3 (parental sIL-13R clone) | 16.6 | 35.4 |
| CXP A | 1.4 | 2.5 |
| CXP B | 4.1 | (nd) |
| CXP C | 2.1 | 1.7 |
| CXP D | 3.1 | 6.2 |

The effect of coexpression on the expression levels of sIL-13R in stable CHO cell lines was also evaluated (Figure 4). Cultures were grown with or without a temperature shift to 31°C (Figure 4, dashed or solid lines, respectively). The expression levels of sIL-13R were significantly improved in CHO cells when IL-13 is coexpressed, at both 37°C and 31°C. All stable CHO cell lines coexpressing IL-13 and sIL-13R tested in this manner (n=4) exhibited improved sIL-13R expression relative to the parental cell line (data not shown). Thus, coexpression of IL-13 and sIL-13R can significantly reduce the formation of high molecular weight aggregates of sIL-13R, and significantly increase the expression of sIL-13R, in both COS cell transient expression systems and in stable CHO cell lines.

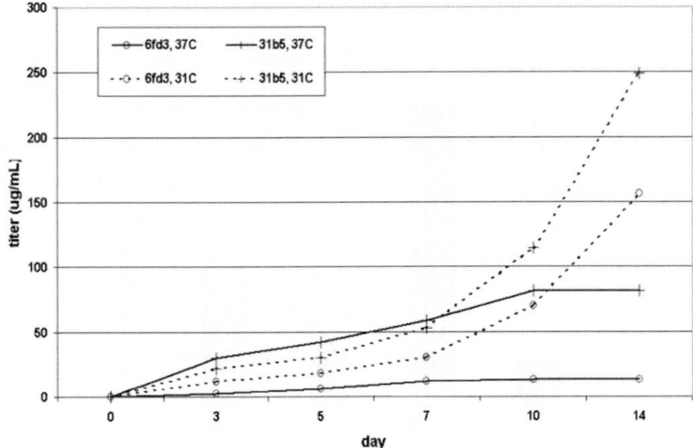


*Figure 4.* Coexpression with IL-13 significantly improves expression of sIL-13R. Cell lines expressing sIL-13R alone (6FD3) or together with IL-13 (31B5) were evaluated in a 14-day fed batch production assay with or without a temperature shift to 31°C on day 3 (dashed or solid lines, respectively). Titers were assessed by Protein A HPLC.

*Mutant variants of IL-13.* Despite the apparent advantages that coexpression of IL-13 confers on the expression of sIL-13R, an important challenge remained. The sIL-13R fusion protein is derived from the high affinity cell surface receptor for IL-13, or IL-13Rα2 (3). The dissociation constant ($k_D$) describing the interaction of IL-13 and IL-13Rα2 is ~$10^{-10}$ M, roughly equivalent to the dissociation constant describing the interaction between Protein A and $IgG_1$. Because sIL-13R was being developed as a potential antagonist of circulating IL-13, it was necessary to demonstrate that IL-13 could be separated from sIL-13R and removed during purification, without disrupting sIL-13R. However, initial attempts to separate IL-13 from sIL-13R during purification proved to be difficult. To facilitate the separation of IL-13 from sIL-13R, mutant variants of IL-13 were created in an effort to lower its interaction affinity for sIL-13R. The variant IL-13 coding sequences were introduced into the parental sIL-13R cell line, and stable IL-13 (mutant or wild-type) coexpressing clones were assessed for expression of sIL-13R, aggregation of sIL-13R (at both 37°C and 31°C) and for ease of separation during purification.

As shown in Figure 5, three mutant variants of IL-13 were compared to wild-type IL-13 for their ability to improve sIL-13R expression when coexpressed with sIL-13R, as compared to the parental cell line. All versions of IL-13 tested were able to increase sIL-13R expression levels in coexpressing cell lines.

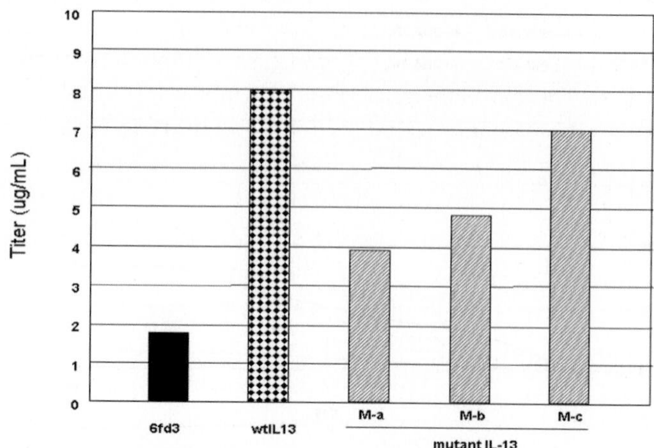


*Figure 5.* Mutant variants of IL-13 can increase expression of sIL-13R in coexpressing cell lines. The parental sIL-13R cell line 6FD3 (black bar) was compared to stable cell lines coexpressing either the wild-type IL-13 (hatched bar) or three different mutant variants of IL-13 (gray bars). Cells were cultured at 37°C for 7 days, and the sIL-13R titer in the conditioned medium was assessed by Protein A HPLC.

The conditioned medium from the parental cell line or the coexpressing cell lines were purified on a Protein A column and analyzed by size exclusion chromatography to assess the levels of high molecular weight (HMW) aggregates of sIL-13R present. As shown in Table 2, the levels of HMW in the parental cell line are 16.4% and 24.8% at 31°C and 37°C, respectively, which is consistent with previous observations. Again, as observed previously, coexpression of wild-type IL-13 significant reduces formation of HMW at both temperatures. Two of the mutant variants of IL-13 tested, "a" and "b", exhibited a HMW profile that was similar to that observed for the wild-type IL-13. However, a third mutant variant, "c", behaved differently when coexpressed with sIL-13R. While the levels of HMW were significantly reduced when cells were cultured at 31°C, the ability to reduce levels of HMW were diminished when cells were cultured at 37°C. This suggested that the interaction between mutant IL-13 "c" and sIL-13R might be more labile as compared to the other versions of IL-13 tested. Indeed, surface plasmon resonance analysis indicated that the $k_D$ of mutant IL-13 "c" and sIL-13R is ~$10^{-8}$ M, or about a 100 fold weaker affinity for sIL-13R than the wild-type IL-13. The mutant IL-13 version "c" was readily separated from sIL-13R using relatively mild disruption conditions during Protein A chromatography, and the levels of residual IL-13 were reduced to extremely low concentrations (data not shown). The presence of an N-terminal 6x-His tag on IL-13 further assisted in the removal of residual IL-13. In contrast, the other variants of mutant IL-13, as well as the wild-type version of IL-13, could be separated from sIL-13R only under conditions that also denatured and destabilizedsIL-13R. These results demonstrated that a mutant variant of IL-13, exhibiting weaker interaction with sIL-13R (particularly at 37°C), could be easily separated from sIL-13R during purification, but can still preserve the ability to increase sIL-13R expression and decrease HMW formation in stable CHO cell lines when coexpressed with sIL-13R.

*Table 2.* Differential ability of mutant variants of IL-13 to reduce formation of sIL-13R HMW species. Cell lines were assessed in a 12-day fed batch production assay, with or without a temperature shift to 31°C on day 3. Conditioned medium harvested at the end of the assay was purified by Protein A chromatography and levels of high molecular weight (HMW) were assessed by SEC-HPLC.

| Clone | %HMW (31°C) | %HMW (37°C) |
|---|---|---|
| 6fd3 (parental sIL-13R clone) | 16.4 | 24.8 |
| Wt CXP | 2.0 | 2.1 |
| Mutant "a" CXP | 2.8 | 3.2 |
| Mutant "b" CXP | 3.9 | 4.6 |
| Mutant "c" CXP | 4.4 | 16.9 |

## 3. CONCLUSION

We have developed a process for the production of a soluble receptor in recombinant CHO cells that is sufficient to supply material to support pre-clinical and Phase I clinical studies. The use of temperature shift during the production phase of the cell culture process was found to have significant impacts on improving productivity and product quality of sIL-13R. Initial studies in transiently transfected COS cells indicated that the molecule was poorly expressed, and poor expression was observed in stably transfected CHO cell lines as well. The poor expression phenotype was found to be partly a consequence of inefficient transit of sIL-13R through the secretory pathway, and not to a defect in transcription or translation of sIL-13R. Lower cell culture temperatures may stabilize a favorable folding conformation and/or the thermodynamic stability of sIL-13R, which may in turn result in more rapid transit through the secretory pathway and a lower likelihood of adopting a folding conformation that is more susceptible to aggregation. Indeed, protein stability has been identified as an important determinant of the probability that a protein will be retained in the ER (10,11). Coexpression of sIL-13R with its natural ligand, IL-13, resulted in an even greater improvement in expression and aggregate reduction than temperature shift. The appearance of sIL-13R in the conditioned medium of coexpressing cell lines with very low levels of HMW aggregates (≤5%), together with the significant increase in sIL-13R expression even at 37°C, strongly suggests that sIL-13R has adopted a more favorable and stable folding configuration in coexpressing cell lines. Further, this effect was not observed when IL-13 was added exogenously to the cell culture, indicating that IL-13 must be acting intracellularly to aid in the folding of sIL-13R. Together, these observations suggest that IL-13 may be acting as a folding or scaffolding chaperone for sIL-13R. Mutant variants of IL-13, designed to weaken the interaction with sIL-13R, preserved their ability to improve expression and reduce aggregation. More importantly, these mutants allowed for ready separation of IL-13 from sIL-13R during purification. It will be interesting to test this approach with other poorly expressed soluble receptors.

## REFERENCES


1. Wills-Karp, M., Luyimbazi, J., Xu, S., Schofield, B., Neben, T.Y., Karp, C.L. and Donaldson, D. (1998) Interleukin-13: central mediator of allergic asthma. Science **282**, 2258-2261.
2. Grunig, G., Warnock, M., Wakil, A.E., Venkayya, R., Brombacher, F., Rennick, D.M., Sheppard, D., Mohrs, M., Donaldson, D.D., Locksley, R.M. and Corry, D.B. (1998) Requirement for IL-13 independently of IL-4 in experimental asthma. Science **282**, 2261-2263.
3. Donaldson, D.D., Whitters, M.J., Fitz, L.J., Neben, T.Y., Finnerty, H., Henderson, S.L., O'Hara Jr., R.M., Beier, D.R., Turner, K.J., Wood, C.R. and Collins, M. (1998) The murine IL-13 receptor α2: molecular cloning, characterization, and comparison with murine IL-13 receptor α1. J.Immunology **161**(5), 2317-2324.
4. Davis, R., Schooley, K., Rasmussen, B., Thomas, J. and Reddy, P. (2000) Effect of PDI overexpression on recombinant protein secreion in CHO cells. Biotechnol Prog. **16**(5), 736-43.
5. Dorner, A.J. and Kaufman, R.J. (1994) The levels of endoplasmic reticulum proteins and ATP affect folding and secretion of selective proteins. Biologicals **22**(2), 103-12.
6. Kaufmann, H., Mazur X., Fussenegger, M. and Bailey, J.E. (1999) Influence of low temperature on productivity, proteome and protein phosphorylation of CHO cells. Biotechnol.Bioeng. **63**(5), 573-582.
7. Yoon, S.K., Song, Y.J. and Lee, G.M. (2003) Effect of low culture temperature on specific productivity, transcription level and heterogeneity of erythropoietin in Chinese hamster ovary cells. Biotechnol Bioeng **82**(3), 289-298.
8. Denning, G.M., Anderson, M.P., Amara, J.F., Marshall, J., Smith, A.E. and Welsh, M. (1992) Processing of mutant cystic fibrosis transmembrane conductance regulator is temperature-sensitive. Nature **358**(6389), 761-4.
9. Berson, J.F., Frank,D.W., Calvo, P.A., Bieler, B.M and Marks, M.S. (2000) A common temperature-sensitive allelic form of human tyrosinase is retained in the endoplasmic reticulum at the nonpermissive temperature. J. Biol.Chem. **275**(16), 12281-9.

10. Ellgaard, L. and Helenius, A. (2003) Quality control in the endoplasmic reticulum. Nat Rev Mol Cell Biol **4**(3), 181-91.
11. Jorgensen, M.M., Bross, P. and Gregersen, N. (2003) Protein quality control in the endoplasmic reticulum. APMIS Suppl. **109**, 86-91.

# Growth, Apoptosis and Functional Genomics Analysis of CHO-K1 Over-Expressing Telomerase

Francesco Crea[1], Donatella Sarti[2], Francesco Falciani[2] and Mohamed Al-Rubeai,[3]

*[1] Department of Chemical Engineering, School of Engineering, and [2]Immuno Genomics Group, School of Biosciences, University of Birmingham, Birmingham, B15 2TT, UK.*
*[3]Department of Chemical and Biochemical Engineering, and Centre for Synthesis and Chemical Biology, University College Dublin, Belfield, Dublin, Ireland*

**Abstract:** The enzyme telomerase plays a crucial role in cellular proliferation. By adding hexameric repeats to the chromosome ends, it prevents telomeric loss and, thus entry into senescence of limited life span cells. It is unclear however, what would be the effect of over-expressing telomerase in an immortalised cell line, characterised by unlimited life span and high levels of apoptosis in sub-optimal growth conditions. In order to address this question, we have transfected the immortal cell line CHO-K1 with the human telomerase reverse transcriptase (hTERT) catalytic sub-unit. Significant difference in the growth profile and apoptosis levels between the cells over-expressing hTERT (Telo) and the cells containing blank vector was found under standard growth conditions. Similarly the Telo cells showed lower levels of apoptosis, greater attachment tendency and higher viable cell density under serum deprived condition compared to the blank cell line, suggesting a major role for hTERT in stressed cultures. Using a mouse cDNA microarray, the collagen type III and V genes were shown to have at least a 10 fold higher expression in the Telo cells than the control cells.

**Key words:** Telomerase, hTERT, CHO-K1, Immortalisation, Apoptosis, Metabolic Engineering.



*R. Smith (ed.), Cell Technology for Cell Products*, 41–45.

# 1. INTRODUCTION

Telomerase is a cellular ribonucleoprotein reverse transcriptase that adds telomeric DNA repeats (TTAGGG) to the ends of chromosomes to compensate for the progressive losses of telomeric sequences inherent to DNA replication. Telomerase has two components required for core enzyme activity: telomerase RNA (TR), which is essential for telomerase activity (Chen *et al.,* 2000; Prescot and Blackburn, 2000), and telomerase reverse transcriptase protein (TERT) that has been implicated as a rate limiting step for telomerase activity (Bodnar *et al.,* 1998; Ouellette *et al.,* 2000). Telomerase activity is present in foetal tissue, germ cells and also in tumour cells but is repressed in most somatic tissue during development (Morales *et al.,* 1999).

The role of telomerase in preventing apoptosis and immortalising primary cells without inducing malignant phenotypic effect is well documented (Ren *et al.,* 2001; Akiyama *et al.,* 2002). However, there is no report thus far regarding the effect of telomerase expression in immortalised cell line, such as CHO cells characterised by unlimited life span and high level of apoptosis when cells are stressed in sub-optimal conditions.

CHO cells are not "genetically optimised" for production-scale bioprocesses. However, recombinant DNA technology is allowing process improvements by the manipulation of cellular characteristics at genetic level. To address this point we produced stable CHO K1 cell clones expressing the hTERT "immortalisation gene" and we measured their apoptotic and proliferative characteristics.

We discovered that the expression of hTERT in CHO K1 markedly increased their resistance to serum-deprivation induced-apoptosis and allowed the serum dependent cell line to survive, attach and divide in un-supplemented basal medium. Additionally this study provides evidence of a new type of interrelation between telomerase and collagen genes, which indicates that telomerase may help cell survival through a mechanism involving the activation of the fibrous proteins that control cell shape and differentiation.

# 2. MATERIALS AND METHODS

CHO-K1 cells were transfected with a pGRN145 vector containing the *hTERT* gene (Telo) (Geron Corporation, Menlo Park, California, USA) and a control empty vector (Blank) using a liposomal transfection. Cells were grown in DMEM-Ham's F12 Medium (1:1) supplemented with 5% FCS. hTERT over-expression was determined using western blot analysis. All batch cultures were done in triplicate. A fluorescence microscope was used for the apoptosis/necrosis count with acridine orange (AO) and propidium iodide (PI) staining. Microarray hybridisation was performed using a two colour competitive hybridisation protocol on glass slides as described by Hegde and collegues (Hegde *et al.*, 2000).

## 3. RESULTS AND DISCUSSION

CHO K1 cells have unlimited cultured life span but the mechanism responsible is different from that normally associated with malignant transformation. These cells are dependent on solid support for growth, however, they are, like tumour cells, do not contact-inhibited at high confluence. The expression of telomerase increased the tumour-associated changes by decreasing serum dependency, thereby leading to altered patterns of growth. It has been suggested that elongation of telomeres and over-expression of telomerase by transfected hTERT are advantageous for stabilising damaged chromosomes and double-stranded DNA break repair (Holt *et al.*, 1999; Akiyama *et al.*, 2002). The results in hTERT-transfected CHO K1 suggest that further increase in telomerase may enhance survival and proliferation by a concentration dependent mechanism, Table 1 provides a summary of the results on the effects of *hTERT* .

*Table 1.* Effects of hTERT over-expression in CHO-K1 cells.

| **3.1 Characteristics** | **Telo** | **Blank** |
|---|---|---|
| Surface attachment (in serum) | high | high |
| Surface attachment (no serum) | high | low |
| Growth rate (in serum) | 0.027±0.002 $hr^{-1}$ | 0.029±0.004 $hr^{-1}$ |
| Growth rate (no serum) | + | - |
| CCT (in serum) | 794.5±27.7x$10^9$ cell h/L | 499.7±108.3x$10^9$ cell h/L |
| Max viable cell number (in serum) | 2.07±0.1x$10^6$ cells/m | 1.37±0.3 x$10^6$cells/ml |
| Max viable cell number (no serum) | 3.8±0.48x$10^5$cells/ ml | 0.88±0.02x$10^5$cells/ ml |
| Apoptosis (in serum) | + | ++ |
| Apoptosis (no serum) | + | +++ |

Serum provides factors that promote cell attachment and spreading, a process essential for the survival of anchorage-dependent cells. Detachment of anchorage dependent cells would normally cause a cessation of growth. Cells are blocked in the G1 phase which is caused by loss of activity of cyclinE/cdk2 complex. The loss of attachment not only stops cell proliferation but also activates the cell apoptotic pathway. To further elucidate a possible link between hTERT and attachment pathways leading to survival, the molecular fingerprints of hTERT transfected cells was studied by microarray analysis. Table 2 summarises the transcriptomic profile of the major genes involved in cell attachment pathway.

*Table 2.* Transriptomic profile of attachment and spreading genes of cells cultivated in serum deprived conditions

| Accession Number | Name | Log ratio T/B |
|---|---|---|
| X65582 | *M.musculus* mRNA for alpha-2 collagen VI | 2.23 |
| M18933 | Mouse alpha-1 type-III collagen mRNA, 5' | 26.50 |
| NM_007737 | *Mus musculus* procollagen, type V, alpha | 3.06 |
| NM_019759 | *Mus musculus* early quiescence protein-1 | 94.52 |
| NM_009242 | *Mus musculus* secreted acidic cysteine rich glycoprotein | 2.43 |
| NM_010577 | *Mus musculus* integrin alpha 5 | 2.61 |
| NM_011058 | *Mus musculus* platelet derived growth factor | 2.42 |
| D10212 | Mouse mRNA for hepatocyte growth factor | 3.11 |
| NM_008005 | *Mus musculus* fibroblast growth factor 18 | 3.11 |

The results strongly suggest an involvement of the telomerase gene in controlling cell adhesion and in remodelling extra cellular matrix proteins such as collagens. Collagen is an integral part of the framework that holds cells and tissues together and has been recognised as a useful matrix for improving *in vitro* culture where it can exert positive effects on the adherence, morphology, growth, migration and differentiation of a variety of cell type (Kleinman *et al.,* 1987).

## 4. CONCLUSION

The telomerase reverse transcriptase catalytic subunit provides resistance to stress-induced cell death in CHO-K1 cells, presumably through a process of healing of DNA breaks by telomerase. The results of this investigation may be potentially useful for the production of recombinant proteins by making cells more robust in large-scale cultures. The work also proved that hTERT is able to modulate key genes involved in producing collagen fibres and that this is sufficient to confer the ability of cells to adhere to a plastic surface in serum un-supplemented medium.

The high cell number and increased resistance to stress-induced apoptosis associated with hTERT expression in CHO offer significant advantages for the *in vitro* production of recombinant proteins.

The present results are the first step in documenting that the expression of hTERT in CHO cells is likely to have many applications and a significant impact on the future of the biotechnology industry.

## REFERENCES


Akiyama, M., Yamada, O., Kanda, N., Akita, S., Kawano, T., Ohno, T., Mizoguchi, H., Eto, Y., Anderson, K.C., Yamada, H., (2002). Cancer Lett., 178:187-97.

Bodnar AG, Ouellette M, Frolkis M, Holt SE, Chiu CP, Morin GB, Harley CB, Shay JW, Lichtsteiner S, Wright WE. (1998). Science 279:349-352.

Chen JL, Blasco MA, Greider CW. (2000). Cell 100:503-514.

Hegde P, Qi R, Abernathy K, Gay C, Dharap S, Gaspard R, Hughes J E, Snesrud E, Lee N, Quackenbush J. (2000). 29:548-550.

Holt SE, Shay JW. (1999). J Cell Physiol. 180:10-18.

Kleinman HK, Luckenbill-Edss L, Cannon FW, Sephel GC. (1987). Anal Biochem. 166:1-13.

Morales CP, Holt SE, Oulette M, Kiran JK, Yan Y, Wilson KS, White MA, Wright WE, Shay JW. (1999). Nature genetics 21:115-118.

Ouellette M M, Liao M, Herbert B S, Johnson M, Holt S E,; Liss H S, Shay J W, Wright W E. (2000). J Biol Chem. 275:10072-10076.

Prescott JC, Blackburn EH. (2000). Mol Cell Biol. 20:2941-2948.

Ren J G, Xia H L, Tian Y M, Just T, Cai G P, Dai Y R. (2001). FEBS Letters 488:133-138.

# Microarray Analysis of Metabolically Engineered NS0 Cell Lines Producing Chimeric Antibody

G. Khoo[1], F. Falciani[2], M. Al-Rubeai[1,3]
*[1]Department of Chemical Engineering and [2]School of Biosciences, University of Birmingham, Edgbaston, Birmingham, B15 2TT, UK, and [3]Department of Chemical and Biochemical Engineering, University College Dublin, Belfield, Dublin 4, Ireland*

**Abstract:** Large scale gene expression provides a powerful approach to the characterisation of cells transcriptional state. Thousand of genes can be monitored in single experiments generating an unprecedented volume of data. In animal cell technology, this information can be used to assign functions to previously unassociated genes, identify potential process variable targets and generate snapshots of transcriptional activity in response to any environmental factor or cellular trigger. We have used a mouse array representing 15000 genes to assess the expression profile of mouse myeloma cell line NS0 and GS-NS0 producing chimeric antibody. Comparisons of gene profiles were also made with proliferation-controlled (over-expressing p21$^{CIP1}$) and apoptosis resistant (over-expressing bcl-2) cell lines. There were 19 genes up regulated and 32 genes down regulated in the apoptosis resistant cell line compared to the parental producing cell line. As for the proliferation-controlled cell line, 54 and 147 genes were up and down regulated respectively. Gene ontology was used to understand the biological relevance of differences in gene expression data. Distinct expression signatures, indicative of observed differences in physiology and productivity between the cell lines, were identified. Our study highlights the potential of microarray technology for the analysis recombinant cell lines as affected by product expression, genetic modification and environmental conditions.

**Key words:** PGS- NS0, microarray, transcriptomics, metabolic engineering, proliferation control, apoptosis, gene ontology



*R. Smith (ed.), Cell Technology for Cell Products*, 47–52.

# 1. INTRODUCTION

As cell culture process technology matures, there is an increasing use of mammalian cells for the production of biologics. This has led to greater interest in cell engineering where research like genetic modification of cell lines and multi-cistronic expression are now gaining attention. In addition, new high throughput technologies like whole genome and proteome analysis are now being harnessed in effort to improve productivity and product efficacy (Korke *et al.*, 2002). These tools are complementary to cell culture engineering as they allow the elucidation of physiological mechanisms within cells.

The use of large scale gene expression analysis in mammalian cell culture have been limited mainly to the study of the consequence of changes in environmental conditions. In this study we examine the effect of over-expression of single gene on the gene expression profile of NS0 cell line. we compare the parental wild type mouse plasmacyoma cell line NS0 to NS0-GS producing chimeric antibody, NS0-GS over-expressing p21$^{CIP1}$ cyclin-dependent kinase inhibitor (cdki) and NS0-GS over-expressing bcl-2 anti-apoptotic protein. Single gene expression can alter physiological events drastically (Mayford *et al.*, 1995; Sapirstein and Bonventre, 2000) but such alteration is a consequence of the manifestation of interacting gene expression at various regulatory hierarchies. To study the transcriptional profile of cells and subsequently use of gene ontology (Ashburner *et al.*, 2000; Lomax and McCray, 2004); www.geneontology.org), and to look for gene pattern alterations we have employed a cDNA array containing 15000 mouse genes.

# 2. METHODOLOGY

## 2.1 Cell culture

Four cell lines were used for gene profiling. The wild type mouse plasmacyoma/myeloma cell line (WT), NS0 6A1 transfected with the glutamine synthetase (GS) expression system and expressing the gene for a human-mouse chimeric cB72.3 IgG4 antibody (Bebbington C.R. *et al.*, 1992) supplied by LONZA Biologics (Slough, U.K), NS0 6A1 bcl-2 (Tey B.T *et al.*, 2000) and NSO 6A1 p21(Watanabe S. *et al.*, 2002). Cells were grown in spinner flask at 37$^0$C before being harvested for analysis.

### 2.2 RNA isolation, labeling and hybridization

Cells were harvested, lysed and the total RNA extracted using 1ml of Trizol (Invitrogen, Rockville MD, USA) before the RNA was cleaned up using the RNeasy Midi kit (Qiagen, Santa Clara, USA). The mRNA from the NS0 cells were then separately reverse transcribed to 1st strand cDNA. The second strand is then generated using the Klenow fragment of DNA polymerase 1 to incorporate dCTP linked fluorescent dyes (Cy3 or Cy5). The differently labelled cDNA were combined Cy3 labeled pellet and/or Cy5 labeled pellet in 25 μl of hybridisation buffer. Hybridisation was done overnight and the arrays were washed.

### 2.3 cDNA description and image analysis

cDNA microarrays of 15247 unique oligo (dT)-primed cDNA clones ("NIA mouse 15K"), from Minoru Ko (NIH)(Tanaka T.S *et al.*, 2000), were generated at HGMP RC Hinxton, UK. Normalization was done using a web-based resource called GEPAS (gepas.bioinfo.cnio.es) (Herrero J *et al.*, 2003), The differentially expressed gene lists were then submitted to FatiGO (fatigo.bioinfo.cnio.es)(Al-Shahrour F *et al.*, 2004) where significant gene associations of Gene Ontology were found. P-values of less than 0.05 as well as false discovery rates (independent adjusted P-value) of less than 0.2 were used as criteria for selection of significant associations(Benjamini Y and Hochberg Y, 1995).

## 3. RESULTS AND DISCUSSION

Clustering genes by their gene ontology allows us to subdivide genes accord to 3 categories as seen in the Figure 1. A summary of the associated is given in Table 1. It is noted that associations within the 6A1 bcl-2 cell line point towards certain biological and metabolic functions being up regulated in order for to maintain the cellular processed. Genes associated with the cell cycle are down regulated as a result and this clearly explains the commonly observed effect of cell cycle arrest in bcl-2 over expressing cell lines(Tey B.T *et al.*, 2000). commonly observed effect of cell cycle arrest in bcl-2 over expressing cell lines(Tey B.T *et al.*, 2000).

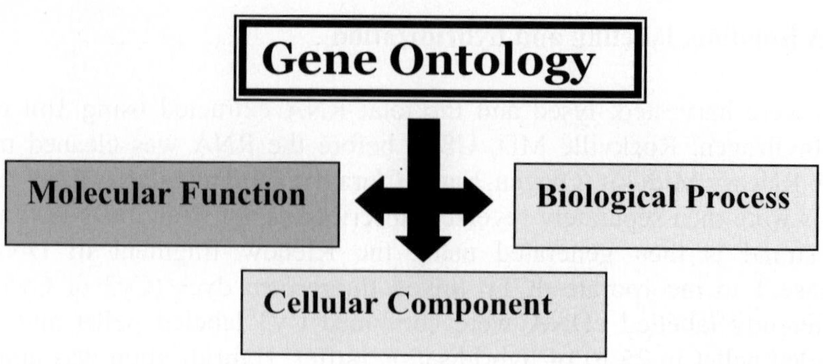


*Figure 1.* Gene Ontology categories.

Both 6A1 p21 and 6A1 bcl-2 have similar associations, showing increased gene expression for cell homeostasis, ion homeostasis and less cell proliferation. However, as stated before, the physiological effect on specific productivity is rather opposite. Despite having a larger number of genes being differentially expressed, the 6A1 p21 cells do not show other significant associations with the cell line. Differentially expressed gene patterns within the WT cell line reveal a higher number of lipid metabolism genes. There are also more genes that are localized in the cell membrane and endoplasmic reticulum while being less associated with the nucleus and chromosome.

Comparing gene changes between cell lines allows us to differentiate genes associated with bcl-2 and p21 and eliminate common features resulting arising from single gene over-expression. Upon closer inspection, it becomes obvious that in 6A1 p21 cells, there are a significant number of genes being down regulated compared to the WT. Most of these genes are associated with various functions including homeostasis, biosynthesis and enzyme activities. Similarly, compared with the 6A1 bcl-2 cells, 6A1 p21 down regulated genes are significantly different. However, 6A1 p21 up regulated genes do not show any significant patterns, possibly due to the fact that as the cell stop proliferating, reducing certain physiogical aspects associated with cell division. The use of associations across cell lines thus allows for a better understanding of the relative levels of differential gene expression. This helps overcome the limitations of associations within cell lines

*Table 1*. Summary of associations derived after statistical analysis.

| Differential gene expression associations based on Gene Ontology | | | | | | |
|---|---|---|---|---|---|---|
| | up regulated // down regulated | down regulated // up regulated | up regulated // up regulated | down regulated //down regulated | up regulated //all genes | down regulated // all genes |
| bcl-2 | Biological process: cell homeostasis, cell ion homeostasis | | | | Biological process: polyol metabolism, respiratory tube development, lactation, acylglycerol metabolism, neutral lipid metabolism, glycerolipid metabolism, epithelial cell differentiation, alcohol biosynthesis Molecular function: glycerol-3-phsophate dehydrogenase activity | Cellular component: chromosome, replication fork Biological process: cell proliferation |
| WT | Cellular component: cytoplasm, integral to membrane | biological process: nucleoside, nucleotide nucleobase and nucleic acid metabolism, regulations of metabolism Cellular component: nucleus, chromosome | | | Biological process: steriod metabolism, sterol metabolism, lipid biosynthesis, Cellular component: endoplasmic reticulum | |
| p21 | Biological process: cell homeostasis, cell ion homeostasis Cellular component: membrane | Biological process:cell proliferation | | | | Molecular function: binding Biological process: cell proliferation |

# REFERENCES

Al-Shahrour F, DÍaz-Uriarte R, and Dopazo J.2004 Bioinfornatics 20, 578-580

Ashburner, M, Ball C.A, Blake J.A, Botstein D, Butler H, Cherry J.M, Davis A.P, Dolinski K, Dwight S.S, Eppig J.T, Harris M.A, Hill D.P, Issel-Tarver L, Kasarskis A, Lewis S, Matese J.C, Richardson J.E, Ringwald M, Rubin G.M, and Sherlock G.2000 Nat.Genet 25, 25-29

Bebbington C.R., Renner G, Thomson S, King D, Abrams D, and Yarranton G.T.1992 Bio/Technology 10, 169-175

Benjamini Y and Hochberg Y.1995 Journal of Royal Statistical Society B 57, 289-300

Herrero J, Al-Shahrour F, DÍaz-Uriarte R, Mateos A, Vaquerizas J.M, Santoyo J, and Dopazo J.2003 Nucleic Acids Research 31, 3461-3467

Korke, R, Rink A, Seow T.K, Wong K, Beattie C, and Hu W.S.2002 Journal of Biotechnology 94, 73-92

Lomax, J and McCray, A. T.2004 Comparative Functional Genomics 5[354], 361

Mayford, M, Abel T, and Kandel E.R.1995. Curr.Opin.Neurobiol. 5, 141-148

Sapirstein, A and Bonventre, J. V.2000 Biochim.Biophys.Acta 1488, 139-148

Tanaka T.S, Jaradat S.A, Lim M.K, Kargul G.J, Wang X, Grahovac M.J, Pantano S, Sano Y, Piao Y, Nagaraja R, Dol H, Wood III W H., Becker K.G, and Ko M.S.2000 Proc Natl Acad Sci USA 97, 9127-9132

Tey B.T, Singh R.P., Piredda L., Piacentini M., and Al-Rubeai M.2000 Journal of Biotechnology 79, 147-159

Watanabe S., Shuttleworth J, and Al-Rubeai M.2002 Biotechnol Bioeng 77, 1-7

# Optimizing Medium for Transient Transfection

Z.-H. Geng, W. Nudson, L. Davis, S. Luo, K. Etchberger,
*JRH Biosciences, Inc. Lenexa, KS, USA*

**Abstract:** During the development of a transient transfection platform, it was observed that several commonly used components were able to support high transfection efficiency and protein production when added at certain steps of the transfection process. Transfection could be inhibited or completely diminished if such critical components were present at other steps of the process. Our data demonstrated that a transfection protocol can be as critical to efficient polyethylenimine (PEI) mediated transient transfection of human embryonic kidney (HEK) 293 EBNA cells as the transfection medium.

**Key words:** Serum-Free medium, process development, PEI, HEK.293 EBNA, transient transfection

## 1. INTRODUCTION

Because components required for cell growth and protein production can inhibit or diminish transfection efficiency when present in transfection medium, development of one medium for the entire transient transfection process can be extremely difficult. A protocol that utilizes different media at different stages of the transfection process not only improves the transfection efficiency but also the viable cell density and total protein production.

## 2. MATERIALS AND METHODS

EX-CELL 60864 (JRH Biosciences, Inc.) cultured HEK 293 EBNA (CRL-10852, ATCC) cells were diluted at log phase growth with variations



*R. Smith (ed.), Cell Technology for Cell Products*, 53–55.

of transfection media (EX-CELL™ 293, EX-CELL 60864 and EX-CELL 65237) at a 1:4 ratio to reach a cell density of (4-6)e$^5$/mL. Cells were seeded at 1mL/well in a 12-well non-tissue culture treated plate (Catalog No. 351143, Falcon, Bedford, MA). Two hours later DNA and PEI complex were added to the cell culture.

To make the DNA (P040400, Gene Therapy Systems Inc) PEI (23966, Polysciences, Inc )complex for 1mL of cell suspension, 3.6ug of PEI was added to 200uL of transfection medium, vortex for 30 seconds, 2.4ug of DNA was then added in the mixture. After 30 minutes, the DNA/PEI complex was added to the cell suspensions.

The level of GFP was quantitated after 72 hours post transfection by fluorescence microscopy using a Spectra MAX Gemini XS plate reader ($\lambda$ex=480 nm $\lambda$em=510 nm).

## 3. RESULTS

The data from Figure 1 showed that the formulations containing either dextran sulfate, a high concentration of hydrolysate or a high concentration of phosphate, yielded zero or significantly reduced transfection efficiencies.

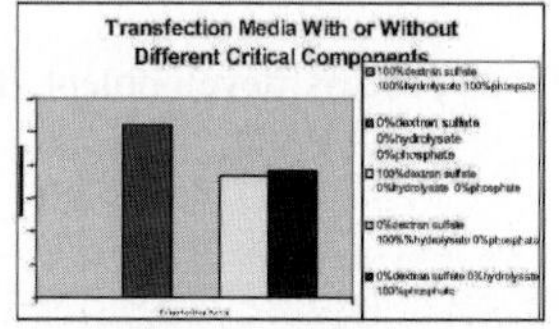


*Figure 1.* GFP fluorescence intensity comparison for different transfection media.

Without these components in the medium viable cell density was decreased (Figures 2 and 3).

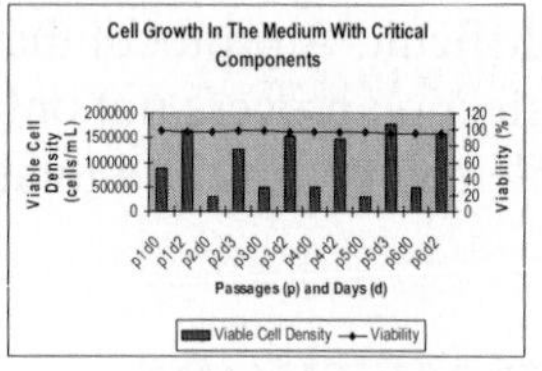


*Figure 2.* Cell growth in complete media.

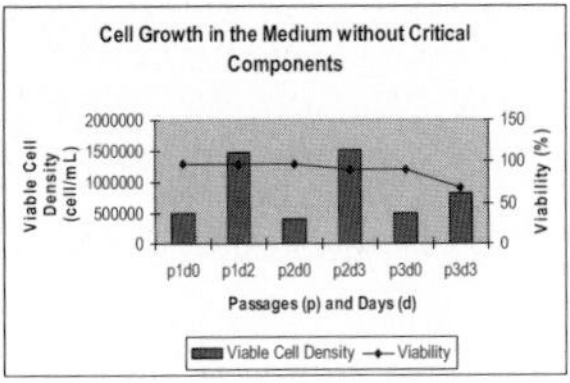


*Figure 3.* Cell growth in deficient medium.

Feeding the cells with complete medium containing dextran sulfate, hydrolysate and phosphate 16 hours after transfection improved total protein productivity as shown in Figure 4.

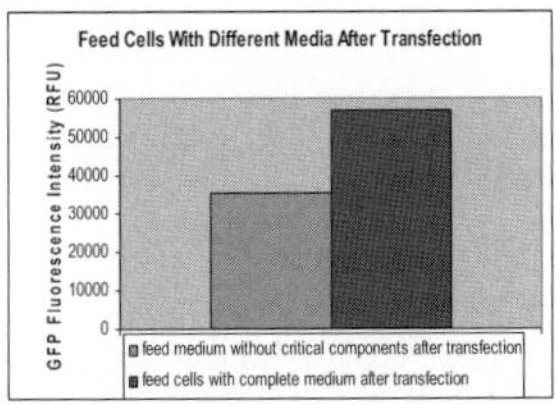


*Figure 4.* GFP fluorescence intensity for the transfected GFP cells with different media.

# 4. CONCLUSION

Dextran sulfate has the most significant negative effect on transfection efficiency. It completely inhibited transfection. At 100% concentration hydrolysate and phosphate also demonstrated significant negative effects on transfection efficiency by causing 20% reduction in efficiencies. However, by performing the transfection process in a deficient medium, followed by feeding cells with a complete medium 16 hours after transfection, 50% transfection efficiency were achieved in combination with good viable cell densities, viabilities and protein production.

# Establishment of Recombinant CHO Cell Lines Under Serum-free Conditions

Maya Kuchenbecker, Bernd Rehberger, Michael Schriek, Judith Gackowski, Ralf Fehrenbach
*Rentschler Biotechnologie GmbH, Erwin-Rentschler-Straße 21, 88471 Laupheim, Germany*

**Abstract:** Due to regulatory requirements today's state-of-the-art cell culture production processes are performed in well defined serum- or protein-free media. Nevertheless, development of recombinant production cell lines is often performed under serum-containing conditions. Accordingly, after clonal selection the cells have to be adapted to serum-free medium, which bears the risk of losing productivity or product quality and significantly extends development time. Through extensive medium screening and testing of multiple transfection methods a CHO dhfr$^-$ host cell based procedure was established, allowing the serum-free performance of all steps being essential for the development of a recombinant cell line

**Key words:** recombinant; CHO dhfr-; serum-free; transfection; selection; amplification; development time; medium screening; single cell cloning; cell-specific productivity; medium optimization; cell line development

## 1. TRANSFECTION

Testing different methods, transfection under serum-free (SF) conditions was enabled using Nucleofection™ technology.

Nucleofection™ was successfully applied to various CHO cell lines, although transfection efficiency differed between the tested cells (Fig. 1A).



*R. Smith (ed.), Cell Technology for Cell Products*, 57–61.

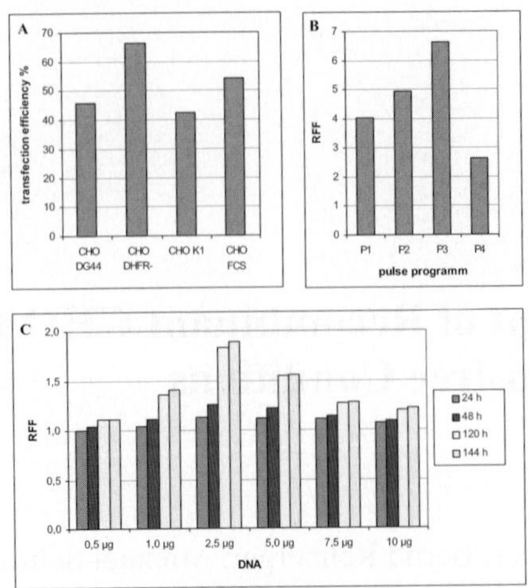


*Figure 1.* Optimization of cell transfection.

CHO cells were transiently transfected with an expression vector comprising a GFP reporter gene. Protein expression was quantified by fluorescence measurement; RFF = relative fluorescence factor; **A** Transfection efficiency for various CHO cell lines; **B** Optimization of the transfection efficiency of a CHO DHFR$^-$ cell line applying different pulse programs and **C** various DNA concentrations.

By optimizing the pulse program (Fig. 1B) as well as cell number / DNA ratio (Fig. 1C) a high efficient method for SF transfection was developed.

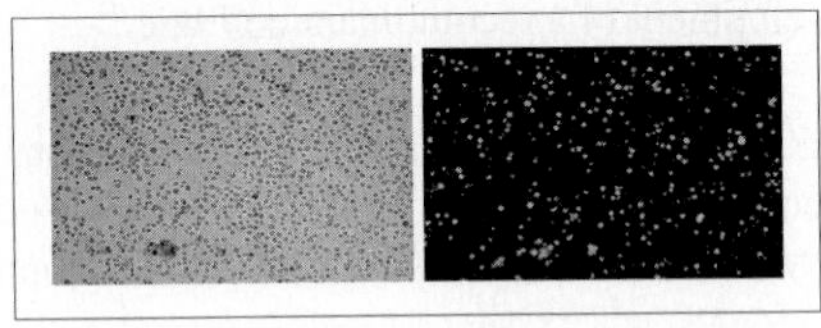

*Figure 2.* CHO DHFR$^-$ cell line transfected under optimized conditions.

Cells transfected with an expression vector comprising a GFP reporter gene are shown with bright field (left) and fluorescence microscopy (right).

## 2. SINGLE CELL CLONING

Under SF conditions the successful generation of clones derived from a single cell strongly depends on the medium used. Substantial medium screening and optimization yielded in a few SF formulations that are suitable to support single-cell cloning by limiting dilution.

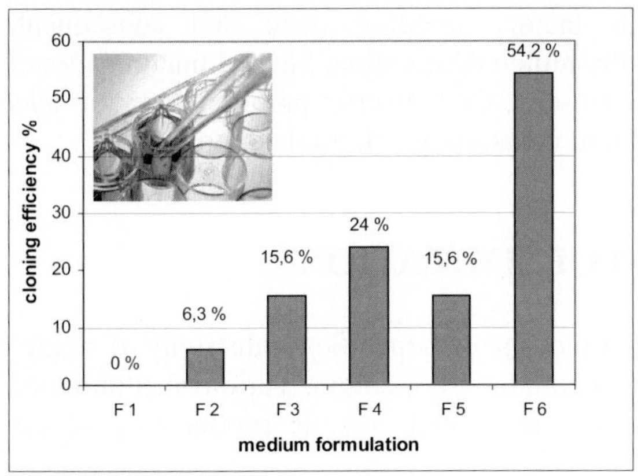


*Figure 3.* Cloning efficiency in different media.

To assess the suitability of different serum-free medium formulations (F1–6) for single cell cloning, in an initial screening 5 cells per well were seeded into 96-well plates and the number of wells showing cell growth were evaluated.

Choosing an appropriate medium single cell cloning, seeding one cell per well into 96-well plates, was enabled.

During the development of a growth factor producing cell line, single cell cloning gave rise to a five-fold increase of cell-specific productivity compared to the cell pool obtained after stable transfection of a CHO DHFR$^-$ cell line.

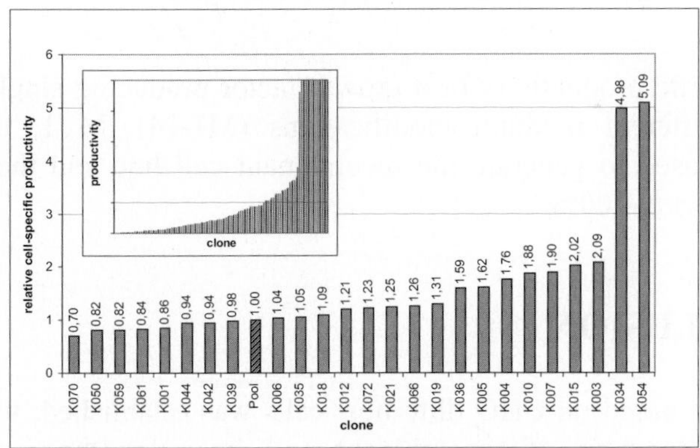


*Figure 4.* Productivity of single cell clones compared to the starting cell pool.

The cell pool was obtained by transfection of a CHO DHFR$^-$ cell line with a growth factor encoding gene and subsequent selection by hypoxanthine / thymidine deprivation. Several hundred clones were screened for productivity (insert). Cell-specific productivities for selected clones are shown in relation to the starting cell pool (set to 1).

## 3. MEDIUM OPTIMIZATION

Further increase of the cell-specific productivity of single cell clones was obtained by optimizing the SF medium. Various medium modifications were tested and an increase of cell specific productivity of up to 400% was reached.

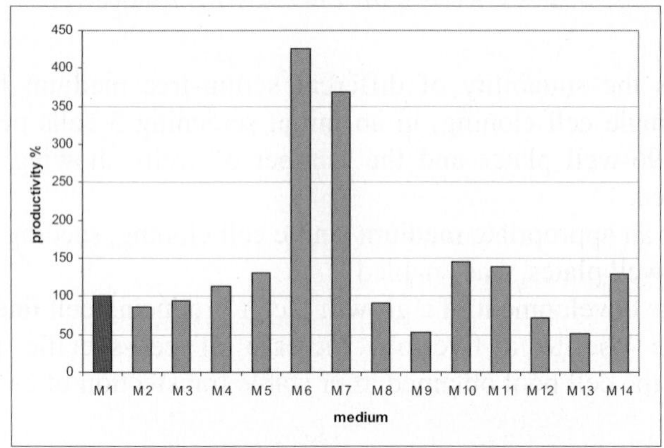


*Figure 5.* Medium optimisation.

Cell-specific productivity of a growth factor producing single cell clone grown in different medium modifications (M1-14). M 1, the medium formulation used to generate the recombinant cell line and the single cell clones, was set to 100%.

## 4. CONCLUSION

A method based on CHO dhfr$^-$ host cells was established, which allows the SF development of recombinant cell lines for the production of biopharmaceutical proteins.

All development steps, including transfection, selection, amplification and single cell cloning were performed under SF conditions.

In addition to regulatory compliance the established procedure significantly shortens the time for cell line development compared to a serum-containing process.

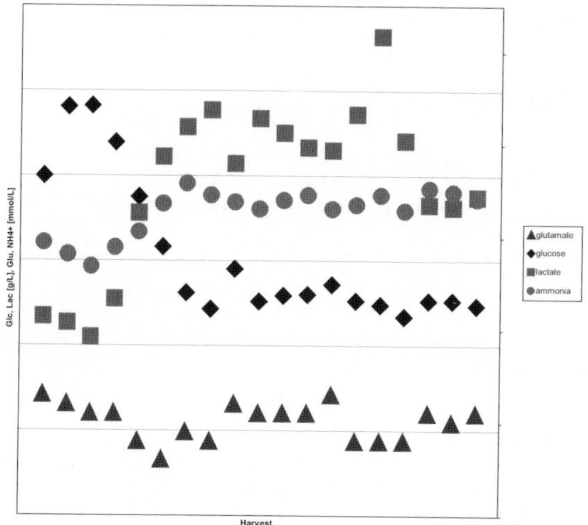

# Novel Cell-based Assay for the Detection of Murine type I IfN

Mariela Bollati-Fogolín, Werner Müller
*Department of Experimental Immunology, German Research Centre for Biotechnology (GBF), Braunschweig, Germany.*

**Abstract:** Type I Interferon (IFN) comprises IFN-alpha and IFN-beta, which are potent biologically active proteins secreted by virus-infected cells. IFN antiviral activity assays were the first type of bioassays developed to measure the relative potency of IFN preparations. Here we report a new virus free, cell-based assay to quantify murine type I IFN. It basically consists of an indicator cell line in which the Cre-recombinase transcription is driven by the IFN-inducible Mx1 promoter and, when expressed, deletes a stop cassette upstream of the eGFP coding region, resulting in the expression of eGFP. The percentage of eGFP expressing cells accurately correlates to the amount of type I IFN added to the culture and can easily be monitored.

**Key words:** reporter gene assay, eGFP, murine type I IFN, Cre-recombinase

# 1. INTRODUCTION

Mx is a typical type I IFN-inducible gene that has been used to generate a mouse strain containing the Cre-recombinase under the control of the Mx1 promotor (Kühn *et al.*, 1995). Indicator mice strain enabling the expression of marker genes upon conditional gene inactivation have recently been created (i.e. the RA/EG mice, see Constien *et al.*, 2001). On this basis, and in order to generate a virus free, cell-based assay, we crossed MxCre mice with RA/EG mice. The novel fibroblastic cell lines proved to be a reliable tool to determine murine type I IFN activity and are expected to be a valuable source to screen for compounds interfering with murine type I signalling.



*R. Smith (ed.), Cell Technology for Cell Products*, 63–65.

# 2. RESULTS

## 2.1 Generation of indicator cell lines: Principle of the IFN-based bioassay

Males homozygous for the RA|EG gene were crossed with MxCre females. Embryonic fibroblasts were isolated according to Torres and Kühn, 1997. Two CRE-positive primary cultures (named as Mx/Rage 4 and Mx/Rage 7) were selected and immortalized with retroviruses (Jat and Sharp, 1989).

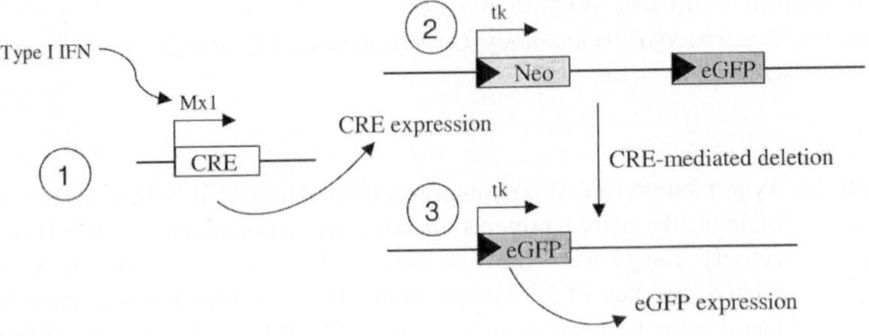


*Figure 1.* Principle of the mIFN-based bioassay (black triangles indicate loxP sites).

In this novel cell line (see Fig. 1), the inducible mouse Mx1 promoter controls the expression of CRE-recombinase gene (1) and is activated upon murine type 1 IFN exposure (2). Consequently, Cre-mediated recombination leads to activation of eGFP transcription by positioning the thymidine kinase (TK) promoter in front of the start site of the promoterless eGFP ORF (3). The percentage of eGFP-expressing cells can be determined by flow cytometry or fluorescence microscopy.

## 2.2 Cell-based assay for the detection of type I IFN

The system was validated using murine type I IFN and measuring the percentage of eGFP-expressing cells by FACS-analysis. Typical dose-response curves were obtained with different amounts of type I mIFN either alpha- or beta-type (Fig. 2-A). A reproducible and linear response was obtained between 20-300 and 10-500 U/ml for mIFN alpha and beta, respectively. Two batches of HEK-293E-derived mIFN alpha were used to test the capacity of this system to quantify murine type I IFN preparations. The titration of the samples displayed a wide range of linearity and paralleled the standard's dose-response curve (Fig. 2-B).

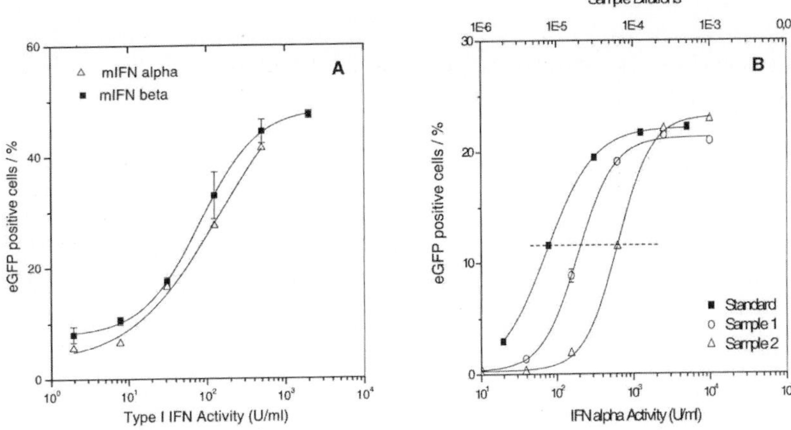


*Figure 2.* Determination of the biological activity of type I IFN: A) Dose-response curve for alpha and beta type IFN. B) Measurement of two HEK-293E derived mIFN alpha samples.

# 3. CONCLUSIONS

The system here described presents the advantage that the induction phase is separated from the readout and, thus, appears suitable for the design of robust high-throughput assays. In comparison with viral-based systems, this method is a biological safe alternative to estimate murine type I IFN activity. In addition, it can also be applied to screen substances able to induce or inhibit murine type I IFN expression. Finally, this novel system offers the potential to be applied for the measurement of human type I IFN upon transfection of these cell lines with a human IFN receptor.

# REFERENCES

Constien, R., Forde, A., Liliensiek, B., Grone, H.J., Nawroth, P., Hammerling, G., Arnold, B. (2001) Characterization of a novel eGFP reporter mouse to monitor Cre recombination as demonstrated by a Tie2 Cre mouse line. Genesis. 30, 36-44.

Jat, P.S., and Sharp, P.A. (1989) Cell lines established by a temperature-sensitive simian virus 40 large-T-antigen gene are growth restricted at the non-permissive temperature. Mol Cell Biol. 9, 1672-1681.

Kühn, R., Schwenk, F., Aguet, M., Rajewsky, K. (1995) Inducible gene targeting in mice. Science 269, 1427-1429.

Torres, R.M., and Kühn, R. (1997). Laboratory protocols for conditional gene targeting. Oxford University Press, Oxford. pp. 1–167.

# Characterisation of the Induction of Cytochrome p450 Enzymes in Primary Cultures of Human Hepatocytes

T.L. Freeman[1], A.P. Chadwick[2], M.S. Lennard[3], C.M. Martin[1], R.G. Turcan[2] and D. Blond[1]

*[1]Biotechnology Department, Covance Laboratories, Harrogate, UK [2]Metabolism Department, Covance Laboratories, Harrogate, UK and [3]Academic Unit of Clinical Pharmacology, Pharmacokinetics and Pharmacogenetics Group, Division of Clinical Sciences South, University of Sheffield, Sheffield, U.K.*

**Abstract:** The induction of drug metabolising enzymes e.g. cytochrome P450s (CYPs), *in vivo*, by drugs, can attenuate the desired clinical effect or can lead to potential pharmacological effects and/or toxicity. The ability to screen for these undesirable characteristics at an early stage of drug development *ex vivo* is clearly an advantage economically and ethically. Validated methods for the determination of the enzyme activities of a range of CYPs have been established at Covance. This work aims to develop a real time quantitative polymerase chain reaction (RT-QPCR) assay to monitor the induction of individual CYPs in human hepatocytes samples by quantitatively measuring mRNA compared to protein. We have designed Taqman primers and probes for human CYP3A4, CYP2B6 and CYP1A2 and the RT-QPCR method has been developed and validated. Using the validated methods described, the relationship between mRNA expression and enzyme activity will be determined in fresh human hepatocytes.

**Key words:** Human hepatocytes; Cytochrome P450; Induction; Quantification; Real-time RT-PCR; TaqMan

## 1. INTRODUCTION

Drug interactions are a major concern both for clinicians and the pharmaceutical industry. A regulatory requirement during the pharmaceutical development process is the *ex vivo* evaluation of the potential of a substance to induce drug metabolising enzymes. *In vivo*



*R. Smith (ed.), Cell Technology for Cell Products*, 67–70.

induction of enzymes which metabolise compounds to inactive products can attenuate the desired clinical effect, whereas increased production of active products or intermediates can lead to potential pharmacological effects and/or toxicity. The ability to screen for undesirable characteristics at an early stage of drug development is clearly an advantage economically and ethically (Bowen *et al.* 2000).

Cytochromes are involved metabolism of various foreign compounds including drugs. It is important to ascertain whether these enzymes are induced or inhibited by a new, potentially therapeutic compound. Various methods have been used in order to measure the expression of cytochromes in biological samples (Rodriguez-Antona *et al.* 1999). However, the use of human hepatocytes has come to be regarded as the 'gold standard' for the *in vitro* quantification of cytochrome P450 (CYP) induction (Roymans *et al.* 2004). Most drugs induce CYP isoforms by upregulating gene transcription, causing increased synthesis of mRNA and enzyme. At present, established enzymatic assays are used at Covance to detect CYP induction. The objective of this study was to develop a real time reverse transcriptase quantitative polymerase chain reaction (RT-QPCR) assay to monitor the induction of individual CYPs in human hepatocyte samples by quantitatively measuring mRNA and comparing this to enzyme activity.

RT-QPCRs to detect the expression of CYP3A4, CYP2B6 and CYP1A2 mRNAs were chosen to be developed initially due to the significance of these CYPs (Plant *et al.* 2003, Faucette *et al.* 2000).

## 2. MATERIALS AND METHOD

TaqMan primers and probes for human CYP3A4, CYP2B6, CYP1A2 and GAPDH mRNA were designed using the Primer Express software version 1.1 (Applied Biosystems). The primers and probes were designed respectively from the NCBI database sequences NM_017460, NM_000767, NM_000761 and NM_002046.

Assays were performed using the Applied Biosystems ABI Prism® 7700 sequence detection system. Using human liver Total RNA (Ambion®) to generate a standard curve, RT-QPCRs for CYP3A4, CYP2B6 and CYP1A2 mRNA relatively efficient to the RT-QPCR for the endogenous control, GAPDH mRNA, were developed following the principles detailed in the ABI Prism® 7700 Sequence Detection System User Bulletin #2. Following this, the RT-QPCR methods for GAPDH, CYP3A4, CYP2B6 and CYP1A2 were then developed and validated in the spirit of the ICH guideline on the validation of analytical procedures, QB2.

## 3. RESULTS

Table 1 summarises the results obtained from the development and validation.

*Table 1.* Results Summary. The results were considered acceptable if less than 6%.

| | GAPDH | CYP3A4 | CYP2B6 | CYP1A2 |
|---|---|---|---|---|
| Relative Efficiency | N/A | $10 - 1x10^6$ pg | $1 - 1x10^6$ pg | $10 - 1x10^5$ pg |
| Specificity | Human<br>Mouse | Human | Human<br>Mouse | Human |
| Linearity and Range | $1 - 1x10^4$ pg | $1 - 1x10^5$ pg | $10 - 1x10^5$ pg | $100 - 1x10^5$ pg |
| Accuracy | N/A | N/A | N/A | N/A |
| Precision | <2.24% | <3.44% | <5.30% | <2.15% |
| Quantification Limit | 1pg | 1pg | 10pg | 10pg |
| Operator Robustness | <1.82% | <3.61% | <3.44% | <2.28% |
| Reagent Robustness | <1.59% | <3.00% | <2.27% | <0.88% |

## 4. DISCUSSION

The methodology used for the amplification and detection of GAPDH, CYP3A4, CYP2B6 and CYP1A2 mRNAs was deemed to be validated.

For measuring cytochrome P450 induction, the range of hepatocyte total RNA to be used will be $10\text{-}1x10^4$, $10\text{-}1x10^4$ and $100\text{-}1x10^4$ pg for the detection of CYP3A4, CYP2B6 and CYP1A2 induction respectively.

Work has begun on in vitro cytochrome P450 induction in human hepatocytes using the inducers rifampicin (CYP3A4), omeprazole (CYP2B6) and phenobarbital (CYP1A2) with enzyme activity quantified using the selective substrates testosterone (CYP3A4), bupropion (CYP2B6) and ethoxyresorufin (CYP1A2). mRNA analysis using the RT-QPCRs described here will be carried out in parallel and the results generated compared with the enzyme activity data.

Total RNA extraction methods have been evaluated.

Preliminary work has shown that the RT-QPCRs have been successful in detecting the cytochrome P450 mRNA extracted from human hepatocytes cultured in our laboratories (results not shown). In addition to detecting mRNA, the RT-QPCRs for GAPDH and CYP2B6 also appeared to detect DNA. This is believed to be due to inadequate DNA removal during the RNA extraction. In these assays, detection of DNA is possible due to the amplicon size, whereas the assays for CYP3A4 and CYP1A2 have larger amplicons and therefore are not detected using the current thermal cycler conditions.

# REFERENCES


Bowen WP, Carey JE, Miah A, McMurray HF, Munday PW, James RS, Coleman RA and Brown AM (2000) Measurement of Cytochrome P450 Gene Induction in Human Hepatocytes using Quantitative Real-time Reverse Transcriptase-Polymerase Chain Reaction. Drug Metab and Dispos **28:**781-788

Rodriguez-Antona C, Jover R, Gomez-Lechon M and Castell JV (1999) Quantitative RT-PCR Measurement of Human Cytochrome P-450s: Application to Drug Induction Studies. Arch Biochem Biophys **376:**109-116

Roymans D, Van Looveren C, Leone A, Brandon Parker J, McMillian M, Johnson MD, Koganti A, Gilissen R, Silber P, Mannens G and Meuldermans W (2004) Determination of Cytochrome P450 1A2 and Cytochrome P450 3A4 Induction in Cryopreserved Human Hepatocytes. Biochem Pharmacol **67:**427-437

Plant NJ and Gibson GG (2003) Evaluation of the Toxicological Relevance of CYP3A4 Induction. Curr Opin Drug Discov Devel **6:**50-56

Faucette SR, Hawke RL, Lecluyse EL, Shord SS, Yan B, Laetham RM and Lindley CM (2000) Validation of Bupropion Hydroxylation as a Selective Marker of Human Cytochrome P450 2B6 Catalytic Activity. Drug Metab Dispos **28:**1222-1230

ABI Prism® 7700 Sequence Detection System User Bulletin #2 on Relative Quantification of Gene Expression

International Conference on Harmonisation of Technical Requirements for Registration of Pharmaceuticals for Human Use (ICH) (1996) Guidance for Industry. QB2 Validation of Analytical Procedures: Methodology. Food and Drug Administration Federal Register: May 19, 1997 (62 FR 27464)

# Effect of Culture Conditions on Glycosylation of Recombinant beta-Interferon in CHO Cells

Maureen Spearman, Jose Rodriguez, Norm Huzel, Kevin Sunley and Michael Butler
*Department of Microbiology, University of Manitoba, Winnipeg, Manitoba, Canada, R3T 2N2*

**Abstract:** CHO cells that produce human recombinant beta-interferon (β-IFN) have been grown under several different culture conditions in an attempt to increase the yields of the therapeutic glycoprotein whilst maintaining consistent glycosylation and minimizing intermolecular aggregation. Supplementation of cultures with either sodium butyrate or sodium chloride increases the productivity of β-IFN. However, sodium butyrate alters the glycosylation profile with an increase in the proportion of more highly branched complex N-linked glycans. A shift in temperature during the culture period from 37°C to 30°C maintains consistent glycosylation of β-IFN for long culture periods with increased overall productivity and reduced aggregation compared to 37°C cultures. The use of Cytopore 1 microcarriers that entrap the cells also increases volumetric productivity of non-aggregated β-IFN while maintaining consistent glycosylation throughout the culture period.

**Key words:** beta-interferon, glycosylation, recombinant protein, temperature, osmolality, sodium chloride, sodium butyrate, Cytopore, microcarrier, bioreactor, aggregation, glycan, sialic acid, glycerol, CHO

## 1. INTRODUCTION

The production of recombinant proteins for biopharmaceuticals has proven to be a major benefit in the treatment of many diseases. The challenges in producing clinically reliable recombinant proteins are maintaining consistency in production and ensuring a high yield (reviewed



*R. Smith (ed.), Cell Technology for Cell Products*, 71–85.

in Wurm 2004, Butler 2005). A large majority of recombinant proteins are glycosylated, adding an extra factor in monitoring product quality.

Many culture conditions have been shown to affect the glycosylation of recombinant proteins such as pH, osmolality, dissolved oxygen, temperature, high shear stress, use of serum and growth on microcarriers. As well, media additives such as sodium butyrate appear to be variable in their affect on glycosylation depending on the cell type, recombinant protein and other culture conditions. The viability and condition of the culture also can affect glycosylation of the product with the accumulation of ammonia (Yang and Butler, 2000 and 2002), nutrient limitation and feeding protocol and the production of glycosidases in later stages of culture. Glycosylation is also dependent on the host cell.

Consistent glycosylation of many recombinant proteins is essential for therapeutic efficacy. Glycosylation can affect the clearance rate of a protein from the patient's circulatory system, immune recognition, biochemical activity, structural stability and aggregation. These problems have been observed for the therapeutic protein, human recombinant beta-interferon (β-IFN), which is used to treat Multiple Sclerosis. A non-glycosylated β-IFN has been produced in E.coli, however, the stability, solubility and biological activity of β-IFN is increased upon the appropriate N-glycosylation at Asn-80 (Karpusas *et al.*, 1998; Runkel *et al.*, 1998). The glycosylated β-IFN produced in mammalian cells also induced fewer neutralizing antibodies in vivo (reviewed in Giovannoni *et al.*, 2002). The hydrophobic protein has a molecular weight of 17.5 kDa which is increased to 25 kDa with the addition of the N-linked oligosaccharide.

The hydrophobicity of β-IFN is a potential problem in production because of its tendency to aggregate in solution. This also reduces the detectability of the β-IFN by ELISA (Rodriguez *et al.*, 2005). The presence of glycan reduces aggregation possibly by shielding the hydrophobic region of the protein (Runkel *et al.*, 1998). Addition of glycerol to the media (Rodriguez *et al.*, 2005) and growth of the β-IFN producing CHO cells on the microcarrier Cytopore (Spearman *et al.*, 2005) stabilized the secreted β-IFN by reducing aggregation and maintaining glycosylation similar to control cultures.

The objective of this research is to continue the study of bioprocess conditions that minimize aggregation and provide consistent glycosylation of β-IFN while at the same time maximizing product yield. This work compares the production and glycosylation of β-IFN under various culture conditions.

# 2. METHODS

## 2.1 Cell Line and Cell Culture

A CHO cell line, transfected with the gene for human β-IFN, was provided by Cangene Corporation (Winnipeg, Canada). Protein expression was constitutive and did not require selective pressure during culture. CHO clone 674 was cultured in serum-free media, CHO-SFM (Biogro Technologies Inc., Winnipeg). Spinner flasks (100 ml) were inoculated at 1 x $10^5$ cells/ml, stirred at 45 rpm and maintained at 37°C in an atmosphere of 10% $CO_2$. For the temperature shift experiments, cultures were grown for 2 days at 37°C prior to a change to 30°C.

Suspension cultures in a controlled bioreactor (3L Applikon) were established with an inoculum of 1 x $10^5$ cells/ml, in a working volume of 2 L and maintained at pH 7.1, dissolved oxygen of 50%, and agitation speed of 100 rpm. Microcarrier cultures were established with Cytopore (1 mg/ml) (Amersham Biosciences) at an initial agitation speed of 45 rpm for one day and then increased to 110 rpm. Sodium butyrate (1mM) was added to selected cultures at 48 hours post inoculation.

## 2.2 β-IFN Determination

Interferon was analyzed in media samples using a specific enzyme-linked immunosorbent assay (ELISA), as previously described (Spearman *et al.*, 2005). The samples values were compared to a standard curve of β-IFN (US Biologicals). Selected samples (100 μl) were denatured by boiling with 1 μl of SDS (10%) and 1 μl of β-mercaptoethanol prior to the ELISA. This treatment ensured disaggregation and denaturation of the protein to maximize the ELISA response.

## 2.3 β-IFN Purification

Culture supernatants were passed through a Hi-Trap Blue column (Amersham Pharmacia Biotech) previously equilibrated with 20 mM sodium phosphate, 0.15M NaCl (pH 7.2) (Buffer A). The flow through was collected, and the column was washed with 35 ml of Buffer A and then further washed with 35 mL of 20 mM sodium phosphate buffer, 2 M NaCl, pH 7.2. The column was eluted with 20mM sodium phosphate, 2M NaCl buffer containing 50% ethylene glycol. The β-IFN containing fraction was dialyzed overnight against phosphate buffer saline (PBS) (Invitrogen) containing 2% glycerol and frozen at -20°C. This preparation was concentrated using Ultrafree-4 Centrifuge filter unit (10K cutoff, Millipore). The Hi-Trap column binds proteins by electrostatic and hydrophobic interaction, and the purification removed background proteins, as evidenced by SDS-PAGE analysis.

### 2.4 Gel Electrophoresis and In-Gel Release of N-glycans

Purified β-IFN samples were run on a 12% SDS-PAGE and stained with Coomassie blue stain. The N-linked glycans were removed from the gel with a scalpel, washed and released from the protein bands with overnight incubation at 37°C with PNGase F (Roche Diagnostics) (Küster *et al.*, 1997). The isolated N-linked glycans were labeled with 2-aminobenzamide (2-AB) according to the method of Bigge *et al.* (1995).

### 2.5 Glycan analysis

2-AB-labelled glycans were separated by normal phase HPLC according to the method of Guile *et al.* (1996). The HPLC instrument consisted of a Waters system with binary pumps, autosampler, and fluorescent detector (excitation wavelength 330 nm and an emission wavelength 420 nm. The glycans were separated using a TSK-GEL Amide-80 column (250 mm x 4.6 mm) (Tosoh Biosep) with a gradient of 50 mM ammonium formate (Buffer A) and acetonitrile (Buffer B) at 30°C. A linear gradient was run from 20% to 58% Buffer A over 150 min (flow rate 0.4 mL/min), followed by another linear gradient to 100% A over 3 min (0.4 mL/min). The elution was calibrated using a 2-AB labeled dextran ladder (glucose homopolymer) and several standard 2-AB labeled N-linked glycans (Prozyme).

### 2.6 Exoglycosidase Digestion

Aliquots of 2-AB labeled glycans were dried in 0.6 ml tubes. A digest array buffered with 50 mM sodium acetate, pH 5.5 consisted of arrays of exoglycosidases (at final concentrations) as follows: Clostridium perfringens sialidase (0.4 U/ml), bovine testes β-galactosidase (0.2 U/ml), Jack bean β-N-acetylhexosaminidase (1.0 U/ml) and bovine kidney α-fucosidase (0.1 U/ml). Digestions were incubated at 37°C for 18 hours. Enzymes were removed by Micropure-EZ enzyme removers (Millipore). The enzymes were from Glyko except the sialidase which was from Sigma.

## 3. RESULTS AND DISCUSSION

### 3.1 Cell Growth and β-IFN Production

Cell growth and volumetric β-IFN production were determined for transfected CHO cells under various culture conditions (sodium butyrate,

NaCl and with macroporous microcarriers, Cytopore 1) in a controlled 2L Applikon bioreactor (Table 1). The extent of molecular aggregation of β-IFN was determined from the difference in ELISA response between untreated samples (nondenatured) and samples treated under conditions to cause protein denaturation by boiling with SDS and mercaptoethanol prior to the ELISA.

A control bioreactor culture with normal culture conditions and no media additives produced 3.2 x $10^6$ cells/ml by day 7 of batch culture. Under non-denaturing conditions the ELISA detected only 0.4 x $10^6$ units/ml β-IFN, however, in the denatured sample 3.0 x $10^6$ units/ml were detected. At day 7 this resulted in 86% of the β-IFN present in an aggregated form which was undetectable by the ELISA. This represents a very high percentage of β-IFN that is potentially not useful as a therapeutic agent. Therefore, various culture conditions and media additives have been tested in an attempt to improve the productivity of monomeric β-IFN.

To improve the yield of β-IFN, sodium butyrate (1 mM) was added to a bioreactor culture at 48 hours post inoculation. At day 5 of culture the volumetric productivity of β-IFN was 5 fold higher as measured under non-denaturing conditions and approximately 2.5 fold higher as measured under denaturing conditions compared to the control bioreactor. However, 73% of the β-IFN was aggregated. By day 6 the extent of β-IFN aggregation in the sodium butyrate culture was so great as to cause precipitation from solution.

The osmolality of culture media was increased by the addition of NaCl (40 mM) prior to inoculation of the culture. This slightly reduced the cell yield compared to the control culture, but increased the volumetric productivity of the culture by over 5 fold under non-denaturing conditions of the ELISA and 1.8 fold increase with denaturing of the β-IFN. This resulted in a reduction in the % aggregation of the β-IFN compared with the control bioreactor, from 86% to 62%.

A bioreactor culture containing the macroporous microcarrier Cytopore 1 was used to increase productivity of the culture. We previously optimized conditions for growth of the β-IFN CHO cells with Cytopore microcarriers (Spearman *et al.*, 2005) and used a 1 mg/ml microcarrier concentration for the bioreactor culture. The volumetric production of β-IFN, measured under non-denaturing conditions, was 5 fold higher than the control culture and 30% higher as measured under denaturing conditions. This was a 48% decrease in the aggregation of the β-IFN compared to control conditions. This suggests that the Cytopore 1 microcarrier culture is an effective alternative to suspension culture for the production of β-IFN in order to reduce the aggregation of the product.

*Table 1.* Effect of culture conditions on cell growth, productivity and aggregation of β-IFN.

| **Culture Condition** | **Culture Day** | **Cell Yield** Cells/ml | **Volumetric Productivity** (nondenatured) Units/ml | **Volumetric Productivity** (denatured) Units/ml | **% Aggregation** |
|---|---|---|---|---|---|
| Control | 7 | $3.2 \times 10^6$ | $0.4 \times 10^6$ | $3.0 \times 10^6$ | 86 |
| Sodium Butyrate (1 mM) | 5 | $1.5 \times 10^6$ | $2.0 \times 10^6$ | $7.5 \times 10^6$ | 73 |
| Sodium Chloride (40 mM) | 7 | $2.3 \times 10^6$ | $2.1 \times 10^6$ | $5.5 \times 10^6$ | 62 |
| Cytopore 1 (1mg/ml) | 7 | $2.5 \times 10^6$ *nuclei/ml | $2.2 \times 10^6$ | $4.0 \times 10^6$ | 45 |

The cultures were established in 2L bioreactors. β-IFN was determined by ELISA from samples at day 7 of culture, except for the sodium butyrate culture which was sampled at day 5.

*Cell yield was determined by trypan blue exclusion counts except for Cytopore 1 which was determined by crystal violet staining of nuclei.

$$\% \text{ Aggregation} = 100 - \left( \frac{\text{non-denatured } \beta\text{-IFN productivity} \times 100}{\text{denatured } \beta\text{-IFN productivity}} \right)$$

## 3.2 Effect of Temperature on Growth and β-IFN Productivity

Spinner flask cultures of β-IFN producing CHO cells were grown under various temperature conditions to determine the effect of reduced culture temperature on cell growth, productivity and aggregation of β-IFN (Table 2). The control culture grown at 37°C had a high cell yield at day 6 of $3.45 \times 10^6$ cells/ml. The total volumetric productivity measured by ELISA under denaturing conditions was $1.7 \times 10^6$ Units/ml with 44% of the β-IFN in the aggregated form.

Decreasing the temperature to 30°C reduced the growth of the cells to 50% of the control cultures. However, the volumetric productivity of the temperature shift culture was enhanced by 3-fold over the control culture and was also accompanied by a 50% reduction in the aggregation of the β-IFN. Initiating the culture at 30°C resulted in much slower growth of the cells resulting in a yield of only $0.69 \times 10^6$ cell/ml by day 10 of culture. The productivity of the culture was only slightly higher than the control culture at day 6 and there was only a small reduction in the aggregation. Addition of 2% glycerol to a 37°C culture increased the productivity of the culture by 2.5 fold, however, the aggregation was increased by 50%. These results show the effectiveness of temperature shift culture for increased productivity of β-IFN with the added benefit of a reduction in β-IFN aggregation.

*Table 2.* Effect of temperature on growth and beta-interferon production in suspension cultures in spinner flasks

| **Culture Condition** | **Cell yield** Cells/ml | **Volumetric Productivity** (non-denatured) Units/ml | **Volumetric Productivity** (denatured) Units/ml | **% Aggregation** |
|---|---|---|---|---|
| 37°C | 3.45 x $10^6$ | 0.96 x $10^6$ | 1.70 x $10^6$ | 44 |
| 37°C to 30°C | 1.80 x $10^6$ | 4.40 x $10^6$ | 5.60 x $10^6$ | 22 |
| 30°C | 0.69 x $10^6$ | 1.50 x $10^6$ | 2.35 x $10^6$ | 37 |
| 37°C Glycerol (2%) | 2.05 x $10^6$ | 1.70 x $10^6$ | 4.50 x $10^6$ | 62 |

The measurements were made at the point of maximum cell yield. This was at day 6 for all cultures except the 30°C culture, which was at day 10. The cell viability was > 90% in all samples.

### 3.3 NP-HPLC analysis of β-IFN glycans produced in bioreactors under different conditions

Altering culture conditions to improve the yield of recombinant proteins with culture additives and temperature and microcarriers has proven effective. However, the glycosylation of β-IFN is important for its stability and quality as a therapeutic agent. We have therefore analyzed the glycans under these altered growth conditions to monitor their effect on the glycan structure of the β-IFN.

The NP-HPLC profile of β-IFN glycans from a control bioreactor (day 6) showed two predominant peaks at 7.9 and 8.2 GU (glucose unit value). Analysis using arrays of exoglycosidase digests (Figure 1) showed these two peaks to be fucosylated biantennary glycans with one (A2G2SF) and two sialic (A2G2S2F) residues, respectively. Also evident on the glycosylation profile are larger molecular weight glycans with more complex structures. Digests indicate the peaks with GU values of 8.6 to 9.5 are triantennary structures with varying amounts of sialic acid and possibly biantennary structures with an extra lactosamine unit with varying sialic acid. Peaks with values of 10 GU and greater represent tetraantennary structures or triantennary structures with an extra lactosamine unit. Analysis of β-IFN glycans by mass spectroscopy (data not shown) have confirmed the presence of these structures. The NP-HPLC profiles of the glycan analysis of β-IFN produced in bioreactor cultures is very similar to profiles of β-IFN produced in spinner flasks (Spearman *et al.*, 2005).

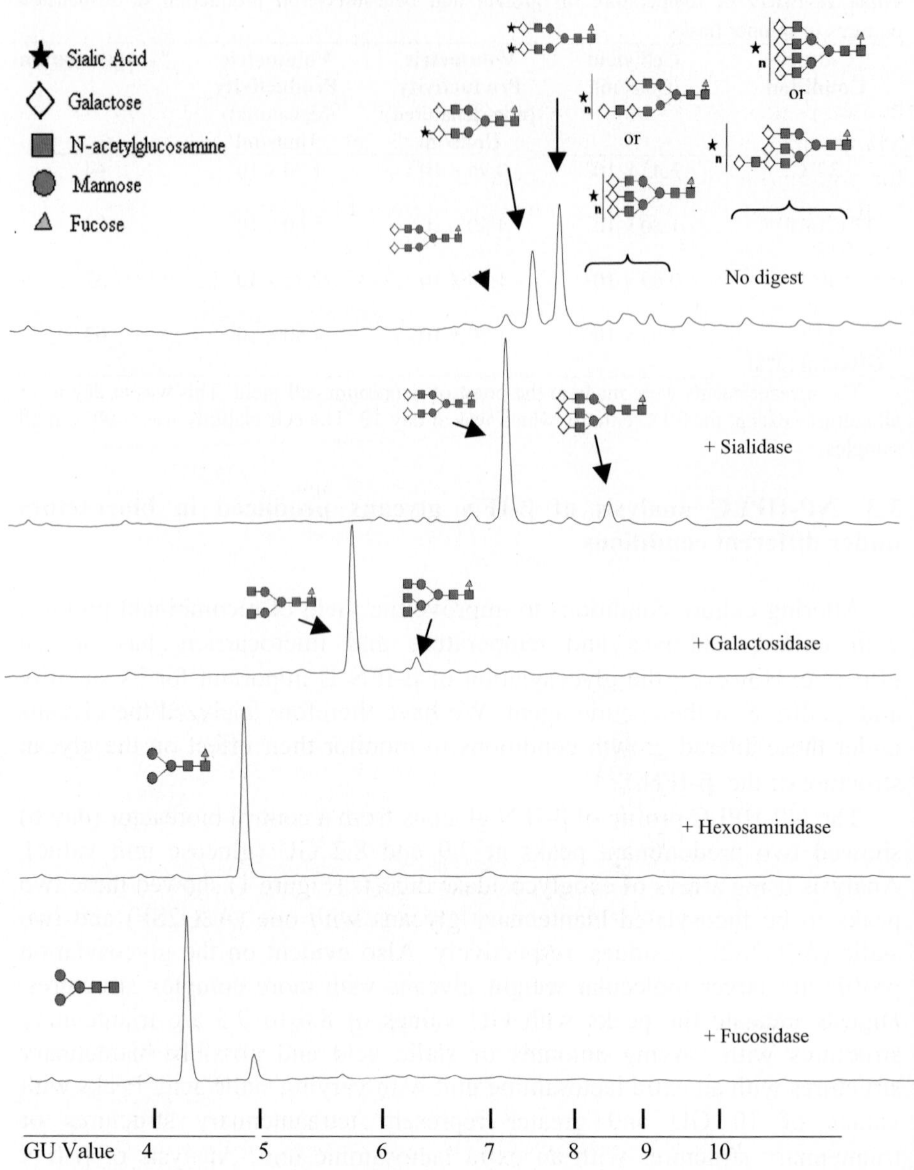


*Figure 1.* NP-HPLC analysis of exoglycosidase digests of β-IFN glycans from a control bioreactor. Glycans were digested with arrays of exoglycosidases and structures assigned based on GU values.

Other studies have found 95% of the human recombinant β-IFN glycan produced in CHO cells to be a core fucosylated biantennary structure with two sialic acid residues with the remaining structures probably tri- or high antennarity (Conradt *et al.*, 1987). Glycosylation of β-IFN is highly cell line dependent and CHO cells produce glycan structures most similar to the native human glycans (Kagawa *et al.*, 1988). They show 74% of natural human β-IFN glycan as a biantennary structure with 1-2 sialic acid residues, 8% as sialylated biantennary structures with an extra lactosamine unit, and the remaining structures as sialylated triantennary structures. β-IFN from CHO cells showed a comparable profile with biantennary structures (68%) and triantennary structures (27%). In our study β-IFN from CHO grown in a control bioreactor had biantennary structures representing approximately 65% of the total glycan with more complex structures representing the remaining structures.

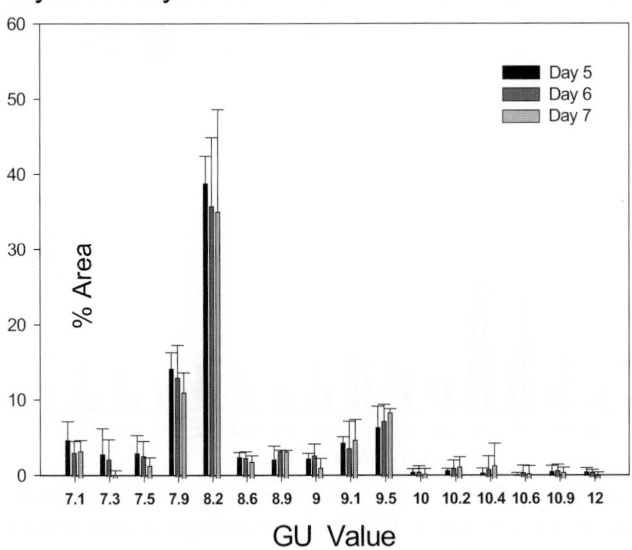


*Figure 2.* NP-HPLC analysis of glycans from β–IFN produced in a control bioreactor. β-IFN was purified from media removed at days 5, 6 and 7 and the glycans isolated and labeled with 2-AB.

Analysis of β-IFN glycans from the control bioreactor cultures showed consistent glycosylation from day 5 to day 7 of culture with no significant change in the glycan profile (Figure 2).

Sodium butyrate is a common media additive for amplifying culture production of recombinant proteins and antibodies. However, sodium butyrate has been found to alter glycosylation of recombinant proteins (Andersen *et al.*, 2000; Sung *et al.*, 2004; Sung *et al.*, 2005; Lamotte *et al.*, 1999). Addition of sodium butyrate (1mM) at 48 hours after inoculation, resulted in a significant increase in β-IFN production (Figure 3). Glycans were analyzed only at day 4 and day 5 of culture because precipitation of the β-IFN in the culture media at day 6 prevented purification. The proportion of the predominant glycan (A2G2S2F) decreased by 50% compared to the control culture. This was associated with an increase in more highly branched glycans with GU values greater than 10. This result indicated that the addition of sodium butyrate increased the proportion of tetraantennary structures or triantennary structures with an extra lactosamine unit. This shift in glycosylation further from the native human form along with the increased aggregation of β-IFN suggested sodium butyrate is not a good media additive in the production of β-IFN.

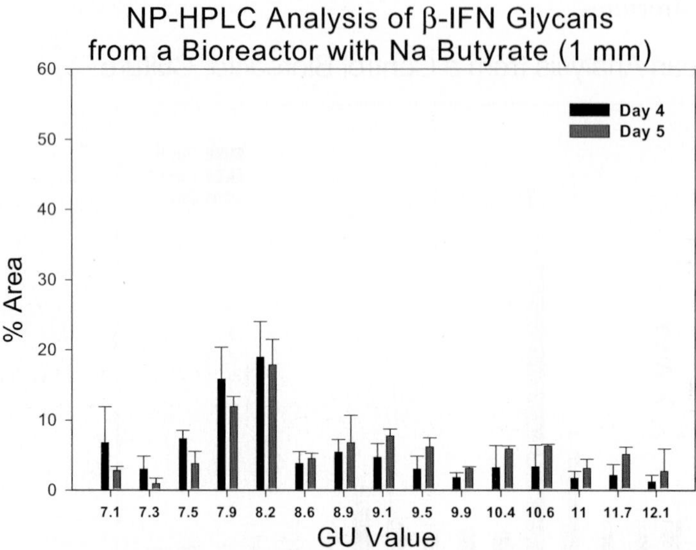


*Figure 3.* NP-HPLC analysis of β-IFN glycans produced in a bioreactor with sodium butyrate (1mM) added at 48 hrs post inoculation. β-IFN glycans were prepared from media removed from the bioreactor on days 4, 5, and 6.

Sodium chloride (40 mM) was added to a bioreactor culture at the time of inoculation and the culture was maintained for 7 days. However, aggregation of the β-IFN at day 7 prevented purification for that day. Glycan analysis of β-IFN produced in this culture at day 4 to 6 showed no significant changes over the course of the culture and no significant differences between the sodium chloride culture and the control culture (Figure 4).

Hyperosomotic pressure has been used to increase antibody production (Kim *et al.*, 2002; Oh *et al.*, 1993) and recombinant protein production (Olejnik *et al.*, 2003). However, this is often accompanied by decreased cell growth. Our results showed that the addition of NaCl (40 mM) to a bioreactor culture to raise the osmolality increased β-IFN production and reduced aggregation to 62%. The glycosylation profile also remained similar to the control cultures. To our knowledge, this is the first report of the effect of increased osmolality on the glycosylation of a recombinant protein. These results suggest that the addition of NaCl for the production of β-IFN and possibly other recombinant glycoproteins can be used to increase production without compromising product quality.

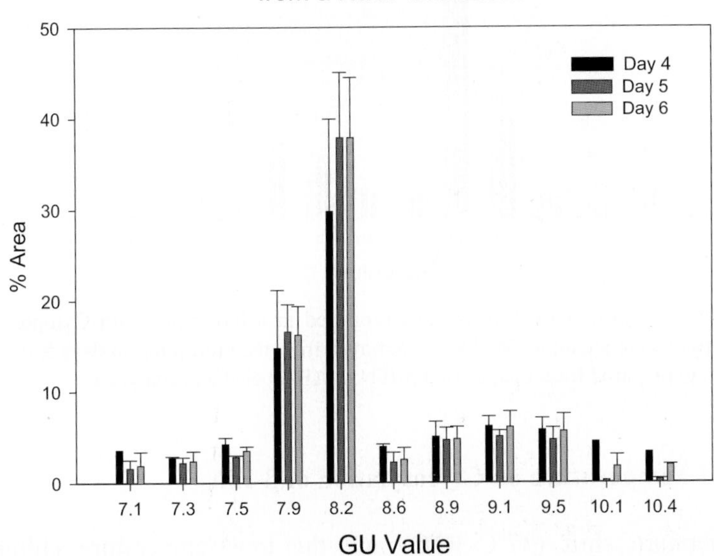


*Figure 1.* NP-HPLC analysis of β-IFN glycans produced in a bioreactor with sodium chloride (40 mM) added prior to inoculation. Media was removed from the bioreactor on days 4, 5 and 6 and glycans were prepared from the purified β-IFN, labeled with 2-AB and analyzed.

We have previously optimized conditions for the growth of β-IFN producing CHO cells in spinner flasks with the microcarriers Cytopore 1 and 2 and analyzed glycosylation of β-IFN produced in these cultures (Spearman *et al.*, 2005). Here, we continue these studies with bioreactor cultures. Glycan analysis of β-IFN produced in a bioreactor culture with Cytopore 1 showed very similar profiles to a control suspension culture and with no significant changes in glycosylation from day 5 to day 7 of culture (Figure 5). At day 7 the A2G2S2F structure was slightly higher than in the control culture. Therefore, the use of Cytopore 1 in bioreactor cultures does not significantly change the glycosylation profile of β-IFN and therefore is an effective means of increasing β-IFN production.

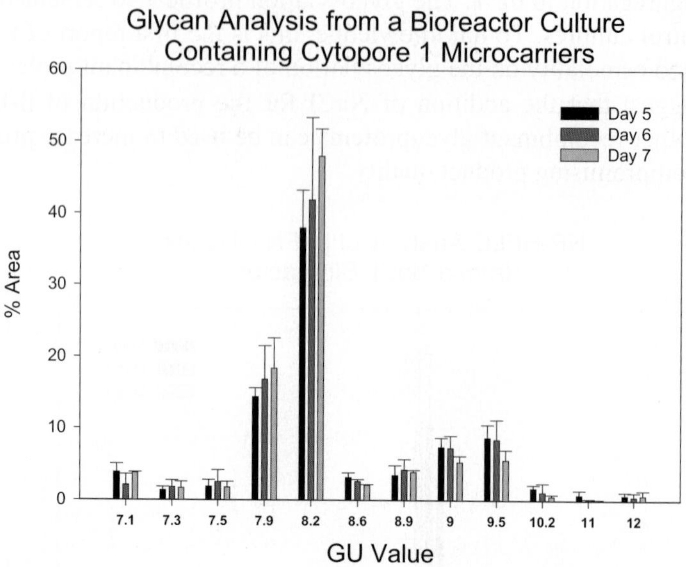


*Figure 2.* NP-HPLC analysis of β-IFN glycans produced in a bioreactor with Cytopore 1 (1mg/ml) added prior to inoculation. Media was removed from the bioreactor on days 5, 6 and 7. The glycans were prepared from the purified β-IFN, 2-AB labeled and analyzed.

## 3.4 Effect of Temperature on Glycosylation of β-IFN

The temperature shift (37°C-30°C) and the low temperature culture (30°C) had similar glycosylation profiles (Figure 6) consistent with glycan analysis in the control bioreactor culture (Figure 2). The results indicated that culturing cells at 30°C will maintain cell viability and the glycosylation profile of the β-IFN for longer culture periods than equivalent cultures at 37°C. Extended spinner flask cultures (8 days) at 37°C had a significantly lower amount of the predominant glycan species at GU 8.2 (A2G2S2F) compared to these cultures. The addition of glycerol to a 37°C culture increased the level of the biantennary glycan but not to the level of the low temperature cultures. The 37°C culture also had slightly higher levels of glycans with GU values 9.1 and 9.5 which are complex glycans, either biantennary glycans with an extra lactosamine unit or triantennary structures. These changes in the glycan profile may be due to reduced cell viability, that is not evident at equivalent time points in the low temperature cultures. Initiating the culture at 37°C and shifting to 30°C did not significantly change the glycosylation of the β-IFN suggesting temperature shift can be used to increase productivity of recombinant protein without affecting glycosylation.

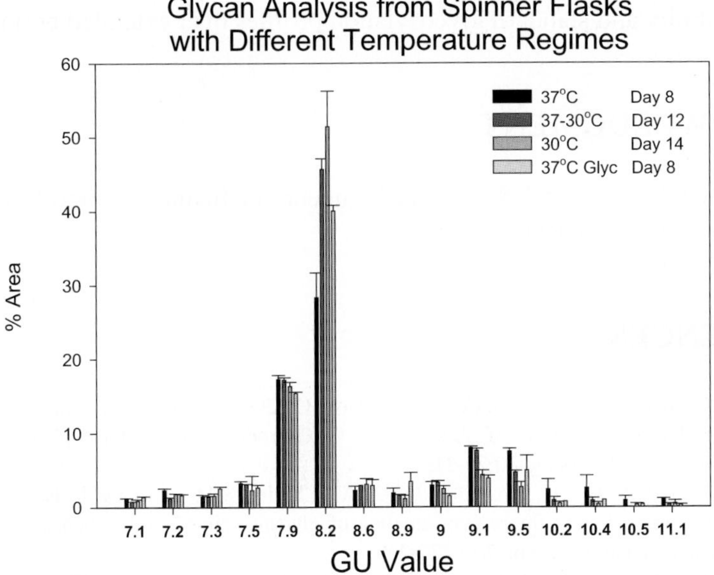


*Figure 3.* NP-HPLC analysis of β-IFN glycans produced in 100 ml spinner flasks cultured at 37oC (harvest day 8), 37°C shifted to 30°C after 48 hours (harvest day 12), 30°C culture (harvest day 14) and 37°C with glycerol (2%) (harvest day 8). β-IFN was purified from pooled media from duplicate flasks for glycan analysis.

## 4. SUMMARY

1. The predominant glycan structures of human recombinant β-IFN produced in CHO cells are fucosylated biantennary glycans with one (A2G2S1F) or two sialic acid residues (A2G2S2F) with smaller amounts of more highly branched glycans.
2. Addition of sodium butyrate (1 mM) increased β-IFN productivity but only slightly reduced aggregation of β-IFN. The glycan profile has reduced A2G2S2F with an increase in more highly branched structures.
3. NaCl (40 mM) addition to a bioreactor culture increased β-IFN production and reduced aggregation while maintaining a glycosylation profile similar to control cultures.
4. Growth of CHO cells on Cytopore 1 microcarriers in a bioreactor increased productivity and decreased aggregation without significantly changing the glycan profile of β-IFN.
5. Cultures at low temperature (30°C) or under a temperature shift regime (37°C to 30°C) showed increased volumetric productivity with

significantly reduced aggregation of β-IFN. These cultures maintained cell viability and standard glycosylation profiles over extended periods.

## ACKNOWLEDGMENT

We wish to acknowledge NSERC and Cangene for financial support of this work in the form of a CRD grant.

## REFERENCES

Andersen, D.C., Bridges, T., Gawlitzek, M., and Hoy, C., 2000, Multiple cell culture factors can affect the glycosylation of Asn-184 in CHO-produced tissue-type plasminogen activator, Biotechnol. Bioeng. 70:25-31.

Bigge, J.C., Patel, T.P., Bruce, J. A., Goulding, P.M. Charles, S.M. and Parekh, R.B., 1995, Nonselective and efficient fluorescent labeling of glycans using 2-amino benzamide and anthranilic acid, Anal. Biochem. 230:229-238.

Butler, M., 2005, Animal cell cultures: recent achievements and perspective in the production of biopharmaceuticals, Appl. Microbiol. Biotechnol. DOI: 10.1007/s00253-005-1980-8 in press.

Conradt, H.S., Egge, H., Peter-Katalinic, J., Reiser, W., Siklosi, T., and Schaper, K., 1987, Structure of the carbohydrate moiety of human interferon-b secreted by a recombinant Chinese hamster ovary cell line, J. Biol. Chem. 30:14600-14605.

Giovannoni, G., Munschauer, F.E. 3rd, and Deinsenhammer, F., 2002, Neutralising antibodies to interferon beta during the treatment of multiple sclerosis, J. Neural Neurosurg. Psychiatry 73:465-469.

Guile, G.R., Rudd, P.M., Wing, D.R., Prime, S.B., and Dwek, R.A., 1996, A rapid high-resolution high-performance liquid chromatographic method for separating glycan mixtures and analyzing oligosaccharides profiles, Anal. Biochem. 240:210-226.

Kagawa, Y., Takasaki, S., Utsumi, J., Hosoi, K., Shimizu, H., Kochibe, N., and Kobata, A., 1988, Comparative study of the asparagine-linked sugar chains of natural human interferon-β1 produced by three different mammalian cells, J. Biol. Chem. 33:17508-17515.

Lamotte, D., Buckberry, L., Monaco, L., Soria, M., Jenkins, N., Engasser, J., and Marc, A., 1999, Na-butyrate increases the production and α2-6 sialylation of recombinant interferon-γ expressed by α2-6-sialyltransferase engineered CHO cells, Cytotechnol. 29:55-64.

Karpusas, M., Nolte, M., Benton, C.B., Meier, W., Lipscomb, W.N. and Goelz, S., 1997, The crystal structure of human interferon-β at 2.2-Å resolution, Proc. Natl. Acad. Sci. 94:11813-11818.

Kim, N.S. and Lee, G.M., 2002, Response of recombinant Chinese hamster ovary cells to hyperosmotic pressure: effect of Bcl-2 overexpression, J. Biotechnol. 95:237-248.

Oh, S.K.W., Vig, P., Chua, F., Teo, W.K. and Yap, M.G.S., 1993, Substantial overproduction of antibodies by applying osmotic pressure and sodium butyrate, Biotechnol. Bioeng. 42:601-610.

Olejnik, A., Grajek, W., and Marecik,R., 2003, Effect of hyperosmolarity on recombinant protein productivity in baculovirus expression system, J. Biotechnol. 102:291-300.

Rodriguez, J., Spearman, M., Huzel, N., and Butler, M., 2005, Enhanced production of monomeric interferon-β by CHO cells through the control of culture conditions, Biotechnol. Prog. 21:22-30.

Runkel, L., Meier, W., Pepinsky, R.B., Karpusas, M., Whitty, A., Kimball, K., Brickelmaier, M., Muldowney, C., Jones, W., and Goelz, S.E., 1998, Structural and functional differences between glycosylated and non-glycosylated forms of human interferon-beta (IFN-beta), Pharm. Res. 15:641-649.

Spearman, M., Rodriguez, J., Huzel, N., and Butler, M., 2005, Production and glycosylation of recombinant β-interferon in suspension and Cytopore microcarrier cultures of CHO cells, Biotechnol. Prog. 21:31-39.

Sung, Y.H. and Lee, G.M., 2005, Enhanced human thrombopoietin production by sodium butyrate addition to serum-free suspension culture of Bcl-2-overexpressing CHO cells, Biotechnol. Prog. 21:50-57.

Sung, Y.H., Song, Y.J., Lim, S.W., Chung, J.Y., and Lee, M.G., 2004, Effect of sodium butyrate on the production, heterogeneity and biological activity of human thrombopoietin by recombinant Chinese hamster ovary cells, J. Biotechnol. 112:323-335.

Wurm, F. M., 2004, Production of recombinant protein therapeutics in cultivated mammalian cells, Nature Biotechnol. 22:1393-1398.

Yang, M. and Butler, M., 2000, Effects of ammonia on CHO cell growth, erythropoietin production, and glycosylation, Biotechnol. Prog. 68:370-380.

Yang, M. and Butler, M., 2002, Effects of ammonia and glucosamine on the heterogeneity of erythropoietin glycoforms, Biotechnol. Prog. 18:129-138.

# Induction of Caspase-3-dependent Apoptosis by Electrolyzed Reduced Water/platinum Nanoparticles in Cancer Cells

T. Hamasaki, S. Aramaki, T. Imada, K. Teruya, S. Kabayama[#], Y. Katakura, K. Otubo[#], S. Morisawa[#], S. Shirahata

*Department of Genetic Resources Technology, Kyushu University, 6-10-1 Hakozaki, Higashi-ku, Fukuoka 812-8581, Japan; [#]Nihon Trim Co. LTD., 34-8-1 Ooyodonaka, Kita-ku, Osaka 531-0076, Japan*

**Abstract:** Electrolyzed reduced water (ERW) supplemented with platinum nanoparticles (Pt Nps)(ERW/Pt Nps) exhibited strong antioxidative and active hydrogen donating activities. ERW/Pt Nps-containing medium caused lowering of cell viability. Analysis of cell cycle and activity of caspase-3 suggested that ERW/Pt Nps induced apoptosis in cancer cells. This apoptosis inducibility ERW/Pt Nps was lost by using degassed ERW and recovered by bubbling of hydrogen gas. These results suggested that active hydrogen produced from molecular hydrogen in ERW by catalysis of Pt Nps induced apoptosis of cancer cells.

**Key words:** Antioxidant, Apoptosis, Platinum, Nanoparticles, Reactive oxygen species

## 1. INTRODUCTION

We previously reported electrolyzed reduced water (ERW), produced near the cathode by electrolysis, has a reductive activity (Shirahata *et al.*, 1997). We also revealed that ERW contains platinum nanoparticles (Pt Npps) derived from Pt-coated titanium electrodes in addition to high concentration of dissolved molecular hydrogen. Pt Nps exhibited strong reactive oxygen species (ROS) scavenger activity and catalysis activity converting molecular hydrogen to active hydrogen. Therefoe, ERW supplemented with Pt Nps (ERW/Pt Nps) is expected to has a strong



*R. Smith (ed.), Cell Technology for Cell Products*, 87–89.

antioxidative activity. A hypothesis has been proposed that high level of intracellular ROS promotes malignant properties of cancer cells such as activation of growth signals, apoptosis tolerance, metastasis, and angiogenesis. Here, we investigated apoptosis inducibility of ERW/Pt Nps on cancer cell lines.

## 2. EXPRIMENTAL PROCEDURE

### 2.1 Cell culture and measurement of cell proliferation

Normal cell lines; human diploid embryonic lung fibroblasts (TIG-1), human diploid fibroblasts (WI-38) and human diploid embryonic lung cell line (MRC-5), and cancer cell lines; cervical carcinoma cells (HeLa) and human hepatocellular carcinoma cell line (HepG2) were obtained from the Japanese Collection of Research Bioresources (JCRB, Osaka, Japan). Adherent cell cells were cultured in Eagle's MEM medium supplemented with 10% fetal bovine serum (FBS) at 37°C in a 95% air/5% $CO_2$ atmosphere. Briefly, cells (floating cells, 1.25 x $10^4$ cells/$cm^2$ and adherent cells, 2.5 x $10^4$ cells/$cm^2$) were seeded in 24-well plates. The cultures were incubated with ERW supplemented with different concentrations of Pt Nps for 24 hours. Cytotoxicity was determined using the 2-(4-iodophenyl)-3-(4-nitrophenyl)-5-(2,4-disulfophenyl)-2H-tetra-zolium,monosodium salt (WST-1) assay.

### 2.2 Measurement of intracellular ROS by flowcytometer

The amount of intracellular ROS, especially the intracellular $H_2O_2$, was determined by using a fluorescent dye, 2',7'-dichlorofluorescin-diacetate (DCFH-DA). Cells were pre-cultured for 10 min in $Ca^{2+}$, $Mg^{2+}$-free HBSS buffer with ERW/Pt Nps. After the removal of the supernatant, 5 μM DCFH-DA was added and incubated for 10 min. After resuspended in PBS, intracellular redox state of cells was analyzed immediately using a flowcytometer.

## 3. RESULT AND DISCUSSION

ERW/Pt Nps exhibited strong antioxidative and active hydrogen donating activities. Pt Nps did not cause lowering of cell viability (Fig. 1). However, ERW/Pt Nps exhibited strong cytotoxicity on cancer cells but not on normal

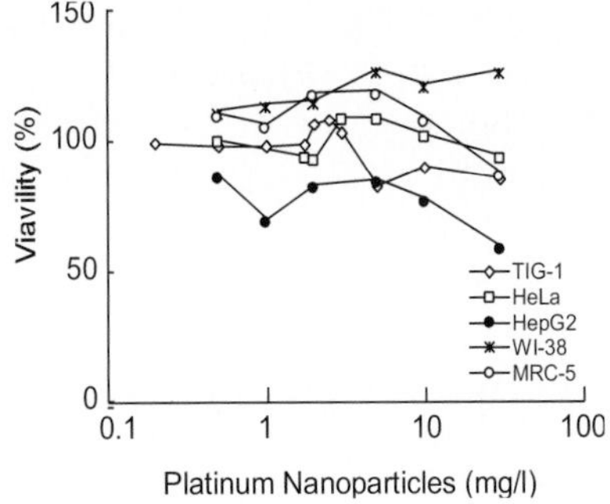


*Figure 1.* Cytotoxicity test of ERW/Pt Nps.

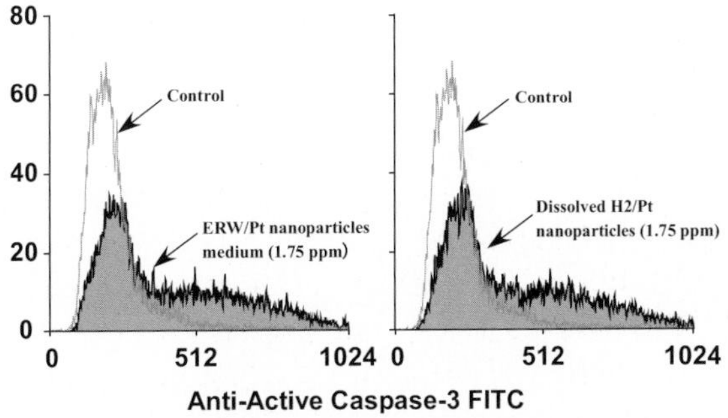


*Figure 2.* Determination of activated caspase-3 in HeLa cells.

cells (data not shown). Analysis of cell cycle and caspase-3 activity suggested that ERW/Pt Nps induced apoptosis in cancer cells. This apoptosis inducibility of Pt Nps/ERW was lost by using degassed ERW and recovered by bubbling of hydrogen gas (Fig. 2). The cytotoxicity of ERW/Pt Nps on cancer cells was rapidly lost in high concentrations of Pt Nps, suggesting that the phenomenon is due to agglutination of Pt Nps. These results suggested that active hydrogen produced from molecular hydrogen in ERW by catalysis of Pt Nps induced apoptosis of cancer cells. ERW/Pt Nps scavenged intracellular ROS in cancer cells. Investigation on the relationship between ROS-scavenging activity and apoptosis is undergoing.

# Molecular Mechanism of TAK1-Induced Repression of hTERT Transcription

*TAK1 represses the transcription of hTERT*

M. Maura, Y. Katakura, T. Miura, T. Fujiki, H. Shiraishi, S. Shirahata

*Department of Genetic Resources Technology, Faculty of Agriculture, Kyushu University, Fukuoka 812-8581, Japan*

**Abstract:** Telomerase activity is tightly regulated by the expression of its catalytic subunit, human telomerase reverse transcriptase (*hTERT*). In most human tumor cells, *hTERT* is expressed to gain immortality. Here we demonstrate that transforming growth factor-β-activated kinase 1 (TAK1), originally identified as an MAPKKK, can repress the transcription of *hTERT* gene in A549 human lung adenocarcinoma cell line through promoter fused to luciferase gene and reverse transcriptase-polymerase chain reaction analysis. Furthermore, Trichostatin A relieved the TAK1-induced repression of *hTERT* transcription. These results suggest that histone deacetylase (HDAC) involves TAK1-induced repression of *hTERT* transcription, resulting in transcriptional silencing of the *hTERT* gene in human cancer cells.

**Key words:** TAK1, *hTERT*, HDAC

## 1. INTRODUCTION

Telomeres are essential elements that contribute to chromosomal stability. Telomerase is a specialized reverse transcriptase that synthesizes telomeres. The absence of telomerase activity in most normal human cells, but immortalized and cancer cells contain detectable telomerase activity and consequently maintain their telomere length and proliferative potential [1]. The human telomerase reverse transcriptase (hTERT) subunit is selectively expressed in most immortalized and cancer cells, indicating that hTERT is the rate-limiting component of telomerase activity. Telomerase activity is



*R. Smith (ed.), Cell Technology for Cell Products*, 91–93.

tightly regulated at the transcriptional level of *hTERT*. In normal human cells, Sp1 and Sp3 are involved in the HDAC-mediated transcriptional repression of the *hTERT* [2]. TAK1 is a member of the MAPKKK family and is activated by various cytokines [3, 4]. TAB1 functions as an activator promoting TAK1 autophosphorylation [5]. In the present study, we demonstrated that TAK1-induced repression of *hTERT* transcription is caused by recruitment of HDAC on the *hTERT* promoter.

## 2. MATERIALS AND METHODS

### 2.1 Cell culture and regent

A549 human lung adenocarcinoma cells were cultured in eRDF medium supplemented with 5% fetal bovine serum (FBS). Trichostatin A (TSA) was dissolved in $CH_3OH$ and added to the culture medium at final concentration of 100 nM.

### 2.2 RT-PCR

Total RNA was prepared by using GenElute Mammalian Total RNA Miniprep Kit (Sigma). One μg of total RNA was used as a template for the cDNA synthesis reaction using M-MLV reverse transcriptase RNase $H^-$ (Promega). Analysis of the expression of *hTERT* and *TEP1* was performed by RT-PCR amplification as described previously [6].

## 3. RESULTS AND DISCUSSION

### 3.1 TAK1 represses the transcription of *hTERT* gene

Previously we reported that TGF-β is able to repress telomerase activity and *hTERT* transcription in A549. Here we attempted whether TAK1 can repress *hTERT* transcription. We confirmed an ability of TAK1 to induce transcriptional repression of *hTERT* gene by using recombinant adenoviruses expressing TAK1/TAB1 and LacZ as an infection control. Three days after the transduction, we investigated the expression of *hTERT* mRNA by RT-PCR. As shown in the result, marked reduction of *hTERT* transcription was observed in A549 cells transducted both with TAK1 and TAB1 (100 moi each) (Table 1), indicating that TAK1 represses the transcription of *hTERT* gene in A549 cells.

*Table 1.* Transcriptional repression of the hTERT gene by TAK1.

| | *hTERT*/*TEP1* Ratio (% of control cells) |
|---|---|
| A549 | 100 |
| LacZ | 108.07 |
| TAK1/TAB1 | 50.21 |

## 3.2 Involvement of HDAC in the TAK1-induced repression of *hTERT* gene

Several studies have shown that HDAC is involved in repression of *hTERT* transcription in normal human cells. Here we examined an involvement of HDAC in the TAK1-induced repression of *hTERT* transcription by using HDAC inhibitor, TSA. Result shows that reduced *hTERT* transcription by the transduction with TAK1 and TAB1 was recovered by the treatment with TSA (Table 2), suggesting that HDAC is involved in the repressed transcription of *hTERT* gene in A549 cells transduced with TAK1 and TAB1.

*Table 2.* Involvement of HDAC in the TAK1-induced repression of hTERT gene.

| | *hTERT*/*TEP1* Ratio (% of control cells) |
|---|---|
| A549 | 100 |
| A549+TSA | 91.72 |
| LacZ | 113.34 |
| LacZ+TSA | 100.30 |
| TAK1/TAB1 | 40.26 |
| TAK1/TAB1+TSA | 139.32 |

## REFERENCES

1. Kim, N. W. *et al.* (1994) *Science* **266**, 2011-2015
2. Won, J. *et al.* (2002) *J. Biol. Chem.* **277**, 38230-38238
3. Yamaguchi, K. *et al.* (1995) *Science* **270**, 2008-2011
4. Ninomiya, J. *et al.* (1999) *Nature* **398**, 252-256
5. Shibuya, H. *et al.* (1996) *Science* **272**, 1179-1182
6. Nakamura, T. M. *et al.* (1997) *Science* **277**, 955-959

# Analysis of Genetic Parameters in Order to Get More Information on High Producing Recombinant CHO Cell Lines

Christine Lattenmayer, M. Loeschel, K. Schriebl, E. Trummer, K. Vorauer-Uhl, D. Mueller, H. Katinger, R. Kunert
*Austrian Center of Biopharmaceutical Technology, Vienna, Austria; Department of Biotechnology, Institute of Applied Microbiology, University of Natural Resources and Applied Life Sciences, Vienna, Austria; Polymun Scientific Vienna, Austria*

**Abstract:** Recombinant mammalian cells for the production of glycoproteins and antibodies are mainly established by gene amplification techniques, using selection and screening methods based on specific productivity and growth rate. In the present study we will show the selection and screening process of a recombinant CHO cell line expressing a Fc fusion protein by analysis of gene copy numbers, mRNA-levels, specific productivity and integration locus.

**Key words:** CHO dhfr$^-$, Selection, Screening, gene copy number, mRNA-level

## 1. INTRODUCTION

Different factors influencing the stability of the exogenous target are only poorly analysed during clonal development. Therefore it is inevitable to control parameters influencing the productivity and growth properties already during clone development. Clonal changes in transcriptional and translational behaviour according to selection and amplification have to be identified and comprised into the screening process. Additionally we have checked the integration locus by Southern Blot analysis for a better characterisation of the clones.



*R. Smith (ed.), Cell Technology for Cell Products*, 95–98.

## 2. MATERIAL AND METHODS

Two plasmids carrying the target epoFc and the selection marker dhfr were cotransfected into CHO dhfr- under serum-dependent conditions (clones were later adapted to protein-free medium). Stable clones were analysed for specific productivity (qp) by ELISA as well as for gene copy number (GCN) and mRNA-level by quantitative PCR using the SYBR-green method. The results are presented as the ratio of epoFc related to a housekeeping gene, ß-actin. Analysis of the integration locus was performed by Southern blot detecting the 3´end of the epoFc cDNA.

## 3. RESULTS

Figure 1 shows specific productivities as well as gene copy numbers of six clones (5 transfectants as well as 13F5, a subclone of 2C10) cultivated in protein-free medium (average over three passages in spinner flasks). 13F5 exhibited the highest specific productivity among the investigated clones, 10D9 and 8C6 showed the lowest titers. Concerning gene copy numbers, an increase of epoFc for the subclone 2C10/13F5 is evident. 10D9 shows highest gene copy numbers of epoFc which is not reflected in a higher specific productivity.

Comparing mRNA-levels to product discharge, a high level of target mRNA also results in high specific productivity. However, subclone 13F5 does not express more mRNA although specific productivity increased. In order to gain more information on the insertion loci, Southern Blot analysis was performed (see Figure 2). Hybridisation with epoFc gave the same restriction pattern (digestion with BamHI) for clone 2C10, 2G6 as well as 2C10/13F5 indication equal integration sites. In contrast, clone 10D9 exhibits a rather complex restriction pattern.

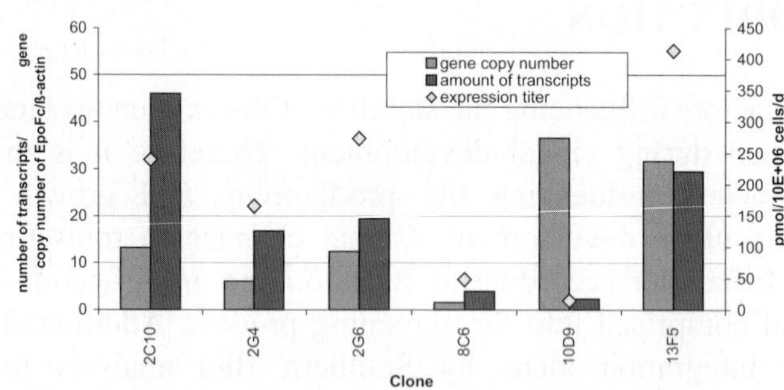


*Figure 1.* Comparison of clones and subclones for gene copy number, amount of transcripts and expression titer.

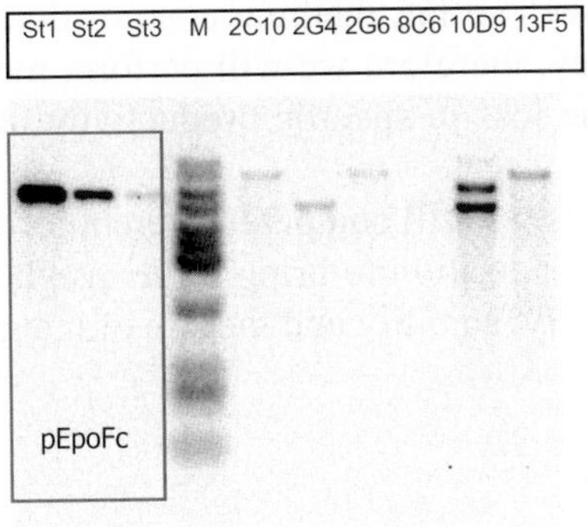


*Figure 2:* Southern blot after digestion of genomic DNA: St1, St2, St3 (digested expression vector $2 \cdot 10^8$, $2 \cdot 10^7$, $2 \cdot 10^6$ copies); λ DNA/EcoRI+HindIII marker; 2C10, 2G4, 2G6, 8C6, 10D9, 2C10/13F5 genomic DNA (1.6 µg per lane).

Furthermore, we analysed expression titers as well as genetic parameters of six 2C10 subclones before and after protein-free adaptation. After switching to protein-free conditions a decrease in specific productivity of about 50% was evident. Concerning the gene copy numbers except for two subclones, where a decrease after protein-free adaptation was observed, no significant differences were evident.

## 4. DISCUSSION AND CONCLUSION

Concerning specific productivity highest titers were obtained for the serum-dependent cultivated subclones of 2C10 (data not shown). After protein-free adaptation a decrease of 50% in specific productivity is observed for all investigated clones. This phenomenon can not be explained by a reduction of the gene copy number as well as the specific mRNA-level, which was found to be unchanged (data not shown). Thus, the reason for the loss in qp must be on the translational level. Medium optimisation will therefore be the next step.

In general a positive correlation of GCN with productivity was observed. However, clone 10D9 showing low expression titers exhibited a high GCN, at least four times higher compared to the other clones. A different restriction pattern in the Southern blot was observed. mRNA level as well as qp were diminished, furthermore the adaptation to protein-free conditions was rather difficult. In contrast, clones 2C10, 2C10/13F5 as well as 2G6 showing high stability in specific productivity during spinner cultivation, have inserted the epoFc into the same locus evident by equal restriction pattern in the Southern blot.

After all we have seen so far, medium composition might have an effect on translation efficiency, therefore we will perform medium optimisation in order to circumvent the loss in specific productivity observed after protein-free adaptation.

In future experiments we will compare different recombinant cell lines by fluorescence in situ hybridisation to bring more insight into the interrelation between high productivity, stability and the site of integration.

## Acknowledgement

This research was kindly funded by the ACBT (Austrian Center of Biopharmaceutical Technology), a competence center supported by the Federal Ministry for Economy and Labour, federal states of Vienna and Tyrol.

# Redox Regulation by Reduced Waters as Active Hydrogen Donors and Intracellular ROS Scavengers for Prevention of type 2 Diabetes

S. Shirahata[1], Y. Li[1], T. Hamasaki[1], Z. Gadek[2], K. Teruya[1], S. Kabayama[3], K. Otsubo[3], S. Morisawa[3], Y. Ishii[4], Y. Katakura[4]

*[1]Department of Genetic Resources Technology, Faculty of Agriculture, Kyushu University, Fukuoka, Japan; [2]Centre for Holistic Medicine and Naturopathy, Nordenau, Germany, [3]Nihon Trim Co.Ltd., Osaka, Japan; [4]Hita Tenryosui Co. Ltd., Hita, Oita, Japan*

**Abstract:** The analysis using the DBNBS reduction method and the DCFH-DA intracellular reactive oxygen species (ROS) determination method revealed that ERW and diseases-improvable natural waters such as Nordenau water in Germany and Hita water in Japan were all reduced waters (RWs) which could function as active hydrogen donors and intracellular ROS scavengers. RWs suppressed the activity of protein tyrosine phosphatase (PTP), which inactivates insulin receptor, suggesting their anti-type 2 diabetes effects via redox regulation. The clinical test of 356 diabetes patients drinking Nordenau water in Germany resulted in the improvement of the relevant tests parameters after 6 days, suggesting the correlation of these changes with the fluctuation of ROS levels in their blood.

**Key words:** reduced water, active atomic hydrogen, reactive oxygen species, diabetes

## 1. INTRODUCTION

Hydrogen-rich electrolyzed reduced water (ERW) scavenged reactive oxygen species (ROS) (Shirahata *et al.*, 1997) and was applied to suppress the oxidative stress of hemodialysis patients (Huang *et al.*, 2003). Some natural mineral waters such as Nordenau water found in Germany in 1992 and Hita water (Hita Tenryousui®) found in Japan in 1997 as well as ERW



*R. Smith (ed.), Cell Technology for Cell Products*, 99–101.

scavenged ROS and protected pancreatic ß-cells from oxidative stress induced by alloxane (Li *et al.*, 2002). Here, we report redox regulation by reduced waters (RWs) for prevention of type 2 diabetes and clinical trials of type 2 diabetes patients.

## 2. MATERIALS AND METHODS

### 2.1 Evaluation of waters as active hydrogen donors and ROS scavengers

A novel colorimetric determination method of active hydrogen in aqueous solution with a spin trap reagent, 3,5-dibromo-4-nitrosobenzene-sulfonic acid sodium salt (DBNBS) was utilized to evaluate the functions of waters as active hydrogen donors. Intracellular ROS levels of rat L6 myotubes were determined using DCFH-DA as described previously (Li *et al.*, 2002).

### 2.2 Determination of glucose uptake, phosphorylated insulin receptor and activity of PTPase

Glucose uptake into L6 cells was determined using $^{3}$H-2-deoxyglucose. Fully differentiated L6 myotubes were incubated with various waters for 72 h. After stimulation with insulin for 20 min, total cell lysates were separated by SDS-PAGE, immunoprecipitated with anti-insulin receptor (IR) β-subunit, and then immunoblotted with anti-phospho-IR (pY1158) antibody. The protein tyrosine phosphatase (PTPase) activity was measured using *p*-nitrophenyl phosphate as substrate.

### 2.3 Clinical trials on type 2 diabetes patients drinking Nordenau water

Changes in the relevant tests parameters of 356 type 2 diabetes patients (average age; 71.5 years old) drinking Nordenau water (2 liter per day) were examined after 6 days. The diagnostic parameters such as blood sugar, HbA1c, cholesterol, LDL and serum creatinine concentration were tested twice – at the beginning (MP1) and at the end of the participants stay (MP2) in Nordenau. ROS in randomly sampled bloods of 81 patients were examined by FORT (Free Oxygen Radical) test.

## 3. RESULTS AND DISCUSSION

The DBNBS active hydrogen determination method and the DCFH-DA intracellular ROS determination method revealed that ERW and diseases-improvable natural waters such as Nordenau water and Hita water, but not most commercial natural mineral waters were all reduced waters (RW) which could function as active hydrogen donors and intracellular ROS scavengers. Protein tyrosine phosphatase (PTPase) is a redox-regulatable enzyme. RWs suppressed the PTP activity, leading to the activation of IR and the enhancement of glucose uptake into L6 myotubes.

The clinical trials of 356 type 2 diabetes patients revealed that blood sugar and HbA1c average values at 74.4% of 356 diabetes patients were significantly improved by drinking Nordenau water in a very short time of six days. Roughly 54% of responders showed the high statistical improvement rate of their cholesterol, LDL and serum creatinine concentration values. The parallel downward trend of ROS concentration in the randomly sampled bloods of the tested persons suggested the ROS scavenging ability of Nordenau water *in vivo*.

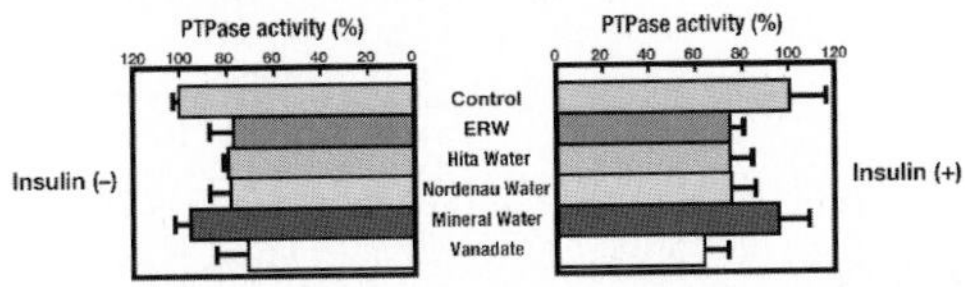


*Figure 1.* Inhibition of PTPase activity by reduced waters.

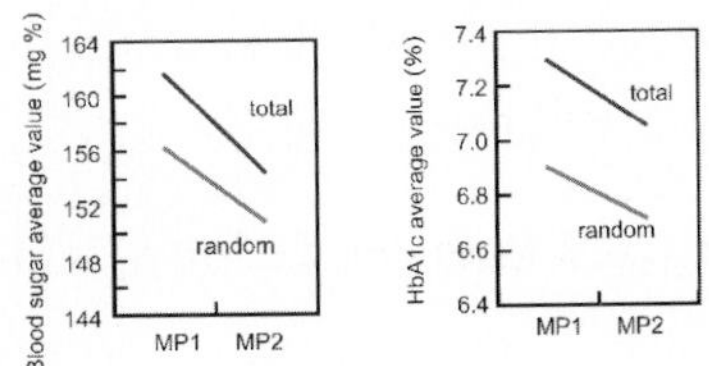


*Figure 2.* Suppressive effect of Nordenau water on blood sugar levels (left) and HbA1c levels (right) of 256 type 2 diabetes patients. Random, randomly sampled 81 patient's data.

## REFERENCES

Huang, K.-C., Yang, C.-C. *et al.*, 2003. Kidney Intern., **64**: 704-714.

Li, Y., Nishimura T. *et al.*, 2002, Cytotechnology, **40**: 139-149.

Shirahata, S., Kabayama, K. *et al.*, 1997, BBRC, **234**: 269-274.

# Transcription Profiling of Different Recombinant CHO Clones Based on Cross-species Microarray Analysis

Evelyn Trummer[1,3], Wolfgang Ernst[1,3], Dethardt Müller[1,3], Kornelia Schriebl[1,3], Christine Lattenmayer[1,3], Karola Vorauer-Uhl[1,3], Renate Kunert[1,3], Hermann Katinger [1,2] and Friedemann Hesse[1,3]

[1] *Institute of Applied Microbiology, Muthgasse 18, 1190 Vienna, Austria* [2] *Polymun Scientific, Immunbiologische Forschung, Nussdorfer Lände 11, 1190 Vienna, Austria* [3] *Austrian Center of Biopharmaceutical Technology, Muthgasse 18, 1190 Vienna, Austria*

**Abstract:** Cross-species analysis was used to monitor the global transcription profile of CHO production cell lines. Mouse and rat-oligoarrays served to compare gene expression signatures of different recombinant CHO clones as well as different conditions in order to point out correlations between the observed phenotype and the transcriptome.

**Key words:** CHO cells, clonal variability, microarray analysis, cross-species hybridisation, mouse-oligo-DNA array

## 1. INTRODUCTION

Since different CHO-clones have been shown to exhibit a large degree of variability regarding recombinant protein production performance, the development of more efficient screening and selection methods is obligatory. Therefore, we combined the phenotypic characterisation of cell cultures with transcription profile analysis using microarray technology to understand the cellular mechanisms that control the phenotype of animal cells and to identify new potential selection parameters.



*R. Smith (ed.), Cell Technology for Cell Products*, 103–106.

# 2. MATERIAL AND METHODS

Experiments were carried out with clones of a serum-free adapted recombinant CHO-dhfr$^-$ cell line producing the fusion protein Epo-Fc. Different CHO clones were cultivated in the multireactor system Sixfors (Infors) during clone development in order to investigate their applicability for technological relevant production systems. The bioreactor cultivation was performed in repeated batch mode (5 batches), whereby total RNA of cells was isolated at the end of each batch. Preparation of RNA samples, reverse transcription, fluorescent dye labeling, hybridization and further array processing was done using the Agilent Low RNA input linear amplification kit. The fluorescently labeled targets were hybridized to mouse and rat 22k Oligo-60mer-Microarrays (Agilent). Agilent´s feature extraction software and Gene Spring 6.2 were used for data analysis. Differentially expressed genes were classified by an experimental threshold, which was determined by self versus self hybridization of equimolar mixtures and alternating dye labeling of all RNAs used. Based on a 99% confidence interval of the dataset, the following thresholds were determined: Mouse (>1.47/<0.64); Rat (>1.33/<0.64). The datasets were further filtered on confidence of replicate data. Changes in gene expression with $p<0.05$ was used as a cut-off to produce preliminary lists of genes. The gene lists were linked to GeneBank accession numbers, GO-terms, Biocarta and KEGG using MASI (Insilico Bioinformatics).

# 3. RESULTS AND DISCUSSION

## 3.1 Experiment 1: Phenotypic characterisation and transcriptional profiling of one clone at two different conditions

In the course of bioreactor cultivation the clone 2G4 showed altered growth characteristics, as the specific growth rate was reduced from 0.5 ($d^{-1}$) (1st and 2nd batch) to 0.37 ($d^{-1}$) (3rd, 4th and 5th batch). In the fifth batch reduced growth was followed by the induction of cell death, as viability declined from 90% to 65%. Specific recombinant protein production rates and degree of product sialylation remained constant in the course of bioreactor cultivation. RNA samples from the 2nd and 5th batch were analyzed for differential gene expression using microarrays. The results obtained from GO-and pathway analysis indicated a strong activation of MAPK-signaling pathway playing an important role in transcriptional activation, cell cycle arrest and apoptosis. These findings were underlined by the induction of genes like Gadd45a (3.08), CHOP (1.71), ATF-2 (1.66),

Tgfbr (1.66), Hmg-14 (1.60) and Mef2C (1.52) in the 5$^{th}$ batch. The overexpression of GADD45 genes is known to result in substantial activation of p38MAP Kinase and apoptosis (Kyriakis *et al.*, 2001), whereas CHOP leads to growth arrest and apoptosis and is known to be responsive to perturbations that culminate in the induction of ER stress (Zinsner *et al.* 1998). The upregulation of several stress responsive genes and UPR signaling mediators was observed, as follows: (1) BIP (1.68) and several co-chaperones (Dnajb9 (2.0), Sec 63 (1.78) and Dnajc1 (1.6)). (2) CHOP (1.71), ATF-2 (1.66), ATF-4 (2.0), GCN-2 (1.79). (3) The ubiquitin dependent protein catabolism and members of the proteasome were tightly regulated (Psma6, Psma4, Psmc2, Pmsd5). (4) The induction of phospholipid biosynthesis could be a consequence of UPR activation, as this correlates with the presence of an elaborate ER [Sriburi *et al.*, 2004]. To verify data obtained with microarrays, a quantitative RT-PCR analysis of four selected genes from clone 2G4 at two different stages was performed. Relative quantification was done by calculating the ratios of the target gene copy number to that of the reference house keeping gene beta-actin. Out of four selected genes, the expression level of three genes could be verified. qRT-PCR results showed a remarkable correlation to microarray data: Gadd45a (3.1-3.08); BIP (1.58-1.68); Gstm5 (0.99-0.96) (-fold induction qRT-PCR-microarray).

## 3.2 Experiment 2: Phenotypic characterisation and transcriptional profiling of two different clones

Transcription profiles of two clones with different production rates and post-translational modification capacities were analyzed to gain insight into their recombinant protein production performance. Clone 2C10 showed ten-fold higher specific Epo-Fc production rates than clone 8C6. The lowproducer 8C6, however, held potential for higher levels of terminal sialylation, as the secreted product contained 50% more sialic acids compared to clone 2C10. Genes involved in glycerolipid and sterol biosynthesis were highly represented in clone 8C6. The upregulation of genes involved in dolichol synthesis (Hmgcr, Hmgcs, Idi-1) could be an indictator for the higher product quality (sialylation of Epo-Fc) of the low producer 8C6. The concentration of dolichol phosphate, the immediate precursor for synthesis of the lipid linked oilgosaccharides used for N-glycosylation, is believed to be an important factor in determining the amount of glycosylation that occurs [Rosenwald *et al.*, 1990]. In addition, a strong induction of BIP (2.04) was observed, what could be responsible for the observed low specific protein production rate, as enhanced BIP expression results in enhanced protein binding to BIP and, therefore, in blocking of protein secretion (Borth *et al.*, 2005). Genes involved in protein biosynthesis and protein metabolism were overrepresented in the high producer clone 2C10 and represented several cellular components involved in protein processing. Ribosomal upregulation, as well as well as the pronounced activation of translation (translation initiation and translation factor activity) may correlate with the almost 10-fold higher specific production rate of clone 2C10 compared to clone 8C6. At least four different proteins which are structural constituents of the ribosome or participate in ribosome biogenesis were found to be up-regulated (Rpl26, Rps19, RpI7a, Rps16).

## 4. CONCLUSION AND PERSPECTIVES

The data reveal, that cross-species transcription analysis of CHO cells using mouse and rat microarrays allows the identification of key genes responsible for the phenotypic behaviour and point out correlations between the phenotype and the transcriptome. The goal of the future work is to generate a dataset which emphasises those genes relevant to the use of CHO cells as a host for recombinant protein production.

## REFERENCES

Borth N, Mattanovich D, Kunert R, Katinger H., 2005, Effect of increased expression of protein disulfide isomerase and heavy chain binding protein on antibody secretion in a recombinant CHO cell line, Biotechnol Prog. **21**(1):106-11.

Kyriakis J.M. and Avruch J., 2001, Mammalian mitogen-activated protein kinase signal transduction pathways activated by stress and inflammation, Physiol Rev. **81**(2): 807-69.

Rosenwald AG, Stoll J, Krag SS., 1990, Regulation of glycosylation. Three enzymes compete for a common pool of dolichyl phosphate in vivo, J Biol Chem. **265**(24):14544-53.

Sriburi R, Jackowski S, Mori K, Brewer JW., 2004, XBP1: a link between the unfolded protein response, lipid biosynthesis, and biogenesis of the endoplasmic reticulum, J Cell Biol. **167**(1):35-41.

Zinsner H., Kuroda M., Wang X.Z., Batchvarova N., Lightfood R.T., Remotti H., Stevens J.L., Ron D., 1998, CHOP is implicated in programmed cell death in response to impaired function of the endoplasmic reticulum, Genes Dev. **12**(7):982-95.

# Adaptation without Cell Division?

*Enhanced adaptation to suspension and serum free culture by the co-expression of p21 and Bcl-2*

K.L. Astley[1] and M. Al-Rubeai[1,2]

[1] *Department of Chemical Engineering, University of Birmingham, Birmingham B15 2TT, UK,* [2] *Department of Chemical and Biochemical Engineering, University College Dublin, Belfield, Dublin 4, Ireland*

**Abstract:** The inducible expression of p21 $^{CIP1}$ cyclin dependent kinase system in CHO cells has been shown to be capable of uncoupling cell growth and productivity enabling higher levels of productivity without increasing the overall cell number, however arrested cultures expressing p21 appear to proceed into the death phase in a relatively similar time span as non arrested cultures despite the differences in overall cell numbers. By combining p21 and Bcl-2 expression it was possible to maintain cells almost indefinitely in a state of cell cycle arrest through which it was possible to adapt cells to suspension and serum free environments at a much quicker and effective rate than that for the non arrested cells.

**Key words:** P21, Bcl-2, Adaptation, Serum free, Proliferation, Apoptosis.

## 1. INTRODUCTION

The effect of p21 activity in terms of its effect on productivity in CHO and NS0 cultures is well documented (Watanabe *et al*. 2002; Ibara *et al*. 2003; Bi *et al*. 2004), however, it is the ability of p21 to prevent cell proliferation that is of particular interest in this study which aims to investigate the possibility that by halting cell division, cells previously grown as a monolayer and serum supplemented cultures may undergo faster adaptation to suspension and serum free conditions without a significant drop in viability. Before this approach was tested p21 cells were transfected with *bcl-2* gene to enhance



*R. Smith (ed.), Cell Technology for Cell Products*, 107–110.

their robustness by reducing apoptosis that may result from stress invoked by the cell cycle arrest and hydrodynamic shear force in suspension culture or growth factors/nutrient limitation in serum free culture. It is therefore postulated that by combining p21 and Bcl-2 over-expression the time required for adaptation to suspension and serum free cultures could be significantly reduced.

## 2. RESULTS AND DISCUSSION

Induced P21 and constitutive bcl-2 Transfectants were cloned in the presence of the selective agents hygromycin B and Geneticin 418 and the resulting clones were assessed for cell cycle arrest by the addition of IPTG and in the apoptotic inducing conditions of serum deprivation and exposure to Staurosporine. The clone that displayed the best cell cycle arrest and resistant to apoptosis was chosen for this investigation.

***The effect of p21$^{CIP1}$ and Bcl-2 expression on the adaptation of anchorage dependent CHO cell lines to suspension and serum free growth environments***

Cells were trypsanized and seeded into two shake flasks and passaged for 21 days. To induce p21 over-expression IPTG was added for 9 days to one of the cultures. Medium was changed every two days and cell number adjusted. The results in Figure 1 show the combined effect of arresting cell division and bcl-2 expression on culture viability measured at 2 time points during the adaptation period. During the adaptation process the viability of arrested cultures remained significantly higher than that of the dividing (control) cultures (99% after 12 days compared to 68% in the control). After 21 days from the start of the adaptation period the control cells showed some evidence of recovery from the initial collapse of viability thereby giving the first signs of adaptation in the form of increasing viability however this took almost double the amount of time to the previously arrested cultures.

The results in Figure 2 show a similar but more dramatic pattern of results to those seen during adaptation to suspension, however, it was apparent that during serum free adaptation previously arrested cultures displayed a significant decrease in aggregate formation resulting in a significant increase in culture viability (99% compared to 46% in non arrested cultures after 21 days). This reduction in aggregate formation appears to continue to be reduced following removal of cell cycle arrest resulting in overall improved culture adaptation state.

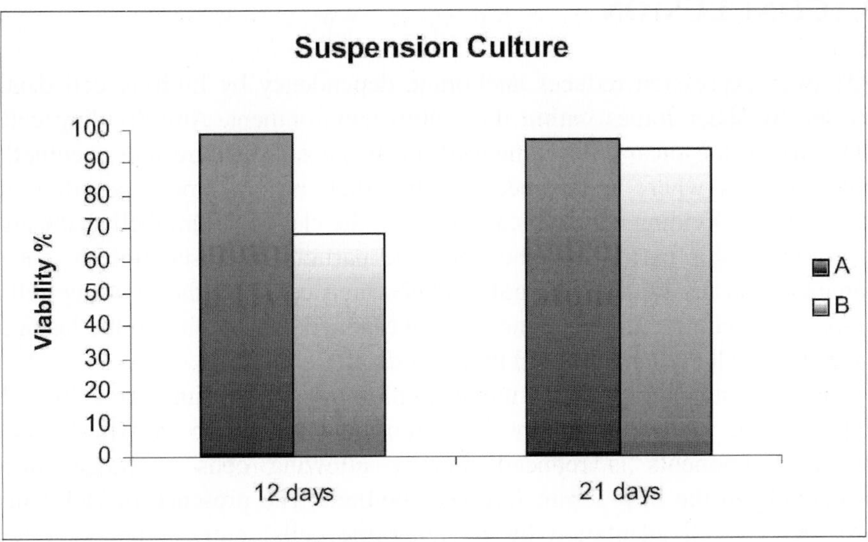


*Figure 1.* Analysis of CHO viability during adaptation to suspension. Arrested (A) and dividing cells (B) were passaged every 3 days.

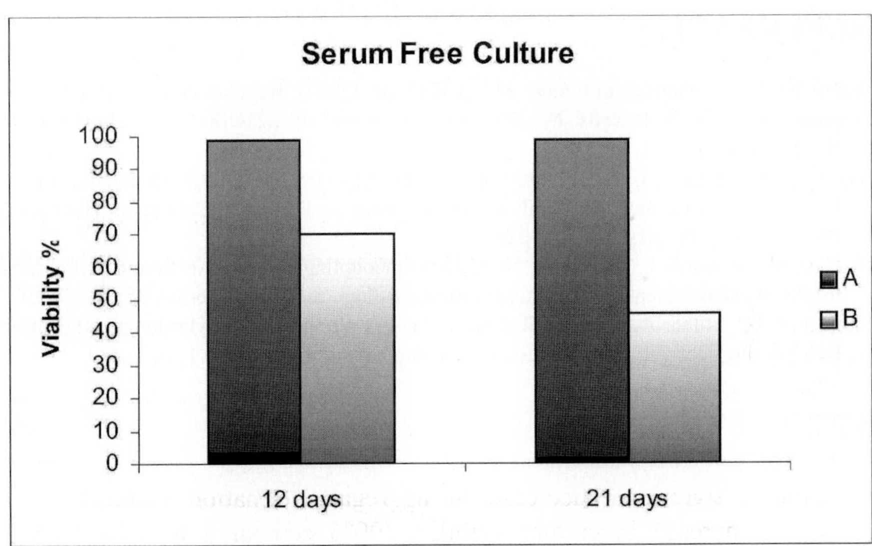


*Figure 2.* Analysis of CHO viability during adaptation to serum free. Arrested (A) and dividing cells (B) were passaged every 3 days.

## 3. CONCLUSION

P21 over expression reduces anchorage dependency by limiting cell death caused by shear forces within the culture environment. Any dividing cells that survive the shear environment of the suspension culture will eventually enter mitosis where it has been shown they become more sensitive to destruction. Dividing cells typically take on the classical dumbbell shape and it is during this period that they become particularly susceptible to shear forces which act on the vulnerable axis which connects the dividing cells, tears occur within the membrane at the weakened axis leading to the leakage of internal cellular contents and thus cell death.

P21 arrest can also reduce the adaptation period to serum free culture. It appears that by preventing proliferation dependence on growth factors/ serum components is reduced thereby allowing cells to adapt more effectively to the new serum free environment. The presence of Bcl-2 has also helped cell adaptation by reducing the cells' nutrient uptake rates. Consumption rates of all essential amino acids have been found to significantly decreased for hybridoma cell line expressing Bcl-2 which provided evidence of an adaptive response to the change in culture environment by down regulation of cellular activity (Simpson *et al.* 1997).

## REFERENCES

Watanabe S., J. Shuttleworth and M. Al-Rubeai (2002) Regulation of cell cycle and productivity in NSO cells by the over-expression of p21CIP1. Biotechnology and Bioengineering. 77, 1-7.

Ibarra N, Watanabe S, Bi,J-X, Shuttleworth J and Al-Rubeai M (2003) Modulation of Cell Cycle for Enhancement of Antibody Productivity in Perfusion Culture of NSO Cells. Biotechnology Progress. 19, 224-228.

Bi J.-X. Shuttleworth J., Al-Rubeai M. (2004) Uncoupling of cell growth and proliferation results in enhancement of productivity. Biotechnology & Bioengineering. 85, 741-749

Simpson N.H., Milner A.E. and Al-Rubeai M. (1997). Prevention of hybridoma cell death by Bcl-2 during suboptimal culture conditions. Biotechnol. Bioeng. 54: 1–16.

# Release of Plasmid-DNC from PEI/DNA Particles

Martin Bertschinger, Arnaud Schertenleib, Gaurav Backliwal, Martin Jordan and Florian M. Wurm

*Laboratory of Cellular Biotechnology, Ecole Polytechnique Federale de Lausanne (EPFL), CH 1015 Lausanne, Switzerland*

**Abstract:** The cationic polymer polyethylenimine (PEI) spontaneously forms complexes with DNA that efficiently transfect mammalian cells. To study the reversibility of the DNA/PEI interaction in vitro, preformed complexes were diluted with varying amounts of plasmid DNA and RNA. The released plasmid DNA was visualized by gel electrophoresis. With this simple assay we determined the kinetics of the release of plasmid DNA. Even at very low concentrations RNA instantaneously replaced most of the plasmid DNA in the complex, whereas DNA-mediated release takes up to one day. The results of these in vitro experiments are important for understanding the transport and release of plasmid DNA from PEI/DNA particles after transfection of cells.

**Key words:** Polyethylenimine, PEI, PEI/DNA complex, PEI/DNA nanoparticle, DNA release

## 1. INTRODUCTION

The cationic polymer polyethylenimine (PEI) spontaneously forms stable complexes with DNA that efficiently transfect mammalian cells. Little is known about the disruption of the DNA/PEI complexes within the cell. Once in the nucleus the plasmid DNA must be free of residual PEI since DNA complexed with PEI shows reduced efficiency to function as a template for RNA and DNA polymerases. DNA/PEI complexes are highly stable in solution, so release of plasmid DNA must be due to interactions with cellular components.



*R. Smith (ed.), Cell Technology for Cell Products*, 111–115.

## 2. MATERIALS AND METHODS

pSP-E1A and pEAK8-EGFP were described previously (Hacker *et al.*, 2004). Plasmid purification was performed with the NucleoBond PC 10000 kit (Macherey-Nagel, Düren, Germany) following the manufacturer's protocol. Total RNA was extracted from CHO DG44 cells with the GenElute Mammalian Total RNA Miniprep kit (Sigma-Aldrich, Steinheim, Germany) and quantified by optical density at 260 nm. Linear 25 kDa PEI (Polysciences, Eppenheim, Germany), JetPEI (PolyPlus Transfection, Illkirch, France) and branched PEIs with molecular weights of 1.8–2 kDa, 10–25 kDa, and 750 kDa (Sigma-Aldrich) were prepared in water at a final concentration of 1 mg/ml, and the pH was adjusted to 7.0 with HCl.

Complexes formed at PEI:DNA ratios of 0.26 (w/w) or more were sedimented by centrifugation at 13,000 rpm for 2 min in a microfuge (Derouazi *et al.*, 2004; Bertschinger *et al.*, 2004). We took advantage of this observation to study PEI/DNA complex disruption by mixing preformed PEI/DNA particles with either excess DNA or RNA. pSP-E1A at 50 μg/ml in 150 mM NaCl was mixed with the same volume of 150 mM NaCl containing PEI at different concentrations. After incubation for 10 min the mature particle was diluted with pEAK8-EGFP or total RNA in 150 mM NaCl at the indicated concentrations. The solution was then centrifuged as described above and the level of pSP-81 in the supernatant was determined by agarose gel electrophoresis.

## 3. RESULTS AND DISCUSSION

At a low PEI:DNA ratio (0.35 (w/w) there was at least a partial release of pSP-81 from preformed complexes in the presence of either pEAK8-EGFP or RNA (Fig. 1). As the PEI/DNA ratio was increased, less pSP-81 was released from preformed complexes in the presence of pEAK8-EGFP (Fig. 1A-D) as compared to RNA (Fig. 1E-H). Even at a low concentration (50 μg/ml) RNA was efficient for the disruption of the preformed complexes (Fig. 1E). PEI seems to have a higher affinity for RNA than DNA as there was a complete release of plasmid DNA at only 50 μg/ml RNA for most PEI:DNA ratios tested (Fig. 1E). Linear 25 kDa PEI and JetPEI® are both very efficient for the transfection of mammalian cells (Derouazi *et al.*, 2004). Plasmid DNA was completely released from the complex with these PEIs in the presence of pEAK8-EGFP (Fig. 2A,B) or total RNA at concentrations of 50 μg/ml (Fig. 2F,G). With complexes formed with branched PEIs, pSP-81 was released at higher DNA or RNA concentrations (Fig. 2C,H) or not at all (Fig. 2 D,E,I,J). This may explain at least some of the differences in transfection efficiencies with these PEIs.

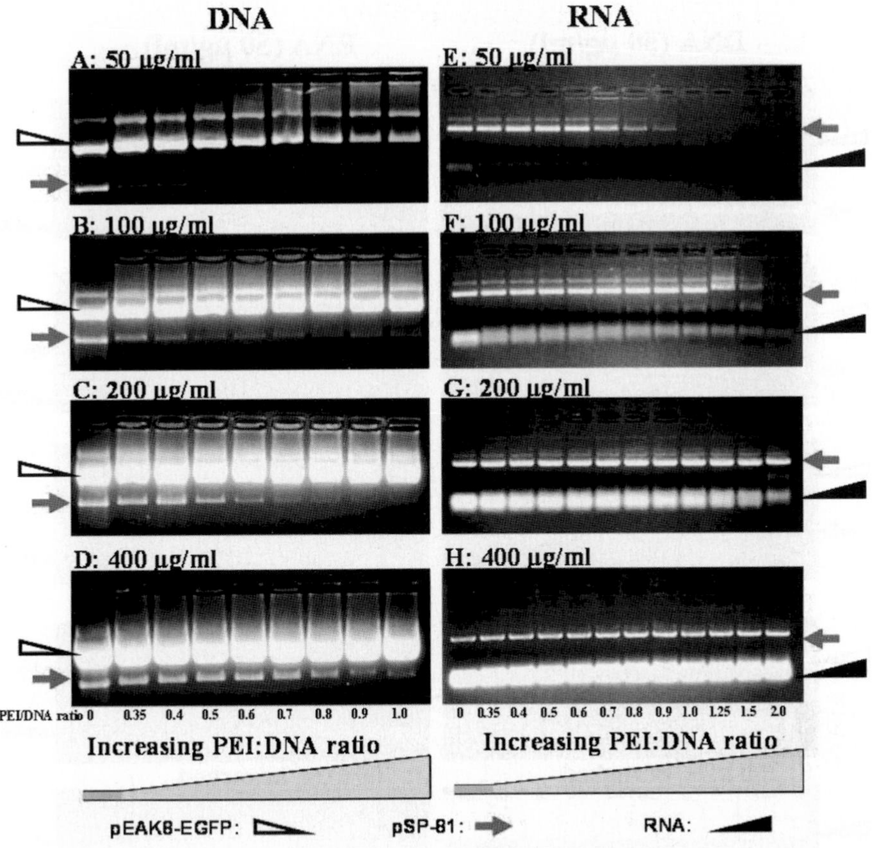


*Figure 1.* Release of pSP-81 from preformed PEI/DNA complexes. PEI/DNA complexes were mixed with increasing concentrations of pEAK8-EGFP (A-D) or total RNA (E-H). After centrifugation, the amount of pSP-81 in the supernatant was determined by agarose gel electrophoresis.

These results may explain why PEI is an efficient transfection agent. The complex of plasmid DNA and PEI may be stable until a cellular compartment with a high RNA or DNA concentration is reached. The plasmid DNA can then be efficiently released and transcribed.

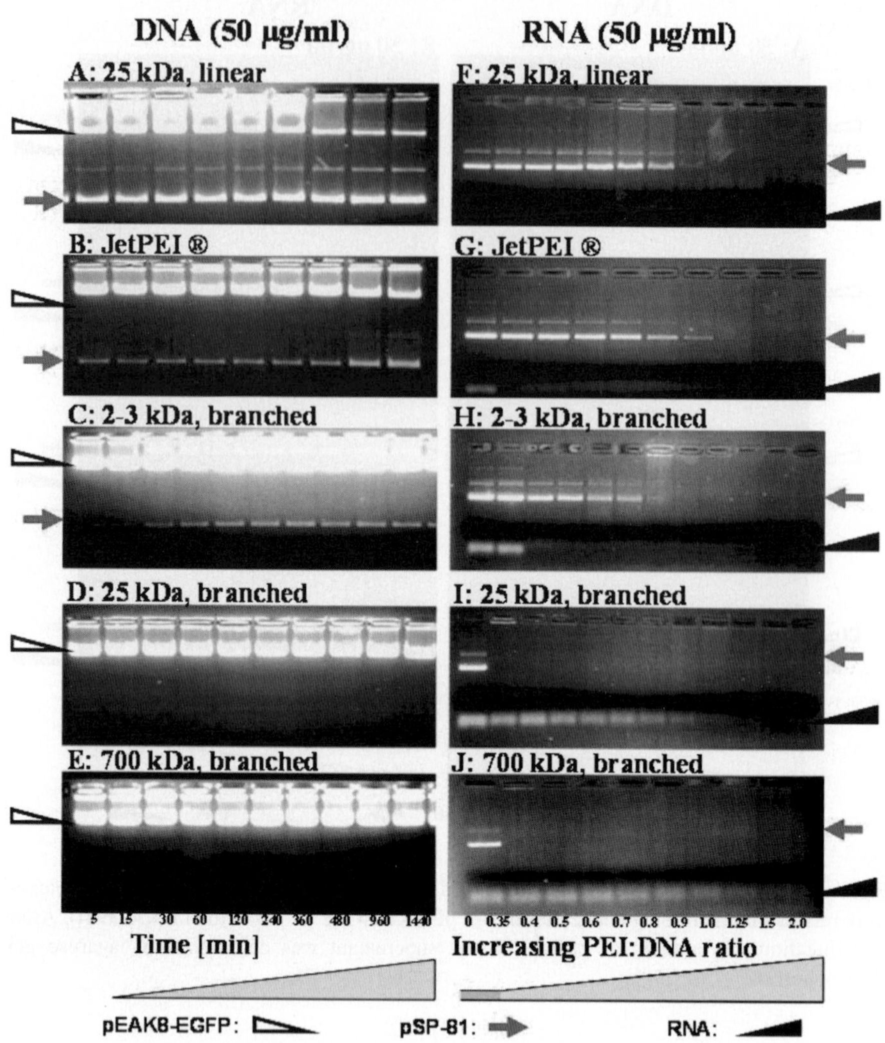


*Figure 2.* Release of pSP-81 from preformed PEI/DNA complexes containing different PEIs. PEI/DNA complexes were mixed with pEAK8-EGFP (A-E) or total RNA (F-J) at 50 µg/ml. After centrifugation, the amount of pSP-81 in the supernatant was determined by agarose gel electrophoresis.

## REFERENCES

Bertschinger, M., Chaboche, S., Jordan, M., Wurm, FM. 2004. A spectrophotometric assay for the quantification of polyethylenimine in DNA nanoparticles. Anal Biochem. 334: 196-8.

Derouazi, M., Girard, P., Van Tilborgh, F., Iglesias, K., Muller, N., Bertschinger, M., Wurm FM. 2004. Serum-free large-scale transient transfection of CHO cells. Biotechnol Bioeng. 87:537-45.

Hacker, DL., Bertschinger, M., Baldi, L., Wurm, FM. 2004 Reduction of adenovirus E1A mRNA by RNAi results in enhanced recombinant protein expression in transiently transfected HEK293 cells. Gene 341: 227-34.

# Efficient Cloning Method for Variable Region Genes of Antigen Specific Human Monoclonal Antibody

Shinei Matsumoto, Yoshinori Katakura, Makiko Yamashita, Yoshihiro Aiba, Kiichiro Teruya, Sanetaka Shirahata

*Graduate School of Systems Life Sciences, Kyushu University, Fukuoka, Japan.*

**Abstract:** We have developed an *in vitro* immunization protocol of human peripheral blood mononuclear cells (PBMC) for generating human antigen-specific antibodies. By using this protocol, B cells producing antigen specific antibody can be propagated within a week. In the present study, we tried to establish an efficient strategy to clone variable region genes of antigen specific human monoclonal antibody by applying *in vitro* immunized PBMC to the phage display method. To evaluate the efficiency of our strategy, heavy and light chain variable region genes were prepared by PCR from PBMC immunized *in vitro* with mite-extract (ME), and the combinatorial library (>$10^5$ members) cloned for display on the surface of a phage. Phage was subjected to a streptavidin magnetic bead panning procedure. After concentrating the ME-specific phage antibody by panning, the phage antibodies specific for ME were detected by ELISA. Results show that rapidly production of antigen-specific human monoclonal antibody from smallish (1.6 x $10^5$) library is possible by using *in vitro* immunized PBL.

**Key words:** *in vitro* immunization, human monoclonal antibody, phage display

## 1. INTRODUCTION

Human monoclonal antibodies (mAbs) have a great potential for diagnosis and treatments of cancer, allergy and other diseases. However, we cannot immunize human with antigen by ethical problems. Monoclonal antibodies from mouse origin are relatively easy to produce, however, their therapeutic availability is restricted by their antigenicity. At present, human monoclonal antibodies are mainly produced by humanizing mouse monoclonal antibodies. But it is difficult to completely remove antigenicity derived from mouse. Therefore, we have developed an *in vitro* immunization



*R. Smith (ed.), Cell Technology for Cell Products*, 117–119.

(IVI) protocol of human peripheral blood lymphocyte (PBL) for generating human antigen specific antibodies. By using this protocol, B cells producing antigen specific can be propagated within a week. In the present study, we tried to establish an efficient strategy to clone variable region genes of antigen specific human monoclonal antibody by applying *in vitro* immunized PBL to the phage display method.

## 2. METHODS

### 2.1 In vitro immunization

Human PBL were cultured for 8 days in ERDF medium containing 10% heat inactivated fetal bovine serum, CpG-ODN, IL-2, IL-4, 2-mercaptoethanol and Mite-Extract (ME). Antibody production in the supernatant of cultured PBL was measured by ELISA.

### 2.2 Construction of phage antibody library

The VH and VL genes of cultured PBL were prepared by RT-PCR. The VH and VL DNA fragments were joined with a linker DNA. The assembled ScFv DNA was digested with *Sfi* I and *Not* I for cloning into the pCANTAB5E phagemid vector. The ligated vector was introduced into competent *E. coli* TG1 cells. Phagemid-containing bacterial colonies were infected with M13KO7 helper phage to yield phage-displayed antibody ScFv fragments.

### 2.3 Detection and production of ME-specific antibody

The phage antibodies were selected by panning against ME. ME-specific phage antibodies bound to biotinylated ME on streptavidin magnetic beads and non-specific phage were washed off. *E. coli* TG1 cells were infected with the binding phage. After panning, 60 individual phage antibodies were picked. Using ME-coated 96 well plates, ME-reactive phage antibodies were detected by ELISA. The VH and VL genes were isolated from the positive clones. The resulting DNA were co-transfected to the CHO cells.

## 3. RESULTS AND DISCUSSION

15 positive clones were obtained from random picked 60 phage clones of *in vitro* immunized library (Figure 1). The VH and VL genes were isolated from the positive clone B11, D12, and E11. The resulting DNA were co-transfected to the CHO cells and secreted IgG in the supernatant were purified and analyzed by Western blotting. ME could be detected by the purified IgG.

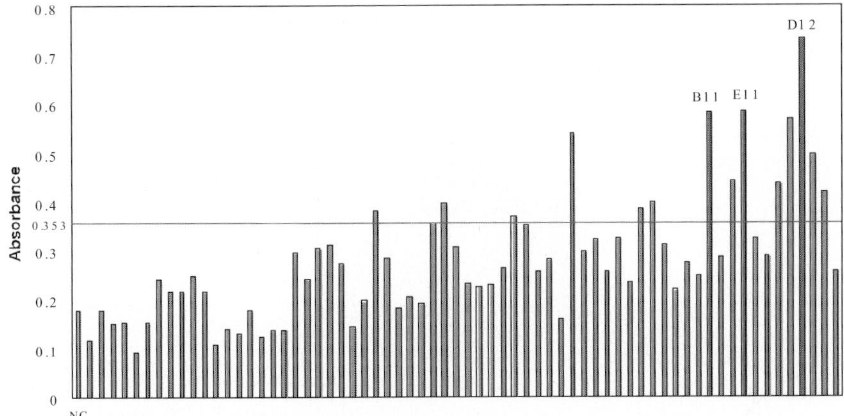


*Figure 1.* Detection of ME-specific scFv (ELISA).

Our results show that it is possible to rapidly produce antigen-specific human monoclonal antibody from smallish (1.6 x $10^5$) library by using *in vitro* immunized PBL. We expect that *in vitro* immunization and phage display method make a strong combination to produce human monoclonal antibody.

## REFERENCES

Ichikawa, A., Katakura, Y., Teruya, K., Hashizume, S. and Shirahata, S. (1999) In vitro immunization of human peripheral blood lymphocytes: establishment of B cell lines secreting IgM specific for cholera toxin B subunit from lymphocytes stimulated with IL-2 and IL-4. Cytotechnology 31, 131.

Marks JD, Hoogenboom HR, Bonnert TP, McCafferty J, Griffiths AD, Winter G. (1991) By-passing immunization. Human antibodies from V-gene libraries displayed on phage. Journal of Molecular Biology. 1991 Dec 5;222(3):581-97.

# Design Approach Regarding Humanization and Functionality of an Anti-CD18 Monoclonal Antibody

Cristina A. Caldas[1,4]; Diorge P. Souza[2]; Maria Teresa A. Rodrigues[3,4]; Andréa Q. Maranhão[2,4]; Ana M. Moro[3,4]; Marcelo M. Brigido[2,4]
[1]*Laboratório de Imunologia, InCor, USP, Brasil;* [2]*Laboratório de Biologia Molecular, UnB,* Brasi; [3]*Laboratório Biofármacos em Célula Animal, Instituto Butantan,Brasil;* [4]*Instituto de Investigação em Imunologia - iii, Institutos do Milênio, Brasil*

**Abstract:** In this work the closest human germline sequence was used as the framework on to which to graft murine CDRs. The proposed fully humanized version displays the same staining pattern to different cell subpopulations like CD4, CD8 and memory CD45RO lymphocytes, when compared with the original anti-CD18 MAb, reinforcing the germline approach as a successful strategy to humanize antibodies with maintenance of affinity and selectivity.

**Key words:** antibody, humanization, CD-18, adhesion, germline, framework, integrin, graft, HAMA, CDR, MAb, FACS, binding, scFv, hypermutation

## 1. INTRODUCTION

Due to HAMA (human anti-mouse antibody) response evoked by the use of murine MAbs, antibody humanization technology was developed for the production of a new molecule that maintains the capacity to recognize the antigen but resemble more like a human antibody. The purpose of this work was to humanize the 6.7 murine anti-human CD18 antibody (David *et al.*, 1991). CD18 is a β2 integrin family component and an antibody that blocks its adhesion property has potential application in a large panel of clinical manifestations through the migration's inhibition of neutrophils and other leucocytes. The huMAb antibody could be used as an anti-inflammatory agent, to decrease post ischemic effects.



*R. Smith (ed.), Cell Technology for Cell Products*, 121–123.

## 2. METHODOLOGY AND RESULTS

The strategy used was to choose the closest human germline sequence (Tomlinson *et al.*,1992) onto which to graft the murine CDRs. The germline sequence does not present somatic hypermutation of rearranged V regions, which can be unique for that specific antibody and thus be seen as immunogenic to patients. The data bank search indicated HG3 as the closest human germline sequence (Rechavi *et al.*, 1983) to receive the anti-CD18 CDRs sequences. The framework residue Ala$^{71}$ was maintained as in the original murine antibody, for the assembly of a correct H2 loop in the humanized antibody (Tramontano *et al.*, 1990). Analysis of the crystallized Fab with sequence most similar to HG3 (1AD9) showed that residue 45 is located in a position of low flexibility near to H-CDR2 and L-CDR3, seen as critical for the maintenance of the H2-H3 loop structure. Based on this assumption, two versions of humanized VH were constructed: Proline version (HP) of murine origin and Leucine version (HL) of the closest germline human sequence (Caldas *et al.*, 2000). The FACS analyses of the variations showed higher binding for the HL version, confirmed by the blocking of the 6.7 original MAb.

<table>
<tr><td></td><td>Median IF</td><td>Blocking (%)</td><td rowspan="4">Hemi-humanized scFvs fragments block the binding of MAb 6.7 FITC to CD18+ cells. scFvs were added to the cells prior the addition of the 6.7 FITC.</td></tr>
<tr><td>6.7 FITC</td><td>40,32</td><td>0</td></tr>
<tr><td>Fv-VH (L)</td><td>9, 65</td><td>76,3</td></tr>
<tr><td>Fv-VH (P)</td><td>12,52</td><td>69,0</td></tr>
</table>

For the light chain humanization the same strategy was followed and the humanized VL gene fragment was expressed together with the selected version Leu$^{45}$ of the humanized VH (Caldas et l., 2003). Two versions were analyzed in detail because of an expected loss of affinity. The version VQ contains glutamine$^{37}$ (present in 78% of both murine and human antibodies) while in the version VL the leucine residue of murine origin was maintained. The loss of affinity effect was shown by FACS analysis (Fig. 1) in comparison with the original 6.7 MAb. The capacity of the two versions' binding to the cell surface was further tested through its blocking effect on the binding of the original anti-CD18 MAb. We found that both the original anti-CD18 and the humanized LL version displayed the same pattern of staining to CD4+, CD8+ and to memory CD45RO+ lymphocytes (Fig. 2).

## 3. DISCUSSION AND CONCLUSION

For the HC, the FACS and further image analyses served as basis to choose the HL (leucine residue) for the position 45. In relation to the LC, a mutation

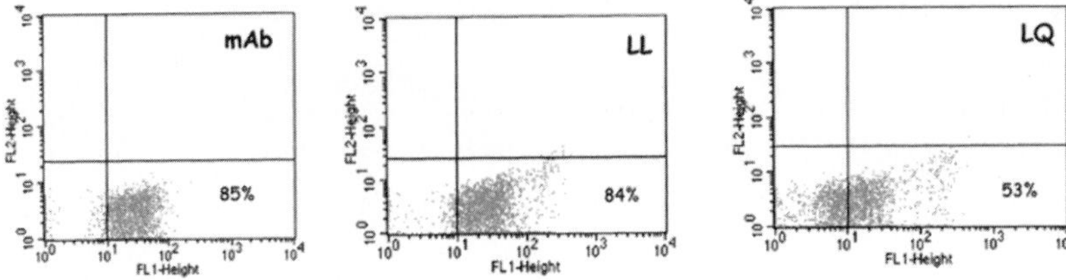


*Figure 1.* Anti-CD18 humanized versions binding to human lymphocyte cell surface; 85, 84 and 53% of the cells were positive when incubated with murine and humanized scFv versions.

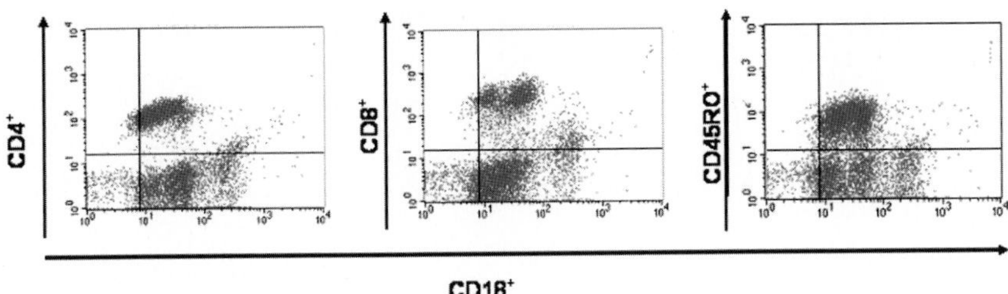


*Figure 2.* - T cell subpopulations stained by the humanized LQ version.

far from the antibody's binding site was identified, resulting in a loss of affinity for the intact CD18 molecule on the cell surface. In the human Vκ germline sequence used in this work, a glutamine residue fills the position 37. This residue is found in many humanized antibodies, including our closest PDB crystal structure, 1AD9, in which the $Gln^{37}$ is mostly buried and it contacts the same set of residues as $Leu^{37}$ in the murine structure. The position 37 is related to the FR2 stabilization and is rarely variable. The only difference is a hypothetical H-bond between the amide group of $Gln^{37}$ and the hydroxyl group of $Tyr^{86}$. $Leu^{46}$ was maintained in the humanized version for its contact with LCDR2 and HCDR3. The construction was able to recognize PMNC harboring CD18, indicating successful humanization with transfer of the original binding capability. We have shown the complete humanization of the 6.7 mAb. The CDR-grafting using the closest germline, although very simple, is a reliable procedure for humanization of antibodies.

## REFERENCES

Caldas C *et al.* (2000) Protein Engineering, 13: 353-360.
Caldas C *et al.* (2003) Mol. Immunol., 39: 341-952.
David D *et al.* (1991) Cel. Immunol., 136: 519-524.
Rechavi G *et al.* (1983). PNAS USA, 80:855-859.
Tomlinson IM *et al.* (1992) J. Mol. Biol., 227: 776-798.
Tramontano A *et al.* (1990). J. Mol. Biol., 215: 175-182.

# Development of a Transient Mammalian Expression Process for the Production of the Cancer Testis Antigen NY-ESO-1

David L. Hacker[1], Leonard Cohen[2], Natalie Muller[1], Huy-Phan Thanh[1], Elisabeth Derow[1], Gerd Ritter[2], and Florian M. Wurm[1]

*[1]Laboratory of Cellular Biotechnology, Ecole Polytechnique Federal de Lausanne (EPFL), 1015 Lausanne, Switzerland, [2]Ludwig Institute of Cancer Research, New York Branch at Memorial Sloan-Kettering Cancer Center, 1275 York Avenue, New York, NY 10021, USA*

**Abstract:** The cancer-testis antigen NY-ESO-1 is expressed in a wide range of tumor types while normal tissue expression is restricted to germ cells. It is highly immunogenic, frequently eliciting spontaneous humoral and cellular immune responses in patients with advanced tumors expressing the protein. A series of different NY-ESO-1 vaccines are currently being explored in early clinical trials worldwide. We have developed a reproducible process for the transient expression of full-length NY-ESO-1 in mammalian cells. Either Chinese hamster ovary (CHO) DG44 or HEK 293 cells were transfected with NY-ESO-1 DNA using polyethylenimine. NY-ESO-1 was expressed intracellularly as a histidine-tagged protein in agitated suspension cultures at volumes to 400 ml at levels up to 30 mg/L as estimated in crude cell lysates by Western blot. The protein was purified by immobilized metal affinity chromatography (IMAC) after detergent lysis and was shown to be reactive with sera from patients with tumors expressing NY-ESO-1.

**Key words:** Cancer-testis antigen, NY-ESO-1, vaccine, recombinant protein, Chinese hamster ovary cells, human embryonic kidney cells, polyethylenimine, transfection, transient gene expression, suspension cells, square bottles, histidine tag, affinity chromatography, antigenicity, ELISA.

## 1. INTRODUCTION

The cancer-testis (CT) antigen NY-ESO-1 is expressed in many tumor types but not in normal tissues except testis (Scanlan *et al.*, 2004). Both humoral and cellular immune responses can occur in patients with tumors



*R. Smith (ed.), Cell Technology for Cell Products*, 125–128.

that express this antigen (Scanlan *et al.*, 2004). In recent clinical trials in which patients with NY-ESO-1-positive tumors were vaccinated with recombinant NY-ESO-1 from *E. coli* in combination with an adjuvant both antibody and T cell responses were observed (Davis *et al.*, 2004). In this report we have demonstrated the feasibility of transient expression of intracellular histidine-tagged NY-ESO-1 (His-NY-ESO-1) in either CHO DG44 or HEK 293 cells in non-instrumented cultivation systems at volumes up to 400 ml. The partially purified His-NY-ESO-1 was shown to be reactive to sera from patients with NY-ESO-1-positive tumors. Transient expression in mammalian cells offers an alternative approach to the production of recombinant proteins for vaccines.

## 2. METHODS

For transfections in 1-L square bottles, suspension CHO DG44 or HEK 293 cells were seeded at 1 x $10^6$ cells/ml in 200 ml of RPMI 1640 medium (Derouazi *et al.*, 2004). The cells were transfected with pcH-NYESO1 in the presence of 25-kDa linear polyethylenimine (PEI) (Polysciences, Eppelheim, Germany) (Muller *et al.*, 2005). At 4 h post-transfection 200 ml of medium was added to each bottle. At 2 days post-transfection, 2 ml aliquots of the culture were collected. The cells were harvested by centrifugation, lysed in Mammalian Protein Extraction Reagent (M-PER) (Perbio Science, Lausanne, Switzerland) according to the manufacturer's recommendations, and the soluble fraction was analyzed by immunoblot using the mouse monoclonal antibody ES121 raised against NY-ESO-1. Antibody binding was detected with horseradish peroxidase-conjugated anti-mouse IgG (Sigma Chemical, St. Louis, MO).

## 3. RESULTS AND DISCUSSION

Histidine-tagged NY-ESO-1 (His-NY-ES0-1) was transiently expressed in both HEK 293E and CHO DG44 cells using PEI for DNA delivery. For both cell lines the highest steady state level of intracellular His-NY-ESO-1 was observed at day 2 post-transfection (data not shown). For production of the recombinant protein, cultures of 200 ml in agitated one-liter square bottles were transfected in serum-free RPMI 1640 (4). As shown in Fig. 1, the production process in 1-L square bottles was reproducible. The transfected cells were harvested at 2 days post-transfection and lysed with detergent. Partial purification of His-NY-ESO-1 was performed by IMAC. Recovery of the protein from CHO DG44 was less efficient than from HEK 293 (data not shown). The antigenicity of His-NY-ESO-1 from HEK 293 cells was determined by ELISA using sera from cancer patients with tumors expressing NY-ESO-1. Importantly, HEK 293-derived His-NY-ESO-1 was equally reactive to *E. coli*-derived NY-ESO-1 (Table 1). Recombinant His-NY-ESO-1 transiently expressed in mammalian cells may become a valuable reagent for the development of NY-ESO-1 cancer vaccines. Potential applications include its use as a vaccine immunogen and as a reagent to monitor vaccine induced "in vitro" and "in vivo" immune responses to NY-ESO-1.

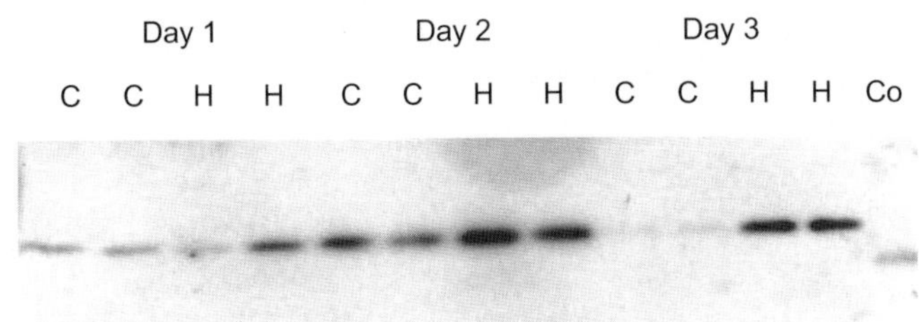


*Figure 1.* Transient His-NY-ESO-1 expression in HEK 293 (H) and CHO DG44 (C) cells. The cells were transfected with pH-NYESO1 and harvested at the times indicated. The lysates were analyzed by western blot using an antibody against NY-ESO-1. Recombinant NY-ESO-1 from E. coli (Co).

*Table 1.* ELISA reactivity of recombinantHis-NY-ESO-1 with sera from cancer patients containing NY-ESO-1 antibodies.

| Patient # | NY-ESO-1 (E. coli) | NY-ESO-1 (HEK) | Control |
|---|---|---|---|
| CA-001 | 1:5000 | 1:5000 | negative |
| CA-002 | negative | negative | negative |
| CA-003 | 1:16000 | 1:8000 | negative |
| CA-004 | negative | negative | negative |
| CA-005 | 1:25000 | 1:60000 | negative |
| CA-006 | negative | negative | negative |
| CA-007 | 1:60000 | 1:60000 | negative |
| CA-008 | negative | negative | negative |
| CA-009 | 1:60000 | 1:60000 | negative |
| Positive control | 1:1600000 | 1:1600000 | negative |
| Negative control | negative | negative | negative |

## REFERENCES

Scanlan, M.J., Simpson, A.J.G., and Old, L.J., 2004, The cancer/testis genes: review, standardization, and commentary. Cancer Immunity **4**:1-15.

Davis, I.D., *et al.*, 2004, Recombinant NY-ESO-1 protein with ISCOMATRIX adjuvant induces broad integrated antibody and CD4+ and CB8+ cell responses in humans. Proc. Natl. Acad. Sci. USA **101**:10697-10702.

Derouazi, M., Girard, P., Van Tilborgh, F., Iglesias, K., Muller, N., Bertschinger, M., and Wurm, F.M., 2004, Serum-free large-scale transient transfection of CHO cells. Biotechnol. Bioeng. **87**:537-545.

Muller, N., Girard, P., Hacker, D.L., Jordan, M., and Wurm, F.M., 2005, Orbital shaker technology for the cultivation of mammalian cells in suspension. Biotechnol. Bioeng. **89**:400-406.

# The Role of the Sumoylation Pathway in Transient and Stable Recombinant Protein Expression in Chinese Hamster Ovary Cells

Martin Bertschinger, David Hacker and Florian Wurm
*Laboratory of Cellular Biotechnology, Institute of Biological Engineering and Biotechnology, Faculty of Life Sciences, Swiss Federal Institute of Technology, 1015 Lausanne, Switzerland*

**Abstract:** One of the essential proteins in the sumoylation pathway is ube2i, one of the subunits of the conjugating complex E2. Transient expression of a short hairpin RNA (shRNA) targeted to the ube2i mRNA in Chinese hamster ovary (CHO) cells increased transient reporter protein expression by 2-3 fold. These results demonstrate the utility of RNAi in studying the role of chromatin modifiers in transgene expression,

**Key words:** SUMO, ubc9, ube2i, transient transfection, RNAi, CHO DG 44, IgG expression

## 1. INTRODUCTION

The modification of histones by acetylation, methylation, phosphorylation, and ubiquitination is known to be important for the regulation of eukaryotic gene expression. Recently, histone sumoylation has also been shown to be important for transcriptional repression (Nahan *et al.*, 2003). Here we show that the reduction of ube2i mRNA by RNA interference results in the enhanced transient expression of recombinant proteins in CHO DG44 cells.



*R. Smith (ed.), Cell Technology for Cell Products*, 129–132.

## 2. MATERIALS AND METHODS

pEGFP-N1 was purchased from Clontech (Palo Alto, CA). The shRNA vector pSP81 was a gift of Richard Iggo (ISREC, Lausanne, Switzerland). To generate pSumo, complementary oligonucleotides (Microsynth; Balgach, Switzerland) with homology to ube2i mRNA were annealed and cloned into pSP81. pControl directs the expression of an shRNA with no known target in CHO DG44 cells. For expression of the anti-human Rhesus D IgG1 antibody the plasmids pEAK8-LH39 and pEAK8-LH41 (Pick *et al.*, 2002) were used.

The polyethylenimine (PEI)-mediated transfection of CHO-DG44 cells in 12-well microtiter plate has been described (Derouazi *et al.*, 2004). For transfections in 250 ml square-shaped bottles, 50 ml of cells were seeded at a density of 2 x 106 cells/ml in ProCHO5 CDM medium. Then 5 ml of 150 mM NaCl containing 12.5 μg DNA and 50 μg PEI (Polysciences, Eppelheim, Germany) were added. At 5 h posttransfection, 50 ml of ProCHO5 CDM medium was added.

The Northern blot was done as described previously (Hacker *et al.*, 2004). For the probe, cDNA of the ube2i mRNA of CHO DG44 cells was amplified with specific primers. The amplicon was cut with BamH1 and Pst1 and cloned into pSP81. The BamH1 and Pst1 fragment was radiolabeled for detection of ube2i mRNA.

## 3. RESULTS AND DISCUSSION

The ube21 knockdown was performed with a plasmid (pSumo) coding for a small hairpin RNA (shRNA) with homology to its mRNA. After transfection with pSumo, ube2i mRNA was decreased to 60% (Fig. 1A,B). Considering a transfection efficiency of 40%, this signifies an efficient knock-down of ube2i mRNA in transfected cells. Co-expression of the ube2i shRNA with a reporter protein resulted in a 1.5 fold increase in green fluorescent protein (EGFP) expression (Fig. 2) and a 2-3 fold incresase in IgG expression (Fig. 3). These results show the utility of RNAi in studying the role of chromatin modifiers in transgene expression, allowing the development of new strategies to increase recombinant protein productivity in mammalian cells.

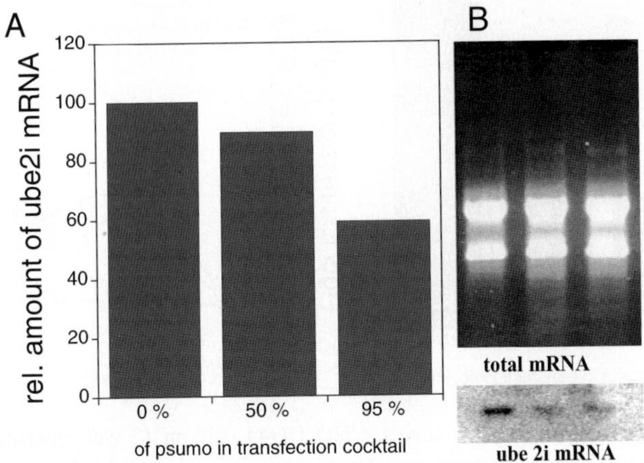


*Figure 1.* Depletion of ube2i mRNA by RNAi. (A) Quantitation of ube2i mRNA from the radiogram shown in B using ImageQuant software. (B) At 3 days post-transfection, total cytoplasmic RNA was recovered and analyzed by northern blot using a radiolabeled probe to ube2i mRNA.

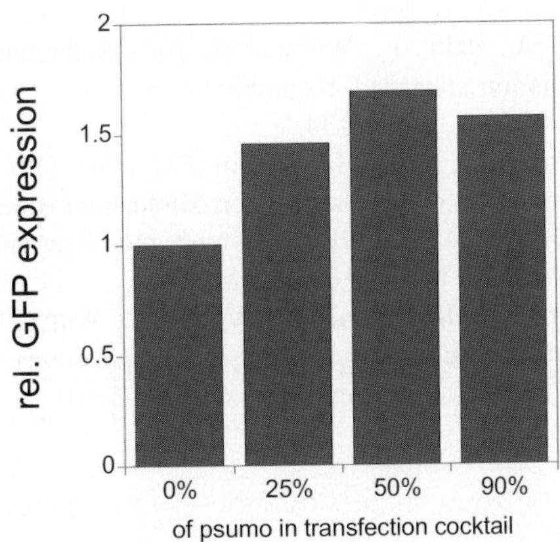


*Figure 2.* GFP expression in 12-well microtiter plates. CHO DG44 cells were transfected with a mixture of 10% pEGFP-N1 and varying amounts of pSumo and pControl (90% of the total DNA). GFP expression was measured at 3 days post-transfection.

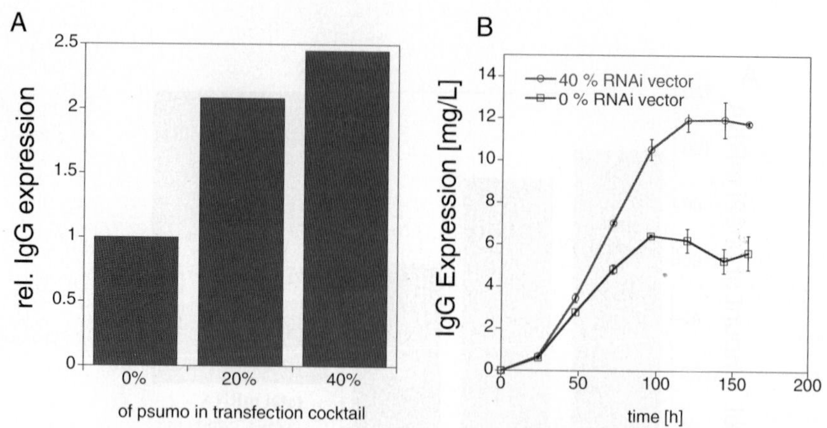


*Figure 3.* IgG expression in different scales. CHO DG44 cells in 12 well microtiter plates (A) or 250 ml square bottles (B) were transfected with 30% pEAK8-LH39, 30% pEAK8-LH41, and 40% pSumo or pControl. The IgG concentration in the medium was measured by ELISA at three days post-transfection (A) or at the times indicated (B).

# REFERENCES

Derouazi, M., Girard, P., Van Tilborgh, F., Iglesias, K., Muller, N., Bertschinger, M., Wurm FM. 2004. Serum-free large-scale transient transfection of CHO cells. Biotechnol Bioeng. 87:537-45.

Hacker, DL., Bertschinger, M., Baldi, L., Wurm, FM. 2004 Reduction of adenovirus E1A mRNA by RNAi results in enhanced recombinant protein expression in transiently transfected HEK293 cells. Gene 341: 227-34.

Muller, N., Girard, P., Hacker, DL., Jordan, M., Wurm, FM. 2005. Orbital shaker technology for the cultivation of mammalian cells in suspension. Biotechnol Bioeng. 89:400-6.

Nathan, D., Sterner, DE., Berger, SL. 2003. Histone modifications: Now summoning sumoylation. Proc Natl Acad Sci U S A. 100:13118-20.

Pick, HM., Meissner, P., Preuss, AK., Tromba, P., Vogel, H., Wurm, FM. 2002. Balancing GFP reporter plasmid quantity in large-scale transient transfections for recombinant anti-human Rhesus-D IgG1 synthesis. Biotechnol Bioeng. 79:595-601

# Effect of Different Materials Used in Bioreactor Equipments on Cell Growth of Human Embryonic Kidney (HEK293) Cells Cultivated in a Protein-free Medium

Peter Aizawa[1], Göran Karlsson, Charlotte Benemar, Vera Segl, Kristina Martinelle, and Elisabeth Lindner
[1]*Biopharmaceuticals, Octapharma AB, SE-11275 Stockholm, Sweden*

**Abstract:** During process development in small-scale bioreactors, different materials used in the bioreactors were found to have a negative effect on cell growth and viability of HEK293 cells grown in protein-free medium. A variety of materials were tested including: stainless steel 316, Duran glass, ethylene propylene diene monomer (EPDM), silicone, C-flex, Viton, and low density polyethylene (LDPE). The most detrimental effect on cell growth and viability was seen with EPDM. Compared to the polystyrene control stainless steel and glass decreased cell growth while Viton, LDPE, and silicone showed the highest biocompatibility.

**Key words:** Human embryonic kidney 293 cell, biocompatibility, culture materials, protein-free medium

## 1. INTRODUCTION

Prior to inoculation (of cells) in a bioreactor, the cultivation medium is equilibrated with the desired physical conditions (e.g. pH, temperature, DO). During the equilibrium phase, the medium may come in contact with many different materials, such as glass, stainless steel, and polymers, which have a potentially detrimental effect on the medium and the cells. The aim of this study was to investigate the effect of different culture materials on cell growth and viability of HEK293 cells cultivated in protein-free medium.



*R. Smith (ed.), Cell Technology for Cell Products*, 133–136.

## 2. MATERIALS AND METHODS

The cell line used was a Human Embryonic Kidney (HEK293) cell line producing a recombinant protein. Cells were preadapted to grow in suspension in FreeStyle 293 expression medium (Invitrogen) or SF5 (a protein-free Octapharma AB proprietary medium). Samples of test materials were incubated in medium for 24 hours at 37°C in 75 $cm^2$ polystyrene (PS) T-flasks. Medium that was incubated in empty PS T-flasks was used as the control. Material-conditioned medium was transferred to new T-flasks and inoculated with cells. Cultivation was thereafter performed on shaker tables at 37°C in a 5%/95% $CO_2$/air incubator. Samples for viable cell density and cell viability were taken every day and analysed by the Trypan blue exclusion method with a CEDEX (Innovatis) cell counter. The ratio between the surface area of the material and the medium volume during incubation was measured and denoted as A/V-ratio [$cm^2/cm^3$ medium]. Analysis of leachables from EPDM was performed by RP-HPLC with a Waters Alliance 2695 HPLC system, including a YMC-Pack ODS-A column.

## 3. RESULTS

Growth was inhibited and viability decreased faster with medium conditioned in stainless steel 316 and glass compared to the polystyrene control (Figure 1). Similar results were obtained when the different culture materials were incubated in SF5 medium (data not shown). Among the tested polymers, C-flex and EPDM decreased cell growth and viability, while silicone, Viton, and LDPE performed similar to the polystyrene control (Figure 2). The A/V-ratio was kept at 0.3 [$cm^2/cm^3$ medium] for all polymers except for the LDPE bag, which, for practical reasons, required an A/V-ratio of 2.1 [$cm^2/cm^3$ medium]. The detrimental effect of EPDM on cell growth was clearly seen when the A/V-ratio was just 0.05 [$cm^2/cm^3$ medium] (Figure 3). With an A/V-ratio of 0.04 [$cm^2/cm^3$ medium] almost all cells were dead within 24 hours. Several leachables, possibly detrimental to cells, are seen in the chromatogram (Figure 4) from water conditioned in EPDM.

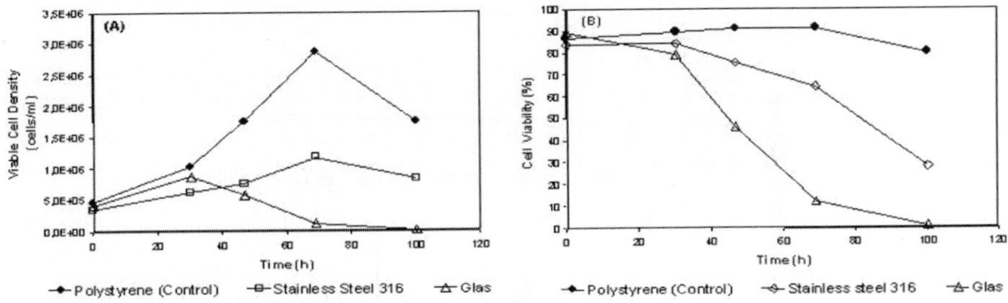


*Figure 1.* Cell growth (A) and cell viability (B) profiles. Stainless steel 316 and glass incubated in FreeStyle 293 expression medium.

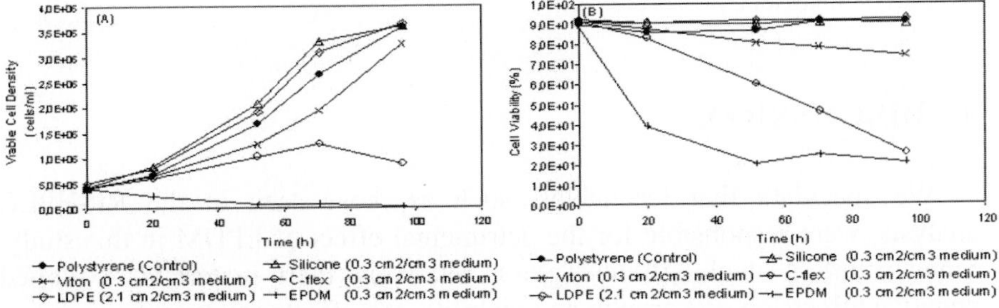


*Figure 2.* Cell growth (A) and cell viability (B) profiles. Different polymers incubated in FreeStyle 293 expression medium.

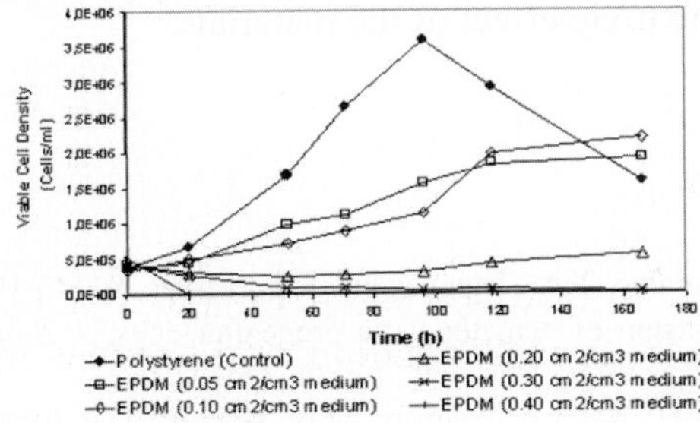


*Figure 3.* Cell growth profiles. EPDM incubated in FreeStyle 293 expression medium. Different ratios between the material surface area and the medium volume (A/V-ratio) were examined.

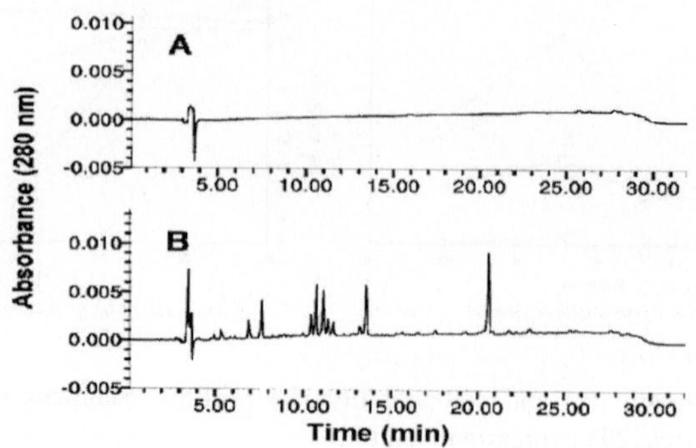


*Figure 4.* RP-HPLC analysis of leachables. Water control (A), an EPDM O-ring (B), incubated in water for 24 hours at 37°C.

# 4. DISCUSSION

We speculate that leachables, such as those seen in the RP-HPLC analysis, were responsible for the detrimental effect of EPDM in this study. Since stainless steel 316 and glass, relatively inert materials, exhibited negative effects on cell growth and viability, the effect may have been due to adsorption of essential medium components. It has been reported elsewhere that the use of serum-containing medium or the preadsorption of a concentrated protein solution (i.e., albumin, IgG, or fibronectin) can sometimes block the toxic effect of the materials.[1, 2]

# REFERENCES

1. LaIuppa, J.A., McAdams, T.A., Papoutsakis, E.T., Miller, W.M., 1997, Culture materials affect ex vivo expansion of hematopoietic progenitor cells, *J. Biomed. Mater. Res.*, **36**, 347-359
2. Ertel, S.I., Ratner, B.D., Kaul, A., Schway, M.B., Horbett, T.A., 1994, In vitro study of the intrinsic toxicity of synthetic surfaces to cells, *J. Biomed. Mater. Res.*, **28**, 667-675

# Antibody Production by GS-CHO Cell Lines Over Extended Culture Periods

A.J. Porter*, A.J. Dickson**, L.M. Barnes** and A.J. Racher*

** Lonza Biologics plc, Cell Culture Process Development, 228 Bath Road, Slough, Berkshire, SL1 4DX, United Kingdom. ** The Michael Smith Building, Faculty of Life Sciences, The University of Manchester, Oxford Road, Manchester, M13 9PT, United Kingdom.*

**Abstract:** When selecting a cell line for the production of a recombinant antibody it is important to assess the suitability of a cell line for the manufacturing process. Any assessment will include examining how the productivity of the cell line changes with increasing population doublings, as this is linked to process robustness and economics. The hypothesis that loss of productivity within antibody producing cell lines is due only to the outgrowth of a low-producer sub-population, with a higher specific growth rate than the high-producer sub-population, was tested.

**Key words:** GS, GS-CHO, stability, suitability, unstable, model, modelled behaviour, mathematical model, sub-population, low-producer, high-producer, non-producer, mixed population, specific growth rate, productivity, productivity kinetics, productivity characteristics, loss of productivity

## 1. INTRODUCTION

Observation of a large decrease in productivity with increasing population doublings (PD) may mean that a cell line is considered economically unsuitable for production of the recombinant antibody.

A possible explanation for loss of productivity with increasing PD is that the decrease in productivity is the result of the outgrowth of a non-producer (NP) or low-producer (LP) cell sub-population (1, 2). It has been suggested that such a sub-population will have a metabolic advantage and hence a growth advantage over a producer (P) or high-producer (HP) sub-population, allowing it to become the dominant sub-population (1, 2). As the proportion of NP/LP cells increases, their contribution to the productivity of the culture increases, resulting in a decrease in the apparent culture productivity.



*R. Smith (ed.), Cell Technology for Cell Products*, 137–140.

A number of models describing the behaviour of cell lines containing a NP/LP cell sub-population and P/HP cell sub-population are reported in the literature (2, 3, 4, 5). However, the authors are aware of only one study (5) that tests a similar model in a commercially relevant system.

## 2. METHODS

Three mixed populations of LP and HP antibody-producing GS-CHO cells were established. Changes in parameters describing the growth and productivity kinetics with increasing PD were trended for the pure cell lines and the mixed populations. The behaviour of the mixed populations was compared against (i) the behaviour of the pure cell lines and (ii) a model developed to predict the behaviour of such a mixed population.

The growth of each population was modelled by the logistic growth equation (a) where t is time, X is the viable cell concentration, μ is the specific growth rate, $X_0$ the initial viable cell concentration and max $X_V$ the maximum viable cell concentration. Given the initial proportions of the two pure cell lines in a population, the equation models the cell concentration of the two sub-populations.

(a) $$X = \mu \, t \, X_0 \left(1 - \frac{X}{\max X_v}\right)$$

## 3. RESULTS

All of the population mixes underwent a significant decrease in product concentration at harvest and specific production rate ($Q_P$). Compared to the pure cell lines, changes observed for the mixed populations were greater and faster (Figure 1).

The model predicts that a faster growing sub-population with a lower $Q_P$ will dominate the population (data not shown). If the model is a good representation of the behaviour in the mixed populations, there should be no substantial difference between the model and the observed data (i.e. linear regression analysis should return a slope of 1 and an intercept of 0), if the LP population has become the dominant population. Values for both the slope and the intercept of the observed data plotted against the model data (Figure 2) were compared with the values 1 and 0 respectively using the regression tool in Minitab® release 14 (Minitab Inc).

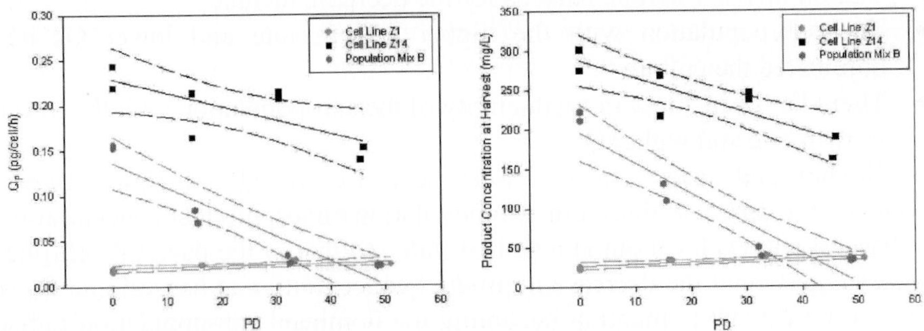


*Figure 1.* Example of the variation in kinetic parameters with increasing PD observed for two pure cell lines and the population mix established from them. Solid line is the regression line; dashed line represents 95% confidence intervals of the regression line.

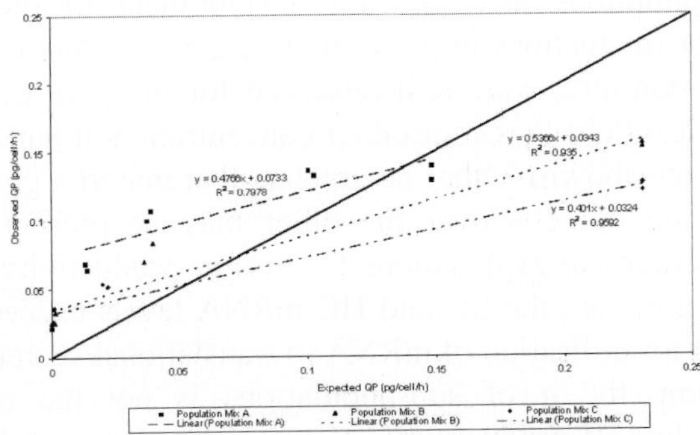


*Figure 2.* Plots comparing the specific production rate obtained from the batch culture assessments with predicted values from the model.

For all population mixes, a significance probability of 0.00 was calculated upon testing the hypotheses that the value of the intercept was zero and that the value of the slope was 1. Therefore, there was very strong evidence to reject both hypotheses. Consequently, the data do not support the initial hypothesis that loss of productivity is entirely due to the outgrowth of a faster growing LP sub-population.

## 4. DISCUSSION

The literature describing the behaviour of a cell line containing a NP/LP sub-population is contradictory. Several reports suggest that the NP/LP sub-population will become the dominant sub-population (1, 2), others report that they do not always become the dominant sub-population (6, 7).

A decrease in product concentration at harvest and $Q_P$ was seen for all population mixes. Possible reasons for the decrease include:

- The sub-population with the higher growth rate and lower $Q_P$ has dominated the culture.
- There has been a loss in productivity of the sub-population with the lower growth rate and higher $Q_P$.

Greater and faster changes were seen for mixed populations when compared to pure cell lines. For two population mixes, product concentration at harvest and $Q_P$ level out at a similar value to that of the pure LP cell line. This suggests that the decrease in product concentration at harvest and $Q_P$ is due to the LP sub-population becoming the dominant sub-population rather than a loss of productivity of the HP sub-population.

Differences between the observed data and the model indicate that the faster growing sub-population with the lower $Q_P$ did not dominate the population mix. Differences between the model and the observed data may be due to assumptions made during development of the model being incorrect. Minor fluctuations in $\mu$ and max Xv, parameters which the model assumed to be constant, were still observed for the pure cell lines – even when no significant changes in product concentration at harvest or $Q_P$ were observed (data not shown). Other parameters that impart a growth advantage to one population of cells over the other may be present, e.g., nutrient utilisation (8). Alternate explanations for loss of productivity may also be a contributing factor, e.g., the LC and HC mRNA levels exceeding a putative saturation point for utilisation of mRNA in translational/secretory events (9).

In conclusion, the $\mu$ of sub-populations is not the only parameter responsible for loss of productivity. A more complex model is required to describe the behaviour of mixed populations of LP and HP cells.

## REFERENCES

1. Frame, K.K. and Hu, W.S., 1990, Biotechnol. Bioeng. 35:469-476.
2. Kromenaker, S.J. and Srienc, F., 1994, Biotechnol. Prog. 10:299-307.
3. Frame, K.K. and Hu, W.S., 1991, Enzyme Microb. Technol. 13:690-696.
4. Lee, G.M., Varma, A. and Palsson, B.O., 1991, Biotechnol. Prog. 7:72-75.
5. Zeng, A.P., 1996, Biotechnol. Bioeng. 50:238-247.
6. Chuck, A.S. and Palsson, B.O., 1992, Biotechnol. Bioeng. 39:354-360.
7. Gardner, J.S., Chiu, A.L.H., Maki, N.E. and Harris, J.F., 1985, J. Immunol. Methods 85:335-346.
8. Kearns, B., Lindsay, D., Manahan, M., McDowall, J. and Rendeiro, D, 2003, BioProcess J. 2:52-57.
9. Barnes, L.M., Bentley, C.M. and Dickson, A.J., 2004, Biotechnol. Bioeng. 85:115-121.

# Growth Promoting Effect of a Hydrophilic Fraction of the Protein Hydrolysate Primatone on Hybridoma Cells

Michael Schomberg[1], Christian Hakemeyer[1], Heino Büntemeyer[1], Lars Stiens[2] and Jürgen Lehmann[1]
[1]*Institute of Cell Culture Technology, University of Bielefeld, P.O. Box 100131, 33501 Bielefeld, Germany;* [2]*Roche Diagnostics GmbH, Nonnenwald 2, 82377 Penzberg, Germany*

**Abstract:** Animal derived protein hydrolysates are commonly used as a complex supplement for cell culture media because of their well known growth promoting properties. In an approach to find components being responsible for this effect we fractionated a protein hydrolysate by solid phase extraction (SPE) on a $C_{18}$-matrix. After lyophilization and further preparation the fractions were tested for their proliferation enhancing ability with a hybridoma cell line in a serum-free DMEM/F12 medium supplemented with human serum albumin and transferrin. We could demonstrate a growth promoting effect caused by a rather hydrophilic fraction. In further investigations preparative High Performance Liquid Chromatography (HPLC) was used to produce growth promoting fractions of less complexity for cell evaluation and analytical HPLC.

**Key words:** Primatone, protein hydrolysate, hybridoma, HPLC, solid phase extraction

## 1. INTRODUCTION

The use of protein hydrolysates is a popular and economical option to enhance cell growth and productivity, especially at large scale cultivation under serum and protein free conditions. Regardless of either plant or animal origin, the latter afflicted with heavy doubts about their biological safety, these hydrolysates are undefined mixtures consisting mainly of amino acids, peptides, other small organic substances and salts. Since we found a strong growth promoting effect of the protein hydrolysate Primatone HS/UF on MF20 hybridoma cells cultivated under insulin free conditions (unpublished



*R. Smith (ed.), Cell Technology for Cell Products*, 141–146.

data), an attempt to elucidate the responsible substances for this effect was made. Primatone HS/UF was fractionated by SPE or HPLC and the fractions were evaluated for their growth promoting property in cell culture under serum and insulin free conditions.

## 2. MATERIALS AND METHODS

### 2.1 Primatone HS/UF

For medium supplementation and SPE or HPLC fractionation a stock solution (20% w/v) of Primatone HS/UF (Quest Intl., Naarden, The Netherlands) was prepared.

### 2.2 Cell Culture

The mouse-mouse hybridoma cell line MF20 (Developmental Studies Hybridoma Bank, University of Iowa) was cultivated in serum-free DMEM/F12 (1:1) with enhanced concentrations of amino acids and pyruvate. The medium was supplemented with human serum albumin and transferrin. For pre-culture, the medium also contained insulin. Parallel cultivations were performed in 250 mL spinner flasks with 60 mL working volume stirred at 50 rpm in a $CO_2$-incubator (37°C, 5% $CO_2$). The cells were supplemented with different SPE or HPLC fractions of Primatone or with Primatone as positive control. Estimation of cell densities was carried out by automatically trypan blue staining with the CEDEX system (Innovatis AG, Bielefeld, Germany).

### 2.3 Fractionation by Solid Phase Extraction

10 mL Primatone were loaded on a $C_{18}$-SPE-cartridge and washed with 12,5 mL water six times followed by stepwise elution with 35 mL of acetonitrile / water mixtures of increasing acetonitrile concentration (10%, 30%, 50% and 95%, respectively). Each washing and elution step was collected as fraction, lyophilized, resuspended in 10 mL medium and sterile filtered for cell evaluation.

## 2.4 Fractionation by preparative HPLC

Preparative HPLC was performed on a Gilson-Abimed HPLC-system. 10 mL Primatone were loaded on a $C_{18}$-column and eluted by a water / acetonitrile gradient. Fractions were collected as indicated in the results, lyophilized, resuspended in 10 mL of medium and sterile filtered for cell evaluation.

## 2.5 Analytical HPLC of Primatone Fractions

Analytical HPLC was carried out using an automated Kontron RP-HPLC-system equipped with a $C_{12}$-column. Analytes were detected by UV at 214 nm.

# 3. RESULTS

## 3.1 Cell Culture Test of $C_{18}$-SPE Fractions

The SPE washing step was collected in six fractions. The impact on proliferation was evaluated for each fraction in a parallel cultivation experiment (Fig. 1). For positive and negative control cells were cultivated in medium supplemented with Primatone (0,8% w/v) and in pure medium, respectively. The first fraction collected in the washing step (designated "Washing Step 1") led to an enhanced proliferation, while other fractions did not influence the cell growth and performed like the negative control.

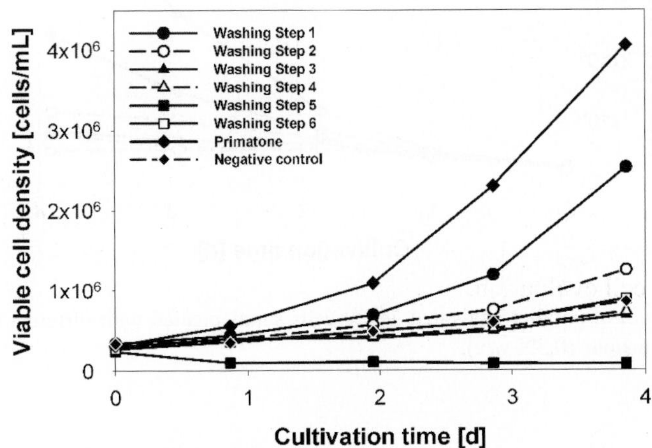


*Figure 1.* Growth of parallel cultivated MF20 cells supplemented with different C18-SPE fractions of Primatone (0,8% w/v).

The more hydrophobic elution fractions of the $C_{18}$-SPE were evaluated in a separate parallel cultivation. None of the fractions showed a better growth performance than the negative control (data not shown).

### 3.2 Cell Culture Test of $C_{18}$-HPLC Fractions

For the first preparative HPLC experiment a rough fractionation strategy was chosen, i.e. the elution volume from 0 to 800 mL was collected in four fractions of 200 mL, which were tested in cell culture (Fig. 2) against Primatone (0,5% w/v) and pure medium. Comparable to the results obtained with the SPE fractions, the fraction eluting in the first 200 mL showed the best growth promoting property, while more hydrophobic fractions collected later had no significant influence on cell growth.

In a subsequent experiment, the initial eluting 50 to 200 mL were collected in three subfractions and evaluated in cell culture. Here, the growth promoting property could be attributed to substances eluting between 100 and 150 mL (data not shown). The proliferation enhancing fraction from 100-150 mL was further subdivided into two smaller fractions (100-125 mL and 125-150 mL). The growth promoting effect could be assigned to the fraction eluting between 125 an 150 mL (Fig. 3).

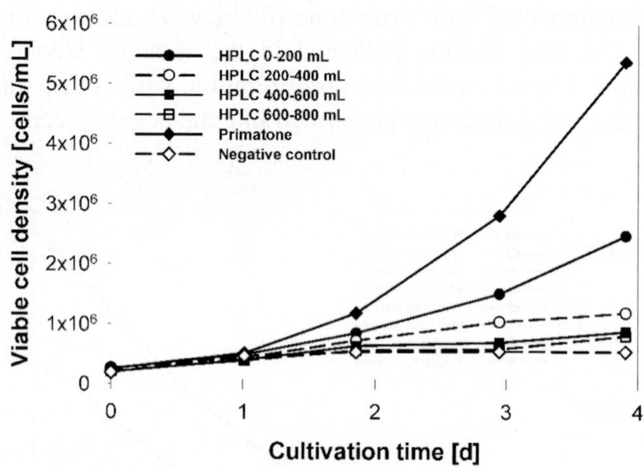


*Figure 2.* Growth of parallel cultivated MF20 cells supplemented with different C18-HPLC fractions of Primatone (0,5% w/v).

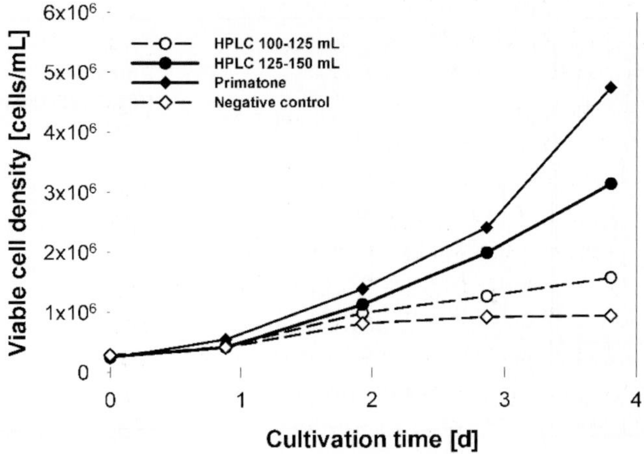


*Figure 3.* Growth of parallel cultivated MF20 cells supplemented with different C18-HPLC fractions of Primatone (1,0% w/v).

## 3.3 Analytical HPLC of Primatone Fractions

As demonstrated in Figures 1 and 4, respectively, the SPE fraction "Washing step 1" and the HPLC fraction "125-150 mL" both showed proliferation enhancing property. HPLC analysis demonstrated that all substances of the proliferation enhancing preparative HPLC fraction "125-150 mL" elute within the first 15 minutes (Fig. 4). While the whole hydrolysate contains a large mixture of different substances, we showed that only a small part of them are responsible for the growth promoting effect on MF20 hybridoma cells. Their early elution time indicates a hydrophilic character.

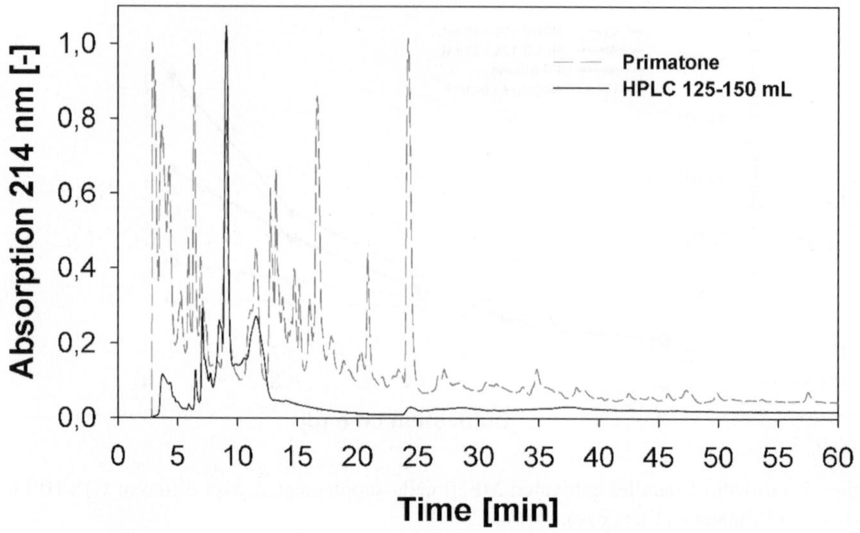


*Figure 4.* Chromatogram of the "125-150 mL" fraction prepared by HPLC (8% w/v). The chromatogram of Primatone (2% w/v) is plotted for reference.

## 4. CONCLUSION AND OUTLOOK

The results obtained with the MF20 hybridoma cells suggest that the complex medium additive Primatone HS/UF can be fractionated into subfractions of less complexity and that one of these fractions contains proliferation enhancing factors of yet undefined character. However, the growth promoting fraction never stimulated growth as well as the whole hydrolysate, so synergistic effects of different beneficial substances can be assumed. Preparation of even smaller fractions by improving the separation methods and use of gas chromatography combined with mass spectrometry as analytical tool may allow to identify single substances being responsible for proliferation enhancing effects of protein hydrolysates.

# Effect of the Scaffold in Maintenance of Suitable Chondrcyte Morphology in Cartilage Regeneration

J. Rodríguez[1], X. Rubiralta[1], J. Farré[1], P.J. Fabregues[2], J.M. Aguilera[3], J. Barrachina[3], F. Granell[3], J. Nebot[2], J.J. Cairó[1], and F. Gòdia[1]

[1]*Departament d'Enginyeria Química. ETSE. Campus Universitat Autònoma de Barcelona, 08193, Bellaterra, Barcelona, Spain,* [2]*Unitat d'Anatomia i Embriologia. Facultat de Medicina. Campus Universitat Autònoma de Barcelona, 08193, Bellaterra, Barcelona, Spain,* [3] *Hospital Asepeyo Sant Cugat. Avda. Alcalde Barnils s/n. 08190, Sant Cugat del Vallés, Barcelona, Spain*

**Abstract:** Using chondrocytes in culture as starting cells, cartilage regeneration and its architectural profile achieved by means of induced extracellular matrix autoregeneration and the use of alginate, collagen and fibrinogen as polimerizable scaffolds is studied.

Alginate allowed to maintain a round cell shape, symilar to that of chondrocytes in cartilage, whereas the autoinduction of the extracellular matrix gave a fibroblastic morphology and the fibrinogen based scaffold yields a neuron-like morphology.

**Key words:** Articular cartilage; Tissue engineering; Biomaterials; Scaffolds

## 1. INTRODUCTION

The repair of defects in cartilage resulting from traumatic or degenerative processes can be envisaged by means of tissue engineering. As it has been previously described (Griffith, 2002), the tissue regeneration process uses specific combinations of cells, scaffolds and cell signalling factors, both chemical (such as growth factors and hormones) and physical (such as mechanic forces and electrical stimulation).There is a great dependence on the three-dimensional structure used as scaffold, and the functional and morphological characteristics of the resultant cell type.



*R. Smith (ed.), Cell Technology for Cell Products*, 147–151.

In the presented work, arthroscopically extracted chondrocytes were used as cell source. *In vitro* cultures were performed with the aim of studing the cartilage regeneration and the architectural profile achieved by means diferent strategies as induced extracellular matrix autoregeneration and also the use of alginate, fibrinogen and collagen as polymerizable scaffolds, in order to generate a healthy tissue for autologous transplant (Figure 1).

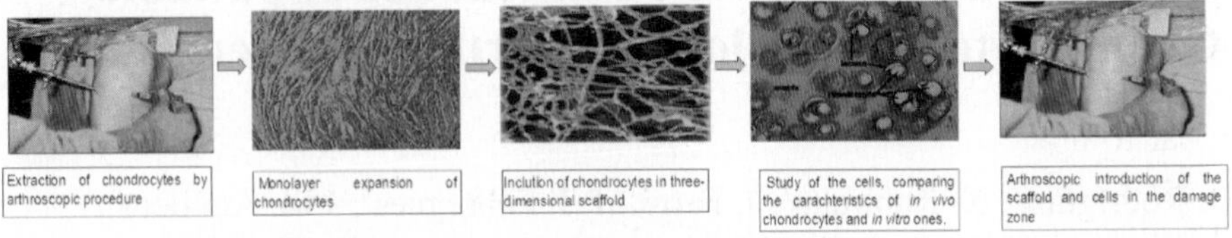
Extraction of chondrocytes by arthroscopic procedure
Monolayer expansion of chondrocytes
Inclution of chondrocytes in three-dimensional scaffold
Study of the cells, comparing the carachteristics of *in vivo* chondrocytes and *in vitro* ones.
Arthroscopic introduction of the scaffold and cells in the damage zone

*Figure 1.* Main goals of the study in order to generate a healthy tissue for autologous transplant.

## 2. MATERIALS AND METHODS

*Cell source and culture conditions.* Chondrocytes were derived from human healthy cartilage. Cartilage harvest was performed under aseptic conditions by arthroscopic procedure. Cells were isolated by digestion with type II collagenase (Roche) and pronase (Roche). They were cultured in Dulbecco's Modified Eagle Medium suplemented with 10% FCS (DMEM; Gibco) in T-75 culture flasks under an atmosphere of 5% $CO_2$ at 37ºC. They were passaged at least five times previous inclusion or culture onto supports.

*Scaffold preparation.* Auntoinduced laminar scaffold: Obtained by centrifugation of $1x10^5$ cell/ml at 500xg for 5 minutes in a 50 mL semicircular shaped tube. Fibrinogen scaffold: Equal volums of *ersta* and *once* (Tissucol duo; Baxter) were mixed with cell suspension to final concentration of $1x10^5$ cell/ml. Alginate scaffold: A solution of alginate (133mg/ml) was mixed with 0,1M $CaCl_2$ and cell suspension to final concentration of $1x10^5$ cell/ml. Collagen scaffold: A solution of rat tail collagen type I (4mg/ml Acetic acid 0,1%)(Sigma) was mixed with cell suspension to final concentration of $1x10^5$ cell/ml.

## 3. RESULTS AND DISCUSSION

The approach used for cartilage tissue regeneration was the inclusion of cultured cells into 3D scaffolds or the generation of an autoinduced matrix. In order to evaluate the performance of the proposed strategies for the obtention of funtional-like tissues, the morphological changes were studied. The obtained morphologies were compared to that observed in healthy cartilage tissues (Figure 3f) and approach is discused.

As starting point we studied cell growth, glucose consumption and lactate production of chondrocytes expanded in monolayer culture (Figure 2).

As we can see in the Figure 3a, chondrocytes expanded under this conditions, present a fibroblastic morphology. Mid-exponential phase monolayer cultures were posteriorly confined in 3D scaffolds or used for autoinduced matrix regeneration. From all studied cases, only alginate included chondrocytes (Figure 3c) pesented a morphological structure similar to the healthy hialine cartilage, with spheroid cellular shape. For the rest of the cases (Figures 3b, 3d, 3e), cultures in fibrinogen, collagen or autoinduced matrix, a fibroblastic morphology was observed, as in monolayer cultured chondrocytes.

Interestingly, in the fibrinogen scaffold it has observed how the cells evolved to a morphology resembling adipose cells (Figure 4), that rises the hypothesis of a probable dediferentiation of chondrocytes to adipose cells and vice versa

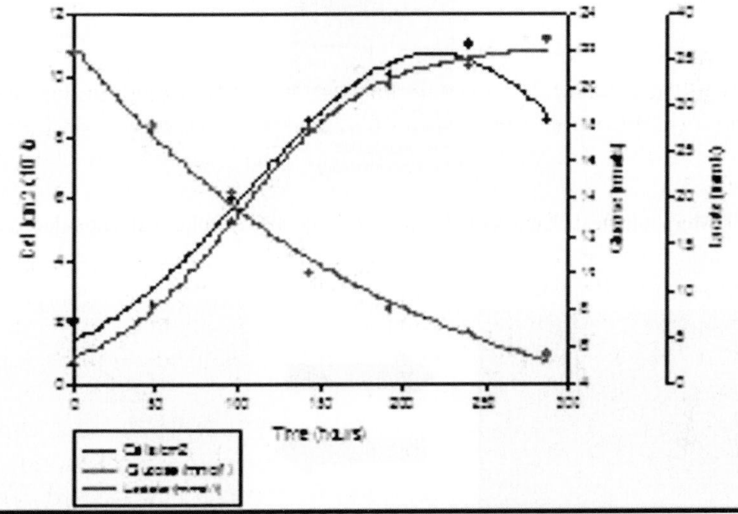

*Figure 2.* Cell growth, glucose consumption and lactate production of chondrocytes expanded in monolayer culture.

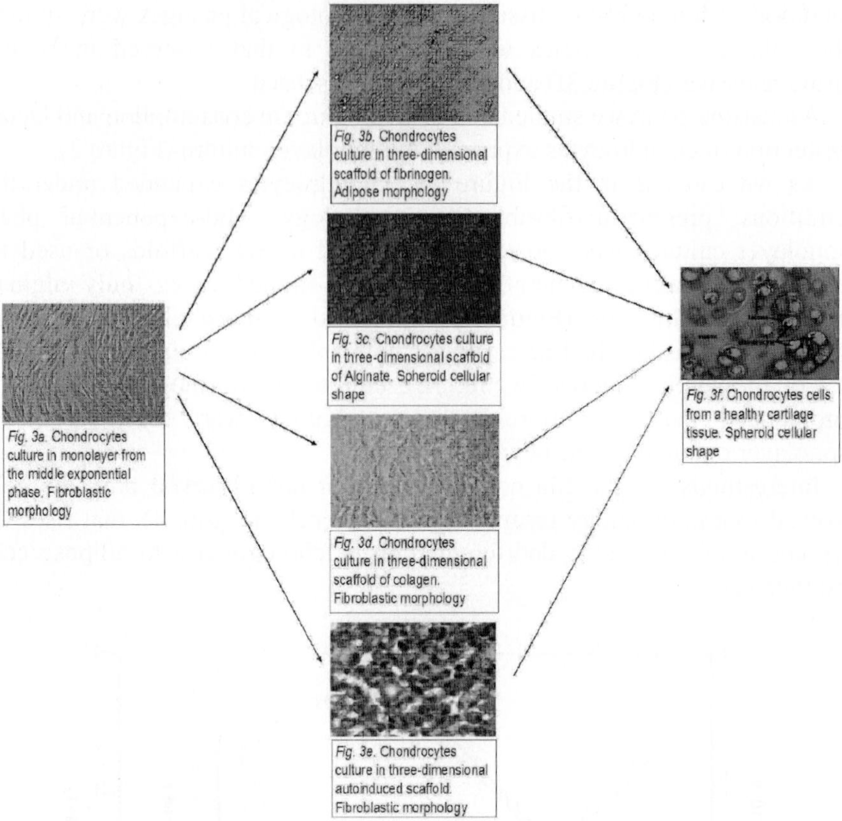

Fig. 3b. Chondrocytes culture in three-dimensional scaffold of fibrinogen. Adipose morphology
Fig. 3c. Chondrocytes culture in three-dimensional scaffold of Alginate. Spheroid cellular shape
Fig. 3a. Chondrocytes culture in monolayer from the middle exponential phase. Fibroblastic morphology
Fig. 3f. Chondrocytes cells from a healthy cartilage tissue. Spheroid cellular shape
Fig. 3d. Chondrocytes culture in three-dimensional scaffold of colagen. Fibroblastic morphology
Fig. 3e. Chondrocytes culture in three-dimensional autoinduced scaffold. Fibroblastic morphology

*Figure 3.* Morphological changes of chondrocytes into 3D scaffolds and autoinduced matrix.

Chondrocyte cells
Adipose cells

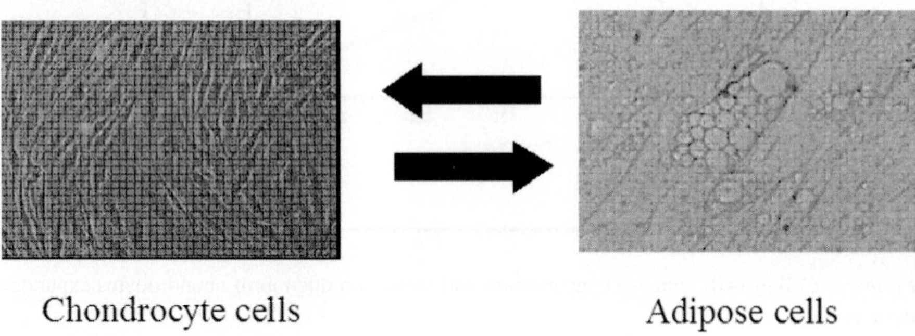

*Figure 4.* Probable deddiferentiation of chondrocytes to adipose cells and vice versa.

# 4. CONCLUTIONS

A great dependence has been observed between the chondrocyte morphology and the type of scaffold used to promote three-dimensional structure.

The case where the obtained three-dimensional structure is more similar to that found in vivo is for chondrocyte inclusion in alginate polymers.

Possible differentiation of chondrocytes to adipose cells will be further studied, since this could open new strategies for cartilage regeneration, with adipose tissue as cell source, enabling less invasive treatments.

# REFERENCES

Mitsuo Ochi, Nobuo Adachi, Hiroo Nobuto, Shinobu Yanada, Yohei Ito, and Muhammad Agung. 2004. Articular. Artificial Organs., **28**(1):28-32

X. Fei, B.-K. Tan, S-T. Lee, C.-L. Foo, D.-F. Sun, and S.-E. Aw. 2000. Effect. Transplantation Proceedings, **32**: 210-217.

Willem J.C.M. Marijnissen, Gerjo J.V.M. van Osch, Joachim Aigner, Henriette L. Verwoerd-Verhoef, Jan A.N. Verhaar. 2000. Tissue-engineered. Biomaterials, **21**: 571-580.

# Metabolic Changes Occuring During Adaptation to Serum Free Media

Sam Denby[1], Frank Baganz[1], David Mousdale[2]
[1]*The Advanced Centre for Biochemical Engineering, University College London, Torrington Place, London WC1E 7JE, UK* [2]*Beocarta,Room 555, Dept. Of Bioscience,Royal College Building,204 George Street,Glasgow,G1 1XW*

**Abstract:** This work aims to investigate the changes occurring in metabolism of a cell population during adaptation of a cell line from serum containing to serum free media. In order to enable growth in a serum free media a supplement was added to DME:F12. Cells grown in 1.25% serum media were used to investigate the effect of the supplement, data for growth, antibody production and key metabolites are presented All data presented here are derived from cells banked at different serum concentrations during the adaptation and subsequently grown in 50mL culture in 250mL shaken ehrlenmyer flasks.

**Key words:** Hybridoma, Metabolism, Serum, Adaptation, Shake Flask

## 1. INTRODUCTION

Recent trends have moved mammalian cell culture away from traditional serum containing media towards the use of more defined media. In addition there are increasing cost pressures on the biochemical engineering insustry as cost of goods becomes a larger consideration. The adaptation to defined, serum and hydrolysate free medium has several benefits, in this instance it simplifies modeling of intracellular fluxes during the course of the culture. Metabolic modeling is being increasingly applied to media development and allows a rational approach to media design.

The cell line used for this work was VPM8 a murine murine hybridoma developed as part of a veterinary PhD project. The cell line in question did not adapt directly into a commercially available serum free medium, DMEF12 or supplemented DMEF12 medium. Over the course of multiple



*R. Smith (ed.), Cell Technology for Cell Products*, 153–156.

passages the cells became able to proliferate in DMEF12 medium supplemented with amino, acids, vitamins, lipids, insulin, transferring and other defined compounds. The aim of this ongoing work is to investigate what occurs during these passages and why they are necessary.

## 2. MATERIALS AND METHODS

All of the data presented here is gathered from 50mL working volume cultures grown in 250ml disposable ehrlenmyer flasks (corning). Cells were grown in DMEF12 media supplemented with serum or an in house chemically defined supplement. All components of the media, including the supplement, were supplied by Sigma. The data presented here is from cells banked at each stage of the adaptation process, revived and used to seed cultures which were grown simultaneously on a shaker (IKA) set to 120rpm in a $CO_2$ incubator (RS biosciences) maintained at 37 °C and 5% $CO_2$. Media was prepared immediately prior to the experiment from the same batch of non-supplemented media.

Viable cell counts were performed using a CASY TTC cell counter. Antibody quantitation was performed via an in house sandwich ELISA developed suring the project. Glutamine analysis was performed by revese phase HPLC along with other amino acid analysis (data not shown here). Lactate, and glucose analysis was carried out using a YSI select 2700 analyser, ammonia was quantified using the indophenol blue assay.

## 3. RESULTS AND DISCUSSION

Figures 1 and 2 show that whilst there are significant differences in the viable cell concentrations achieved during the cultures, particularly between the serum containing and serum free conditions. In addition it shows that there is a real but small difference in the final antibody titre (mg/L) between cultures, ranging from 45 to 80 mg/L.

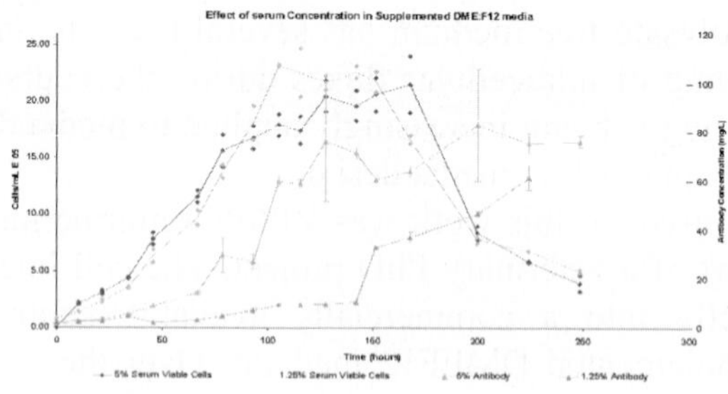


*Figure 1.* Comparison of 5% and 1.25% serum concentrations.

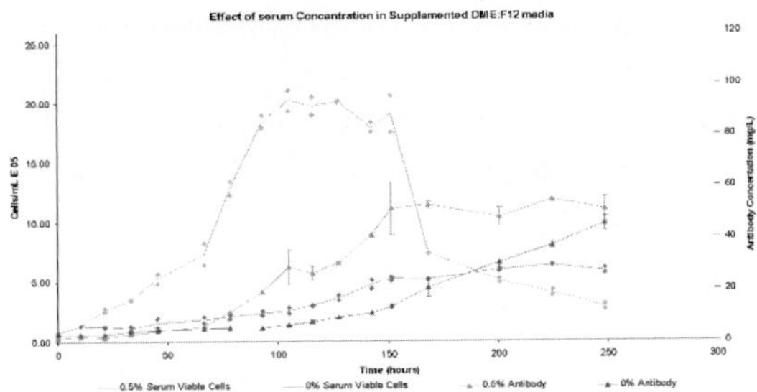


*Figure 2.* Comparison of 0.5% and 0% Serum concentrations.

Figures 3, 4 and 5 show the effect that the supplement added to DMEF12 is having on viable cell count, antibody concentration, ammonia, lactate, glutamine and glucose concentrations during the early phases of the culture. It can be seen that whilst the supplement leads to significantly higher viable cell counts it also leads to significantly lower antibody titres. There are no striking differences in glutamine or glucose consumption or lactate and ammonia accumulation that would account for this in the early stage of the fermentation and further investigation is being carried out.

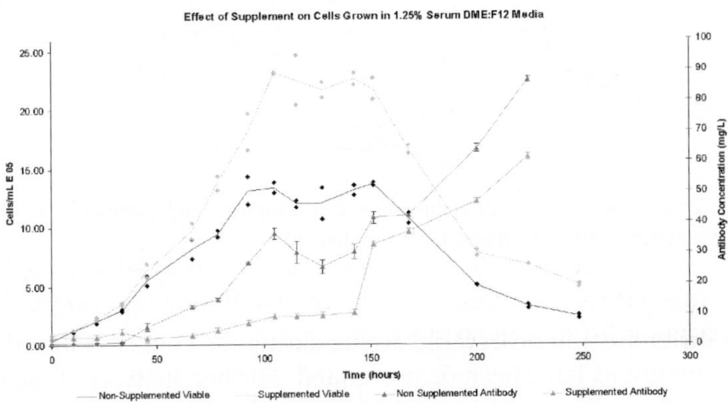


*Figure 3.* Comparison of Growth and Antibody Production in 1.25% Serum Media with and without CDS supplement.

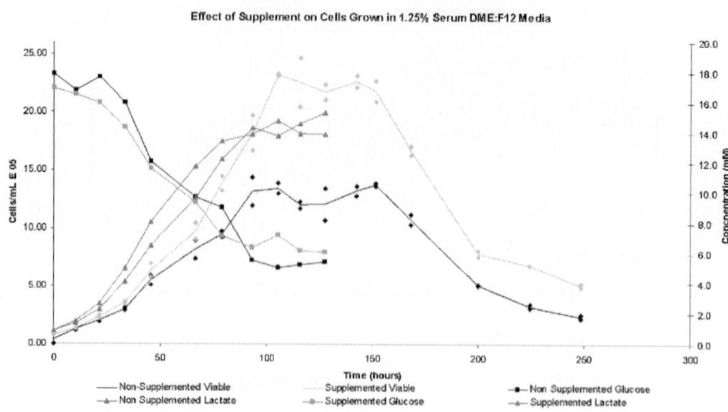


*Figure 4.* Comparison of Growth, Glucose Consumption and Lactate Production in 1.25% Serum Media with and without CDS supplement.

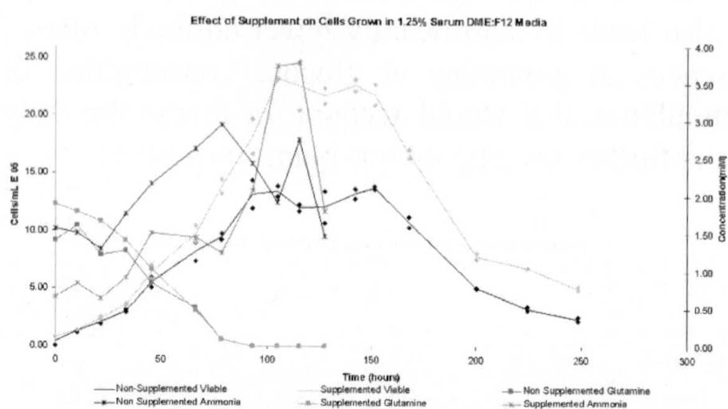


*Figure 5.* Comparison of Growth, Glutamine Consumption and Ammonia Production in 1.25% Serum Media with and without CDS supplement.

Clear differences in observable events during the course of the cultures in different conditions have been demonstrated. Further analysis of the external metabolites from these cultures is being performed and will be used to determine intracellular fluxes. A greater understanding of the physiology behind them can only improve process control and optimisation

# The Expression of the *neo* gene in GS-NS0 Cells Increases their Proliferative Capacity

Farlan Veraitch[1,2] and Mohamed Al-Rubeai[1,3]
[1]*Department of Chemical Engineering, University of Birmingham, Birmingham B15 2TT, UK.* [2] *Present address: Biochemical Engineering, University College London, Torrington Place, London, UK.* [3]*Present address: Department of Chemical and Biochemical Engineering, University College Dublin, Belfield, Dublin 4, Ireland*

**Abstract:** Prior work has reported that cotransfecting a gene of interest with the selectable marker *neo* can seriously perturb a number of cellular processes. In this study the influence of the *neo* gene on the growth, death and metabolism of a murine myeloma NS0 cell line, expressing a chimeric antibody, was investigated. A pool of *neo* transfectants, 6A1-NEO, was selected with 500 μg/ml G418. Batch cultivation of 6A1-NEO showed that there was a 36% increase in maximum viable cell concentration, a 20% increase in the maximum apparent growth rate and a 134% increase in cumulative cell hours as compared with the parent, 6A1-(100)3. Batch cultivation of five randomly selected clones illustrated that 6A1-NEO's advantage over the parent was not due to clonal variation. Neither the use of G418 during the selection process nor the cultivation of cells in the presence of G418 were responsible. This implied that the *neo* gene product, APH(3')-II, was causing the changes in proliferative capacity. These results show that there was an increase in growth rate and proliferative capacity caused by the expression of recombinant APH(3')-II.

**Key words:** *neo*, aminoglycoside phosphotransferase, G418, NS0 myeloma cells, cell cycle, apoptosis, Bcl-2, metabolism



*R. Smith (ed.), Cell Technology for Cell Products*, 157–162.

# 1. INTRODUCTION

The *neo* gene has been commonly used as a selectable marker for the introduction of genes into mammalian cells. The aminoglycoside antibiotic G418 is used to select for cells which have successfully integrated *neo* into their chromosomal DNA (Colbere-Garapin *et al.*, 1981; Southern and Berg, 1982). However, this selection system assumes that the *neo* gene product is neutral and that G418 selection does not affect the cellular properties under investigation (Von Melchner and Housman, 1988). In eukaryotes, G418's primary function is to kill cells by binding to the 80S ribosome thereby blocking protein synthesis (Bar-Nun *et al.*, 1983). The *neo* gene codes for a bacterial aminoglycoside phosphotransferase (APH(3')-II) which inactivates G418 by phosphorylation (Davis and Smith, 1978; Shaw *et al.*, 1993).

Increasingly, immortalised mammalian cell lines such as CHO, NS0 and baby hamster kidney (BHK) are being used to produce recombinant proteins for clinical applications (Wurm, 2004). A number of attempts have been made to optimise cell growth and suppress cell death using genetic engineering techniques (reviewed in Fusseneger *et al.*, 1999; Arden and Betenbaugh, 2004). The majority of these studies have utilised the *neo* selection system for the over-expression of recombinant proteins such as Bcl-2, Bcl-$x_L$, XIAP and p21 (Itoh *et al.*, 1995; Mastrangelo *et al.*, 2000; Sauerwood *et al.*, 2002; Watanabe *et al.*, 2001). However, the authors are not aware of any studies which have compared the cell growth of a *neo* transfectant with the parent. The objective of this work was to investigate whether transfection with the *neo* gene altered key cellular processes such as proliferation, cell death, metabolism and cell cycle. The host was a NS0 cell line commonly used for the production of recombinant proteins. The effect of the *neo* transfection on the proliferative capacity of a pool of transfectants and five randomly selected clones was monitored in static, serum-supplemented batch cultures. We found that the pool of *neo* transfectants were capable of reaching a higher maximum viable cell concentration than the parent in batch cultures. In addition all five *neo* clones outgrew five randomly selected parent clones.

# 2. MATERIALS AND METHODS

## 2.1 Cell line

The parent cell line, 6A1-(100)3, was generated by transfecting a murine myeloma NS0 cell line with a GS expression vector (Lonza Biologics, Slough, UK) encoding the human–mouse chimeric antibody, cB72.3.

### 2.2 Transfection

The expression vector pEF-MC1*neo*pA encodes the *neo* gene under the control of a human thymidine kinase promoter (Cory, 1995). This vector does not contain any other the genes apart from *neo*. 6A1-(100)3 was transfected with pEF-MC1*neo*pA using lipfectamine (Gibco) to create the 6A1-NEO cell line

### 2.3 Cloning

5 random clones of 6A1-(100)3 and 6A1-NEO were derived using the limited dilution technique

## 3. RESULTS AND DISCUSSSION

### 3.1 Serum-supplemented batch cultures

6A1-(100)3 was transfected with pEF-MC1*neo*pA and selected with 500 μg/ml G418. The control cells, which had been treated with lipofectamine in the absence of any DNA, were less than 5% viable 4 days after the addition of the antibiotic. The pool of *neo* transfectants, named 6A1-NEO, was passaged until its growth was reproducible. The viable cell numbers of 6A1-(100)3 and 6A1-NEO were monitored in serum-supplemented batch cultures. 6A1-NEO grew faster than the parent and reached a higher maximum viable cell concentration (Figure 1(a)). After 144 hours there were 36% more viable 6A1-NEO cells than the parent. In addition 6A1-NEO's cumulative cell hours (CCH) at the end of the batch was $410 \pm 9 \times 10^9$ cell h/l, 134% higher than the parent's final CCH. This increase in proliferative capacity could have been caused by the expression of APH(3')-II, G418 acting as a growth enhancing supplement or the inadvertent selection of cells influenced by one or more of G418's secondary properties. 6A1-NEO was cultured for four weeks in the presence and absence of 500 μg/ml G418. Subsequent batch cultivation of 6A1-NEO with and without G418 revealed that the viable cell concentrations of both cultures were very similar (Figure 1(a)). This confirmed that G418 was not enhancing the growth of 6A1-NEO by acting as a growth additive. Furthermore, untransfected 6A1-(100)3 cells were allowed to recover from 4 days of treatment with 500 μg/ml G418. This population, 6A1-(100)3-GP1, was passaged in the absence of G418 until its viability had recovered. There were no significant differences between the growth of 6A1-(100)3-GP1 cells and the parent during batch cultivation (Figure 1(b)), indicating that the G418 selection process did not cause an increase in proliferative capacity. These results indicate that the expression of APH(3')-II was causing the increase in proliferative capacity exhibited by 6A1-NEO.

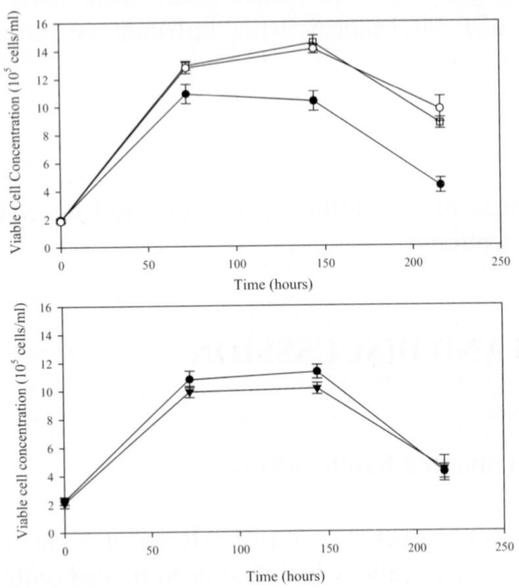


*Figure 1.* (a) shows serum-supplemented batch cultivation of 6A1-(100)3 (●), 6A1-NEO with G418 (○). 6A1-NEO was cultivated for four weeks without G418 before batch cultivation in the absence of the antibiotic (≤). (b) shows the batch cultivation of 6A1-(100)3 (●) and G418 treated parent cells, 6A1-(100)3-GP1 (▼). Viable cell concentration was measured using the trypan blue exclusion assay.

## 3.2 Clonal variation

6A1-(100)3 was transfected with the *neo* gene to create a mixed population of transfectants, 6A1-NEO. This population was cloned using the serial dilution technique and five clonal cell lines were selected at random. Control cell lines were generated by randomly selecting five 6A1-(100)3 clones. The viable cell concentration of all ten cell lines was monitored in serum-supplemented batch cultures. After 96 hours of cultivation all five 6A1-NEO clones had reached a higher viable cell concentration than any of the parent clones (Figure 2). The fastest growing clone, 6A1-NEO-4D5, reached a maximum viable cell concentration of 12.05 x $10^5$ cells/ml, whilst the fastest growing parent clone, 6A1-(100)3-1B6, only reached 8.25 x $10^5$ cells/ml. The parent clones progressed into the death phase whilst all five 6A1-NEO clones remained in the stationary phase. After 240 hours the

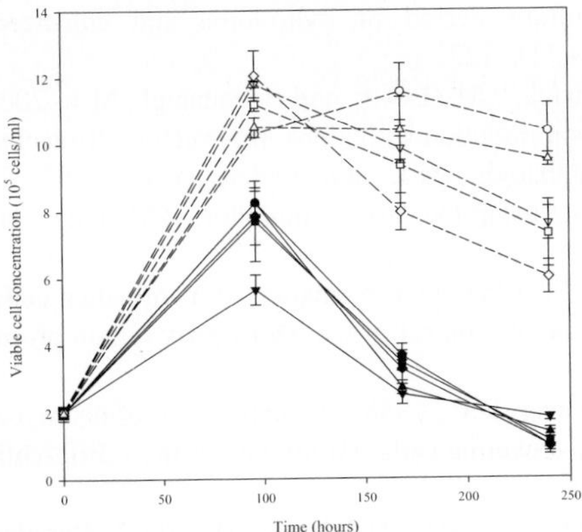


*Figure 2.* Batch cultures of five clones randomly selected from the pool of neo transfectants (dashed lines). The viable cell concentration of 6A1-NEO-1D6 (○), 6A1-NEO-2B10 (σ), 6A1-NEO-2F7 (≤), 6A1-NEO-4D5 (↓) and 6A1-NEO-4F4 (ρ) were measured using the trypan blue exclusion method. 6A1-(100)3 was cloned to create five control cell lines: 6A1-(100)3-1B6 (●), 6A1-(100)3-2C7 (▼), 6A1-(100)3-2F10 ('), 6A1-(100)3-3C7 (↵) and 6A1-(100)3-E10 (▲).

average viable cell concentration of the *neo* clones was 8.15 x $10^5$ cells/ml, 5.6 fold higher than the average of the parent clones. Although there was some variability between the individual clones all five randomly selected clones exhibited the increase in proliferative capacity shown by the pool of transfectants, 6A1-NEO, in Figure 1(a). These results confirm that 6A1-NEO's advantage over the parent was not due to clonal variability.

# REFERENCES


Arden N. and Betenbaugh, M.J., 2004, Life and death in mammalian cell culture: Strategies for apoptosis inhibition, Trends Biotechnol 22:174-180.

Bar-Nun, S., Shneyour, Y. and Beckmann, J.S., 1983, G-418, an elongation inhibitor of 80S ribosomes, Biochim Biophys Acta 741:123-127.

Colbere-Garapin, F., Horodniceanu, F., Kourilsky, P. and Garapin, A.C., 1981, A new dominant hybrid selective marker for higher eucaryotic cells, J Mol Biol 150:1-14.

Cory, S., 1995, Regulation of lymphocyte survival by the bcl-2 gene family, Annual Rev Immunol 13:513-543.

Davies, J. and Smith, D.I., 1978, Plasmid-Determined Resistance to Antimicrobial Agents, Ann Rev Microbiol 32:469-508.

Fussenegger, M., Bailey, J.E., Hauser, H. and Mueller, P.P., 1999, Genetic optimisation of recombinant glycoprotein production by mammalian cells, TIBTECH 17: 35-42.

Itoh, Y., Ueda, H. and Suzuki, E., 1995, Overexpression of bcl-2, apoptosis suppressing gene: prolonged viable culture period of hybridoma and enhanced antibody production, Biotechnol Bioeng 48:118–122.

Mastrangelo, A.J., Hardwick, J.M., Bex F. and Betenbaugh, M.J., 2000, Part I. Bcl-2 and Bcl-xL limit apoptosis upon infection with alphavirus vectors, Biotechnol Bioeng 67:544-554.

Sauerwald, T.M., Betenbaugh, M.J. and Oyler, G.A., 2002, Inhibiting apoptosis in mammalian cell culture using the caspase inhibitor XIAP and deletion mutants, Biotechnol Bioeng 77:704-716.

Southern, P.J. and Berg, P., 1982, Transformation of mammalian cells to antibiotic resistance with a bacterial gene under control of the SV40 early region promoter, J Mol Appl Genet 1:327-341.

Von Melchner, H., Housman, D.E., 1988, The expression of neomycin phosphotransferase in human promyelocytic leukemia cells (HL60) delays their differentiation, Oncogene 2:137-140.

Watanabe, S., Shuttleworth, J. and Al-Rubeai, M., 2002, Regulation of cell cycle and productivity in NS0 cells by the over-expression of p21CIP1, Biotechnol Bioeng 7:1–7.

# CHAPTER III: THERAPEUTIC CELL ENGINEERING

# The Therapeutic Potential of Human Embryonic Stem Cells

S.H. Cedar[1, 3] M.J. Patel[1] S.J. Pickering[2] P.R. Braude[2] S.L. Minger[1]
[1] *Stem Cell Biology Laboratory, Wolfson Centre for Age Related Disease, King's College, London SE1 1UL UK,* [2] *Department of Women's Health, GSKT School of Medicine, St. Thomas' Hospital, London SE1 7EH UK,* [3] *Faculty of Health & Social Care, London South Bank University, London SE1 0AA UK*

**Abstract:** Embryonic stem cells can be maintained in their undifferentiated state over long periods in culture. They are pluripotent, able to form all the cell types of the body. These two findings open the way for the expansion and differentiation of embryonic stem cells for use in therapeutic transplantation for a myriad of diseases. Here we look at the hurdles that must be overcome for them to reach their potential use in the clinical arena.

**Key words:** Embryonic stem cells (ESC), Somatic stem cells (SSC), pluripotent, multipotent, transplantation.

## 1. INTRODUCTION

A stem cell is a cell that is capable of self-renewal to an extent that it can perpetuate itself indefinitely. Each division in vivo is thought to be unequal, providing a cell that takes the role of a stem cell and a cell that takes the role of progeny. The progeny cell is capable of proliferation and differentiation. The stem cell is thought of as undifferentiated in that it carries out limited functions beyond self-perpetuation while the progeny cells become increasingly differentiated during each cell cycle and less able to self-renew. Many of these notions of what constitutes a stem cell and how cells become restricted in their potential may be questioned and the physiological role of stem cells in vivo may differ from their behaviour in vitro (Joseph & Morrison 2005).



*R. Smith (ed.), Cell Technology for Cell Products*, 165–184.

The fertilised egg is seen as the ultimate totipotent cell capable of forming all the types of cells in the body and all the structures of the embryo, foetus and thus adult. It is highly proliferative and of short life-span existing only for the first three divisions in humans until the embryo becomes eight cells. From the eight-cell stage onwards the cells become more restricted in their potency (see Smith 2001for a review). Up until the blastocyt stage many cells, if not all, are highly plastic, able to form many cell types. Even during these stages, however, the cells are becoming more restricted. Amongst the blastocyst cells are pluripotent stem cells of the inner cell mass that can form all the cell types of the body, but not the entire organism. As development continues, each compartment of the body gains its own somatic stem cell, a multipotent cell capable of forming progeny that can repopulate the compartment in which it exists. Recently, evidence has suggested that these multipotent, somatic stem cells may be more plastic and able to cross compartments or to provide progeny of other compartments (see Alison *et al.*, 2004 for a review). Thus as proliferation proceeds during early development differentiation gives rise to all the various structures of the body. Concomitant with this is the loss of potency with stem cell characteristics such as self-renewal separated from the ability to differentiate to form functional end cells with limited life-spans. Cell numbers, cell differentiation, cell death and cell replacement are kept under strict homeostatic control. Throughout life repair, replacement and regeneration compete in the community of cells that form the multicellular organism. Throughout life, therefore, stem cells exist, perhaps more or less compartment-restricted, to replace cells as needed. The totipotent fertilised germ cells represent one end of a spectrum of stem cell potentials with the unipotent stem cell, present in some compartments, at the other. Each species possesses different ranges of stem cells for various life-cycles able to regenerate to a greater or lesser extent.

## 2. MILESTONES IN STEM CELL BIOLOGY AND REGENERATIVE MEDICINE

Stem cells are increasingly seen as potential therapies for organ and tissue failure. This potential takes its lead from four major scientific developments:

- The discovery of stem cells
- Immunological and surgical advances in organ transplantation
- In vitro fertilisation
- Isolation of embryonic stem cell.

## 2.1 The discovery of stem cells

Stem cells have been of interest to biologists and have been studied for decades. Until the late $17^{th}$ century theories in embryogenesis and development proposed suggested the expansion of the pre-formed human. With the advent of better microscopes and the discovery of the cell these early theories were gradually replaced by theories of morphogenesis. The presence of cells that can replenish other cells was first demonstrated in the haemopoietic compartment. Colony forming cells from donors were introduced into irradiated mice and were able to colonise and form all the blood cells that had been ablated (Ford *et al.*, 1959, Till and McCulloch 1961). These cells were shown to be present in low numbers in the bone marrow and divide slowly (Metcalf 1979). Their discovery led to the first bone marrow transplants in identical twins in 1968 by a team at the University of Minnesota and allogeneic recipients in 1973 at the Memorial Sloan Kettering Hospital.

## 2.2 Immunological and surgical advances in organ transplantation

With increasing knowledge of blood groups, immunological compatibility and surgical skills transplantation has became more successful. From the 1970's onwards the way forward for organ failure was seen as organ, rather than cell transplant. There is now a shortfall between organ demand and organ supply (see NHS UK Transplants (2005) for UK and United Network for Organ Sharing (2004) for USA organ needs) and cell transplants are seen as a potential area for meeting these demands.

## 2.3 In vitro fertilisation

In 1978 the first IVF baby was born in the UK (Steptoe and Edwards, 1978). The ability to isolate oocytes, fertilise them in vitro and re-implant them was a major step forward in the ability to manipulate and maintain early embryos. This, coupled to pre-natal genetic diagnosis, opened the door to the extraction of the material from which human embryonic stem cells could be isolated.

## 2.4 Isolation of embryonic stem cells

In 1981 mouse embryonic stem cells were isolated and grown in the UK (Evans and Kaufman 1981). This has been followed by the isolation and growth of human embryonic stem cells (Thompson *et al.*, 1998 Shamblott *et al.*, 1998).

Transplantation technologies and stem cell biology has been brought together in the human foetal brain transplants seen for the treatment of Parkinson's Disease in the 1970 and 1980s (Sladek *et al.*, 1987). Autologous transplants of differentiated cells, which avoids immunlogical problems, have been used to replace defective or degenerative tissue. Autologous progenitors, mainly from the bone marrow or peripheral blood compartment, are used in cancer treatments and are now being used in regenerative medicine (see Fodor 2003 for a review). These are mainly mesenchymal cells from bone marrow which may have pluripotency (Pittinger *et al.*, 1999). Trials are currently underway to test their efficacy in a variety of disorders, serving as repair tissue for myocardium (Orlic *et al.*, 2001) liver (Terai *et al.*, 2002) and diabetes (Ianus *et al.*, 2003). Embryonic stem cells offer greater pluripotency and indefinite growth in culture (see Stolkovic *et al.*, 2004 for review).

# 3. HURDLES IN STEM CELL RESEARCH

The hurdles that hold stem cell research back are many. These include:

- Stem cell sources
- Regulatory hurdles
- Technological hurdles.

## 3.1 Stem cell sources

In vitro studies of stem cells started with cells that were accessible, blood. Somatic stem cells have since been isolated from various tissues including hepatic (Michaelopoulos & DeFrances 1997), skin epidermis (Jensen *et al.*, 1999) epithelium of respiratory and digestive tracts (Potten *et al.*, 1997) mesodermal derivatives and muscle stem cells (Hansen-Smith & Carlson 1979, Ham & Cormack 1978) and neural (Ray *et al.*, 1993). These somatic stem cells (SSC) have varying degrees of ease of isolation and have varying degrees of developmental potential. Bone marrow and peripheral blood as well as transformed cell lines (Horton *et al.*, 1983) have been the tissue of choice for the investigation of the behaviour of SSC in vitro and their isolation due to the ease of accessibility and some evidence for pluripotency. Stem cells are thought to constitute a very low percentage of cells in a body compartment. In bone marrow they constitute about 1 in 10,000 cells. This rarity, together with their limited growth and lifespan, contributes to the difficulty in accessing sufficient quantities for expansion of numbers in vitro.

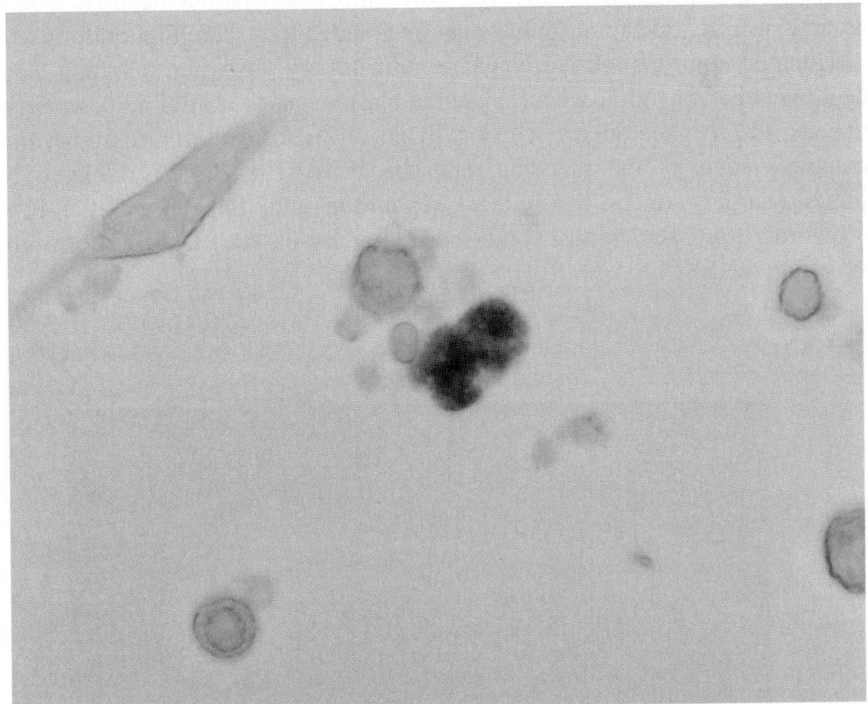

*Figure 1.* Adult stem cell numbers are low within a tissue.

Cord blood has also proved a useful and accessible source of stem cells, mainly of the haematopoietic compartment. Cord blood for autologous use and for allogeneic donation can be stored. Foetal tissue has also proved a source of SSC, but presents many ethical problems in its isolation and use.

Embryonic Stem Cells (ESC) are isolated from the inner cell mass (ICM) of a day 5-6 pre-implantation blastocyst These blastocysts are gathered from people undergoing in vitro fertilisation (IVF). During these procedures occytes are harvested from super-ovulated females and fertilised in vitro with freshly collected sperm. The fertilised egg is then incubated in vitro until the eight cell stage (day 3-4) and one cell may be removed for prenatal genetic diagnosis (PDG) in cases where genetic abnormalities are known. The healthy embryos can then be implanted or spare healthy embryos cryopreserved in liquid nitrogen for future use. Unhealthy embryos are generally discarded. Alternatively, spare embryos, whether healthy, frozen or carrying a genetic abnormality may be donated to research laboratories.

Donated embryos are cultivated in vitro to day 5-6 when hatching from the zona pelucida should occur. Hatched blastocysts consist of an outer layer of cells, the trophectoderm which will go on to form the extra-embryonic structures such as the placenta, and the ICM. The ICM is isolated by immuno-surgery or by mechanical separation and the cells of the ICM grown in vitro tissue culture conditions (Thompson *et al.*, 1998). These cells constitute the ESC.

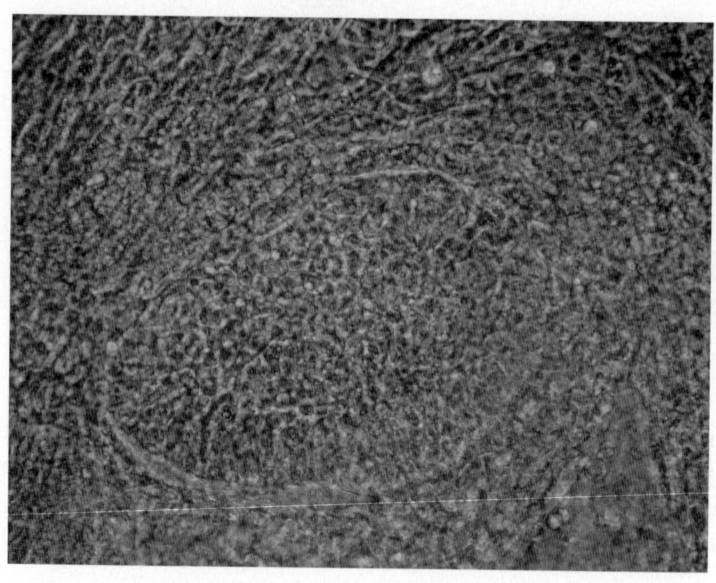

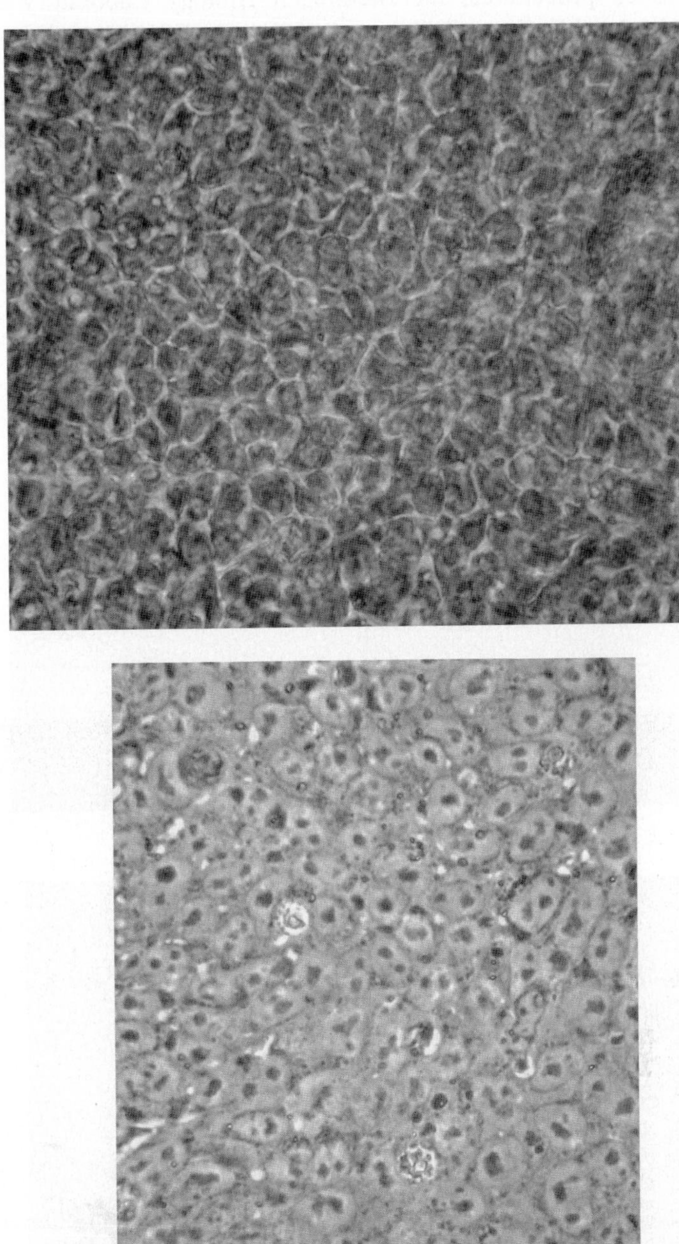

*Figure 2.* Generation of a second cell line, WT3, at King's College from a healthy embryo.

Using such procedures, the Stem Cell Biology Laboratory at Kings College, London has derived three unique human ESC lines. Two of these lines, WT3 (Pickering *et al.*, 2003) and WT4 (Pickering & Minger unpubl.) are genetically normal human ESC lines, whilst CF1 (Pickering *et al.*, 2005) was derived from an embryo screened by pre-implantation genetic diagnosis and shown to be homozygous for the delta F508 mutation in the CFTR that accounts for approximately 70% of all Cystic Fibrosis.

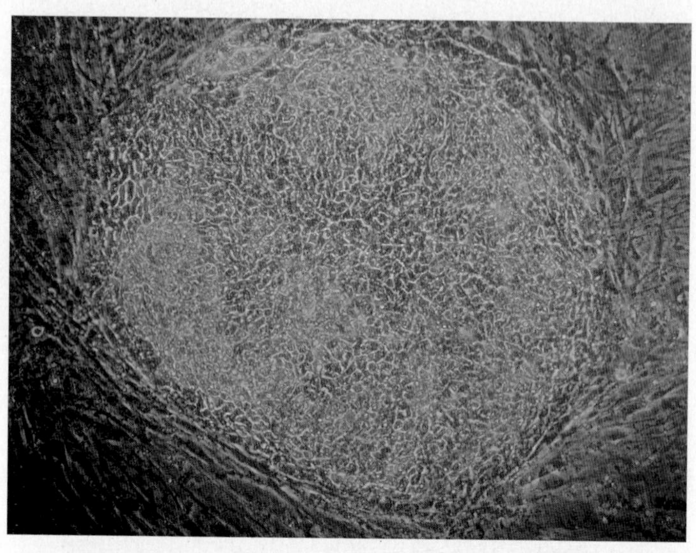

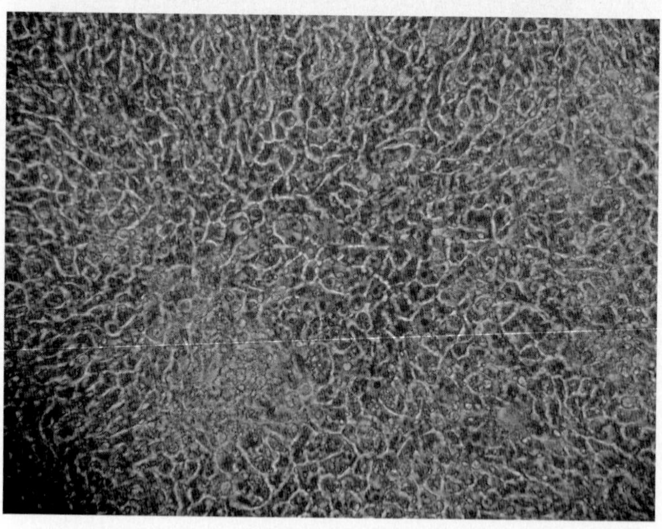

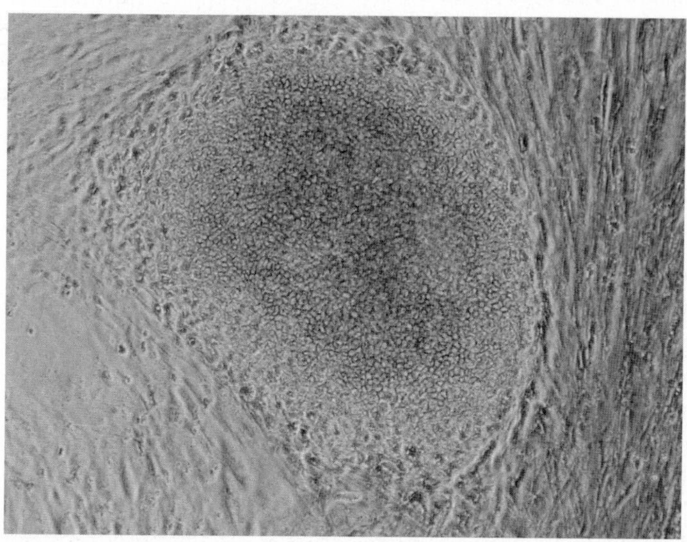

*Figure 3.* Generation of ECS from a healthy embryo, WT4 and from a CF homozygous embryo at King's College.

## 3.2 Regulatory hurdles

Somatic stem cell isolation presents some ethical hurdles. For allogeneic transplants isolation of stem cells from those not capable of giving consent raises the same issues that are seen in organ donation under similar circumstances. Isolation of cells from cord blood, for instance, obviously cannot be consensual on the part of the foetus. For autologous transplants the ethical problems of isolating stem cells is reduced.

The isolation of human ESC raises many ethical issues. The embryos that are used have been created for IVF. Surplus embryos are usually created in this procedure. These can be cryopreserved for future use should they be needed. They can be destroyed or they can be donated for research purposes. The embryos that are donated for research are then rendered incapable of being re-implanted or of producing a foetus due to the isolation of the ICM to culture ESC.

This use of embryos from IVF is regulated according to each country's legislature. In Europe there is variation between countries. In the USA there is variation between states. The UK has a highly regulated embryonic stem cell research programmed currently licensed by the Human Fertilisation and Embryology Authority (HFEA). This was created in 1990 (HFEA 1990) and licenses all reproductive medicine and human embryo research in the UK. From 1990-2002 HFEA had five main research areas:

- Treatment of infertility
- Investigation of congenital disease
- Investigation of the causes of miscarriage
- Development of more effective conception
- Improvement in pre-implantation genetic diagnosis.

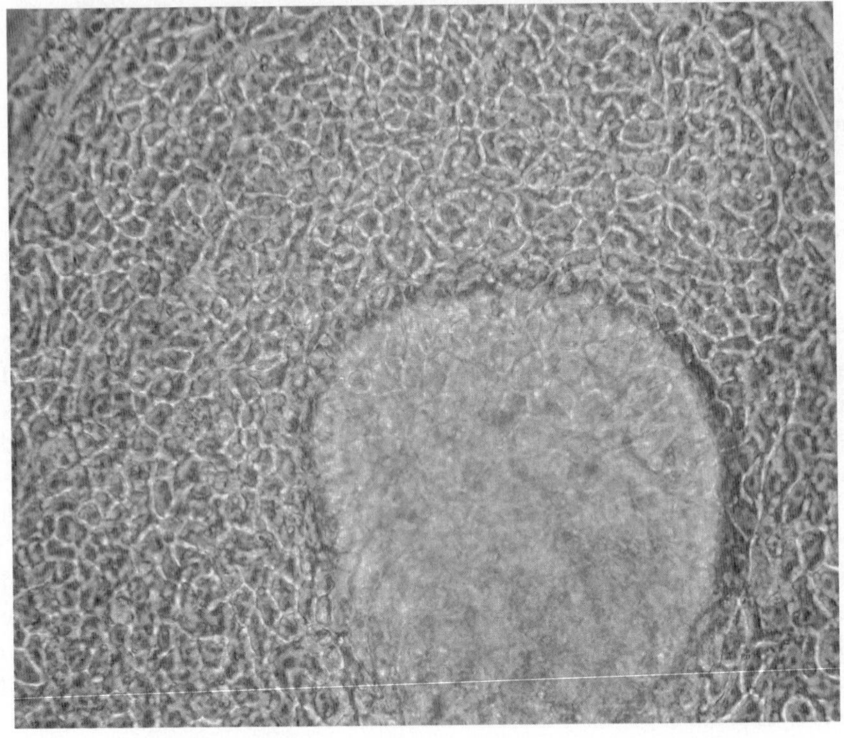

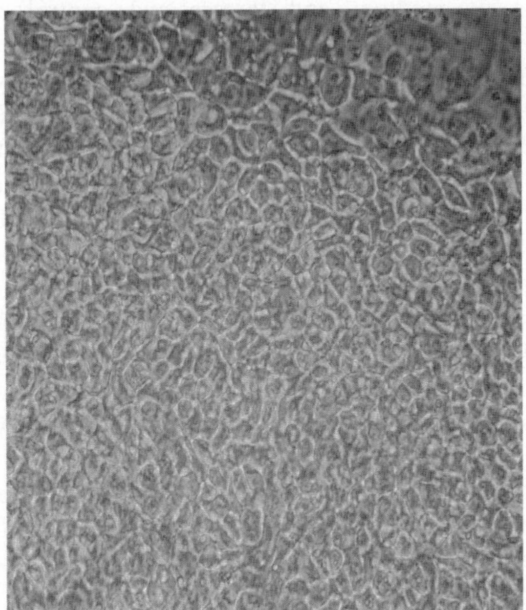

*Figure 4.* Licenses to grow human ESC result in new stem cell lines in the UK.

Later acts have been passed to allow embryonic research (HFEA 2001)

Now HFEA has added to the list to include regulation and licensing allowing:

- the donation of embryos for research with informed consent from regulated IVF clinics,
- the growth of embryos in vitro up until day 14,
- the isolation of the ICM
- the culture of ESC from the ICM in vitro
- the use of somatic cell nuclear transfer (SCNT) procedures using oocytes, which are enulceated and nuclei from somatic cells introduced to form ESC with the genetic material of the donor somatic nucleus (HFEA 2005).

## 3.3 Technological hurdles

Growth of any cells in vitro is beset with many technological hurdles. Aseptic techniques are needed to prevent contamination of cells. Cell cloning is required to ensure a pure population of cells. Culture conditions, batch variations of medium and factors and control of the growth environment all present problems to be overcome.

Stem cells from human sources have specific hurdles to overcome. Somatic stem cells represent a rare group of cells in each compartment of the body and are thus hard to isolate. They are also hard to identify among the 206 cell types of the body. Their main physiologic characteristic, their ability to divide indefinitely, does not represent a useful marker for identification. A battery of cell surface antigens unique to stem cells would be useful. Currently the main cell surface markers for SSC isolation come from the haemopoietic lineage such as the cell differentiation (CD) markers CD 34 and CD133. Whether there are unique cell surface markers for stem cells is still an area of debate as those found so far can appear, disappear and reappear and have different expression between human ESC lines (Rosler *et al.*, 2004).

ESC are isolated from the Inner Cell Mass (ICM) of an embryo and are thus a purer population of cells. The main antigen markers for human ESC are Oct 4 and some stage specific embryonic antigens (SSEA- 3 and 4, TRA-1-60 and 81) which vary between those found on the mouse ESC and those found on human ESC (Amit and Itskovitz-Elder 2002). Also expressed by human ESC are alkaline phosphatase and high levels of telomerase (Thompson *et al.*, 1998, Reubinoff *et al.*, 2000)

The numbers of donated embryos, their physiological fitness and the ability to isolate a productive ICM which then grows into ESC varies. The maintenance of the pluripotential, undifferentiated state in vitro also varies from embryo to embryo and laboratory to laboratory. Whether ESC lines represent a true stem cell or a system artifact is also questionable. The control of differentiation and the control of growth are yet to be established.

## 4. THE CLINICAL USE OF STEM CELLS

For stem cells to be able to be used in the clinic some minimal requirements are needed.

The stem cell or progenitor cell used must be able to proliferate, perhaps indefinitely, and, at least for extended periods in culture, be able to self-renew and expand in numbers to form a clone of cells. Embryonic stem cells appear to be able to do this while somatic stem cells seem to be more limited in their lifespan.

The stem cell or progenitor cell must be phenotypically stable over time (Minger *et al.*, 1996). This includes a stable karyotype, shown in ESC (Carpenter *et al.*, 2004). They must also be able to be maintained in their undifferentiated stem cell state in vitro over time. The ESC must be able to remain pluripotent in vitro. The SSC must be able to remain multipotent. ESC have been shown to remain pluripotent over time. Somatic stem cells are also being investigated for their potential to be plastic and pluripotent.

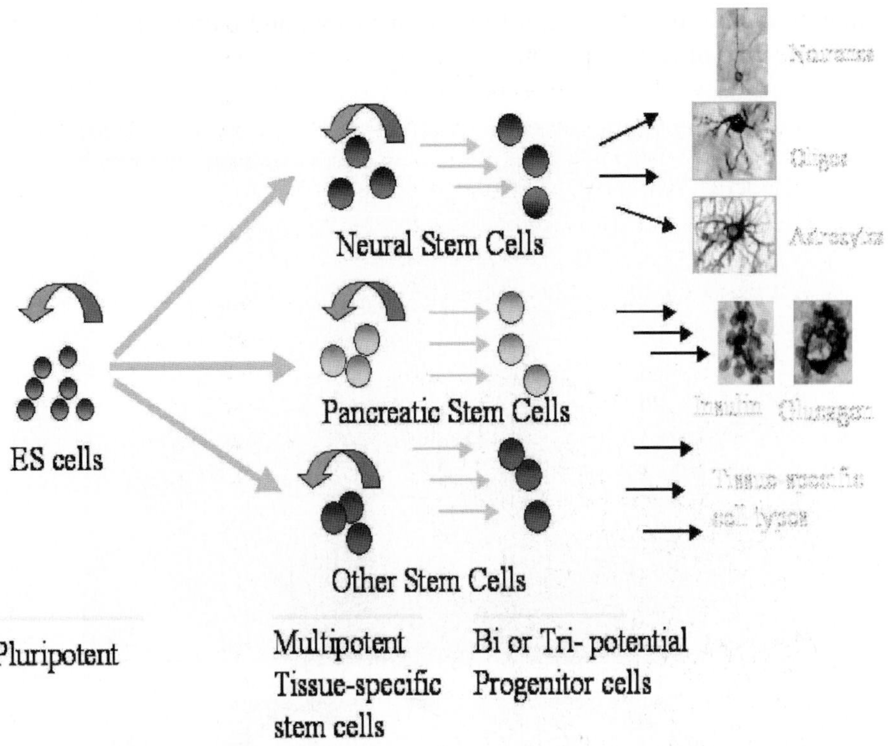


*Figure 5.* Pluripotent embryonic stem cells must be able to become multipotent stem cells.

The stem cell or progenitor cell must be able to generate the desired cell type during differentiation. This will be the next hurdle for all stem cell research. Each stem cell line varies genetically and immunologically and this may affect its growth characteristics in vitro. The culture conditions are also variable from cell line to cell line. Each laboratory has its preferred procedures for cultivating stem cells. The UK Human Embryonic Stem Cell Initiative has brought together many ESC laboratories to share a common practice. After establishment of a cell line the UK Stem Cell Bank holds a stock of each line for distribution to other stem cell laboratories The UK ESC groups also meet regularly to share information. Controlling gene expression and growth through the use of factors and culture conditions is the future goal of ESC research.

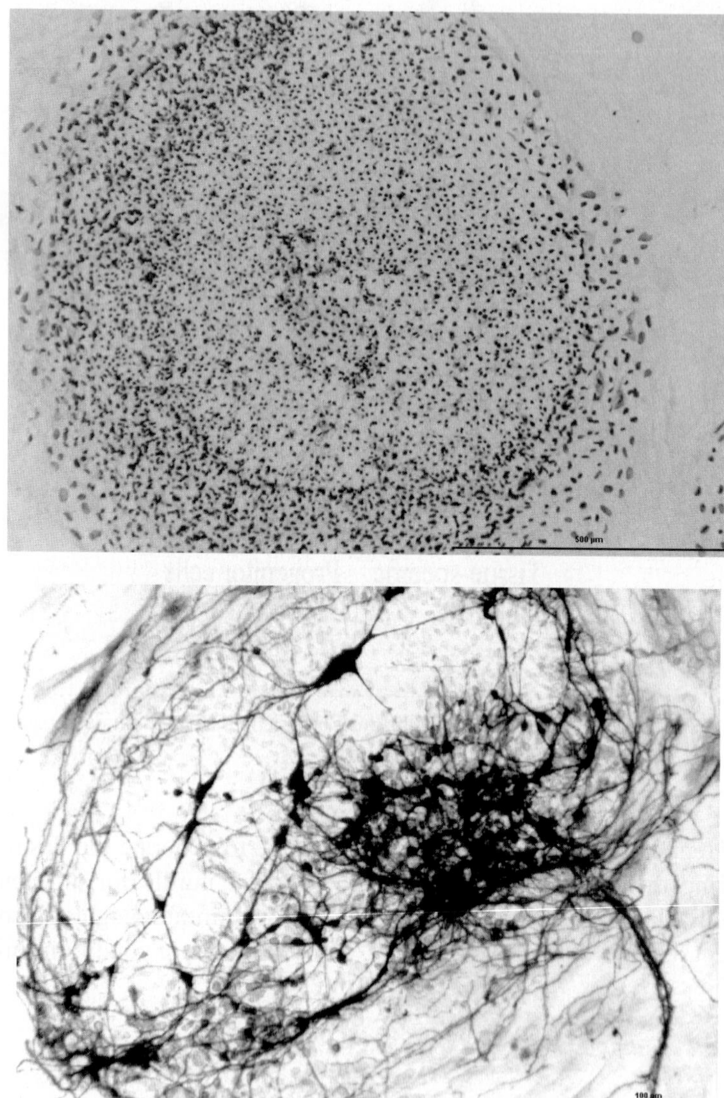

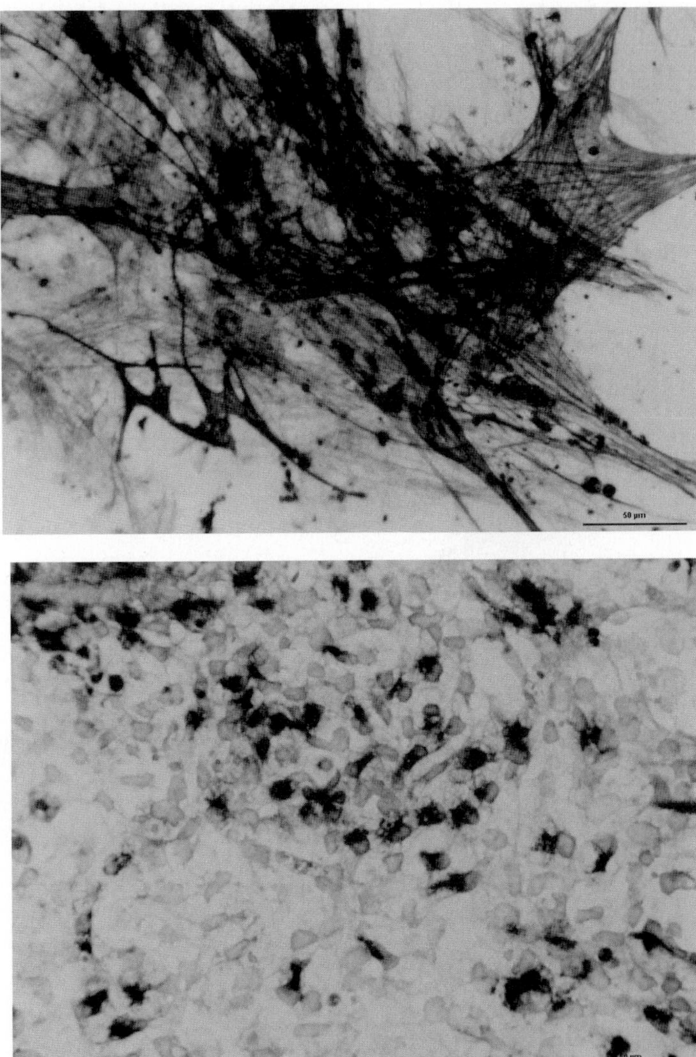


*Figure 6.* ESC undifferentiated (Oct 4) or differentiated along the ecotdermal (b- tubulin), endodermal (albumin) or mesodermal (actin) lineages.

Stem cells or their differentiated progeny must survive implantation. While an in vitro environment attempts to mimic an in vivo environment the former is more controlled than the latter and the variation will affect the behaviour of transplanted cells.

The transplanted cells must functionally integrate into the host tissue. While some problems are reduced using autologous somatic stem cells, their rarity for isolation may prove an obstacle in their use. Currently (2005) autologous somatic stem cells are being tested in clinical trials for their ability to integrate into host tissue and show functional improvement *(personal communication)*. The differentiated progenitors of ESC are yet to be transplanted in to humans, but human ESC have been transplanted into animals to test for pluripotency by formation of benign teratomas. Their differentiated progeny have been transplanted into animals (see Taylor & Minger for transplants in Parkinson's). They must also integrate with host cells, avoid immunological rejection and provide functional improvement while avoiding being tumourigenic.

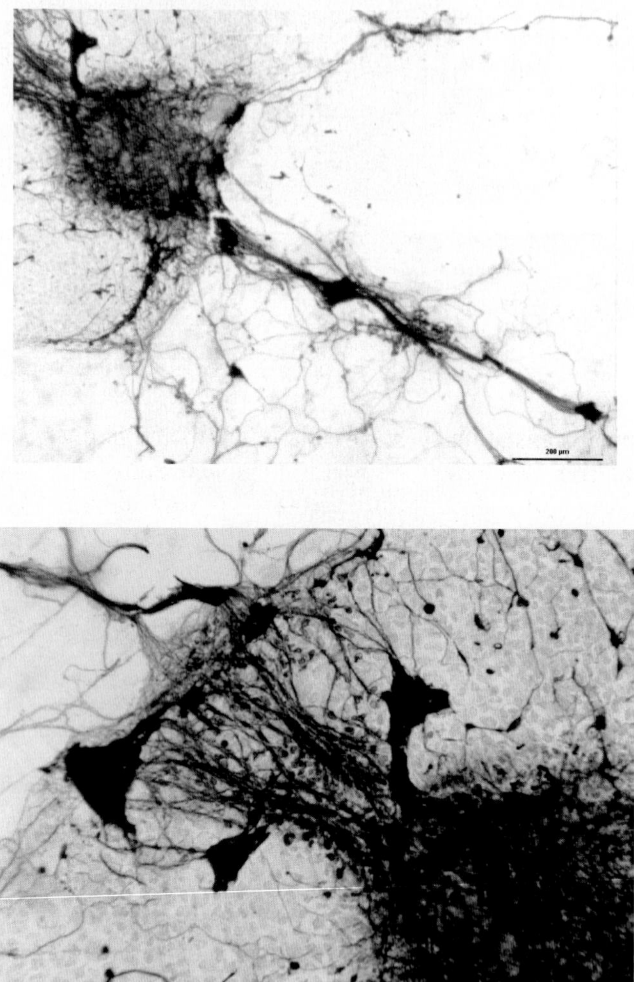

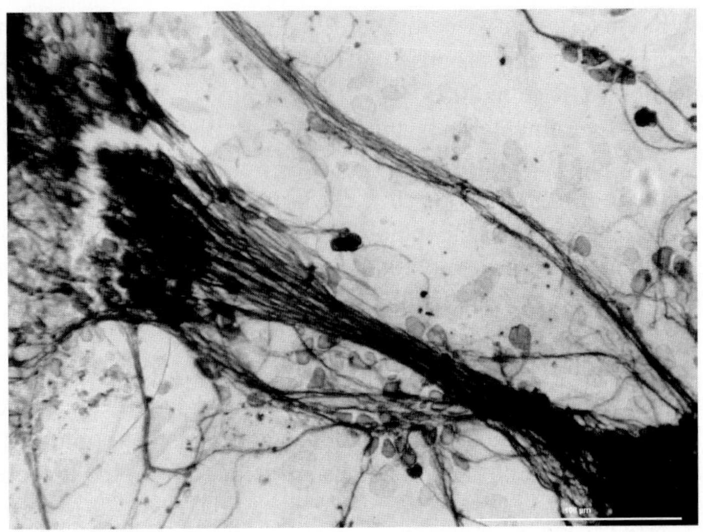

*Figure 7.* Human ESC generate complex neural aggregates when Embryoid Bodies (EBs) are generated in IGF-1.

**(Figure- generation of multipotent stem cells from pluripotent ESC)**

For clinical translation of stem cells much of the biology remains to be elucidated. This includes a greater understanding of their growth characteristics and needs, the control of their growth and of their differentiation and the transition from stem cell to progenitor cell and reduction of potency. Identification of unique surface markers to aid isolation of rare stem cells is also needed as well as identification of growth factors for cell expansion of specific populations.

## 5. THE UK FACTOR

The UK has been at the forefront of stem cell research. This is due to a number of factors:

- There are world-class academic research centres already established in the UK.
- Tight regulation on Assisted Reproduction (AD) and human embryo research by HFEA allows public confidence and support as well as academic validity

- The UK Government is committed to Stem Cell research having given £45M in 2003-4 and proposed £100M for a 10 year strategy from the Chancellor's Office in 2005
- The UK Government funds the Stem Cell Bank foe foetal, adult and embryonic stem cell lines
- The Human Embryonic Stem Cell Forum encourages the sharing of knowledge between all the major UK human ESC laboratories
- The London Regenerative Medicine Network integrates a research-led drive towards clinical applications.

The generation of therapeutic grade lines under GMP (Good Manufacturing Procedures) will require specialised facilities and expertise. For GMP the processes must be consistent, there must be control of supply of the materials used, and the processes must be safe, reliable, robust and well documented.

With strong regulations and financial incentives the UK has become a major centre for stem cell research. In the field of human ESC many technical hurdles need to be overcome to produce cells for GMP. Clinical translation requires GMP as a minimum standard. The UK Stem Cell Bank will be at the forefront of setting these standards. Humans ESC provide a very useful source of pluripotential cells which, under conditions gradually being elucidated, can be controlled to expand cell numbers and to produce all the cell types of the body. In the field of regenerative medicine with a lack of whole organ donors, this could prove to be the way forward. While the use of human ESC in regenerative medicine may be some way off, the potential therapeutic use of ESC cannot be ignored.

## ACKNOWLEDGEMENTS

This work was supported by The Charitable Foundation of Guy's and St. Thomas's Hospital, The MRC and London South Bank Research Fellowship Programme.

## REFERENCES

Alison M.R, Poulson R, Otto W.R, Vig P. Britton M, Dirokzo N.C, Lovell M, Fang T.C, Preston S.L, Wright N.A 2004, Recipes for adult stem cell plasticity:fusion cuisine or readymade? J.Clin. Pathol . **57** 113-120

Amit M, Itskovitz-Elder J, 2002, Derivation and spontaneous differentiation of human embryonic stem cells. J of Anatomy **200** 225-232

Carpenter M.K, Rosler E.S, Fisk G.J, Bradenberger R, Ares X, Miura T, Lucero M, Rao M.S 2004 Properties of four human embryonic stem cell lines maintained in feeder-free culture system. Developmental Dynamics **229** 243-258

Evans M.J, Kaufman M.H 1981 Establishment in culture of pluripotent cells from mouse embryos. Nature **292** 154-6

Fodor W.L, 2003 Tissue engineering and cell based therapies, from the bench to the clinic: The potential to replace, repair and regenerate. Reprod Biol and Endocrinol **1** 102-122

Ford C.E, Hamerton J.L, Barnes D.W.H, Loutit J.F 1956, Cytological identification of radiation chimaeras. Nature **117** 452

Ham A.W and Cormack D.H 1978, Histology. 8$^{th}$ ed. Philadelphia, JB Lippincott p377-462

Hansen-Smith F.M and Carlson B.M 1979, Cellular responses to free grafting of the exteriror digitorum longus muscle of the rat J Neurol Sci 41, 149-173

Human Fertilisation and Embryology Act, 1990 c37. HMSO Stationary Office, London, UK

Human Fertilisation and Embryology Act (Research Purposes). Human Reproductive Cloning Act 2001; c23. HMSO Stationary Office. London UK

Human Fertilisation and Embryology Authority HEFA (2005) Facts and Figures. www.hfea.gov.uk

Horton M.A, Cedar S.H, Maryanka D, Mills F.C, Turberville C, 1983 Multiple differentiation programs in the K562 erythroleukemia cells and their regulation. Prog Clin. Biol. Res **134** 305-22

Ianus A, Holz G.G, Theise N.D, Hussain M.A, 2003, In vivo derivation of glucose-competent pancreatic endocrine cells from bone marrow without evidence of cell fusion. J Clin. Invest **111** 843-850

Jensen U.B, Lowell S and Watt F 1999, The spatial relationship between stem cell and their progeny in the basal layer of human epidermis: a new view based on whole mount labelling and lineage analysis. Develoment 126' 2409-2418

Joseph N.M and Morrison S.J 2005, Toward an understanding of the physiological function of mammalian stem cells. Devlop. Cell **6** 173-183

Metcalf D 1979, detection and analysis of human granulocyte-monocyte precusors using semi-solid culture. Clinics in Haematol **8** 263

Michaaelopoulos G.K and DeFrances M.C 1997 Liver regeneration. Science 276, 60-66

Minger S.L, Fisk V.L, Ray J, Gage F.H, 1996 Long-term survival of transplanted basal forebrain cells following in vitro propagation with fibroblast growth factor 2. Exp. Neurol. **141**(1) 12-24

Pickering SJ, Braude P.R, Patel M, Burns C.J, Trussler J, Bolton V, Minger S, 2003, Preimplantation genetic diagnosis as a novel source of embryos for stem cell research. Reprod. Biomed. Online **7**(3) 252-64

NHS UK Transplants (2005). Statistics. http://www.uktransplants.org.uk

Orlic D, Kajstura J, Chimenti S, Bodine D.M, Leri A, Anversa P, 2001, Transplanted adult bone marrow cells repair myocardial infarcts in mice. Ann N Y Acad Sci USA **938** 221-9

Pickering S.J, Minger S.L, Patel M, Taylor H, Black C, Burns C.J, Ekonomou A, Braude P.R 2005, Generation of a human embryonic stem cell line encoding the cystic fibrosis mutation delta F508 using preimplantation genetic diagnosis. Reprod Biomed. Online **10**(3) 390-7

Pittinger M.F, Mackay A.M, Beck S.C, Jaiswal R.K, Douglas R, Mosca J.D, Moorman M.A, Simonetti D.W, Craig S, Marshak D.R, 1999 Multilineage potential of adult human mesenchymal stem cells. Science **284** 143-147

Potten C.S, Booth C and Pritchard D.M 1997, The intestinal epithelial stem cell: the mucosal govenor. Int J Path. 78 219-243

Ray J, Peterson DA, Schinstine M, Gage F.H, 1993 Proliferation, differentiation and long-term culture of primary hippocampal neurons. Proc. Ntal. Acad. Sci. USA **90** 2602-06

Reubinoff BE, Pera M.F, Fong C-Y, Trounson A, Bongso A 2000 Embryonic stem cell lines from human blastocysts: somatic differentiation in vitro. Nature Biotechnol.**18** 399-404

Rosler E.S, Fisk G.J, Ares W, Irving J, Miura T, Rao M.S, Carpenter M.K 2004 Long-term culture of human embryonic stem cells in feeder free conditions Developmental Dynamics **229** 259-274

Shamblott M.J, Axelman J, Wabg S, Bugg E.M, Littlefield J.W, Donovan P.J, Blumenthal D, Huggins G.R, Gearhart J.D, 1998, Derivation of pluripotent stem cells from cultured human primordial germ cells. Proc. Natl. Acad. Sci. USA **95** 13726-13731

Smith A 2001, Embryonic stem cells. In Stem Cell Biology: Marshak D.R, Gardner R, Gotlieb D, eds. Cold Spring Harbor: Cold Spring Harbor Laboratory Press pp205-230

Sladek J.R, Redmond D.E Jr, Collier T.J, Haber S.N, Elsworth J.D, Deutch A.Y, Roth R.H 1987 Transplantation of fetal dopamine neurons in primate brain reverses MPTP induced parkinsonism. Prog. Brain Res. **71** 309-23

Steptoe P.C, Edwards R.G 1978 Birth after the reimplnatation of a human embryo . Lancet **2**(8085) 366

Stojkovic M, Lako M, Strachan T, Murdoch A, 2004 Derivation. Growth and applications of human embryonic stem cells. Reproduction **128** 259-267

Taylor H, Minger S.L 2005 Regenerative medicine in Parkinson's disease: generation of mesencephalic dopaminergic cells from embryonic stem cells. Current Opinion in Biotechnol. **16** 1-6

Terai S, Yamamoto N, Omori K, Sakaida I, Okita K, 2002, A new cell therapy using bone marrow cells to repair damaged liver. J Gastroentrol. **37.** Suppl 14. 162-163

Till J.E and McCulloch E.A 1961 A direct measurement of the radiation sensitivity of normal mouse bone marrow cells. Radiation Res **14** 213

Thompson JA, Itskovitz-Elder J, Shapiro S.S, Waknitz M.A, Swiergiel J.J, Marshall V.S, Jones J.M, 1998, Embryonic stem cell lines derived from human blastocysts. Science **282** 1145-1147

United Network for Organ Sharing (2004) www.unos.org

# Investigations of Murine Embryonic Stem Cell Maintenance by Analyses of Culture Variables and Gene Expression

James M. Piret[1,2], Clive H. Glover[1], Michael Marin[1,3], Mohammed A. S. Chaudhry[1,2], Connie J. Eaves[4,7], Bruce D. Bowen[2], R. Keith Humphries[5,7], Cheryl D. Helgason[6,8] and Jennifer Bryan[1,3]

*[1]Michael Smith Laboratories, 2185 East Mall, Departments of [2]Chemical & Biological Engineering, [3]Statistics, [4]Medical Genetics, [5]Medicine and [6]Surgery, University of British Columbia, Vancouver, BC, Canada; [7]Terry Fox Laboratory and [8]Cancer Endocrinology, BC Cancer Agency.*

**Abstract:** The clinical realization of many gene and cell therapies requires robust, scalable methods for expanding stem cells *ex vivo* without compromising their developmental potential. This requirement is usually complicated by the low frequency of stem cells in most tissues (< 0.1%) and the lack of assays for quantifying stem cells. Murine embryonic stem cells (ESC) provide a useful model for stem cell bioprocess research since ESC can be propagated indefinitely under defined conditions and assayed functionally. Nevertheless, conventional ESC cultures require daily medium exchange and an understanding of their environmental tolerance ranges is still lacking. We have now begun to explore these using the embryoid body (EB) assay to quantify R1 ESC integrity as a function of culture variables. We are also exploring the feasibility of using mRNA-based assays to assess the effect of altered culture conditions on ESC maintenance. Analysis of microarray data has revealed a subset of genes whose change in expression was consistently correlated with loss of ESC pluripotency following the induction of differentiation by LIF removal, DMSO or retinoic acid treatment. Regression models were fitted to the data and their parameter estimates used to quantify the differential expression of genes, providing new candidates for future investigation of the mechanisms by which ESC may lose pluripotency.

**Key words:** embryonic stem cell, culture variables, pH, osmolality, gene expression profiling



*R. Smith (ed.), Cell Technology for Cell Products*, 185–189.

# 1. INTRODUCTION

Pluripotent embryonic stem cells (ESC) have been derived from the inner cell mass of blastocyst-stage embryos from mouse, primate and human sources [1-3]. The clinical potential of human ESC arises from their ability to generate a variety of cell types found in the body. For this reason they have been proposed as a raw material for cellular therapies [4]. Extensive stem cell expansion will be required to meet requirements for ESC-based therapies. Maintaining a large pool of cells in an undifferentiated state requires an understanding of the pH and osmolality effects.

To understand the status of stem cells, the developmental potential must be monitored. We have previously shown in the mouse ESC system that SSEA-1 and Oct-4 expression, both commonly used markers of undifferentiated mouse ESC, are lagging indicators of functional output and thus do not accurately demonstrate developmental potential [5]. This problem could be overcome, if it were possible to identify other genes whose expression change during differentiation is consistently correlated with loss of functionality independent of how loss of pluripotency was induced. Such genes could then provide the basis of an mRNA-based assay for the proportion of undifferentiated ESC in a given test population.

# 2. MATERIALS AND METHODS

## 2.1 ESC Maintenance Cultures

R1 ESC (p17) [6] were maintained as described previously [5]. ESC were thawed and maintained on irradiated primary embryonic fibroblasts (PEF) for 2-3 passages (96-144 h) prior to initiation of pH, osmolality and differentiation experiments. Residual PEF were removed by preplating cells on tissue culture treated plastic in maintenance medium for 20 – 60 min at 37°C. Cells were centrifuged and resuspended in relevant media before replating on tissue culture dishes (Sarstedt) coated with 0.1% porcine gelatin (Sigma) at a density of 5,500 cells/cm$^2$ (pH/osmolality experiments) or 80 – 1500 cells/cm$^2$ (differentiation experiments). The pH of the maintenance medium was adjusted by the addition of either 1 N HCl or 1 N NaOH. The osmolality of the medium was adjusted by addition of appropriate volumes of 1M NaCl to a custom made DMEM formulation with low NaCl concentration. The pH was measured on a Bayer Rapidlab 348 pH/Blood Gas Analyser (Bayer, Toronto, ON) and osmolality was measured using an osmometer (Advanced Instruments, Norwood, MA).

The differentiation media were based on the maintenance medium with the following differences: (1) DMSO – maintenance medium with no LIF and 1% DMSO and (2) RA – maintenance medium with 2 μM retinoic acid. Cells were sampled at the end of culture and the embryoid body (EB) assay used as previously described [5]. Alternatively, cytoplasmic RNA was extracted using the RNeasy mini kit (Qiagen, Mississauga, ON) and standard protocols (Affymetrix, Santa Clara, CA) were used for the generation of the probe RNA from 5 μg of sample RNA before hybridization to Affymetrix MOE430 Genechips according to manufacturer's instructions. All experiments were performed in duplicate.

### 2.2 Microarray data analysis

Two distinct data sets were analyzed. The generation of the DMSO/RA data set is described here. The LIF removal data set is described in [5]. All statistical analyses were performed using the statistical analysis environment R (www.r-project.org). Probe intensities were corrected for background, normalized and summarized within probe sets using the RMA algorithm [7]. A linear model was then used to estimate expression changes and genes were labeled as differentially expressed if they exhibited at least a 2-fold change relative to the control condition (+LIF in the DMSO/RA data set, 0 h in the LIF removal data set) and a fold-change:error ratio greater than 2. In addition, genes were labeled as differentially expressed in the DMSO/RA data set if they exhibited changes in the same direction and, in the LIF removal data set, if they showed a monotonic expression change throughout the duration of the experiment.

## 3. RESULTS AND DISCUSSION

### 3.1. Osmolality and pH studies

To determine the range of osmolality that ESC were exposed to during standard maintenance culture, R1 ESC were cultured on PEF for 48 h with daily medium exchange. Fresh medium had an osmolality 304 ± 4.0 mOsm/kg. After growth for 24 h, the medium had an osmolality of 329 ± 8.6 mOsm/kg and at 48 h this had risen to 405 ± 10.5 mOsm/kg, despite the exchange of medium at 24 h (due to the exponential growth of cells during this period following a lag phase during the first 24 h). To explore the influence of variation of osmolality and pH on the proliferation and differentiation of R1 ESC, dose-response experiments were performed where the osmolality and pH of the medium were varied. The highest growth rate and EB formation per initial cell were obtained for pH 7.3 (not shown) and an average osmolality of 340 mOsm/kg (Figure 1). The yield of EB decreased by 60% ($p<0.05$) for R1 ESC cultured at pH 7.0 or 400 mOsm/kg osmolality (compared to a pH 7.3 or 300 mOsm/kg). This was due to both relative and absolute decreases in the rate of EB-forming cell expansion. Thus, modest variations in both pH and osmolality had a large impact on culture performance.

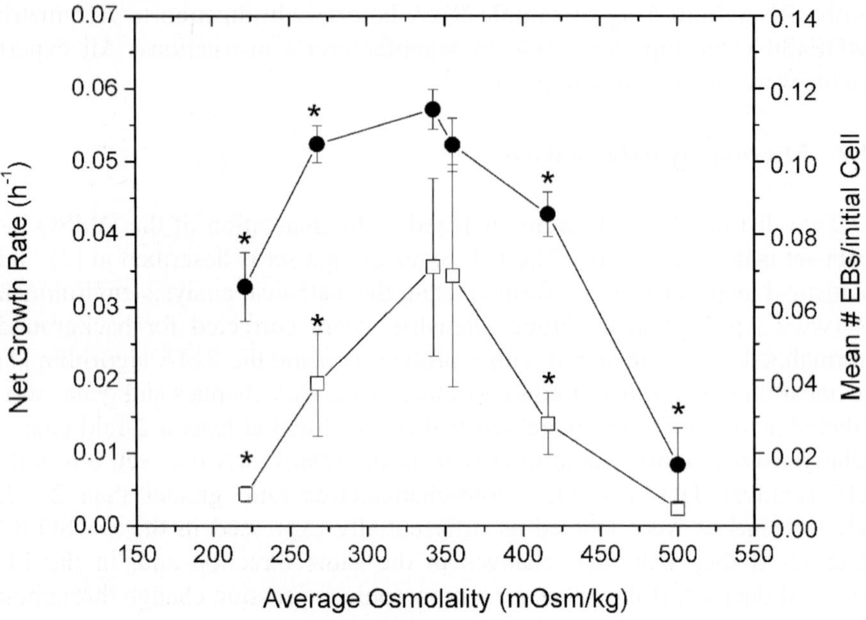


*Figure 1.* Variation of mean number of EBs per initial cell (ϒ) and growth rate (•) as a function of average osmolality. (* - $p<0.05$ vs. average osmolality of 340 mOsm/kg by paired t-test).

### 3.2. Gene expression studies

R1 ESC were cultured for 96 h in maintenance medium without LIF in 1% DMSO or in maintenance medium with 2 μM RA. Cells in standard maintenance conditions served as a positive control. Inoculum cells had 7.8 ±1.9% EB forming cells. After 96 h, DMSO and RA conditions exhibited 0.1 ± 0.1% and 0.2 ± 0.1% EB forming cells respectively while cells cultured in +LIF had 6.8 ± 1.9% EB forming cells.

Gene expression profiling was initiated on ESC grown in DMSO and RA for 96 h using the Affymetrix MOE430 platform using +LIF as a positive control. To define a set of differentially expressed genes whose expression change was independent of the treatment applied, analysis of this data set was combined with previously published data [5] from gene expression changes 18 and 72 h following the removal of LIF. 42 genes showed a consistent downregulation in all conditions and included many previously identified markers of ESC (Table 1). Approximately 15 of the 42 genes have no known biological function. These genes provide a starting point from which to develop mRNA based assays to understand the impact of culture conditions on the maintenance and differentiation potential of ESC.

*Table 1.* Gene expression markers of undifferentiated ESC. Expression of genes is shown as a percentage of the levels in undifferentiated control cells.

| Gene | -LIF (72 h) | DMSO (96 h) | RA (96 h) |
|---|---|---|---|
| Nanog | 55 | 36 | 19 |
| Sox2 | 36 | 16 | 7 |
| Esrrb | 20 | 9 | 9 |
| Zfp42/Rex1 | 14 | 9 | 10 |
| Dppa3 | 36 | 37 | 32 |
| Dppa5 | 44 | 17 | 25 |
| Fbx15 | 10 | 11 | 33 |
| Nr0b1 | 10 | 6 | 8 |
| Ptch1 | 31 | 6 | 17 |
| Nmyc1 | 58 | 30 | 25 |
| Klf2 | 41 | 11 | 11 |
| Klf4 | 7 | 9 | 33 |
| Cobl | 32 | 26 | 29 |
| Leftb | 46 | 31 | 18 |

# REFERENCES


1. Evans, M.J. and M.H. Kaufman, *Establishment in culture of pluripotential cells from mouse embryos.* Nature, 1981. **292**(5819): p. 154-6.
2. Thomson, J.A., *et al.*, *Isolation of a primate embryonic stem cell line.* Proc Natl Acad Sci U S A, 1995. **92**(17): p. 7844-8.
3. Thomson, J.A., *et al.*, *Embryonic stem cell lines derived from human blastocysts.* Science, 1998. **282**(5391): p. 1145-7.
4. Zandstra, P.W. and A. Nagy, *Stem cell bioengineering.* Annu Rev Biomed Eng, 2001. **3**: p. 275-305.
5. Palmqvist, L., *et al.*, *Correlation of murine embryonic stem cell gene expression profiles with functional measures of pluripotency.* Stem Cells, 2005. **23**: p. 663-80.
6. Nagy, A., *et al.*, *Derivation of completely cell culture-derived mice from early-passage embryonic stem cells.* Proc Natl Acad Sci U S A, 1993. **90**(18): p. 8424-8.
7. Irizarry, R.A., *et al.*, *Summaries of Affymetrix GeneChip probe level data.* Nucleic Acids Res, 2003. **31**(4): p. e15.

# Generation of Cardiomyocytes Derived from Mouse ESC in a Scale-up System

S. Niebruegge[1], H. Ebelt[2], M. Jungblut[3], H. Baer[4], M. Schroeder[5], R. Zweigerdt[6], T. Braun[3], J. Lehmann[1]

*[1] Inst. of Cell Culture technology, University of Bielefeld, Germany, [2] Department of Medicine III and [3] Inst. of Physiol. Chemistry, Martin-Luther-University, Halle, Germany, [4] Department of Cell Biology, German Cancer Research Center Heidelberg, Germany, [5] Inst. of Biotechnology 2, Res. Center Juelich GmbH, Germany, [6] ITZ, University of Duesseldorf, Germany*

**Abstract:** The cultivation of embryonic stem cells (ES cells) in a standard stirred tank bioreactor makes it possible to generate large numbers of therapeutically usable cells such as cardiomyocytes. We have used this technique to expand genetically engineered ES cells transfected with a fusion gene that allows an efficient enrichment of cardiomyocytes via antibiotic addition.

**Key words:** stem cells, cardiomyocytes, scale-up, transplantation

## 1. INTRODUCTION

In this work a draw-and-fill perfusion process in a controlled stirred tank reactor at 2L scale was developed which allows a long-term cultivation under stable, reproducible conditions. Throughout the culture period flow cytometry, immunohistochemistry and fluorescence microscopy of cardiac specific antigens were performed. About 9e8 cardiomyocytes in 2e5 cardiac bodies were generated. The purity of the population was about 99% measured by immunohistochemistry and fluorescence microscopy. To analyze the functionality of ES-cell derived cardiomyocytes *in vivo* cardiac bodies were transplanted into female mice 7 days after myocardial infarction induced by ligation of a major blood vessel. Both echocardiographic



*R. Smith (ed.), Cell Technology for Cell Products*, 191–195.

follow-up and histological analysis 28d after transplantation showed an improved left ventricular function of hearts that received a transplant.

Here we demonstrate that it is feasible to generate functional cardiomyocytes derived from embryonic stem cells in a bioreactor at a large-scale that improve cardiac functions of infarcted hearts.

## 2. MATERIAL AND METHODS

The used cell line was the J1 ES cell clone CM7/1 carrying the neomycin resistance gene driven by the cardiac myosin heavy chain (MyHc) promotor. The addition of G418 makes cardiomyocyte enrichment possible. Expansion and differentiation via Embryoid body (EB) formation was established in a standard 2L stirred tank bioreactor (B. Braun) which allows the control of culture parameters like pH, $pO_2$ and shear force. The process was inoculated with $2x10^5$ undifferentiated CM7/1 cells/mL in 1L volume of DMEM medium supplemented with 10% FCS. We chose a stirrer speed of 65 rpm using a blade impeller generating axial medium flow (Fig. 1). The resulting EB-suspension is shown in Figure 2.

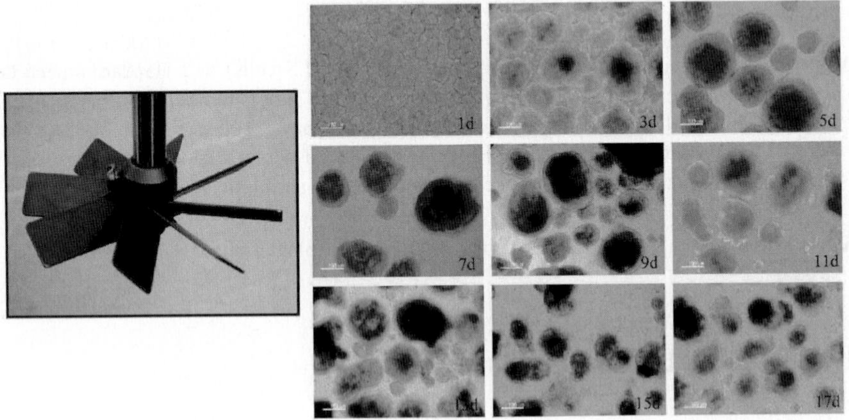


*Figure 1.* Pitched-blade turbine used for stirring.

*Figure 2.* EB-Formation during the production process. Start of selection on day 9.

For cardiomyocyte enrichment the process was divided in 2 phases. The expansion and differentiation phase via EB formation lasted 9 days, followed by the selection phase and the addition of G418. At the end of the cultivation all EBs were beating and termed Cardiac bodies (CBs). Figure 3 gives an overview of the experiment.

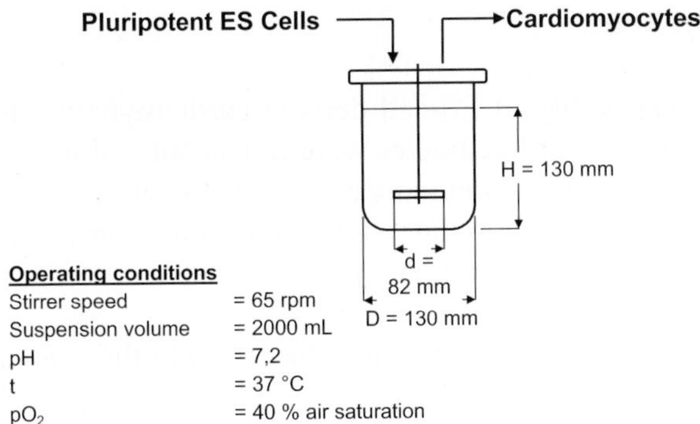


*Figure 3.* Overview of the experiment conditions for the production of cardiomyocytes at 2L scale.

# 3. RESULTS

After 9 days of expansion and differentiation via EBs the highest cell concentration was achieved on day 9. The selection of cardiomyoctes by addition of G418 resulted in a reduction of total cell numbers (Fig. 4). EB concentration dropped from approx. 261 EBs/mL at day 9 (before initiating selection) to approx. 86 CBs/mL at the end of cultivation. All of these generated CBs were beating. In order to characterize the selected cell population fluorescence staining and confocal laser scanning microscopy of cardiac specific antigens was performed (data not shown).

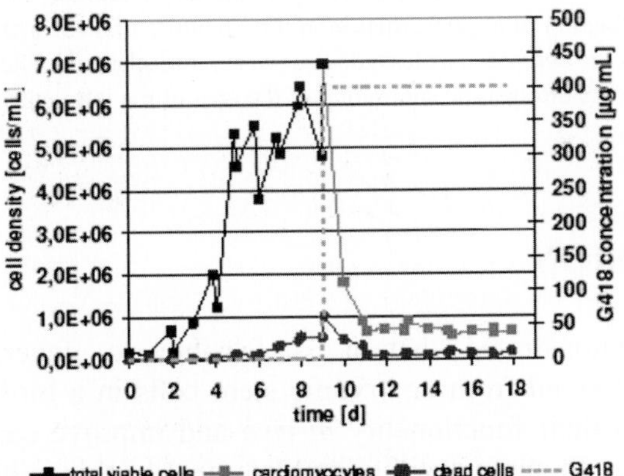


*Figure 4.* Cell densities during the process. The antibiotic addition on day 9 resulted in a drop of viable cells. The culture was inoculated with $2x10^5$ cells/mL. After 48 h the culture volume was adjusted form 1 to 2 liter. Thereafter half of the medium was changed daily.

### 3.1 Transplantations

To analyze the ability of ES-cell derived cardiomyocytes to reconstitute damaged heart tissue, cardiac bodies were transplantated into female mice 7 days after myocardial infarction induced by LAD-ligation.

Echocardiographic follow-up 28d after transplantation showed an improved left ventricular function of hearts that received a transplant in comparison to sham-treated mice (Fig. 5).

Integration of the transplanted cardiomcytes into the scared tissue could be detected.

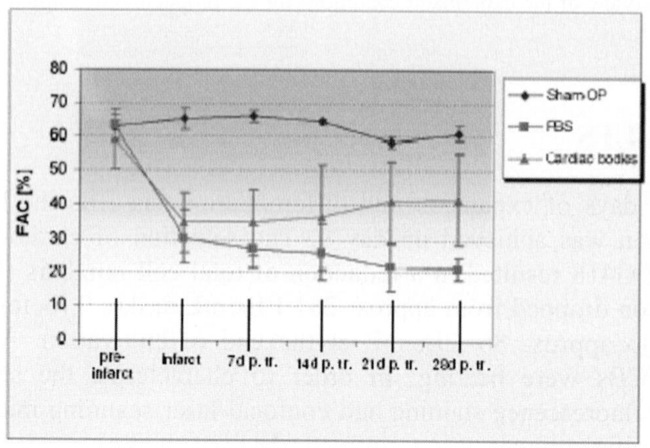


*Figure 5.* Echocardiografic follow-up after transplantations of Cardiac Bodies into infarcted mice. 7 days after induction of a myocardial infarct by ligature of a coronary artery of the left ventricle the bioreactor-derived cardiomyocytes were transplanted. The fraction of area change (FAC) represents the systolic shortening of the area of the left ventricle.

## 4. CONCLUSION

We have demonstrated that it is feasible to generate functional cardiomyocytes derived from embryonic stem cells in a bioreactor at large scale. They retain their functionality *in vivo* and improve cardiac functions of infarcted hearts. In summary the establishment of this cultivation system can serve as a tool for the production of therapeutically usable cells and as an appropriate cell supply for regenerative medicine applications.

## ACKNOWLEDGMENT

The author is supported by the Max-Buchner-Forschungsstiftung, DECHEMA. H. Ebelt was supported by the Wilhelm-Roux-Program for Research, Martin-Luther-University, Halle.

## REFERENCES

Schroeder, M., et al., 2005, Embryonic stem cell differentiation and lineage selection in a stirred bench scale bioreactor with automated process control, Biotech. Bioeng. (reviewed)

# Expansion of Undifferentiated Human Embryonic Stem Cells

Andre B.H. Choo[1], Jeremy Crook[2] and Steve K.W. Oh[1]
*[1]Stem Cell Group, Bioprocessing Technoloogy Institute, 10 Biopolis Way, #06-01, SINGAPORE. [2]ES Cell International, 11 Biopolis Way, #05-06, Helios, SINGAPORE. E-mail: steve_oh@bti.a-star.edu.sg*

**Abstract:** Human embryonic stem cells (hESC) are pluripotent cells derived from the inner cell mass of blastocysts. They have the potential to proliferate indefinitely in culture, but still retain their capacity for differentiation into a wide variety of cells. hESC are currently maintained either on feeder layers or on matrigel with conditioned-medium (CM) from primary mouse embryonic fibroblasts (MEFs).

We describe our group's strategy and results to date in establishing defined culture conditions for the continuous expansion of undifferentiated hESC in a feeder-free and serum-free culture system. This includes the development of a robust hESC passaging method based on enzymatic dissociation and culture on immortalized mouse and human feeders for the generation of consistent batches of conditioned-media, and the development of an alternative serum free culture platform based on sphingosine-1-phosphate and platelet derived growth factor.

**Key words:** Human embryonic stem cells, pluripotency, immortal feeders, serum-free, feeder-free cultures

## 1. INTRODUCTION

Human embryonic stem cells (hESC) are a gift of nature for potentially healing, replacing or restoring the function of tissues in the human body which would otherwise lack the capacity to do so in adults. Added to this is the fact that they have a natural proliferative capacity in vitro which potentially enables them to be manufactured to large quantities for eventual



*R. Smith (ed.), Cell Technology for Cell Products*, 197–205.

therapeutic applications. This key feature is what distinguishes hESC from other stem cells.

While hESC were first derived on primary mouse feeders [Thomson *et al.*, 1998], researchers have since attempted to grow them on human [Choo *et al.*, 2004] and immortal feeders [Choo *et al.*, 2005]. However, recognizing that for successful therapeutic applications, a defined serum-free, feeder-free culture system is necessary, we describe our progress towards establishing this defined culture platform. This entailed creating an immortal feeder cell line which could be used for producing conditioned media for analysis and eliminating the use of serum replacer with defined components from serum. A summary of the Stem Cell Group's research areas is presented in Figure 1.

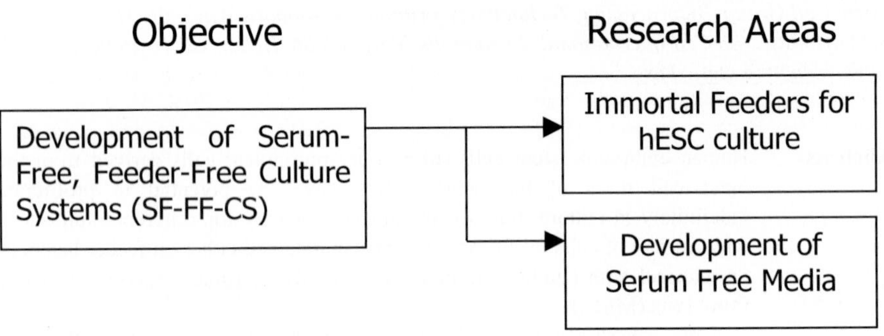


*Figure 1.* Development of serum-free, feeder-free culture systems for hESC.

## 2. CULTURING HUMAN EMBRYONIC STEM CELLS

Traditionally, hESC have been grown in basal media with serum and on human or mouse feeders [Thomson *et al.*, 1998] which produce both the matrix as well as the conditioned media (CM). We have generated immortal feeders which produce a consistent source of CM that can support hESC cultured on either matrigel, fibronectin or laminin. The eventual aim is to transit from this system to a fully defined culture with growth factors and supplements replacing serum replacer and CM as shown in Figure 2.

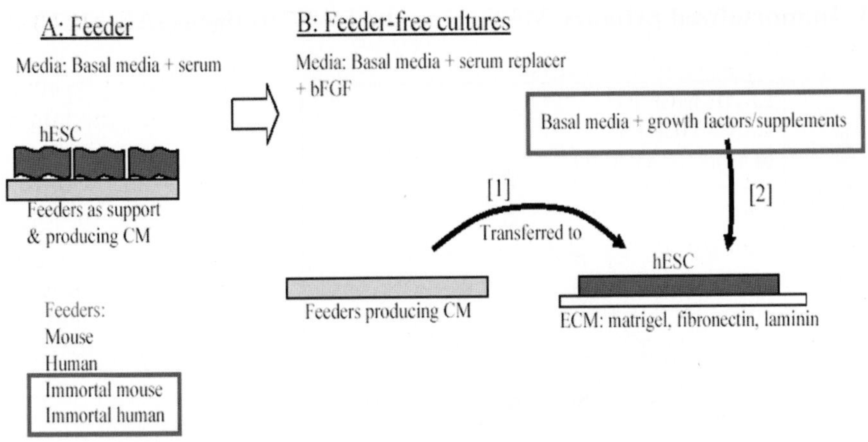


*Figure 2.* HESC are traditionally cultured on mouse or human primary feeders. Immortal mouse and human feeders were created; these cells produce consistent conditioned media for analysis to enable us to eventually develop defined conditions with known growth factors, supplements and extracellular matrices for hESC expansion.

# 3. IMMORTAL FEEDERS FOR HESC CULTURE

Immortal feeders were generated by transfecting mouse feeders with E6/E7 viral antigens which allowed the feeders to be passaged for an indefinite period of time. Figure 3.1 shows these immortal feeders, ΔE-MEF growing at passage 7 and 45 compared to the primary MEF which senesced at passage 7. HESC were enzymatically digested with collagenase and physically broken into smaller clusters of cells by repeated pipetting. These clusters were then seeded onto mitomycin-C treated ΔE-MEF at $10^5$ cells/organ culture dish. Alternatively hESC clusters were seeded onto matrigel coated plates and fed with conditioned media (CM) from a separate plate of mitomycin-C treated ΔE-MEF. Figure 3.2 also shows that these feeders were able to support the undifferentiated hESC phenotype of tight clusters of cells with high nuclear to cytoplasm ratio in both feeder and feeder free conditions. Figure 3.3 illustrates the FACS data of stable and high Oct-4 expression from 3 hESC lines cultured on the immortal ΔE-MEF as well as with CM produced from ΔE-MEF.

**1. Immortalized primary MEF using the E6/E7 antigens (ΔE-MEF)**

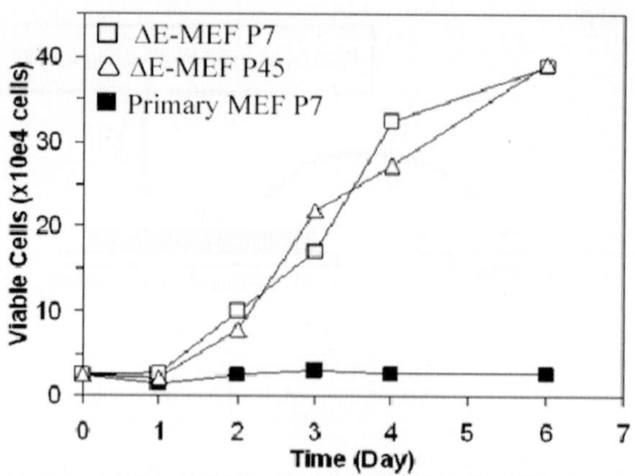


**2. hESC proliferation for >40 passages**

HES-2, HES-3 and HES-4

Maintained undifferentiated morphology

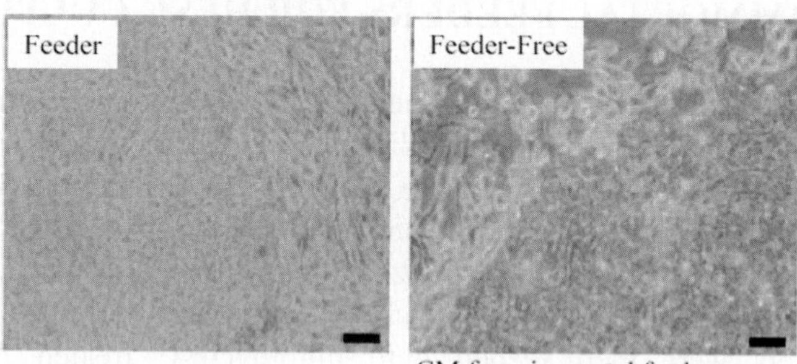


CM from immortal feeders + Matrigel

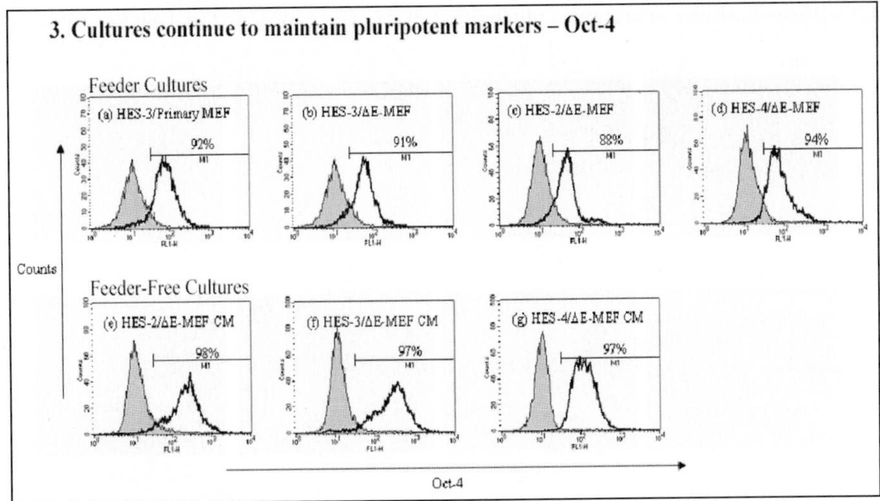


*Figure 3.* 1) Immortal feeders (ΔE-MEF) cultured for passage 7 and 45 vs. primary feeders which senesced at passage 7; 2) 3 hESC lines, HES-2, HES-3, and HES-4 were grown on immortal feeders or with conditioned media from immortal feeders and continue to maintain an undifferentiated morphology; 3) Three hESC lines HES-2, HES-3, HES-4 showing high Oct4 pluripotent marker expression (88% to 98%) on immortal feeder cultures and in feeder free conditions compared to culture on primary feeders.

Furthermore, hESC on feeders or feeder-free with CM from ΔE-MEF also expressed the other pluripotent surface markers, SSEA-4, TRA-1-60, TRA-1-81 and the enzyme alkaline phosphatase (Figure 4). These expanded hESC were also able to form teratomas with structures representation of all 3 lineages from ecto-, meso- and endoderm when injected into SCID mice (results not shown.); indicating that they have extensive differentiation capacity. The ΔE-MEF feeders have been routinely used in our lab to support hESC lines for more than 50 passages.

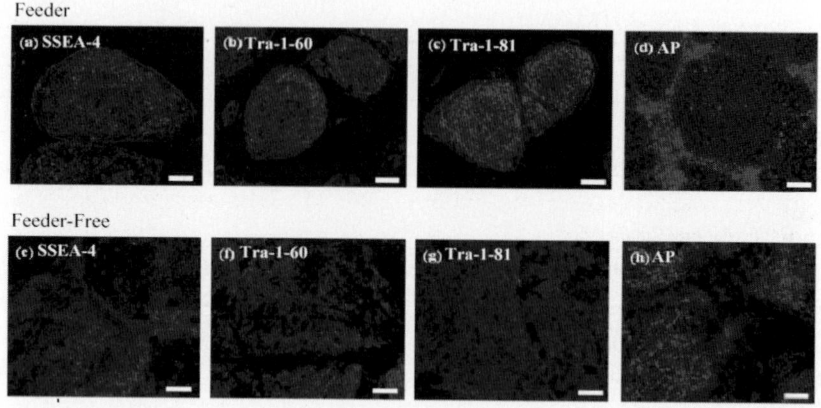


*Figure4.* hESC culture expressing pluripotent markers, SSEA-4, Tra-1-60, Tra-1-81 and alkaline phosphatase in feeder and feeder-free conditions.

Table 1 shows that the ΔE-MEF were able to support hESC expansion at increasing scales from the an organ culture dish, to a T-flask, a triple flask and finally a Nunc cell factory with a surface area of over 600cm$^2$. The final condition gave a total yield of almost 300 million cells while maintaining similar cell densities/cm$^2$ as the smaller culture devices. Thus in principle they are capable of being used to support the scale up of hESC to therapeutic quantities for animal studies.

*Table 1.* Expansion of hESC at increasing scales from organ culture dish, tissue culture flask, triple flask and cell factory on immortal feeders. Up to 300 million cells can be produced in a cell factory while retaining the cell density per unit surface area.

| Culture Platform | Surface Area ($cm^2$) | Total Cell Yield (x $10^6$ cells) | Cell Density (x$10^6$ cells/$cm^2$) |
|---|---|---|---|
| Organ Culture Dish | 2.4 | 1.0 | 0.42 |
| Tissue Culture Flask | 75.0 | 29.4 | 0.39 |
| Triple Flask | 500.0 | 187.5 | 0.37 |
| Cell Factory | 632.0 | 295.3 | 0.47 |

# 4. SERUM FREE CULTURE OF HESC

In collaboration with Embryonic Stem Cell International Pte. Ltd. (ESI), we attempted to remove serum replacer from the culture and replace it with sphingosine-1-phosphate (S1P) and platelet derived growth factor (PDGF) [Pebay *et al.*, 2005]. HESC were transferred onto feeders in the presence of this media along with a parallel control without S1P and PDGF. As shown in Figure 5.1, after 6 continuous passages, the one without the 2 supplements had differentiated into more fibroblastic cells and lost the marker TRA-1-60; whereas culture with these supplements retained the pluripotent morphology and marker. Figure 5.2 shows that Oct-4 transcription factor is partially down-regulated in the absence of the supplements but with S1P and PDGF, expression is similar to the control culture grown with serum replacer. More obviously, the marker SSEA-4 is significantly down-regulated without the supplements but its expression is retained in the presence of S1P and PDGF. Cell densities of the culture with S1P and PDGF were similar to controls, at about $10^6$ cells / organ culture dish. Work is progressing to culture hESC in serum-free conditions on fibronectin or matrigel in place of feeders.

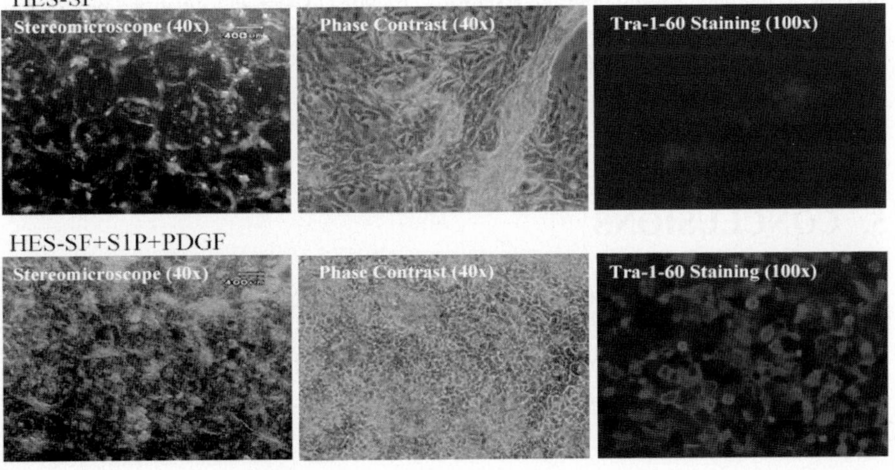

2. FACS: Oct-4 and SSEA-4 Expression

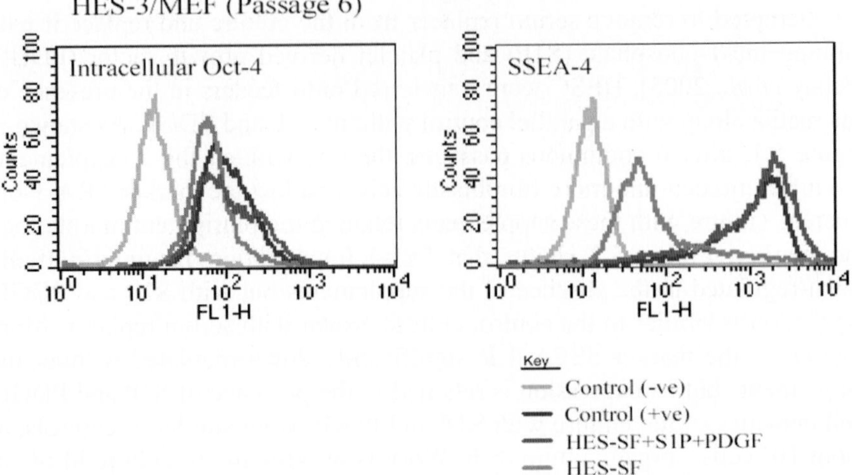


*Figure 5.* 1) Serum free culture of hESC without S1P and PDGF and with these 2 supplements after 6 passages, the latter continues to show compact cell morphology and pluripotent marker Tra-1-60 whereas the serum free culture without supplement has a fibroblastic morphology and does not stain for Tra-1-60; 2) Oct-4 and SSEA-4 expression is high in serum free culture with S1P and PDGF, being similar to the feeder control culture, compared to the culture without the supplements which has down-regulated both pluripotent markers.

# 5. CONCLUSIONS

In conclusion, we have derived immortal mouse and human feeders which can support undifferentiated hESC cell expansion and established a culture platform that does not require the addition of serum or serum replacer in the culture media.

# ACKNOWLEDGEMENTS

We thank ASTAR, Singapore for generous funding of this research and the excellent technical help of Jayanthi Padmanabhan, Angela Chin, Fong Wey Jia, Ang Sheu Ngo, and Tan Heng Liang.

# REFERENCES

Choo, A. B., Padmanabhan J., Chin, A. C., Fong, W. J., and Oh, S. K. W. Immortal feeders for the scale up of human embryonic stem cells in feeder and feeder-free conditions. J.Biotechnol. 2005.

Choo AB, Padmanabhan J, Chin AC, Oh SK (2004) Expansion of pluripotent human embryonic stem cells on human feeders. *Biotechnol Bioeng* 88: 321-331.

Pebay A, Wong RC, Pitson SM, Wolvetang E, Peh GS, Filipczyk A, Koh KL, Tellis I, Nguyen LT, Pera MF (2005) Essential roles of sphingosine-1-phosphate and platelet-derived growth factor in the maintenance of human embryonic stem cells. *Stem Cells.*

Thomson JA, Itskovitz-Eldor J, Shapiro SS, Waknitz MA, Swiergiel JJ, Marshall VS, Jones JM (1998) Embryonic stem cell lines derived from human blastocysts. *Science* 282: 1145-1147.

# New Electrofusion Devices for the Improved Generation of Dendritic Cell-tumour Cell Hybrids

Janina Schaper[1], Hermann Richard Bohnenkamp[2], Thomas Noll[1]
[1] *Research Center Juelich GmbH, Institute of Biotechnology 2, Cell Culture Technology, 52425 Jülich, Germany;* [2] *Cancer Research UK, Guy's Hospital, London, UK*

**Abstract:** Hybrids of two different cell types are of particular interest for a couple of applications (*e.g.* hybridoma cells for the production of monoclonal antibodies). In recent years, cancer immunotherapy using hybrid from antigen-presenting cells (especially dendritic cells) and tumour cells has been shown to efficiently induce anti-tumour immune responses. However, the available processes for generating such hybrids by chemically induced or electro-fusion are inefficient. Therefore, the clinical application of this technology is severely limited. In this study, we try to overcome poor fusion efficiencies and low yield of viable hybrids by an alternating arrangement of the fusion partners inside the fusion chamber, thus avoiding homologous fusion events. Two different devices have been developed using either microfluidics or micropatterning for cell arrangement introducing a new and reliable electrofusion procedure. Clinical requirements have been considered throughout process development and both approaches can be dimensioned to produce a sufficient amount of hybrid cells at once.

**Key words:** electrofusion, cancer vaccines, hybrid vaccines, dendritic cell-tumour cell vaccines, cancer immunotherapy, hybrids, fusion yield, electrofusion chamber, microfluidics, cell immobilisation

## 1. INTRODUCTION

Hybrid cells generated by the fusion of dendritic cells with tumour cells have been shown to induce anti-tumour immunity in several cancer models and clinical studies (Orentas *et al.*, 2001, Hayashi *et al.*, 2002, Rosenblatt *et al.*, 2005). Their application as whole cell vaccines in cancer immunotherapy requires a reliable method, which allows the reproducible production of sufficient amounts of heterologous hybrid cells. Especially in types of cancer



*R. Smith (ed.), Cell Technology for Cell Products*, 207–216.

where autologous tumour material is limited (*e.g.* breast cancer) a high fusion efficacy is of paramount importance. Unfortunately, most electrofusion procedures use random conditions without paying attention to statistical distribution or cell specific parameters.

The electrofusion procedure (also called electric field-mediated fusion) consists of two crucial steps. First of all, the fusion partners have to be brought into close contact (dielectrophoresis or cell alignment). This is followed by a short fusion pulse resulting in hybrid formation (Sugar *et al.*, 1987). The theoretical maximum fusion efficiency of 50% for heterologous hybrid formation (using a 1:1 mixture of the fusion partners) cannot be obtained in commercially available electroporation or electrofusion chambers. The low yield of 15-20% hybrid formation achieved has in our hands been due to the formation of cell clusters of homologous cells (see Fig. 1), which are typically formed during cell alignment owing to similar electrophysical parameters (such as cell size, shape and transmembrane potential). Ergo the prevention of these clusters ranks first.

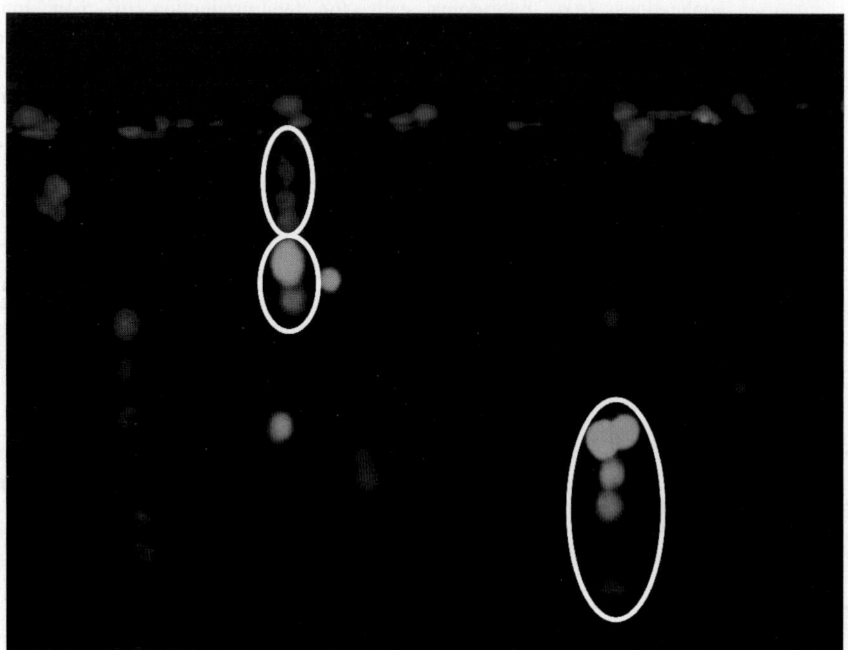

*Figure 1.* Cluster formation between MCF-7 (stained in red) and T47D (green) breast cancer cell lines during cell alignment in commercially available fusion chamber.

# 2. MATERIAL AND METHODS

## 2.1 Cells

Dendritic cells have been generated from mononuclear cells from peripheral blood of healthy donors as been published previously (Bohnenkamp*et al.*, 2003). Furthermore, two breast cancer cell lines, MCF-7 and T47D, were used. Both were cultured in Iscove´s modified Dulbecco´s medium (IMDM; Biochrom, Berlin, Germany) supplemented with 10% fetal calf serum (FCS; Biochrom).

## 2.2 Fusions

Electrofusions were performed in hypoosmolar fusion buffer (Eppendorf, Hamburg, Germany), which enhances hybrid formation due to a slight increase in membrane permeability by osmotic stress (Schmitt *et al.*, 1989).

The vital fluorescent dyes used for staining the cells were 5-chloromethylfluorescein diacetate (CMFDA; Molecular Probes, Eugene, OR) and 5-(and 6-)-(((4-chloromethyl)benzoyl)amino)tetramethylrhodamine (CMTMR; Molecular Probes).

## 2.3 Immobilisation

CD45-antibodies (BD Bioscience, San Jose, CA, USA) were immobilised following a standard protocol (Xinglong *et al.*, 2005).

# 3. RESULTS

## 3.1 Process Development

### 3.1.1 Aim

The aim of process development was to establish a method to improve fusion efficiency. This can be done by prevention of homologous fusion events, which in random fusion conditions occur at least in the same ratio as heterologous fusion events, but are presumably favoured due to cluster formation and more similar cell parameters.

Thus, two different electrofusion chambers for arranged fusion were developed. The underlying ultimate principle is the alternating arrangement of the fusion partners inside the fusion chamber by either antibody-based immobilisation of the cells in micropattern or parallel microfluidic single-cell flows. For an explanation of these methods see Fig. 2 for the immobilisation approach and Fig. 3 for the microfluidic approach.

### 3.1.2 Chamber design

The designed microdevices were fabricated using a combination of photolithography, metal-film deposition and glass etching technologies. They include up to 9 alternating cell lanes preventing cluster formation and homologous fusion events by this special cell arrangement. Direct visualization of the fusion procedure is permitted in both lab-on-a-chip approaches, thus giving more insight into electrofusion. Furthermore, all dimensions have been chosen to correlate sizes of cells and to allow fusion of large numbers of cells simultaneously.

The fabrication of the immobilisation chamber required the design of the chamber, the immobilisation of antibodies in small lines of 10 to 15 μm and the implementation of electrodes.

At first, CD45-antibodies were immobilised on glass using reverse microcontact printing. Therefore, the glass surface had to be functionalised with 3-aminotriethoxysilane and crosslinked with glutardialdehyde. The immobilisation of the antibody is then carried out by a Schiffsche base reaction between a primary amino group of the antibody and the carboxyl group of the cross-linker.

Later on, the surface chemistry has been switched due to problems with the integration of the electrodes. For the final design, the antibodies were immobilised on small gold lanes. Thus, a self-assembled monolayer (SAM) of 11-mercaptoundecanoic acid (MUA) was used. The carboxyl groups of MUA were activated with a mixture of N-hydroxysuccinimide (NHS) and 1-ethyl-3-(dimethylamino-propyl)carbodiimide (EDC). The resulting succinimide ester-activated monolayer then reacts with a primary amino group on the CD45-antibody to give a peptide bond.

Simultaneously, the chamber has been designed, which is build up around the immobilised antibody lanes. A gold structure, comprising electrodes and small lanes for antibody immobilisation is evaporated onto glass. On top of it the chamber is constructed. It consist of a bonding layer, made from medical double sided tape, with a flow channel, filling holes for the channel and holes for contacting the electrodes (all structures were cut

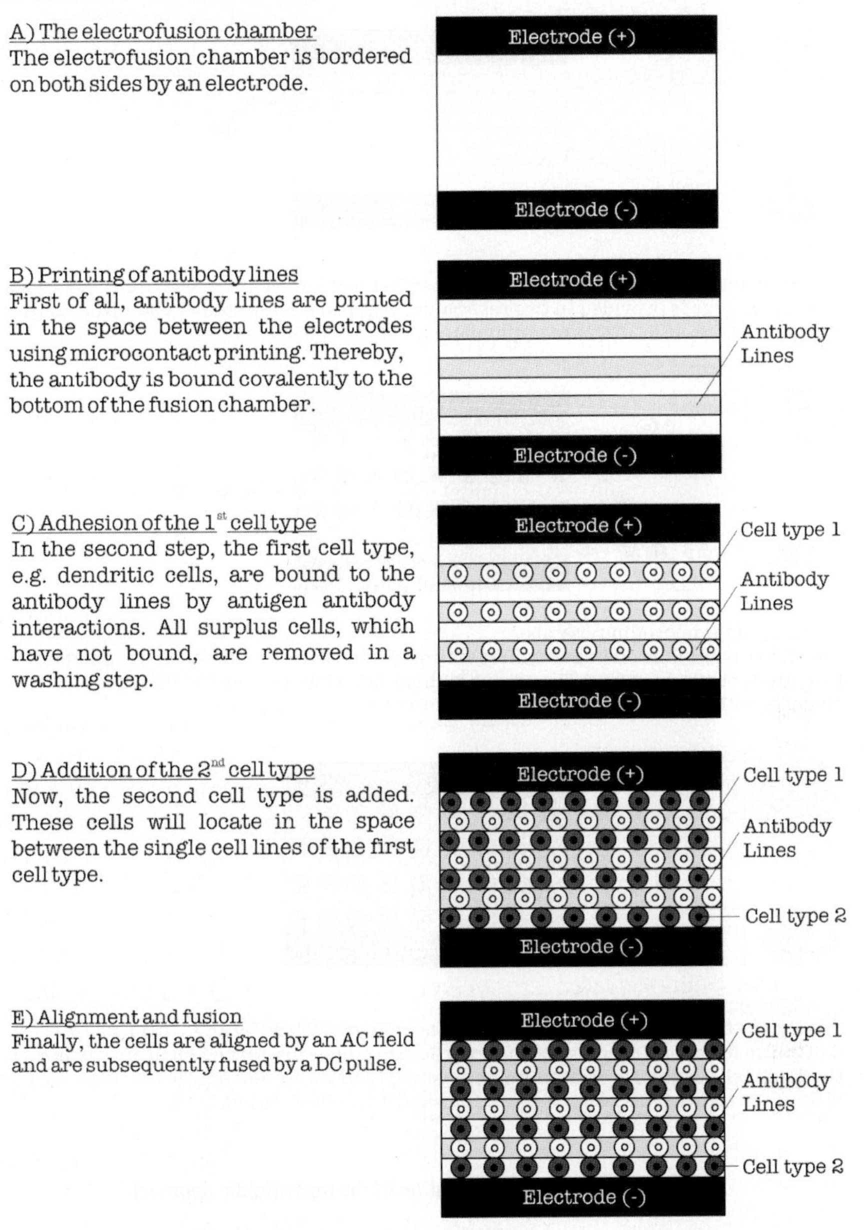
A) The electrofusion chamber
The electrofusion chamber is bordered on both sides by an electrode.
Electrode (+)
Electrode (-)
B) Printing of antibody lines
First of all, antibody lines are printed in the space between the electrodes using microcontact printing. Thereby, the antibody is bound covalently to the bottom of the fusion chamber.
Electrode (+)
Antibody Lines
Electrode (-)
C) Adhesion of the 1st cell type
In the second step, the first cell type, e.g. dendritic cells, are bound to the antibody lines by antigen antibody interactions. All surplus cells, which have not bound, are removed in a washing step.
Electrode (+)
Cell type 1
Antibody Lines
Electrode (-)
D) Addition of the 2nd cell type
Now, the second cell type is added. These cells will locate in the space between the single cell lines of the first cell type.
Electrode (+)
Cell type 1
Antibody Lines
Cell type 2
Electrode (-)
E) Alignment and fusion
Finally, the cells are aligned by an AC field and are subsequently fused by a DC pulse.
Electrode (+)
Cell type 1
Antibody Lines
Cell type 2
Electrode (-)

*Figure 2.* Schematic representation of the immobilisation approach.

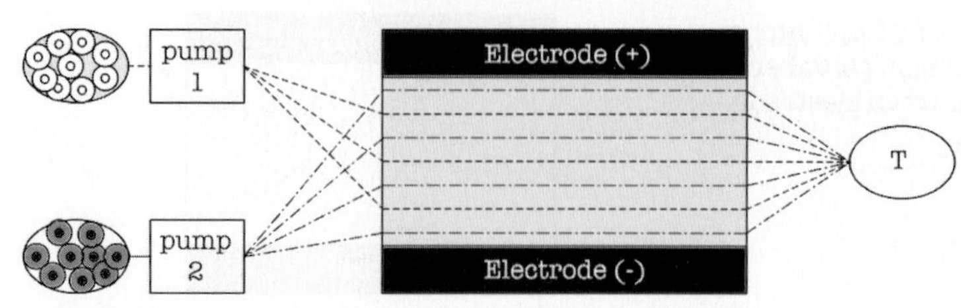


A) The electrofusion chamber
Two electrodes narrow the electrofusion chamber on two sides. A single cell suspension of each cell type is provided in two reservoirs, which are connected via pumps with the chamber. At the outlet of this microfluidic chamber an collecting tank (T) is placed.

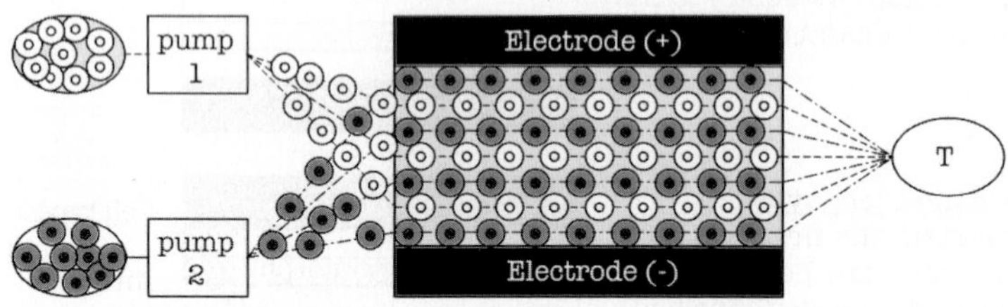


B) Filling of the microfluid chamber
The empty microfluid chamber is filled with single cell flows by dint of the two pumps due to hydrodynamic focussing. The arising laminar flows stream over the total length of the chamber without mixing, delivering the different cell types in lanes.

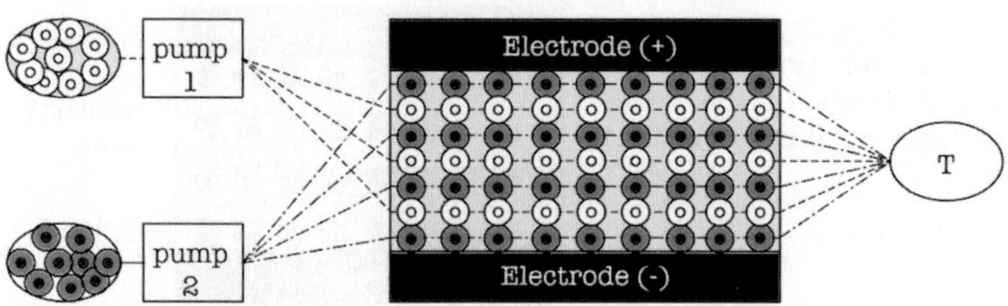


C) Alignment and fusion
Finally, the flow is stopped or reduced and the cells are aligned by an AC field, followed by a DC pulse for hybrid formation. Afterwards, the pumps are turned on again, replacing the fused cell suspension containing the desired hybrids, by new single cell suspensions of the fusion partners. The hybrids are collected in the intercepting tank.

*Figure 3.* Schematic representation of the microfluidic approach.

with a $CO_2$ laser) and a top layer made of acetate foil, sealing the flow channel (see Figs. 4 and 5 for design and realisation).

The cell arrangement has been approved within the completed chamber. Further research has been done into the width of the antibody lanes, the cell immobilisation and their detachment afterwards (see Figs. 6-8).

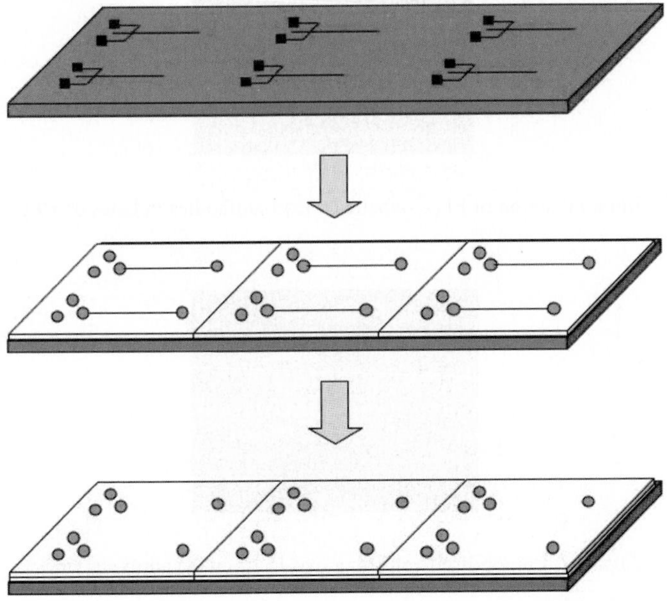

*Figure 4.* Final design of the immobilisation chamber.

*Figure 5.* Realisation of the immobilisation chamber.

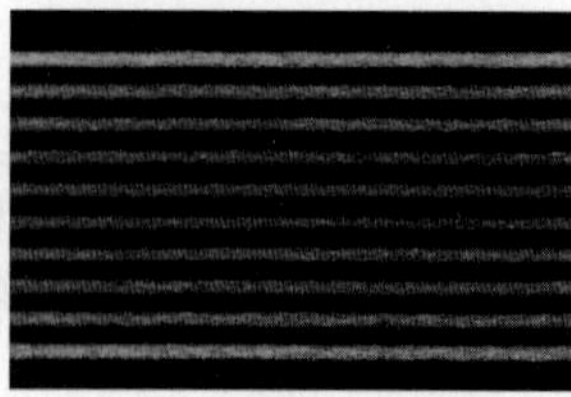

*Figure 6.* Immobilisation of FITC-labelled CD45-antibodies in lanes of 10 μm width.

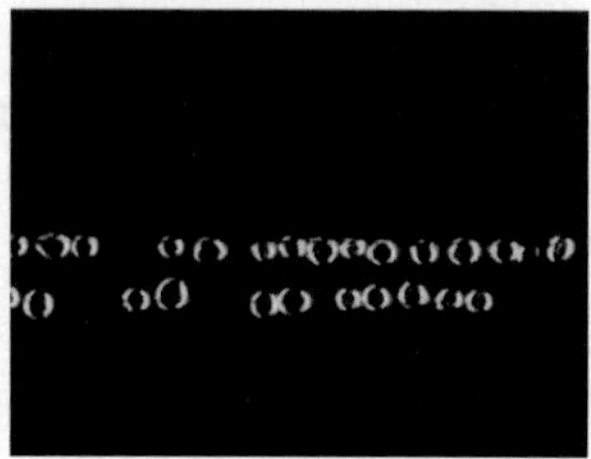

*Figure 7.* Immobilised cells on a two 15 μm wide antibody lanes.

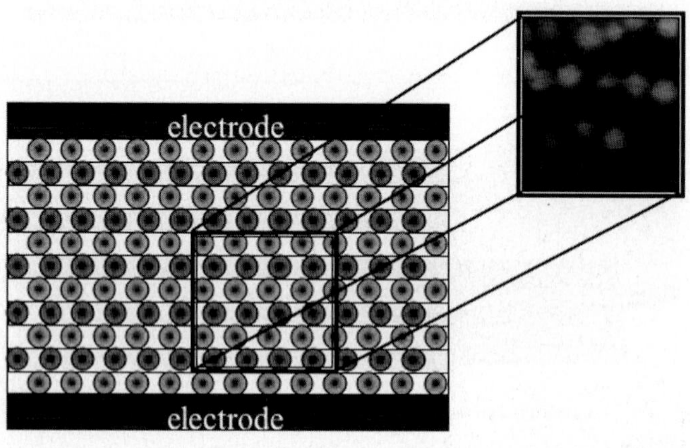


*Figure 8.* Arrangement of two different cell types within the fusion chamber.

Similar inquires have been done for the microfluidic approach. After the design (see Fig. 9) and its realisation (Fig. 10), the chamber has been tested for laminar flow conditions (Fig. 11) and the arrangements of cells inside this chamber.

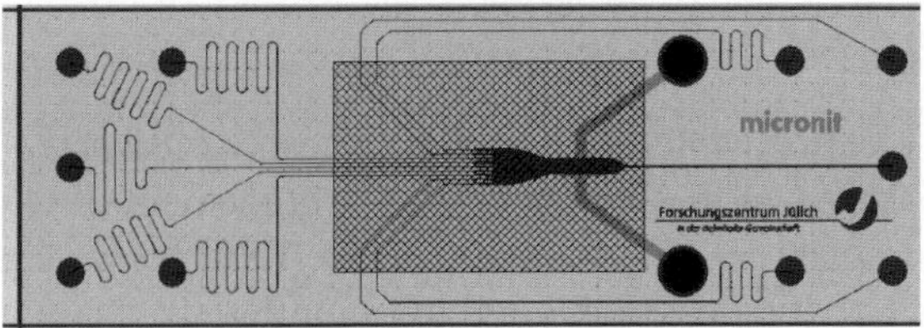


*Figure 9.* Final design of the microfluidic chamber.

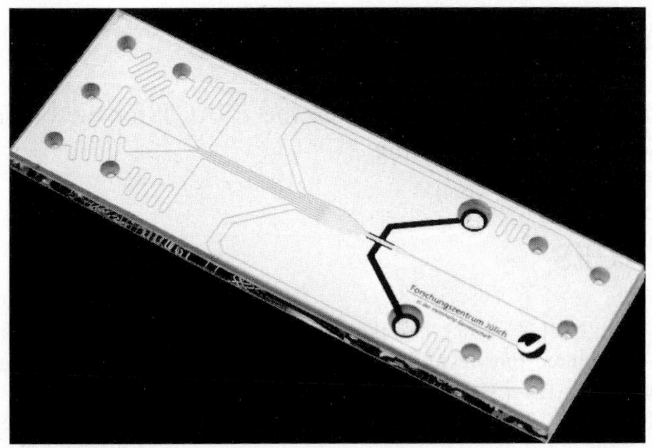


*Figure 10.* Realisation of the microfluidic chamber.

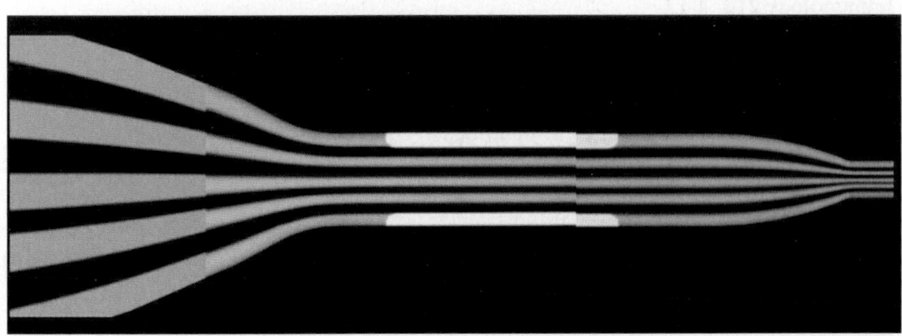

*Figure 11.* Laminar flow inside the microfluidic fusion chamber visualised by a fluorescent tracer.

## 4. CONCLUSION

Two new electrofusion devices have been constructed and put into operation. It was shown that an alternating arrangement of two different cell types is possible in both chambers. They are now used for fusion experiments to investigate which yield of homologous fusion can be reached. The resulting hybrids will subsequently be utilised for an immunological evaluation, *e.g.* for the induction of tumour-specific T cells.

## ACKNOWLEDGEMENTS

The authors gratefully acknowledge the ESACT for the bursary and the European Commission for funding.

## REFERENCES

Bohnenkamp, H.R., Noll, T., 2003, Development of a standardized protocol for reproducible generation of matured monocyte-derived dendritic cells suitable for clinical application, Cytotechnology **42**: 121-131

Hayashi, T., Tanaka, H., Tanaka, J., Wang, R., Averbook, B.J., Cohen, P.A., Shu, S., 2002, Immunogenicity and therapeutic efficacy of dendritic-tumor hybrid cells generated by electrofusion, Clinical Immunology **104**:14-20

Orentas, R.J., Schauer, D., Bin, Q., Johnson, B.D., 2001, Electrofusion of a weakly immunogenic neuroblastoma with dendritic cells produces a tumor vaccine, Cellular Immunology **213**: 4-13

Rosenblatt, J., Kufe, D., Avigan, D., 2005, Dendritic cell fusion vaccines for cancer immunotherapy, Expert. Opin. Biol. Ther. **5(5)**: 703-715

Schmitt, J.J., Zimmermann, U., 1989, Enhanced hybridoma production by electrofusion in strongly hypo-osmolar solutions, Biochimica et Biophysica Acta **983**: 42-50

Sugar, I.P., Förster, W., Neumann, E., 1987, Model of cell electrofusion, Biophysical Chemistry **26**: 321-335

Xinglong, Y., Dingxin, W., Xing, W., Xiang, D., Wei, L., Xingsheng, Z., 2005, A surface plasmon resonance imaging interferometry for protein micro-array detection, Sensors and Acutators B **108**: 765-771

# Influence of c-Myc and bcl-2 CO-expression on Proliferation, Apoptosis and Survival in CHO Cultures

Vasiliki Ifandi[1] and Mohamed Al-Rubeai[2]

*[1]Biomedical Research Institute, University of Warwick, Coventry, CV4 7AL, UK, [2]Dept. of Chemical and Biochemical Engineering, and Centre for Synthesis and Chemical Biology, University College Dublin, Dublin 4, Ireland*

**Abstract:** Under the constitutive control of c-Myc over-expression the CHO cultures showed an increase in growth rate and maximum cell number accompanied by a similar decrease in specific glucose consumption rate. Additionally over-expression of c-Myc appeared to induce morphological transformation and partial anchorage-independence. Although, the *c-myc* transfected cell line exhibited apoptosis at much lower rates than it is widely reported and associated with the over-expression of c-Myc, it was nevertheless apparent that c-Myc was responsible for the induction of higher apoptotic rates when compared with the control cell line. Hence, the anti-apoptotic gene *bcl-2* was also used to transfect the c-Myc CHO cell line, in order to reduce cell death. Over-expression of both oncoproteins resulted in a cell line that exhibited higher proliferation rates and maximum cell numbers, with a decrease in apoptosis when compared to the parental cell line.

**Key words:** Bcl-2 c-Myc, CHO-K1, proliferation, apoptosis, metabolic engineering

## 1. INTRODUCTION

The use of metabolic engineering for the positive control of cell proliferation is a rapidly developing field in biotechnology for the production of genetically modified cell lines. Great research has been directed towards the development of proliferation and apoptosis controlled cell lines with high cell density, controlled proliferation, apoptosis resistance, and easy adaptation into cultures of serum free media. These are



*R. Smith (ed.), Cell Technology for Cell Products*, 217–220.

some of the desirable characteristics for the cost effective production of biopharmaceuticals, mainly because genetically modified cell lines can afford greater efficiency and control. Some of the strategies employed in cell culture for the management of cell proliferation include among others the over-expression of important regulators of proliferation and apoptosis pathways such as growth factors and cell cycle genes.

*c*-myc in its role as a transcription factor that is involved in regulation of cell cycle and *bcl*-2 that is involved in cell survival are two prime candidates from a selection of genes, that regulate cell proliferation and cell death in such a manner as to consider the advantages of their introduction in cell lines (Hockenbery *et al.,* 1990; Singh, *et al*., 1995; Simpson *et al*., 1997).

As previously reported ectopic c-Myc over-expression has been successfully achieved in a CHO-K1 cell line (Ifandi and Al-Rubeai, 2003). The result was the development of a new cell line, namely cmyc-cho which exhibited a significant increase in growth rate and maximum cell number, in various cultures conditions. While interestingly, the very large percentages of apoptosis, as widely reported in previous studies, were not as profound in this cell line, nevertheless, apoptosis was the main form of cell death observed in the cultures. Therefore, the anti-apoptotic gene *bcl-2* was introduced in the cmyc-cho cell line, in order to reduce the apoptotic effect of c-Myc over-expression.

## 2. MATERIALS AND METHODS

Cmyc-cho cells were transfected with a *bcl-2* vector (pEFY28Abcl-2pGKpuro and a control empty vector (pEFY28ApGKpuro) (all plasmid were kindly donated by Dr David Huang, Melbourne, Australia), using a liposome mediated-transfection. Cells were grown in Ham's F12 supplemented with 5% FCS. Protein over-expression was determined using western blotting. All batch cultures were performed in triplicate.

## 3. RESULTS

Having characterized the influence of c-Myc over-expression on CHO-K1 cell growth morphology and death kinetics it was decided to introduce Bcl-2 on the cmyc-cho cell line, to address the apoptosis levels that the cell line was exhibiting. Following transfection, and having established stable clones, it was determined that c-Myc and Bcl-2 were stably over-expressed in the new cell line, cmyc-cho-bcl-2. Our results show that constitutive co-expression of c-Myc and Bcl-2 results in an increase of proliferation rate, with a marked decrease in the rate of decline for the cultures. Cmyc-cho-bcl-2 cells reached higher cell numbers and exhibited higher growth rates (Fig. 1) when compared to the control cell line, even under conditions of absent growth factors (serum). Although the cell numbers for the cmyc-cho-bcl-2 cultures were significantly higher in the presence of serum, the positive effect of Bcl-2 on cell death was more obvious in the cultures grown in the absence of serum.

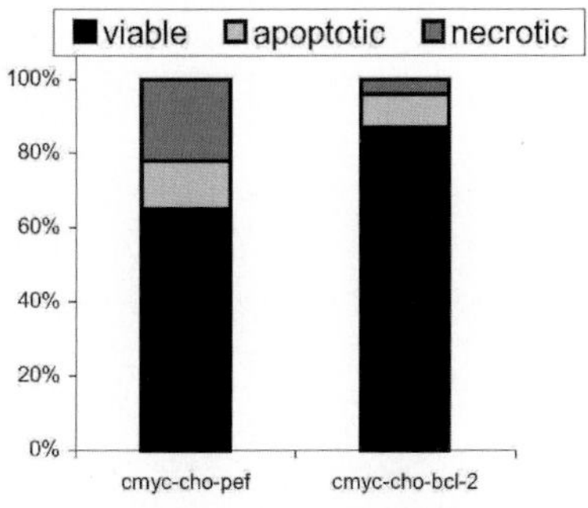


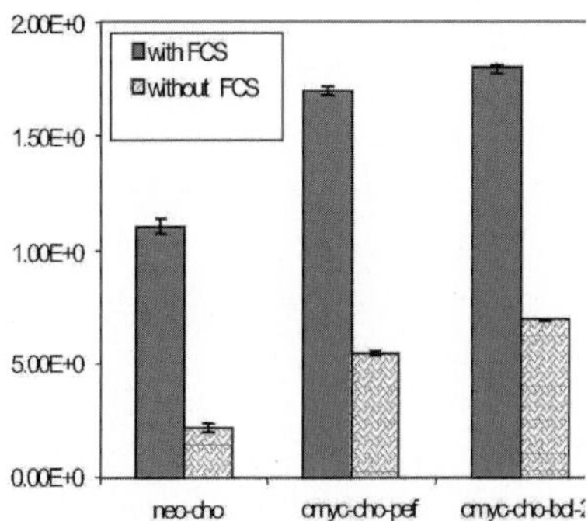


*Figure 1.* 1a) Viable cell density of neo-cho (control), cmyc-cho-bcl-2 (c-Myc and Bcl-2 over-expressed) and cmyc-cho-pef (c-Myc over-expressed) on day 6, in static batch culture (error bars represent standard deviation, n=3). 1b) Percentages of viable, apoptotic, and necrotic cells of cmyc-cho-pef and cmyc-cho-bcl-2 on day 9 of batch culture in the presence of FCS.

## 4. CONCLUSION

The results of this study have shown that the co-expression of c-Myc and Bcl-2 in CHO cells leads to the development of a cell line that comprises the robust nature of CHO-K1 with the high proliferative and transforming nature of c-Myc along with the anti-apoptotic function of Bcl-2. This cell line has enhanced survivability and it shows that Bcl-2 and c-Myc can cooperate in promoting cell yield and survival.

## REFERENCES

Hockenbery, D., Nunez, G., Milliman, C., Schreiber, R.D., and Korsmeyer, S.J., (1990). Bcl-2 is an inner mitochondrial membrane protein that blocks programmed cell death. *Nature*, **348**:334-336

Ifandi, V.; Al-Rubeai, M. Stable transfection of CHO cells with the c-myc gene results in increased proliferation rates, reduces serum dependency, and induces anchorage independence. Cytotechnology, 2003, 41, 1-10.

Simpson N, A. E. Milner, M. Al-Rubeai (1997) Prevention of hybridoma cell death by bcl-2 during sub-optimal culture conditions. *Biotechnology and Bioengineering* **54**, 1-16.

Singh, R.P., Al-Rubeai, M., and Emery, A.N., (1996). Apoptosis: exploiting novel pathways to the improvement of cell culture processes. *Genet. Eng, Biotech*, **16**: (4): 227-251.

# Ex Vivo Expansion of Hematopoietic Stem Cells Using Defined Culture Media

J. Hartshorn,[2] A.S.-H. Chan,[1] Z.-H. Geng,[2] L. Davis,[2] S. Luo,[2] J. Ni,[1] K. Etchberger,[2]

[1] *R&D Systems, Inc., Minneapolis, Minnesota, USA;* [2] *JRH Biosciences, Inc., Lenexa, Kansas, USA*

**Abstract:** *Ex vivo* expansion of hematopoietic stem cells (HSCs) is frequently exploited to overcome the limited availability of HSCs and progenitors for clinical applications. The use of conventional basal cell culture medium supplemented with bovine serum often has resulted in unpredictable HSC expansion and presents potential exposure to adventitious agents (AVAs). For this reason, we have investigated the development of defined cell culture reagents capable of supporting optimal HSC *ex vivo* expansion and differentiation. One of the developed formulations was capable of expanding and maintaining the CD34+ population to levels comparable or greater than that of commercially available media. In summary, preliminary tests of a panel of formulations have confirmed that CD34+ HSCs can be expanded and maintained in a defined medium for up to 14 days, and extensive proliferation and expansion into hematopoietic lineages are achievable with administered cytokine cocktails.

**Key words:** Hematopoietic stem cell, *ex vivo* expansion, CD34+, serum free cell culture media, defined cell culture media

# 1. INTRODUCTION

Hematopoietic stem and progenitor cells are frequently characterized by the surface expression of CD34. Down-regulation of CD34 has been shown to correlate with the loss of self-renewal property and reflect the differentiated fate of HSC in cell culture. Here, the growth and development of HSCs cultured in various serum free media were monitored by flow cytometry of the CD34 expression. The overall expansion capacity of the medium was determined by the fold expansion of total nucleated cells (TNC) and of CD34+ cells obtained after 14 days in culture.



*R. Smith (ed.), Cell Technology for Cell Products*, 221–224.

The benefits of using a serum free medium for the expansion of HSCs include minimizing risk of AVAs in addition to minimizing performance effects due to lot-to-lot variability in serum. Therefore, the use of serum free media for the expansion of CD34+ populations is the preferred method for *ex vivo* expansion of cells.

## 2. MATERIALS AND METHODS

### 2.1 Cells

Granulocyte colony stimulating factor (GCSF)-mobilized CD34+ cells were obtained from Cell Therapy Core at the University of Minnesota Cancer Center.

### 2.1 Cell Culture Media

Control medium (StemSpan #09600) was purchased from StemCell Technologies (Vancouver, BC). Cell culture test media were prepared by JRH Biosciences ImMEDIAte Advantage Program (Lenexa, KS).

### 2.2 Reagents

Low density lipoprotein (LDL, part number L7914) and human holo-transferrin (T0665) were purchased from Sigma Chemical Company (St. Louis, MO); human holo-transferrin (part number 4455) was purchased from Serologicals (Norcross, Georgia). Recombinant human SCF (rh-SCF), rhTPO, and Flt-3 (kit number SMPK-8); bovine serum albumin (BSA), and human holo-transferrin were provided by R&D Systems (Minneapolis, MN).

### 2.3 Cell Culture

Cell culture was performed at R&D Systems. Cells were seeded in wells of a 24-well tissue culture plate in medium containing 100ng/mL of each of the following cytokines: rhSCF, rhTPO, rhFlt-3. LDL was supplemented where appropriate. Cells were maintained for a total of 14 days, and fed on days 4, 8, and 12 of culture. On day 14, cells were counted to obtain the total nucleated cell (TNC) count.

### 2.4 Flow Cytometry

FACS analysis of the expanded cell cultures was performed at R&D Systems. On day 14, samples were labeled with CD34-conjugated phycoerythrin (PE) antibody (BDIS; San Jose, CA). Cells were analyzed by flow cytometry; data was acquired using a FACSCalibur and CellQuest software.

## 3. RESULTS AND DISCUSSION

A total of twenty-six different base formulations were screened. Results were calculated as the fold increases in total nucleated cells (TNC) from day 0 to day 14 as well as the fold expansions of CD34+ staining cells. One test formulation performed the best and was selected for further optimization.

Through screening the initial base formulations, it was learned that both albumin and transferrin were critical for growth and expansion of CD34+ stem cells in serum free media. Results from screening BSA lots indicated that one lot did not support TNC expansion and that another lot supported CD34+ proliferation equivalent to that in control medium. Human holo-transferrin also was tested for CD34+ expansion capability. Two lots supported equivalent or better specific expansion than the control medium. The best combination of BSA and transferrin yielded 70% greater CD34+ cell populations as compared to controls (Figure 1). Because of the improved selectivity of the desired cell populations, faster *ex vivo* expansion capabilities are achievable using this optimized medium.

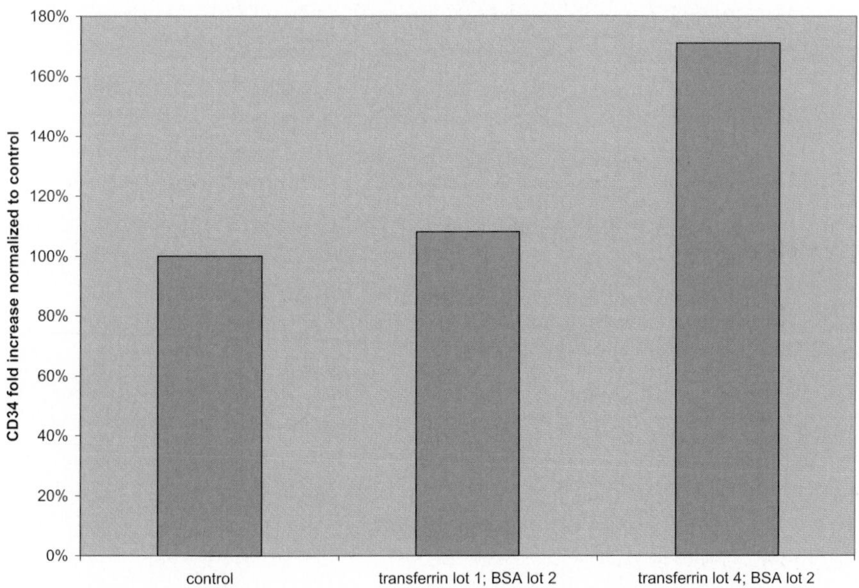


*Figure 1.* CD34+ expansion normalized to control in optimized medium.

## 4. CONCLUSIONS

A serum free medium formulation has been developed which supports the expansion of HSCs and, more importantly, maintains the CD34+ phenotype as well or better than the benchmark product. Optimal combinations of critical components, including BSA and transferrin, can yield populations of CD34+ cells up to 170% of that cultured in the benchmark product. This targeted expansion capability of the optimized medium will enable a faster scale-up of CD34+ cell populations for use in *ex vivo* therapies. Additionally, the use of a defined serum free medium significantly reduces risk of AVAs for the patients undergoing these therapies, as compared to the use of a basal medium supplemented with serum.

# Development of A Novel Serum-free Cryopreservative Solution

T. Toyosawa[1], M. Sasaki[2], Y. Kato[2], H. Yamada[2] and S. Terada[1]
[1]*Dept. of Applied Chem. and Biotech, University of Fukuii, 3-9-1 Bunkyo, Fukui 910-8507, Japan,* [2]*Technology Department, Seiren Co., Ltd.., 1-10-1 Keya, Fukui 918-8560, Japan*

**Abstract:** There has been a rapid development of cell therapy or regenerative medicine and of the production of bio-medicine with mammalian cell cultures. The cryopreservation of cells used for therapy is vitally important for a stable and continuous supply. Currently, fetal bovine serum (FBS) supplemented with 5-10% DMSO is predominantly used as a cryopreservative, but because bovine spongiform encephalopathy (BSE) or other infections are of serious concern, serum-free cryopreservative solutions are preferable. In this study, we found that the silk protein sericin was effective for the cryopreservation of mammalian cells and successively constructed a novel serum-free cryopreservative solution comprising 1% sericin, 0.5% maltose, 0.3% proline, 0.3% glutamine and 10% DMSO in PBS. The cells cryopreserved in this solution survived as well as in the conventional FBS supplemented with DMSO, and survival was better than in two commercially available solutions.

**Key words:** sericin, serum-free, cryopreservation, cryoprotectant, cryopreservative solution, viability, freeze, thaw, hybridoma, myeloma, BSE, DMSO

## 1. INTRODUCTION

Recently, mammalian cell cultures have been applied to many fields. A variety of bio-medicines, such as erythropoietin, are produced by mammalian cell cultures because of the bioactivity derived from glycosylation. Various cell types, including skin and lymphocytes, are cultured *ex vivo* for regenerative medicine and cell therapy. For these applications, cryopreservation is vitally important because cryopreservation of cells adds great flexibility to clinical transplant programs. Currently, fetal



*R. Smith (ed.), Cell Technology for Cell Products*, 225–231.

bovine serum (FBS) supplemented with dimethyl sulfoxide (DMSO) is extensively used as cell cryopreservative solution. However, FBS should be avoided and an alternative to FBS is eagerly desired. Bovine spongiform encephalopathy (BSE) and other infections, including viruses, are of serious concern and the cells to be used as transplants and in bio-medicines should not be infectious.

The present study aimed to develope a novel serum-free cryopreservative solutions by using sericin as a supplement. Sericin is a protein derived from cocoon silk. We previously reported that sericin accelerated the proliferation of mammalian cells (1) and an insect cell line and found that sericin successfully protected cells from toxicity of DMSO (S. Terada *et al.*, unpublished). Therefore, it can be expected that sericin could be used as an alternative to FBS and to develop serum-free cryopreservative solutions by preventing cell death induced by DMSO.

Previous studies revealed that several biomolecules including sugars (2, 3) and amino acids (4) were potent cryoprotectants when supplemented to cryopreservative solutions. Therefore, we searched for the best biomolecules and their optimum concentrations for the development of an effective serum-free cryopreservative solution.

## 2. MATERIALS AND METHODS

### 2.1 Cell lines and culture conditions

A murine hybridoma cell line 2E3-O (5) was cultured in ASF103 serum-free medium (Ajinomoto, Tokyo, Japan) and murine myeloma P3U1 and rat insulinoma RIN5F were cultured in RPMI 1640 medium supplemented with 10% FBS.

### 2.2 Determination of cell number and viability

The viable and dead cell densities were determined by the trypan blue exclusion method using a Neubauer improved hemocytometer.

### 2.3 Construction of cryopreservative solutions

Novel serum-free cryopreservative solutions were developed by supplementing cryoprotectants into phosphate-buffered saline (PBS) consisting of 0.137 M NaCl, 2.7 mM KCl, 1.4 mM $KH_2PO_4$ and 10 mM $Na_2HPO_4$. The cryoprotectants used were: DMSO, glucose and fructose (monosaccharides), maltose, trehalose, sucrose, lactose (disaccharides), maltitol, mannitol, xylitol, sorbitol, inositol (sugar alcohols), hydroxyethylcellulose, methyl cellulose, dextran, carboxymethyl cellulose, xanthane gum, sodium alginate (polysaccharides), proline, glutamine (amino acids) and sericin. As a positive control a conventional solution comprised of 90% FBS and 10% DMSO was used.

### 2.4 Cryopreservation

Cells in the logarithmic growth phase were collected and resuspended at $1*10^6$ cells / ml (except for Fig. 5; $5*10^5$ cells / ml) and 1 ml of each solution was dispensed into cryovials. The cryovials were ice-chilled for 5 minutes and then placed at -80°C in a deep freezer for 1 day. After thawing, the cells were cultured for 21 hours (Fig. 1) or 1 hour in their respective culture medium (Figs. 2, 3, 4) before counting viable cell numbers. For RIN5F, the viable cell number was immediately assessed after thawing.

## 3. RESULTS AND DISCUSSION

### 3.1 Sericin improved cell survival during cryopreservation

First of all, we investigated the effect of sericin as a cryoprotectant on cell survival during cryopreservation. Figure 1 shows that sericin improved hybridoma cell survival after crypreservation, as shown by a higher viable cell density in the sericin cryopreservative solution than in the basal solution, although the sericin solution was inferior to the FBS solution. This result indicates that sericin is an effective cryoprotectant and encouraged us to develop a novel serum-free, sericin-based cryopreservative solution supplemented with other cryoprotectants.

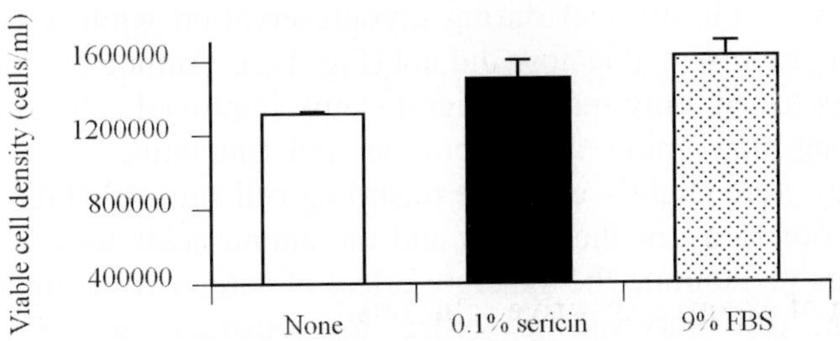


*Figure 1.* Protective effect of sericin on cell viability after cryopreservation
RPMI supplemented with 10% DMSO was used as a basal cryopreservative solution, shown as "none". Sericin or FBS was supplemented to the basal solution. The experiment was performed on a hybridoma cell line.

### 3.2 Effects of sugars and amino acids on cell survival

We tested the cryoprotective effect of monosaccharides, disaccharides and amino acids and the results are shown in Fig. 2.

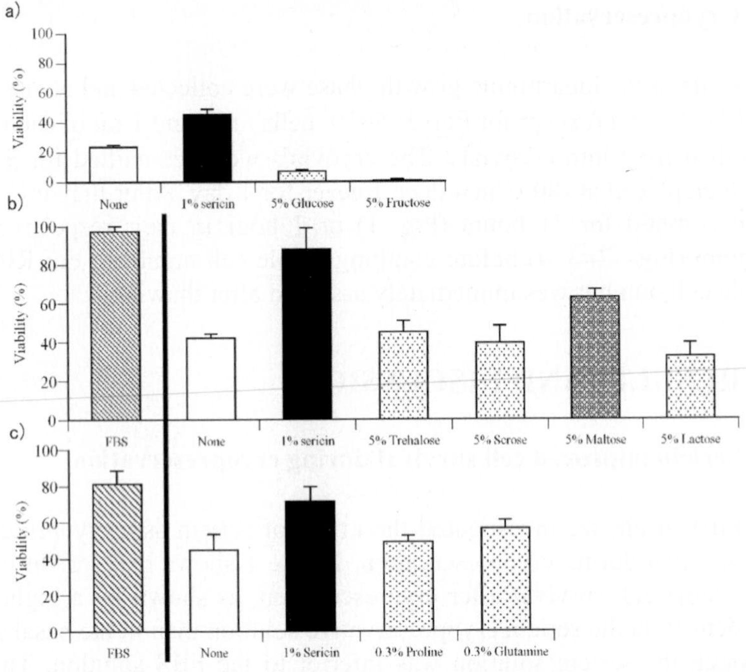


*Figure 2.* Effect of sericin, sugars and amino acids on cell survival
PBS supplemented with 10% DMSO was used as a basal cryopreservative solution, shown as none. Experiment was performed on a hybridoma cell line.

Sericin improved cell survival during cryopreservation while glucose, fructose, trehalose, sucrose and lactose did not (Fig. 1-a). Among the mono- and di-saccharides tested, only maltose significantly improved cell survival (Fig. 1-b). Among the amino acids, proline and glutamine were also examined and they were slightly effective regarding cell survival (Fig. 1-c). Sericin was superior to all of the sugars and the amino acids tested. This result prompted us to examine the synergic effect of sericin with sugars or amino acids on cell survival in order to construct an effective cryopreservative solution.

### 3.3 Synergic effect of sericin with other cryoprotectants on cell survival

We examined synergic effect of sericin with sugars or amino acids and the result is shown in Fig. 3. Of the disaccharides tested, maltose alone improved viable cell numbers, while the others did not (Fig. 3-a). Maltitol of the tested sugar alcohols and dextran, HEC, and methyl cellulose (MC) of the tested polysaccharide had synergically improved cell survival with sericin (Fig. 3-b). Subsequently, a screening of sugars and amino acids was performed to determine the optimum cryoprotectants in a cryopreservative solution including sericin that showed the highest viability. Out off all the solutions, the one containing sericin and maltose was superior (Fig. 3-c). Further experiments revealed that the optimum concentration of maltose was 0.5% (data not shown).

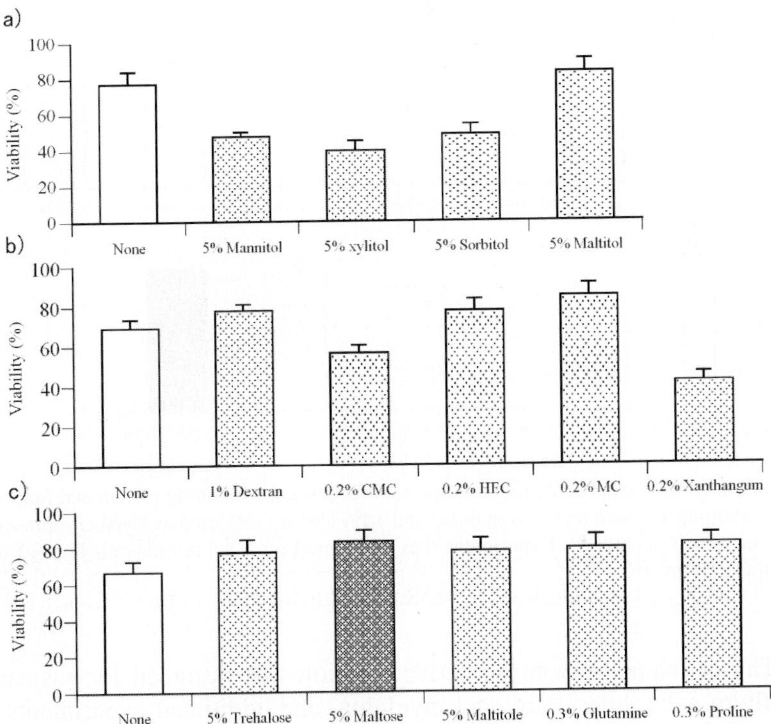


*Figure 3.* Synergic effect of sericin with sugar alcohols (a), polysaccharide (b) or sugars and amino acids (c) on cell survival during cryopreservation

PBS supplemented with 10% DMSO was used as a basal cryopreservative solution, shown as none. Experiment was performed on a hybridoma cell line.

### 3.4 Improvement of the solution containing sericin and maltose

The newly developed cryopreservative solution containing 1% sericin, 0.5% maltose and 10% DMSO was effective. To further improve this solution, we searched for another cryoprotectant to improve this solution. The supplementation of dextran to the solution slightly improved cell survival (Fig. 5-a), but the supplementation of proline or glutamine to the solution significantly improved cell survival and co-supplementation of both proline and glutamine showed the greatest improvement (Fig. 5-b).

From these results, a novel cryopreservative solution was designed; PBS consisting of 1% (v/w) sericin, 0.5% (v/w) maltose, 0.3% (v/w) proline, 0.3% (v/w) glutamine and 10% (v/v) DMSO.

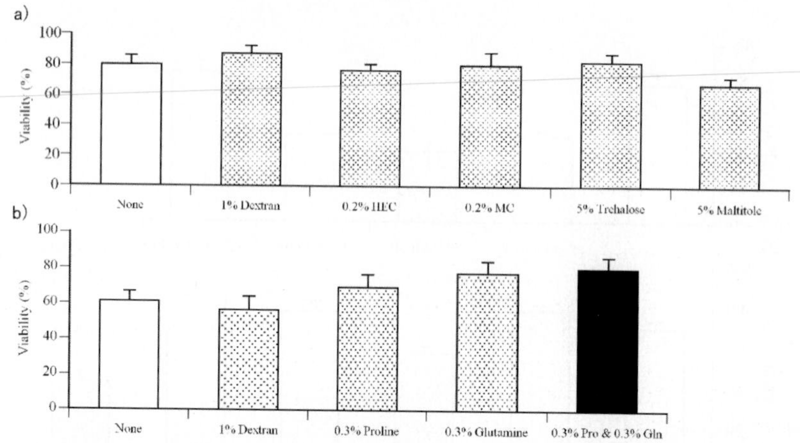


*Figure 4.* Cryoprotectant screening to improve the solution containing sericin and maltose PBS containing 1% sericin, 0.5% maltose and 10% DMSO was used as basal cryopreservative solution, shown as "none". Experiment was performed on a hybridoma cell line (a) or on a myeloma cell line (b).

The developed cryopreservative solution was designed for suspensions of culture cells, hybridoma and myeloma, and additional experiments were performed to see if this solution was effective for adhesion culture cells. For adhesion cells, rat insulinoma RIN5F were used. Figure 5 shows that the sericin-based solution was equal to the conventional FBS solution, and superior to the two commercially available solutions.

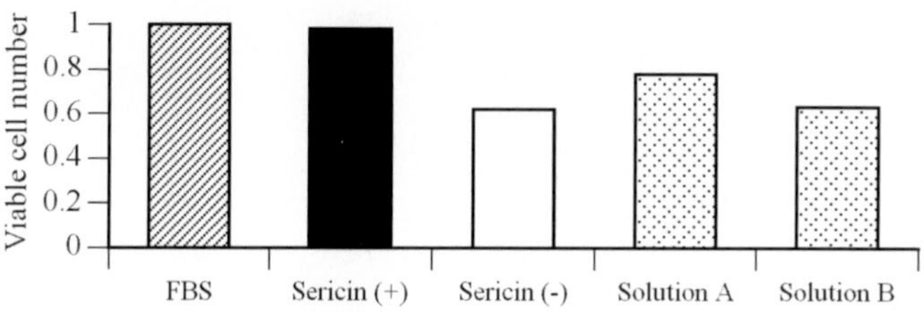


*Figure 5.* Performance of the designed solution
Viable cell numbers were standardized against viable cell number in FBS solution. FBS solution consists of 90% FBS and 10% DMSO. Sericin (+) solution was our designed solution. Sericin (-) solution was our designed solution without the sericin. Solution A and B were commercially purchased serum-free solutions.

# REFERENCES

(1) S. Terada, T. Nishimura, M. Sasaki, H. Yamada and M. Miki, 2002, Sericin, a protein derived from silkworms, accelerates the proliferation of several mammalian cell lines including a hybridoma, Cytotechnology. **40**: 3-12.

(2) C. Koshimoto and P. Mazur, 2002, The effect of osmolality of sugar-containing media, the type of sugar, and the mass and molar concentration of sugar on the survival of frozen-thawed mouse sperm, Cryobiology. **45**: 80-90.

(3) G. Erdag, A. Eroglu, R. Jefferey, Morgan and M. Toner, 2001, Cryopreservation of fetal skin is improved by extracellular trehalose, Cryobiology. **42**: 218-228.

(4) J. Kruuv and D. J. Glofcheski, 1992, Protective effect of amino acids against freeze-thaw damage in mammalian cells, Cryobiology. **29**: 291-295.

(5) S. Terada, E. Suzuki, H. Ueda and F. Makishima, 1996, Cytokines involving gp130 in signal transduction suppressed growth of a mouse hybridoma cell line and enhanced its antibody production, Cytokine. **8**: 889-894.

# Hepatic Cell Lines CultureD on Different Scaffolds and in Different Stages for Bioartificial Liver Systems

Kozue Kaito, Yumi Narita and Satoshi Terada
*Department of Applied Chemistry and Biotechnology, Faculty of Engineering, University of Fukui,* 3-9-1 Bunkyo, Fukui, Japan 910-8507

**Abstract:** Because of a lack of livers being donated for transplantation, the use of bio-artificial livers (BALs) using hepatic cells is a valid alternative. There are two major problems: one is the source of potent hepatic cells for BALs and the other is that BAL's working period is still too short because of apoptosis of the hepatic cells. In order to prolong the working period of BALs, we previously generated an anti-apoptosis hepatic cell line, Hep-bcl2, which maintained cell viability and improved albumin productivity for longer periods. In this study we tested the effect of collagen on the proliferation and liver function, albumin synthesis, of the anti-apoptosis hepatic cell line Hep-bcl2 to improve the culture. Hep-bcl2 cultured on collagen showed increased proliferation, while albumin productivity was decreased. This result implies that a two-step culture condition should be developed: First, to expand the pre-culture population, Hep-bcl2 should be cultured on collagen-coated dishes because of the rapid proliferation. Second, for BAL culture, Hep-bcl2 should be cultured without collagen because of higher liver function and slow proliferation, which would avoid cell death due to over-growth.

**Key words:** bio-artificial liver, apoptosis, hepatoma, HepG2, *bcl-2*, Hep-bcl2, proliferation, albumin, type collagen, adhesion, scaffold

## 1. INTRODUCTION

Liver transplantation is the most effective remedy in serious liver disease, but currently there is a critical problem since the number of donors is much smaller than the number of patients in need of a transplant. This is why



*R. Smith (ed.), Cell Technology for Cell Products*, 233–238.

bioartificial liver (BAL) support systems composed of artificial materials and living liver cells were developed. If a temporary replacement for a failing liver could be achieved, liver regeneration occurred in patients with liver failure and many patients could be saved.

Although a variety of BAL systems have been investigated by many researchers, their working period is too short to allow a full liver recovery in patients [1]. The strategies for improving BAL systems could be classified into two categories. The first is the development of a BAL module such as hollow fibers, radial flow, and so on. The second is the improvement of cells used in BALs, their viability and liver-specific function. We focused on the second strategy and tried to generate novel human hepatic cell lines that do not undergo apoptosis and introduced the anti-apoptotic gene *bcl-2* into hepatoma HepG2 cells. The generated cell line was named Hep-bcl2 and it maintained higher viability for a longer period and had improved liver-specific functions such as albumin productivity and metabolic activity (CYP1A1) [2]. In order to achieve improved BALs, hepatic cells should be cultured under the best possible conditions. For this purpose, we focused on collagen, an effective scaffold, during the pre-culture stage before putting BAL's module.

Collagen is one of the most ubiquitous proteins in the animal world and occupies one third of total protein in human. Collagen constitutes connective tissue and functions as a matrix between a large number of cells forming animal body for tissue morphology, influencing cell adhesion and proliferation and enhancing the repair response after various damages [3]. Recently, hepatic cells cultured between layers of collagen gel, termed collagen sandwich [4], have been applied to BAL systems.

In this study, we cultured hepatic cell lines on 35 mm dish coated with type I collagen and measured the liver-specific function through proliferation and albumin synthesis and compared it with the cells cultured on uncoated dishes.

# 2. MATERIALS AND METHODS

## 2.1 Cell lines and culture conditions

The human hepatoma cell lines used were HepG2 and Hep-bcl2. The latter was established by transfection of HepG2 with the vector BCMG-bcl-2-neo. The medium was Dulbecco's modified Eagle medium (DMEM) containing 0.2% sodium bicarbonate, 10 mM HEPES, 2 mM L-glutamine, 0.06 mg/ml kanamycin, and 10% fetal bovine serum. The cells were grown in 35 mm culture dishes (Sumitomo Bakelite, Japan) at 37℃ in humidified air containing 5% $CO_2$.

### 2.2 Morphology

Morphology of HepG2 and Hep-bcl2 cells was observed under phase-contrast microscopy (Olympus, Japan).

### 2.3 Measurement of growth curves

HepG2 and Hep-bcl2 cells were seeded at $3*10^4$ and $4*10^4$ cells / 35 mm dish, respectively, on wells coated with collagen or on uncoated wells. Cells were maintained in DMEM medium and cell number was assessed using a hemocytometer. Viable and dead cell densities were determined by the trypan blue exclusion method.

### 2.4 Determination of albumin productivity

The albumin concentration in the culture supernatant was determined by ELISA. The 96-well plate used for ELISA was first coated with goat anti-human albumin polyclonal antibody and blocked with skim milk. Subsequently, the standard wells were incubated with purified human serum albumin. The rest of the wells were incubated with the experimental samples. Finally, the wells were incubated with horseradish peroxidase-conjugated rabbit anti-human albumin polyclonal antibody and added of citric acid buffer containing *o*-phenylenediamine. The absorbance was read at 490 nm.

## 3. RESULTS

### 3.1 Morphology

In order to investigate the morphology of the hepatic cell lines HepG2 and Hep-bcl2 cultured on 35 mm dishes coated with collagen and uncoated dishes, the cells were observed at 100x magnification using a phase-contrast microscope. Morphologically, HepG2 (Fig. 1A) and Hep-bcl2 cells (Fig. 1C) on collagen appeared fuzzily and their hand with adhesion plaque extended between the cells, forming tough adhesion. On uncoated dishes, both HepG2 (Fig. 1B) and Hep-bcl2 cells (Fig. 1D) clearly formed spherical clusters probably due to a weak attachment of the cells to the culture dish. The marked difference between the morphological shape of HepG2 and Hep-bcl2 cells were not appeared. These observations showed that collagen coated dishes seemed to provide a tough adhesion for the cells so the area of dish could be used fully.

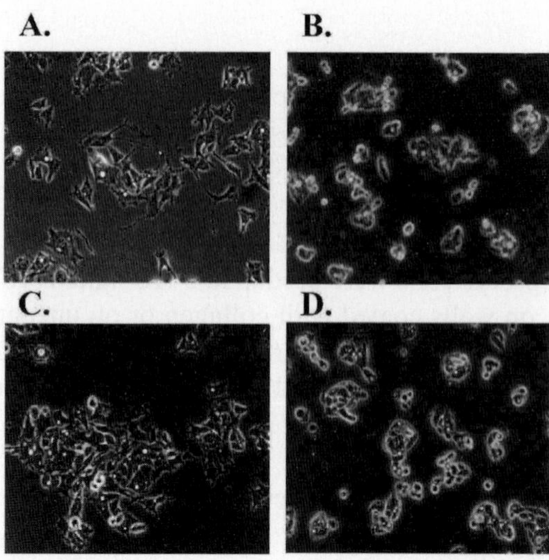


*Figure 1.* Phase-contrast microscopic observation of hepatic cell lines on collagen or on uncoated dishes. (A, B) HepG2; (C, D) Hep-bcl2. (A, C) The hepatic cell lines on collagen and (B, D) on uncoated. Cells were observed at 100x magnification.

### 3.2 Proliferation

To investigate whether culturing on collagen coated dishes altered the growth characteristics or not, the hepatic cell lines on the dishes coated with collagen and on the uncoated dishes were separately batch-cultured. As shown in Fig. 2A, viable cell density of HepG2 cells on uncoated dishes was not increased after 2 days because of a weak attachment of the cells to the culture dish, while HepG2 cells on collagen coated dishes showed an increase in viable cell number because of a tough adhesion induced by collagen. During the exponential growth phase, the growth rate of HepG2 cells on collagen was similar to that of untreated HepG2 cells. Untreated HepG2 cells started dying at day 8 because of a depletion of nutrients or growth factors, while HepG2 cells on collagen coated dishes seemed to have a prolonged life-span and higher maximum viable cell density. Hep-bcl2 cells also showed similar results (Fig. 2B). Proliferation of hepatic cell lines enhanced by collagen seemed to be dependent on cell adhesion immediately after seeding. These findings suggest that cells cultured on collagen might be useful at the precondition stage to rapidly expand the population of hepatic cells for BALs.

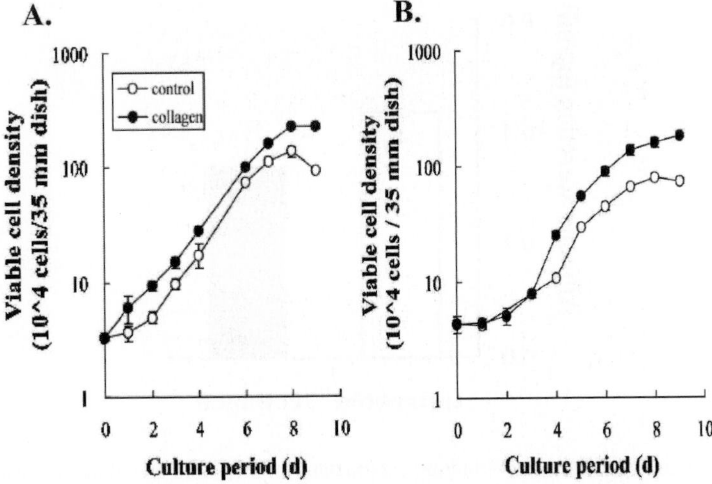


*Figure 2.* Proliferation of hepatic cell lines on collagen or untreated. (A) 3 * $10^4$ cells HepG2 were batch-cultured in 35mm dishes coated with collagen closed circle and on uncoated dishes (open circle for 9 days. (B) Similarly, 4 * $10^4$ cells Hep-bcl2 was batch-cultured. Viable cell densities were determined by the trypan blue exclusion method.

### 3.3 Albumin productivity

In order to estimate liver-specific functions of the hepatic cell line, human serum albumin concentrations in the culture supernatant were determined (Fig. 3). 0.44 and 0.34 μg/ml albumin was secreted by 3 days into the supernatant of Hep-bcl2 on collagen coated dishes and on uncoated dishes, respectively. The hepatic cell lines on uncoated dishes presented with increased albumin production compared to cells grown on collagen coated dishes. These results suggest that cells cultured without collagen might be useful to maintain liver-specific functions of hepatic cell lines in BALs. Albumin productivity per cell and per culture period (day) was calculated and shown in Table 1. The productivity per Hep-bcl2 cell was five times higher than that of HepG2 cells. In both cell lines, albumin productivity on untreated dishes was three times higher than that of cultures on collagen coated dishes.

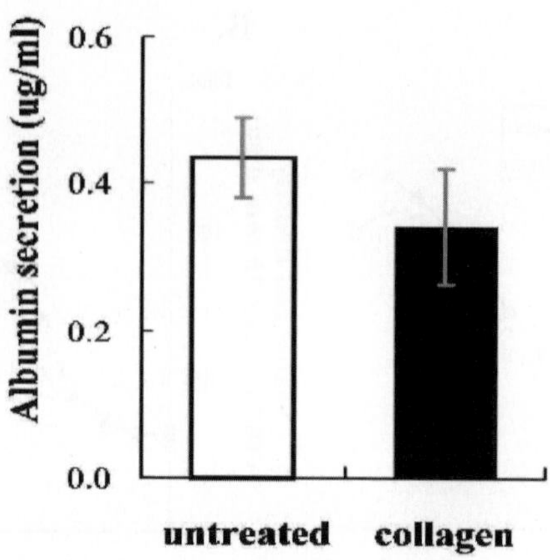


*Figure 3.* Albumin concentration of culture supernatants of Hep-bcl2 on collagen (closed bar) or uncoated dishes (open bar) at day3.

*Table 1* Calculated albumin productivity (pg/cell/day).

| | Uncoated | coated with collagen |
|---|---|---|
| HepG2 | 1.69 ± 0.14 | 0.55 ± 0.15 |
| Hep-bcl2 | 9.75 ± 1.22 | 2.87 ± 0.66 |

## 4. DISCUSSION

These results suggest that a two-step culture strategy for BAL systems would be effective. At the first pre-culture stage for expansion of the population of cells, Hep-bcl2 cells should be cultured on collagen-coated dishes because of the rapid proliferation. At the second stage within the BAL, Hep-bcl2 cells should be cultured without collagen because of higher liver function and slower proliferation that could reduce cell death due to over-growth.

## REFERENCES

1. Park, J. K. *et al.*, 2005, Bioartificial Liver Systems; culture status and future perspective, J. Biosci. Bioeng. **99(4)**: 311-319.
2. Terada, S. *et al.*, 2003, Generation of a novel apoptosis-resistant hepatoma cell line, J. Biosci. Bioeng. **95(2)**: 146-151
3. Pachenc,e J. M. *et al.*, 1996, Collagen-based device for soft tissue repair, J. Biomed Mater. Res. **33(1)**: 35-40.
4. Dun, n J. C. *et al.*, 1991, Long-term in vitro function of adult hepatocytes in a collagen sandwich configuration, **7(3)**: 237-245.

# Comparison of Flask and Bag Systems in Cultivation of Dendritic Cells for Clinical Application

L. Macke[b], B. Wrobel[a], W. Meyring[a], H.S.P. Garritsen[b], B. Wörmann[b], M. Rhode[a], W. Lindenmaier[a], K.E.J. Dittmar[a]

*[a]GBF, Department of Molecular Biotechnology, Braunschweig, Germany,[b]Department of Transfusion Medicine and Haematology/Oncology, Klinikum Braunschweig gGmbH, Germany*

**Abstract:** Dendritic cells (DC) represent the antigen-presenting cells which play a decisive role in the regulation of immune responses. Large numbers of monocyte-derived DC can be generated from leukapheresis cells for clinical applications by positive selection of CD14+ cells or by negative selection (removal of CD2+ and CD19+ cells). The positive selection showed a lower yield but higher purity of monocytes. Additionally, we compared the culture flask system (as "gold standard") with our integrated-closed-bag system designed for clinical use. Bags resulted in slightly lower yields of viable cells on day 7 compared to flasks. Structural and functional properties of DC generated by these cultivation conditions showed no significant difference. A high donor variability was observed. In regard to clinical application and GMP compliance CD14+ selection in combination with our closed-bag system is a promising approach to generate defined dendritic cells.

**Key words:** bag, cell selection, cell separation, dendritic cell, flask, human, immunology, immunotherapy, leukapheresis, system, vaccines.

# 1. INTRODUCTION

The induction of an immune response is crucial dependent on the activation status of the dendritic cells (DC). Interaction of antigen presenting cells with T-lymphocytes determines an immunogenic or tolerogenic outcome of immune response. Therefore the choice of antigen presenting cell is critical. Monocyte derived dendritic cells are preferentially used for immunotherapy. Several cultivation protocols have been established to



*R. Smith (ed.), Cell Technology for Cell Products*, 239–242.

generate sufficient DC numbers from leukapheresis. In previous clinical trials different isolation and cultivation methods had been used for DC based vaccines resulting in a poor comparability. However, due to this lack of standardization there is an urgent need for standardized production protocols.

## 2. COMPARISON OF SELECTION METHODS FOR GENERATION OF DC

### 2.1 Selection strategies

Positive (pos.) selection (CliniMACS®) and negative (neg.) selection (Isolex 300i®) were compared in one large scale leukapheresis from donor LK85. The pos. technique selects $CD14^+$-cells as starting population for DC generation. The neg. selection procedure is based on the depletion of $CD2^+$/$CD19^+$-cells after enrichment of leukocytes by a ficoll gradient. In this setting the $CD2^-$/$CD19^-$-population is used as starting population for DC generation. Results from donor LK85 are listed in Table 1 (see appendix) for both separation methods. Starting with the same leukocyte sample, the recovery of CD14+ cells was 80% with pos. selection vs. 94% with neg. selection. In contrast, with only a contamination of 1%, pos. selected monocytes were essentially pure compared to only 55% CD14+ cells in the neg. selection.

### 2.2 Cultivation of selected monocytes

Both starting populations were cultivated in the presence of GM-CSF/IL-4. On day 5, maturation was induced by addition of TNF-$\alpha$, IL-1$\beta$, IL-6 and $PGE_2$. After cell harvest on day 7, yield of DC from neg. selection was 34% (appendix: Figure 2, right). The proportion of mature DC was independent of the selection process with 91% and 94% of CD83+ cell, respectively. The cultivation of pos. selected cells resulted in a yield of 11% for this donor, but with only 0.5% contaminating lymphocytes, whereas the lymphocyte population in the neg. selected cultivation was still 27.0% (appendix: Figure 2, left). We found high induction of chemotactic cytokines for monocytes and T-lymphocytes in the neg. selected culture, especially RANTES, MCP-1, IP-10, MIG and IL-10 (data not shown)

## 3. COMPARISON OF DC GENERATION FROM CD14 MONOCYTES IN BAG VS. FLASK SYSTEMS

Positively selected monocytes were used for comparison of DC generation in bags from 4 different manufacturers vs. flasks. Generation of DC from the donors LK68, LK85 and LK101 show a high variation in yield of viable cells on day 7, i.e. between 11% and 33% in flask cultivation. Monocyte yield after bag cultivation from different manufacturers varied from 1.4% to 7.6%. Cultivation bags from M4 suitable for the closed system were optimized and give results from 4.4% to 27.8% (data not shown). The maturation of DC on day 7, indicated as percentage of CD83+ cells was measured by flow cytometry. In bag cultivation, the proportion of mature DC was approx. 84%, whereas in flasks variations could be observed between the different donors (appendix: Figure 2, right). The stimulatory capacity of peptide loaded DC was additionally tested in an IFN-γ ELISpot assay. Tyrosinase pulsed DC from day 7 either cultured in bags or in flasks were capable of stimulating a tyrosinase specific CTL clone (data not shown).

## 4. DISCUSSION

The critical point of DC generation for clinical trials is the standardization of isolation and cultivation protocols. Since we are working on the development of an integrated closed bag cultivation system suitable for cGMP compliant production of defined autologous therapeutic DC vaccines, we investigated the questions: Should pos. or neg. selection for monocytes be used? Are bags suitable cultivation containers to generate reliable DC numbers? In our study the neg. selection resulted in higher yield on day 7 but lower purity of monocytes compared to pos. selection. Contaminating lymphocytes seemed to interact with DC during cultivation. The effects of this interaction are difficult to control, therefore for clinical use, the high purity of DC of pos. selected cells is advantageous. DC generated in bags or flasks showed no significant difference in maturation and stimulatory capacity at day 7. The yield of viable DC was higher in flasks, but sufficient numbers of DC for the proposed clinical protocol were generated from bag cultivation. The completely closed-bag system in combination with pos. selection is a promising approach to generate dendritic cells for clinical applications under standardized conditions.

# 5. APPENDIX

*Table 1.* Overview of results from selection processes.

| *Leukapheresis LK85* | *positive selection* | *negative selection* |
|---|---|---|
| Starting cell number | $10.0x10^9$ | $6.3x10^9$ |
| CD3+ | 72% | |
| CD19+ | 7% | |
| CD14+ | 16% | |
| Theoretical number of CD14+ | $1.6x10^9$ | $1.1x10^9$ |
| total number of isolated cells | $1.3x10^9$ | $1.8x10^9$ |
| purity CD14+ | 99% | 55% |
| recovery CD14+ | 80% | 94% |

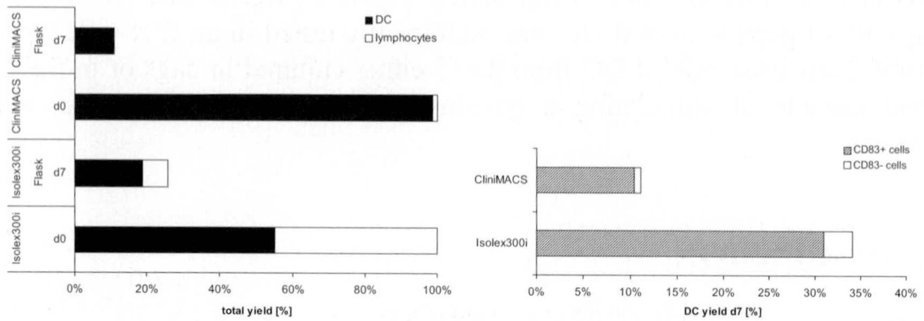


*Figure 1.* Characteristics of cultured DC: left, total yield of viable cells at the indicated days of cultivation; right, yield of viable DC on day 7.

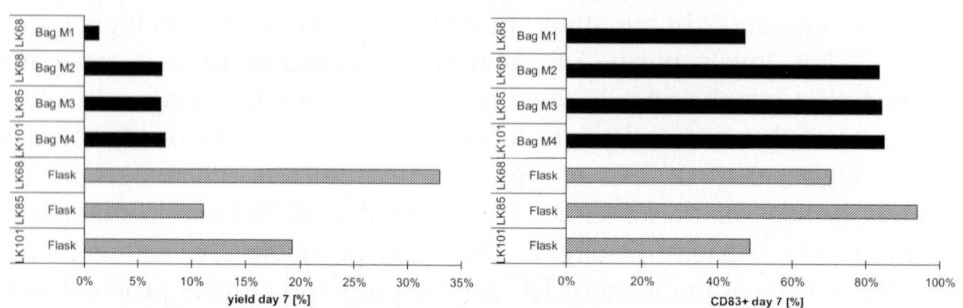


*Figure 2.* Cultivation details of DC: left, yield of viable DC on day 7; right, maturation of viable DC on day 7 measured as CD83+ cells Different bag producers are indicated as manufacturer M1 – M4.

# T-Lymphocytes Transduced with SFCMM-3 Vector in a Closed System

Claudia Benati, Roberto Sciarretta Birolo, Simona La Seta Catamancio, Marina Radrizzani, Cecilia Sendresen, Salvatore Toma
*MolMed S.p.A., Milano, Italy*

**Abstract:** Safety, yield and scalability represent critical issues in production of gene-modified cells (GMC) for clinical applications. Currently open systems, based on culture flasks, are used for production of TK-transduced T cells. Such systems may be used, without significant risk of contamination, only if limited numbers of cells (up to 108-109) are required. Thus a method for cell manipulation and culturing in a closed system has been developed.

Lymphocytes are cultured in bags, activated with OKT3, then transduced with the vector SFCMM-3, carrying the HSV-TK and ΔLNGFR genes. After a few days, lymphocytes are collected and the percentage of transduced cells is evaluated by FACS analysis. Transduced cells are sorted by antibody conjugated magnetic beads and then expanded in culture. The full process lasts 10 days. The following instruments are used: CytoMateTM Cell Processing System (Nexell Therapeutic Inc.), for cell thawing, washing and harvest, and IsolexTM 300 Magnetic Cell Separation System (Baxter), for selection of the transduced cells.

The closed-system is very efficient and the cell yield is higher than that obtained in the open-system (final cell number/initial cell number: 0.64 versus 0.39). A detailed analysis of the process indicates that the steps in which the closed system shows major advantages are transduction (23% versus 17%) and selection (recovery of the cells 67% versus 36%). The purity of the final product is very similar and shows the homogeneity of the final population of GMC (94% and 95% LNGFR positive cells, respectively). Only the proliferation rate is lower in the closed than in the open system, especially post-transduction (2.5 versus 4.7) and post-selection (3.6 versus 5.6). However, the development of cell culturing using different bags or different culture media could improve cell growth.

These results demonstrate the feasibility of a method for transduction and selection of GMC in a closed system that is safe and efficient.

**Key words:** T-Lymphocyte, transduction, closed system



*R. Smith (ed.), Cell Technology for Cell Products*, 243–250.

## 1. INTRODUCTION

Safety, yield and scalability represent critical issues in the production of gene-modified cells (GMC) for clinical applications. Current processes for production of TK-transduced T cells used in suicide gene therapy protocols in the context of allogeneic bone marrow transplantation are open systems, based on culture flasks.

Aim of the present work was the development of a new method for cell manipulation and culturing in a closed system.

## 2. MATERIALS AND METHODS

### 2.1 Vectors

Transductions were performed with the retroviral vector SFCMM-3 (S. Verzeletti *et al.*, Human Gene Therapy 9:2243, 1998) or with a variant containing a single base, silent mutation within the HSV-TK gene (F. Salvatori *et al.*, manuscript in preparation). SFCMM-3 vectors carry both the suicide gene HSV-tk and the marker gene DLNGFR.

### 2.2 Transduction and Culture Process

Schematic representations of the current open system process and the new closed system process for production of TK-transduced T cells are shown in Tables 1 and 2, respectively.

In the experiments with the closed system, human peripheral blood lymphocytes (PBL) were washed by CytoMate$^{TM}$ device, then cultured at an initial cell concentration of $10^6$ cell/ml in X-VIVO 15 + 3% human plasma + 2 mM glutamine + 600 IU/ml IL2 with soluble anti-CD3 mAb (OKT3, 30 ng/ml).

At day 2, OKT3-activated cells were transduced with the SFCMM-3 vector in RetroNectin® coated bags. Cell density was $1x10^6$ cells/ml of retroviral supernatant diluted 1:2 with medium, corresponding to an MOI of 0.5-2. The day after, cells were washed and seeded at $0.2x10^6$ cells/ml of culture medium in VueLife$^{TM}$ Culture Bags.

At day 6, the transduction efficiency was determined by LNGFR immunofluorescence and FACS analysis; the cells were immunomagnetically selected using anti-LNGFR mAb 20.4 (1 μg/$20x10^6$ cells) and Dynabeads anti-Mouse IgG ($5x10^6$ beads/$10^6$ positive cells). Results were expressed as % of cells positive for LNGFR.

At day 8, the beads were removed by magnet separation and the number of cells recovered was determined by trypan blue cell counting. Cells were then seeded at a density of $0.2x10^6$ cells/ml.

The purity was evaluated at day 10 and day 14 by LNGFR immunofluorescence and FACS analysis.

At the end of the process, cells were extensively washed, then resuspended in infusion medium (saline solution containing 4% human serum albumin) at a cell density of $10\text{-}25x10^6$ cells/ml. Aliquots were withdrawn for quality control analysis, while the infusion bag was stored at 4°C for stability analysis.

The entire process was performed using VueLife™ Culture Bags of different size depending cell numbers. In the closed system, CytoMate™ Cell Processing System (Nexell Therapeutic Inc.) (Figure 2) was used for cell thawing, washing and harvest steps and the Isolex™ 300 Magnetic Cell Separation System (Baxter) (Figure 3) for selection of the transduced cells and removal of magnetic beads from positive selected cells.

The connections between different bags were performed in sterile conditions using a Terumo Sterile Connecting Device.

# 3. RESULTS

## 3.1 Production of TK-transduced T cells in a closed versus a open system: cell expansion

Thirty independent open system experiments were conducted with peripheral blood lymphocytes (PBL) from 12 different donors, and 9 closed system experiments were performed with PBL from 9 donors. Mean starting cell numbers were 3.6x109 (range: 3.0-5.5x109) for open system experiments and 1.7x106 (range: 1.0-3.1x109) for closed system experiments, respectively.

Experiments performed in the closed and the open system were compared step by step for cell expansion (Figure 1).

Relative cell numbers (mean ±sd) represent ratios between viable cell numbers (determined by trypan blue cell counting) at the indicated days.

No statistically significant differences were detected between the closed and the open system ($p>0.01$, T test, for all culture phases).

The production of TK-transduced T cells in the closed system allowed consistent cell expansion during all the culture phases between 3 and 14 days, similar to the open system process.

### 3.2 Production of TK-transduced T cells in a closed versus a open system: transduction and selection efficiency

Experiments performed in the closed and the open system were compared step by step for transduction efficiency, selection yield and final purity (Figure 2).

% LNGFR+ cells before selection indicates the percentage of transduced cells, genetically modified with the SFCMM-3 vector, as determined by immunofluorescence and FACS analysis at day 6.

% selection yield is the ratio between the number of cells collected after the selection step and the number of LNGFR+ cells before selection.

Final % LNGFR+ cells indicates the percentage of genetically modified cells in the final product, as determined by immunofluorescence and FACS analysis at day 10-14.

No statistically significant differences were detected between the closed and the open system ($p>0.01$, T test, for all parameters).

The production of TK-transduced T cells in the closed system allowed consistent transduction efficiency, selection yield and final product purity, similar to the open system process.

### 3.3 Production of TK-transduced T cells in a closed system: final product stability

At the end of the closed system production process, the stability of the final product was evaluated (Figure 6). The final bag, containing TK-transduced T cells in infusion buffer, was incubated at 4°C. At the indicated time points, cell number and viability were determined by trypan blue cell counting.

No significant decrease in cell viability was observed until 72 hours ($p>0.01$, T test, for all time-points respect to T0 value). At 120 hours, cell viability dropped to 66%, which is below the minimum value (70%) required for infusion in patients, according to current regulatory guidelines for gene therapy and cell therapy products.

Thus, TK-transduced T cells produced in closed system are stable at 4°C for up to 72 hours.

## 4. CONCLUSIONS

In summary, these results indicate that the closed system process for production of TK-transduced T cells gives similar results in comparison to the open system process currently used for clinical preparations: no significant differences were detected in terms of cellular expansion, efficiency of transduction, selection yield and final product purity.

TK-transduced T cells produced by the closed system process are stable for up to 72 hours when stored at 4°C without significant decrease in cell viability, which is an important parameter for product release for infusion.

The shift from an open to an easily applicable closed system for production of TK-transduced T cells represents a relevant improvement in terms of product safety.

Moreover, it is a necessary pre-requisite for the scale-up of the process. In fact, in our experience an open system may be used, without significant risk of contamination, if number of cells are required up to9x109 final cells, but it could be very critical for the production of higher cell numbers.

Finally, the possibility to store the final product for up to 3 days at 4°C before infusion could facilitate the clinical application of TK-transduced T cells.

*Table 1.* Production of TK-transduced T cells in open system.

| Step | Day | Process step and conditions |
|---|---|---|
| 1 | 0 | Thawing and stimulation<br>106 cells/ml; RPMI 1640 + 2mM glutamine +3% autologous plasma + 600 IU/ml IL-2 +30 ng/ml OKT3T ; 162 flasks |
| 2 | 2 | First transduction round<br>5x106 cells/ml; SFCMM-3 supernatant; 4 mg/ml protamine; Spinoculation: T75 flasks, 2 hours, 1000xg; Seeding as at day 0, w/o OKT3 |
| 3 | 3 | Second transduction round<br>5x106 cells/ml; SFCMM-3 supernatant; 4 mg/ml protamine; Spinoculation: T75 flasks, 2 hours, 1000xg; Seeding as at 0.5x106 cells/ml, w/o OKT3 |
| 4 | 6 | LNGFR+ cell selectionEvaluation of LNGFR expression (IF + FACS)<br>Incubation with anti-LNGFR mAb (20 mg/108 total cells) (2x107 cells/ml in 250 ml tubes); Washing; Incubation with Dynabeads anti-mouse IgG (5x106/106 LNGFR+ cells) (2.5x107 cells/ml in IsolexTM tube); Magnet: IsolexTM 300Cell density: 0.5 x106 cells LNGFR + /ml |
| 5 | 8 | Dynabeads removal Magnet:MPC-1<br>Seeding as at 0.5x106 cells/ml, w/o OKT3 |
| 6 | 10 | Cell harvestConcentration and washing<br>Filling for infusion:10-50x106 cells/mlsaline solution + 4% HSAtransfer bag |

*Table 2.* Production of TK-transduced T cells in closed system.

| Step | Day | Process step and conditions |
|---|---|---|
| 1 | 0 | Thawing and stimulation<br>Cell washing using CytoMateTM<br>X-VIVO 15, + 3% human plasma + 2 mM glutamine + 600 IU/ml IL2; Stimulation: 30 ng/ml of anti-CD3 Ab |
| 2 | 2 | Transduction cycle<br>Bags coated with RetroNectin®Cell; washing using CytoMateTM; Final cell density: 1x106 cells/ml sup diluted 1:2 with medium; Viral supernatant (SFCMM-3) |
| 3 | 3 | Cell washing using CytoMateTM and seeding<br>Final cell density: 0.25+0.05 x106 cells/ml in culture medium |
| 4 | 6 | LNGFR+ cell selection using CytoMateTM and IsolexTM 300 SA<br>40x106 cells/ml in selection buffer + 0.5% human Ig<br>15' at room temperature in constant agitation; 20x106 cells/ml in selection buffer + 1mg of anti-LNGFR Ab/5x106 cells 20' at room temperature in constant agitation, washing 20-40x106 cells/ml in selection buffer + Dynabeads anti-Mouse IgG (5x106 beads/106 of positive cells); Selection using IsolexTM 300 SA<br>Cell seeding after selection<br>Cell density: 1 x106 cells/ml |
| 5 | 8 | Dynabeads removal using IsolexTM 300 SA<br>Cell seeding after beads detachment; Cell density: 0.25+0.05 x106 cells/ml |
| 6 | 10 | Cell harvestConcentration and washing<br>CytoMateTMFinal cell density: 10-25x106 cells/mlInfusion medium: saline solution + 4% HSA |

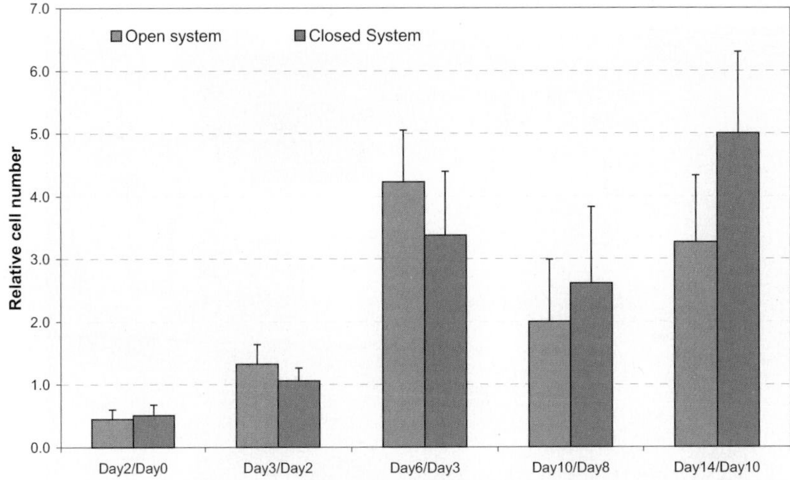


*Figure 1.* Cell expansion.

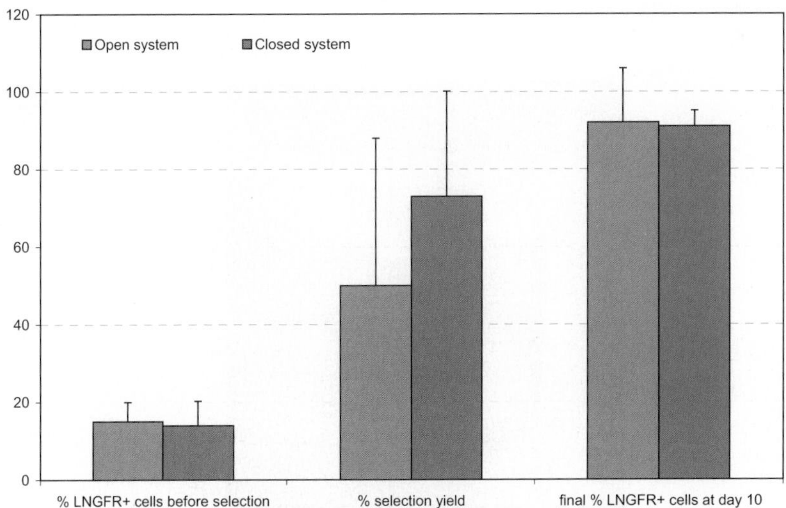


*Figure 2.* Transduction and selection efficiency.

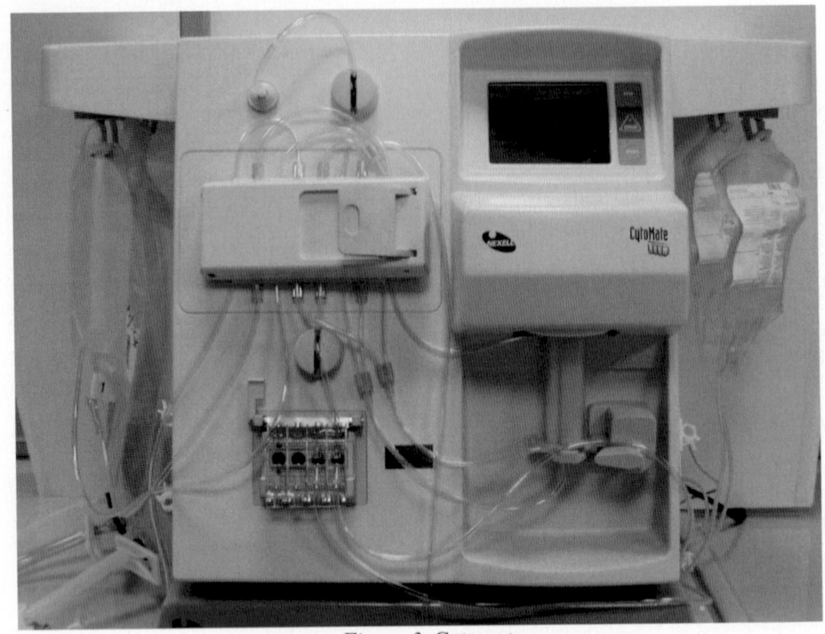

*Figure 3.* Cytomate.

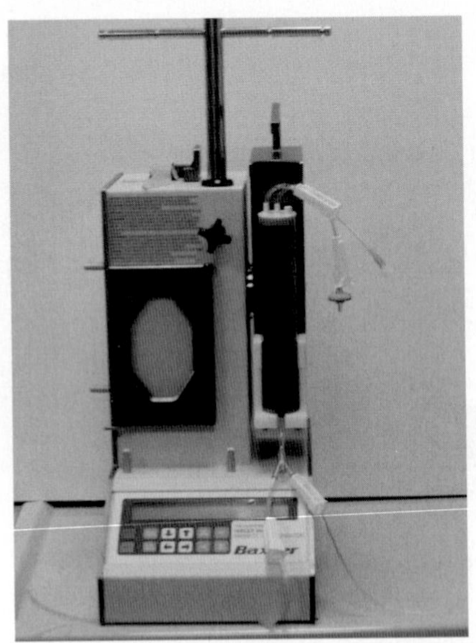

*Figure 4.* Isolex 300A.

# Influence of Down-regulation of Caspase-3 and CASPASE-7 by Sirnas on Sodium Butyrate-induced Apoptotic Cell Death of Chinese Hamster Ovary Cells Producing Thrombopoietin

Yun Hee Sung, Soon Hye Park, Sung Hyun Kim, Gyun Min Lee
*Department of Biological Sciences, KAIST, 373-1 Kusong-Dong, Yusong-Gu, Daejon 305-701, Republic of Korea*

**Abstract:** Sodium butyrate (NaBu) can enhance the expression of foreign protein in recombinant Chinese hamster ovary (rCHO) cells, but it can also inhibit cell growth and induce cellular apoptosis. Thus, the beneficial effect of using a higher concentration of NaBu on foreign protein expression is compromised by its cytotoxic effect on cell growth. To overcome this cytotoxic effect of NaBu, we down-regulated caspase-3 by expressing caspase-3 siRNA. However, it was not sufficient for the inhibition of apoptotic cell death induced by NaBu. When subjected to NaBu treatment, the rCHO cells expressing caspase-3 siRNAs compensated for the lack of caspase-3 by increase of active caspase-7 level.

For the successful inhibition of apoptosis of rCHO cells, we down-regulated both caspase-3 and caspase-7 using the siRNA expression vector system. The rCHO cell line (FR26 cell) with reduced levels in both caspase-3 and caspase-7 was developed by transfecting F21 cells with the caspase-7 siRNA expression vector.

**Key words:** CHO cell, Cell Engineering, Anti-apoptosis, Productivity, siRNA, Caspase 3,7

## 1. INTRODUCTION

Active caspase-7 has been shown to be associated with the mitochondria and the Endoplasmic reticulum (ER) membranes, whereas active caspase-3 remains cytosolic.1 Although they act on distinct substrates in different cellular compartments2, their activations were interconnected during the



*R. Smith (ed.), Cell Technology for Cell Products*, 251–258.

apoptosis induced by NaBu addition. They are highly homologous and structurally alike3 and both of them prefer DEXD-based cellular substrates during apoptosis. The hTPO is a potential therapeutic glycoprotein for the amelioration of thrombocytopenia caused by chemotherapy, irradiation and bone marrow transplantation.4

In this study, we downregulated both caspase-3 and caspase-7 using the siRNA expression vector system for the successful inhibition of apoptosis of rCHO cells producing human TPO (hTPO). We evaluated the effect of downregulation of caspase-3 and caspase-7 on inhibition of NaBu-induced apoptosis, enhancement of culture longevity, and hTPO productivity.

## 2. RESULTS

### 2.1 Selection of Caspase-7 SiRNA Target Site

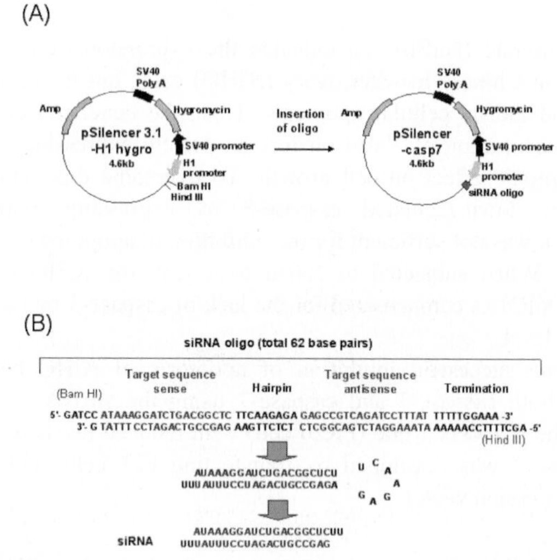


*Figure 1.* (A) The procedure of vector construction and (B) the expression of caspase-7 siRNAs in cells. Amp, ampicillin resistance; Hygro, hygromycin resistance. pSilencer 3.1-H1 hygro contains H1 RNA polymerase III promoter for the expression of siRNAs and hygromycin resistant gene for the antibiotic selection of transfected cells.

## 2.2 Development of Cell Line with Reduction in Both Caspase-3 and Caspase-7 Levels

The rCHO cells expressing a low level of caspase-3 (F21 cells) were transfected with pSilencer-casp7 vector containing siRNA oligo of target site R and cultivated in the selection medium with hygromycin, as described earlier. The C1 cells were R-2-3-2 cells transfected with two negative control vectors, pSuper-neo and pSilencer 3.1-H1 hygro. The C2 cells were F21 cells transfected with a negative control vector, pSilencer 3.1-H1 hygro. To screen the clone expressing low levels of caspase-3 and caspase-7, 29 hygromycin-resistant clones were assayed through Western blot analysis. FR26 cell clone showed the significant low levels in caspase-7 as well as caspase-3 and was selected for further study (Fig. 2B). C2 cells maintained a low level of caspase-3 during hygromycin selection (Fig. 2B).

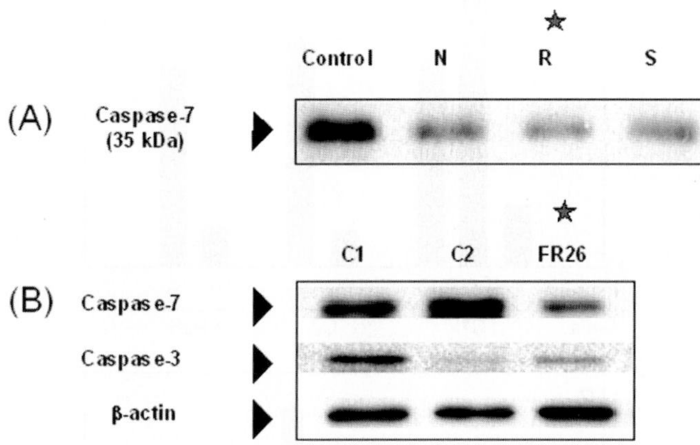


*Figure 2.* Selection of siRNA oligo sequence with reduced caspase-7 level by Western blot analysis. (B) Reduced caspase-3 and caspase-7 level in selected clone(FR26) relative to controls by Western blot nalysis. C1,containing null vectors of caspase-3 and 7; C2,containing caspase-3 siRNA and caspase-7 null vector ; FR26,containing caspase3 and 7 siRNAs.

## 2.3 Caspase-3/7 Activity

Figure 3A shows relative caspase-3/7 activity in the cells sampled just before or 5 days after replacement of serum-medium with serum-free medium (SFM-YH). The relative caspase-3/7 activity was estimated as a ratio of the activity in cells just before or after medium replacement to that in C1 cells just before medium replacement. In serum-free medium, C2 and FR26 cells exhibited considerably lower activities compared to that of C1 cells. It indicates that expression of caspase-3 siRNAs or expression of both caspase-3 and -7 siRNAs decreased caspase-3/7 activities in serum-free medium.

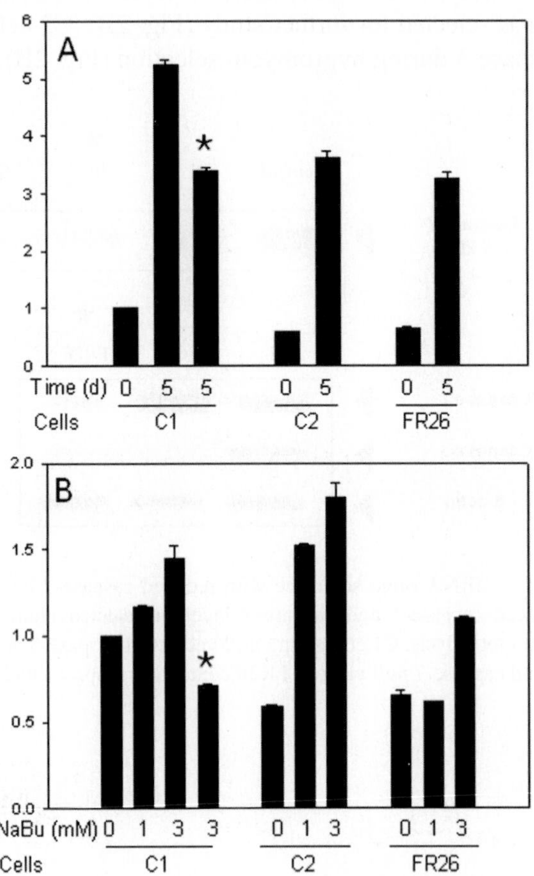


*Figure 3*. Relative Caspase 3,7 activity in C1, C2 and FR26 cells. (A) culture in serum-free media and (B) 0,1,3 mM NaBu treatment. (*) caspase inhibitor was treated as a control of assay.

Figure 3B shows relative caspase-3/7 activity in the cells sampled just before or one day after NaBu addition. The relative caspase-3/7 activity was estimated as a ratio of the activity in cells just before or after NaBu addition to that in C1 cells just before NaBu addition. In C2 cells, the caspase-3/7 activity in presence of NaBu was not decreased by caspase-3 siRNA expression. This may be caused by compensation of caspase-7 for the lack of caspase-3 activity in the presence of NaBu, as shown in our previous study (Sung and Lee, 2005). However, in FR26 cells, it was reduced by expression of both caspase-3 and -7 siRNAs. This suggests that downregulation of both caspase-3 and -7 by siRNAs can efficiently inhibit total enzymatic activity of caspase-3 and -7 in the presence of NaBu.

## 2.4 Cell Culture with NaBu Addition

In all cell lines, the highest hTPO concentration was achieved in the culture with 1 mM NaBu. This was resulted from enhanced $q_{hTPO}$ by NaBu addition. Compared to cultures without NaBu addition, more than 3-fold

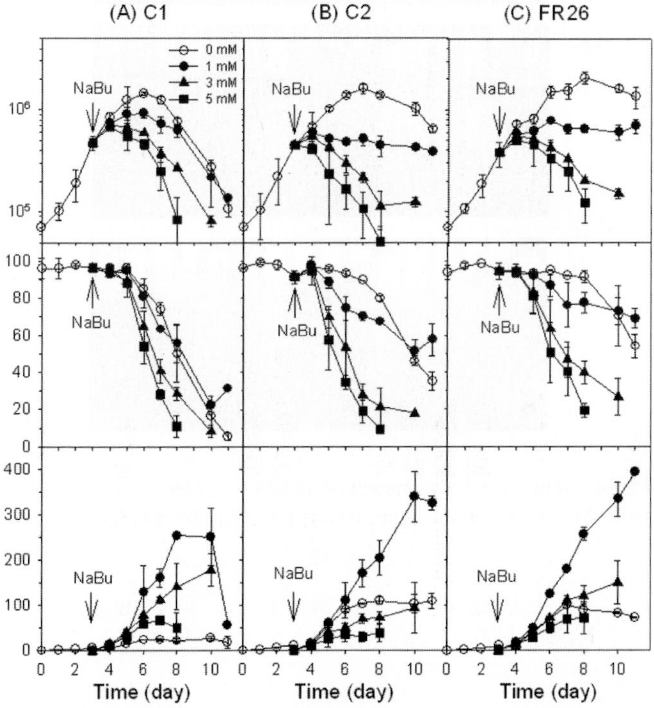


*Figure 4.* Cell growth, cell viability, and hTPO production of (A) C1 control cells, (B) C2 control cells and (C) FR26 cells during the cultivation using NaBu. Arrows indicate the time of NaBu addition. Error bars represent the standard deviation determined in duplicate experiments.

increases of the $q_{hTPO}$ were obtained in the cultures with 1 mM NaBu addition (Table 1). In the cultures with 1mM NaBu addition, the hTPO concentration in FR26 cells was slightly higher than or similar to that in control cells (C1 and C2 cells) (Table 1). This suggests that the dramatic enhancement of hTPO concentration can not be obtained in the cultivation of rCHO cells with reduced levels of caspase-3 and caspase-7.

## 2.5 Apoptosis Assay

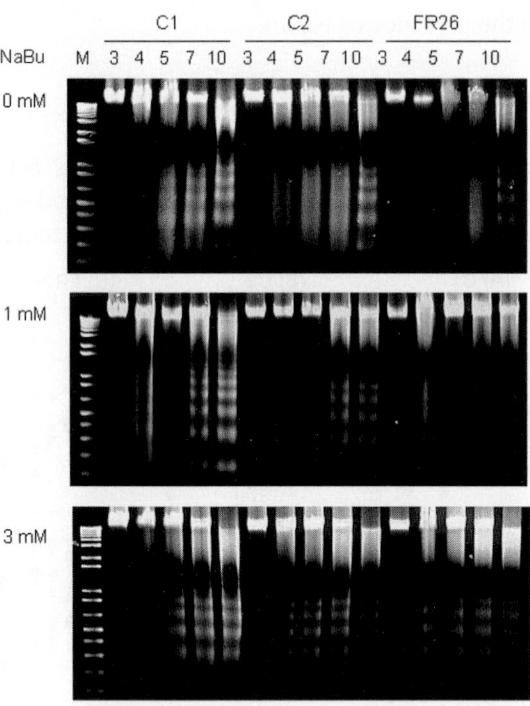


*Figure 5*. Chromosomal DNA fragmentation in (A) C1 control cells, (B) C2 control cells and (C) FR26 cells. M: a molecular weight marker, 1-kb plus DNA ladder.

## 2.6 Detection of Mitochondrial Membrane Potential

When C1 cells were subjected to 3 mM NaBu, the percentage of population with low mitochondria membrane potential increased by 11.2%. This reduction by 3 mM NaBu could not be overcome by downregulation of caspase-3 or/and caspase-7. The percentage of population with low mitochondria membrane potential was also increased in C2 and FR26 cells by 3 mM NaBu addition. However, overexpression of Bcl-2 maintained high mitochondria membrane potential even in the presence of 3 mM NaBu. The percentage of population with low mitochondria membrane potential was not increased in R-bcl2-14 cells by 3 mM NaBu addition.

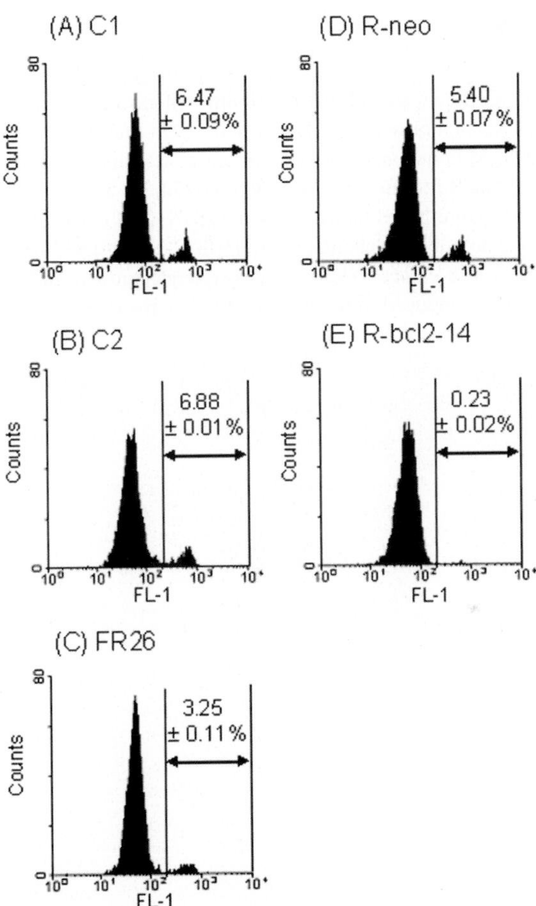


*Figure 6.* Mitochondrial membrane potentials were examined by flow cytometry in (A)C1, (B)C2, (C)FR26,(D)R-neo and (E)R-bcl2-14 the population with high green fluorescence (FL1) and low red fluorescence (FL2) represents the population with low mitochondria membrane potential.

## 3. CONLUSION

1. Expression of caspase-3 and 7 siRNAs could down-regulate efficiently and specifically the intracellular caspase-3 and 7 level in rCHO cells producing hTPO.
2. Down-regulation of both caspase-3 and 7 didn't yield highly positive effect as we expected, and it needs to be investigated further.
3. NaBu-induced apoptosis was slightly prevented in FR26 cells, but anti-apoptotic effect of caspase-3 and 7 down-regulation was not sufficient enough when compared with that of bcl-2 overexpression.

## REFERENCES

Chandler *et al.*, 1998, Different subcellular distribution of caspase-3 and caspase-7 following Fas-induced apoptosis in mouse liver. J Biol Chem. 1998 May 1;273(18):10815-8.

Earnshaw *et al.*, 1999, Mammalian caspases: structure, activation, substrates, and functions during apoptosis. Annu Rev Biochem. 1999;68:383-424. Review.

Sulpizi *et al.*, 2003, Structure-based thermodynamic analysis of caspases reveals key residues for dimerization and activity. Biochemistry. 2003 Jul 29;42(29):8720-8.

Hokom *et al.*, 1995; Ulich *et al.*, 1995 , Megakaryocyte growth and development factor ameliorates carboplatin-induced thrombocytopenia in mice. Blood. 1995 Aug 1;86(3): 971-6.

# CHAPTER IV: GENE MEDICINE

# UTRtech™: Exploiting mRNA Targeting To Increase Protein Secretion From Mammalian Cells

Beate Stern[1,2], Christine Gjerdrum[1], Stian Knappskog[2], Laura Minsaas[1], Stian Nylund[1], Lene C. Olsen[1,2], Litta Olsen[1], Eystein Oveland[1,2], Hanne Ravneberg[1], Christiane Trösse[1,2], Anja Tveit[1], Endre Vollsund[1,2], John E. Hesketh[1,3], Albert Tauler[1,4] and Ian F. Pryme[1,2]

*[1]UniTargetingResearch AS, Bergen, Norway; [2]University of Bergen, Norway; [3]University of Newcastle upon Tyne, UK; [4]University of Barcelona, Spain*

**Abstract:** The technology of UniTargetingResearch AS (UTRtech™) is based on the finding that the efficiency of directing mRNA to the endoplasmic reticulum is influenced by targeting signals. Using selected signals, genetically engineered mammalian cells are generated from which a protein of interest can be efficiently secreted. An industrial collaboration has revealed that UTRtech™ has the potential to significantly enhance the production of therapeutic proteins.

**Keywords:** CHO cells, IgG supersecretion, recombinant protein, seamless cloning, signal peptide, 3'UTR.

## 1. INTRODUCTION

In order to make mammalian cell systems competitive as cell factories it is of extreme importance to overcome the problem of low levels of protein production. Numerous approaches have been applied during recent years where attention has been paid mainly to modifying the cell's growth conditions (media composition and process control) and increasing the transcriptional activity of the recombinant gene (e.g. utilisation of strong promoters/enhancers in the expression vector, amplification of gene copy number) (Wurm, 2004). A new approach is currently being developed in our laboratory where the focus is on aspects of post-transcriptional events. We



*R. Smith (ed.), Cell Technology for Cell Products*, 261–268.

have earlier shown that polysomes in CHO cells can be fractionated into free, cytoskeletal-bound and membrane-bound populations (Pryme *et al.*, 1996). Based on our finding that the efficiency of directing mRNA to membrane-bound polysomes and thus to the endoplasmic reticulum, is dependent on the presence of targeting signals (specific signal peptides and 3'UTRs) (Partridge *et al.*, 1999), we have addressed our efforts to the improvement of protein synthesis/secretion by exploiting these results. We have compared the relative efficiencies of different signal peptides and signal peptide/3'UTR "doublets" from various sources on the production of a model protein (a marine luciferase). This led to the development of UTRtech™. In collaborative experiments with an industrial partner we have tested the technology for its ability to promote increased immunoglobulin (IgG) production. Using the same approach we have investigated the possibility to secrete an intracellular model protein (EGFP). The results from these studies are reported here.

## 2. Materials And Methods

### 2.1 Cell culture and transfection

CHO cells were grown in monolayer culture using DMEM medium containing 10% FBS where appropriate, and 100 U/ml penicillin. Cells were incubated in a humified atmosphere of 5% $CO_2$ at 37°C. Transfection was performed using lipofectamine and stable cell populations were established within 4 weeks using a selection medium.

### 2.2 Seamless cloning

All vector constructs were made using a seamless cloning strategy. This PCR-based method was developed in our laboratory for the directional insertion/substitution of DNA, independent of restriction enzyme sites and avoiding the incorporation of linker sequences.

### 2.3 Bioluminescence assay, ELISA, Western and Northern blotting

These were carried out according to routine laboratory procedures.

# 3. Results

## 3.1 Defining Utrtech™ Using A Model Protein

Using a naturally secreted luciferase from the marine copepod *Gaussia princeps* as a model protein, a number of signal peptides have been tested for their ability to promote synthesis and secretion of this protein in transfected CHO cells. Figure 1A shows the relative efficiencies of selected signal peptides with respect to the amounts of luciferase produced and recovered in the growth medium. The amounts differ by factors of up to 50.

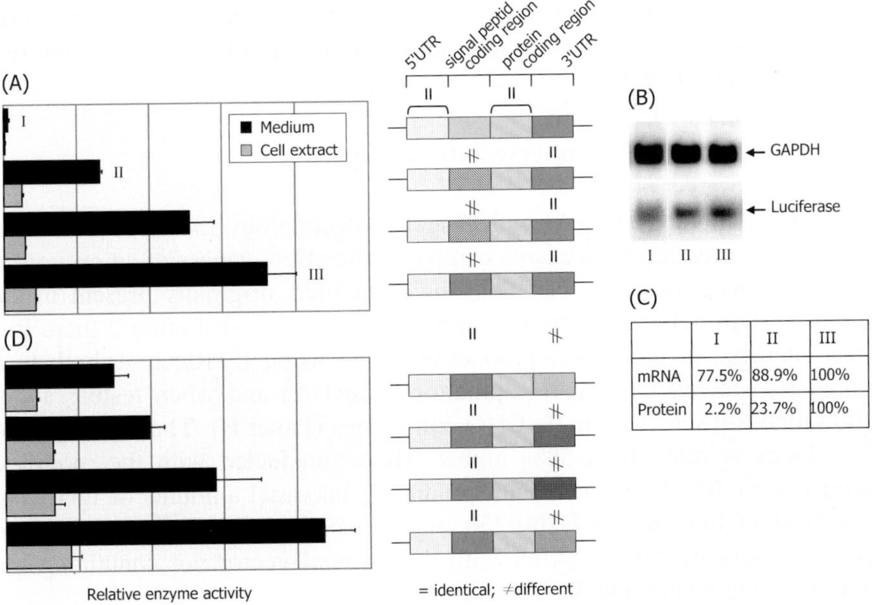


| | I | II | III |
|---|---|---|---|
| mRNA | 77.5% | 88.9% | 100% |
| Protein | 2.2% | 23.7% | 100% |

*Figure 1.* A) Effect of different signal peptides on synthesis/secretion of Gaussia luciferase. Luciferase activity in culture medium and cell extracts of stably transfected CHO cell populations was measured with a standard bioluminescence assay and corrected according to cell number. The signal peptides used are derived from the following proteins (top to bottom): human albumin, human interleukin-2, human trypsinogen, Gaussia luciferase. B) Northern blot analysis of mRNA isolated from selected CHO cell populations in (A), designated I-III. DNA probes against GAPDH (control) and luciferase coding region were used. C) Quantification of luciferase mRNA bands in (B) using QuantityOne software (Bio-Rad). Both mRNA measurements and luciferase activity measurements (medium values from panel A) are related to the maximal values achieved using the Gaussia-luciferase signal peptide (set as 100%). D) Effect of various 3'UTRs combined with the Gaussia-luciferase signal peptide on synthesis/secretion of Gaussia luciferase.

In order to eliminate the possibility that the striking differences observed were due to varying levels of mRNA, Northern blot analysis was performed (Figure 1B). The levels of mRNA and secreted luciferase were quantified (Figure 1C). Interestingly, although the amounts of luciferase mRNA where the albumin signal peptide was used, is about 80% of that found with the *Gaussia*-luciferase signal peptide, the level of secreted luciferase is only 2%.

In a further set of experiments the *Gaussia*-luciferase signal peptide was tested in combination with various 3'UTR sequences derived from genes coding for secreted proteins. As exemplified in Figure 1D, again a significant variation in the levels of luciferase secretion was observed. Efficient signal peptide/3'UTR doublets were identified. The doublets together with the seamless cloning technology to be used for the construction of "secretion cassettes" containing the coding region of a protein of interest is defined as UTRtech™.

### 3.2 Proof of Principle: Supersecretion of Igg

In a collaborative project with Angel Biotechnology Ltd aiming to boost human IgG production, we have compared the effectiveness of an optimised signal peptide/3'UTR doublet with those doublets originally present in the vector system determined by the company. As shown in Figure 2 there is a clear increase in intact IgG production when using UTRtech™, both in a stably transfected CHO cell population (Panel A) and when testing about 200 randomly selected single CHO cell clones (Panel B). The mean titre of the clones is more than 50% higher when transfected with the modified *versus* unmodified vector. Clones producing maximal amounts of IgG were observed in the range of 200-250 mg/ml using UTRtech™ compared to clones producing 80-90 mg/ml with the original vector not containing the optimised signal peptide/3'UTR doublet.

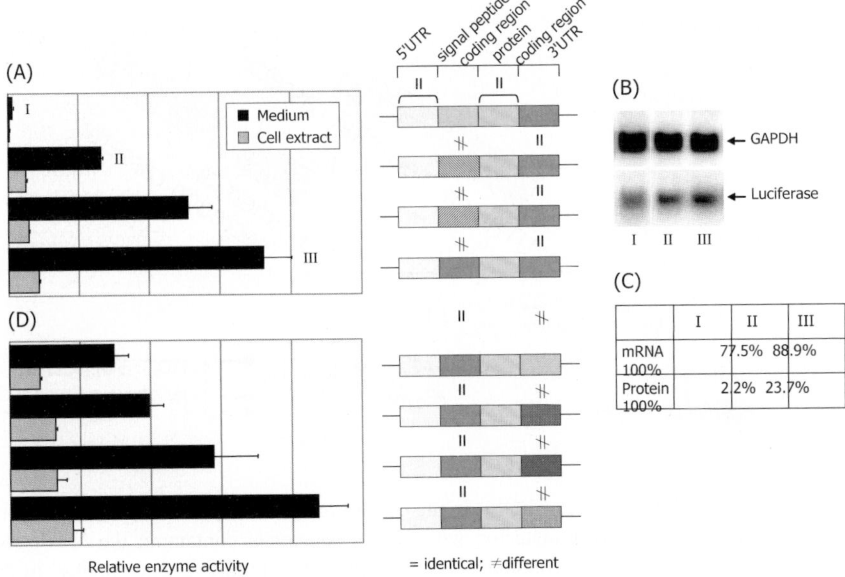


*Figure 2.* A) Assembled IgG identified in the medium of stably transfected CHO cell populations. B) Assembled IgG identified in the medium of randomly selected single clones from transfected CHO cells (work performed by Angel Biotechnology Ltd). In (A) two vectors encoding either the IgG heavy or light chain were cotransfected, whereas in (B) both chains were encoded by the same vector. In (A) medium samples were adjusted according to total protein concentration and subjected to Western blot analysis. In (B) medium samples were subjected to ELISA. "Partner" indicates the original and "UTRtech" the modified vector(s).

## 3.3 Secretion of an Intracellular Protein

UTRtech™ has been employed to successfully secrete an intracellular model protein, namely EGFP (Figure 3). Further, the experiment demonstrates the importance of the seamless cloning technology in order to produce authentic protein. When comparing the result obtained with our secretion vector and the result obtained with a commercially available secretion vector where cloning requires the use of restriction enzymes, it is evident that the presence of linker sequences is disadvantageous.

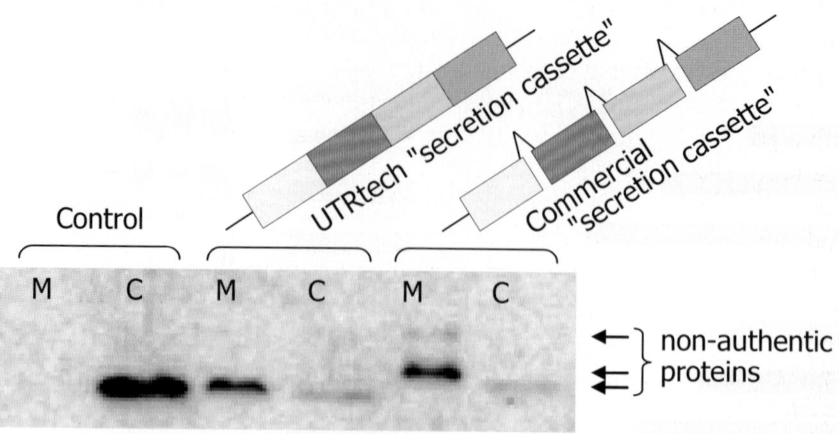


*Figure 3.* Western blot analysis of EGFP in the culture medium (M) and cell extracts (C) of CHO cell populations stably transfected with vector constructs containing the "cassettes" specified. The different genetic components are designated in Figure 1. The “hooks” in the commercial secretion vector indicate linker sequences.

## 4. Discussion

The observations we made when testing individual signal peptides were very surprising in that we had expected that the signal peptide of albumin, a protein produced and secreted constitutively in large quantities by the liver, would be extremely effective with regards to production of recombinant protein in mammalian cells. The amount of luciferase recovered in the culture medium when using the albumin signal peptide, however, was only 2% of the amount achieved using the signal peptide from *Gaussia* luciferase. Interestingly, similar results as with CHO cells were obtained with HepG2 cells. Also other signal peptides of human origin expected to perform well (e.g. interleukin-2, trypsinogen) did not reach the same high level of recombinant protein secretion achieved using the marine signal peptide. Since the levels of mRNA measured in the samples from CHO cells transfected with the albumin/interleukin/*Gaussia*-luciferase signal peptide constructs vary to a much lesser extent than the corresponding levels of the secreted luciferase, the results would strongly suggest that the effect is at the post-transcriptional level. This is supported by preliminary results obtained from a collaboration with Selexis, a company focusing on transcriptional enhancement by employing specific DNA elements binding to nuclear scaffolding. Combining UTRtech™ with their technology (MARtech™) resulted in a 100% increase in luciferase secretion, demonstrating that

UTRtech™ is complementary to approaches aiming at improved recombinant protein production through elevated mRNA levels.

Since the results had been obtained using a model reporter protein, it was important to ascertain whether or not similar observations could be made with a protein of commercial interest. Collaborative experiments performed together with Angel Biotechnology Ltd indeed showed that incorporating UTRtech™ into their expression system significantly enhanced IgG production on the laboratory scale.

Further collaborative experiments performed with academic/industrial partners have demonstrated that in addition to recombinant protein production, UTRtech™ has the potential to be adaptable for use in other applications such as gene and cell therapy where efficient protein secretion is imperative.

Regarding the successful application of UTRtech™ to secrete an authentic intracellular protein, this has important implications e.g. for the production of certain "difficult-to-express" proteins which exhibit toxicity when accumulating in the host cell. Their secretion would circumvent the problem of cell damage/death.

The results described here clearly illustrate the fact that mammalian signal peptides hitherto thought to be extremely effective with regard to recombinant protein production (e.g. the signal peptide from albumin), can be replaced by far more efficient sequences originating from non-mammalian sources. Further, they demonstrate the importance that the nature of the 3'UTR is also taken into consideration such that the most effective signal peptide/3'UTR doublet is chosen for the recombinant protein whose production rate is to be boosted.

## 5. Summary

1. The choice of signal peptide/3'UTR doublet is imperative when one has the goal of achieving improved yields of recombinant protein in mammalian cells.
2. The adaptation of seamless cloning technology results in the production of authentic protein.
3. UTRtech™ can be used to enhance the production of naturally secreted proteins and to secrete intracellular proteins.
4. Industrial collaborations have demonstrated the incremental nature of UTRtech™.
5. UniTargetingResearch AS (www.unitargeting.com) is the sole supplier and IP holder of UTRtech™.

## ACKNOWLEDGEMENTS

We would like to thank our collaborators at Angel Biotechnology Ltd (www.angelbio.com) and at Selexis (www.selexis.com) for their participation in this work. The Norwegian Research Council is acknowledged for financial support.

## REFERENCES

Partridge K., Johannessen A.J., Tauler A., Pryme I.F., Hesketh J.E.; Competition between the signal sequence and a 3'UTR localisation signal during redirection of beta-globin mRNA to the endoplasmic reticulum: implications for biotechnology. *Cytotechnol* 1999, 30:37-47.

Pryme I.F., Partridge K., Johannessen A.J., Jodar D., Tauler A., Hesketh J.E.; Compartmentation of the protein synthetic machinery of CHO cells into free, cytoskeletal-bound and membrane-bound polysomes. *Gen Eng Biotechnol* 1996, 16:137-144.

Wurm F.M.; Production of recombinant protein therapeutics in cultivated mammalian cells. *Nat Biotechnol* 2004, 22:1393-1398.

# Semi-Synthetic Mammalian Gene Regulatory Networks

B.P. Kramer, M. Fischer and M. Fussenegger
*Institute for Chemical and Bio-Engineering, Swiss Federal Institute of Technology, ETH Hoenggerberg HCI F115, Wolfgang-Pauli-Strasse 10, CH-8093 Zurich, Switzerland. E-mail: fussenegger@chem.ethz.ch*

**Abstract:** While synthetic biology was until recently restricted to network assembly and testing in prokaryotes, decisive advances have been achieved in eukaryotic systems based on current availability of different human-compatible transgene control technologies. The majority of transgene control networks available to date are fully synthetic. Yet, in order to develop their full anticipated therapeutic potential, synthetic transgene control circuits need to be well interconnected with the host cell's regulatory networks in order to enable physiologic control of prosthetic molecular expression units. We have designed three semi-synthetic transcription control networks able to integrate physiologic oxygen levels and artificial antibiotic signals to produce expression readout with NOT IF or NOR-type Boolean logic or discrete multi-level control of several intracellular and secreted model product proteins. Subtle differences in the regulation performance of the endogenous oxygen-sensing system in CHO-K1 and human HT-1080 switched the semi-synthetic network's readout from a classic four-level (high, medium, low, basal) regulatory cascade to a network enabling six discrete transgene expression levels.

**Key words:** Synthetic Biology, Gene Networks, BioLogic Gates, Hypoxia, Erythromycin, Tetracyclin

# 1. INTRODUCTION

In this work we pioneer several semi-synthetic gene networks able to integrate endogenous (oxygen levels) and artificial (clinically licensed antibiotics) signals and produce NOT IF- or NOR-type BioLogic as well as



*R. Smith (ed.), Cell Technology for Cell Products*, 269–272.

multi-level transgene transcription responses in rodent and human cell lines. We expect semi-synthetic mammalian gene networks to generate decisive impact on future gene therapy and tissue engineering initiatives.

## 2. RESULTS AND DISCUSSION

### 2.1 Design of hybrid promoters

In order to design chimeric promoters which are induced by low oxygen concentrations (hypoxia, $H_{OX}$) and repressed by the transcriptional silencer E-KRAB (Weber *et al.*, 2002a), hypoxia-responsive promoters were equipped with E-KRAB operator modules placed between hypoxia-responsive promoter and luciferase reporter gene (*luc*).

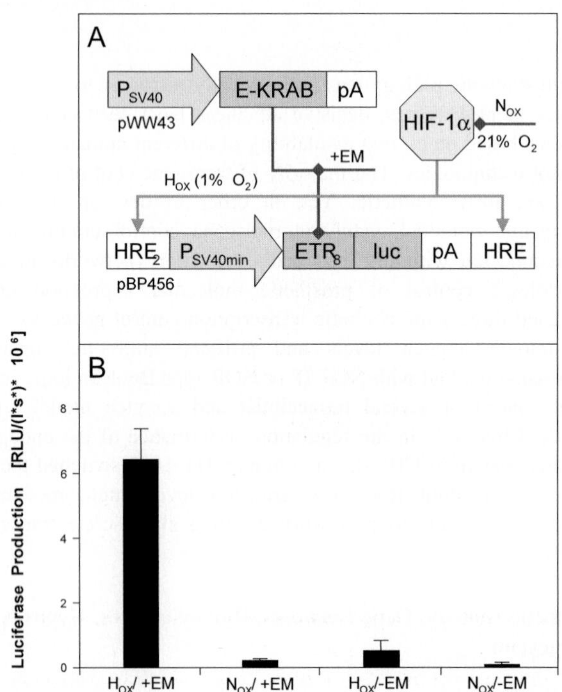


*Figure 1.* Diagram of oxygen- and erythromycin-responsive hybrid promoters, assembled in a semi-synthetic regulatory network, enable transgene expression with NOT IF Boolean logic (A). Luciferase production of HT-1080 transfected with pWW43 (PSV40-E-KRAB-pA) and pBP456 (HRE2-PSV40min-ETR8-Luc-pA-HRE) (B) and cultivated for 48h in the presence or absence of erythromycin (±EM) and under hypoxic (HOX, 1% O2) or normoxic (NOX, 21% O2) conditions. (U/L, units per liter).

## 2.2 Discrete four-level transgene expression control by a semi-synthetic transcription cascade in CHO-K1

Interconnection of a two-step sequential antibiotic cascade with endogenous hypoxia-controlled regulatory networks results in a semi-synthetic three-step regulatory cascade (Figure 2), characterized by (i) oxygen-responsive $HRE_2$-$P_{SV40min}$-driven PIT expression triggered by endogenous HIF-1α ($HRE_2$-$P_{SV40min}$-PIT-pA-HRE; pBP445), (ii) PIT controlling tTA (tetracycline-dependent transactivator) transcription in a PI-responsive $P_{PIR}$-driven manner ($P_{PIR}$-tTA-pA; pBP134) and (iii) tetracycline (TET)-repressible tTA-dependent reporter gene production driven by the tetracycline-responsive promoter $P_{hCMV*-1}$ ($P_{hCMV*-1}$-SAMY-pA; pBP99 SAMY, heat-stabile *Bacillus stearothermophilus*-derived secreted α-amylase).

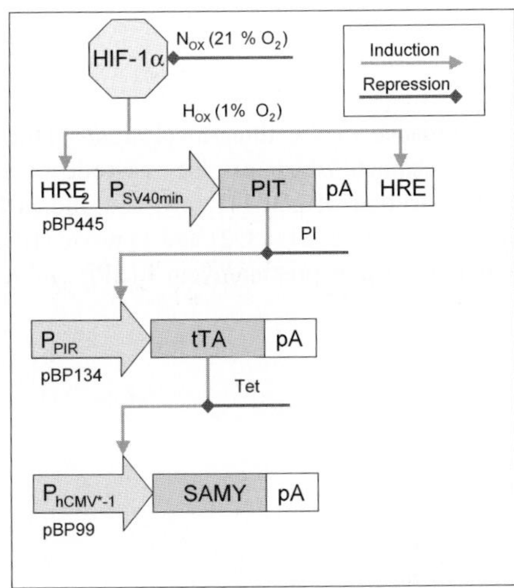


*Figure 2.* Design of a three-level semi-synthetic regulatory cascade. At hypoxic conditions ($H_{OX}$) HIF-1α induces $P_{SV40min}$-mediated expression of the pristinamycin-dependent transactivator PIT encoded on pBP445 ($HRE_2$-$P_{SV40min}$-PIT-pA-HRE). In the absence of pristinamycin (-PI), PIT drives $P_{PIR}$-mediated expression of the tetracycline-dependent transactivator tTA encoded on pBP134 ($P_{PIR}$-tTA-pA). tTA finally triggers $P_{hCMV*-1}$ (tetracycline-responsive promoter) -mediated SAMY (*Bacillus stearothermophilus*-derived secreted α-amylase) production from pBP99 ($P_{hCMV*-1}$-SAMY-pA) whenever transgenic cells are cultivated in tetracycline-free (-TET) medium.

Following cotransfection of pBP445, pBP134 and pBP99 into CHO-K1, the semi-synthetic network enabled four distinct SAMY expression levels correlating with four different culture conditions (Figure 3).

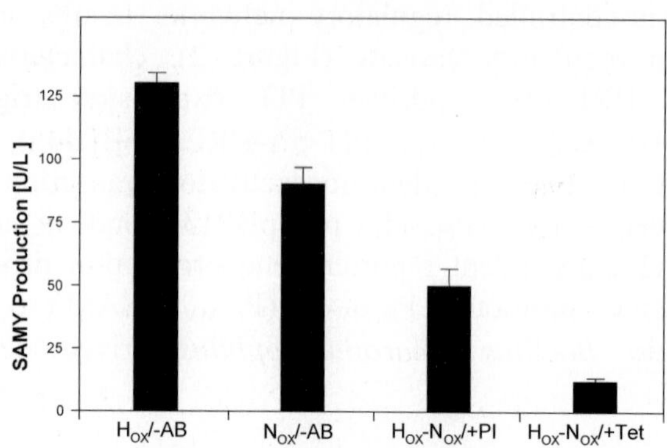


*Figure 3.* Regulation performance of the three-level semi-synthetic regulatory cascade in CHO-K1. The regulation characteristics of the cascade were assessed by SAMY quantification following transfection of pBP445, pBP134 and pBP99 into CHO-K1. Cells were cultivated under normoxic (NOX, 21% O2) and hypoxic (HOX, 1% O2) conditions as well as in the presence and absence of pristinamycin I (±PI) and tetracycline (±TET). (POI, protein of interest).

# Acoustic Cell Processing for Viral Transduction or Bioreactor Cell Retention

V.M. Gorenflo[1,2*], P. Beauchesne[1,2], V. Tayi[1,2], O. Lara[1], H. Drouin[1,2], J.B. Ritter[1], V. Chow[1,2], C. Sherwood[1], B.D. Bowen[2], J.M. Piret[1,2]

*[1]Michael Smith Laboratories and Department of [2]Chemical & Biological Engineering, University of British Columbia, 2185 East Mall, Vancouver, BC, V6T 1Z4, Canada*

*[*]Sanofi Pasteur Limited, Toronto, ON, Canada*

**Abstract:** A major limitation of retrovirus gene therapy technology is the often low percentage of target cells transduced. This is in part due to the low diffusivity of retroviruses, as well as their short half-life (~5 h). An approach to overcome these limitations has been developed using the forces in an acoustic standing wave field. An air-backflush mode of operation obtained up to 8-fold increases in TF-1 cell transduction compared to static controls and this was sustained from 2 to 24 h. The transduction increased as a function of power input, but at elevated power levels the acoustic transducer generated excessive heat. A new design with improved heat dissipation allowed continuous acoustic treatment over 2 days with no decrease in cell viability. This acoustically increased transduction reduces the need for additives and avoids the complications of recovering anchored cells. While acoustic separators can be used for bioreactor volumes ranging from hundreds of mL to >100 L, it is also important to define operational settings that avoid negative thermal influences on the cells. Additional cell culture experiments with CHO cells were performed to determine the acceptable temperature variations.

**Key words:** separator, gene therapy, retroviral transduction, mammalian cell retention, separation, gene transfer, standing wave, acoustic field, perfusion



*R. Smith (ed.), Cell Technology for Cell Products*, 273–278.

# 1. INTRODUCTION

Recombinant retroviruses are widely used for scientific research and are being developed for gene therapy applications, due to their ability to efficiently deliver transgenes (Andreadis *et al.,* 1999; Chan *et al.,* 2001). A major limitation of this technology is the often low percentage of cells transduced, even when high-titre viral preparations are available. This is, in part, due to the low diffusivity of retroviruses (~6 x $10^{-8}$ cm/s), as well as their short ~5 h half-life (Higashikawa *et al.*, 2001). Various colocalization strategies to enhance virus-cell contact, such as centrifugation (Kuhlcke *et al.*, 2002) and membrane flow-through transduction (Chuck *et al.*, 1996), have increased transduction efficiencies by factors of 3- to 10-fold. Other protocols have relied on additives, including fibronectin (Bajaj *et al.*, 2001) and polycations, such as protamine sulfate and Polybrene (Cornetta *et al.*, 1989; Hennemann *et al.*, 2000), to increase viral-cell interaction by enhancing adsorption, or by reducing the electrostatic repulsion between these negatively charged particles. However, many of these techniques are not readily scaleable for the large numbers of cells often required in clinical protocols. Therefore, it is valuable to develop technologies that can increase the effective viral transduction rates despite low and variable virus titres.

We are currently developing a strategy using acoustic forces to overcome the diffusion limitations associated with viral transduction protocols. The diameter of retroviruses is ~130 nm, about two orders of magnitude smaller than that of mammalian cells. The primary radiation forces in the standing wave field (Kilburn *et al.*, 1989; Woodside *et al.*, 1997) are opposed by drag forces, such that the acoustically driven particle velocities are roughly proportional to the particle radius squared. Thus, forces in a resonant acoustic field displace cells to the pressure node planes at much greater velocities than those of viral particles. Mammalian cells are retained in the pressure node planes while the viral particles are more readily displaced by fluid flow. These phenomena result in relative motion between virus particles and target cells, greatly increasing their encounter frequency. This use of acoustic forces is distinct from other reported uses of ultrasound for non-viral gene transfer (Koch *et al.*, 2000; Newman *et al.*, 2001) since the latter systems rely on cavitation-induced shear forces in order to generate transient permeabilization of cell membranes.

# 2. MATERIAL AND METHODS

Acoustically enhanced retroviral transduction was performed using a modified 10L BioSep acoustic separator (Applikon Biotechnology) mounted on a 100 mL spinner flask (Bellco Glass) containing $8 \times 10^5$ TF-1 cells/mL

(ATCC CRL-2003) resuspended in a mixture of 75 mL of DMEM supplemented with 10% FBS (Gibco) and 6 ng/mL GM-CSF, and 75 mL of MSCV-IRES-GFP virus-containing medium (VCM). The retroviruses were obtained from vector-transduced PG13 cells (gift from the Terry Fox Laboratories, B.C. Cancer Agency) engineered to produce a vector with a green fluorescent protein (GFP) reporter gene (Hennemann *et al.*, 1999). The use of an air-backflush mode of operation (Gorenflo *et al.*, 2003) eliminated the need to circulate the virus-cell suspension through a pump. Using a SC3010 controller (SonoSep Technologies), the resonant frequency was adjusted to 1.95 MHz and the power input to the piezoelectric transducer was set to 5 W. The transducer was cooled by air flow at 10 L/min. Every 4 min, a total of 25 mL from this culture flowed through the separator at 7 mL/min and then, by back flushing, was returned to the spinner. This process was repeated continuously over a period of 24 h. The temperature of the spinner was maintained at 37 °C with a water bath and the pH controlled using 10% $CO_2$ in air. Viability was measured by a Trypan blue (Sigma) assay and using a Cedex automated cell counter (Innovatis). Samples were diluted 1:5 using DMEM supplemented with 10% FBS and 6 ng/mL GM-CSF to minimize post-sampling transduction. The cells were expanded in tissue culture treated 6-well plates (Nunc) for 2 days (Klein *et al.*, 1997) and then analyzed by flow cytometry for GFP expression using a FACScan cytometer (BD Biosciences). In order to exclude non-viable cells from analysis, 5 μg/mL propidium iodide was added to samples prior to analysis. Results from the acoustically treated spinner were compared to that of transduction performed in a static control consisting of a T-75 flask (suspension cells, Sarstedt), a 100 mL spinner flask (Bellco Glass) and the acoustic system without power supplied to the piezoelectric transducer. Chinese Hamster Ovary (CHO) cell experiments were performed in 100 mL working volume spinner flasks (Bellco Glass). The spinners were modified to allow aeration, sampling and automated pumping in/out to a heat exchanger to obtain varying temperature exposures.

## 3. RESULTS AND DISCUSSION

A 3.5-fold increase in transduction efficiency, compared to static controls, was obtained within the initial 2 h of acoustic treatment and was sustained for a period of up to 24 h (Figure 1). After 24 h of treatment 23% of viable cells expressed GFP while only 6.7% were transduced in the controls. Additional trials (not shown) yielded 3- to 8-fold increases in transduction. No significant variations in transduction were observed between the static control, the spinner flask alone and the acoustic system without power supplied to the transducer. Viability remained above 91%

during the initial 6 h and dropped to 86% after 24 h of treatment compared to 97% for the control. The transduction increased as a function of power input, but at elevated power levels the acoustic transducer generated excessive heat resulting in decreased cell viability. A new design with improved heat dissipation allowed continuous acoustic treatment over 2 days with no significant decrease in cell viability.

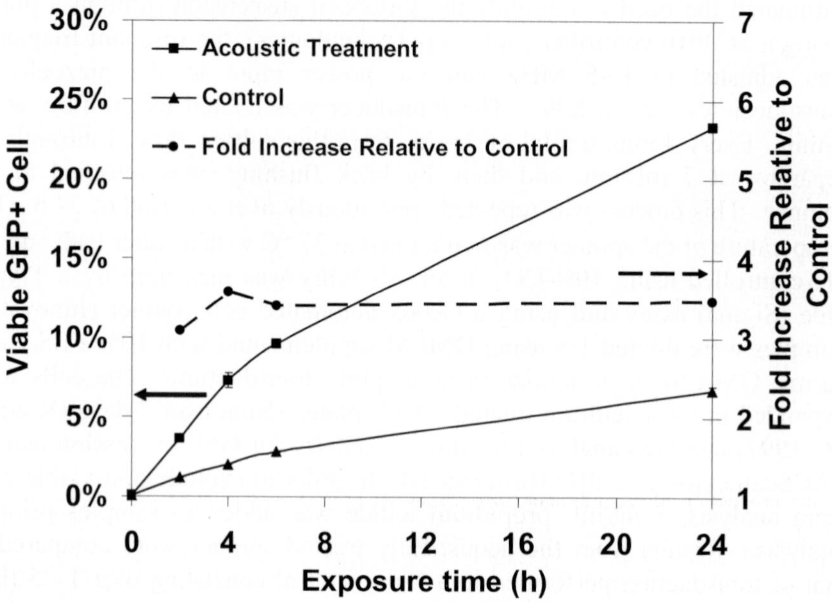


*Figure 1.* Viable TF-1 cells transduced after acoustic treatment (squares) compared to control (triangles). Transduced cells were assessed after a 2-day expansion based on GFP expression analyzed by flow cytometry. (Error bars: SEM, sample replicate, n=2).

Additional experiments were performed with CHO cells, where 4% of the culture volume was repeatedly exposed to 1-min excursions from 37 °C to between 27 and 40 °C. Remarkably, it was found that cycling to 27 °C, as would be the case for many external cell retention devices, significantly decreased the growth rate of the whole culture. In the range of temperature excursions from 31.5 to 38.5 °C, there was minimal impact on cell growth or metabolism. Overall, these studies define conditions that maintain temperatures within acceptable limits while maximizing bioprocess performance for both perfusion and viral transduction applications.

These results demonstrate the potential for increased retroviral transduction efficiency through the use of acoustic forces. This method should reduce the time needed for transduction while increasing the yield. Furthermore, the ease of cell inoculation and recovery from such acoustic systems could make them far simpler to use than other systems. This approach could also offer other advantages particularly relevant to safety in clinical applications. For instance, it could eliminate the need for animal derived protein additives, such as fibronectin or protamine sulfate. In addition, the ease of aseptic cell manipulations in this closed system would be advantageous for cell and gene therapy applications.

## ACKNOWLEDGEMENTS

The authors would like to thank Felix Trampler (SonoSep Technologies Inc., Vienna, Austria) for his technical assistance and Nancy Chan (University of British Columbia). This work was supported by a grant from the Natural Sciences and Engineering Council, NSERC); fellowships from Deutsche Forschungsgemeinschaft for V. Gorenflo and Stem Cell Network for P. Beauchesne.

## REFERENCES

Andreadis, S. T., C. M. Roth, J. M. Le Doux, J. R. Morgan and M. L. Yarmush 1999. "Large-scale processing of recombinant retroviruses for gene therapy." Biotechnol Progr **15**(1): 1-11.

Bajaj, B., P. Lei and S. T. Andreadis 2001. "High efficiencies of gene transfer with immobilized recombinant retrovirus: kinetics and optimization." Biotechnol Prog **17**(4): 587-96.

Chan, L. M. C., C. Coutelle and M. Themis 2001. "A novel human suspension culture packaging cell line for production of high-titre retroviral vectors." Gene Ther **8**(9): 697-703.

Chuck, A. S. and B. O. Palsson 1996. "Consistent and high rates of gene transfer can be obtained using flow-through transduction over a wide range of retroviral titers." Hum Gene Ther **7**(6): 743-50.

Cornetta, K. and W. F. Anderson 1989. "Protamine Sulfate as an Effective Alternative to Polybrene in Retroviral-Mediated Gene-Transfer - Implications for Human-Gene Therapy." J Virol Methods **23**(2): 187-194.

Gorenflo, V. M., S. Angepat, B. D. Bowen and J. M. Piret 2003. "Optimization of an acoustic cell filter with a novel air-backflush system." Biotechnol Prog **19**(1): 30-6.

Hennemann, B., J. Y. Chuo, P. D. Schley, K. Lambie, R. K. Humphries and C. J. Eaves 2000. "High-efficiency retroviral transduction of mammalian cells on positively charged surfaces." Hum Gene Ther **11**(1): 43-51.

Hennemann, B., E. Conneally, R. Pawliuk, P. Leboulch, S. Rose-John, D. Reid, J. Y. Chuo, R. K. Humphries and C. J. Eaves 1999. "Optimization of retroviral-mediated gene transfer to human NOD/SCID mouse repopulating cord blood cells through a systematic analysis of protocol variables." Exp Hematol **27**(5): 817-25.

Higashikawa, F. and L. J. Chang 2001. "Kinetic analyses of stability of simple and complex retroviral vectors." Virology **280**(1): 124-131.

Kilburn, D. G., D. J. Clarke, W. T. Coakley and D. W. Bardsley 1989. "Enhanced sedimentation of mammalian cells following acoustic aggregation." Biotechnol Bioeng **34**: 559-562.

Klein, D., S. Indraccolo, K. vonRombs, A. Amadori, B. Salmons and W. H. Gunzburg 1997. "Rapid identification of viable retrovirus-transduced cells using the green fluorescent protein as a marker." Gene Ther **4**(11): 1256-1260.

Koch, S., P. Pohl, U. Cobet and N. G. Rainov 2000. "Ultrasound enhancement of liposome-mediated cell transfection is caused by cavitation effects." Ultrasound in Medicine and Biology **26**(5): 897-903.

Kuhlcke, K., B. Fehse, A. Schilz, S. Loges, C. Lindemann, F. Ayuk, F. Lehmann, N. Stute, A. A. Fauser, A. R. Zander and H. G. Eckert 2002. "Highly efficient retroviral gene transfer based on centrifugation-mediated vector preloading of tissue culture vessels." Mol Ther **5**(4): 473-8.

Newman, C. M., A. Lawrie, A. F. Brisken and D. C. Cumberland 2001. "Ultrasound gene therapy: On the road from concept to reality." Echocardiography **18**(4): 339-347.

Woodside, S. M., B. D. Bowen and J. M. Piret 1997. "Measurement of Ultrasonic Forces for Particle-Liquid Separations." AIChE Journal **43**(7): 1727-1736.

# Comparison of Different Transient Retroviral Transfection Systems

V. Kleff, I. Rattmann, C. Ludwig, M. Flasshove, T. Moritz and B. Opalka
*Dept. of Internal Medicine (Cancer Research), University of Duisburg-Essen, Medical, School. Essen. Germany*

**Abstract:** Oncoretroviral preparations were produced by PEI and $CaPO_4$ precipitation-mediated transient transfection of producer cells. For optimizing the latter transfection method, the doses of different supporting agents were varied and incubation times were altered. Best results were achieved with a modified $CaPO_4$ protocol. For the production of foamy and lentiviral preparations different transient transfection agents were tested. Foamyviral supernatants had best titers after transfection with Polyfect and $CaPO_4$, while lentiviral preparations generated with $CaPO_4$ achieved the best results.

**Key words:** Oncoretrovirus, lentivirus, foamy virus, transfection, titers of viral preparations

## 1. INTRODUCTION

Oncoretroviral vectors represent an established technology for an efficient and stable gene transfer into hematopoietic $CD34^+$ stem/progenitor cells and their daughter cells. The full potential of the technology has recently been highlighted by the successful treatment of children with severe combined immunodeficiency (SCID) disease. Here functional reconstitution of the immune system following transfer of the missing gene encoding the common gamma chain of interleukin receptors into hematopoietic stem cells has been shown. The most common oncoretroviral vectors are derived from viruses such as the Moloney murine leukemia virus (MMLV) or the spleen focus-forming virus (SFFV). Because of their unique biological characteristics lentivruses (LV) and foamyviruses (FV) offer promising new



*R. Smith (ed.), Cell Technology for Cell Products*, 279–282.

alternatives and have recently been employed successfully for hematopoietic stem cell gene transfer.

However, for successful gene transfer robust virus production protocols are required. Therefore lenti, foamy, and oncoretroviral supernatants were produced using various agents for transient transfections of producer cells. The viral titers were used as a measure for the efficiency of the initial transfection step.

## 2. PRODUCTION OF ONCORETROVIRAL SUPERNATANTS

For the preparation of oncoretroviral supernatants a three plasmid system was used. One plasmid carries the gene of interest. For this purpose two vectors expressing the $MGMT^{P140K}$ gene in the 3' and 5' position and the EGFP marker gene in 5' and 3' positions with respect to an IRES were used. Both retroviral vectors share a hybrid backbone containing elements of the murine embryonic stem cell virus (MESV) and the spleen focus forming virus (SFFV) *[Baum et al., J Virol, 1995]*. In order to allow the production of virus particles for gene therapy, the genes encoding the retroviral structure proteins (*gag*), the reverse transcriptase (*pol*) and the envelope protein (*env*) are supplied *in trans* by two other plasmids which are cotransfected into producer cells.

For producing high titer preparations a transient $CaPO_4$ transfection system was used. In this system 293T or PNXg/p cells were transfected with three plasmids. The vectors were mixed and added to the transfection buffer by bubbling. Transfections were carried out in the presence of chloroquine (25 μM). After 48 h of cultivation, retroviral supernatants were harvested and subsequently titered on 3T3 cells.

From the different modifications of the $CaPO_4$ transfection system tested, the $CaPO_4$ protocol with sodium butyrate and without heat inactivated FCS achieved the best results. In the absence of sodium butyrate and heat inactivated FCS PEI-produced supernatants showed higher transduction rates than supernatants produced by $CaPO_4$-mediated transfection (Table 1).

The best titers were achieved with supernatants produced with 293T cells treated with the modified $CaPO_4$ protocol with sodium butyrate and without heat inactivated FCS. The titers were about 2 x $10^6$ virus particles per milliliter, one log higher in comparison to the same conditions with PNXg/p cells.

*Table.* Transduction efficiency of supernatants produced with different transfection protocols on PNXg/p cells.

| | | |
|---|---|---|
| $CaPO_4$<br>**without** sodium butyrate<br>**without** incubation after bubbling<br>**without** heat inactivated FCS | 2 x $10^4$ | 1 x $10^4$ |
| $CaPO_4$<br>**without** sodium butyrate<br>**without** incubation after bubbling<br>**with** heat inactivated FCS | 6 x $10^4$ | 1 x $10^4$ |
| $CaPO_4$<br>**with** sodium butyrate (10 mM)<br>**without** incubation after bubbling<br>**without** heat inactivated FCS | 1 x $10^4$<br>(same sample **without** sodium butyrate: 8 x $10^3$) | --- |
| $CaPO_4$<br>**with** sodium butyrate (10 mM)<br>**with** incubation after bubbling<br>**without** heat inactivated FCS | 3 x $10^5$ | --- |
| PEI<br>**without** sodium butyrate<br>**without** heat inactivated FCS | 2 x $10^5$ | 2 x $10^4$ |

# 3. GENERATION OF FOAMY- AND LENTIVIRAL PREPARATIONS

For the production of the foamy viral and the lentiviral preparations two plasmid systems were used. The transfer vector carries the *gag* and *pol* genes as well as the transgene (here eGFP). The second plasmid contains the corresponding envelope gene. For the generation of the foamy viral particles the foamy viral wild type *env* gene under the control of a CMV promoter was used. The lentiviral based vectors were pseudotyped with the vesicular stomatitis virus glycoprotein G (VSV-G).

For the production of foamy viral supernatants, five different transfection agents were tested: Polyfect®, $CaPO_4$, FuGene6®, jet PEI® and Lipofectamine® 2000. 293T cells were transfected with the two plasmids carrying the transgene (eGFP), the structure genes *gag-pol* and the gene encoding for the envelope protein and one of the reagents. 24 h after transfection, the medium was removed and the cells cultured in fresh medium containing sodium butyrate for eight hours. Sodium butyrate can act to increase the efficiency of transfection and expression for both transient and stable transfections. It increases enhancer-dependent transcription of inserts driven by the CMV promoter. After another medium change and 24 h of cultivation, retroviral supernatants were harvested and subsequently titered on HT1080 cells. The preparations that were generated with Polyfect® and $CaPO_4$ achieved the highest titers (3 x $10^4$ to 5 x $10^4$ virus particles / ml).

Lentiviral supernatants were generated by transfecting 293T cells with the two plasmids using FuGene6® and $CaPO_4$ as transfection agent. The lentiviral preparations generated with $CaPO_4$ achieved higher titers (about 3 x $10^6$ virus particles per ml) compared to FuGene6® (about 8 x $10^5$ virus particles / ml).

## REFERENCE

Baum C, Hegewisch-Becker S, Eckert HG, Stocking C, Ostertag W. Novel retroviral vectors for efficient expression of the multidrug resistance (mdr-1) gene in early hematopoietic cells. J Virol, 1995, 69:7541-7547.

# CHAPTER V: CELLS TO TISSUES

# Artificial Human Lymphnode

*A device for Generation of Human Antibodies and Testing for Immune Functions in vitro*

C. Giese and U. Marx,
*ProBioGen AG, Goethestr. 54, D-13086 Berlin, Germany*

**Abstract:** Complex cultures of immune-organoids supported by the microperfusion bioreactor HIRIS can be used as an in vitro-model of secondary human lymphatic organs such as lymph nodes or spleen. Various functions of the human immune system can be engineered for the induction of humoral or cellular immune responses:

1. Generation of fully human antibodies against any therapeutical antigens
2. Testing of immune modulating effects for drug screening
3. Testing for immunogenicitiy or immunotoxicity

HIRIS provides the effective interaction of antigen-presenting DC, antigen specific regulatory T-cells and antibody-producing B-cells by combination of matrix embedded semi-stationary cell phase and a circulating mobile phase of suspended cells. The matrix assisted micro environment of cytokine- and chemokine gradients induces directed cell migration, formation of cell-cell contacts, secretion of mediators and at least induced proliferation, differentiation and antibody secretion.

HIRIS can be used for the generation of B-cells expressing affinity maturated antibodies angainst defined antigens. In advantage of established in vitro assays using suspended cells such as mixed lymphocyte reaction HIRIS-culture allows investigations of complex interactions, e.g. TH1/TH2-Shifts and B-cell induction.

**Key words:** immune, organoid, bioreactor, perfusion, dendritic cell, lymphocyte, antibody, micro environment

## 1. INTRODUCTION

Therapeutic antibodies play an increasingly important role in modern medicine. Apart from their highly specific effectiveness, these antibodies have to be extremely safe and must show excellent tolerability in patients.



*R. Smith (ed.), Cell Technology for Cell Products*, 285–290.

The ideal active substance is a fully human antibody derived from human B lymphocytes matured during a natural immune response to antigens in vitro. The primary goal of the research & development activities is the generation of high affinity human antibodies for therapeutic use in a organoid system.

Using tissue engineering technology, a complex tissue culture of antigen presenting dendritic cells, regulatory T-cells, and antibody producing B-cells allows the understanding and experimental modeling of the immunologic occurrences within a lymph node in vitro.

Furthermore, human immunofunctional organoids will offer the opportunity for immunogenicity and immunotoxicological testing and screening in vitro.

A suitable bioreactor system is needed for the essentials of the lymph node environment regarding cellular composition, microanatomy and fluidics.

In vivo, the lymphnode directs blood based lymphocytes to interact with the stream of antigens and antigen presenting cells transported by the lymphatic fluid from all peripheral compartments of the body. In terms of technical equivalency the lymph node can be described as a biological cross flow filter system for blood and lymphatic fluid. The lymphatic tissue of stroma cells entraps antigens, antigen presenting cells and lymphocytes.

For a successfull bioreactor development matrix assisted compartments for immobilized cells have to be combined with a perfusion system for mobile, suspended cells. Effective interaction, by cell adhesion, migration and stimulation can be realized under a well-balanced relation of a minimum of perfusion and a maximum of self conditioned and self adjusted biochemical microenvironment.

In contrast to already established tissues like skin or mucosal epithelia or liver parenchym the lymph node is characterized by a highly dynamic behaviour during infection and inflammation. In vivo, a new immune response of a lymph node is induced by infiltration of antigen presenting cells, e.g. dendritic cells (DC) via lymphatic flow, the activation and proliferation of blood circulating T-lymphocytes followed by B-lymphocytes and their growth, resulting in antibody-secretion and the swarming of effector and memory cells. Finally, the lymph node reaction ist termined by controlled tissue regression.

Two pivotal steps must have to be realized: The antigenspecific activation of T-cells and the induction of B-cells towards secretion, affinity maturation and class switch of antibodies.

# 2. MATERIALS AND METHODS

## 2.1 Cell preparation and primary culture

The complex co-culture of human dendritic cells, T-and B-lymphocytes is realized in an autologuous way. Cells are prepared from leucapheresis material of adult, healthy donors. All cell populations are isolated from PBMC-preparations by immune magnetic bead separations (Miltenyi Biotech) for CD14, CD4 or CD19. Dendritic cells are generated in vitro using stardard differentiation protocols by cytokines and growthfactors (GM-CSF, IL-4, TNF-$\alpha$). Aliqots of lymphocytes can be kryopreserved and revitalized on demand.

## 2.2 Perfusion bioreactor HIRS ™

The HIRIS ™ bioreactor is a hollow-fiber based miniaturized device for longterm cultivation of immune competent human micro-organoids. Culture volume and geometry is designed for multiparallel runs with respect on the availability of the cells by single donors.

Sheets of microporous matrices containing immobilized cells are mounted into the central part of the culture space. They are supplied for oxygen and pH by planar sets of hollow-fibers using a defined, humidified gas mix. The continous but moderate perfusion of culture medium provides local microenvironment in the matrix.

A mobile cell phase of suspended lymphocytes can be infused in circular through the matrix to initiate cell-cell interaction of mobile and immobile cells. The given perfusion rates and geometry of the culture space should have to ensure sufficient cellular orientation in the matrix. Migration and cell interaction is driven by local microgradients of cytokines and chemokines which will be formed in a process of self conditioning of the involved cells.

Matrix sheets are formed by hygrogels of agarose or alginate, non woven polyamid fibrics or macroporous collagene sponges.

A system of disposable optical sensors (PreSens GmbH) is intergrated for on line montoring of dO2 and pH.

## 2.3 Organoid culture

The inoculation of antigen loaded dendritic cells and T-lymphocytes into the matrix sheets is performed in multiwell plates. Sheets are mounted into the culture compartment and equilibrated in situ by regular perfusion. Cultitivation time ist about 14 days, recently. For enhanced stimulation mitogenic supplementation of bacterial lipopolysaccharid (LPS) is useful. At the end of the bioreactor runs matrix sheets are removed and prepared for histological characterisation.

### 2.4 Analytics of metabolites and cytokines

Samples of culture supernatants are collected daily and analyzed for glucose and lactate concentration. Frozen aliquots (-80°C) are stored for later analyse of cytokines. The cytokines were quantified by Cytometric Bead Array (CBA, BD Biosciences) for a given set of cytokines (IL-2, IL-4, IL-5, IL-10 TNF-α and INF-γ).

## 3. RESULTS AND DISCUSSION

The HIRIS ™ bioreactor device allows coninuously perfused longterm cultivation of primary human immune competent cells in complex matrix assisted co-culture for about 14 days. In co-culture DCs are forming a dendritic network (Fig. 1). T-cells migrate to the denendritic network and are activated by direct contact. The following T-cells proliferation leads to the generation of T –cell aggegates. For the given biomass and perfusion rates meatobolite and cytokine concentrations (IFN-γ, IL-5, IL-2, IL-10, IL-4 and TNF-α) are measurable and can be monitored to describe the process of antigenspecifc T-cell activation. Remarkable peaks of enhanced secretion

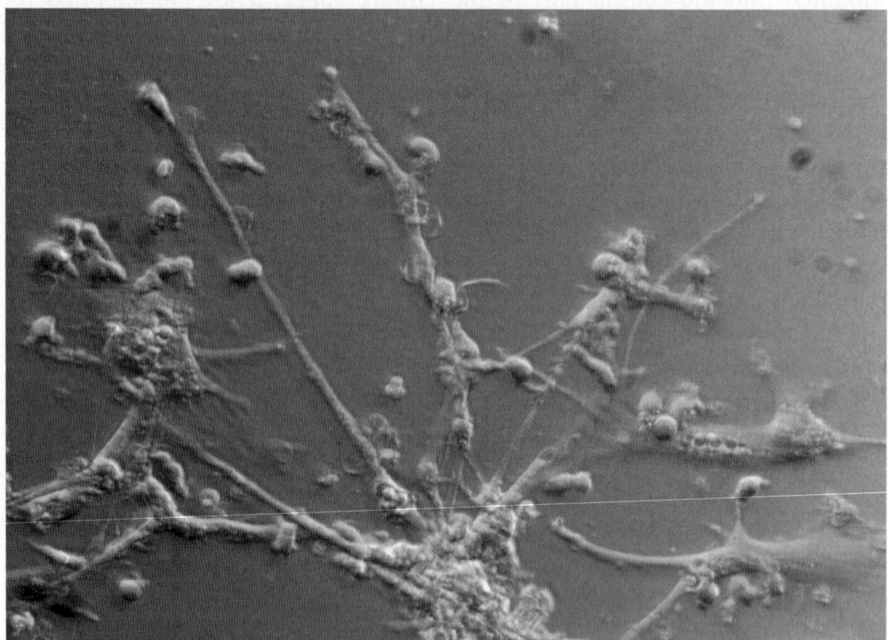

*Figure 1.* A dendritic network is formed in a antigen stimulated co-culture of DC and T-cells (microscop. magn. 20x).

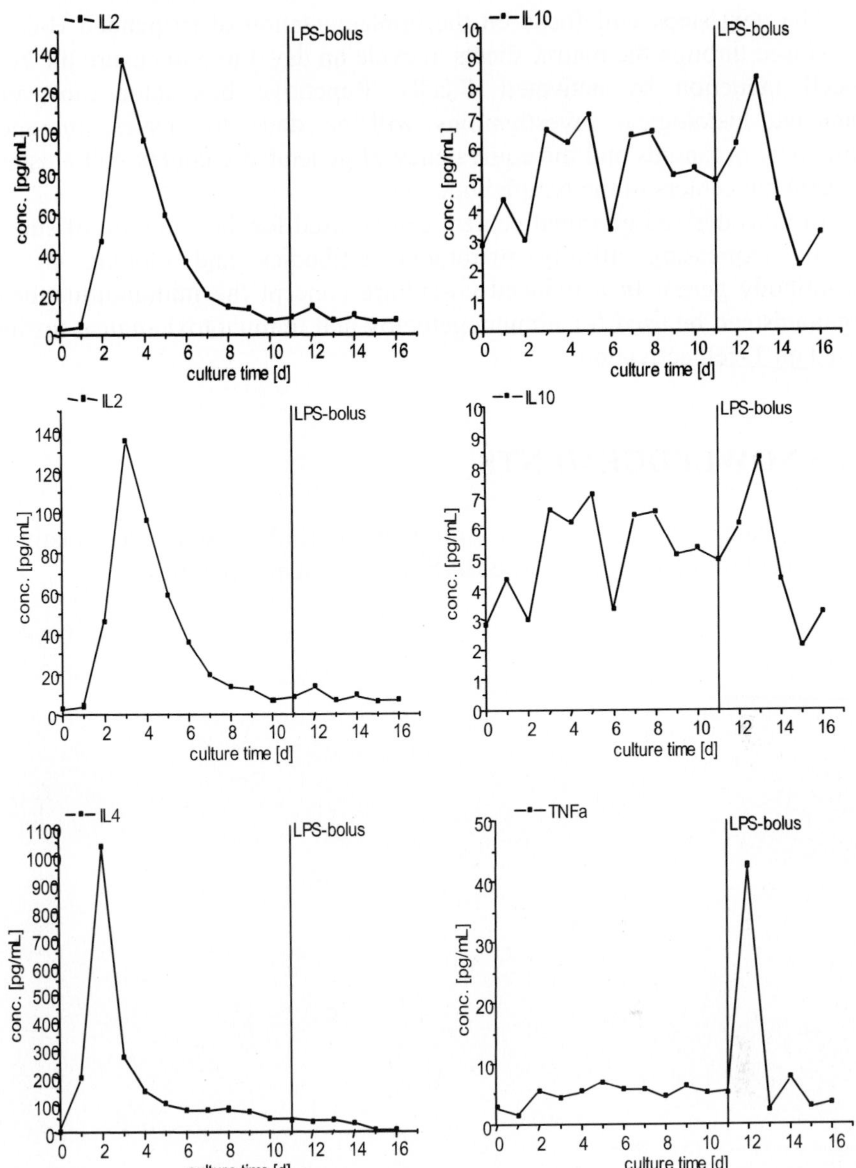


*Figure 2.* Dynamics of secreted cytokines of a HIRIS™ bioreactor run (IFN-γ, IL-5, IL-2, IL-10, IL-4 and TNF-α). Remarkable peaks of enhanced secretion can be observed after mitogenic re-stimulation on day 11.

can be observed after mitogenic re-stimulation on day 11 (Fig. 2). Recently the antigenspecific activation of a co-culture of DCs and T-cells could be shown.

The next steps will focus on the implementation of suspended B-cells transfused through the matrix sheets in cycle on day 4 to 5 to ensure the best B-cell induction by activated T-cells. Repetitive bioreactor runs and intensive histological investigations will be done to ensure immuno-functional organoids and the equivalency of generated complex cell clusters to germinal centers of the lymph node in vivo.

In vitro derived germinal centers can be used for the isolation of single B-cells expressing affinity maturated antibodies and cloning of of theantibody genes. In a reduced co-culture concept the immunofunctional organoids can be used for immunogenicitiy and immunotoxicolgical testing based on T-cell activation.

## ACKNOWLEDGEMENTS

We thank Prof. Hans-Dieter Volk and Dr. Kathrin Schmolke (Institute for Clinical Immunology, Charité Berlin) for cytokine analytics.

# Single Cell Study in a Hydrogel

T. Braschler, R. Johann, U. Seger, H. Van Lintel, P. Renaud
*Swiss Federal Institute of Technology of Lausanne (EPFL), Lausanne, Switzerland*

**Abstract:** A single cell study is performed in a microfluidic device employing the versatility of a newly developed technique of controlled cell entrapment in a calcium alginate hydrogel. A Jurkat cell captured in a way that half of its body is held in the gel, the other half exposed to flowing medium, shows an unexpected behaviour: it elongates and moves out of the gel into the medium. We speculate, that the directed migration is stimulated by shear stress exerted by the fluid flow, creating an effect in our device, that is similar to diapedesis.

**Keywords:** Microchip, cell immobilization, hydrogel, jurkat, diapedesis

## 1. INTRODUCTION

Cell immobilization is fundamental for cell studies. Cell entrapment in hydrogels on chip is a means to achieve cell immobilization in a microfluidic environment, and is an area of intense research. Casting gels in liquid precursor form (Tan and Desai), photolithography of PEG-hydrogels (Koh and Pishko) and photothermal etching of agar (Kojima *et al.*) are some of the more current approaches. We use the binary reaction of $Ca^{2+}$ with alginate to form a solid hydrogel from two liquid precursors under precise microfluidic control (Johann *et al.*, Braschler *et al.*). It has the advantage of full biocompatibility and enbles to precisely place a selected cell before hydrogel formation.

Cellular morphology and function is dependent on the local microenvironment. The chemical composition of the extracellular matrix and communication with neighbouring cells are important factors, but also spatial microheterogeneity: chemotaxis and polar cell organisation with apical and basal membranes show that spatial gradients are detected and reacted upon by cells. In microtechnology, structures at the cellular scale or even smaller are routinely produced. However, most of the fabrication



*R. Smith (ed.), Cell Technology for Cell Products*, 291–295.

technologies are not biocompatible, because of the aggressive chemicals and/or UV radiation used in the process. In PEG-hydrogel photolithography, for instance, the UV doses used are sufficiently low to allow cell survival, although the possibility of mutations and UV stress remain unresolved issues. By adapting the alginate hydrogel technology to microchips, UV and toxic chemicals are avoided, while spatial control of the reaction is maintained using microfluidics. The basic techniques as well as a method for positioning single cells before hydrogel formation has been published recently (Braschler *et al.*). Here we extend the technique to mammalian cells by demonstrating viability of immobilized jurkat and HEK cells.

An interesting aspect of microfluidics is the precise spatial control at dimensions comparable to the size of a single cell. For instance, cells can be half-trapped: a defined fraction of the cell protrudes into the culture medium, while the other part is immoblized in the gel. As a first application of the alginate entrapment technology on chip, we use the partial entrapment of jurkat cells to demonstrate shear-induced changes in the cell morphology. Lymphocytes are capable of crossing the endothelium by diapedesis. Although still under investigation, the role of chemical factors such as selectins during rolling, chemokines and integrin action for lymphocyte activation and attachment are quite well understood (Roitt *et al.*). Shear stress induced by blood flow is currently under investigation as a supplementary factor modulating diapedesis, as show for neutrophils in (Cinamon *et al.*). It is in this context that our hydrodynamical model system may be of interest.

## 2. EXPERIMENTAL

### 2.1 Device Fabrication, Interfacing and Operation

Standard micromachining techniques were used for fabricating fluidic channels, which were closed by reversibly sealing a PDMS piece with access holes on top of a glass chip. The assembly was interfaced with a PMMA block containing reservoirs for the liquids used in the experiment. Pneumatic pressure generated by a pressure-dividing box described by Braschler *et al.* was used to drive and control the different fluid streams in the microchip.

### 2.2 Chemicals

PDMS (Sylgard 184) was obtained from Omya, Sodium Alginate, $CaCl_2$-dihydrate, EDTA, Tris buffer and fluoresceine diacetate from Sigma-Aldrich, DMEM culture medium from Gibco; the jurkat and CHO cells were a courtesy of Prof. Wurm's group (EPFL), HEK 293 a courtesy of Prof. Vogel's group (EPFL).

For gel formation, 4 solutions were prepared: low concentration $CaCl_2$ solution (1mM $CaCl_2$, 50mM Tris) and high concentration $CaCl_2$ solution (15mM $CaCl_2$, 50mM Tris); the cells were contained in an alginate solution (0.8% w/w) in DMEM culture medium, supplemented with 15mM EDTA; the final culture medium was DMEM.

# 3. RESULTS AND DISCUSSION

## 3.1 Viability of immobilized mammalian cells

Viability of included HEK 293 and jurkat cells were assessed by loading them with fluoresceine diacetate and observing leakage of the dye. This indicates membrane integrity. Using flows on the order of 100μm/s and stable gel position, no cells were observed to die during immobilization and the minutes thereafter, indicating no immediate toxicity or harmful effect. However, violent displacements of the gel by experimentally introduced brutal pressure changes rapidly killed the cells, especially the HEK 293. This indicates that while controlled slow flow and stable inclusion into the gel are not problematic, large deformations and shear forces are harmful to the cells. The immobilized HEK 293 are probably more sensitive to such mechanical stress because they are not only caught in the gel but often establish also a firm contact with the glass surface, giving thus rise to conflicting forces when the gel is displaced. For HEK 293 cells adhering to the border of the gel, intentional displacement of the gel sometimes resulted in displacement of the cell, sometimes in detaching the cell from the gel, corroborating the hypothesis of conflicting adhesion forces leading to cell disruption.

## 3.2 Positioning Cells at the gel border

For controlled entrapment of a cell at the border of the gel, the following protocol was developed (Figure 1).

In an initialization phase (Figure 1-A), fluid flow was initiated and the pressures adjusted. In this phase, contact of the alginate with concentrated calcium solution is avoided, so no gel formation takes place. Gel formation is then initiated by bringing the concentrated $CaCl_2$-solution (15mM) into contact with the alginate solutions containing the cells (Figure 1-B). Rapid gel growth leads to full inclusion of the cells into the gel. By changing back from high to low calcium concentration, gel dissolution is initiated (Figure 1-C). When a cell is exposed to the desired extent, the alginate-EDTA supply is replaced with culture medium. As a consequence, the gel is bordered by a solution containing $Ca^{2+}$ (1mM) and culture medium, and is thus stabilized for an indefinite period of time (Figure 1-D). This protocol exploits both the possibility of controlling the gel width by varying the composition of the liquids next to the gel and the possibility of maintaining the gel indefinitely in solutions containing physiologic calcium concentrations. However, it presents the inconvenience of a transient contact of the cells with EDTA. Although such transient contact with EDTA is routinely used for detaching cells in cell culture passaging, conditions could be made more biocompatible by using citrate or EGTA instead.

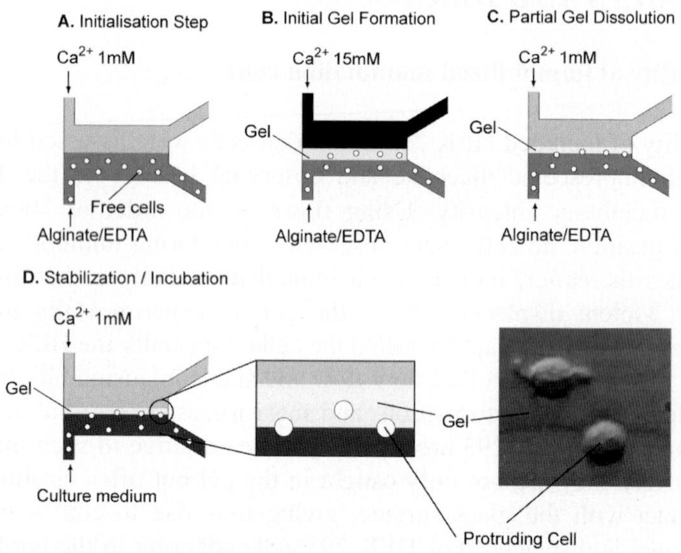


*Figure 2.* Microfluidic protocol for positioning cells at the gel border. The procedure is detailed in the text. The cells shown in 1-D are HEK 293.

### 3.3 Cellular responses to shear stress

When jurkat cells were partially entrapped such that about half of the cell protruded into the free flowing culture medium (flow speed 100-500μm/s, channel height 30μm), while the other half was contained in the hydrogel, characteristic morphological changes were observed. The cells elongate and constrict in a way that is reminiscent of diapedisis through endothelium (Figure 2). Migration is towards the culture medium. HEK293 or CHO cells, in similar conditions, show no directed migration, although changes in cellular shape (CHO, HEK 293) and exploration of the surroundings by filopodia (HEK293) were observed on the scale of 30 minutes. These results are preliminary, as precise conditions for migration are currently under investigation.

*Figure 3.* Artificial diapedesis. A jurkat cell was partially entrapped at the gel border and exposed to shear for 20 minutes. The micrograph shows the resulting cell morphology.

# REFERENCES

Braschler, T., Johann, R., Heule, M., Metref, L., Renaud, Ph., 2005, Gentle Cell Trapping and Release on a Microfluidic Chip by in situ Alginate Hydrogel Formation, Lab on a Chip 5: 553-559

Cinamon, G. *et al.*, 2004, Chemoattractant Signals and beta 2 Integrin Occupancy at apical endothelial Contacts combined with Shear Stress Signals to promote transendothelial neutrophil Migration, J Immunol. 173:7282-91

Johann, R., Braschler, T. Renaud, Ph. 2004, Reversible Cell Trapping in a Hydrogel on Chip, Nanotech-Montreux, Conference Proceeding

Koh, W.G., Pishko, M., 2003, Photoreaction Injection Molding of Biomaterial Microstructures, Langmuir 19: 10310-10316

Kojima, K. *et al.*, 2003, Two-dimensional Network Formation of Cardiac Myocytes in Agar Microculture Chip with 1480 nm infrared Laser photo-thermal Etching, Lab on a chip 3: 292-296

Roitt, I. *et al.*, Immunology, Fourth Edition, London, 1996

Tan, W., Desai, T.A., 2003, Microfluidic patterning of Cells in extracellular Matrix Biopolymers: Effects of Channel Size, Cell Type, and Matrix Composition on Pattern Integrity, Tissue engineering 9: 255-261

# Predicting Tissue Assembly of Prostate Cancer Spheroids

Kim O'Connor[1,3], Cristina Vidulescu[2], Sanda Clejan[2,3], Hong Song[1,3] and Mark Venczel[1]

[1]*Department of Chemical and Biomolecular Engineering, Tulane University, New Orleans, LA 70118, USA.* [2]*Department of Pathology, Tulane Health Sciences Center, New Orleans, LA 70112, USA.* [3]*Interdisciplinary Molecular and Cellular Biology Program, Tulane Health Sciences Center, New Orleans, LA 70112, USA.*

**Abstract:** Computational methods that predict tissue assembly aid in the production of biological substitutes that mimic native tissue. In particular prostate cancer cells self-assemble on an attachment-limiting substrate into spheroids that resemble micrometastases and have application to in vitro drug testing. Two mathematical models of spheroid formation have been developed using the population-balance and Monte Carlo method. The models accommodate a variety of size populations: single cells and spheroids of different sizes. The population-balance model predicts spheroid size distributions over a 5-fold range of cell concentrations in the inoculum. Monte Carlo simulations predict long-range interactions between aggregating cells on the order of several cell diameters. This study provides evidence of intercellular bridges between the cancer cells that contain alpha-tubulin and can extend at least 100 microns in length. The computational methods presented here are robust in predicting spheroid assembly and underlying biological phenomena. Since spheroid composition is size-dependent, the models may be able to predict both spheroid size and composition from properties of the inoculum.

**Key words:** aggregation, assembly, cancer, computation, filopodia, intercellular bridges, kinetics, micrometastases, model, Monte Carlo, population balance, prostate, spheroid, tissue, tubulin.



*R. Smith (ed.), Cell Technology for Cell Products*, 297–302.

# 1. INTRODUCTION

Tissue engineering has produced a remarkable variety of metabolic and structural tissue substitutes for implantation, as extracorporeal devices and for in vitro drug testing (Langer and Vacanti, 1993). The production of a living tissue substitute from cell culture involves the assembly of vast numbers of cells into three-dimensional structures in a controlled culture environment. Computer simulations of this complex process aid in the rational design of engineered tissue by facilitating the comparisons of different production strategies with virtual cell cultures. For example, mathematical models and computational analysis have been helpful in the design of a soft-tissue implant (Song *et al.*, 2002), bioartificial nerve graft (Rutkowski and Heath, 2002) and liver assist device (Allen and Bhatia, 2003).

Our research group has recently developed mathematical models of the self-assembly of multicellular spheroids using liquid-overlay cultures of prostate cancer cells as a representative system (Enmon *et al.*, 2002; Song *et al.*, 2003). Prostate cancer spheroids mimic micrometastases and have application to in vitro drug testing and, in particular, drug development and delivery, and the design of patient-specific therapies (O'Connor, 1999). Our models are unique in their ability to account for the aggregation of attachment-dependent cells at the resolution of one-cell increments. A population-balance model was designed to investigate the effect of cell concentration on spheroid size distribution (Enmon *et al.*, 2002), and another using the Monte Carlo method was developed to resolve intercellular phenomena during spheroid assembly (Song *et al.*, 2003).

The present study demonstrates the predictive capacity of the population-balance model with respect to cell concentration and the Monte Carlo model with respect to long-range intercellular interactions.

# 2. MATERIALS AND METHODS

## 2.1 Cell Cultivation

DU 145 human prostate cancer cells (ATCC HTB 81, Manassas, VA) were cultivated at 37 °C, 5% $CO_2$ and 95% relative humidity in GTSF-2 complete medium (Lelkes *et al.*, 1997) containing 7% fetal bovine serum (pH 7.4). Multicellular spheroids were prepared in liquid-overlay cultures on an attachment-limiting substrate of 1% agar (Difco, Detroit, MI) as previously described (Enmon *et al.*, 2002; Vidulescu *et al.*, 2004). Liquid-overlay cultures were inoculated with 5.0 x $10^3$ cells/cm$^2$ to 2.5 x $10^4$ cells/cm$^2$ at a constant volumetric density of 1.6 x $10^5$ cells/ml.

The concentration of single cells and aggregates in liquid-overlay culture was determined by automated analysis of digital images captured with light microscopy as described by Enmon *et al.* (2002). Image analysis was performed with Image-Pro Plus software (Media Cybernetics, Silver Springs, MD). Between 14,000 to 36,000 aggregates per time period were examined.

## 2.2 Fluorescence Microscopy

Filopodial dynamics was monitored with fluorescence microscopy by labeling cell membranes with the lipophilic stain Vybrant® DiI (Molecular Probes, Eugene, OR). DU 145 cells were suspended at a density of 1 x $10^6$ cells/ml in serum-free GTSF-2 medium and incubated with 5 µL/ml stain at 37 °C in the dark for 20 min. After centrifugation and suspension in complete medium, labeled cells were plated in liquid-overlay cultures as described above.

DU 145 cells were subjected to in situ immunofluorescence to detect alpha-tubulin. Samples were fixed for 10 min with 3.7% methanol-free formaldehyde in PBS and permeabilized for 5 min with 0.1% Triton X-100 in PBS containing 1% bovine serum albumin. Immunostaining was performed with anti-alpha-tubulin mouse $IgG_1$ monoclonal antibody and fluorescent anti-mouse IgG antibody according to manufacture's instructions (Molecular Probes).

## 2.3 Mathematical Models

The development of the population-balance and Monte Carlo model is described in detail in Enmon *et al.* (2002) and Song *et al.* (2003), respectively. Briefly, the accumulation of cell aggregates of a given size $i$ during spheroid self-assembly was modeled as a population balance between the formation of aggregates of size $i$ from smaller aggregates and their depletion for the formation of larger aggregates. The model accounts for all size classes of aggregates in culture at a resolution of one-cell increments. The rate of aggregation is described by a set of kinetic rate constants $k_{ii}$ and $k_{ij}$ that are a function of aggregate size. The former represents aggregation between aggregates of the same size $i$; the latter, dissimilar sizes $i$ and $j$.

The Monte Carlo model is based on the cluster-cluster aggregation model developed by Meakin (1984) for particle flocculation. It was modified to mimic cell aggregation by introducing a radius of influence, adhesion probability, initial aggregate size distribution, Brownian-like cell movement and convergence of aggregated cells. The radius of influence is defined as the maximum distance between cell centers in which aggregation can occur.

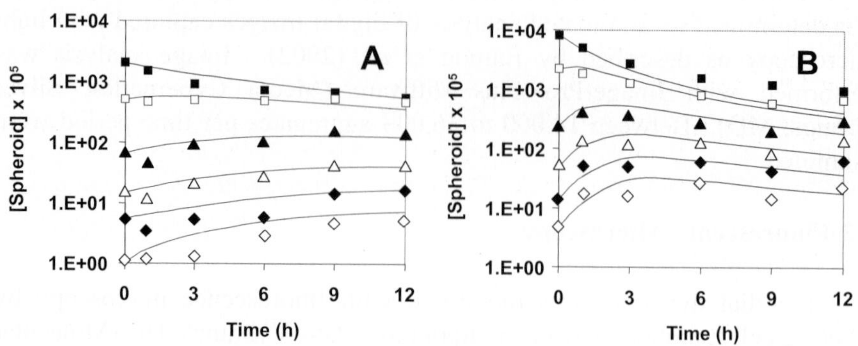


*Figure 1.* Kinetics of DU 145 cells aggregating on agar in liquid-overlay cultures at 5 x $10^3$ cells/$cm^2$ (A) and 2.5 x $10^4$ cells/$cm^2$ (B). Symbols: ■, single cells; •, dimers; ▲, 4-mers; △, 6-mers; •, 8-mers; and •, 10-mers.

# 3. RESULTS AND DISCUSSION

## 3.1 Cell Concentration

Figure 1 describes the depletion of single cells and formation of cell aggregates during the self-assembly of DU 145 spheroids on agar in liquid-overlay culture over 12 h when inoculated with 5.0 x $10^3$ cells/$cm^2$ (Figure 1A) and 2.5 x $10^4$ cells/$cm^2$ (Figure 1B). The symbols depict experimental aggregate concentrations that are expressed as dimensionless values, aggregates/single cell; whereas, the curves represent simulated aggregate concentrations from the population-balance model. Data are presented for aggregate size $i$ = 1 to 10. Model simulations were obtained with a single set of kinetic rate constants as model input [$k_{ii}$ ($h^{-1}$) = 1.7 + 1.9$i$ - 5.2 x $10^{-1}i^2$ + 6.7 x $10^{-2}i^3$ and $k_{ij}$ = 0.5($k_{ii}$ + $k_{jj}$)]. There is good fit between simulated and experimental concentrations over a 5-fold range of inoculum densities. The population-balance model is capable of predicting the concentrations of single cells through 10-mers as a function of time for the entire culture period.

There is some error in model predictions particularly for large aggregates at 5.0 x $10^3$ cells/$cm^2$ (Figure 1A). Error between model simulations and experimental data is more uniformly distributed over all size classes at 2.5 x $10^4$ cells/$cm^2$ (Figure 1B). We attribute this error to the uncertainty of measuring large aggregates that appear infrequently in culture, intrinsic changes in the rate of aggregation as a function of inoculum density, and difficulty in resolving aggregate boundaries in concentrated cultures.

The predictive capacity of the population-balance model may be useful in predetermining spheroid composition. The composition of spheroids is highly dependent on their size. For example, the fraction of proliferating cells is the largest on the spheroid surface and rapidly declines in the interior of static spheroids over a distance of four cell diameters (Song *et al.*, 2004). With this size-composition relationship, model predictions could identify inoculum concentrations to produce spheroids with a desired composition.

## 3.2 Intercellular Interactions

Our research group observed during development of the Monte Carlo model that the best fit of simulated to experimental aggregation data occurred when the radius of influence was significantly greater that the cell diameter (Song *et al.*, 2003). In other words, the model solution suggests that long-range intercellular interactions are involved in spheroid self-assembly. These interactions could be physical and/or chemical in nature.

As shown in Figure 2A, we have detected filopodial extensions and intercellular bridges between DU 145 cells that have diameters from 100 nm to 5 μm and lengths from a few microns to at least 50 μm to 100 μm. These reach lengths are on the order of the radius of influence predicted by Monte Carlo simulations. Immunofluorescence detected alpha-tubulin in filopodia, indicative of microtubule bundles (Figure 2B). Filopodial extension is likely an early step in cell aggregation and spheroid self-assembly.

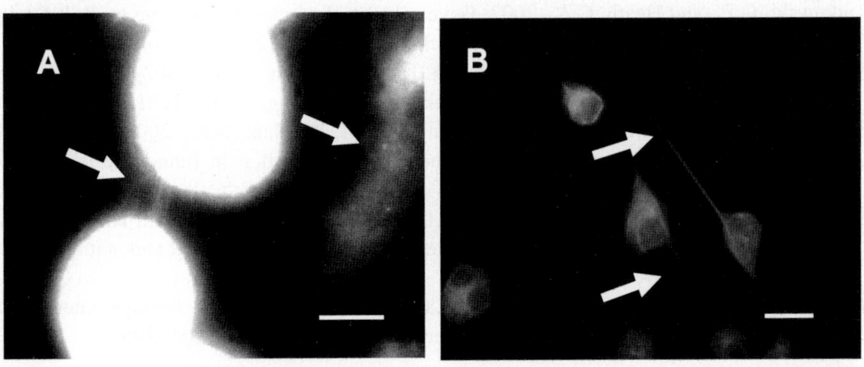


*Figure 2.* Extensions and intercellular bridges of filopodia from DU 145 cells labeled with DiI (A, 10 μm scale) and anti-alpha-tubulin antibody (B, 25 μm scale).

In a related study, our research group discovered that filopodia are also involved in intercellular communication. Specifically, filopodial extensions can function as conduits for the transport of membrane vesicles between adjacent DU 145 cells (Vidulescu *et al.*, 2004). For DU 145 and many other cancer cells that are deficient in gap junctional intercellular communication, flow of cellular material across intercellular bridges may partially compensate for impeded transport through defective gap junctions.

## ACKNOWLEDGMENTS

This research was funded with grants from NASA (NAG-9-1351) and the Tulane Cancer Center.

## REFERENCES

Allen, J.W., and Bhatia, S.N., 2003, Formation of steady-state oxygen gradients in vitro: application to liver zonation, Biotechnol. Bioeng. 82:253-262.

Enmon, R.M., O'Connor, K.C., Song, H., Lacks, D.J., and Schwartz, D.K., 2002, Aggregation kinetics of well and poorly differentiated human prostate cancer cells, Biotechnol. Bioeng. 80:580-588.

Langer, R., and Vacanti, J.P., 1993, Tissue engineering, Science 260:920-926.

Lelkes, P.I., Ramos, E., Nikolaychik, V.V., Wankowski, D.M., Unsworth, B.R., and Goodwin, T.J., 1997, GTSF-2: a new, versatile cell culture medium for diverse normal and transformed mammalian cells, In Vitro Cell. Dev. Biol. Anim. 33:344-351.

Meakin, P., 1984, Diffusion-controlled aggregation on two-dimensional square lattices: results from a new cluster-cluster aggregation model, Phys. Rev. B 29:2930-2942.

O'Connor, K.C., 1999, Three-dimensional cultures of prostatic cells: tissue models for the development of novel anti-cancer therapies, Pharm. Res. 16:486-493.

Rutkowski, G.E., and Heath, C.A., 2002, Development of a bioartificial nerve graft. I. Design based on a reaction-diffusion model, Biotechnol. Prog. 18:362-372.

Song, H., O'Connor, K.C., Papadopoulos, K.D., and Jansen, D.A., 2002, Differentiation kinetics of in vitro 3T3-L1 preadipocyte cultures, Tissue Eng. 8:1071-1081.

Song, H., O'Connor, K.C., Lacks, D.J., Enmon, R.M., and Jain, S.K., 2003, Monte Carlo simulation of LNCaP human prostate cancer cell aggregation in liquid-overlay culture, Biotechnol. Prog. 19:1742-1749.

Song, H., David, O., Clejan, S., Giordano, C.L., Pappas-LeBeau, H., Xu, L., and O'Connor, K.C., 2004, Spatial composition of prostate cancer spheroids in mixed and static cultures, Tissue Eng. 10:1266-1276.

Vidulescu, C., Clejan, S., and O'Connor, K.C., 2004, Vesicle traffic through intercellular bridges in DU 145 human prostate cancer cells, J Cell. Mol. Med. 8:388-396.

# Cultivating Cells of Different Origin for 3d Bone Constructs Considering Physiological Conditions

K. Suck[1], C. Kasper[1], C. Hildebrandt[1], S. Diederichs[1], M. Fischer[1], T. Scheper[1], M. van Griensven[2]
[1]*Institut für Technische Chemie, Callinstr. 3, 30167 Hannover, Germany*
[2]*Ludwig Boltzmann Institut für experimentelle und klinische Traumatologie, Donaueschingenstr. 13, 1200 Wien, Austria*

**Abstract:** One major objective in bone tissue engineering is the construction and application of precise matrices in order to support and guide cell growth and differentiation. These matrices are supposed to replace the extracellular matrix (ECM), which naturally provides cells with a supportive framework of structural proteins, carbohydrates and signaling molecules. The ideal matrix has to mimic these ECM characteristics and should form the desired structure with similar physical properties as bone tissue. Moreover it is known that mechanical strain is an essentiell stimulus for proper function of tissue like bone and cartilage. The mechanical strain mimics the physiological environment and thus supports the differentiation process. In this work MC3T3-E1 cells (mouse) were seeded onto macroporous ceramic scaffolds and cultured with and without differentiation conditions. Cell viability was monitored via MTT test. The differentiation status of MC3T3-E1 cells was investigated using the alkaline phosphatase assay and von Kossa staining of the extracellular matrix. Furthermore BMSCs were subjected to mechanical strain. Strained cells were compared to non strained cells with regard to proliferation, differentiation and repair mechanism by performing standard BrdU assay and collagen I and III specific radio immuno assays (RIA).

**Key words:** bone tissue engineering, Sponceram®, Biostat® RBS, rotating bed system, bioreactor, differentiation, mineralisation, BMP-2, MC3T3-E1 cells, BMSCs, mechanical strain, alkaline phosphatase, collagen I, collagen III, von Kossa, MTT assay



*R. Smith (ed.), Cell Technology for Cell Products*, 303–311.

# 1. INTRODUCTION

One approach in tissue engineering is to isolate cells or a tissue from a patient and to generate a construct of matrix and cells *ex vivo*, before reimplantation into the patient.

These matrices have to be biocompatible, should support cell attachment, growth and differentiation towards the desired phenotype.

The matrix for bone engineering has to fulfil requirements concerning their mechanical stability, biodegradability and porosity. Currently, the most frequently used materials for bone tissue engineering are ceramics like calciumphosphates, hydroxyapatites, degradable polymers such as poly (glycolic acid)/poly(lactid acid) or composite materials (Burg *et al.*, 2000, Hutmacher, 2000, Logeart-Avramoglou *et al.*, 2005). Also zirconium oxide has been investigated as a possible suitable implant material (Hentrich *et al.*, 1971). Zirconia is often applied in composite materials to increase the strength of ceramics (Lee *et al.*, 2004). In this study we used Sponceram® as biomaterial for bone engineering. It is a highly porous ceramic compound, consisting of doped zirconium oxide ($Zr0_2$). Moreover, special proteins, such as growth factors play an important role in tissue engineering. Bone morphogenetic proteins are the main inductors for bone and cartilage formation (Reddi 1992). They belong to the TGF-β superfamily (Sebald *et al.*, 2004). Besides BMP-7, BMP-2 is the most prominent used cytokine for the differentiation process into bone tissue (Hollinger *et al.*, 1998, Yamaguchi *et al.*, 2000).

Recently, the field of tissue engineering has utilized mechanical stimulation *in vitro* as a tool for promoting the development of a number of tissue types including bone (Wakitani *et al.*, 1994; Yang *et al.*, 2002), cartilage (Davisson *et al.*, 2002), ligament (Altman *et al.*, 2002), skeletal muscle, and cardiac muscle (Kim & Mooney 2000; Zimmermann *et al.*, 2002).

For bone tissue, according to the general physiological strain *in vivo,* the most frequently used systems elongate, compress or deflect cells grown on different substrates (Neidlinger-Wilke *et al.*, 1994; Ngan *et al.*, 1988; Pender & McCulloch 1991; Schaffer *et al.*, 1994). While in circular devices cells are strained radially which is adequate for cells from dermal tissues or muscle cells, bone cells should better be strained longitudinally on rectangular substrates, where a homogenous strain distribution is given. Such a strain mimics a physiologic strain as present in bone tissue *in vivo*.

Cellular reaction to mechanical strain *in vitro* is dependent on substrate material and strain parameters like elongation amplitude, frequency, or strain duration. In this study, we chose rectangular elastic silicone dishes which are simple to deal with and do not influence cell morphology, total protein content and alkaline phosphatase activity (Neidlinger-Wilke *et al.*, 1994). A frequency of 1 Hz was obtained as optimum for the proliferation of human osteoblastic cells on silicon dishes (Kaspar *et al.*, 2002). In contrast to the frequency, for elongation amplitude and duration, no optimal values seem to exist yet. Therefore, different strain durations were tested in this study.

## 2. MATERIALS AND METHODS

### 2.1 Scaffold production

Under *static conditions* Sponceram® (Zellwerk, Germany) matrices (3 x 4 mm) were incubated for 24 h in standard medium (DMEM, 10% FCS, antibiotics) at 37°C, 5% $CO_2$ in 96-well dishes.

1.5 x $10^4$ MC3T3-E1 cells were seeded on each matrix in 96-well dishes for 30 min at gentle stirring at 37°C. Non attached cells were removed and the wells were filled up with 200 µl medium: 1. standard medium; 2. differentiation medium (1 µM dexamethasone, 10 mM β-glycerolphosphate, 50 µg/ml ascorbic acid); 3. BMP-2 medium (differentiation medium + 10 ng/ml BMP-2) and for each of the following tests the matrices were placed into a new 96-well dish. Scaffolds were cultured for up to 20 days. Since it was not possible to estimate the number of attached cells, the first measurements were performed directly after cell seeding.

Cell viability was assayed using MTT test (Sigma, Germany). DNA concentration was measured using the Pico Green assay (Molecular Probes, USA). Alkaline phosphatase (ALP) was determined by an assay based on the hydrolysis of p-nitrophenyl phosphate to p-nitrophenol.

Under *rotating conditions*, MC3T3-E1 cells were injected by a sterile canula onto dry Sponceram® disc for 30 min. Two different types of cultivation were performed in the Biostat® RBS (rotating bed system) bioreactor (Sartorius BBI Systems, Germany). One cultivation was performed for 28 days under standard medium conditions (seeding cell number: 25 x $10^6$/disc). The second one was cultivated for 10 days under standard medium followed by 11 days cultivation under differentiation medium + 10 ng/ml BMP-2 (seeding cell number: 55 x $10^6$/disc).

At the end of the cultivation period, the cells were briefly washed with PBS and fixed in ice-cold 100% ethanol for 20 min at room temperature.

For von Kossa staining, fixed cells were washed with deionized water and incubated for 30 min in 5% $AgNO_3$ in the dark, washed with deionized water and exposed to ultraviolet light for 2 min.

Scaffolds for scanning electron microscopy were fixed in Karnovsky´s buffer (Ito and Karnovsky 1968) and subsequently dried in increasing acteone solutions.

### 2.2 Mechanical strain experiments

For the strain experiments, stem cells obtained from bone marrow were used. Human bone marrow aspirates were obtained during routine orthopaedic surgical procedures needing exposure of the iliac crest. Bone Marrow derived Stromal Cells (BMSCs) were isolated using Percoll density gradient centrifugation. Isolated cells were subsequently transferred to T-75 $cm^2$ flasks in DMEM containing 10% FCS and antibiotics. Adhering cells are believed to be BMSC. Cells in suspension were discarded. For strain experiments, $1.5 \times 10^5$ second passage BMSCs were seeded onto silicone dishes (11.5 $cm^2$) prepared of a two component silicone resin. Serum concentration was reduced to 1% for 24 h to align the cells into the G0 cell cycle phase. The cells on the silicone dishes were strained longitudinally with a frequency of 1 Hz, an amplitude of 5% for 15 min and 60 min, respectively. In order to investigate the proliferation characteristics of the cells, BrdU was added to the culture medium two hours before strain initiation. BrdU, being a thymidin analogue, is incorporated in the DNA of proliferating cells during the S-phase of the mitosis cycle.

For proliferation assay, cells were fixed with 70% ethanol in 0.5 M HCl at 20°C for 30 min and washed three times with DMEM. Cell proliferation was performed with a standard BrdU assay (Hoffmann La Roche, Germany).

Collagen I and III production was determined with a competitive Radio Immuno Assay (RIA) in the supernatant against the C-terminal propeptide of collagen I and the N-terminal propeptide of collagen III using $^{125}I$ radiolabeled antigens (DPC Biermann, Germany). A standard curve was used to determine the exact concentrations.

## 3. RESULTS AND DISCUSSION

### 3.1 Scaffold production under controlled conditions

To test the applicability of Sponceram® for the cultivation of MC3T3-E1 cells, a first screening in cell culture dishes under static conditions was performed. The cells were seeded in 96-well plate onto the matrices and cultured for up to 20 days. Several analyses for cell proliferation and differentiation into bone were carried out.

Cultivation of MC3T3-E1 cells in BMP-2 medium showed that the cells grew well inside the macroporous structure of Sponceram® with a typical flat morphology of osteoblatic like cells (Fig. 1A). The presence of single cells is rare. The cells grow as an interconnecting network having intercellular contacts to the surrounding cells (Fig. 1B).

*Figure 1.* Scanning electron microscopie image of MC3T3-E1 cells cultured for 8 days in BMP-2 medium on Sponceram®.

### 3.1.1 Cell proliferation and alkaline phosphatase activity

Cell viability of MC3T3-E1 cells cultured under static conditions was analysed with standard MTT assay. Cultured MC3T3-E1 cells on Sponceram® in standard medium, differentiation medium and BMP-2 medium showed a fast cell growth during the first ten days. Due to high confluence on the matrix, the proliferation decreased after 10 days of culture (Fig. 2, left). To evaluate the differentiation process of MC3T3-E1 cells cultured on Sponceram®, the activity of the early osteogenic marker alkaline phosphatase was determined. Cells were seeded onto Sponceram® matrices in 96-well dishes and cultured in standard medium, differentiation medium and BMP-2 containing differentiation medium. Due to the differentiation induction of BMP-2 the alkaline phosphatase activity had a maximum at day 5 of cultured MC3T3-E1 cells in the cytokine containing medium. These results clearly indicate the beginning differentiation process into bone tissue on the matrix (Fig. 2, right).

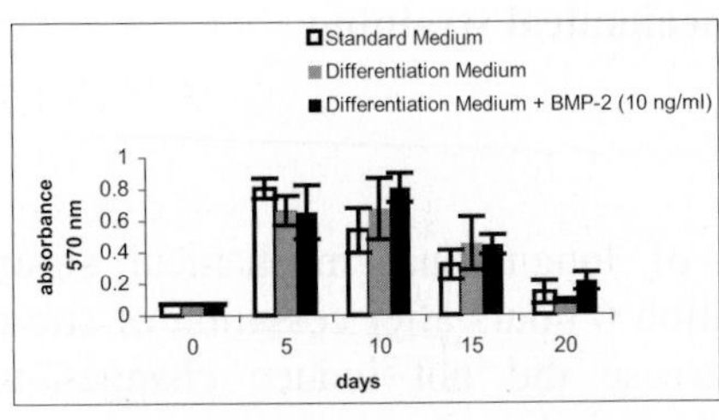


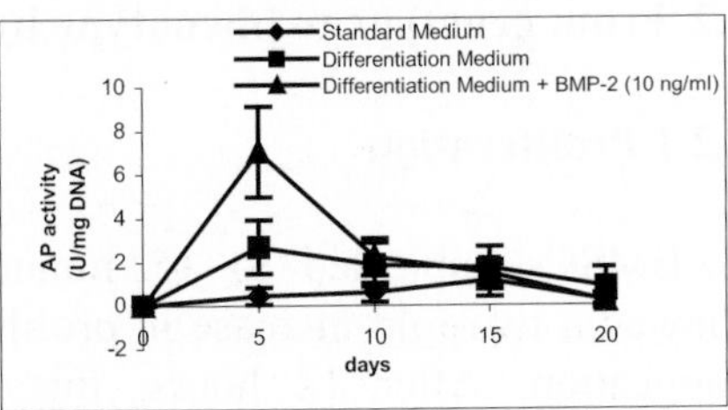


*Figure 2.* Cell viability (left) and alkaline phosphatase activity (right) of MC3T3-E1 cells cultured on Sponceram® over a time period of 20 days. Values represent the mean of 5 samples of cultured scaffolds +/- SD.

### 3.1.2 Mineralisation of ECM

MC3T3-E1 cells were cultured for 4 weeks in the Biostat® RBS bioreactor. One cultivation was performed in standard DMEM medium, in an additional cultivation, cells were cultured for two weeks in standard medium followed by two weeks in BMP-2 medium.

The production of calcified extracellular matrix (ECM) by MC3T3-E1 cells cultured under described conditions in the Biostat® RBS was qualitatively determined by histochemical staining with von Kossa.

Cells cultivated under differentiation conditions with BMP-2 showed a more intense von Kossa black positive staining, which indicates a higher mineralisation (Fig. 3, upper right). The same results were obtained by Alizarin red staining (data not shown).

*Figure 3.* Results of mineralisation after cultivation in the Biostat® RBS. Von Kossa staining of non seeded Sponceram® (middle), cultured in standard medium (upper left) and in differentiation medium (upper right).

However, also MC3T3-E1 cells cultured under standard conditions showed mineralisation of the ECM, which means, that the Sponceram® matrix itself induced the bone differentiation process (Fig. 3, upper left). Thus differentiation medium with BMP-2 potentiate this to a large extent.

## 3.2 From genotype to phenotype by mechanical straining

### 3.2.1 Proliferation

BMSCs subjected to 15 minutes of longitudinal mechanical strain showed a two-fold increase in proliferation 6 hours after cessation of strain application. After 12 hours, this increase did not induce changes in proliferation during the entire observation period.

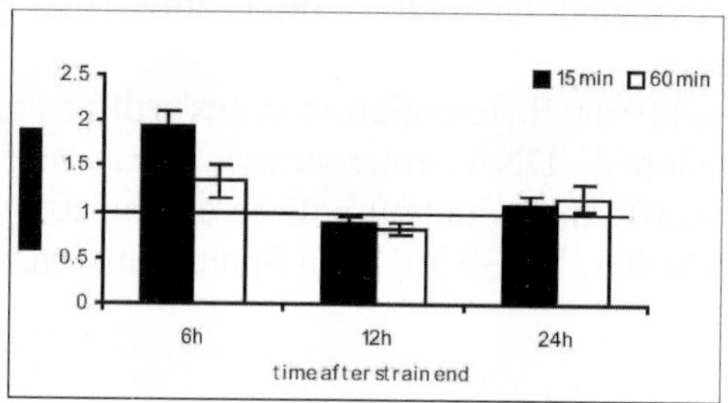


*Figure 4.* Cell proliferation measured by BrdU incorporation. Values represent the mean of 10 patients +/- SD. The baseline at 1 shows the control of non strained cells.

### 3.2.2 Differentiation and repair mechanism

15 minutes duration of strain application resulted in an increase of collagen typ I production. The maximum production was observed 12 hours after cessation of strain application (Fig. 5, left). This strain duration did not result in any change in collagen typ III production. Conversely, 60 minutes of strain induced increases of collagen III already 6 hours after termination of strain application (Fig. 5, right). This increased production remained until the end of the observation period. No significant deviation from baseline values were observed for the collagen I production using this strain duration.

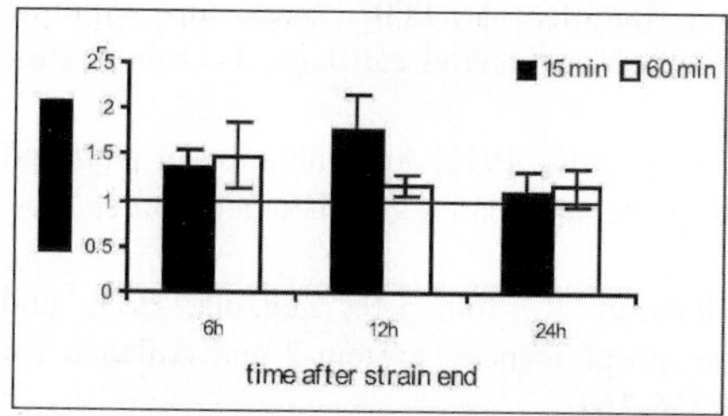


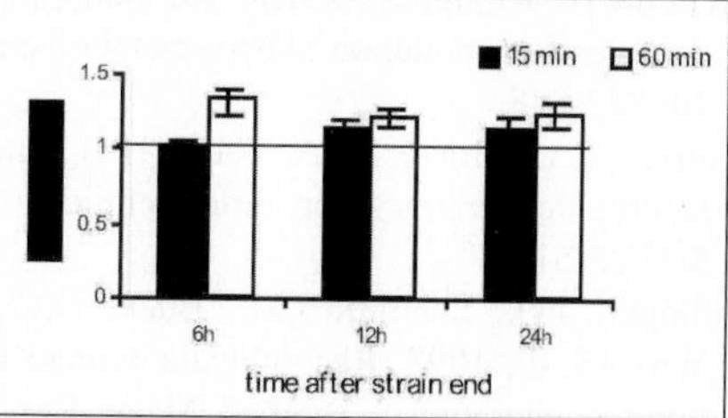


*Figure 5.* Collagen I (left) and III (right) production measured with a RIA. Values represent the mean of 10 patients +/- SD. The baseline at 1 shows the control of non strained cells.

## 4. CONCLUSION AND OUTLOOK

The results of cultivating MC3T3-E1 cells on macroporous Sponceram® showed, that it represents a suitable matrix for bone engineering. For a more controlled cultivation of functional bone tissue the Biostat® RBS was successfully applied. Our mechanical strain experiments showed that a longitudinal strain accelerates the differentiation of human bone marrow stromal cells towards bone tissue. Moreover, choosing an optimal duration leads to physiologic extracellular matrix composition (like collagen I) and increases proliferation.

In further experiments the differentiation status will be analyzed in more detail by self developed DNA microarrays. The mechanical strain experiments and the scaffold production will be combined by cultivation of preconditions cells (BMSCs, MC3T3-E1) on Sponceram® matrices.

## ACKNOWLEDGEMENTS

We thank Prof. W. Sebald (University of Würzburg, Germany) for kindly donating BMP-2 and Zellwerk GmbH (Oberkraemer, Germany) for the support during the cultivation. The SEM images were kindly performed by Dr. A. Feldhoff and F. Steinbach (University of Hannover, Germany).

## REFERENCES

Altman, G.H., Horan, R.L., Martin, I., Farhadi, J., Stark, P.R., Volloch, V., Richmond, J.C., Vunjak-Novakovic,G., and Kaplan, D.L., 2002, Cell differentiation by mechanical stress, Faseb J. **16:** 270-272.

Burg, K.J., Porter, S., and Kellam, J.F., 2000, Biomaterial developments for bone tissue engineering, Biomaterials **21**(23): 2347-59.

Davisson,T., Kunig, S., Chen, A., Sah, R., and Ratcliffe, A., 2002, Static and dynamic compression modulate matrix metabolism in tissue engineered cartilage, J. Orthop. Res. **20:** 842-848.

Hentrich, R.L., Graves, G.A., Stein, H.G., and Bajpai, P.K., 1971, An evaluation of inert and resorbable ceramics for future clinical orthopedic applications, J Biomed Mater Res. **5**(1):25-51.

Hollinger, J.O., Schmitt, J.M., Buck, D.C., Shannon, R., Joh, S.P., Zegzula, H.D., and Wozney, J., 1998, Recombinant human bone morphogenetic protein-2 and collagen for bone regeneration, J. Biomed. Mater. Res. **43**:356-364.

Hutmacher, D. W., 2000, Scaffolds in tissue engineering bone and cartilage, Biomaterials **21**(24): 2529-43.

Ito, S., Karnovsky, M.J., 1968, Formaldehyde-glutaraldehyde fixatives containing trinito compounds, J. Cell Biol. **39**: 168a (abstr).

Kaspar, D., Seidl, W., Neidlinger-Wilke, C., Beck, A., Claes, L., and Ignatius, A., 2002, Proliferation of human-derived osteoblast-like cells depends on the cycle number and frequency of uniaxial strain, J. Biomech. **35:** 873-880.

Kim, B.S. and Mooney, D.J., 2000, Scaffolds for engineering smooth muscle under cyclic mechanical strain conditions, J. Biomech. Eng **122:**210-215.

Lee, T.M., Yang, C.Y., Chang. E., and Tsai, R.S., 2004, Comparison of plasma-sprayed hydroxyapatite coatings and zirkonia-reinforced hydroxyapatite composite coatings: in vivo study, J Biomed Mater Res A. **71**(4):652-60.

Logeart-Avramoglou D., Anagnostou, F., Bizios, R., and Petite, H., 2005, Engineering bone: challenges and obstacles, J. Cell. Mol. Med. **9** (1):72-84.

Neidlinger-Wilke, C., Wilke, H.J., and Claes, L., 1994, Cyclic stretching of human osteoblasts affects proliferation and metabolism: a new experimental method and its application, J. Orthop. Res. **12:** 70-78.

Ngan, P.W., Crock, B., Varghese, J., Lanese, R., Shanfeld, J., and Davidovitch, Z., 1988, Immunohistochemical assessment of the effect of chemical and mechanical stimuli on cAMP and prostaglandin E levels in human gingival fibroblasts in vitro, Arch. Oral Biol. **33:** 163-174.

Pender, N. and McCulloch, C.A., 1991, Quantitation of actin polymerization in two human fibroblast sub-types responding to mechanical stretching, J. Cell Sci. **100 ( Pt 1):** 187-193.

Reddi, A.H., 1992, Regulation of cartilage and bone differentiation by bone morphogenetic proteins, Curr. Opin. Cell Biol., **4**:850-855.

Schaffer, J.L., Rizen, M., L'Italien, G.J., Benbrahim, A., Megerman, J., Gerstenfeld, L.C., and Gray, M.L., 1994, Device for the application of a dynamic biaxially uniform and isotropic strain to a flexible cell culture membrane, J. Orthop. Res. **12:** 709-719.

Sebald, W., Nickel, J., Zhang, J.L., and Mueller, T.D., 2004, Molecular recognition in bone morphogenetic protein (BMP)/receptor interaction, Biol. Chem., **385**:697-710.

Wakitani, S., Goto, T., Pineda, S.J., Young, R.G., Mansour, J.M., Caplan, A.I., and Goldberg, V.M., 1994, Mesenchymal cell-based repair of large, full-thickness defects of articular cartilage, J. Bone Joint Surg Am. **76:** 579-592.

Yamaguchi, A., Komori, T., and Suda, T., 2000, Regulation of osteoblast differetiation mediated by bone morphogenetic proteins, hedgehogs, and Cbfa1, Endocr Rev., **21**(4):393-411.

Yang, Y., Magnay, J.L., Cooling, L., and El, H.A., 2002, Development of a 'mechano-active' scaffold for tissue engineering, Biomaterials **23:** 2119-2126.

Zimmermann, W.H., Schneiderbanger, K., Schubert, P., Didie, M., Munzel, F., Heubach, J.F., Kostin, S., Neuhuber, W.L., and Eschenhagen, T., 2002, Tissue engineering of a differentiated cardiac muscle construct, Circ. Res. **90:** 223-230.

# Guidance on Good Cell Culture Practice

## *A Report of the Second ECVAM Task Force on Good Cell Culture Practice*

Sandra Coecke,[1] Michael Balls,[2] Gerard Bowe,[1] John Davis,[3] Gerhard Gstraunthaler,[4] Thomas Hartung,[1] Robert Hay,[5] Anna Price,[1] Otto-Wilhelm Merten,[6] William Stokes,[7] Leonard Schechtman[8] and Glyn Stacey[9]
*[1]ECVAM, European Centre for the Validation of Alternative Methods, Institute for Health & Consumer Protection, European Commission Joint Research Centre, 21020 Ispra (VA), Italy; [2]FRAME, Russell & Burch House, 96-98 North Sherwood Street, Nottingham NG1 4EE, UK; [3]Research and Development Department, Bio-Products Laboratory, Herts. WD6 3BX, UK; [4]Department of Physiology, Innsbruck Medical University, 6010 Innsbruck, Austria; [5]ATCC, Manassas, VA 20110-2209, USA; [6]Généthon, 91000 Evry, France; [7]National Toxicology Program Interagency Center for the Evaluation of Alternative Toxicological Methods Environmental Toxicology Program, National Institute of Environmental Health Sciences, Research Triangle Park, NC 27709, USA; [8]National Center for Toxicological Research, Food and Drug Administration, Rockville, MD 20857, USA; [9]Division of Cell Biology and UK Stem Cell Bank, National Institute for Biological Standards and Control, Blanche Lane, South Mimms, Potters Bar, Herts. EN6 3QG, UK*

**Abstract:** The history of cell culture teaches us that there is a need to attend to certain fundamental aspects of best practice when preparing cells for routine use. As cell culture techniques and applications become more complex it becomes increasingly necessary to be aware of the impact of poorly controlled or suboptimal cell culture procedures. The Good Cell Culture Practice guidance has been prepared to promote an awareness of a broad range of important issues in cell culture in workers coming to use cells for their work for the first time and to remind others of the fundamental aspects of good practice in cell and tissue culture.

**Key words:** cell culture, best practice

Following the proposal for publication of minimum guidelines for cell and tissue culture at the 3rd World Congress on Alternatives and Animal Use in the Life Sciences (Bologna, Italy, 1999): outline guidance on Good



*R. Smith (ed.), Cell Technology for Cell Products*, 313–315.

Cell Culture Practice was published by Hartung *et al.*, in 2002. In October 2003, a new task force was convened in Ispra, Italy, with a broader range of expertise in cell and tissue culture, in order to produce a more-detailed GCCP guidance document which could be of practical use in the laboratory. This Guidance is required to serve the rapidly expanding use of *in vitro* systems: in basic research, to meet regulatory requirements for chemicals and products of various kinds; in the manufacture of various products; in medical diagnostics; and in therapeutic applications such as tissue engineering, and cell and gene therapy. The new GCCP document, now published in full in ATLA (Coecke *et al.*, 2005), is intended to support best practice in all aspects of the use of cells and tissues *in vitro*, and to complement, but not to replace, any existing guidance, guidelines or regulations.

This GCCP Guidance is based upon the following six operational principles:

1. Establishment and maintenance of a sufficient understanding of the *in vitro* system and of the relevant factors which could affect it.

2. Assurance of the quality of all materials and methods, and of their use and application, in order to maintain the integrity, validity, and reproducibility of any work conducted.

3. Documentation of the information necessary to track the materials and methods used, to permit the repetition of the work, and to enable the target audience to understand and evaluate the work.

4. Establishment and maintenance of adequate measures to protect individuals and the environment from any potential hazards.

5. Compliance with relevant laws and regulations, and with ethical principles.

6. Provision of relevant and adequate education and training for all personnel, to promote high quality work and safety.

This guidance is important for research work, to avoid poor reproducibility of data and the invalidation of results from cell culture processes, and it is especially valuable for the establishment of new cell lines. Critical testing procedures and manufacturing procedures where cells are used have well developed guidelines and regulations, but will benefit from the basic guidance in GCCP which is not given in current quality standards and guidance.

The full text of the GCCP guidance is available on the ECVAM website (http://ecvam.jrc.it) as a 'task force publication' and has been published in the journal ATLA (Coecke *et al.*, 2005).

## REFERENCES

Coecke, S., Balls, M., Bowe, G., Davis, J., Gstraunthaler, G., Hartung, T., Hay, R., Merten, O-W., Price, A., Shechtman, L., Stacey, G.N. and Stokes, W., 2005, Guidance on Good cell Culture Practice. A report of the second ECVAM Task Force on Good Cell Culture Practice, ATLA, **33**: 1-27.

Hartung, T., Balls, M., Bardouille, C., Blanck, O., Coecke, S., Gstraunthaler G. and Lewis, D., 2002, Good Cell Culture Practice Task Force Report 1. ATLA, **30**: 407-414.

# In Vitro Vascularization of Human Microtissues

*VEGF profiling and angiogenesis in human connective micro tissues*

J. M. Kelm, C. Diaz Sanchez-Bustamante and M. Fussenegger
*Institute for Chemical and Bioengineering, Swiss Federal Institute of Technology, ETH Hoenggerberg HCI F115, Wolfgang-Pauli-Strasse 10, CH-8093 Zurich, Switzerland. E-mail: fussenegger@chem.ethz.ch*

**Abstract:** Owing to its dual impact on tissue engineering (neovascularization of tissue implants) and cancer treatment (prevention of tumor-induced vascularization), management and elucidation of vascularization phenomena remain clinical priorities. Using a variety of primary human cells and (neoplastic) cell lines assembled in microtissues (MTs) by gravity-enforced self-aggregation in hanging drops we (i) studied VEGF production of MTs in comparison to isogenic monolayer cultures, (ii) characterized the self-organization and VEGF-production potential of mixed-cell spheroids and (iii) analyzed VEGF-dependent capillary formation of HUVEC (human umbilical vein endothelial cells) cells coated onto human primary cell spheroids.

**Key words:** Tissue Engineering, Vascularization, Angiogenesis, Regenerative Medicine, Self-organization, Endostatin, VEGF, Drug Screening, Cell-based assay system, Blood vessel formation

## 1. INTRODUCTION

An artificial tissue with more than a few cubic millimeters cannot survive by simple diffusion and requires formation of new capillaries to supply essential nutrients/oxygen and enable connection to the host vascular system following implantation. A variety of strategies for therapeutic angiogenesis have been designed including (i) delivery of recombinant angiogenic molecules through controlled-release devices and (ii) functionalized matrices or (iii) transfection/transduction of (engineered) angiogenesis-modulating cDNAs. However, the use of growth factors such as the vascular endothelial growth factors (VEGF) bears some risks. The biological effects of VEGF are extremely dose dependent. Loss of even a single allele results in fatal



*R. Smith (ed.), Cell Technology for Cell Products*, 317–320.

vascular defects in the embryo and insufficient levels of VEGF lead to post-natal angiogenesis and ischemic heart disease, whereas uncontrolled VEGF expression may lead to angioma-genesis.

Based on our previous observations that VEGF production in myocardial microtissue is strictly correlated to cell number and microtissue size we have established an entirely human cell-based microtissue format to provide new insight into VEGF production, angiogenesis and blood vessel formation.

# 2. RESULTS AND DISCUSSION

## 2.1 VEGF production of human microtissue cultures

ELISA-based technology was used to quantify vascular endothelial growth factor (VEGF) production by human cell lines and primary cells grown assembled as microtissues. Based on previous observations suggesting cell type-specific cell number – microtissue size correlations we have used tailored cell concentrations to obtain microtissues of 350 μm in diameter following a 3-day cultivation period.

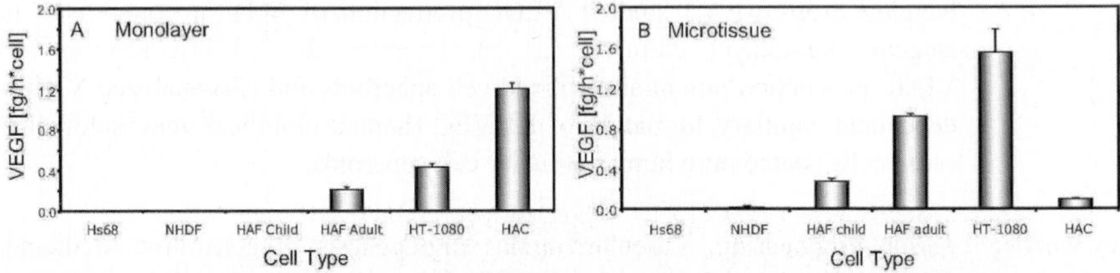


*Figure 1.* Vascular endothelial growth factor (VEGF) production profiling in monolayer and microtissue cultures (350 μm in diameter). HAC: human aortic chondrocytes; HAF child: human aortic fibroblasts from a newborn donor; HAF adult: human aortic fibroblasts from an adult donor; Hs68: embryonic dermal fibroblast cell line; HT-1080: fibrosarcoma cell line; NHDF: normal human dermal fibroblast.

## 2.2 Self-organization potential of different cell phenotypes in a microtissue format

In order to investigate self-organization forces underlying cell migration during angiogenesis we assembled HAF:HUVEC and HepG2:HUVEC suspension cocultures to mixed microtissues in hanging drops. According to the differential adhesion hypothesis HAF/HepG2:HUVEC populations resulted in concentric HAF/HepG2-inside:HUVEC-outside structures, reminiscent of blood vessel cross-sections.

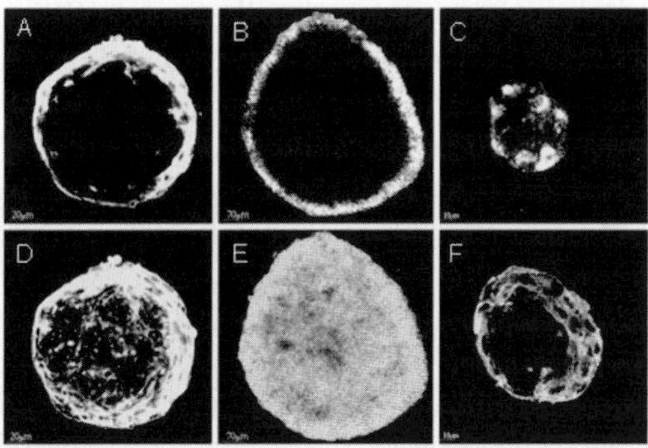

*Figure 2.* Mixed populations of HUVECs, stained for von Willbrand factor (A-C) and HAFs, stained for F-Actin (D), HepG2, stained for F-Actin (E), as well as human umbilical aortic smooth muscle cells, stained for smooth muscle alpha-actin (F).

## 2.3 In vitro vascularization

To vascularize it was not sufficient to produce mixed HUVEC microtissue cultures. Therefore we coated pure HAF microtissues of different size (125 μm – 335 μm; produced by 2-day gravity-enforced assembly) by cocultivation with monodispersed HUVECs in hanging drops. Transmission electron microscopy revealed that the migrated endothelial cells are characterized by lumen formation, with endothelial-periendothelial cell contacts. Thus, lower VEGF production was measured in the vascularized microtissues.

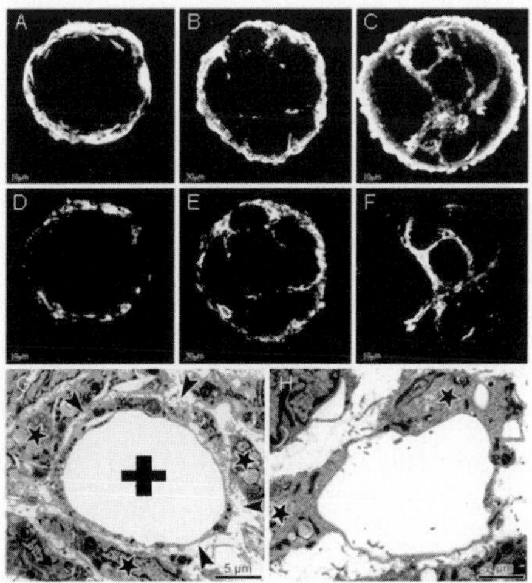

*Figure 3.* Migration of HUVECs, stained for von Willbrand factor (A-C) and CD31 (D-F) in different sized microtissues, initiated with 500 HAFs coated with 600 HUVEC (A, D); 5,000 HAFs coated with 900 HUVECs (B, E); 10,000 HAFs coated with 1,200 HUVECs (C, F), 6 days post coating. Endothelial cells are typically characterized by intracellular lumen formation (+) and are tightly covered by human aortic fibroblasts (asterisks) (G, H).

The C-terminal cleavage product of collagen XVIII known as endostatin is a key anti-angiogenic factor. We evaluated endostatin action on HAF microtissues coated with HUVECs.

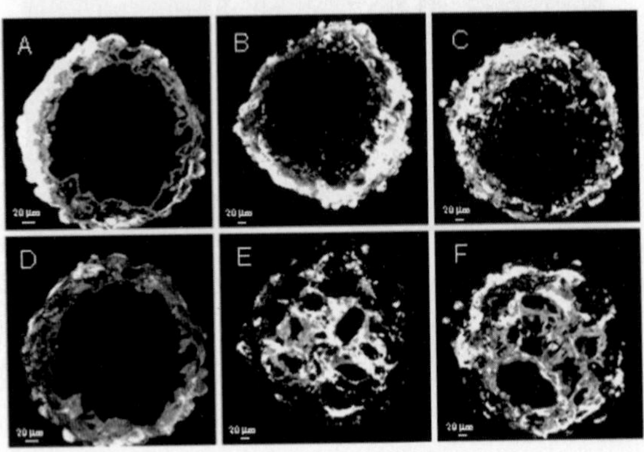


*Figure 4.* Endostatin induced inhibition of HUVEC-based capillary formation. HUVECs were stained for von Willebrand factor (A-C) and CD31 (D-F). Endostatin was supplemented with HUVECs (A, D), two days postcoating (B, E) and the control culture, grown without addition of Endostatin, is shown in (C, F).

# Preparation of Salmon Atelocollagen Fibrillar Gel and Its Application to the Cell Culture

Nobuhiro Nagai[1], Shunji Yunoki[2], Takeshi Suzuki[3], Yasuharu Satoh[1], Kenji Tajima[1], and Masanobu Munekata[1]

*[1]Division of Molecular Chemistry, Graduate School of Engineering, Hokkaido University, N13-W8, Kita-ku, Sapporo, Hokkaido 060-8628, Japan [2]National Institute for Materials Science Namiki 1-1 Tsukuba, 305-0044 Japan [3]Ihara & Company Ltd., 3-263-23 Zenibako, Otaru, Hokkaido 047-0261, Japan*

**Abstract:** Salmon atelocollagen (SAC) has not been used as biomaterials due to its low denaturation temperature (19 degrees C). In the present study, we succeeded in preparation of SAC fibrillar gel stable at an actual physical temperature of human by cross-linking during fibril formation, and in cultivating human periodontal ligament cells on the cross-linked SAC fibrillar gel.

**Key words:** collagen, salmon, fibrillar gel, denaturation temperature, cell proliferation

## 1. INTRODUCTION

Generally, collagens for biomaterials are prepared from mammalian sources, such as bovine and porcine skins. However, the use of mammalian sources has to be reconsidered and limited because of the risks of pathogens such as bovine spongiform encephalopathy (BSE). Although fish collagen is thought to be safe (1), it has not been widely used for biomaterials due to its low denaturation temperature.

Large quantities of fish skin are discarded as waste in the food industry. Collagens are easily extracted from wasted fish skins with high yield. The use of fish collagens, therefore, could contribute to the recycling of natural unutilized resources. Fish collagens have the potential to be used as a novel source for collagen biomaterials if the thermal stability could be improved.



*R. Smith (ed.), Cell Technology for Cell Products*, 321–323.

Recently, we have reported the fabrication of collagen sponges from salmon atelocollagen (SAC) using UV irradiation, dehydrothermal treatment (2), and chemical cross-linking (3). It was found that the thermal stability of SAC sponges was comparable to that of bovine atelocollagen sponges. Next, we tried to prepare the SAC fibrillar gel stable at an actual physical temperature of human.

Collagen molecules self-assemble into cross-striated fibrils in tissues. The aggregation and alignment of molecules improves the thermal stability and mechanical strength of collagen matrix. The collagen fibrils, therefore, provide the major biomechanical scaffold for cell attachment, allowing the shape and form of tissues to be maintained. We hypothesized that the introduction of cross-linking among collagen fibrils during fibril formation would result in a further increase in the thermal stability of SAC fibrillar gel.

As a cross-linking reagent for SAC, 1-ethyl-3-(3-dimethylaminopropyl)-carbodiimide (EDC), a water-soluble-carbodiimide, was used. EDC has become popular as a cross-linking reagent for collagens due to ease of handling and potentially low cytotoxicity (4). Here, we show the increase in the thermal stability of SAC fibrillar gel and its application to the cell culture.

## 2. EXPERIMENTAL

SAC was prepared from fresh skin of chum salmon (*Oncorhynchus keta*) (2) and the purified SAC was found to be mainly composed of type I collagen. The introduction of EDC cross-linking during fibril formation was performed by mixture of acidic SAC solution and EDC solution in pH 6.8 at 4 degrees C (5). Porcine atelocollagen (PAC) was purchased from Nitta Gelatin (Cellmatrix Type I-P, Japan) and the PAC fibrillar gel was prepared (5). The thermal stability, the mechanical strength, and the fibril structure of the gels were evaluated (5). The cell proliferation on the collagen fibrillar gels was measured *in vitro* using human periodontal ligament cells (HPDL cells). The cell number was directly counted by hemocytometer after digestion of gels with collagenase and trypsin. Alkaline phosphatase (ALP) activity, a differentiated cell function of HPDL cells, was measured (3).

## 3. RESULTS AND DISCUSSION

The introduction of EDC cross-linking during fibril formation resulted in further increase of thermal stability of SAC fibrillar gel, and the denaturation temperature was found to be 55 degrees C at an EDC concentration of 60 mM. At an EDC concentration of more than 60 mM, the fibril formation was inhibited, and the denaturation temperature decreased. These results indicate that the maximum synergistic effect of the EDC cross-linking and the aggregation of collagen molecules is obtained at 60 mM. We have reported that the maximum denaturation temperature of SAC fibrillar gel was 47 degrees C at an EDC concentration of 50 mM (5). In the present study, we investigated the condition of EDC concentration in more detail, and it was found that the denaturation temperature reached maximum values at 60 mM.

The mechanical strength of SAC fibrillar gel was five times higher than that of PAC fibrillar gel. Scanning electron microscopy observation showed that SAC fibrillar gel had thin fibrils and well-developed fibril network compared with PAC fibrillar gel. The fibril formation rate of SAC was found to be faster than that of PAC. It was considered that the higher the fibril formation rate the thinner the fibril size. The well-developed fibril network and the cross-links could contribute to the mechanical strength of SAC fibrillar gel.

The proliferation rates of HPDL cells cultured on SAC fibrillar gel were faster than that cultured on PAC fibrillar gel. The mechanism of high cell proliferation rate was unclear. The high mechanical strength and well-developed fibril network could influence the cell proliferation. On the other hand, ALP activities of HPDL cells cultured on PAC fibrillar gel were higher than those on SAC fibrillar gel. The differences of cell activities between SAC and PAC fibrillar gel are under consideration.

SAC fibrillar gel could have the potential being used for biomaterials such as a scaffold for tissue engineering. We have reported the application of SAC fibrillar gel to cell-sheet preparation using collagenase digestion (6). Cell-sheet engineering is a novel method to develop the artificial tissues by layering cell-sheets. We are studying to prepare the layering cell-sheets using SAC fibrillar gel.

## REFERENCES

1. Swatschek, D., Schatton, W., Kellermann, J., Muller, W. E., and Kreuter, J., 2003, Marine., Eur. J. Pharm. Biopharm., 53, 107-113
2. Yunoki, S., Suzuki, T., and Takai, M., 2003, Stabilization., J. Biosci. Bioeng., 96, 575-577
3. Nagai, N., Yunoki, S., Suzuki, T., Sakata, M., Tajima, K., and Munekata, M., 2004, Application., J. Biosci. Bioeng., 97, 389-394
4. Lee, C. R., Grodzinsky, A. J., and Spector, M., 2001, The effects., Biomaterials, 22, 3145-3154
5. Yunoki, S., Nagai, N., Suzuki, T., and Munekata, M., 2004, Novel., J. Biosci. Bioeng., 98, 40-47
6. Nagai, N., Yunoki, S., Satoh, Y., Tajima, K., and Munekata, M., 2004, A method., J. Biosci. Bioeng., 98, 493-496

# Propagation of USSC for Therapeutic Approaches

J. Lamlé [1], H. Büntemeyer [1], A. Reinel[2], T. Menne[2], C. van den Bos[2], J. Lehmann[1]
[1] *Institute of Cell Culture Technology, University of Bielefeld, Germany,* [2]*Kourion Therapeutics AG, Langenfeld, Germany*

**Abstract:** Human adult stem cells isolated from umbilical vein cord blood show promising properties for therapeutic applications. But as the amount of stem cells extracted from an umbilical cord is insufficient to transplant into an adult, it is necessary to expand the undifferentiated stem cells in such a way to enable them for therapeutic issues.

Therefore approaches have been made to cultivate the unrestricted somatic stem cells (USSC) on microcarriers in a stirred bioreactor system with gentle mixing and bubble free aeration ("SuperSpinner"). An establishment of the cultivation of USSC in a bioreactor system could lead to a simplified scale-up process which allows monitoring of all critical parameters and provides larger surfaces for cell attachment and propagation due to the use of microcarriers.

As the use of animal derived serum has to be avoided in terms of therapeutic applications, it is important to develop a serum free cultivation process. For this reason all essential serum ingredients have to be replaced by defined substitutes. Different available serum replacements were tested as well as the supplementation with single substances. Specific growth parameters of USSC cultivated in serum-reduced conditions were measured and compared with those of USSC growing in media containing standard serum concentration in order to identify the best culture conditions. The influence of serum-free cultivation on the differentiation status of USSC is not yet analysed.

**Key words:** cord blood, *in vitro* expansion, stem cells, undifferentiated



*R. Smith (ed.), Cell Technology for Cell Products*, 325–329.

# 1. INTRODUCTION

Human somatic stem cells derived from placental cord blood could serve as a highly valuable resource for the development of cellular therapeutics as they can be propagated in large quantities while retaining their ability to differentiate into different tissue cell types (1). The use of human embryonic stem cells for therapeutic applications is excluded for several reasons, although these cells have the broadest differentiation potential (2). The fact that adult tissue-specific stem cells possess an intrinsic differentiation potential to other tissues is not proven yet, but could be an alternative choice to embryonic stem cells. This so-called "plasticity" appears to be an extremely rare event (3). Subsequent data have even shown that certain results may be the consequence of cell fusion (4). Recently multipotent adult progenitor cells (MAPC) could be isolated from rodent bone marrow which differentiated *in vitro* into cells of all three germ layers. A phenotypically identical cell was isolated from human bone marrow (5). In contrast to adult bone marrow, stem cells derived from cord blood are less mature, have a higher prolific potential associated with an extended life span (6). Furthermore there is plenty of cord blood available, it is harvested without risk to the donor and infections are rare (7).

Recently, a CD45 and HLA class II-negative stem cell could be isolated which shows robust in vitro proliferation capacity without spontaneous differentiation but with an intrinsic potential to develop into mesodermal, ectodermal and endodermal cell types (8). These cells are termed unrestricted somatic stem cells (USSC).

# 2. METHODS AND MATERIALS

Unrestricted somatic stem cells (USSCs) from three different donors were used during this work, termed KCB 330, KCB 554 PM and KCB 704 B. General culture media was composed of DMEM and 20% FBS, supplemented with dexamethasone ($10^{-7}$ M), glutamine (2mM) and antibiotics. For the reduction of serum in culture medium different commercial serum replacements were tested as well as a conditioned medium. During the SuperSpinner cultivation the standard medium composition had been retained unchanged except the addition of pluronic F68 to absorb occurring forces (e.g. shearing) which may damage the microcarriers.

The three-dimensional cultivation system consisted of a 500 ml Duran glass bottle, a flask for inoculation and harvest, and a sterile sampling system. The aeration rates were kept constant at 49.3% $N_2$, 33.2% air and 14.0% $CO_2$, respectively, with a volume flow of 12.4 Nl/h. As a result, the oxygen tension corresponded to the *in vivo* conditions of the *V. umbilicalis* just before birth. The chosen $CO_2$ aeration rate served to maintain the physiological pH value of 7.2 during cultivation. Cultispher-G microcarriers at a concentration of 1 g/l were used as the surface area for cell cultivation. The cell density was about 60 cells/microcarrier at the time of inoculation.

To determine the concentration of glucose and lactate, the biosensor YSI 2700 Select was used. Ammonium was detected fluorometically after derivatisation, whereas the amino acids were analysed by RP-HPLC (9).

For immunohistological staining a monoclonal mouse anti-human CD49e IgG I (X-6, Dianova) and a polyclonal Cy3-conjugated donkey anti-mouse IgG (Dianova) as second antibody were used.

## 3. SUPERSPINNER CULTIVATION ON CULTISPHER-G

After a resting time of 24 hours in order to enhance cell attachment, the system was stirred at intervals at 35 rpm. Afterwards, the system was stirred continuously at 40 rpm. The culture medium was changed regularly at a ratio of 1:3. The growth behaviour of the USSCs on the microcarriers was documented using MTT (3-(4,5-dimethylthiazol-2-yl)-2,5-diphenyl-tetrazoliumbromide) and DAPI (4'-6-Diamidino-2-phenylindole) staining. Two methods were used to determine the cell concentration on the microcarriers. First, the total number of cells was determined after crystal violet staining. Second, the number of viable cells was determined following treatment of the microcarriers with trypsin and subsequent staining with trypan blue. In this context, it should be pointed out that the samples taken had different microcarrier concentration levels due to an inhomogeneous distribution of microcarriers in the culture fluid. For this reason, none of the two methods allowed a reliable determination of cell density. During the first 24 hours, the microcarriers formed aggregates of up to 30 microcarriers. It was not possible to break the aggregates by increasing the stirring speed. In the process of development, it became apparent that the cell density on the microcarriers in the beginning of the cultivation was an important parameter. Single cells on individual microcarriers hardly proliferated, whereas cell growth on microcarriers that showed a higher cell density was much more pronounced. Aggregates in particular were densely colonised after a relatively short period of time. In addition, MTT and DAPI staining revealed that the cells were not just growing on the surface, but migrating into the porous structure of the microcarriers.

Although an exact determination of the cell density was not possible, it seemed as if cells proliferate on densely covered microcarriers whereas on rarely covered microcarriers cell growth could not be observed. The process was not limited all along the cultivation.

## 4. REDUCTION OF SERUM IN CULTURE MEDIUM

Different serum replacements as well as a conditioned medium were tested over a period of five to seven passages.

MarrowGrow is a ready-to-use medium which was originally developed for optimizing the culture of bone marrow cells. It contains a small percentage of FBS, glutamine, growth factors and gentamycin. NuSerum was used in a concentration of 20%. It contains a 25% serum base along with a standardized, consistent formulation of serveral growth factores and hormones, amino acids, vitamins and trace elements. The actual serum concentration in the beginning of these two cultivations was about 5%. The Conditioned Medium was produced by cultivating pimary mouse fibroblasts into a serum-free culture media for two weeks. The complex mixture of secreted components in the conditioned media enhance adhesion and growth of cord blood derived stem cells.

At the end of each cultivation a fluorescence staining against CD49e was made to show that the stem cells are still undifferentiated. CD49e is a antigen which only undiffentiated cells present on the cell surface. To visualize the ratio of differentiated to undifferentiated cells, all cell nuclei were stained with DAPI.

## 5. RESULTS AND CONCLUSIONS

The present work shows that it is possible to cultivate human unrestricted somatic stem cells isolated from umbilical cord blood over a period of at least 20 days without differentiation. Although the determination of the cell density was not conclusive, a slight increase in cell density could be observed during the cultivation on Cultispher-G microcarriers. However, additional cultivations are needed to reproduce the results. One way to solve the problem of aggregate formation could be the use of a bioreactor that can be operated at higher speeds without strong shearing forces. This would result in a more homogeneous distribution of microcarriers in the culture medium and a reproducible determination of cell density. Moreover, successful cultivation in a bioreactor would offer the opportunity of a scale-up so that cells could be cultivated in larger quantities.

In comparison to standard culture conditions used in the SuperSpinner cultivations, the cell growth under MarrowGrow is highly increased during the first passage, but decreases subsequently until it reaches values comparable to standard culture levels. In comparison to 20% serum containing culture conditions, the metabolism of the cells is reduced when NuSerum or conditioned medium is used, which leads to a generally decreased cell growth in the course of cultiviation. Nevertheless a propagation of human unrestricted somatic stem cells in the absence of serum was possible over a period of at least five passages. Cultivation under reduced serum containing conditions was possible even longer.

## REFERENCES

Kuehnle, I. and Goodell, M.A., 2002, The therapeutic potential of stem cells from adults, BMJ. **325**:372-376.

Thomson, J.A. *et al.*, 1998, Embryonic stem cell lines derived from human blastocysts, Science, **282**:1145-1147.

Wagers A.J. *et al.*, 2002, Little evidence for developmental plasticityof adult hematopoietic stem cells, Science, **297**:2256-2259.

Wang, X. *et al.*, 2003, Cell fusion is a principle source of bone-marrow-derived hepatocytes, Nature, **422**:897-901.

Reyes, M. *et al.*, 2001, Purification and ex vivo expansion of postnatal human marrow mesodermal progenitor cells, Blood, **98**:2615-2625.

Vaziri, H. *et al.*, 1994, Evidence for a mitotic clock in human hematopoietic stem cells : loss of telomeric DNA with age, Proc. Natl. Acad. Sci. USA, **91**:9857-9860.

Rubinstein, P. *et al.*, 1993, Stored placental blood for unrelated bone marrow reconstruction, Blood, **81**:1679-1690.

Kögler, G. *et al.*, 2004, A new human somatic stem cell from placental cord blood with intrinsic pluripotent differentiation potential, J. Exp. Med., **200**:1-13.

Büntemeyer H., 2000, Off-line analysis in animal cell culture, methods, Encyclopedia of Cell Technology Ed. R.E. Spier,Wiley, New York: 945-959.

# Bioreactor Based Production of Functional Conditioned Medium for the Propagation of Undifferentiated Human Embryonic Stem Cells

A. Bauwens[1], H. Büntemeyer[1], S. Terstegge[2], T. Nottorf[1], O. Brüstle[2] and J. Lehmann[1]

[1] *Institute of Cell Culture Technology, University of Bielefeld, Germany,* [2] *Institute of Reconstructive Neurobiology, University of Bonn Medical Centre, Germany*

**Abstract:** Embryonic stem (ES) cells are characterized by their capacity for self-renewal and their pluripotency. A crucial prerequisite for a therapeutic use of human ES cells is the ability to expand them to large numbers without affecting their pluripotency. For many murine ES cell lines, leukaemia inhibitory factor suffices to prevent spontaneous differentiation. In contrast, the human ES cell lines available depend on feeder cells, i.e., coculture with mitotically inactivated mouse embryonic fibroblasts (MEFs) or treatment with conditioned medium (CM) derived from such cultures. This dependency on feeder cells is associated with batch-to-batch variations, difficulties in implementing large scale suspension culture of ES cells and potential carry-over of pathogens from the feeder to the ES cell population.

**Key words:** human embryonic stem cells, conditioned medium, feeder cells, bioreactor

## 1. INTRODUCTION

The bioreactor based production of conditioned medium (CM) has shown advantages: Completely controlled conditions minimize the batch-to-batch variations while the amount of generated CM is high enough to enable large scale ES cultivation processes with pathogen screened CM. Different cell types, as STO (ATCC CRL-1503) and primary murine fibroblasts, were cultivated on porous microcarriers in a 2 litre stirred and bubble-free aerated perfusion process. The produced CM was tested on human ES cells (H 9.2)



*R. Smith (ed.), Cell Technology for Cell Products*, 331–334.

over multiple passages while the differentiation and proliferation of these cells were used as a marker for the evaluation of the CM´s functionality.

## 2. MATERIAL AND METHODS

Feeder cells (primary mouse fibroblasts, CD1) were isolated from 13 days old mouse embryos. In order to cultivate these adherent cells in a continuously stirred bioreactor, micro- and macroporous CultiSpher-G microcarriers were used. Due to a lack of attachment and growth factors in the ES cell medium (Table 1) the carriers were colonized under serum containing conditions (10% FCS). Therefore the microcarriers (MCs) have to be washed several times before inoculation. The growth of the MEFs on the carriers was inspected with MTT & DAPI staining periodically.

*Table 1.* Composition of 2,5 litre ES cell medium.

| | |
|---|---|
| Knockout serum replacement | 500 mL |
| Knockout DMEM with pyruvate | 2000 mL |
| Non essential aminoacids (1:100) | 25 mL |
| L – glutamin | 1 mM |
| 2 – Mercaptoethanol | 0,1 mM |

During the cultivation in the bioreactor (Fig. 1) several parameters, e.g. concentration of glucose, lactate, ammonium and lactate-dehydrogenase were examined at regular intervals. The cell density was determined with trypan blue staining after enzymatic degradation of the carriers and with a crystal violet staining. After a short period of batch cultivation, the perfusion was started. The harvest containing dead cells and cell debris was sterile-filtered and screened for mycoplasma contamination using PCR. Afterwards the CM was stored frozen. Before the CM was used as full-media for ES cells growing on Matrigel™ coated petri dishes it was supplemented with basic fibroblast growth factor (4ng/mL). The differentiation level of the ES cell culture was proven over at least three passages. Functionality of the CM was determined by staining for alkaline phosphatase (AP), a cellular marker for differentiation.

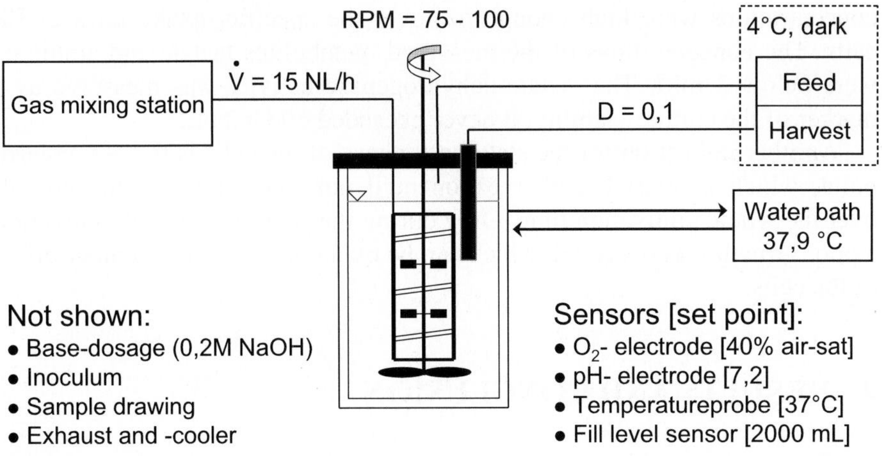


*Figure 1.* Schematic diagram of the 2 liter bioreactor used for the production of CM.

## 2.1 Inoculum

MEFs proliferate in the serum containing preculture-medium only. Because the CM should not contain complex animal components, the colonization of the MCs had to take place outside the bioreactor. Another consequence is, that no bead-to-bead transfer will occur in the bioreactor. To archive a cell density of at least $2 * 10^5$ cells/mL, a preparation protocol for the preculture (Table 2) was developed. The colonization of the carriers was validated with MTT & DAPI staining.

*Table 2.* Protocol for the generation of the inoculum.

| Day | Action |
|---|---|
| 1 | Thaw cells for 4 confluent T-175 flasks |
| 2 | Split cells to 12 T-175 flasks |
| 4 | Detach cells, add 5g (dry mass) prepared carrier |
| 7-8 | Detach carrier mechanically<br>Wash carrier until serum concentration is less than 0,01%<br>Inoculate the bioreactor, which should be in steady state |

# 3. ARCHIEVED PROCESS DATA

Although there is no cell growth in the bioreactor the perfusion was started at day 4 in order to satisfy the maintenance demand of the MEFs. During the cultivation the concentrations of glucose and amino acids remained at a constant level. Because the pyruvate concentration decreased during the process we can assume that murine embryonic fibroblasts utilize pyruvate and glucose as carbon source. However, the substrate concentrations were high enough to cover the specific uptake rates of ES cells. The concentrations of the measured metabolites lactate and ammonia were below 2 mM. The lactate-dehydrogenase activity was measured as a marker of the culture viability, it never exceeded 60 Units/L.

Another indication for the stationary phase of the cells is the $pO_2$ control point, which increased neither exponential nor linear but stayed constant over the whole cultivation time. Only during the first two days of cultivation a poor growth was observed, which can be explained with the memory effect of the cells.

## 4. RESULTS AND CONCLUSION

The goal to establish an easy to use, reproducible bioreactor process for generating great amounts of functional conditioned medium is nearly reached. The tests of the produced CM showed that the pluripotency of the ES cells can be maintained. In comparison, the functionality of the CM produced during the bioreactor process is at least equal to the functionality of CM achieved from mitotic inactivated feeder cells in t-flasks. Three bioreactor cultivatons had been done under same conditions. As well as the cultures showed similar behaviour during the process the harvested medium had the same quality concerning the functionality.

Comparable with inactivated feeder cells the mouse embryonic fibroblasts in the bioreactor remain in a stationary phase. Besides glucose the feeder cells utlilize pyruvate as carbon source, like early human embryonic cells do. As shown by a test of metabolite depleted medium, the small amounts of built metabolites does not affect the functionality of the CM.

A future goal is the establishment of a process with human feeder cells.

# Large-scale Expansion of Cells for Use as Therapeutic Agents

Juan Melero-Martin[1], Shanmugapriya Santhalingam[1],
Mohamed Al-Rubeai[1,2]
*1Department of Chemical Engineering, The University of Birmingham, Edgbaston, B15 2TT, UK. 2Department of Chemical and Biochemical Engineering and Centre for Synthesis and Chemical Biology, University College Dublin, Belfield, Dublin, Ireland*

**Abstract:** Expansion of cell population in vitro has become an essential step for the tissue engineering processing of articular cartilage and there is an urgent need to develop optimal culture conditions for chondrocytes. This presentation focuses on using a mathematical approach coupled with small-scale experiments to determine the optimal process parameters for cell expansion of chondroprogenitor cells and comparing the process to CHO cells expansion.

**Key words:** Tissue Engineering, Chondroprogenitor, CHO, Expansion, Growth Rate, Exponential Growth Kinetic.

## 1. INTRODUCTION

A relatively new approach to the treatment of damaged articular cartilage is the direct use of cells as therapeutic agents. The expected demand from the millions of afflicted individuals, coupled with the expected demand from biotechnology companies creating therapies, has fuelled the need to develop large-scale culture methods for these cells. There is an urgent need to develop optimal culture conditions for chondrocytes. The analysis of seeding density, passage length and operation cost is considered crucial to optimise the expansion process, and the correct selection of these parameters should be taken as a pre-requisite to establish a successful culturing system. The determination of the optimal seeding density and the corresponding passage length for cell expansion in a serial passaging operation is found to be a compromise between growth kinetics and process time. This work focuses



*R. Smith (ed.), Cell Technology for Cell Products*, 335–338.

on using a mathematical approach coupled with small-scale experiments to determine the optimal process parameters for cell expansion in large-scale processes. Moreover, expansion studies of mammalian cells for inoculation of industrial scale bioreactors for the production of antibody are examined and compared to chondroprogenitors, including studies that characterise important scale-up parameters and lessons for reducing costs are drawn.

## 2. MATERIAL AND METHODS

Bovine chondroprogenitor cells isolated from the superficial zone of articular cartilages of metatarsophalangeal joints [1] and cultured in T-Flasks at different seeding densities using DMEM+ (DMEM, 2 mM L-Glutamine, 1% Non-essential Amino Acids, 50 IU/ml Penicillin - 50 mg/ml streptomycin) , 40% FCS and 1 ng/ml TGF-β1 [2]. For any given initial cell number, the optimal passage length was found by maximizing the final total viable cell number expressed as Max(F)=Max(LN(Ep)/tp), where F, Ep and tp correspond to the total cell number, the passage expansion factor and passage length respectively. CHO cells producing gamma interferon were cultured in suspension at different cell densities in shaker flasks and agitated at 125 rpm. The medium used was CHO DHFR-Medium Animal Component-Free (Sigma) + 2% L-Glutamine + 1μm MTX + 1% P/S.

## 3. RESULTS AND DISCUSSION

### 3.1 Chondroprogenitor cells: optimal cell density

Two points were considered essential for the optimisation of chondroprogenitor cultures. Firstly, it was necessary to determine the seeding density for optimal cell expansion. Secondly, for any selected seeding density, it was necessary to establish the value of the passage length that makes the expansion process optimal in a serial operation. Although lower seeding densities obviously led to superior multiplication of the cells by the end of a single passage, the slower growth kinetics imposed by low seeding densities significantly increased the time necessary to achieve the desired expansion factor. The determination of an optimal seeding density is a compromise between expansion factor achievable and growth kinetics, which also imposes a significant role for the length of each monolayer passage. The value of $10^4$ cell/cm$^2$ for seeding density with 73 hours of passage length were found to be the optimal conditions for cell expansion in a serial passaging operation (Table 1).

*Table 1.* Optimal expansion of chondroprogenitor cells at different seeding densities.

| Seeding Density (cell/cm$^2$) | Expansion Factor (-) | LN(E)/tp ($h^{-1}$) | Optimal passage length (h) |
|---|---|---|---|
| 1000 | 97 | 0.02362 | 149 |
| 5000 | 35 | 0.02411 | 108 |
| 10000 | 29 | 0.02974 | 73 |
| 50000 | 9 | 0.02462 | 62 |
| 100000 | 5 | 0.01928 | 55 |

## 3.2 Cost considerations

Although the optimal passage length gives the desired expansion factor in the minimum process time, longer values of the passage length could create a situation where the same desired expansion factor is achieved with a lower cost by reducing the total number of passages required. This analysis led to the selection of an alternative optimal value of 120 h for the length of each passage. Although this value lead to a sub optimal proliferation criteria and process time, the running cost of the process will nevertheless be reduced by more than 60% (Fig. 1).

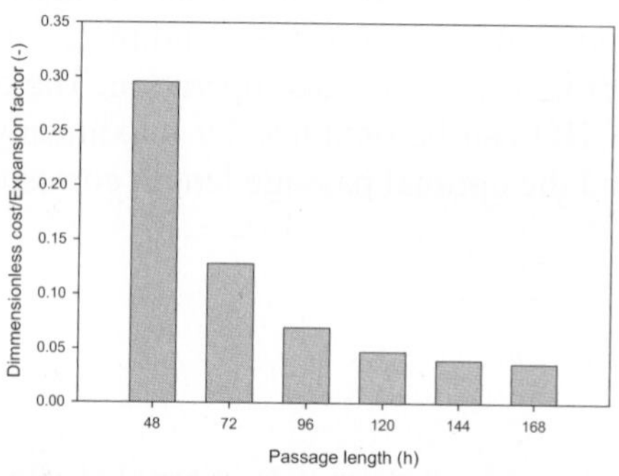


*Figure 1.* Effect of passage length on process cost for chondroprogenitor cell cultures.

## 3.3 CHO-320: optimal cell density

The majority of the cell lines exhibits non-exponential growth patterns in vitro [3] including chondroprogenitor cells [2]. However, an example of exponential growth kinetic was found for the industrial cell line CHO-320. In this case, the growth curves follow true exponential equations characterised by constant values of the specific growth rates (Table 2). The optimisation approach for this cell line is necessarily different from that shown for chondroprogenitor cells and the optimal conditions will always correspond to that at the end of the exponential phase.

*Table 2.* Growth kinetics of CHO320 at different seeding densities.

| **Seeding cell density (cells/ml)** | **Specific growth rates ($h^{-1}$)** |
|---|---|
| $5x10^4$ | 0.0215 |
| $1x10^5$ | 0.0190 |
| $2x10^5$ | 0.0185 |
| $4x10^5$ | 0.0152 |
| $6x10^5$ | 0.0137 |
| $1x10^6$ | 0.0139 |

## 4. CONCLUSIONS

Mathematical expressions of the growth curves provide a detailed analysis of the effects that the seeding density has on both cell growth kinetics and expansion process performance. The kinetics of chondroprogenitor cells consist of an initial period of growth acceleration followed by a later phase of deceleratory growth, and therefore cell growth in culture is predominantly non-exponential. On the other hand, CHO cell growth follows an exponential growth kinetics. The culture optimisation of these cells with different growth kinetics necessarily imposes different approaches. The value of $10^4$ cell/cm$^2$ for seeding density with 73 hours of passage length were found to be the optimal conditions for cell expansion of chondroprogenitor cells in a serial passage operation. The optimal conditions for the expansion of CHO can be obtained by maximising the value of the specific growth rate and the optimal passage length corresponds to the end of the exponential phase.

## REFERENCES

[1] Dowthwaite, G.P., Bishop, J.C., Redman, S.N., Khan, I.M., Rooney, P., Evans, D.J., Haughton, L., Bayram, Z., Boyer, S., Thomson, B. *et al.*, 2004, The surface of articular cartilage contains a progenitor cell population, *J. Cell Sci,*. **117**, 889-897

[2] Martin, J.M.M., Smith, M. and Al-Rubeai, M., 2005, Cryopreservation and in vitro expansion of chondroprogenitor cells isolated from the superficial zone of articular cartilage, *Biotechnol. Prog.,* **21**, 168-177

[3] Skehan, P., and Friedman, S.J., 1984, Non-exponential growth by mammalian cells in culture, *Cell Tissue Kinet.,* **17**, 335-343

# *EX VIVO* Expansion of CD34+ Cells in Stemline™ II Hematopoietic Stem Cell Expansion Medium Generates a Large Population of Functional Early and Late Progenitor Cells

Stacy L. Leugers[1] , Daniel W. Allison[1] , Ying Liang[2] , Carol Swiderski[2] , Gary Van Zant[2] & Laurel M. Donahue[1]

*[1] Cell Culture R&D, Sigma-Aldrich, St. Louis, MO, United States, [2] University of Kentucky, College of Medicine, Markey Cancer Center, Lexington, KY, United States*

**Abstract:** Hematopoietic stem cell replacement therapy is an area of research lacking an optimal culture system that allows for the *ex vivo* expansion of CD34+ cells for transplant. The necessity for expansion is due to the lack of sufficient material from umbilical cord blood (UCB), the preferred source of CD34+ cells. The final composition of the expanded cells is also very important to a successful transplant. The material must contain both early and late progenitor cells to ensure the long-term engraftment that is required in patients with genetic disorders and those that have gone through high dose chemotherapy.

Stemline™ II Hematopoietic Stem Cell Expansion Medium was developed to facilitate the development of this replacement therapy. Culturing CD34⁺ cells in Stemline™ II consistently yields higher levels of expansion than seen with other commercially available serum-free media. Analysis of the expanded material for lineage indicating markers by flow cytometry demonstrates that the material contains the required early and late progenitors. Furthermore, NOD/SCID studies confirm that the expanded cells are capable of long-term engraftment and are therefore functional. Together, this data supports the conclusion that culturing CD34⁺ cells in Stemline™ II Hematopoietic Stem Cell Expansion Medium is a key step in developing the optimal culture system.

**Key words:** Hematopoietic, Stem Cell, Expansion, Serum Free, Medium



*R. Smith (ed.), Cell Technology for Cell Products*, 339–341.

# 1. INTRODUCTION

Hematopoietic stem cells (HSC) have the ability to repopulate the hematopoietic system by differentiating into all of the necessary erythroid, lymphoid, and myeloid lineages. Due to this rare ability, HSCs are used as therapeutic agents in the treatment of malignant and benign diseases of the blood forming and immune systems. There have been many advances in the area of clinical HSC research, but the availability of suitable cells for transplantation still remains a major limiting factor. In order to expand these very specific cell types, an optimized serum-free medium and cytokine cocktail are needed. To this end, Stemline™ Hematopoietic Stem Cell Expansion Media were developed for the expansion of HSCs. They are serum-free media that allow for expansion of both differentiated and undifferentiated HSCs. The original medium, Stemline™, expands $CD34^+$ cells better than or equal to other commercially available serum-free HSC media. Stemline™ II is a newer version of the medium that has an increased expansion potential for $CD34^+$ cells.

# 2. RESULTS

In bench-scale experiments, Stemline™ media expand $CD34^+$ cells from all three sources better than other commercially available serum-free media. The expanded material contains the proper cell types, both early progenitors and late progenitors. Figure 1 has two representative graphs demonstrating the superior expansion characteristics of the Stemline™ media when using $CD34^+$ cells from cord blood. Similar results were seen in mobilized peripheral blood and bone marrow (data not shown).

The expansion capabilities of the Stemline™ media were also tested on a clinical scale. The expansion of $CD34^+$ cells cultured in both media was performed in clinical bags and the final material analyzed by flow cytometry. The assays demonstrated that both media are able to support the expansion of these cells, but Stemline™ II was able to achieve a 377-fold increase in cell number compared to a 238-fold increase achieved by Stemline™. The majority of the cells expanded in Stemline™ remained undifferentiated ($CD34^+$, $CD38^-$), while cells expanded in Stemline™ II contained both early and later progenitors ($CD34^+$, $CD38^+$) (data not shown).

A sample of cells from the Stemline™ and Stemline™ II clinical cultures were prepared for transplantation into NOD/SCID mice. A high percentage of the mice survived transplantation with cells expanded from both media (higher with Stemline™ II), all of which contained a small number of $CD45^+$ human cells as proof of engraftment (an even smaller number of which were also $CD34^+$). A summary of NOD/SCID results is in Table 1.

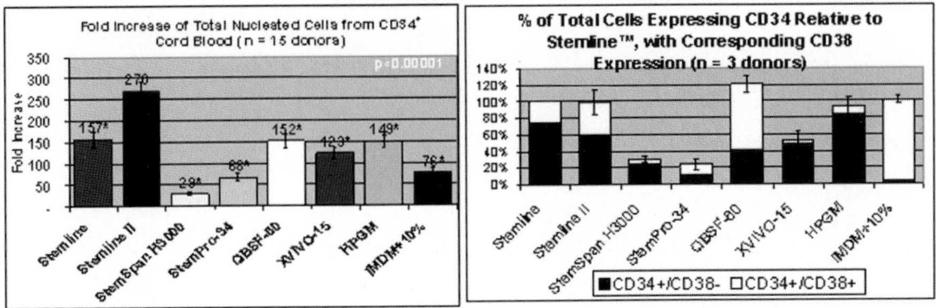


*Figure 1.* Expansion and characterization of CD34+ cells from cord blood in Stemline™ and Stemline™ II compared to other commercially available, serum-free, HSC media.

*Table 1.* Summary data from the transplantation of expanded cells in the NOD/SCID mouse model.

| | *Stemline ™* | | | *Stemline ™ II* | | |
|---|---|---|---|---|---|---|
| *Injected Cells* | *600,000* | *1,800,000* | *5,400,000* | *600,000* | *1,800,000* | *5,400,000* |
| *Survival Rate* | *5/10 50%* | *3/10 30%* | *3/7 43%* | *7/10 70%* | *6/10 60%* | *6/7 86%* |
| *Average % CD45+* | *0.064 ± 0.061* | *0.017 ± 0.006* | *0.143 ± 0.081* | *0.036 ± 0.013* | *0.018 ± 0.019* | *0.108 ± 0.162* |
| *Average % CD45+/ CD34+* | *0.000 ± 0.000* | *0.003 ± 0.006* | *0.007 ± 0.006* | *0.011 ± 0.009* | *0.002 ± 0.004* | *0.010 ± 0.000* |

## 3. DISCUSSION

The excellent performance of the Stemline™ media for the expansion and maintenance of functionality of CD34+ cells makes both media superior to all other commercially available, serum-free, HSC media. Both media expanded enough functional, early progenitors to achieve long-term engraftment in NOD/SCID mice. The higher survival rate in Stemline™ II may be explained by the higher levels of the late progenitors required for early engraftment and amelioration of the post-transplant nadir in mature myeloid cells. Additionally, Stemline™ media have Device Master Files (DMF) and are formulated in a state-of-the-art cGMP facility making them well suited for clinical applications.

# In Vitro Bone Formation of Multipotential Human Mesenchymal Cells in Fixed Bed Perfusion Bioreactors

Nadia Zghoul[1], Martijn van Griensven[2], Johannes Zeichen[3], Kurt E.J. Dittmar[1], Manfred Rohde[1], Volker Jäger[1]

*[1]German Research Centre for Biotechnology, Mascheroder Weg 1, 38124 Braunschweig, Germany, [2]Ludwig Boltzmann Institute for Experimental and Clinical Traumatology, Donaueschingenstr. 13, 1200 Vienna, Austria, [3]Hannover Medical School, Carl-Neuberg-Str.1, 30625 Hannover, Germany*

**Abstract:** The objective of this study was to characterize growth and differentiation of human bone marrow-derived mesenchymal cells and trabecular bone-derived cells regarding their expansion and their multipotential capacity as well as their ability, to eventually form bone tissue in vitro in combination with a PLGA/CaP scaffold.

**Key words:** Tissue engineering, Trabecular Bone, Bone Marrow, Biomaterial, Bioreactor, Flow Cytometry, Adipogenesis, Osteogenesis

## 1. INTRODUCTION

Tissue engineering approaches for promoting the repair of skeletal tissues have focused on cell-based therapies involving multipotent bone marrow-derived mesenchymal stem cells (MSCs). There is evidence to suggest that other tissues such as muscle, skin, fat and periosteum also contain multipotent progenitor cells (Dragoo *et al*). Our purpose was to isolate, expand and characterise human mesenchymal cells from small trabecular bone specimens and to investigate whether human trabecular bone explants contain multipotential progenitors. Further, the capabilities of these cells were tested in combination with a PLGA/ CaP scaffold to produce bone tissue in vitro in a continuously perfused bioreactor.



*R. Smith (ed.), Cell Technology for Cell Products*, 343–346.

## 2. CELL ISOLATION AND EXPANSION

Bone marrow-derived cells: Bone marrow aspirates were centrifuged over a density gradient and suspended in ZKT-I medium with 15% human serum. Phase contrast microscopy showed typical star-shaped colonies. After 30-40 days in culture first confluent cell populations were harvested.

Trabecular bone-derived cells: Cells were allowed to migrate from adult human bone fragment placed in ZKT-I medium supplemented with 15% human serum (see Fig. 1).

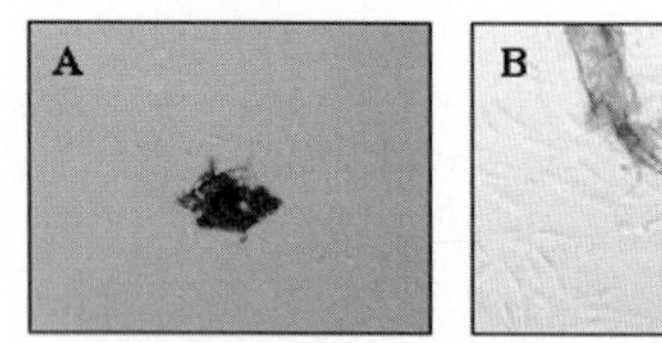

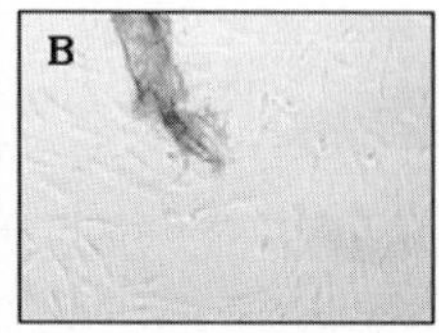

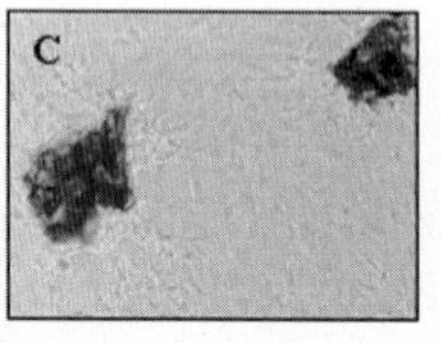

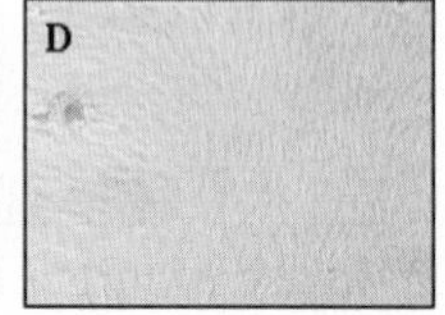


*Figure 1.* Trabecular bone explantation and cell outgrowth within 4-6 weeks (x100). Appearance of bone fragments (A). Cells migrating from the bone fragments after 14 (B) and 25 (C) days of culture reaching a confluent monolayer after 32 days of explant culture (D).

### 2.1 Cell Characterisation and Multilineage Potential

FACS analysis of a panel of cell surface markers reported to be specific for MSCs (Barry et al) was shown to be positive for both the bone marrow-derived as well as for the trabecular bone cells.

*Table 1.* Flow Cytometry.

| | CD 73 | CD 90 | CD 105 |
|---|---|---|---|
| Bone Marrow Cells | 97% | 94% | 85% |
| Trabecular Bone Cells | 99% | 91% | 76% |

To confirm the MSC characters of trabecular bone-derived cells their adipogenic (stimulated by isobutylmethylxanthin, indomethacin, insulin, hydrocortisone) and osteogenic (stimulated by β-glycerophosphate, ascorbic acid, dexamethasone) potential was compared to that of bone marrow-derived cells. Adipogenic differentiation was demonstrated by staining lipid vacuoles (A, B) with Sudan III and osteogenic differentiation was shown by both alkaline phosphatase (C, D) and von Kossa staining (E, F).

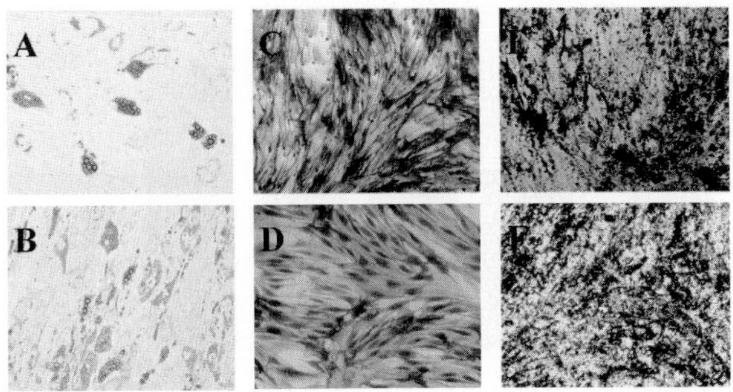


*Figure 2.* Adipogenesis (A) and osteogenesis (C, E) of bone marrow-derived cells compared to both the adipogenic (B) and the osteogenic potential of trabecular bone cells (D, F) (x100).

# 3. THREE-DIMENSIONAL CULTIVATION IN A FIXED BED PERFUSION BIOREACTOR

5E6 trabecular bone cells were seeded onto PLGA/CaP scaffolds (Osteoscaf TM, 100-700μm pore size)and cultivated in a continuously perfused bioreactor (Barthold *et al*). Differentiation was analysed by determining the expression of alkaline phosphatase and collagen type I (Fig. 3) as well as by immunofluorescence of osteogenic markers (not shown). The viability of the cells was determined with the WST-1 test throughout the culture period (Fig. 4). The seeded scaffolds were also assayed using scanning electron microscopy (Fig. 5).

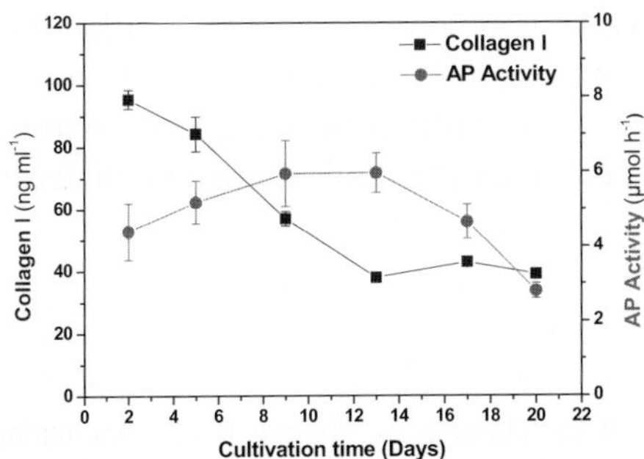


*Figure 3.* Osteogenic differentiation of trabecular bone cells on OsteoscafTM cultured in the bioreactor system. Alkaline phosphatase activity reached a maximum after 13 days coinciding with the decline in collagen I levels and indicating the start of mineralisation.

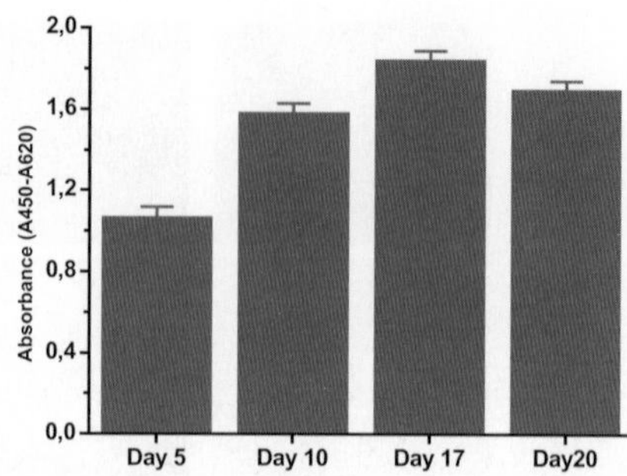


*Figure 4.* Cell proliferation as measured by the WST-1 assay was observed to increase throughout the first two weeks followed by a gradual plateau of cell numbers.

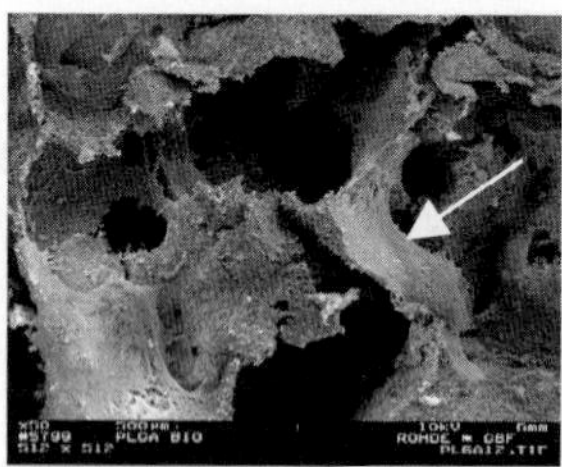

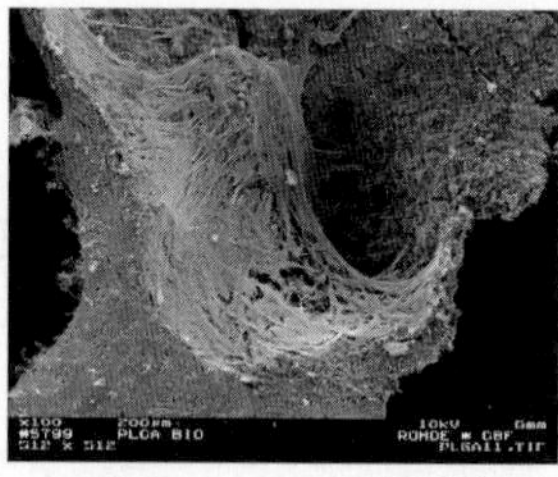

*Figure 5.* (A) SEM of the interior scaffold 20 days post seeding showing a line of cells between the pores. (B) Higher magnification of spread cells with deposited collagen fibrils.

# 4. CONCLUSION

Our data of isolated and characterised cells support the hypothesis that mesenchymal stem cells are present in trabecular bone. Given their accessibility, they may prove to be a source of cells for the engineering of bone tissue. A bioreactor for cultivating these cells within a porous matrix was shown to support cell spreading and osteogenic differentiation.

# REFERENCES

Dragoo J.L., Samimi B., Zhu M., Hame S.L., Thomas B.J., Liebermann J.R., Hedrick M.H., Benhaim P., 2003, Tissue-engineered cartilage and bone using stem cells from human infrapatellar fat pads, J Bone Joint Surg Br. **85**:740-7.

Barry F.P., Murphy J.M., 2004, Mesenchymal stem cells : clinical applications and biological characterization. Int J Biochem Cell Biol., **36**:568-84.

Barthold M., Majore I., Fargali S., Stahl F., Schulz R., Lose S., Mayer H., Jäger V., 2005 3-D Cultivation and characterization of osteogenic cells for the production of highly viable bone tissue implants. F. Godia, M. Fusenegger (eds.) 'Animal Cell Technology Meets Genomics', Springer, Dordrecht, Netherlands, 199-205.

# Influence of Basal Media Formulations and Calcium on Growth and Directed Differentiation of Human Osteogenic Cells *In Vitro*

Ingrida Majore, M. Morris Hosseini, Hubert Mayer, Volker Jäger
*Gesellschaft für Biotechnologische Forschung mbH, Mascheroder Weg 1, D-38124 Braunschweig, Germany*

**Abstract:** For the generation of human bone tissue *in vitro*, we were searching for suitable methods to control the ongoing differentiation process of osteogenic cells with a possibility to switch between the expansion of immature progenitor cells and the differentiation into cells of the mature osteogenic phenotype producing mineralised extracellular matrix. This study is focussing on the influence of different basal media formulations which were tested using primary human cells derived from either sites of heterotopic ossification or trabecular bone. The different basal media formulations DMEM, αMEM, RPMI 1640 and our in-house formulation ZKT-I revealed a substantial impact on cellular growth and differentiation. Cells in DMEM were growing relatively slow and differentiated rapidly into mature osteoblasts with strong matrix mineralisation, whereas cells in αMEM were growing more rapidly with slightly slower differentiation and mineralisation kinetics. A strong mineralisation was always accompanied by apoptotic processes as indicated by higher levels of LDH in the supernatant. In contrast, cells in ZKT-I medium showed a sufficient growth with markedly retarded differentiation kinetics which made it most suitable for cell expansion. To induce cell-induced matrix mineralisation using ZKT-I, a moderate increase of the calcium concentration was shown to be a sufficient trigger. With this simple supplementation, ZKT-I proved to be highly suitable for both phases of bone tissue generation *in vitro*.

**Key words:** calcium, osteoblast, bone, extracellular matrix, cell proliferation, cell differentiation, matrix mineralisation, DNA microarray, lactate dehydrogenase, alkaline phosphatase



*R. Smith (ed.), Cell Technology for Cell Products*, 347–351.

# 1. INTRODUCTION

In order to generate human bone tissue in vitro, it is highly desirable to control the ongoing differentiation process of osteogenic cells with the possibility to make a switch between a phase of expansion of immature progenitor cells and a phase of differentiation into cells of the mature osteogenic phenotype producing mineralised extracellular matrix. This study is focussing on the influence of different basal media formulations which were tested using primary human cells derived from either trabecular bone or sites of heterotopic ossification.

# 2. RESULTS

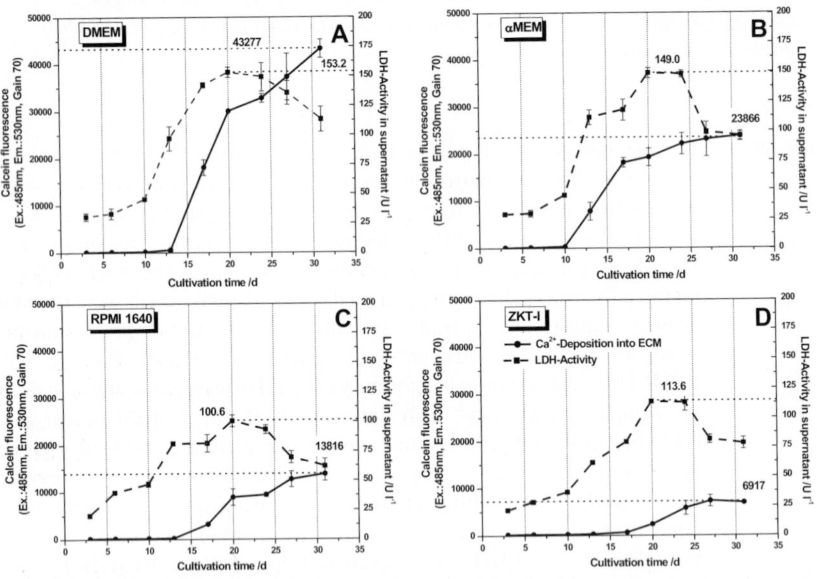


*Figure 1.* Mineralisation of ECM as measured by calcein assay and Von Kossa staining (not shown) as well as LDH activity kinetics during long-term cultivation of human osteogenic cells. Cells were cultured in three frequently used basal media DMEM (A), αMEM (B), RPMI 1640 (C) as well as in our in-house formulation ZKT-I (D). All media were supplemented with 10% human serum and osteogenic factors (ascorbic acid-2-phosphate, β-glycerophosphate, dexamethasone). Different cell culture media were shown to have a severe impact on the mineralisation of ECM. These differences are based on the ongoing differentiation of osteogenic cells and - as measured by LDH activity in the supernatant - are accompanied by an increased level of dead cells undergoing apoptosis due to locally increasing concentrations of inorganic phosphate ($P_i$) in the mineralisation nodules (Meleti *et al.*, 2000). $P_i$ is formed at a high level in the mineralization centres from β-glycerophosphate under the catalysis of alkaline phosphatase (AP) (Bellows *et al.*, 1992; Coelho *et al.*, 2000). Osteoblasts producing much AP are found in a particularly high number within the nodules (Bellows *et al.*, 1992; Schecroun & Deloye, 2003).

*Table 1.* Evaluation of cell culture media with regard to their potential in supporting cell proliferation, differentiation, and mineralisation of ECM during long-term cultivation.

| | DMEM | αMEM | RPMI1640 | ZKT-I |
|---|---|---|---|---|
| Proliferation Capacity | ++ | ++++ | ++++ | +++ |
| Differentiation Velocity | ++++ | +++ | ++ | + |
| Mineralisation Level | ++++ | +++ | ++ | + |

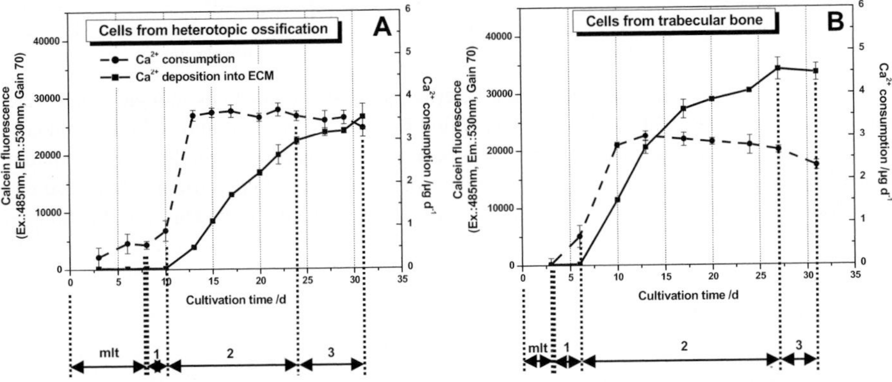


*Figure 2.* 'Mineralisations lag time' (mlt) (osteoid maturation period) and the three phases of matrix mineralisation: 1 - Initiation phase, 2 - Progression phase, 3 - Plateau phase of mineralisation. Human osteogenic cells isolated either from a heterotopic ossification (A) or trabecular bone (B) were cultivated in 48-well plates using αMEM supplemented with 10% human serum and osteogenic factors. Medium was replaced twice (A) or three times (B) a week. The incorporation of calcium can be subdivided into several phases (Bellows et al., 1992). During progression phase calcium consumption reached rapidly a plateau which corresponds to a complete depletion of calcium in the culture medium. Ossification in contrast to calcification is actively controlled by bone cells, respectively collagens and non-collagenous proteins produced by those cells (Boskey, 2003). The most sensitive index for detecting proper mineralisation is the determination of the so called "mineralisation lag time" (Zhang et al., 2002) which also allows to distinguish between ossification and premature calcification.

*Table 2.* $Ca^{2+}$ concentrations and osmolarity in ZKT-I media with normal and increased $Ca^{2+}$ levels. Complete medium contains 10% human serum, osteogenic factors and antibiotics. Osmolarity was measured with a Knauer automatic osmometer. The calcium content of serum was 73.9 mg $L^{-1}$ as measured with the calcium kit (bioMerieux).

| Nr. | Non-supplemented ZKT-I medium | | Complete ZKT-I medium | | Complete medium as measured with the calcium kit | | Osmolarity in complete medium |
|---|---|---|---|---|---|---|---|
| | mg $L^{-1}$ | mmol $L^{-1}$ | mg $L^{-1}$ | mmol $L^{-1}$ | mg $L^{-1}$ | mmol $L^{-1}$ | mosmol $kg^{-1}$ |
| 1 | 35.79 | 0.89 | 39.24 | 0.98 | 37.45±4.58 | 0.93 | 332.7±2.1 |
| 2 | 71.58 | 1.78 | 71.65 | 1.79 | 72.64±3.10 | 1.78 | 331.7±0.6 |
| 3 | 143.16 | 3.56 | 134.8 | 3.36 | 145.38±6.32 | 3.56 | 335.3±0.6 |

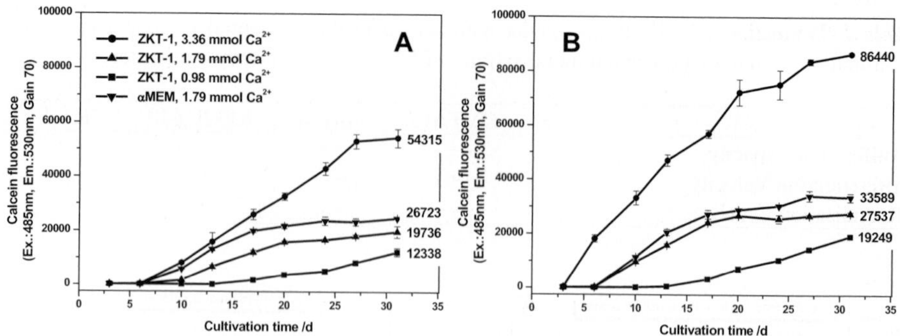


*Figure 3.* Osteogenic cells isolated from a heterotopic ossification (A) or spongiosa (B) were cultivated in either αMEM or ZKT-I media with different $Ca^{2+}$ concentrations. Matrix mineralisation was measured by a calcein fluorescence assay. The data clearly indicate that calcium concentration is a major (but not the only) factor of ECM matrix mineralisation. It is also demonstrated that very high $Ca^{2+}$ concentrations initiate an unspecific premature calcification.

## 3. CONCLUSIONS

Basal media represent an important factor for osteogenic cell cultivation. They substantially influence the differentiation of human osteoprogenitor cells, especially the mineralisation of extracellular matrix. ZKT-I proved to be most suitable cell culture medium for propagation of osteoprogenitor cells whereas αMEM proved to be the best medium for initiation of a directed differentiation process with matrix mineralisation.

Gene expression profiles as indicated by a DNA microarray (data not shown) revealed that there is no significant shift in the expression of collagens and non-collagenous proteins involved in ECM synthesis and regulation of mineralisation. Instead, calcium appears to be the major factor responsible for the initiation of matrix mineralisation. It is known, that the plasma membrane of osteogenic cells is endowed with a sodium-dependent $P_i$ transport system as well as with ion channels and receptors for calcium (Caverzasio & Bonjour, 1996; Dvorak & Ricardi, 2004). Because of this the bone cells can react very sensitively to changes of extracellular calcium and inorganic phosphate concentrations.

Very high calcium concentrations result in an undesirable, unspecific premature calcification. On the other hand, calcium at lower concentrations is rapidly depleted from the culture medium suggesting a continuous feeding of calcium (as it is accomplished in our continuously perfused biorector system) as the method of choice to achieve optimal matrix mineralisation.

# REFERENCES

Bellows, C.G., Heersche, J.N., Aubin, J.E., 1992, Inorganic phosphate added exogenously or released from beta-glycerophosphate initiates mineralization of osteoid nodules in vitro, J. Bone Miner. Res. **17**: 15-29.

Boskey, A.L., 2003, Biomineralization: an overview, Connect Tissue Res. **44 (Suppl. 1)**: 5-9.

Caverzasio, J., Bonjour, J.P., 1996, Characteristics and regulation of $P_i$ transport in osteogenic cells for bone metabolism, Kidney Int. **49**: 975-980.

Coelho, M.J., Fernandes, M.H., 2000, Human bone cell cultures in biocompatibility testing. Part II: effect of ascorbic acid, β-glycerophosphate and dexamethasone on osteoblastic differentiation, Biomaterials **21**: 1095-1102.

Dvorak, M.M., Riccardi, D., 2004, $Ca^{2+}$ as an extracellular signal in bone, Cell Calcium **35**: 249-255.

Meleti, Z., Shapiro, I.M., Adams, C.S., 2000, Inorganic phosphate induces apoptosis of osteoblastic-like cells in culture, Bone **27**: 359-366.

Schecroun, N., Delloye, C., 2003, Bone-like nodules formed by human bone marrow stromal cells: comparative study and characterization, Bone **32**: 252-260.

Zhang, M., Xuan, S., Bouxsein, M.L., von Stechow, D., Akeno, N., Faugere, M.C., Malluche, H., Zhao, G., Rosen, C.J., Efstratiadis, A., Clemens, T.L., 2002, Osteoblast-specific knockout of the insulin-like growth factor (IGF) receptor gene reveals an essential role of IGF signalling in bone matrix mineralization, J. Biol. Chem. **277**: 44005-44012.

# CHAPTER VI: PROTEIN PRODUCTS

# BHRF-1 as a Tool for Genetic Inhibition of Apoptotis in Hybridoma Cell Cultures

S. Juanola[1], J. Vives[2], J. Gálvez[1], P. Ferrer[1], R. Tello[1], J.J. Cairó[1], F. Gòdia[1]

[1] *Dept. d'Eng. Química, UAB/U. d'Eng. Bioquímica, CeRBa, Edifici C, 08193 Bellaterra, Barcelona, Spain,* [2] *ISCR, Uni. Edinburgh, King's Buildings, West Mains Road, EH9 3JQ Edinbugh, Scotland*

**Abstract:** Cell death is a critical factor in hybridoma cell cultures. At high cell densities, the limitation of nutrients, mitogens and oxygen leads to elevated death rates and to the accumulation of cell debris affecting the quality of the product and downstream processing, therefore compromising the productivity and cost-effectiveness of the bioprocess. In the present work, we engineered a hybridoma cell line to express either the antiapoptosis gene Bcl-2 or its viral homologue BHRF-1 (isolated from Epstein Barr virus) and analyse their effects on the improvement of cell viability and productivity under different cell culture strategies as batch, continuous and perfusion.

**Key words:** Hybridoma, Apoptosis protection, Metabolic engineering, BHRF-1, Bcl-2

## 1. INTRODUCTION

Our previous studies on the behaviour of the hybridoma KB26.5 under conditions of glutamine depletion evidenced the success of caspase-inhibitor treatment in the delay of the apoptosis programme and the capacity to recover those cultures when they were brought to normal growing conditions, i.e. complete medium (Tintó *et al.*, 2002). However, the high cost of these molecules bars them from their use in large-scale cultures. For this reason, genetic strategies directed to control regulatory points upstream of effector caspases are desirable for batch, continuous and perfusion processes at industrial scale. Consequently, a particular attention has been given to the need to interfere in the apoptotic pathways at the level of the mitochondria, which play a central role in apoptosis integrating death signals through Bcl-2 family members and co-ordinating caspase activation through the release of apoptogenic factors that are normally sequestered in the mitochondrial intermembrane space (Figure 1). The involvement of several caspases in channelling



*R. Smith (ed.), Cell Technology for Cell Products*, 355–361.

the apoptotic signal indicates that potential genetic modifications should either consider more than one single target or be directed upstream to their activation, which is at the mitochondrial level.

## 2. MATERIALS AND METHODS

### 2.1 Cell line, medium and culture conditions

The KB26.5 murine hybridoma was cultured as described (Sanfeliu *et al.*, 1997).

### 2.2 Plasmid constructs

Sequences encoding cDNA from Bcl-2 and BHRF-1 were isolated as described (Vives *et al.*, 2002) and cloned into pcDNA3.1 (Invitrogen) and pIRES-neo/pIRES-puro2 (Clontech).

### 2.3 DNA delivery into hybridoma cells

Three methods were used: (1) lipofection with DMRIE-C reagent (Life Technologies); (2) electroporation as described in the protocol provided by BTX; and (3) calcium phosphate-mediated transfection by inclusion of adenovirus in coprecipitates (Lee and Welsh, 1999). Adenoviral particles (Ad2-CMV-GFP) were supplied by CBATEG (UAB, Bellaterra, Spain) at titers of approximately $5{,}2 \times 10^{10}$ IU (Infectious Unit)/ml. In the optimisation process, we used pEYFP-1 (Clontech) to assess the efficiencies.

### 2.4 Hemagglutination assay

The presence of antibodies was determined by hemagglutination test (Sanfeliu, 1995).

### 2.5 Apoptosis detection

Cell viability, Annexin-V-Fluos positive cells and DNA fragmentation were measured as described (Tintó *et al.*, 2002).

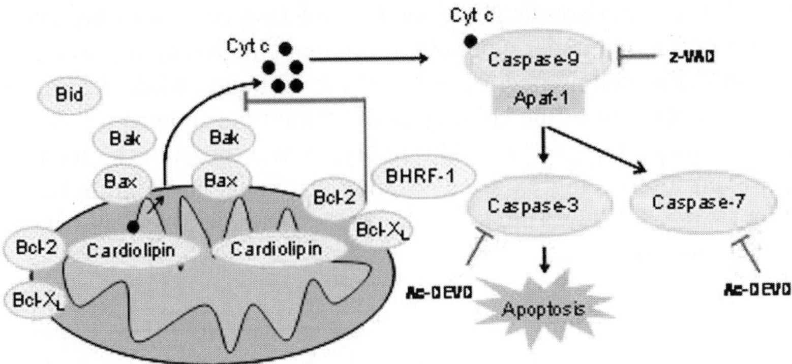


*Figure 1.* Model for the control of apoptosis and location of the Bcl-2 family members and caspase-inhibitors (z-VAD-fmk and Ac-DEVD-cho). Cytochrome c release is inhibited by anti-apoptotic Bcl-2 family members such as Bcl-2, Bcl-XL, and BHRF-1, while it is directly enhanced by the pro-apoptotic members such as Bak, Bax, and Bid. Once released from the mitochondria, Cytochrome c activates pro-caspase-9 via the apoptosome.

# 3. RESULTS AND DISCUSSION

## 3.1 Genetically engineered hybridoma cells

For most of animal cell lines, the two principal difficulties in the genetic engineering endeavour are their modest transfection efficiencies and the decline in the expression levels after several passages. We evaluated alternative approaches, such as the improvement of gene delivery methods and the use of plasmids based on bicistronic technology. The inclusion of adenovirus in DNA:CaPi coprecipitates (8%) provides by far the best results in comparison with eletroporation (1%) and lipofection (3%). Moreover, the maintenance of initial expression levels, and therefore antiapoptosis protection, was achieved with bicistronic vectors. In fact, our results showed at least 3 months stability of the transfectants, even in absence of selectable marker, allowing them to be employed in perfusion systems for long-lasting operational times.

## 3.2 Apoptosis protection in batch cultures

As reported previously, the use of protective members of the Bcl-2 family from viral origin, such as BHRF-1, not containing the $Asp^{34}$ cleavage site of caspase-3, confer enhanced protection against apoptosis when compared to *bcl-2-* (Figures 2a and 2c) and *bclxl*-expressing cells (Vives *et al.*, 2003). Moreover, the use of bicistronic vectors harbouring *bhrf-1* substantially improved such protection in glutamine-deficient medium

(Figure 2d in comparison to 2b). *bhrf-1*-expressing cells not only exhibited the highest protection under glutamine-deficient culture conditions but also a decrease in viable cell number of only 15% at 72 hours, while the number of viable cells in the rest of the cultures at the same time showed a reduction in viability by more than 60%. Additionally, it was possible to recover the cultures even after being exposed to apoptosis-inducing conditions during significant time window, up to 72 hours, as demonstrated in continuous cultures (Figure 3).

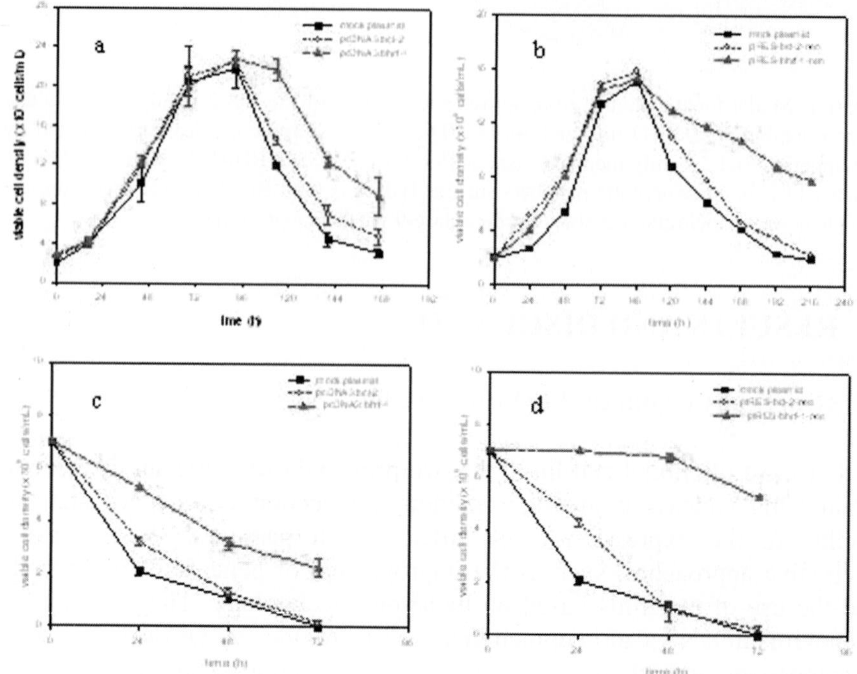


*Figure 2.* Effect of the expression of the antiapoptotic genes bcl-2 and bhrf-1 in batch cultures of hybridoma KB26.5. Cell growth profile of cells engineered using (a) monocistronic and (b) bicistronic vectors. Protection against apoptosis induced by glutamine deprivation in cultures of cells engineered with (c) monocistronic and (d) bicistronic vectors.

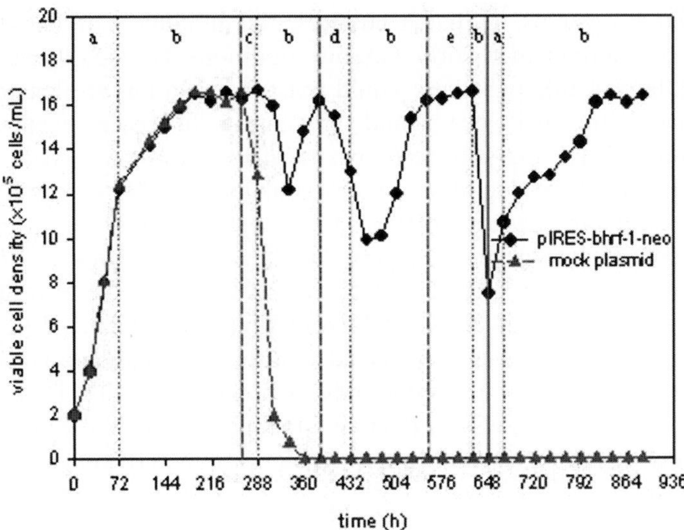


*Figure 3.* Continuous cultures of mock plasmid transfected (control) and bhrf-1-expressing hybridoma cells to evaluate the effect of the limitation of nutrients and the capacity to rescue cell viability. When the continuous culture was operated at a dilution rate of 0.16 h-1, the growth of both cell cultures was similar and reached the highest density at 264 h (a). At that point, continuous was suspended for 24 h and apoptosis was irreversibly triggered in the control culture coinciding with a fall in culture viability. By contrast, the specific growth rate of the bhrf-1-transfected cells could be quickly recovered when the system was operated again with elevated levels of cell survival after a momentary decline phase. As soon as cell densities were restored, another perturbation was applied to the system, for 48 h (d). This time, cell viability fell to 9.9 x 105 cells/mL, and they required a longer time to be recovered. Finally, a third perturbation was applied, for 72 h (e), and once more cells were recovered showing the protective effect conferred by the expression of bhrf-1.

## 3.3 Perfusion

High-density hybridoma perfusion cultures are extensively used for producing large amounts of MAbs. Although perfusion culture technology allows the maintenance of high cell concentration for extensive operational times, cell death is an important limitation: whereas viable cell density remains steady, accumulation of dead cells affects the quality of the product, downstream processing and clogging the perfusion system (Figure 4a). We performed perfusion cultures using mock plasmid transfected (control) and *bhrf-1*-expressing hybridoma cells and compared viable and dead cell numbers. Results show a dramatic reduction of dead cell number in *bhrf-1*-transfected cell cultures, 12.1 x $10^5$ cells/mL vs 55.4 x $10^5$ cells/mL (Figure 4b) in the control culture. Both batch and perfusion culture strategies are the preferred choice for the production of MAbs. However, they are totally different systems. In batch cultures, the limitation of nutrients or oxygen triggers apoptosis in fairly extreme conditions. On the other hand, in perfusion cultures, although there could exist a limitation at elevated cell densities, triggering cell death and preventing to reach even higher concentrations, there are still resources available for every cell in the bioreactor. Under these conditions, mock transfected cultures sustain high percentage of cells in phase S, 50.1% at 192 h vs 47% at 96 h (in the middle of the exponential growth phase, Figure 4a). In contrast, cultures of *bhrf-1*-expressing cells show 36.3% at 192 h vs 53.9% at 96 h. Regarding the antibody production, *bcl-2*- and *bhrf-1*-expressing cells generate 85.2% and 87.5% with respect of mock transfected control, respectively. This slight reduction (less than 15%) in productivity is compensated by the fact that *bhrf-1*-expressing cells can be effective in long operational cultures and therefore it doesn't represent a drawback in their use in perfusion systems.

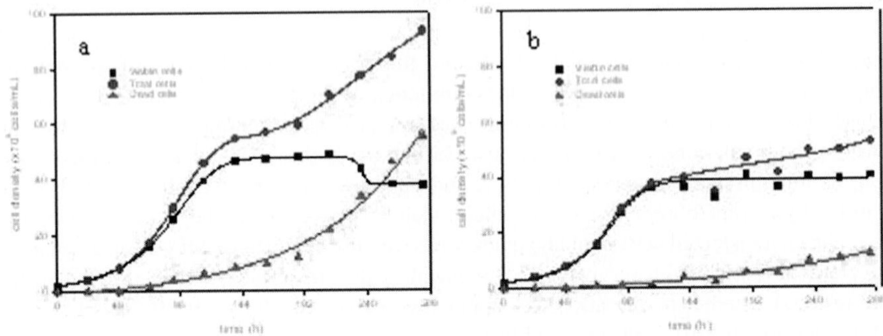

*Figure 4.* Perfusion cultures of mock plasmid transfected (control) (a) and bhrf-1-expressing hybridoma (b).

## 4. CONCLUSIONS

The performance and robustness of *bhrf-1*-expressing cells were tested under glutamine limitation in order to induce apoptosis as a consequence of starvation of this amino acid in batch cultures, interruption of continuous feeding in continuous cultures and depletion in perfusion operations. Under these conditions *bhrf-1*-expressing cells showed a high protection in front of the apoptotic process when compared to controls. It has been observed an increase in life span of 72 hours in batch and continuous experiments, and a reduction of almost 80% in cell death in perfusion operations. In addition, the protection conferred by BHRF-1 allowed the recovery of the cultures even after prolonged periods of time under apoptosis-inducing conditions, therefore evidencing the potentiality of this technology for long operation periods in animal cell bioprocesses.

## ACKNOWLEDGEMENTS

The present work has been developed in the framework of the Centre de Referència en Biotecnologia (Generalitat de Catalunya) and supported by the Plan Nacional de Biotecnologia (BIO97-0542).

## REFERENCES

Lee J., Welsh M. "Enhancement of calcium phosphate-mediated transfection by inclusion of adenovirus in coprecipitates." *Gen.Ther.* **6** (1999), 676-682.

Sanfeliu A. "Producció d'anticossos monoclonals mitjançant el cultiu in vitro d'hibridomes en bioreactors: anàlisi de la fisiologia i metabolisme cel·lulars". PhD Thesis (1995). UAB.

Sanfeliu A., Paredes C., Cairó J.J., Gòdia F. "Identification of key patterns in the metabolism of hybridome cells in culture." *Enz. Micr. Tech.* **21** (1997), 421-428.

Tintó A., Gabernet C., Vives J., Prats E., Cairó J.J., Cornudella L., Gòdia F. "The protection of hybridoma cells from apoptosis by caspase inhibition allows culture recovery when exposed to non-inducing conditions". *J. Biotechnol.* **95** (2002), 205-214.

Vives J., Juanola S., Cairó J.J., Prats E., Cornudella L., Gòdia F. "Protective effect of viral homologues of *bcl-2* on hybridoma cells under apoptosis-inducing conditions". *Biotechnol. Prog.* **19** (2003), 84-89.

# Karyotype of CHO DG44 cells

D. Martinet[1], M. Derouazi[2], N. Besuchet[1], M. Wicht1, J. Beckmann[1] and F.M. Wurm[2]

*[1]Laboratoire de Cytogénétique, Service de Génétique Médicale, Centre Hospitalier Universitaire Vaudois CHUV, 1011 Lausanne Switzerland, [2]Laboratory of Cellular Biotechnology, Faculty of Life Science, Ecole Polytechnique Fédérale de Lausanne EPFL, 1015 Lausanne Switzerland*

**Abstract:** The CHO cell line was mutagenized with radiation and by chemical treatment, and a CHO DG44 mutant carrying a double deletion for the dihydrofolate reductase gene located on chromosome 2 was isolated and characterized. Among the different sublines, CHO DG44 cells are widely used for the stable production recombinant proteins since they exhibit the DHFR-selection and amplification system. To our knowledge, the karyotype of CHO-DG44 cells has not been studied. After analysis of more than 100 metaphases, we consistently found 20 chromosomes whereas the normal diploid Chinese hamster genome was characterized with 22 chromosomes. The quasi-diploid CHO DG44 cells were found to have a karyotype altered from the original Chinese hamster. Only seven of the chromosomes were normal including the two chromosomes 1, one chromosome 2, 4, 5, 8 and 9. The four chromosomes Z1, Z4, Z8 and Z13 were identified as described by Deaven and Peterson (Deaven and Petersen 1973). The remaining chromosomes were 7 derivative chromosomes (rearrangement within a single chromosome or involving two or more chromosomes) and 2 marker chromosomes (structurally abnormal chromosome in which no part can be identified). We believe that insights from these studies can be of value for a genetic characterization of CHO DG44 cells expressing a recombinant protein as well as for work on targeted or homologous integration into the genome.

**Key words:** CHO DG44, karyotype, recombinant proteins



*R. Smith (ed.), Cell Technology for Cell Products*, 363–366.

# 1. INTRODUCTION

Primary diploid cells from the Chinese hamster (*Cricetulus griseus)* have 22 chromosomes, and the karyotype of these cells serves as the basis for the banding nomenclature of all CHO-derived cell lines (Ray and Mohandas 1976). The parental CHO cell line (CHO-K1) has only 21 chromosomes with 9 of these designated the "Z group" chromosomes that are structurally different than the normal chromosomes of *C. griseus* [3]. The deletions, pericentric inversions, and translocations of the Z chromosomes have been fully described (Deaven and Petersen 1973). The CHO DG44 strain described here was generated from a praline deficient CHO cell lines (Pro-3) by a round of chemical mutagenesis followed by exposure to gamma irradiation (Urlaub and Chasin 1980; Urlaub *et al*., 1983). Since both alleles of the *dhfr* gene in CHO DG44 cells are deleted, selection of cells cotransfected with an exogenous *dhfr* gene and a transgene in medium without hypoxanthine and thymidine allows the efficient recovery of DHFR-positive recombinant cell lines (Kaufman R.J. *et al*., 1985). Here we report the first karyotype of CHO DG44 cell line.

# 2. RESULTS

Initially, one hundred metaphase chromosome spreads from CHO DG44 cells were analyzed. In most cells, a total of 20 chromosomes were observed (Fig. 4). Based on the karyotype of the near-euploid Chinese hamster cell line DEDE and the subline TGA 102a, the karyotype of CHO DG44 cell line was established. The chromosome number and giemsa banding pattern were defined according to the system proposed by Ray and Mohandas (Ray and Mohandas 1976). All the descriptions of chromosomal rearrangements followed the conventions established by the International Standard Committee on Human Cytogenetic Nomenclature (ISCN 1995). A derivative chromosome was defined as a structurally altered chromosome generated by one or more rearrangements within a single chromosome or rearrangements involving two or more chromosomes. Derivative chromosomes were defined as a structurally rearranged chromosome generated by more than one rearrangement within a single chromosome or involving two or more chromosomes. Those derivative chromosome(s) had an intact centromere. Marker chromosomes were defined as structurally abnormal chromosomes in which no part was matched to a region(s) of a parental chromosome.

The quasi-diploid CHO DG44 cells were found to have a altered karyotype compared to that of diploid Chinese hamster cells (Ray and Mohandas 1976). Twenty rather than 22 chromosomes were observed, and only seven of these appeared to be normal; these included the two chromosome 1 and one copy each of chromosomes 2, 4, 5, 8 and 9 (Fig. 1). In some cells, however, a chromosome 4 with additional genetic material in the long arm was seen. A normal chromosome 4 (chromosome 4A) was observed in 47% of the metaphase spreads, while 44% of the spreads had a chromosome 4 of intermediate length (chromosome 4B) and another 9% had an even longer chromosome 4 (chromosome 4C) (Fig. 2). In addition to the normal chromosomes, four Z-group chromosomes (Z1, Z4, Z8 and Z13), 7 derivative chromosomes (der), and 2 marker chromosomes (mar) were identified (Fig. 1) [5].

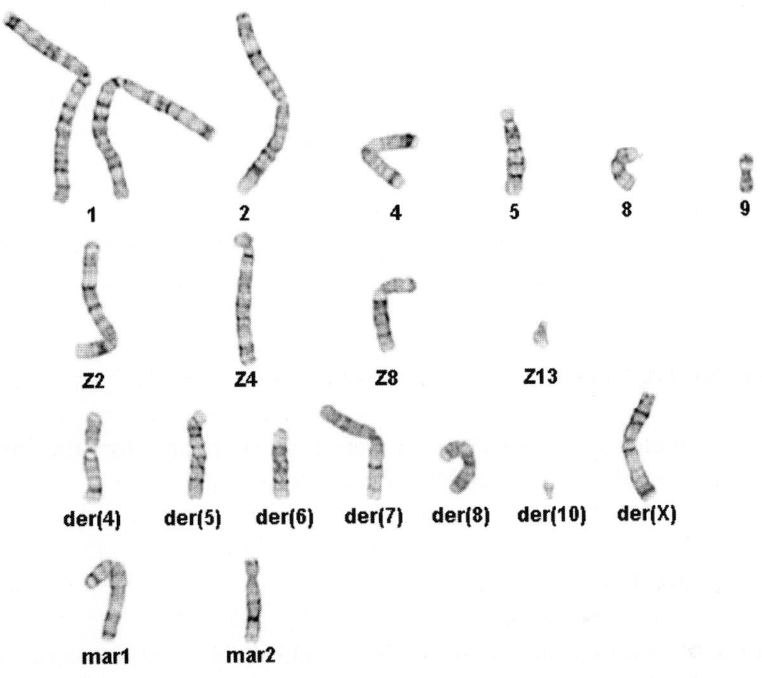


*Figure 1.* G-banded karyotype of CHO DG44 cell line.

For the Z group and derivative chromosomes, the high resolution observed with giemsa banding allowed a precise description of each structural rearrangement (Fig. 4). The deletion observed in Z2 occurred on the long arm from q11 to q22. The pericentric inversion of chromosome 3 in Z4 occurred on the short arm at p33 and on the long arm at q12. Der(4) resulted from the joining of the long arm of chromosome 4 and a part of the short arm of chromosome 3, from p21 to pter. Der(5), a derivative of chromosome 5, has a duplication in theist long arm, probably from q22 to q25. Der(6) was identified as a derivative of chromosome 6 without the short arm. Der(7) was formed from the long arms of chromosome 7 and the X chromosome. Der(8) was derived from the short arm of chromosome 8 and the long arm of chromosome 6 with a pericentric inversion between q17 and q29. Der(10) was identified as a derivative of chromosome 10 without a long arm. Der(X) was formed from the shorts arms of the X and Z3 chromosomes. The origins of the two mar chromosomes were not determined since they were not related to either the euploid Chinese hamster or to to the Z-group chromosomes.

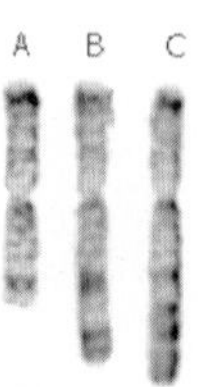


*Figure 2.* Variability of the chromosome 4 within the CHO DG44 cell line. A: normal chromosome 4. B: add(4) and C: add add (4), contained additional genetic material at the end of the long arm.

## 3. CONCLUSION

These studies are expected to serve as standard for the genetic characterization of recombinant DG44 cell lines.

## REFERENCES

Deaven LL, Petersen DF. 1973. The chromosomes of CHO, an aneuploid Chinese hamster cell line: G-band, C-band, and autoradiographic analyses. Chromosoma 41(2):129-44.

ISCN. 1995. An International System for Human Cytogenetic Nomenclature. Mitelman F, editor: Karger.

Kaufman R.J., Wasley L.C., Spiliotes A.J., Gossels SD, Latt S.A., Larse GR, M. KR. 1985. Coamplification and coexpression of human tissue-type plasminogen activator and murine dihydrofolate reductase sequences in chinese hasmter ovary cells. Molecular and Cellular Biology 5(7):1750-1759.

Ray M, Mohandas T. 1976. Proposed banding nomenclature for the Chinese hamster chromosomes (Cricetulus griseus). Cytogenet Cell Genet 16(1-5):83-91.

Urlaub G, Chasin LA. 1980. Isolation of Chinese hamster cell mutants deficient in dihydrofolate reductase activity. Proc Natl Acad Sci U S A 77(7):4216-20.

Urlaub G, Kas E, Carothers AM, Chasin LA. 1983. Deletion of the diploid dihydrofolate reductase locus from cultured mammalian cells. Cell 33:405-412.

# From Biopsy to Cartilage-carrier-constructs by Using Microcarrier Cultures as Sub-process

Stephanie Nagel-Heyer, Svenja Lünse, Christian Leist, Roman Böer, Christiane Goepfert, Ralf Pörtner
*Technische Universität Hamburg-Harburg, Bioprozess- und Bioverfahrenstechnik, Denickestr. 15, 21071 Hamburg, Germany*

**Abstract:** The production of cartilage-implants includes several steps of propagation accopamied by an unwanted dedifferentiation of the cells. The aim of this work was to investigate the propagation of human chondrocytes on microcarriers in a specially designed conical bioreactor as a sub-process for the production of cartilage-carrier-constructs. This technique for increasing the number of chondrocytes without losing their characteristic phenotype and ability to produce cartilage matrix can improve the development and quality of hyaline-like cartilage implants.

**Key words:** tissue engineering, cartilage, bioreactor, microcarrier, proliferation, cartilage-carrier-construct, matrix formation

## 1. INTRODUCTION

For the production of artificial cartilage, several million chondrocytes are needed. Therefore, chondrocytes harvested from a biopsy have to be expanded by a factor of 100-1000. The propagation in T -flasks comprises a couple of trypsination steps leading to a dedifferentiated phenotype of the chondrocytes (reduction of collagen type II expression and production of collagen type I). In this work, an alternative propagation technique was developed, which may avoid these drawbacks. The propagation of chondrocytes on microcarriers (Figure 1) has the following advantages: (i) bioreactors can be used for better process control and easier handling, (ii)



*R. Smith (ed.), Cell Technology for Cell Products*, 367–369.

repeated treatment with proteolytic enzymes (trypsin) can be avoided, (iii) detachment of chondrocytes after the expansion period for further use is possible.

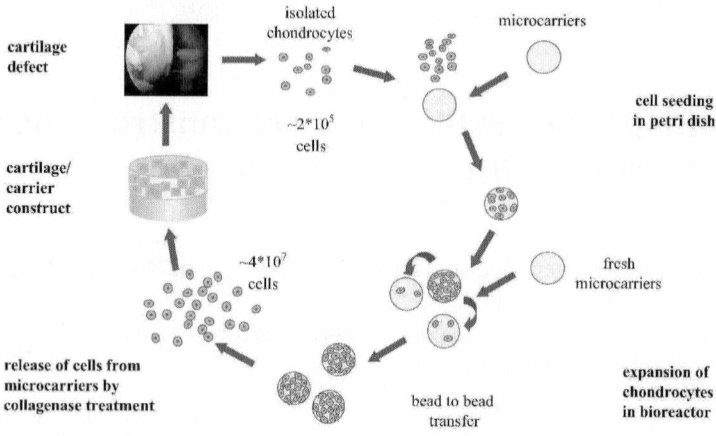


*Figure 1.* Scheme of the cultivation principle for propagation of chondrocytes on microcarriers in a stirred bioreactor.

## 2. MATERIALS AND METHODS

Chondrocytes of a human donor were isolated from articular cartilage and seeded on Cytodex 3 microcarriers (Amersham Bioscience) at a cell density of $5 \cdot 10^4$ or $2 \cdot 10^5$ per 3 mg of the microcarriers. The culture medium (DMEM hG) was supplemented with 10% FCS and 10 ng/ml bFGF.

In parallel, chondrocytes were expanded in T-flasks for two passages (P1, P2). Afterwards, the cells were seeded on microcarriers and further expanded in the bioreactor. Primary, P1 and P2 chondrocytes were compared for their ability to attach to the microcarriers and to reach cell numbers relevant for *in vitro* production of cartilage.

The cultivation process was started in petri dishes with 3 mg of the microcarriers using 2 mL of culture medium. When reaching confluence, the cultures were scaled up to 30 mg and 300 mg of the microcarriers, respectively. Cultivations with 300 mg microcarriers were done in a conical bioreactor (working volume 100 mL).

Finally, the chondrocytes were harvested from the microcarriers using collagenase type Ia (SIGMA) and used to establish cartilage cultures on top of ceramic cylinders. To stimulate cartilage matrix production, TGF-ß and IGF-I were added to the culture medium.

## 3. RESULTS

The increase in cell numbers for the primary cells and for cells in P1 and P2 was as follows:

| | primary chondrocytes | | passage no 1 | passage no 2 |
|---|---|---|---|---|
| initial cell number | $5 \cdot 10^4$ | $2 \cdot 10^5$ | $5 \cdot 10^4$ | $5 \cdot 10^4$ |
| cultivation time | 58 days | 43 days | 46 days | 58 days |
| final cell number | $43 \cdot 10^6$ | $46 \cdot 10^6$ | $29 \cdot 10^6$ | $20 \cdot 10^6$ |

Primary human chondrocytes attached more slowly to the microcarriers than passaged cells (P1 and P2). On the other hand, primary cells reached higher cell yields and cell densitites per area than expanded chondrocytes. The cells were successfully used to engineer cartilage-carrier-constructs.

## 4. CONCLUSIONS

The cultivation of primary human chondrocytes on Cytodex 3 microcarriers was successful. Expansion times and final cell numbers of human articular chondrocytes on Cytodex 3 depend on the initial cell numbers. Treatment with proteolytic enzymes during expansion can be avoided. Recovery of viable chondroctes from microcarriers after collagenase treatment was about 80%.

## REFERENCES

Nagel-Heyer, S., Goepfert, Ch., Adamietz, P., Meenen. N.M., Petersen, J.-P., and Pörtner, R., 2005, Flow-chamber bioreactor culture for generation of three-dimensional cartilage-carrier-constructs. Bioproc. Biosyst. Eng., in press

Nagel-Heyer, S., Goepfert, Ch., Morlock, M.M., and Pörtner, R., 2005, Relationship between gross morphological and biochemical data of tissue engineered cartilage-carrier-constructs. Biotechnol. Lett. **27**: 187-192

# Effect of Peptones and Study of Feeding Strategies in a CHO Based Fed-batch Process for the Production of a Human Monoclonal Antibody

Véronique Chotteau, Caroline Wåhlgren, Helena Pettersson
*Biovitrum, Biopharmaceuticals, Process Development, Process Sciences, Stockholm, Sweden*

**Abstract:** Eight commercial peptones, derived from plants, were studied for their ability of improving the cell growth and the productivity of a CHO cell line producing a human monoclonal antibody. They were also compared to yeast, lactalbumin and meat derived peptones. Seven plant peptones were selected and further studied in combination by Design of Experiment. The best three peptones were then tested in combinations in fed-batch cultivation. The fed-batch process was based on low concentrations of glucose and glutamine with feeding of amino acids, peptones and feed medium including vitamins, metal traces and biosynthesis precursors. This process was based on Biovitrum protein-free proprietary medium for the base medium and the feeding medium. Different feeding strategies, different peptone combinations and phosphate feeding were studied for their ability to improve the cell density, the cell specific productivity and the cultivation longevity.

**Key words:** peptone, protein hydrolysate, CHO, Chinese Hamster Ovary, fed-batch, feed, DoE, Design of Experiment, antibody, MAb, soy, wheat, rice, pea, yeast, cotton, Primatone, lactalbulmin, phosphate



*R. Smith (ed.), Cell Technology for Cell Products*, 371–374.

# 1. RESULTS AND DISCUSSION

## 1.1 Screening of various peptones, dosage study and combination study

The effect on the cell growth and the productivity of 8 plants peptones (HyPep 1510 soy, HyPep 4601 wheat and HyPep 4605 wheat, HyPep 5603 rice, HyPep 7401 pea, HyPep 7504 cotton seed, all Kerry, and P-0521 soy Sigma, Soy Protein Acid Hydrolysate Sigma) and 1 yeast peptone (HyPep 7455 Kerry) was studied including a comparison with Primatone meat Kerry and HyQ lactalbumin cow milk Hyclone. The results indicated that 5 plant peptones resulted in the best cell growth and productivity at levels comparable to the ones obtained by using Primatone and that the best results were obtained between 5 and 10 g/L peptone with worse results at 1 g/L and 11 g/L peptone and above. Combinations of 2 or 3 peptones of the 5 selected peptones plus 2 additional peptones were studied by Design of Experiment; the total peptone concentration being 5 g/L. From this, 3 plant peptones, coded PEP D, E and G, were selected for their effect on the cell growth, the viability or the productivity. No synergy effect or detrimental effect of combining the peptones were observed in this experiment.

## 1.2 Fed-batch cultivation with peptone addition

Diverse combination of peptones PEP D, E and G in the base medium and the feed were studied in fed-batch spinners as well as feeding of phosphate. Comparing feeding or not the feed medium while glucose, glutamine and amino acids were fed showed that the feed medium addition was beneficial. It was showed that the presence of peptone PEP G resulted in better productivity and growth and that feeding phosphate did not improve the process. An early unexplained decline in viability was observed reproducibly in two run series despite the fact that the by-product accumulation or the osmolarity were not too high. A screening was then performed in 6-wells plate system showing that the early viability decline was due to the simultaneous feeding of feed medium, peptones D and E and amino acids. The feed medium composition was also slightly modified and a new series of fed-batch runs in 100 ml spinners was performed to compare the effect of different combinations of 2 peptones from PEP D, E and G in the base medium (BM), the effect of feeding frequency of feed medium (FM) every day (1/1), every $2^{d}$ (1/2) or $3^{d}$ day (1/3) of the same total feed medium volume and the effect of diluting 1.5 (concentrated FM) versus 2.5 times over the whole fed-batch cultivation while adding the same amount of nutrients. It was observed that the combination of PEP G and E gave higher productivity than the combination of PEP D and G in the base medium, feeding every 3d days was better than feeding every 2d day for the cell growth and the productivity, feeding every day seemed to help maintaining the viability but the productivity was similar as feeding every 2d day and finally higher cultivation dilution 2.5 versus 1.5 was more favourable to the viability and the productivity, see Figures 1 to 3.

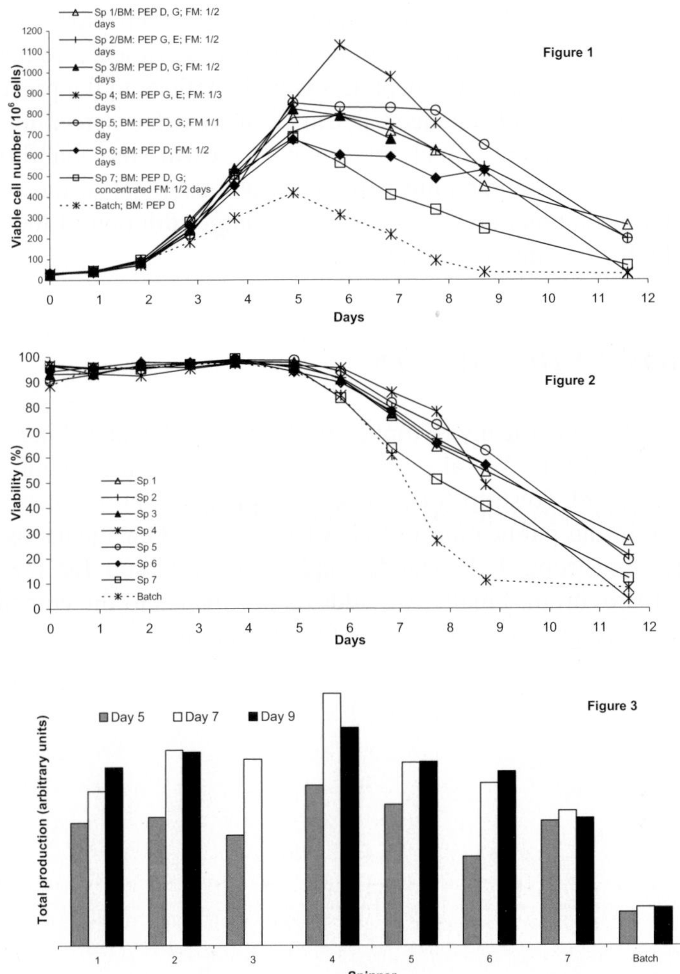


*Figure 1-3.* Viable cell number, viability and product accumulation in fed-batch spinners.

## 2. CONCLUSIONS

A series of 8 plant peptones and 1 yeast peptones was studied together with a meat- and milk-derived peptones as references. 3 plant peptones were selected for further fed-batch process investigation due to their positive effect on the cell density, viability or productivity. The fed-batch strategy was improved by the use of plant peptones for this cell line. It was found that the amino acids should not be fed simultaneously with peptones D and E and the feed medium. This peptone combination was detrimental to the fed-batch process even if the individual peptones were beneficial to the process and the same combination had not been detrimental in batch screening. Different peptones had different effect on the cell growth, the viability and the productivity so that combining peptones was beneficial to the process; the best peptone combination was identified. A larger dilution of the cultivation brought by the feeding was beneficial.

## ACKNOWLEDGEMENTS

We thank the company Symphogen, Denmark, for their kind permission to use their cell line. We also want to thank Dr. Sören Bregenholt and Dr. Anne Bondgaard Tolstrup for their kind support to get this permission and helping us in the publication release of this work. Our thanks as well to J. Mühr, K. Engström, J. Kinnander and U. Oswaldsson for the analytical support at Biovitrum. Finally, we address our thanks to the company Kerry, which provided graciously the Kerry/Quest peptones.

# Retinoic Acid Induced Up-regulation of Monoclonal Antibody Production

Yuichi Inoue[1]*, Hiroharu Kawahara[2], Sanetaka Shirahata[3] and Yasushi Sugimoto[4]

*[1]Faculty of Agriculture, and [4]The United Graduate School of Agricultural Sciences, Kagoshima University, 1-21-24 Korimoto, Kagoshima 890-0065, Japan; [3]Division of Life Engeering, Graduate School of Systems Life Sciences, Kyushu University, 6-10-1 Hakozaki, Higashi-ku, Fukuoka 812-8581, Japan; *Present address: [2]Kitakyushu National College of Technology, 5-20-1, Shii, Kokuraminami-ku, Kitakyushu 802-0985, Japan*

**Abstract:** The expression of retinoic acid receptor (RAR) and retinoid X receptor (RXR) was compared between RA responsive and unresponsive hybridoma cell lines. RT-PCR analysis showed that most of cell lines treated with RA expressed the RAR-alpha, -beta and -gamma, and the RXR-alpha and -beta. Interestingly, RXR-alpha was little expressed in the unresponsive cell lines. In addition, the transfection of RXR-alpha siRNA to the responsive cell line significantly suppressed the increasing effect of RA, suggesting that RXR-alpha may be associated with MAb up-regulation by RA. The RXR or RAR selective ligand alone and their combinations all could increase MAb production. Their abilities to increase the MAb production were almost paralleled with those to induce the expression of the RAR/RXR-mediated gene, RAR-beta, suggesting that MAb production may be up-regulated through, in part, the RAR/RXR heterodimers. Taken together, our results suggest that RXR-alpha may play an important role in MAb up-regulation through the RAR/RXR heterodimers

**Key words:** antibody production, hybridoma, monoclonal antibody, retinoic acid, retinoid receptor.

## 1. INTRODUCTION

The vitamin A metabolite, retinoic acid (RA), has been shown to increase monoclonal antibody (MAb) production in several human hybridoma cell



*R. Smith (ed.), Cell Technology for Cell Products*, 375–379.

lines (Aotsuka *et al.*, 1991; Inoue *et al.*, 2000). However, not all human hybridoma cell lines show increased production by RA. The understanding of the molecular mechanism underlying RA action leads to the development of high MAb expression systems. RA was known to act through two kinds of retinoid nuclear receptors, the RA receptor (RAR) and the retinoid X receptor (RXR) (Ballow *et al.*, 2003). Each receptor consists of three types, alpha, beta, and gamma. All three RARs bind both all-trans RA and 9-cis RA, while all three RXRs bind only 9-cis RA. In this study, we examined the expression of RARs and RXRs and the effects of their selective ligands on MAb production in human hybridoma cell lines.

## 2. MATERIALS AND METHODS

### 2.1 Hybridoma cell line and culture conditions

The hybridoma cell lines, AE6, BD9, HB4C5, and HF10B4 were used for this study (Kawahara *et al.*, 1992; Murakami *et al.*, 1985). Cells were maintained in the serum-free ERDF medium (Kyokuto Pharmaceutical Industrial Co., Tokyo, Japan) supplemented with ITS-X Supplements (GIBCO BRL, Tokyo, Japan) which contains 10 μg/ml of insulin, 5.5 μg/ml of transferrin, 2 μg/ml of ethanolamine and 6.7 ng/ml of sodium selenite, at 37°C in humidified 5% $CO_2$/95% air. All experiments were done in triplicate and the average value was used for analysis.

### 2.2 Retinoic acid and selective ligands

A RAR and RXR ligand 9-cis RA (Wako, Osaka, Japan), a selective RAR ligand AM-580 or TTNPB (BIOMOL, PA, USA), and a selective RXR ligand methoprene acid (BIOMOL, PA, USA), were dissolved in ethanol at a concentration of 1 mM, and stored at –20°C in small aliquots under light protection. For each experiment, stock aliquots were diluted with ethanol, and immediately added to the culture medium at a given concentration.

### 2.3 Measurement of antibody concentration and viable cell number

The MAb concentrations in culture medium were measured by an enzyme-linked immunosorbent assay (ELISA) as described previously (Shoji *et al.*, 1994), using anti-human immunoglobulin (Ig) antibody (IgM #AH1601, IgG # AH1301; Biosource International, Inc., USA) as the first antibody, and anti-human Ig peroxide conjugate antibody (IgM #AH1604, IgG # AH1304; Biosource International, Inc., USA) as the second antibody.

Cell number was counted by using a hemacytometer, and viability was determined by the trypan blue dye exclusion method.

### 2.4 RT-PCR analysis

RT-PCR analysis was done as reported previously (Inoue *et al.*, 2002). Briefly, total RNA was recovered from cells using the RNA extraction reagent. Reverse transcription was done using oligo-dT primers as described in the manufacturer's protocol (Amersham Pharmacia Biotech Inc., USA). PCR amplification reactions were done in 50 μl reaction volumes containing 5 μl of 10 x PCR buffer, 4 μl of 2.5 mM deoxynucleoside triphosphates, 3 μl of first strand cDNA, 10 pmol of each primer and 1.25 units of *Taq* DNA polymerase (Takara Biomedicals, Osaka, Japan). The mixture was denatured at 94°C for 2 min, followed by 35 cycles at 94°C for 30s, at 55°C for 30s, and at 72°C for 30s. The final elongation step was extended for an additional 5 min. The amplified products were analyzed by electrophoresis on a 1.5% agarose gel and stained by ethidium bromide.

## 3. RESULTS AND DISCUSSION

### 3.1 Expression of retinoid receptors

RT-PCR analysis showed that the hybridoma cell line AE6, BD9, HB4C5 and HF10B4 treated with 9-cis RA expressed the RAR-alpha, -beta and -gamma, and the RXR-beta. Interestingly, RXR-alpha was little expressed in the RA unresponsive cell line HB4C5 and HF10B4 (Fig. 1). In addition, the transfection of RXR-alpha siRNA to the RA responsive cell line AE6 and BD9 significantly suppressed the increasing effect of 9-cis RA (Fig. 2). These findings suggest that RXR-alpha may be associated with MAb up-regulation by RA.

### 3.2 Effects of selective retinoid receptor ligands on antibody production

The RAR selective ligand AM-580 or TTNPB, the RXR selective ligand methoprene acid (MA), and their combinations AM-580+MA and TTNPB+MA could increase MAb production in the human hybridoma cell line BD9 (Fig. 3). The abilities of these ligands to increase the MAb production were almost paralleled with those to induce the expression of the RAR/RXR-mediated gene, RAR-beta (data not shown). These findings were similar in the human hybridoma cell line AE6, suggesting that MAb production may be up-regulated through, in part, the RAR/RXR heterodimers. Taken together, our results suggest that RXR-alpha may play an important role in MAb up-regulation through the RAR/RXR heterodimers.

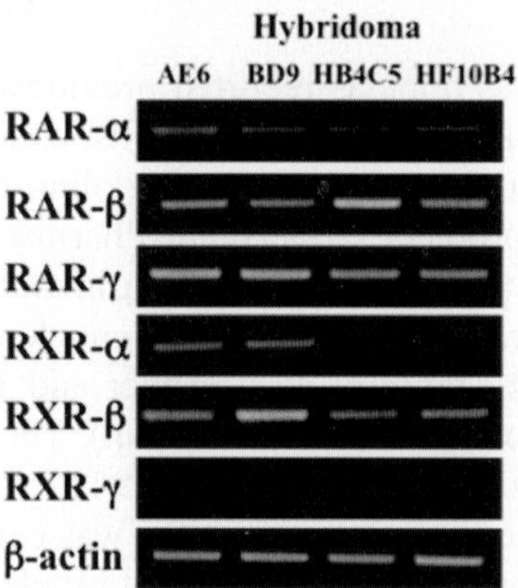


*Figure 1.* RAR and RXR expression in human hybridoma cell lines. The hybridoma cell lines (1 x 105 cells/ml) were cultured in ITS-X-ERDF medium supplemented with 100 nM 9-cis RA. After 2 days, RAR and RXR expression were examined by RT-PCR analysis. Beta-actin expression was also examined as a control.

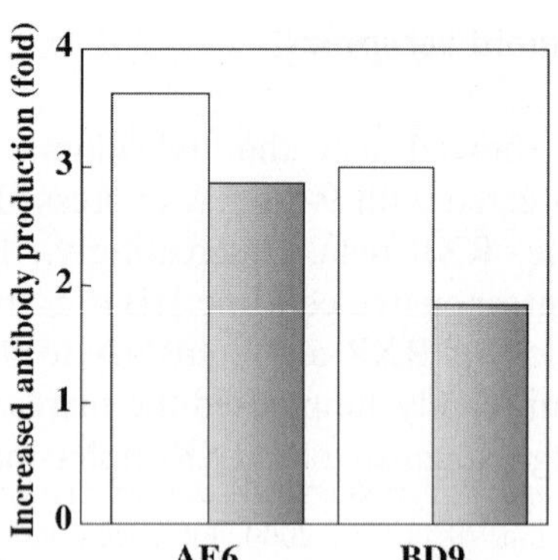


*Figure 2.* Effect of RXR-alpha siRNA transfection on antibody production increased by RA. siRNA transfection was performed according to manufacturer's instructions. After transfection, The hybridoma cell lines AE6 and BD9 (1 x 105 cells/ml) were treated with 10-7 M 9-cis RA for 2 days. Increased antibody production by RA was determined by ELISA. Open and shadow bars indicate control and test, respectively.

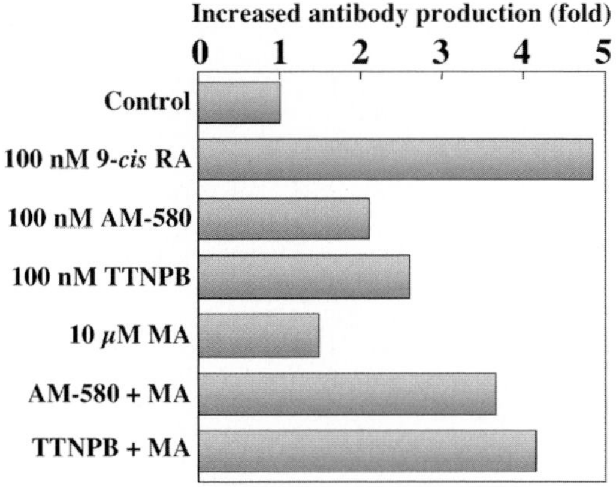


*Figure 3.* Effect of the selective retinoid receptor ligands on antibody production. The hybridoma cell line BD9 (1 x 105 cells/ml) were cultured in ITS-X-ERDF medium supplemented with or without retinoid receptor ligand(s). After 2 days, IgG production was measured by ELISA. 9-cis RA: RAR and RXR selective; AM-580 or TTNPB: RAR selective; MA: RXR selective.

## ACKNOWLEDGEMENTS

This work is supported in part by a Grant-in-Aid for Young Scientists from Japan Society for the Promotion of Science.

## REFERENCES

Aotsuka, Y. and Naito, M., 1991, Enhancing effects of retinoic acid on monoclonal antibody production of human-human hybridomas, *Cell. Immunol.* **133**: 498-505.

Ballow, M., Wang, X., Xiang, S. and Allen, C., 2003, Expression and regulation of nuclear retinoic acid receptors in human lymphoid cells, *J. Clin. Immunol.* **23**: 46-54.

Inoue, Y., Fujisawa, M., Shoji, M., *et al.*, 2000, Enhanced antibody production of human-human hybridomas by retinoic acid, *Cytotechnology* **33**: 83-88.

Inoue, Y. and Shirahata, S., 2002, Increased antibody production by retinoids is related to the fusion partner of human hybridomas, *Biosci. Biotechnol. Biochem.*, 66 (1): 215-217.

Kawahara, H., Shirahata, S., Tachibana, H. and Murakami, H., 1992, In vitro immunization of human lymphocytes with human lung cancer cell line A549, *Hum. Antibod. Hybridomas* **3**: 8-13.

Murakami, H., Hashizume, S., Ohashi, H., *et al.*, 1985, Human-human hybridomas secreting anibodies specific to human lung carcinoma, *In Vitro Cell. Dev. Biol.*, **21**: 593-596.

Shoji, M., Kawamoto, S., Sato, S., *et al.*, 1994, Specific reactivity of human antibody AE6F4 against cancer cells in tissues and sputa from lung cancer patients, *Hum. Antibod. Hybridoma* **5**: 116-122.

# Media and Cell Manipulation Approaches to the Reduction of Adaptation Time in GS NS0 Cell Line

Paul Clee[1] and Mohamed Al-Rubeai[1,2]
[1]*Department of Chemcial Engineering, University of Birmingham, Birmingham, B15 2TT Email:pxc235@bham.ac.uk and* [2] *Department of Chemcial and Biochemcial Engineering, University College Dublin, Belfield, Dublin 4, Ireland. Email:m.al-rubeai@ucd.ie*

**Abstract:** Initial attempts to grow GS NS0 cells in the serum-free media provided by Cambrex failed owing to a lack of certain additions. It was decided to evaluate the addition of cholesterol, cholesterol encapsulated with a cyclodextrin carrier and Glutamine Synthetase Expression Supplements (GSES). Aim was to evaluate if the adaptation time could be reduced, while still maintaining equal to higher viable cell density and IgG antibody production. It was found that GS NS0 cells required the addition of cholesterol to sustain growth in the medium. The addition of cholesterol encapsulated with cyclodextrin with the GSES additives reduced adaptation time to 50 days from 127 days for cholesterol alone.

**Key words:** GS NS0, CHOLESTEROL, GSES.

## 1. INTRODUCTION

Serum-supplemented media are normally used to culture mammalian cells, however, growing concerns over serum supply and the potential for exogenous serum contamination has driven the transition to serum-free media. Some cell lines (e.g., cholesterol auxotrophic cell line, NS0) have unique requirements for specific media supplementation to be effective, as demonstrated by Keen and Steward (1). However, the production of purified recombinant proteins has created a demand for media capable of supporting growth without relying on these supplements. The use of serum-free media requires some adaptation effort to ensure the most viable cell population is



*R. Smith (ed.), Cell Technology for Cell Products*, 381–383.

selected for long-term cell culture studies or the continuous production of desired cell derived proteins. Adaptation protocols are tedious, time consuming and vary according to cell lines and production methods. We have demonstrated that with the addition of cholesterol and glutamine synthetase expression media additives (GSEM) adaptation time can be significantly reduced.

## 2. RESULTS AND DISCUSSION

The GS NS0 cell line in Figure 1, was not adapted to the serum-free media. Initial attempts to get the cell line to grow without the addition of cholesterol failed. When cholesterol was added the media , the NS0 cell line was able to proliferate and sustain growth in the medium. The addition of cholesterol and GSES additives gave an increase in the viable cell density over cholesterol alone. The use of cholesterol encapsulated with cyclodextrin with GSES additives gave the largest increase in viable cell density. The NS0 cells were then adapted using a weaning protocol.

*Table 1.* Showing the cholesterol requirement of NS0.

| Days | Serum-Free Media Only (Cells/mL) | Serum-Free Media + Cholesterol (Cells/mL) | Serum-Free Media + Cholesterol & GSES (Cells/mL) |
|---|---|---|---|
| 0 | $2 \times 10^5$ | $2 \times 10^5$ | $2 \times 10^5$ |
| 1 | $1.7 \times 10^5$ | $2.2 \times 10^5$ | $2.2 \times 10^5$ |
| 2 | $0.9 \times 10^5$ | $2.9 \times 10^5$ | $3.0 \times 10^5$ |
| 3 | $0.3 \times 10^5$ | $3.3 \times 10^5$ | $3.4 \times 10^5$ |
| 4 | 0 | $3.8 \times 10^5$ | $4.2 \times 10^5$ |
| 5 | 0 | $4.3 \times 10^5$ | $4.5 \times 10^5$ |

Mean values plotted. n=3

*Table 2.* Adaptation time for NS0 cell line.

| Compounds Used | Adaptation Time (Days) |
|---|---|
| Serum-Free Media with Cholesterol | 127 |
| Serum-Free Media with Cholesterol and GSES | 90 |
| Serum-Free Media with Cholesterol:Cyclodextrin and GSES | 50 |

Table 2, shows the results for adaptation time with the selected components from Table 1. It was observed to take 127 days for the NS0 cell line to adapt to the serum-free media with the addition of cholesterol. Adding GSES additives to the media with cholesterol improved the adaptation time to 90 days for the NS0 cell line. The quickest time observed was 50 days, which included the addition of cholesterol encapsulated with cyclodextrin and GSES additives.

*Table 3.* Comparison of adapted NS0 serum-free cells in serum-free media, verses NS0 cells in basal media.

| Days of Batch | Serum-free Cell Count (Cells/mL) | Basal Media Cell Count (Cells/mL) | Serum-Free IgG Production (μg/mL) | Basal Media IgG Production (μg/mL) |
|---|---|---|---|---|
| 0 | $2.5 \times 10^5$ | $2.5 \times 10^5$ | | |
| 1 | $3.4 \times 10^5$ | $3.2 \times 10^5$ | | |
| 2 | $4.9 \times 10^5$ | $4.6 \times 10^5$ | | |
| 3 | $7.8 \times 10^5$ | $6.7 \times 10^5$ | 37 | 35 |
| 4 | $10.7 \times 10^5$ | $8.8 \times 10^5$ | 48 | 42 |
| 5 | $9.8 \times 10^5$ | $9.1 \times 10^5$ | 62 | 57 |
| 6 | $7.3 \times 10^5$ | $7.2 \times 10^5$ | 64 | 55 |
| 7 | $5.4 \times 10^5$ | $5.5 \times 10^5$ | 50 | 37 |
| 8 | $2.8 \times 10^5$ | $3.1 \times 10^5$ | 32 | 21 |
| 9 | $1.6 \times 10^5$ | $1.2 \times 10^5$ | | |
| 10 | $0.9 \times 10^5$ | $0.4 \times 10^5$ | | |

Mean values plotted. n=3.

The adapted NS0 cell line in Table 3, was tested to see if the adaptation process had been successful. The serum-free media out-performed the basal media in respect of viable cell density and antibody production. A common observation observed when adapting cell lines to serum-free media is the actual reduction of antibody production in comparison to basal media, even though viable cell density may be proportional to basal media.

## 3. CONCLUSION

It has been demonstrated in Figure 1, that the NS0 cell line required the addition of cholesterol to maintain growth in the serum-free media used. The addition of GSES additives also aided in maintaining growth. Figure 2 showed that with the additions observed in Figure 1, that the adaptation time can be reduced. To adapt the NS0 cells in the shortest time period it was observed that the additions of cholesterol encapsulated with cyclodextrin and GSES additives gave a 50 day period. The slowest time observed was from only adding cholesterol alone leading to a 127 day period. Subsequent performance testing, Figure 3, showed that the NS0 cell line had a higher viable cell density and equal to higher IgG productivity in comparison to basal media.

## REFERENCES

Keen, M. J.; Steward, T. W. Adaptation of cholesterol requiring NS0 mouse myeloma cells to high density growth in a fully defined protein-free and cholesterol-free culture medium. *Cytotechnology* 1995, 17, 203-211.

# The Production of Human Growth Hormone

Fabienne Anton[1], Alexander Tappe[1], Cornelia Kasper[1], Alexander Loa[2], Bernd-Ulrich Wilhelm[3], Thomas Scheper[1]
[1]*Institut für Technische Chemie, Callinstr. 3, 30167 Hannover, Germany*
[2]*CCS CellCulture Sevice GmbH, Falkenried 88, D-20251 Hamburg, Germany*
[3]*Sartorius BBI Systems GmbH, Schwarzenberger Weg 73-79, 34212 Melsungen, Germany*

**Abstract:** For the production of recombinant proteins for clinical use animal cell cultivation is used. Only these cells are able to perform correct folding and glycosylation of the desired protein. As the production process is expensive and long, cheaper and/or swifter cultivation routines are required. Chinese hamster ovary (CHO) cells were used to examine the expansion of cells and the production of recombinant human growth hormone in different cell culture systems which are supposed to achieve higher cell densities and product concentrations compared to conventional cell culture systems. The CHO cells were grown in suspension in serum-free, low-protein medium. Five different culture systems were used for batch-cultivation: Biostat B, BelloCell 500, spinner flask, RCCS-D and miniPERM. The systems differed in oxygen supply and medium agitation. While cells are agitated by stirrers in Biostat B and spinner flask, the whole medium is revolved in BelloCell 500, RCCS-D and miniPERM. Unlike the other systems the BelloCell 500 retains the CHO cells on a matrix. The aim was to maximize cell growth and productivity, which was achieved best in BelloCell and RCCS-D. In a second step the influence of temperature on growth and product formation was examined.

**Key words:** animal cell culture, protein production, Chinese hamster ovary CHO cells, human growth hormone hGH, bioreactor, bioprocess engineering, Biostat B, BelloCell 500, miniPERM, RCCS-D, serum-free medium, batch cultivation



*R. Smith (ed.), Cell Technology for Cell Products*, 385–391.

# 1. INTRODUCTION

Human growth hormone (hGH) is a polypeptide consisting of 191 amino acids with a molecular weight of 22 kDa (Ribela, 2003). It is produced in the pituitary gland of every healthy human being and is responsible for the growth process of human beings and the stimulation of protein, lipid and bone metabolism. Since microsomia is one of the most consequences of lacking hGH, the protein is of pharmaceutical significance.

The progress of recombinant DNA technology in the 1970s made it possible to express recombinant genes via microorganismen like *Escherichia coli.* (Simonsen, 1994). Nowadays the production of pharmaceutical proteins for clinical use often depends on the use of mammalian cell culture in order to ensure the correct folding and posttranslational modifications of the desired protein (Hauser, 1997, MacDonald, 1990). The most widely used mammalian cell line for the production of recombinant biopharmaceuticals are derived from the Chinese Hamster Ovary (CHO) cell line (Simonsen, 1994).

Due to long and expensive cultivation processes it is necessary to optimise the process. The aim is to increase cell growth and the production of complex proteins to the maximum. This can be achieved by using different bioreactors or by the optimisation of the cultivation parameters like stirring speed, pH-regulation or temperature (Birch, 1994, Bloemkolk, 1992, Hu, 1992, Sureshkumar, 1991). A decrease of temperature from 37 °C to lower temperatures can have a positive influence on cell productivity (Rössler, 1996).

The aim of this study was to examine growth and productivity of the recombinant $CHO^{SFS}$ hGH cells cultivated in different culture systems which shall result in a higher cell density and product concentration. Furthermore the influence of temperature on cell growth and hGH production was examined.

# 2. MATERIAL AND METHODS

All experiments were carrierd out with the $CHO^{SFS}$ hGH cell line which was provided by CCS Cell Culture Service GmbH (Hamburg, Germany). The cells were cultivated under serum free conditions and low protein concentrations.

Cell count was carried out with the trypan blue method and the amount of hGH in the supernantant was measured with a specific sandwich ELISA (Roche, Diagnostik GmbH, Germany). Cell densities in BelloCell 500 were calculated from glucose uptake rates.

### 2.1 Cultivation in different cell culture systems

In the first part five different cultivation systems were tested towards their suitability for batch-cultivation in order to achieve better cell growth and higher productivity. Therfore the experiments were carrierd out in:

- 250 ml spinner flask (Techne, England): 100 ml medium, 20 rpm
- 2 l Biostat B (Sartorius BBI Systems, Germany): 1.5 l medium, 100 rpm
- miniPerm (Vivascience, Germany): 35 ml medium, 30 rpm
- RCCS-D (Synthecon, USA): 45 ml medium, 8 rpm
- BelloCell 500 (CESCO Bioengineering Co., Taiwan): 300 ml medium, up/down 1mm/s, top/bottom delay 10 s

The systems differed in airiation and agitation. All cultures held at 37 °C and 5% $CO_2$. In Biostat B and spinner flasks the cells were mixed by mechanical stirring while in miniPerm, RCCS-D and BelloCell 500 the whole medium is revolved. Only the BelloCell 500 retains the CHO cells in a matrix of fibers.

### 2.2 Temperature experiments

The influence of temperature on cell growth and product formation was determined. The experiments were carried out in 250 ml spinner flasks. The suspension batch-cultures were grown at 37 °C, 34 °C and 31 °C, in an incubator at 5% $CO_2$ and 20 rpm. The spinner flasks were inoculated with an initial cell density of $1.5 \cdot 10^5$ cells per ml.

## 3. RESULTS AND DISCUSSION

### 3.1 Effect of culture system on cell growth and hGH production

Figure 1 shows the cell density of CHO cells in different culture systems measured with the trypan blue method. The highest cell density of $9.3 \cdot 10^6$ cells per ml is achieved in the BelloCell 500, but the cells die rapidly afterwards and cell density decreases to $2 \cdot 10^6$ cells per ml due to lack of glucose. Cell density reaches about $2.2 \cdot 10^6$ cells per ml in the RCCS-D and miniPerm, operating with revolving medium, whereas the continious stirred systems resulted in only $1 \cdot 10^6$ cells per ml.

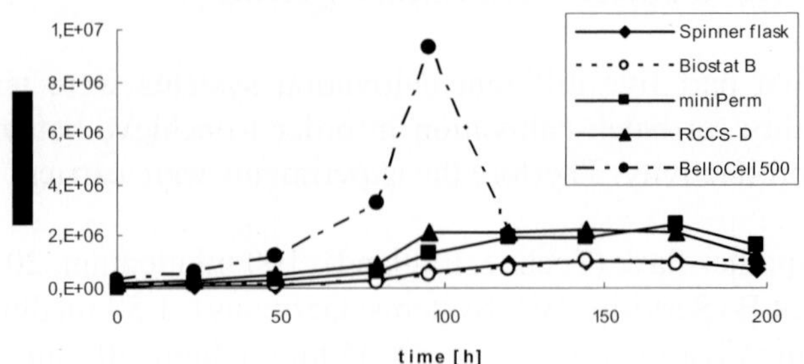


*Figure 1.* Number of living cells during the cultivation of CHOSFS hGH cells at 37 °C at an atmosphere of 5% CO2 in different culture systems.

Figure 2 shows the hGH concentration in the supernatant. The hGH concentration in the supernatant was measured with a specific sandwich ELISA.

Spinner flask, Biostat B and miniPerm show similar concentrations of 23.9-26.4 µg/ml. The highest hGH concentration of 49 µg/ml is observed in the RCCS-D, the lowest concentration of 15.9 µg/ml in the BelloCell 500. While cell growth in BelloCell is fastest, the production rate of hGH is slowest by far with only 1.7 $\mu g/(day \cdot 10^6$ cells). The hGH concentration of 23.9-26.4 µg/ml are the result of either a good growth rate and a mediocre production rate (4.2-7.7 $\mu g/(day \cdot 10^6$ cells)) or vice versa. Only in the RCCS-D, a fast growth rate correlates to a sufficient production rate of 9 $\mu g/(day \cdot 10^6$ cells), leading to a high hGH concentration in the supernatant.

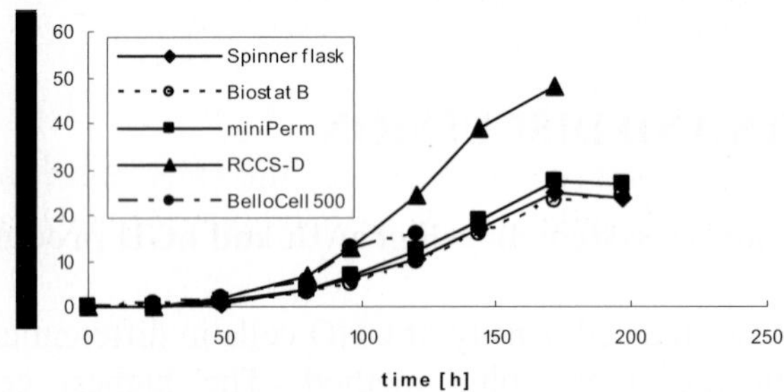


*Figure 2.* hGH concentration during the cultivation of CHOSFS hGH cells at 37 °C at an atmosphere of 5% CO2 in different cultur systems.

## 3.2 Effect of temperature on cell growth and hGH production

Figure 3 illustrates the effect of temperature on cell density. Cultured cells at 37 °C and 34 °C showed a fast cell growth during the first 100 hours while cells cultured at 31 °C showed a slower cell growth. In comparison to the cultivation at 37 °C, the cultivations at temperatures of 34 °C and 31 °C reach higher cell densities. The highest cell density of about $2.6 \cdot 10^6$ cells per ml was achieved at 34 °C.

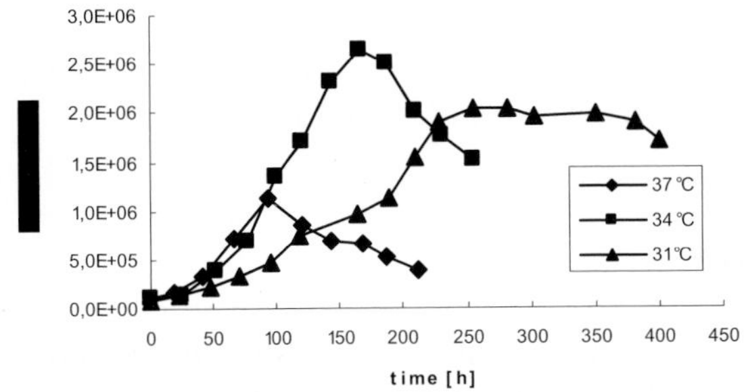


*Figure 3.* Number of living cells during the cultivation of CHOSFS hGH cells in spinner flasks at 37 °C, 34 °C and 31 °C at an atmosphere of 5% CO2 and 20 rpm.

The reduction of the temperature from 37 °C to 34 °C and 31 °C had a positive influence on cell growth and resulted in elongated cultivation times. But the prolonged cultivation times cause also an increase of process costs.

Figure 4 illustrates the effect of temperature on hGH production in the supernatant during the cultivation in spinner flask measured with a specific sandwich ELISA.

In the first 50 hours a slow increase of hGH concentration at all temperatures could be observed. At 34 °C, the hGH concentartion increases up to 47.2 µg per ml while at 37 °C a maximum of hGH concentration of only 10 µg per ml is achieved with the slowest production rate of 3.23 µg/(day$\cdot 10^6$ cells). While cell growth at 31 °C is slowest, the hGH concentration of 62.8 µg per ml is highest. The production rate of 3.88 µg/(day$\cdot 10^6$ cells) at this temperature is lower than at 34 °C. Only at 34 °C a fast growth rate correlates to a high production rate of 5.14 µg/(day$\cdot 10^6$ cells), resulting in the highest possible product yield in best time. The results show, that a decrease of temperature from 37 °C to 31 °C has a positive influence on product formation.

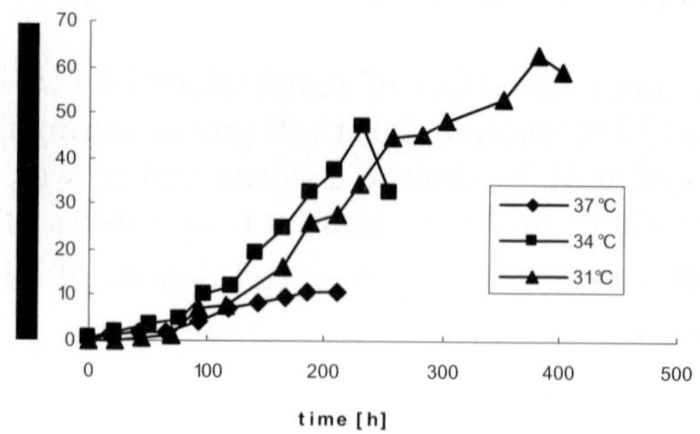


*Figure 4.* hGH concentration during the cultivation of CHOSFS hGH cells in spinner flasks at 37 °C, 34 °C and 31 °C at an atmosphere of 5% CO2 and 20 rpm.

## 4. SUMMARY

Both experiments show that cell growth and product formation is influenced both by the culture system and the temperature. Compared to traditional cultivation systems like spinner flask and Biostst B the highest cell density is achieved in the BelloCell 500 but the maximum of hGH amount is achieved in the RCCS-D. The reduction of cultivation temperature in CHO batch cultures from 37 °C to 34 °C and 31 °C has a positive influence on cell growth and productivity.

## ACKNOWLEGEMENT

The CHO cell line was provided by CCS Cell Culture Service GmbH (Hamburg, Germany).

The BelloCell 500 was donated by CESCO Bioengineering Co. (Taiwan).

## REFERENCES

Birch, J.R., Froud S.J., 1994, Mammalian cell culture systems for recombinant protein production, Biologicals, **22**:127-133

Bloemkolk, J.W., Gra, M.R., Merchant, F., Mosmann, T.R., 1992, Effect of temperature on hybridoma cell cycle and mAb production, Biotechnol.Bioeng., **42**:427-431

Hauser, HJ., Wagner, R., 1997, Mammalian cell biotechnology in protein production, Walter de Gruyter Berlin New York, **3.1**:279

Hu, W.-S., Piret, J.M., 1992, Mammalian cell culture processes, Current Opinion in Biotechnology, **3**:110-114

MacDonald, C., 1990, Development of new cell lines for animal cell biotechnology, Critical reviews in Biotechnology, **2** (10):155-178

Ribela, M.-T. C.P., Gout, P.W., Bartolini, P., 2003, Synthesis and chromatographic purification of recombinant human pituitary hormones, Journal of Chromatography B, **790**:285-316

Rössler, B., Lübben, H., Kretzmer, G., 1996, Temperature: a simple parameter for process optimization in fed-batch cultures of recombinant Chinese hamster ovary cells, Enzyme and Microbial Technology, **18**:423-427

Simmonsen, C.C, McGrogan, M., 1994, The molecular biology of production cell lines, Biologicals **22**:85-94

Sureshkumar, G.K., Mutharasan, R., 1991, The Influence of temperature on a mouse-mouse hybridoma growth and momoclonal antibody production, Biotechnol. Bioeng., **7**:292-295

# Preparation of Membrane Receptor Proteins Utilising Large Scale Micro-carrier Culture and a "Pearl Freezing Factory" to Finalise the Product

Irma Jansson, Thomas Fröberg, Anette Danielsson and Elke Lullau
*Global Protein Science & Supply, DECS, AstraZeneca R&D Södertälje, Building 841, S-151 85 Södertälje*

**Key words:** Micro-carrier, pearl freezing

## 1. INTRODUCTION

Bioreactor culture of adherent cells offers clear advantages for scale-up of cells for membrane preparations. However, harvesting cells from micro-carriers is time-consuming. In addition, enzymatic treatment may lead to loss of ligand binding and functional activity. To address this, we investigated scalable procedures to streamline cell harvesting and membrane preparation avoiding the use of enzymatic treatment. The concept is based on wash of cell-loaded micro-carriers in the fermentor, harvest and freezing of the cells directly without enzymatic treatment. Starting from the thawed carrier-cell material we compared enzymatically treated with non treated membrane preparations with respect to yield of functionally active receptors.



*R. Smith (ed.), Cell Technology for Cell Products*, 393–396.

## 2. MATERIAL AND METHODS

### 2.1 Cell line maintenance and preparation of seeding cultures

CHO cells (AstraZeneca proprietary) expressing membrane receptor B was adapted to our in-house, amino acid enriched DMEM/F12 (Invitrogen) medium for static cultures, supplemented with 5% FCS (Invitrogen), 1mM $Ca^{2+}$, 4 mM l-Gln, 1 mg/ml G418 (Invitrogen), and maintained in T225 flasks (Nunc).

The fermentation inoculum was grown in 10 layer Cell factories(Nunc).

### 2.2 Production procedure

Large-scale micro-carrier cultivation was performed according to a generic in-house protocol. The fermentation inoculum was grown in 10 layer cell factories seeded at a density of $2x10^4$cells/$cm^2$ and harvested when 80% confluent.

The cells were enzymatically removed from the cell factories by the use of Accutase (TCS Cellworks). After removal, cells and CultiSpher G™ (Percell Biolytica.) were pooled into an inoculum flask at a density of $0.3x10^6$ cells/ml in one third of the final working volume.

Cells were grown at 37°C, 5% $CO_2$ and pH 7.2, in a 50L stirred bioreactor (Sartorius BBI) with direct sparging aeration. Intermittent stirring for the first 12 hours after inoculation allowed the cells to attach to the carriers.

24h prior to harvest, when the cell culture had reached the desired density, the medium was renewed, sodium butyrate (5mM) was added and the temperature was decreased to 31° C.

### 2.3 Harvesting procedure

For the non-enzymatic method the micro-carrier and cell suspension was left to settle in the bioreactor vessel for 20 min. The supernatant was removed via a dip tube; the carrier slurry was washed with PBS directly in the bioreactor and harvested through the bottom valve. The wet carrier sediment was frozen at – 70°C.

The harvest procedure for the enzymatic method was different in that the carriers and cells were removed from the bioreactor and collected by centrifugation and washed once with PBS. The gelatin carriers were then enzymatically dissolved as described in the supplier's instructions and cells were collected by centrifugation. The material was then stored at –70°C for further membrane preparation according to standardised procedures for enzymatic treated membrane preparation.

### 2.4 Large scale membrane preparation

The frozen cell containing micro-carriers were thawed on ice in protease inhibitor containing homogenisation buffer (50 mM Tris/HCl (pH 7, 4°C), 2,5 mM EDTA. Subsequently the suspension was homogenized using an Ultraturrax at medium speed in short bursts of 60 seconds. The suspension was then filtered through a 75 μm metal mesh. This procedure was repeated three times in order to ensure complete removal of the cell lysate from the micro-porous carriers. All filtrates were pooled and kept on ice. Membranes were recovered through a series of centrifugation and disintegration steps prior to freezing in liquid-nitrogen using an in-house pearl freezing device.

## 3. RESULTS

This novel improved large-scale membrane preparation method was successfully applied to 3 x 50 L carrier campaigns (Figure 1). The comparison of specific binding between the different non-enzymatic membrane preparation methods showed that the streamlined preparation method resulted in as much as 12 times higher specific activity compared to the control sample (Figure 1).

The obtained functionality in the binding assay was higher in #3 than the initial evaluation due to reduced process time (Figure 1).

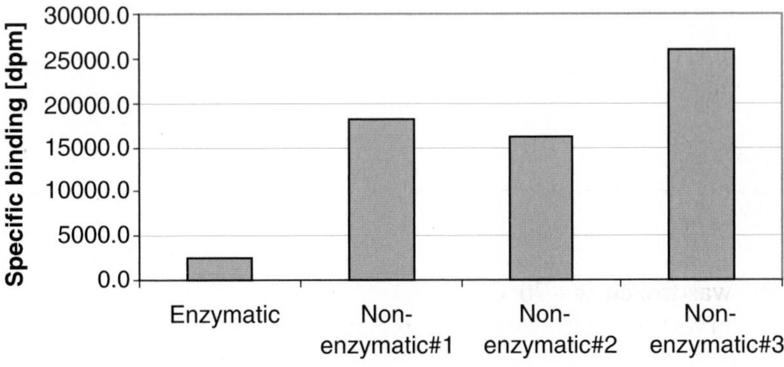


*Figure 1.* Comparison of specific binding between enzymatic and non-enzymatic (3 x 50L campaigns) membrane preparations.

The process time was reduced both through shorter handling times during membrane preparations and through the use of pearl freezing as a fast final formulation.

## 4. CONCLUSIONS

We have developed a simplified, straightforward and scalable procedure for membrane preparations from large-scale micro-carrier fermentations. The increase in specific binding is due to the fact that we have left out the use of proteases in the recovery steps and that the process time had been shortened several hours.

## REFERENCES

Jansson, Kanu (2004) Instruction for MEMBRANE PREPARATION from animal cells on micro-carriers (CultiSpher) for membrane receptor assays. Proprietary protocol within AstraZeneca.

Fenge *et al.* (2001) Process Development for Functional Membrane Receptor Production in Mammalian Cells. In: Animal Cell Technology: From Target to Market 17th ESACT meeting.

# Silk Protein Sericin Improves Mammalian Cell Culture

*Sericin As a Mitogenic Supplement to Culture Media*

Satoshi Terada, Naoki Takada, Kazuaki Itoh, Takuya Saitoh, Masahiro Sasaki* and Hideyuki Yamada*
*Department of Applied Chemistry and Biotechnology, University of Fukui, 3-9-1 Bunkyo, Fukui 910-8507, Japan, *Technology Department, Seiren Co., Ltd., 1-10-1 Keya, Fukui 918-8560, Japan*

**Abstract:** Mammalian cell cultures generally require supplementation with fetal bovine serum (FBS), or its replacement, into the culture media. Sera contain various unidentified and unknown factors and the risk of infections, including bovine spongiform encephalopathy (BSE), is of serious concern. Therefore, the supplementation of sera into culture media is a major obstacle for purification to recover cell products and this limits pharmaceutical acceptance of products. In this study, we examined whether the sericin protein, derived from the silkworm cocoon, can be effectively used as a substitute for FBS in mammalian cell culture. Together with fibroin, sericin is a major component of raw-silk and is removed from raw-silk by a treatment called degumming to make the silk lustrous and semitransparent. In order to investigate the effect of sericin on the proliferation and the productivity of mammalian cells, sericin was added to cultures of various mammalian cell lines, such as murine hybridoma 2E3-O. Sericin successfully accelerated the proliferation of the cells. Moreover, the production of MoAb by the hybridoma cells was also improved in the presence of sericin. Although heat easily denatures and inactivates most proteins, sericin maintained mitogenic activity after conventional autoclaving (20 minutes) and longer (60 minutes) autoclaving.

**Key words:** sericin, hybridoma, monoclonal antibody, serum-free medium



*R. Smith (ed.), Cell Technology for Cell Products*, 397–401.

# 1. INTRODUCTION

A variety of mammalian cells, including CHO and BHK, are industrially cultured to produce biomaterials such as proteins and gene therapy vectors and are used as transplants for cell therapy. Most mammalian cells need serum, or a replacement, in the culture medium, and fetal bovine serum (FBS) is used most frequently. However, FBS is frequently contaminated with viruses, and even the risk of bovine spongiform encephalopathy (BSE) is of major concern and serum also contains numerous factors outside the operator's control. Thus, the supplementation of serum to culture medium is a serious obstacle in the purification of products. Therefore, an alternative to FBS as a supplement to mammalian cell culture is eagerly desired.

We focused on sericin as an alternative to FBS and we reported that sericin accelerated the proliferation of various mammalian cells (Terada *et al.*, 2002). In this study, we examined the optimal concentration of sericin when supplemented to hybridoma cultures and the condition of sterilization of sericin.

Sericin and fibroin are the two major components of raw silk with fibroin being the predominant component. Sericin, a gummy coating on raw-silk filaments, is removed by a treatment called degumming to make the silk lustrous and semitransparent. The degumming treatment is essentially an alkaline scouring operation and is carried out at boiling temperatures. During this treatment, sericin is degraded and solubilized in water and abolished from silk. Various functions of sericin have been revealed and novel applications have been proposed. Sericin inhibits tyrosinase and lipid peroxidation (Kato *et al.*, 1998), and so sericin is utilized in cosmetics. Dietary supplementation of sericin in mice successfully suppressed colon carcinogenesis induced by 1,2-dimethylhydrazine (Sasaki *et al.*, 2000a). Enhancement of the bioavailability of several metal ions during consumption of sericin was also indicated in rat (Sasaki *et al.*, 2000b). A recombinant sericin peptide protected E. coli from freezing stress (Tsujimoto *et al.*, 2001). Sericin also improved mammalian cell survival during cryopreservation (Sasaki *et al.*, in press)

# 2. MATERIALS AND METHODS

The hybridoma cell line 2E3-O was cultured in serum-free ASF104 medium (Ajinomoto, Japan) in 24 well plates (Sumitomo Bakelite, Japan) at 37°C in humidified air containing 5% CO2.

The numbers of viable and dead cells were determined by counting in a hemocytometer under a phase contrast microscope using trypan blue exclusion. MoAb concentration in the culture supernatant was determined by ELISA.

## 3. RESULTS

In order to determine the optimum concentration of sericin, the proliferation of the hybridoma cells in the presence of 0.05%, 0.10% and 0.15% sericin was measured and shown in Fig. 1. At day 2, the viable cell number of the cultures with 0.15% sericin was the highest, but at day 3, viability with 0.1% sericin was the highest. Further experiments were performed to determine the optimal concentration of sericin for hybridoma cultures.

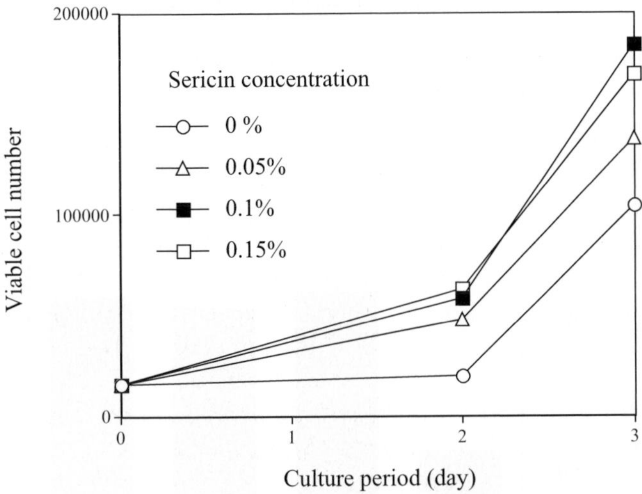


*Figure 1.* Growth curve of hybridoma 2E3-0 in the presence of sericin.

As shown in Fig. 2-a, the highest cell number was found in cells that were cultured in the presence of 0.1% sericin. Figure 2-b shows that the culture with 0.075% sericin produced the highest amount of MoAb, and the cultures with 0.1% produced the second highest amount. These results indicate that 0.075 - 0.1% is the optimum concentration of sericin to be supplemented to the hybridoma cultures.

Furthermore, we investigated the effect of sterilization on the mitogenic activity of sericin. Sericin was sterilized under various conditions and supplemented to the hybridoma cultures. As shown in Fig. 3, the mitogenic activity of sericin was not compromised, even after longer (60 minutes) autoclaving. These results indicate that sericin is suitable as a supplement for mammalian cell culture.

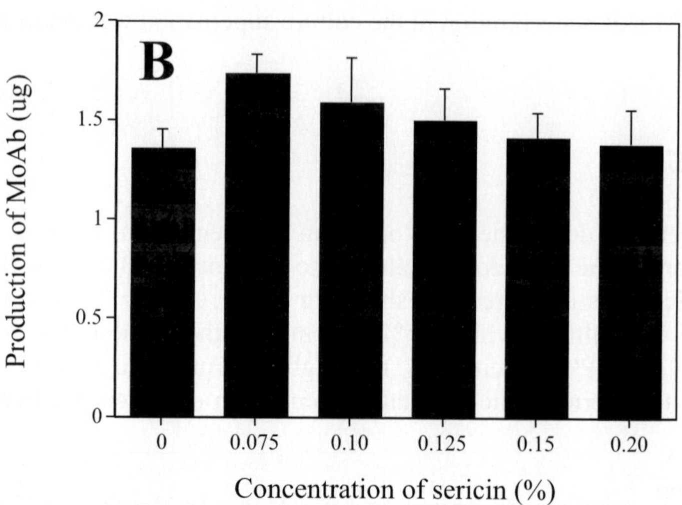


*Figure 2.* Effect of sericin on the proliferation (A) and MoAb, and (B) production of hybridoma 33000 cells were seeded and cultured for 24 hours (N=4).

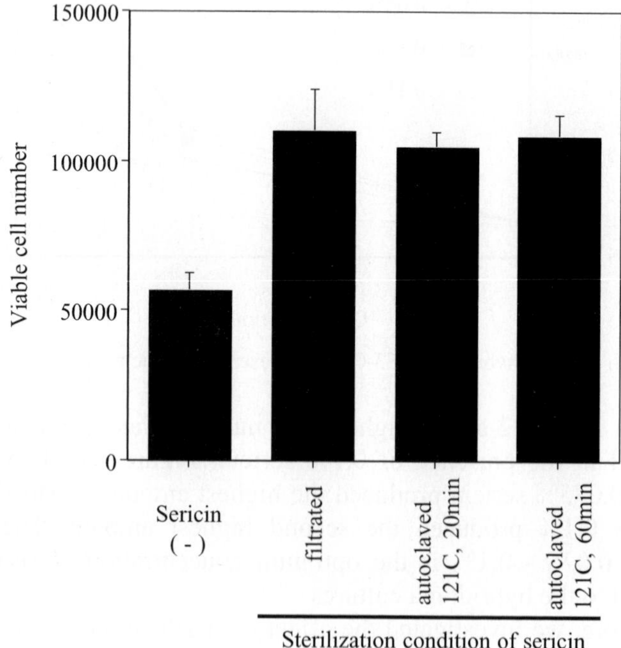


*Figure 3.* Mitogenic effect of sericin sterilized by different conditions on the hybridoma 29,000 hybridoma cells were seeded and cultured for 2 days (N=4).

# REFERENCES

Terada, S., T. Nishimura, M. Sasaki, H. Yamada, M. Miki, 2002, Sercin, a protein derived from silkworms, accelerates the proliferation of several mammalian cell lines including a hybridoma, Cytotech. **40**: 3-12

Kato, N., S. Sato, A. Yamanaka, H. Yamada, N. Fuwa and M. Nomura ,1998, Silk protein, sericin, inhibits lipid peroxidation and tyrosinase activity, Biosci. Biotechnol. Biochem. **62**: 145-147

Sasaki, M., N. Kato, H. Watanabe and H. Yamada, 2000a, Silk protein, sericin, suppresses colon carcinogenesis induced by 1,2-dimethylhydrazine in mice, Oncol. Rep. **7**: 1049-1052

Sasaki, M., H. Yamada and N. Kato, 2000b, Consumption of silk protein, sericin elevates intestinal absorption of zinc, iron, magnesium and calcium in rats, Nutr. Res. **20**: 1505-1511

Takahashi, M., K. Tsujimoto, H. Yamada, H. Takagi and S. Nakamori, 2003, The silk protein, sericin, protects against cell death caused by acute serum deprivation in insect cell culture, Biotech. Lett. **25**: 1805-1809

Tsujimoto, K., H. Takagi, M. Takahashi, H. Yamada and S. Nakamori, 2001, Cryoprotective effect of the serine-rich repetitive sequence in silk protein sericin, J. Biochem. **129**: 979-986

Sasaki, M., Y. Kato, H. Yamada and S. Terada, Development of a novel serum-free freezing medium for mammalian cells using silk protein sericin, Biotechnology and Applied Biochemistry, in press

# Design of Culture Media for Drosophila Melanogaster S2 Cells Producing Recombinant G Glycoprotein from Rabies Virus

A.L.L. Galesi,[1] F.R.X. Batista,[1] R.Z. Mendonça,[2] C.A. Pereira,[2] and A.M. Moraes,[1]

[1]*Department of Biotechnological Processes, School of Chemical Engineering, State University of Campinas - CP 6066, 13083-970, Campinas, SP, Brazil. E-mails: allima@feq.unicamp.br; fbatista@feq.unicamp.br; ammoraes@feq.unicamp.br.,* [2]*Laboratory of Viral Immunology, Butantan Institute, Av. Vital Brasil, 1500, 05503-900, São Paulo, SP, Brazil. E-mails: grugel@butantan.gov.br; zucatelli@butantan.gov.br.*

**Abstract:** The synthesis of recombinant proteins represents one of the main challenges in molecular biology as well as in preventive and therapeutic medicine. Many expression systems such as plant, bacterial, mammalian and insect cells have been developed with success for the expression of heterologous proteins, and the use of the dipteran embrionary cells *Drosophila melanogaster* Schneider 2 (S2) is one of them. The goals of this work were to design culture media with reduced concentration of fetal bovine serum and totally free of animal protein-derived supplements, assuring the target protein safety requirements, the G glycoprotein from rabies virus. The reduction or even elimination of bovine serum by the addition of yeast extract was evaluated, as well as the effects of the addition of glucose, fructose, glutamine, methionine, tyrosine, Pluronic F68, and a lipid emulsion on transfected S2AcGPV2 cells metabolism, concentration and viability. Preliminary studies of the cell growth were performed. The obtained results were compared to those achieved with cultures incubated with traditional culture media such as the three basal media supplemented with 10% (v/v) of fetal bovine serum, TC100, IPL-41 and Grace, and also the chemically defined media SF900II. It was observed that the behavior of the tested transfected cells is strongly affected by medium composition and, for this cell line, the use of media with reduced fetal bovine serum percentage is feasible.

**Key words:** S2 insect cells, G glycoprotein, yeast extract, lipid emulsion, culture media, media formulation



*R. Smith (ed.), Cell Technology for Cell Products*, 403–413.

# 1. INTRODUCTION

During the last decade, insect cells have been used for recombinant proteins production. Many types of expression systems have been used to produce these proteins: plant, bacterial, mammalian and insect cells. The *Drosophila* Expression System (DES) has several advantages that make it adequate for protein expression at laboratory and industrial scales. It is a non-lytic insect expression system that uses simple plasmid vectors and easy-to-use *Drosophila* S2 cells for proven high–level protein production. Currently, the system utilizes *Drosophila* S2 cells that grow rapidly, do not require $CO_2$ incubation, and grow in culture media that do not require expensive supplements.

Rabies is a serious public health problem and a cause of mortality in many regions of the world. The annual number of deaths worldwide caused by rabies is estimated to be between 40,000 and 70,000, and it is estimated that 10 million people receive post-exposure treatments each year after being exposed to rabies suspect animals (WHO, 2004). The first rabies vaccine successfully employed in humans was developed by Louis Pasteur and colleagues in 1885 (Pérez and Paolazzi, 1997). Since then, several types of vaccines have been developed.

One of the key factors for obtaining high recombinant proteins yield is adequate nutrient provision (Palomares *et al.*, 2004). For many years, insect cell culture has been routinely performed in basal media, such as Grace, TNM-FH and TC100 media supplemented with fetal bovine serum (FBS). This component not only increases media cost but is also associated to quality control problems and, due to its protein content, FBS can interfere with the downstream processing of the target product. Therefore, the search for chemically defined culture media in recent years culminated in the development of optimized commercial serum-free media such as SF900II, which is very effective on stimulating cell growth, but considerably expensive (Marteijn *et al.*, 2003; Ikonomou *et al.*, 2003).

Many studies have been performed with the aim of improving insect cell culture media formulation (Wilkie *et al.*, 1980; Mitsuhashi, 1989; Wang *et al.*, 1993; Drews *et al.*, 1995; Vaughn and Fan, 1997; Maranga *et al.*, 2003, among others, as reviewed by Schalaeger, 1996). Supplements such as yeast extract, lipid emulsion, milk, colostrum, and protein hydrolysates have been evaluated (Taticek *et al.*, 2001, Ramírez *et al.*, 1990, Ikonomou *et al.*, 2003). However, specifically for *Drosophila melanogaster* S2 cells there are few data in literature about cell growth parameters in different media.

Therefore, in this work, the effects of the addition of glucose, fructose, glutamine, methionine, tyrosine, Pluronic F68, and a lipid emulsion to Grace's and IPL-41 basal media on transfected S2 cells concentration and viability will be discussed, as well as media formulations including hydrolysed lactoalbumin, yeastolate and low concentrations of FBS. The obtained results are compared to those achieved with cultures performed with the chemically defined media SF900II and also to several basal media supplemented with FBS.

# 2. MATERIALS AND METHODS

## 2.1 Materials

The insect cell line S2 (donated by the Laboratory of Viral Immunology, Butantan Institute, São Paulo, SP, Brazil) transfected with pAc 5.1/V5-His A (S2AcGPV2) was used in this study. The transfected S2AcGPV2 cells were maintained in SF900II (Gibco) and in TC100 (Cultilab) media. The media supplements tested were glucose (Gibco), fructose, lactose, methionine and tyrosine (Sigma Chem. Co.), FBS (Cultilab), yeastolate (Gibco), lipid emulsion (Gibco) and hydrolysed lactoalbumin (Becton, Dickinson and Co.).

## 2.2 Cell Culture in Different Commercially Available Media

S2AcGPV2 cells were cultured in SF900II, TC100, Grace's and IPL-41 media, the last three media supplemented with 10% FBS, to evaluate cell behavior in these different media formulations. The cells were adapted before inoculation in each media, with exception of Grace's medium. All experiments were carried out in 100 mL shake flasks inoculated with $7.5x10^5$ cells/mL, with working volumes of 20 mL, and incubated in a rotary shaker at 100 rpm and 28°C.

## 2.3 Evaluation of Media Supplements

The first set of experiments was performed employing TC100 as the basal medium. The effects of FBS percentage (1 to 3% v/v), yeastolate concentration (1 to 8 g/L) and a lipid emulsion percentage (1 a 5% v/v) were studied to evaluate the influence of the culture medium composition on cell growth and viability. The analysis of these effects was performed employing a $2^3$ factorial experimental design, in a total of 12 experiments. Besides that, hydrolysed lactoalbumin (3.3 g/L), glucose (to reach 10 g/L), glutamine (to reach 3.5 g/L) and Pluronic F68 (0.1% w/v) were added to the cultures.

In another set of experiment, the effects of the variables FBS percentage (1 to 3% v/v) and yeastolate concentration (1 to 3 g/L) were further evaluated employing a $2^2$ factorial design, in a total of 7 experiments, employing as controls TC100 medium containing 10% FBS (minimal culture medium) and SF900II medium (rich chemically defined serum-free medium). Glucose, glutamine, lipid emulsion, and Pluronic F68 were added to these cultures to reach, respectively, 10 g/L, 3.5 g/L, 1% (v/v) and 0.1% (w/v). For both sets of experiments, cells were maintained in TC100 medium with 10% (v/v) FBS containing 10 g/L glucose and 3.5 g/L glutamine before inoculation.

A set of assays was also performed with media presenting totally synthetic formulations. Grace´s medium was used as the basic formulation after supplemented with glucose (to reach 10 g/L), glutamine (3.5 g/L), fructose (3.75 g/L), lactose (0.75 g/L), methionine (0.4 g/L), tyrosine (0.4 g/L) and lipid emulsion (3% v/v). The effects of also adding FBS (1% v/v) and yeastolate (2 g/L) to the enriched Grace's medium formulation were assessed, as described in the previous set of experiments for TC100 medium.

Finally, another set of assays was performed with media presenting totally synthetic formulations using Grace's and IPL-41 media as basic formulations after supplemented with glucose (to reach a final concentration of 10 g/L), glutamine (3.5 g/L), fructose (3.75 g/L), lactose (0.75 g/L), methionine (0.4 g/L), tyrosine (0.4 g/L) and lipid emulsion (3% v/v). The effects of also adding FBS (1% v/v) and yeastolate (2 g/L) to the enriched Grace's and IPL-41 media were assessed. In this set of experiments, cells maintained in TC100 supplemented with 10% (v/v) FBS were inoculated in the tested media and the results were evaluated after four subsequent passage (individual formulations listed in Table 1). After that, a $2^2$ factorial design study in star configuration was performed, in a total of 13 experiments, to evaluate the effects of FBS (1 to 3% v/v) and yeastolate (4 to 8 g/L) on cell growth and viability during the stationary phase.

*Table 1.* Media formulation based on supplemented IPL-41 medium. The minimum formulation was IPL-41 medium with 10g/L glucose, 3.5g/L glutamine, 3.75g/L fructose, 0.75g/L lactose, 0.4g/L methionine, 0.4g/L tyrosine and 3% (v/v) lipid emulsion.

| ***Medium formulation*** | ***Yeastolate (g/L)*** | ***FBS (%)*** |
|---|---|---|
| 20 | 0 | 0 |
| 21 | 0 | 1 |
| 22 | 2 | 0 |
| 23 | 2 | 1 |

All experiments were carried out in 100 mL shake flasks inoculated with $7.5x10^5$ viable cells/mL, with work volumes of 20 mL, incubated in a rotary shaker at 100 rpm and 28ºC. The contrasts of the independent variables on the response variables were calculated using the software Statistica.

### 2.4 Analytical Methods

Cell concentration and viability were determined by optical microscopy (Olympus, model CK2) with Trypan Blue.

## 3. RESULTS AND DISCUSSION

### 3.1 Cell Growth Kinetic in Different Commercially Available Media

The first step in the development of the reduced serum medium was the selection of the basal formulation among different commercially available media. The behavior of the S2AcGPV2 cells was evaluated in IPL-41, Grace's and TC100, all supplemented with 10% (v/v) FBS, and compared to that observed in SF900II, as shown in Figure 1. Cells were adapted in these media before inoculation, with exception of Grace's medium.

It can be observed in Figure 1A that TC100 medium containing 10% FBS could efficiently support cell growth, as well as SF900II medium. In TC100 medium, the maximum viable cell concentration was around $3.5 x 10^6$ cells/mL and, in SF900II medium, it was about $9.1 x 10^6$ cells/mL. S2 cells maintained in Grace's and IPL-41 media showed low growth rates.

Figure 1B shows that cell viability was higher than 90% in SF900II, whereas in IPL-41 medium it was above 80% during the entire culture period. In TC100 medium, cells maintained viability up to 90% until the $10^{th}$ day of culture. The initial cell viability in Grace's medium was 76%, but it decreased and remained between 40 and 60% during the culture period.

As shown in Figure 1, TC100 supplemented with 10% FBS presented the most adequate results on viable cell concentration ($3 x 10^6$ cells/mL) among the basal media evaluated. Cells cultured in Grace's medium containing 10% FBS presented low growth, attaining a maximum cell density of $1 x 10^6$ cells/mL, and a viability around 50%. A similar behavior was observed in IPL-41 medium with 10% FBS, with exception of cell viability that was around 90% during all culture period. Based on these results, TC100 was selected as the first basal medium in which the experiments would be performed, since the behavior of the cells in this medium was satisfactory, however, IPL-41 and Grace's media were also further evaluated.

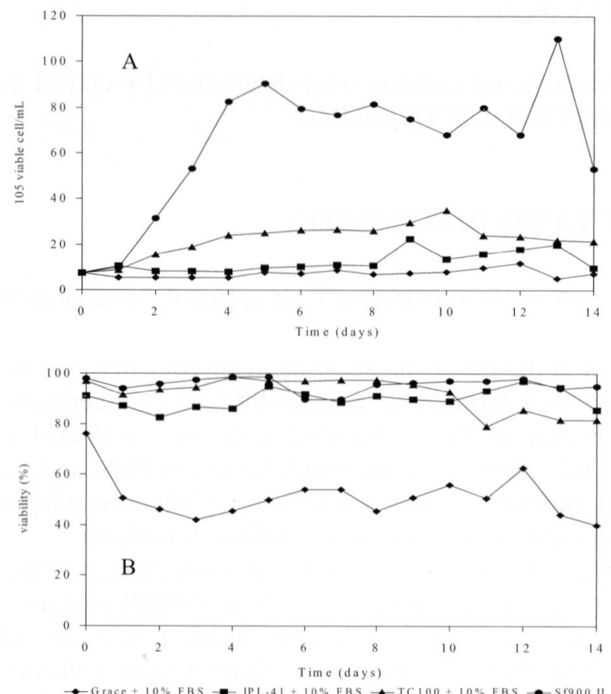


*Figure 1.* Viable cell concentration (A) and viability (B) of S2AcGPV2 cells in IPL-41, Grace's, TC100 and SF900II media (first three media supplemented with 10% v/v FBS).

## 3.2 Cell Behavior in Supplemented TC100 Medium

In a first experiment, the influence of yeastolate concentration, FBS percentage and a lipid emulsion on cell growth and viability were analyzed through a $2^3$ experimental design. The results obtained showed that cell growth occurred only in two of the culture media (data not shown). An increment in lipid concentration caused a decrease in cell concentration and viability, and the addition of lipid emulsion in percentages higher than 1% (v/v) resulted in cell growth inhibition. This result is not in agreement with previously published data for media supplementation with a mixture of lipids and Pluronic F68 (Donaldson and Shuler, 1998).

A further study in which lipid emulsion percentage was maintained at low level (1% v/v) was performed. The range of yeastolate concentration was reduced to 1 to 3 g/L, and the range of FBS percentage was maintained at 1 to 3% v/v. A $2^2$ experimental design study was carried out, totaling 7 experiments, and the achieved results are shown in Figure 2 and in Table 2. Lactoalbumin hydrolysate was not added in this experiment to prevent excessive osmolality increase; only glucose, glutamine and Pluronic F68 were further added.

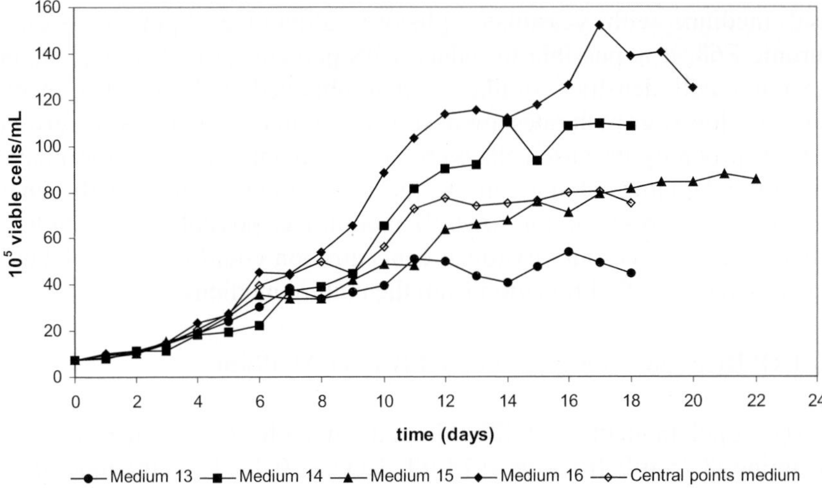


*Figure 2.* Viable cell concentration of S2AcGPV2 cells obtained in the experiment employing the 22 factorial experiment design, in which the effects of yeastolate and FBS were analyzed.

*Table 2.* Maximum cell concentration (Xs) and viability in the different media tested through the 22 experimental design, in which the effects of FBS and yeastolate were evaluated on cell concentration and viability.

| *Medium formulation* | *Yeastolate (g/L)* | *SFB (%)* | *$X_s$ max (cells/mL)* | *Viability (%)* |
|---|---|---|---|---|
| 13 | 1 | 1 | $5.1 \times 10^6$ | 94.1 |
| 14 | 3 | 1 | $11.0 \times 10^6$ | 96.9 |
| 15 | 1 | 3 | $8.5 \times 10^6$ | 94.4 |
| 16 | 3 | 3 | $15.2 \times 10^6$ | 95.9 |
| 17 | 2 | 2 | $8.6 \times 10^6$ | 97.7 |
| 18 | 2 | 2 | $8.4 \times 10^6$ | 93.3 |
| 19 | 2 | 2 | $7.8 \times 10^6$ | 92.3 |

Cell fast adaptation and growth was observed in all tested media. In cultures in which the yeastolate concentration was higher, cells attained the highest cell densities. Higher cell concentrations were obtained for media 14 and 16, in which yeastolate concentration was higher. The analysis of the effects on cell growth showed that yeastolate and FBS had positive effects on maximum cell density at 95% confidence level (data not shown).

Maximum cell concentration attained in media 14 and 16 were higher than in TC100 containing 10% FBS, however, in SF900II, cell concentration was higher than in formulation 16 and similar to that obtained in medium 14.

The achieved results show that, through the supplementation of TC100 basal medium with yeastolate, glucose, glutamine, lipid emulsion and Pluronic F68, it is possible to reduce FBS percentage to 1 to 3%, attaining maximum cell densities similar to that obtained with SF900II medium. However, lower growth rates are observed in comparison to those verified in Sf900II, probably because cells were not yet adapted to these formulations. With the adaptation of the cells to the tested media, cell growth behavior similar to the observed for SF900II medium is possible. This study also indicates that increasing yeastolate concentration could result in higher cell densities and inFBS elimination from the culture medium.

### 3.3 Cell Behavior in Supplemented IPL-41 Medium

The supplementation of IPL-41 medium with 10 g/L glucose, 3.5 g/L glutamine, 3.75 g/L fructose, 0.75 g/L lactose, 0.4 g/L methionine, 0.4 g/L tyrosine, and 3% (v/v) lipid emulsion, as well as with 2 g/L yeastolate and/or 1% FBS was performed. As shown in Figure 3, satisfactory cell growth was noticed only in media formulations 22 and 23. In general, cell viability was maintained at 99% during cell culture, but a decay to around 20% was observed in media formulations 20 and 21 after the $10^{th}$ culture day. A similar study was carried out with Grace's medium, not resulting in cell growth.

Since yeastolate and FBS have shown to be relevant supplements to S2 cells growth in enriched IPL-41, their effects on maximum cell concentration (viable cell concentration in stationary phase) and viability were further evaluated through a $2^2$ factorial design in star configuration. The achieved results are shown in Table 3.

Viable cell concentration varied from 2.5 to $11 \times 10^6$ viable cells/mL. Cell viability was around 97% or above in all experiments and media formulation with 6 g/L of yeastolate and 0.59 or 3.4% of FBS (v/v) showed the best results for cell maximum concentration at the stationary phase.

Statistically significant effects were observed only for viable cell concentration. According to non linear multiple regression and analysis of variance, the calculated F value was 32.5, being, therefore, more than 7.3 times higher than the listed value, equal to 4.46. The mathematical model generated is presented in Equation 1, which was used to plot the response surface and contour diagram shown in Figures 4A and 4B, respectively.

$$X_S = 95.06 + 26 * C_{yeastolate} + 26.59 * C_{yeastolate}^2 \qquad (1)$$

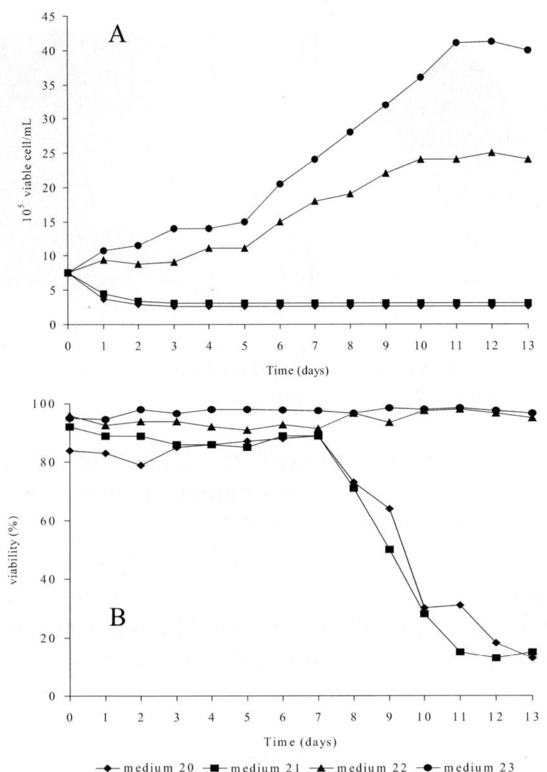


*Figure 3.* Comparison of cell concentration (A) and viability (B) in media 20 to 23 with cultures performed in IPL-41 medium containing 10 g/L glucose, 3.5 g/L glutamine, 3.75 g/L fructose, 0.75 g/L lactose, 0.4 g/L methionine, 0.4 g/L tyrosine, and 3% (v/v) lipid emulsion, as well as 2 g/L yeastolate and/or 1% FBS.

*Table 3.* 22 experimental design in star configuration, in which were evaluated the effects of FBS and yeastolate on cell concentration (Xs) and viability at the stationary phase.

| ***Medium formulation*** | ***Yeastolate (g/L)*** | ***FBS (%)*** | ***$X_s$ max ($10^5$ cells/mL)*** | ***Viability (%)*** |
|---|---|---|---|---|
| 24 | 4 | 1 | 25 | 97 |
| 25 | 8 | 1 | 90 | 97 |
| 26 | 4 | 3 | 37 | 97 |
| 27 | 8 | 3 | 100 | 97 |
| 28 | 3.2 | 2 | 18 | 97 |
| 29 | 8.8 | 2 | 79 | 97 |
| 30 | 6 | 0.59 | 110 | 97 |
| 31 | 6 | 3.4 | 110 | 98 |
| 32 | 6 | 2 | 91 | 97 |
| 33 | 6 | 2 | 80 | 97 |
| 34 | 6 | 2 | 95 | 97 |

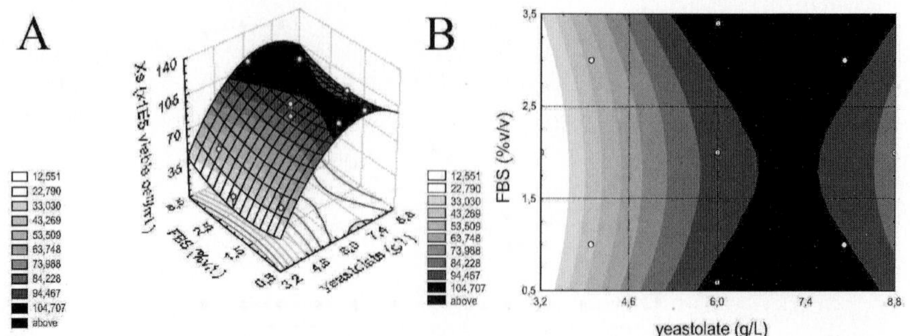


*Figure 4.* Response surface (A) and contour (B) diagrams of maximum cell concentration as a function of yeastolate concentration and FBS percentage.

The response surface was plotted to define the optimal working conditions for maximum cell concentration at the stationary phase. Considering the optimal conditions for $X_S$, a yeastolate concentration of 6 g/L represent the best condition to obtain high density of S2AcGPV2 cells at 95% of confidence. This result can be compared with those obtained by Drews *et al.* (1995), in which Sf9 cells reached high density when yeastolate was employed at range from 4 to 8 g/L. FBS at percentages varying from 1 to 3, on the other hand, did not show relevant statistical effects on cell growth.

# 4. CONCLUSION

When comparing the achieved results for the several media formulations developed with cultures incubated with the traditional culture media TC100, IPL-41 and Grace's supplemented with FBS at 10% (v/v), and also with the chemically defined media SF900II, it was observed that the best performance was obtained in SF900II medium. However, TC100 and IPL-41 basal media were capable of supporting cell growth and, with the addition of a set of supplements, these media could provide adequate cell concentrations. The attempts of fetal bovine serum reduction in TC100 medium were successfully performed with the addition of the supplements glucose, glutamine, lipid emulsion and, mainly, yeastolate. Similar cell concentrations to the usually obtained employing the rich medium SF900II were already attained, and the results showed that higher cell densities could be achieved with the augmentation of yeastolate concentration. On the other hand, media based on IPL-41 medium supplemented with glucose, glutamine, fructose, lactose, methionine, tyrosine, a lipid emulsion, as well as with yeastolate, provided high cell growth even when low percentages of FBS were used.

## ACKNOWLEDGEMENTS

The authors are grateful to the Fundação de Amparo à Pesquisa do Estado de São Paulo (FAPESP) and to Coordenação de Aperfeiçoamento de Pessoal de Nível Superior (CAPES) from Brazil, for the financial support to this research.

## REFERENCES

Drews, M., Paalme, T., Vilu, R. The growth and nutrient utilization of the insect cell line *Spodoptera frugiperda* Sf9 in batch and continuous culture. *Journal of Biotechnology*, 40, 187-198, 1995.

Donaldson, M. S., Shuler, M. L. Low-cost serum-free medium for the BTI-Tn5B1-4 insect cell line. *Biotechnology Progress*, 14, 573-579, 1998.

Ikonomou, L., Schneider, Y. J., Agathos, S. N. Insect cell culture for industrial production of recombinant proteins. *Applied Microbiology and Biotechnology*, 62, 1-20, 2003.

Maranga, L., Mendonça, R.Z., Bengala, A., Peixoto, C.C., Moraes, R.H.P., Pereira, C.A., Carrondo, M.J.T. Enhancemernt of cell growth and longevity through supplementation of culture medium with hemolymph. Biotechnol. Progr., v. 19, p. 58-63, 2003

Marteijn R.C.L., Jurrius O., Dhont J., de Gooijer C.D., Tramper J., Martens D.E. Optimization of a feed medium for fed-batch culture of insect cells using a genetic algorithm, Biotechnology and Bioengineering, v. 81, n. 3, p. 269-278, 2003

Mitsuhashi, J. In: *Invertebrate Cell System Applications* (Mitsuhashi J. ed.), p. 3-20, CRC Press, 1989.

Palomares, L.A., López, S., Ramírez, O.T. Utilization of oxygen uptake rate to assess the role of glucose and glutamine in the metabolism of insect cell cultures. *Biochem. Eng. J.*, v. 19, p. 87-93, 2004.

Pérez, O., Paolazzi, C.C. Production methods for rabies vaccine. *Journal of Industrial Microbiology & Biotechnology*, v. 18, p. 340-347, 1997

Ramírez, O.T., Sureshkumar, G.K., Mutharasan, R. Bovine colostrum or milk as serum substitute for the cultivation of a mouse hybridoma. *Biotechnol. Bioeng.*, v 35, p.882-889, 1990.

Schlaeger, E.-J. Medium design for insect cell culture. *Cytotechnology*, v. 20, p. 330-336, 1996.

Taticek, R.A., Choi, C., Phan, S-E., Palomares, L.A., Shuler, M.L. Comparation of growth and recombinant protein expression in two different insect vell lines in attached and suspension culture. *Biotechnoogy Progress*, 17, p. 676-684, 2001.

Vaughn, J.L., Fan, F. Differential Requirements of two Insect Cell Lines for Growth in Serum-Free Medium. *In Vitro; Cellular and Developmental Biology*, 33, 479-482, 1997.

Wang, M.Y., Kmong, S., Bentley, W.E. Effects of oxygen/glucose/glutamine feeding on insect cell baculovirus protein expression: a study on epoxide hydrolase production. *Biotechnol. Progr*, v. 9, p. 355-361, 1993.

Wilkie, G. E. I.; Stockdale, H.; Pirt, S. J. Chemically defined media for production of insect cells and viruses in vitro. *Dev. Biol. Stand.*, v. 46, p. 29-37, 1980.

WHO (World Health Organization) http://www.who.int/mediacentre/factsheets/fs099/ en/, accessed in December 2004.

# Comparison of the Cultivation of Wild and Transfected *Drosophila Melanogaster* S2 Cells in Different Media

K. Swiech,[1] A.L.L. Galesi,[2] A.M. Moraes,[2] R.Z. Mendonça,[3] C.A. Pereira,[3] and C.A.T. Suazo[1]

*[1]Department of Chemical Engineering, Federal University of São Carlos, [2]Department of Biotechnological Processes, School of Chemical Engineering, State University of Campinas - CP 6066, 13083-970, Campinas, SP, Brazil, [3]Laboratory of Viral Immunology, Butantan Institute, Av. Vital Brasil, 1500, 05503-900, São Paulo, SP, Brazil.*

**Abstract:** Insect cells have been increasingly employed for the production of recombinant proteins. One of the most widely used dipteran cells in transfection studies is the Schneider 2 (S2) cell line, established from *Drosophila melanogaster* embryonic tissue. In this work, the growth and proliferation of wild and transfected S2 cells expressing the G glycoprotein from rabies virus (GPV) were compared employing different culture media. For the transfected (S2AcGPV2) cell contruction, the plasmid pGPV encoding the sequence of interest and the vectors pAc 5.1/V5-His A under the control of the *Drosophila* actin promoter were utilized. The selection vector pCoHygro carrying genes coding for hygromicin-inactivating enzymes was also employed and the cells were transfected using lipofectin. Due to wide utilization of TNM-FH and TC100 media (both requiring supplementation with fetal bovine serum) for insect cells, their low cost and the low protein content of SF900II medium, these three media and the mixture TNM-FH-SF900II (1:1) were evaluated for the choice of a suitable media for S2 cells. The results indicated that the wild and the transfected cells presented different growth characteristics in the distinct media. In SF900II medium, larger accumulation and consumption of lactate were observed for the wild cell culture. In TC100, however, S2AcGPV2 cells did not produce lactate. In TNM-FH, the S2 cells presented lower growth rate ($\mu_{max}$=0.0078 $h^{-1}$) when compared to the other media ($\mu_{max}$=0.0375 $h^{-1}$ and $\mu_{max}$=0.0112 $h^{-1}$ for SF900II and TC100, respectively), with an accentuated viability drop during the first days of culture. The mixture of TNM-FH and SF900II at a 1:1 volume ratio resulted in cell growth ($\mu_{max}$=0.0377 $h^{-1}$) similar to that observed in SF900II medium only, allowing significant culture media cost reduction. Transfected



*R. Smith (ed.), Cell Technology for Cell Products*, 415–423.

cells could not be adapted to TNM-FH medium. Cell growth of S2AcGPV2 cells in TC100 medium ($\mu_{max}$=0.0151 $h^{-1}$) was lower than in SF900II medium ($\mu_{max}$=0.0407 $h^{-1}$).

**Key words:** S2 insect cells, G glycoprotein, TNM-FH, TC100, SF900II

## 1. INTRODUCTION

Rabies or hydrophobia is an acute viral infection of the central nervous system that can occur in animals and humans. The virus is transmitted from an animal to a person, or from one animal to another, through infected saliva, most often by biting, however, the contact of torn skin with infected saliva can also cause the disease. The virus travels from the contact location to the spinal cord and brain, and once the disease symptoms develop, death (caused by convulsions, exhaustion, or paralysis) is usually inevitable. The only treatment after symptoms appear is rest and sedation.

Despite being a vaccine-preventable disease, rabies is still a public health problem in several developing countries. Around 50.000 deaths from rabies are reported annually around the world, and most of the victims are children under 15 years of age bitten or scratched by infected dogs. Therefore, vaccination of dogs is a cost-effective means for eliminating rabies in humans and studies on less expensive and more efficient vaccines are of relevance.

Animal vaccine against rabies can be produced through animal cell culture (using Vero cells), but Fuenzalida & Palácios vaccine, produced with virus purificated from infected mouse neural tissue, is still widely used in developing countries.

In the last decade, insect cells have been used for the abundant expression of recombinant proteins and also to produce wild-type baculoviruses-based biopesticides. *Drosophila melanogaster* S2 cells, for instance, can be employed for the expression of heterologous proteins such as dopamine, gamma-amino butyric acid (GABA), opioid receptores, and human plasminogen, among others. The surface antigen of hepatitis B virus can be also produced using stably transfected S2 cells (Deml *et al.*, 1999) and this cell line seems to be an interesting system for the expression of antirabies vaccine components.

Insect cells present several advantages when compared to mammalian cells, such as ease of culture, higher tolerance to osmolality and by-product concentration. In addition, insect cells can be cultured at lower costs than mammalian cells, since larger cell densities can be achieved at room temperature under regular air atmosphere. Except for N-glycosylation, insect cells perform similar post- and cotranslational modifications as compared to mammalian cells.

Insect cell culture can be performed in basal media, such as Grace's medium, TNM-FH and TC100 supplemented with 5% or 10% (v/v) serum, most usually fetal bovine serum (FBS). However, problems such as lot-to-lot inconsistency, possible contamination with foreign proteins, prions, viruses and mycoplasma, and complications in product downstream processing normally limit its widespread use in animal cell culture for the production of biopharmaceutics. Therefore, it is observed an increased tendency on the use of serum-free media in animal cell culture. When FBS is not added to animal cells culture media, its supplementation with several different components is required.

Several commercial serum-free media are now available for insect cell culture, such as Sf900II and Express Five SFM from Life Technologies, Insect-XPRESS from Biowhittaker, EX-Cell 400, 405 and 420 from JHR Biosciences and HyQ SFX-Insect from Hyclone (Ikonomou *et al.,* 2003). These media are capable to support high cell densities, however, their cost is too high when large scale heterologous protein production is desired.

Therefore, the purpose of this work was to compare the cultivation of wild and transfected *Drosophila melanogaster* S2 cells in the commercially available TNM-FH, TC100 and SF900II media and their mixtures.

## 2. MATERIALS AND METHODS

### 2.1 Cell transfection and preservation

S2 cells donated by the Laboratory of Viral Immunology, Butantan Institute, São Paulo, Brazil were used through out the study. For S2AcGPV2 cell contruction, the plasmid pGPV encoding the sequence of interested and the vectors pAc 5.1/V5-His A (Invitrogen) under the control of the *Drosophila* actin promoter were utilized. For cell selection, the selection vector pCoHygro carrying genes coding for hygromicin-inactivating enzymes was employed.

S2 cells were transfected with pAc/GPV + pCoHygro vectors using lipofectin (Invitrogen). For selection of the cells with higher GPV expression, confluent transfected cultures were subcultured in 96 well plates at a cell concentration of $10^5$ cells/well and subsequently cultured in 24, 12 and 6 well plates. The selected cells were then cultured at $10^6$ cells/mL in 25 $cm^2$ T-flasks (Nunc), in 3 mL of SF900II medium with 200 μg/mL of hygromicin.

Wild and recombinant cells S2AcGPV2 were preserved in liquid nitrogen, activated and cultured at 28ºC in 100 and 250 mL Schott flasks in shaker at 100-120 rpm.

### 2.2 Culture media and culture conditions

TC100 culture medium (Cultilab) and TNM-FH medium (prepared from Grace's medium provided by Gibco BRL, with 3.33 g/L of lactalbumin hydrolysate and 3.33 g/L of yeastolate ultrafiltrate, Life Technologies), both supplemented with 10% v/v fetal bovine serum (Cultilab), were utilized. The serum free medium SF900II (Gibco BRL) and a 1:1 mixture of these two cultures media (formulation MM) were also employed.

For cultivation of wild S2 cells, cell concentration for inoculating SF900II, MM and TC100 media was equal to $5x10^5$ cells/mL, while for TNM-FH, it was $1x10^6$ cells/mL, otherwise the cells would not satisfactorily develop.

For the recombinant cell culture (S2AcGPV2), the SF900II and TC100 media were used. Inoculum concentration was $5x10^5$ cells/mL.

### 2.3 Analytical methods

*Cell viability and concentration:* Cell concentration was determined by optical microscopy with a hematocytometer and the viability via trypan blue exclusion.

*Metabolites and substrates analysis:* Glucose, lactate and glutamine concentrations were determined enzymatically using a YSI Biochemical Analyzer model 2700 (Yellow Spring Instruments) after sample centrifugation at 1500 rpm for 5 minutes.

## 3. RESULTS AND DISCUSSION

### 3.1 Cultivation of wild S2 cells culture in distinct media

Due to wide utilization of the TNM-FH medium for insect cells cultivation, it's low cost and the low protein content of SF900II medium, this two media and one 1:1 mixture of both (MM formulation) were utilized for the choice of a suitable growth culture medium for S2 cells. These experiments were performed in 100mL Schott flasks with a working volume of 20 mL agitated at 100rpm and at 28ºC.

Even though the TNM-FH and TC100 media are widely utilized for insect cell culture (Yamagi *et al.*, 1999; Drews *et al.*, 1995 and Mendonça *et al.*, 1999), the S2 cell line employed presented slow growth and accentuated viability drop during the first days of culture in these media ($\mu_{max}$=0.0078 $h^{-1}$ for TNM-FH and 0.0112 $h^{-1}$ for TC100) when compared to the other two formulations ($\mu_{max}$=0.0375 $h^{-1}$ and $\mu_{max}$=0.0377 $h^{-1}$ for SF900II and MM, respectively), as shown in Figure 1. Cells cultured in TC100 medium reached an adequate level of viability later than cultures in TNM-FH medium. There are not data in literature with regard to *Drosophila melanogaster* culture in these media. The most utilized media for the culture of this cell are HyQSFX serum free insect cell medium from Hyclone (Valle *et al.*, 2001); Schneider's medium from Life Technologies (Sondergaard, 1996) and M3 medium (Shield and Sang M3 insect medium, Park *et al.*, 1999 and Shin *et al.*, 2003). Cell growth in SF900II and MM was very similar, allowing process cost reduction, without significant reduction in cell growth, since TNM-FH and TC100 are low cost media.

The poor cell growth in TNM-FH and TC100 media may be associated to their low glucose concentration (0.7 g/L and 1.0 g/L, respectively, as compared to around 10 g/L in SF900II) and also low amino acid content. Lactate production in TNM-FH medium was low, due to the low glucose content, since lactate is formed from the piruvate originated from glucose metabolism. In TC100, lactate accumulation was not observed. Lactate production in SF900II and in MM was higher as a consequence of higher glucose availability in these media. In the MM formulation, lactate consumption was noticed when glucose concentration reached low levels, as also observed by other groups (Kamen *et al.*, 1991; Bédard *et al.*, 1993 and Palomares & Ramírez, 1996).

The end of the exponential phase and of stationary phase in the SF900II and MM media are not associated to glucose exhaustion, since glucose concentration at this time (approximately 120 hours of culture) is still high in both media (around 6.5 and 2.0 g/L for SF900II and MM, respectively).

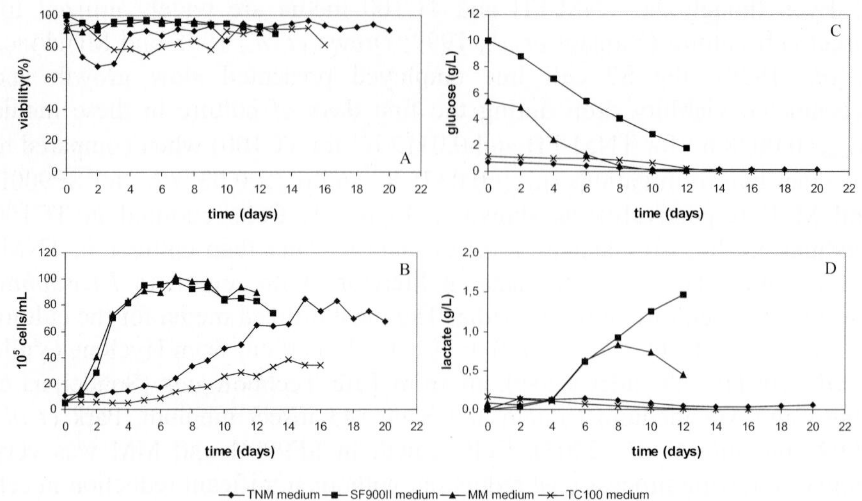


*Figure 1.* Comparison of growth (A and B), glucose consumption (C) and lactate production (D) of wild S2 cells in four culture media in Schott flasks at 28ºC and 100rpm.

## 3.2 Cultivation of transfected S2AcGPV2 cells culture in distinct media

In this study, the growth and the glucose consumption and lactate production of the S2AcGPV2 cells were evaluated in TC100 and SF900II media. Transfected cells could not be adapted in TNM-FH medium. In Figure 2, the results obtained for an inoculum concentration of $5.0x10^5$ cells/mL can be observed. Similarly to S2 cells, S2AcGPV2 cells in TC100 medium presented a lower growth rate ($\mu_{max}$=0.0151 $h^{-1}$) when compared to the cells in SF900II medium ($\mu_{max}$=0.0329 $h^{-1}$). The lower maximum cell concentration and growth rate in TC100 medium were probably due to the lower content of glucose and amino acids in this medium. Cell viability in SF900II medium was higher than 90% during most of the culture period, and, in TC100 medium, viability dropped at the first days and, after that, it increased and remained above 80% until the $9^{th}$ day.

In SF900II medium, glucose was totally consumed by the end of the culture period, and lactate concentration remained low, differently from the behavior observed for wild S2 cells in this medium, in which lactate was produced. In TC100, glucose was exhausted in 6 days and the accumulated lactate was consumed after that. The cessation of the growth phase and the beginning of the stationary phase were associated to the glucose exhaustion.

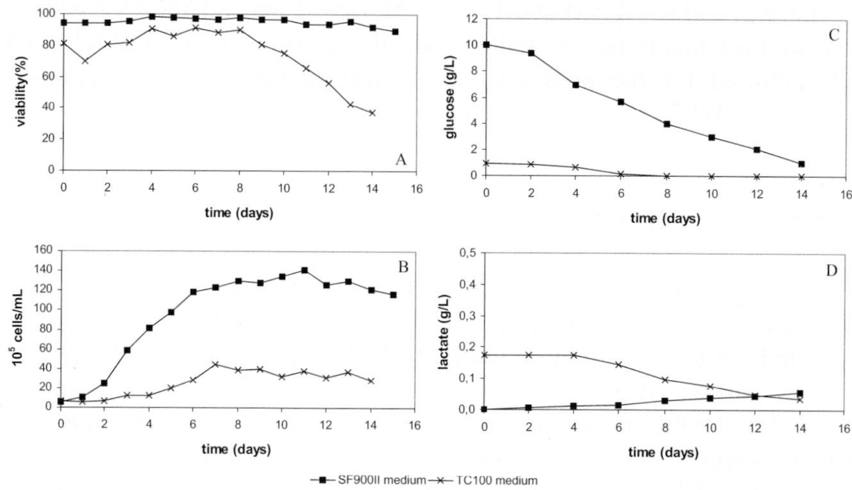


*Figure 2.* Comparison of growth (A and B, glucose consumption (C) and lactate production (D) of transfected S2 cells in two culture media in Schott flasks at 28°C and 100rpm.

## 3.3 Comparison of wild and transfected cells behaviour

Despite TNM-FH medium is widely used for the culture of many insect cell lines, the wild S2 cell line employed presented slower growth in this medium when compared to the other three media (Table 1). In TC100, specific cell growth rate was higher than in TNM-FH, but much lower than those obtained in MM and SF900II media. Despite that, maximum cell concentration obtained in TNM-FH was higher than that obtained for TC100. Maximum cell density in TNM-FH was similar to the values achieved for MM and SF900II media. The performance of the transfected cells in SF900II medium was also better than in TC100 medium. It was not possible to culture the transfected cells in TNM-FH since the cells could not be adapted in it. In TC100 and SF900II media, maximum cell concentrations and specific growth rates were higher for transfected cells.

*Table 1.* Values of specific growth rates (□max) and maximum cell concentrations (Xmax) of wild S2 and transfected S2AcGPV2 cells for in the evaluated media. Inoculum: 5,0x105 cells/mL (with exception of wild S2 cells cultured in TNM-FH medium – 1,0x106 cells/mL).

| ***Medium*** | ***$\mu_{max}$ ($h^{-1}$)*** | | ***$X_{max}$ (cells/mL)*** | |
|---|---|---|---|---|
| | **S2** | **S2AcGPV2** | **S2** | **S2AcGPV2** |
| TNM-FH + 10% FBS | 0.0078 | - | $8.5 \times 10^6$ | - |
| TC100 + 10% FBS | 0.0112 | 0.0151 | $3.8 \times 10^6$ | $4.4 \times 10^6$ |
| MM | 0.0377 | - | $10.2 \times 10^6$ | - |
| SF900II | 0.0375 | 0.0407 | $9.8 \times 10^6$ | $14.1 \times 10^6$ |

The $\mu_{max}$ values for SF900II and MM for S2 and S2AcGPV2 cells are close to those mentioned by Valle *et al.* (2001), from 0.03 to 0.04 $h^{-1}$, for S2 cells cultured for the production of human menin using the serum-free medium HyQSFX, from Invitrogen.

# 4. CONCLUSION

Wild and transfected S2 cells presented differences in distinct media concerning to cell growth and metabolism. The wild S2 cell line employed presented the lowest growth rate in TNM-FH supplemented with 10%FBS, whereas transfected S2 cells could not be adapted to TNM-FH containing media. In SF900II medium, the growth results for both cell types were similar, however the lactate evolution profiles were different. Whereas S2 cells produced high quantities of this metabolite, lactate concentration in S2AcGPV2 culture was low during all culture period. For the wild cells, the mixture of TNM-FH and SF900II media at 1:1 volume ratio resulted in cell growth similar to that observed in SF900II medium alone, allowing significant culture media cost reduction.

# ACKNOWLEDGEMENTS

The authors are grateful to the Fundação de Amparo à Pesquisa do Estado de São Paulo (FAPESP), Coordenação de Aperfeiçoamento de Pessoal de Nível Superior (CAPES), Conselho Nacional de Desenvolvimento Científico e Tecnológico (CNPq), and Butantan Foundation, from Brazil, for the financial support to this research.

# REFERENCES

Bédard, C., Tom, R., Kamem, A. Growth, nutrient consumption, and end product accumulation in Ff9 and BTI-EAA insect cell cultures: insights into growth limitation and metabolism. *Biotechnology Progress*, 9, 615-624, 1993.

Deml, L., Wolf, H. e Wagner, R.. High level expression of hepatitis B virus surface antigen in stably transfected *Drosophila* Schneider-2 cells. Journal of virological methods, 79, 191-203, 1999.

Drews, M., Paalme, T., Vilu, R., The growth and nutrient utilization of the insect cell line *Spodoptera frugiperda* Sf9 in batch and continuous culture. *Journal of Biotechnology*, 40, 187-198, 1995.

Ikonomou, L., Schneider, Y. J., Agathos, S. N., Insect cell culture for industrial production of recombinant proteins. *Applied Microbiology and Biotechnology*, 62, 1-20, 2003.

Kamen, A., Tom, R., Caron, A., Chavarie, C., Massie, B. and Archambault, J.. Culture of insect cells in a helical ribbon impeller bioreactor. Biotechnology and Bioengineering 38, 619-628, 1991.

Mendonça, R. Z., Palomarez, L. A., Ramírez, O. T., An insight into insect cell metabolism through selective nutrient manipulation. Journal of Biotechnology, 72, 61-75, 1999.

Palomares, L.A. and Ramírez, O.T.. The effect of dissolved oxygen tension and the utility of oxygen uptake rate in insect cell culture. Cytotechnology 22, 225-237, 1996.

Park, J.H., Lee, J.M. and Chung, I.S.. Production of recombinant endostatin from stably transformed *Drosophila melanogaster* S2 cells. Biotechnology Letters 21, 729-733, 1999.

Shin, H.S., Lim, H.J. and Chao, H.J.. Quantitative monitoring for secreted production of human interleukin-2 in stable insect *Drosophila* S2 cells using a green fluorescent protein fusion partner. Biotechnology Progress 19, 152-157, 2003.

Sondergaard, L., *Drosophila* cells can be grown to high cell densities in a bioreactor. *Biotechnology Techniques*, 10 (3), 161-166, 1996.

Valle, M. A., Kester, M. B., Burns, A. L., Marx, S. J., Spiegel, A. M., Shiloach, J., Production and purification of human menin from *Drosophila melanogaster* S2 cells using stirred tank reactor. *Cytotechnology*, 35, 127-135, 2001.

Yamagi, H., Tagai, S. and Fukuda, H.. Optimal production of recombinant protein by the baculovirus-insect cell system in shake-flask culture with medium replacement. Journal of Bioscience and Bioengineering 87, 636-641, 1990.

# Low Temperature Culture Effects on sCR1 Productivity by rCHO Cells

Hiroshi Matsuoka, Chie Shimizu, Toshiya Takeda
*Department of Biosciences, Teikyo University of Science & Technology*

**Abstract:** Recombinant CHO, *r*CHO, cells are usually cultured at 37 °C. Unlike the specific growth rate, which decreases at low temperature, effects of low temperature culture on specific productivity rate are not so clear. We studied low temperature culture effects on *s*CR1 productivity in *r*CHO cells and reported here. The specific growth rate of *r*CHO increased according to increase of culture temperature. The specific rate of *s*CR1 at 37 °C was similar to that at 35 °C, however that at 33 °C increased by 1.3 compared with that at 37 °C. These results suggest that the productivity of *s*CR1 could become higher by using the shift of culture temperature.

**Key words:** *r*CHO, CHO, sCR1, CR1, complement receptor, cell cycle, specific growth rate, specific production rate, low temperature culture, batch culture, serum-free medium, yield, glucose, lactate, viable cells

## 1. INTRODUCTION

The human complement receptor type 1, CR1, is a polymorphic membrane glycoprotein expressed on human erythrocytes, monocytes, macrophages, granulocytes, follicular dendritic cells, glomerular podocytes, and B- and T-lymphocytes. Since CR1 can serve several regulatory functions such as co-factor activity with factor I, inactivation of bursting oxidization of neutrophil cells, hemolytic reaction with complement, and processing of C3 to C3a and C5a, it is widely used for clinic application. The CR1 plasmid was transfected to CHO cells, and 220 kD of soluble CR1, *s*CR1, could be expressed and secreted by CHO cells. The purified *s*CR1 is confirmed to



*R. Smith (ed.), Cell Technology for Cell Products*, 425–429.

retain its well-known functions similar to those of cell-bound receptor CR1 in vitro. On the other hand, temperature is a key environmental parameter that influences cell growth and recombinant protein production. Most mammalian cells are cultured at 37 °C. Lowering culture temperature below 37 °C decrease specific growth rate. In many cases, the specific production rate, $q$, of CHO cells was not enhanced by lowering the culture temperature. On the other hand, some reports say that the $q$ of CHO cells increase at lower temperature (Yoon *et al.* (2003), Hendrick *et al.* (2001)). In the present study, we investigated the effect of low temperature culture on CHO cell growth and *s*CR1 production rates.

## 2. MATERIALS AND METHODS

The cell line of CHO (CRL-10052) was used. It was originally an adherent cell, however it was changed to a floating one. Cells were cultured in a 2 L fermentor with a 1.2 L working volume at various temperatures. pH and DO were maintained at 7.2 and 40% of air saturation by $CO_2$ and $O_2$, respectively. Agitation speed was 100 rpm. A serum-free medium on the basis of IMDM with 1% Penicillin-Streptomycin-Neomycin antibiotics mixture was used. Concentrations of glucose, glutamine, glutamate, lactate, and ammonia were measured by Bioprofile 400 (NOVA). *s*CR1 concentration was measured by using the modified method by Wang *et al.* (Kato *et al.* (2002)). Briefly the supernatant of medium sample by centrifuge was dialyzed by using a seamless cellulose tube and then injected to HPLC gel filtration column chromatography (TSK gel G3000SWXL, TOSOH). As elution buffer, the Tris buffer (pH=7.4) containing 0.05% CHAPS was used. To analyze the cell cycle distribution, flow cytometric analysis was carried out by using a FACSAria flow cytometer (BD).

## 3. RESULTS AND DISCUSSION

To determine the effect of low culture temperature on *s*CR1 productivity, batch cultures of *r*CHO were carried out at 4-different temperatures such as 37, 35, 33 and 30 °C (Figure 1). The specific growth rate of *r*CHO increased according to the culture temperature (Table 1). Figure 2 shows the relationship between *s*CR1 concentration and time integrated cell concentration. Since the relationship had linearity, the slope becomes to the $q_{sCR1}$. Other specific rates were obtained in the same way and shown in Table 1. The $q_{sCR1}$ at 37 °C is similar to that at 35 °C. However the $q_{sCR1}$ at 33 °C increased by 1.3 compared with that at 37 °C

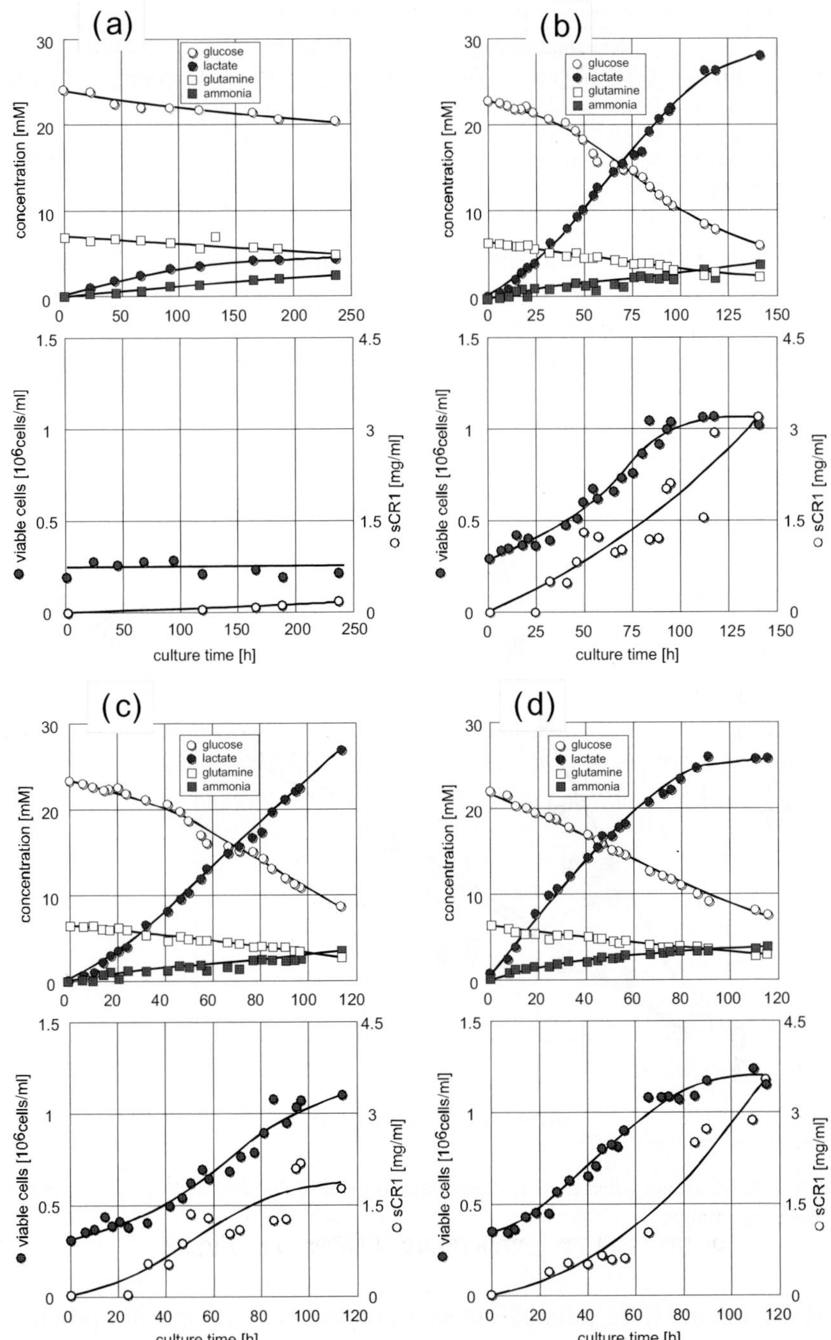


*Figure 1.* Time courses of rCHO culture at various temperatures.

*Table 1.* Effect of culture temperature on specific rates and yields.

| *Temp* [°C] | $\mu$ [$h^{-1}$] | *Specific rate (q)* [μ mol $10^6$cells$^{-1}$ $h^{-1}$] | | | | | *Yield* [-] | |
|---|---|---|---|---|---|---|---|---|
| | | *Gluc* | *Lact* | *Gln* | *Amm* | *sCR1* | *Lac/Gluc* | *Amm/Gln* |
| 37 | 0.0174 | 0.150 | 0.252 | 0.031 | 0.032 | 1.50e-4 | 1.68 | 1.03 |
| 35 | 0.0106 | 0.180 | 0.302 | 0.041 | 0.047 | 1.53e-4 | 1.68 | 1.13 |
| 33 | 0.0082 | 0.170 | 0.253 | 0.035 | 0.036 | 1.90e-4 | 1.48 | 1.03 |
| 30 | 0.0000 | 0.060 | 0.060 | 0.095 | 0.037 | 1.77e-4 | 1.58 | 1.15 |

The $q_{\mathrm{Gluc}}$ and $q_{\mathrm{Lac}}$ were almost same without influence of temperature difference except at 30 °C. The $q_{\mathrm{Gluc}}$ and $q_{\mathrm{Lac}}$ were almost same in all temperature range. However the yields of glucose to lactate change a little in all temperature range as well as the yields of glutamine to ammonia does.

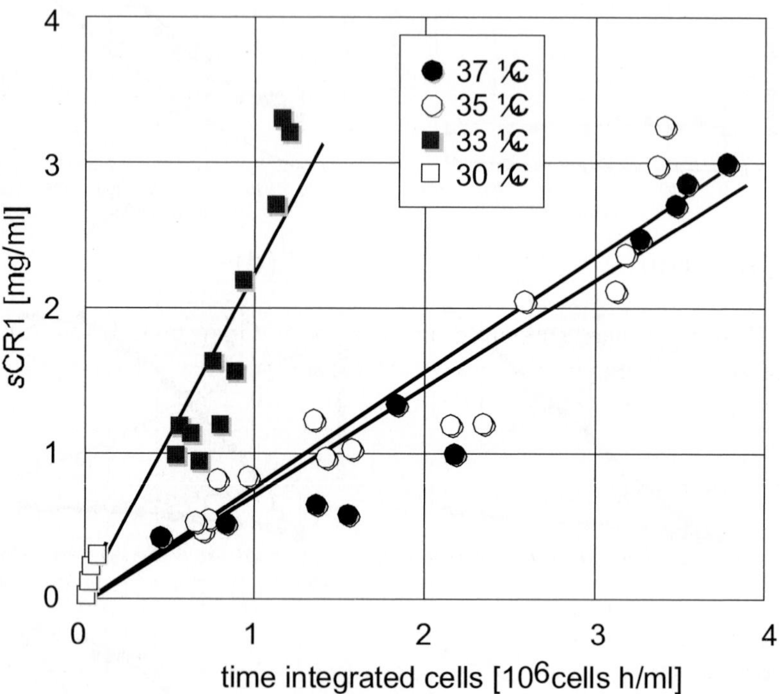


*Figure 2.* Relationship between sCR1 concentration and time integrated cell concentration at various temperatures.

Many reports have pointed out that enhancement of specific production rate is observed by arresting cells in G0/G1 phase and the accumulation of cells in G0/G1 phase could be achieved by lowering culture temperature. In this experiment the ratio of G0/G1 phase and the specific production rate of *s*CR1 increased in lower temperature culture.

*Table 2.* Distribution of cells in different phases on the cell cycle of rCHO cells.

| *Temp* [°C] | *Culture time* [h] | *G0/G1* [%] | *S* [%] | *G2/M* [%] |
|---|---|---|---|---|
| 37 | 55 | 62.3 | 15.7 | 22.3 |
| | 74 | 63.9 | 17.3 | 18.8 |
| | 90 | 67.3 | 15.1 | 17.6 |
| 35 | 57 | 66.6 | 11.5 | 21.9 |
| | 71 | 61.6 | 13.9 | 24.7 |
| | 114 | 58.9 | 17.0 | 24.1 |
| 33 | 115 | 81.5 | 2.7 | 14.7 |
| | 144 | 81.0 | 4.3 | 15.8 |
| 30 | 237 | 78.3 | 5.6 | 16.1 |

# 4. CONCLUSION

Lower temperature culture decreases the specific growth rate of *r*CHO, however it significantly increases the productivity of *s*CR1. It was suggested that the productivity of *s*CR1 could become higher by using the shift of culture temperature.

# ACKNOWLEDGMENTS

This work was supported in part by a Grant-in-Aid for Advanced Scientific Research on Bioscience/Biotechnology Areas from the Ministry of Education, Science, Sports and Culture of Japan.

# REFERENCES

Hendrick, V. *et al.* (2001) Increased productivity of recombinant tissular plasminogen activator (*t*-PA) by butyrate and shift of temperature: a cell cycle phase analysis, *Cytotechnology*. **36**; 71-83

Kato, H., Inoue, T., Ishii, N., Murakami, Y., Matsumura, M., Seya, T., Wang, P.-C. (2002) A nobel simple method to purify recombinant soluble human complement receptor type 1 (*s*CR1) from CHO cell culture, *Biotechnol. Bioprocess Eng.* **7**; 67-75

Yoon, S. K., Song, Ji Y., Lee, G. M. (2003) Effect of low temperature on specific productivity, transcription level, and heterogeneity of erythropoietin in chinese hamster ovary cells, *Biotechnol. Bioeng.* **82**; 289-298

# Proteomic Characterization of Recombinant NS0 Cell Clones Obtained by an Integrated Selection Strategy

Adolfo Castillo[1], Kathya R. De La Luz[1], Svieta Victores[1], Luis Rojas[1], Yamilet Rabasa[1], Simon Gaskell[2] and Rolando Perez[1]
*[1]Research and Development Division, Center of Molecular Immunology, Havana, Cuba,*
*[2]Michael Barber Center for Mass Spectrometry, Manchester University, United Kingdom.*

**Abstract:** In this work several clones of a recombinant myeloma NS0 cell that were isolated in protein-free medium and selected by an integrated methodology that takes in account specific production and growth rates, stability of expression and resistance to apoptosis induced by nutrient limitation were characterized by two-dimensional gel electrophoresis in the pH range of 3-10. We have determined a group of 5 spots that correlated well with the values of selection indexes calculated for different clones.

**Key words:** NS0, proteomics, cell line selection, production rate, stability, apoptosis.

## 1. INTRODUCTION

One of the main goals during production process development of mammalian cell based products is the selection of the 'right' cell line before to start with the cell banking procedures. The selected cell line should have not only good growth properties, but also high and stable expression levels of the product of interest with the desired physico-chemical and biological characteristics. The lack of high throughput clone selection procedure that could render with high efficiency clones with desired characteristics for further scale-up process (i.e. high specific production rates) have been pointed out. However the detailed molecular basis of these desired characteristics have not been well established yet. For these aims new postgenomic technologies, as proteomics could be of great impact to identify



*R. Smith (ed.), Cell Technology for Cell Products*, 431–434.

and study the mechanisms involved in the generation of particular cell phenotype.

Previously several clones of a recombinant myeloma NS0 cell line producing a humanized monoclonal antibody were isolated in protein-free medium and an integrated methodology was developed for the final cell line selection. In order to characterise the molecular basis of obtained phenotypes we have carried out a comparative study of protein expression patterns by two-dimensional gel electrophoresis (2DE) in the pH range of 3-10 between clones with different values of Ks. We have determined 5 spots that correlated with the Ks values for different clones. In summary these results offer useful information about the complex mechanisms and protein expression patterns that are related with improved phenotypic characteristics needed for industrial cell lines.

# 2. RESULTS

## 2.1 Characterization of phenotypic differences

Several clones of a recombinant myeloma NS0 cell line producing a humanised Mab were obtained in protein-free medium (PFM) and those with IgG concentration values higher than 3 mg/mL in 96-well culture plates were selected. Antibody concentration and cell proliferation were measured in 24-well culture plates by ELISA and colorimetric assay respectively. Samples of cells population at the end of exponential phase in spinner flasks batch cultures were taken and apoptosis measured using the Anexin V Assay kit. The integrated selection index (K*s*) was calculated as reported previously (1).

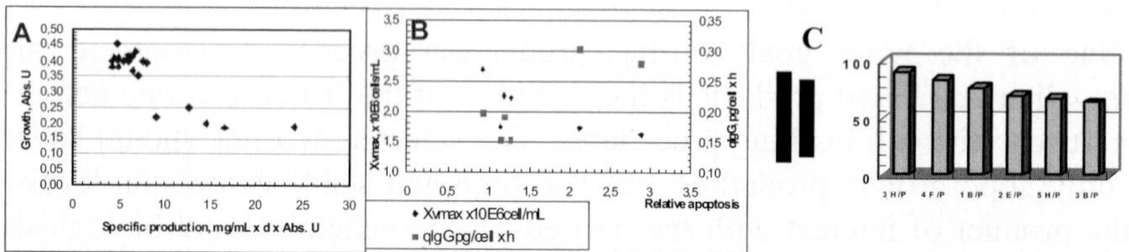


*Figure 1.* Correlation between: (A) specific production rates qab and growth values measured in 24-well plates and (B) maximum viable cell density (Xvmax) and qab measured in spinner flasks versus percent of apoptotic population relative to non apoptotic cell population (positive control), C) Selection indexes calculated for different clones in protein free medium.

## 2.2 Proteomic characterization of different phenotypic behaviours expressed in calculated selection index

Bidimensional gels for each clone was carried out by triplicate employing 17 cm, pH 3-10 linear IPG gel strips for IEF first dimension employing a total of 75 000 V*h. Second dimension SDS-PAGE was performed using 12,5% acrylamide gels and afterward gels were fixed and silver stained.

Obtained gels were analysed using the Melanie IV software: 1550 spots were detected, but only 878 were matched across all replicates for all clones.

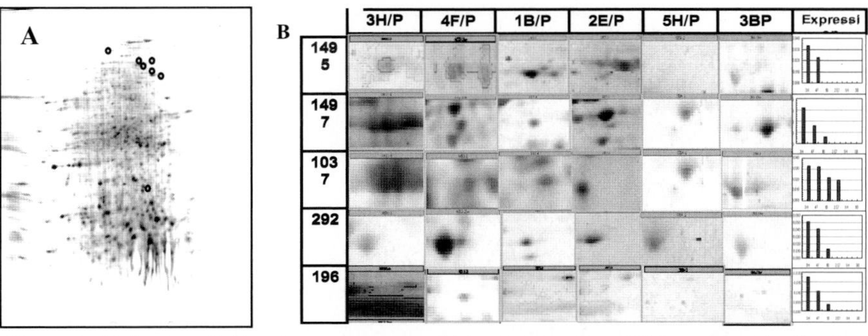


*Figure 4.* A) Spots with lineal correlation coefficient higher than 0.8 between selection index and values of percent of total volume intensity. B) Expression profile for five of selected spots.

# 3. CONCLUSIONS

During cell line selection process we observed the generation of clones with phenotypic differences. We have observed an inverse correlation between proliferation and production rates measured in 96-wells plates.

There exists also an inverse correlation between maximum viable cell number at the end of exponential phase and percent of cell population in apoptosis at this stage. The clones with the highest production rates showed also the highest values of apoptosis induction, that suggests that this phenotype is more sensible to nutrient limitation. We have considered the influence of these factorss in order to select cell lines for process scale-up.

We have found five spots that showed a high correlation coefficient between their expression intensities and calculated selection index, that could be related with the given phenotype for each analysed clone. Interesting to note that in all cases the expression of these spots decrease for clones with lower selection indexes.

# REFERENCE

Castillo A. J., Víctores S., Faife E., Rabasa Y., de la Luz K., Development of an integrated strategy for recombinant cell line selection. Animal Cell Technology

# IFN-γ Glycosylation Macroheterogeneity and Intracellular Nucleotide and Sugar Nucleotide Contents of CHO Cells Cultivated in Batch Culture

N. Kochanowski[1] F. Blanchard[1], E. Guedon[1], A. Marc[1], R. Cacan[2], F. Chirat[2] and J.-L. Goergen[1]
[1] *Laboratoire des Sciences du Génie Chimique, UPR CNRS 6811, ENSAIA-INPL - 2, avenue de la Forêt de Haye, 54 505 Vandoeuvre-lès-Nancy, France,* [2] *Unité de Glycobiologie Structurale et Fonctionnelle, UMR CNRS 8576, 59655 Villeneuve d'Ascq, France.*

**Abstract:** We developed an ion-pair RP-HPLC method to measure variations of intracellular nucleotide and sugar nucleotide pools throughout CHO cell cultures producing IFN-γ as a model glycoprotein. The data suggest that UDP-glucose and nucleotide triphosphate availabilities as well as the cell energetic status could influence the macroheterogeneity of IFN-γ glycosylation.

**Key words:** CHO cells, glycosylation, energy charge, sugar nucleotides,

## 1. INTRODUCTION

IFN-γ bears two N-glycosylation sites resulting in three possible glycoforms: 2N, 1N and 0N. IFN-γ glycosylation is not constant over the course of CHO cell batch cultures: the proportion of 2N IFN-γ declines with a concomittant increase in the 0N glycoform. Nucleotide and nucleotide sugars could be responsible for that phenomenon as they are carrier molecules of carbohydrates in protein glycosylation processus and energetic molecules. Therefore, an ion-pair RP-HPLC assay, allowing to measure thirteen nucleotides and nucleotide sugars, was developed and applied to monitor these molecules in CHO cell extracts during batch cultures.



*R. Smith (ed.), Cell Technology for Cell Products*, 435–438.

## 2. MATERIALS AND METHODS

CHO cells were grown either in serum free RPMI (BSA: 5g.L$^{-1}$; insulin: 5mg.L$^{-1}$; transferrin: 5mg.L$^{-1}$) or RPMI + 10% FCS or PF-BDM, a rich protein-free medium. Batch cultures were performed in 4-L bioreactors (Incelltech SGI), at pH 7.2, 37°C, 50% air saturation and 50 rpm. Cell viability was measured by trypan blue assay. IFN-γ production was determined by ELISA (R&D Systems) and macroheterogeneity was assessed by Western-Blot (Amersham Biosciences). Intracellular nucleotides and nucleotide sugars were analysed by ion-pair RP-HPLC (C18 column Alltech, 254nm, 40°C) after perchloric acid extraction (Ryll *et al.*, 91).

## 3. IFN-γ PRODUCTION AND GLYCOSYLATION MACROHETEROGENEITY

Maximal cell and IFN-γ titers were maximal in rich media, *i.e.* RPMI + 10% FCS and PF-BDM. Maximal IFN-γ concentration was two-fold higher in PF-BDM than in RPMI with or without FCS. In RPMI with or without FCS, 2N IFN-γ proportion decreased during the culture, contrary to PF-BDM.

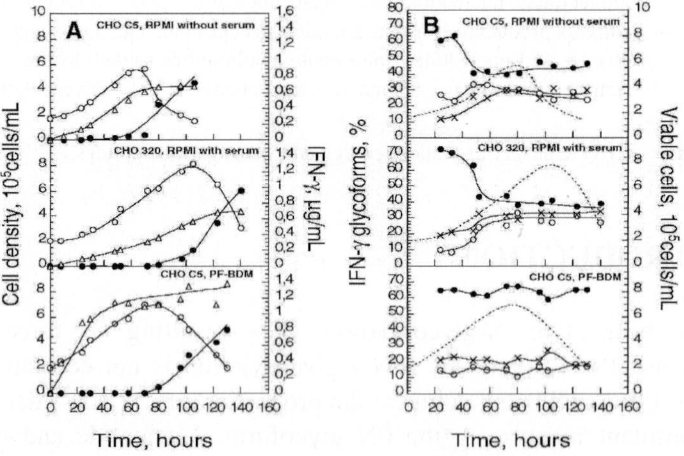


*Figure 1.* Kinetics of viable cells (A,○), dead cells (A,•), IFN-γ production (A, △), and IFN-γ macroheterogeneity (B, IFN 2N : •, IFN 1N: ○ , IFN 0N: ×, viable cells: dotted line).

## 4. SUGAR NUCLEOTIDE CONTENTS

UDP-GalNAc and UDP-GlcNAc contents increased with time in the three cultures probably as a consequence of ammonia accumulation (Ryll *et al.*, 1994). UDP-Glucose titer was maximal simultaneously to the cell specific growth rate, and then decreased. Compared to PF-BDM, UDP-Gal content remained relatively low in RPMI with or without FCS. GDP-Man accumulated throughout the three cultures, meaning that this molecule is not limiting for glycosylation.

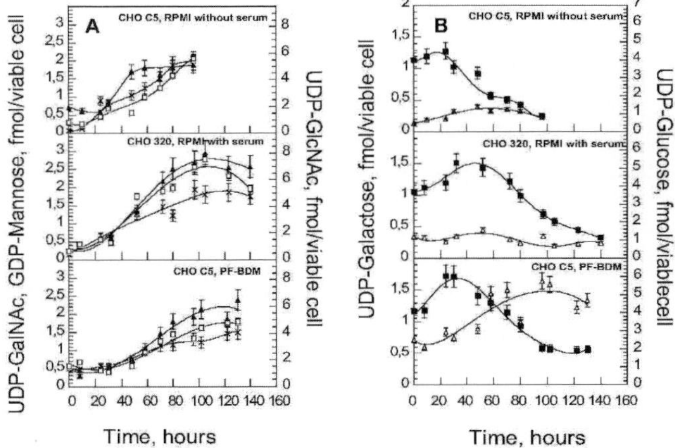


*Figure 2.* Kinetics of sugar nucleotides during CHO cell batch culture (UDP-Glc: ■, UDP-Gal:△ , UDP-GlcNAc:× , UDP-GalNAc: ▲, GDP-Man: □).

# 5. NUCLEOTIDE TRIPHOPHATE CONTENTS AND ADENYLATE ENERGY CHARGE

Adenylate energy charge, defined as (ATP+0.5ADP/ATP+ADP+AMP) reflecting the physiological state of the cells, is maintained at high values (around 0.95) as long as IFN-γ is produced in PF-BDM. Therefore, it can be assumed that this medium may supply precursors that are limiting in RPMI with or without FCS, particularly nucleotides and nucleotide sugars, which are crucial for N-glycosylation reactions.

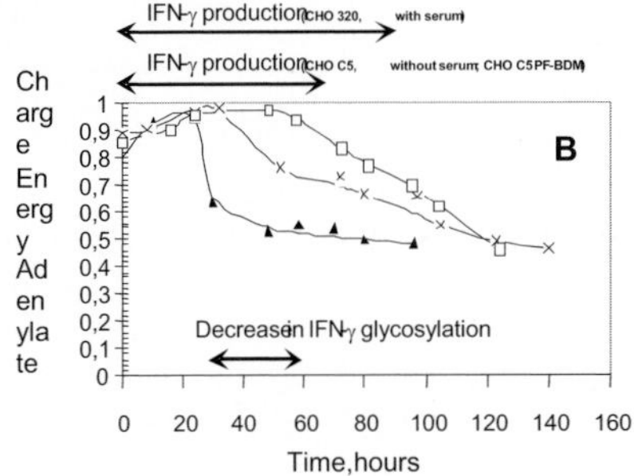


*Figure 3.* Kinetics of adenylate energy charge during CHO cell batch culture in three different media ( RPMI without serum: ▲, RPMI with serum :× , PF-BDM : □).

## REFERENCE

Ryll, T., Valley, U., and Wagner, R., 1994, Biochemistry of growth inhibition by ammonium ions in mammalian cells, Biotechnol. Bioeng. **44**:184-193.

# Influence of the Adaptation Procedure and the Serum-free Conditions on the N-glycosylation Pattern of EPO Produced by CHO Cells

F. Le Floch[1], B. Tessier[1], J.-M. Guillaume[2], P. Cans[2], A. Marc[1], J.-L. Goergen[1]
[1] *Laboratoire des Sciences du Génie Chimique, UPR CNRS 6811, ENSAIA-INPL - 2, avenue de la Forêt de Haye, 54 505 Vandoeuvre-lès-Nancy, France,* [2] *Aventis Pharma, 13, quai Jules Guesde, 94400 Vitry sur Seine, France*

**Abstract:** To discriminate between the influence of the adaptation procedure itself and the proper effects of the absence of serum, EPO production and glycosylation were comparatively studied in various culture conditions. Serum was identified as responsible for EPO desialylation whereas cell adaptation procedure did not affect the protein desialylation that occurred in the same way when adapted and non-adapted cells were cultivated with serum.

**Key words:** glycosylation, sialylation, serum free adaptation, capillary electrophoresis

## 1. INTRODUCTION

The necessity to perform serum-free cultures to produce recombinant glycoproteins generally requires an adaptation procedure of the original cell line to new environmental conditions, which may therefore affect cellular physiology and protein glycosylation (Le Floch *et al.*, 2004).

To assess this phenomenon, we analysed m-EPO production and glycosylation (by HPCE-LIF) of non-adapted and adapted CHO cells to serum-free conditions either in presence or absence of serum.



*R. Smith (ed.), Cell Technology for Cell Products*, 439–442.

# 2. MATERIALS AND METHODS

CHO cells producing recombinant mEPO were grown in serum-free commercial medium (SFX-CHO, HyClone), in 4-L bioreactors (Incelltech SGI), at pH 7.2, 37°C, 50% air saturation and 50 rpm. mEPO was immuno-purified by coupling a monoclonal anti-hEPO antibody to CNBr activated sepharose gel. HPCE-LIF analysis were performed according to a procedure derived from the method described by Chen and Evangelista (1998) with an eCAP N-linked oligosaccharide profiling kit (Beckman). Acidic desialylation was performed with 0.5M acetic acid. Sialidase activity was determined by fluorescence according to the method described by Gramer (2000).

# 3. PROFILE EVOLUTION OF N-LINKED GLYCANS OF EPO PRODUCED BY CELLS ADAPTED OR NON-ADAPTED TO SERUM-FREE CONDITIONS

## 3.1 Non-adapted cells, in presence of serum

HPCE-LIF pattern of EPO N-glycans produced by non-adapted cells in presence of serum (Fig. 1) showed that species with migration times superior to 10 glucose units (GU) appeared with time, and that a concomitant decrease in the size of faster migrating chains occurred. Because of the negative charges brought by sialyl residues, sialylated carbohydrates have migration times much shorter than 10 GU (Chen and Evangelista, 1998). Therefore EPO collected at the beginning of the culture consisted mainly in sialylated glycoproteins; then a desialylation occurred during the course of the culture. On the other hand, mild acidic hydrolysis showed that desialylated glycans were very similar over the culture.

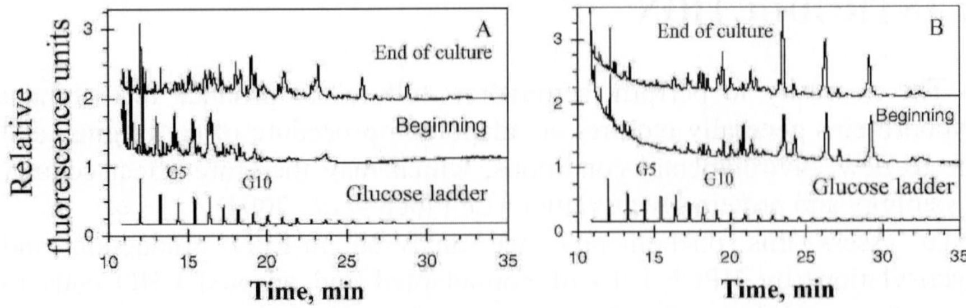


*Figure 1.* HPCE analysis of the N-linked glycans of EPO produced by non-adapted cells in presence of serum; before(A) and after (B) mild acidic hydrolysis.

### 3.2 Adapted cells, in presence of serum

Glycans of EPO collected during the cell growth phase and at the end of the culture performed in presence of serum, but with cells previously adapted to serum-free conditions showed that the N-glycans of EPO were still greatly sialylated at the middle of the culture (major peaks were located before 10 GU), whereas significant amounts of non sialylated species appeared at the end of the process. Therefore, a desialylation of EPO occurred when serum was used, whatever the cells were used. As for the non-adapted cells, the desialylated pattern (upon acidic treatment) remained very stable over the process.

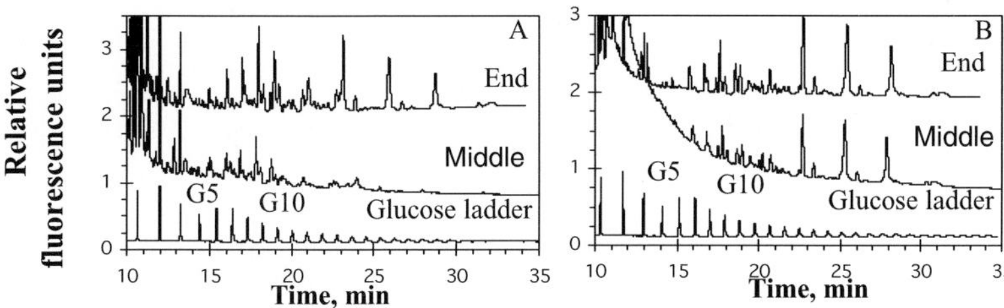


*Figure 2.* HPCE analysis of the N-linked glycans, before (A) and after (B) mild acidic hydrolysis, from EPO produced by adapted cells in presence of serum.

### 3.3 Adapted cells, without serum

The glycosylation patterns of serum-free produced EPO collected at the end of the cultures of adapted cells were very similar to those of EPO collected at the beginning of the two previous experiments. The complete glycan pattern did not show any peak after 10 GU and the major species appearing after acidic desialylation had similar migration times to those obtained in presence of serum. Thus, EPO collected at the end of this serum-free batch culture mainly consisted in sialylated glycoproteins.

An enzymatic degradation may explain the phenomenon of EPO desialylation during the serum-containing cultures. Clearly, a significant sialidase activity was found in the supernatants of serum-supplemented cultures, around 430 nmol.$h^{-1}.l^{-1}$, while a negligible level was observed in serum-free media. The adaptation of the cells did not change the level of released sialidase.

# REFERENCES

Chen, A.F.T. and Evangelista, R.A., 1998. Profiling glycoprotein N-linked oligosaccharide by capillary electrophoresis. Electrophoresis 19, 2639-2644.

Gramer, M.J., 2000. Detecting and minimizing glycosidase activities that can hydrolyze sugars from cell culture-produced glycoproteins. Mol. Biotechnol. 15, 69-75.

Le Floch, F., Tessier, B., Chenuet, S., Guillaume, J.M., Cans, P., Marc, A. and Goergen, J.L., 2004. HPCE-monitoring of the N-glycosylation pattern and sialylation of murine erythropoietin produced by CHO cells in batch processes. Biotechnol. Prog. 20, 864-871.

# Stability and Cytogenetic Characterization of Recombinant CHO Cell Lines Established by Microinjection and Calcium Phosphate Transfection

M. Derouazi[1], D. Martinet[2], N. Besuchet[2], R. Flaction[1], M. Wicht[2], M. Bertschinger[1], D. Hacker[1], J. Beckmann[2] and F.M. Wurm[1]

*[1]Laboratory of Cellular Biotechnology, Faculty of Life Science, Ecole Polytechnique Fédérale de Lausanne EPFL, 1015 Lausanne Switzerland, [2]Laboratoire de Cytogénétique, Service de Génétique Médicale, Centre Hospitalier Universitaire Vaudois CHUV, 1011 Lausanne Switzerland*

**Abstract:** Chinese hamster ovary cells (CHO) are widely used for the stable production of recombinant proteins. Typically, recombinant cell lines are characterized for the stability of protein expression over a period corresponding to the time needed to scale-up the culture and harvest the product (*e.g.* 2 to 3 months), for the number of plasmid copies integrated into the host genome, and for the quality and quantity of the recombinant protein. In this study we extended the characterization to the cytogenetic level. Sixteen recombinant CHO cell lines were established using calcium phosphate transfection and microinjection as DNA transfer methods. For each cell line we observed by fluorescence in situ hybridization a single integration site regardless of the gene delivery method, the topology of the DNA (circular or linear), or the integrated plasmid copy number (between 1 and 50). Integration was not targeted to a specific chromosome. Chromosomal rearrangements were observed in about half of these cell lines. This phenomenon occurred independently of the gene transfer method. Interestingly the rearrangements were not on the chromosome where the plasmid integrated. We observed rearrangements between chromosomes and chromosomal imbalances.

**Key words:** CHO DG44, karyotype, recombinant proteins, calcium phosphate, microinjection, GFP



*R. Smith (ed.), Cell Technology for Cell Products*, 443–446.

# 1. INTRODUCTION

The main approach to the expression of recombinant proteins in CHO and other mammalian cells is the establishment of a cell line in which the transgene(s) is integrated into the host genome and stably expressed over time. The process of generating a recombinant cell line involves the isolation and expansion of a clonal cell line from a pool of cells cotransfected with the transgene and a selectable gene. The stability of a recombinant cell line is determined by monitoring cell growth and protein production over several months. For some cell lines, however, protein productivity diminishes over time usually as the result of changes in the regulation of transgene expression (Strutzenberger *et al.*, 1999). Unfortunately, there are currently no methods available to predict the stability of recombinant protein expression in a given cell line.

Here we report the generation and analysis of several CHO DG44–derived cell lines expressing the green fluorescent protein (GFP). The cell lines were characterized in terms of the long-term stability of recombinant protein expression, the number and location of the transgene integration sites, the transgene copy number, and the karyotype.

# 2. RESULTS

Sixteen recombinant CHO DG44 cell lines expressing the green fluorescent protein (GFP) were generated either by CaPi-mediated transfection or by microinjection of linear or circular pMYKpuroEGFP. At 2 days after gene transfer the cells were exposed to either a low (3 µg/ml) or high (6 µg/ml) concentration of puromycin for 22-28 days. After removal of the selective pressure the GFP expression of each clonal cell line was measured by flow cytometry. Cell lines exhibiting a homogenous level of GFP expression with a coefficient of variation below 20 were only observed in only 7 cases (Fig. 1). Some cell lines such as 10, 12 and 15 clearly had two subpopulation expressing GFP at different levels (Fig. 1). The number of integrated copies of the GFP gene in the 16 cell lines ranged from 0-51 as determined by Southern blot using the full-length GFP coding sequence as a probe (Fig. 2).

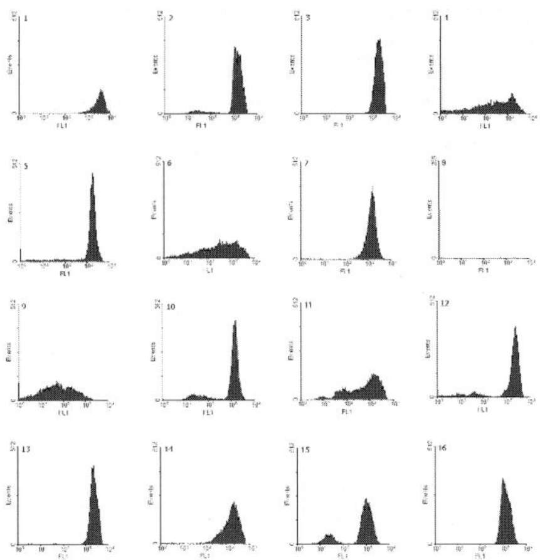

*Figure 1.* Flow cytometry analysis of the 16 recombinant CHO G44 cell lines expressing the GFP.

For each recombinant cell line, the site(s) of pMYKpuroEGFP integration was determined by FISH using the full-length GFP gene as a probe. For all of the GFP-positive cell lines except 7, only one integration site was observed (Fig. 2). The karyotype of each recombinant cell line was also determined and compared to that of the parental CHO DG44 strain. Ten to twenty metaphase spreads were analysed for each of the recombinant cell lines. In most of the recombinant cell lines, 20 chromosomes were observed (Fig. 2). Strong aneuploidy, however, was seen in cell line 7 with 36-41 chromosomes and in cell line 16 with 21, 31, 40, or 41 chromosomes (Fig. 2). Among the recombinant cell lines, only 37% had the same karyotype as the parental CHO DG44 strain (Fig. 2). Interestingly, a single species of chromosome 4, either the normal one or one with additional genetic material, was found in 13 of the 16 cell lines (data not shown). All chromosomal rearrangements detected were specific to each cell line, and these genetic abnormalities were clonal with two exceptions: cell line 11 and 14. For cell line 11, two subpopulations were identified. Most of the cells (75%) had the same karyotype as CHO DG44 and 25% had a deletion in the short arm of der(7) (del der(7p)). For cell line 14, 85% of the cells had additional genetic material on the short arm of chromosome 8 (add(8p)) and in the second subpopulation a deletion was observed on the short arm of der(8) (del(der8p)). Interestingly, among the cell lines characterized for stable recombinant protein expression there was no obvious correlation between genetic rearrangements and the stability of expression (Fig. 2). A cell line with a genetic rearrangement but having stable recombinant protein expression (cell line 16) and a cell line with a normal karyotype but unstable

| Cell lines | Transfection method[a] | GFP copy Number | Protein expression | Chromosomes number | Integration site[b] | Rearrangement | Chromosomal rearrangement[c] |
|---|---|---|---|---|---|---|---|
| 1 | nuclear MI | 1 | stable | 20 | 4 | none | |
| 2 | cytoplasmic MI | 13 | stable | 20 | der(5) | none | |
| 3 | nuclear MI | 3 | stable | 20 | 1 | balanced | t(4q;der(5)) |
| 4 | nuclear MI | 51 | not stable | 19 | der(6) | unbalanced | t(Z13;der(6)),-Z13p |
| 5 | CaPi circular | 2 | not stable | 20 | 2 | none | |
| 6 | CaPi circular | 15 | not done | 20 | 2 | none | |
| 7 | CaPi circular | 3 | not done | 36-41 | 8 and Z2 | unbalanced | iso der(7q) + aneuploidy |
| 8 | CaPi circular | 0 | not done | 20 | - | unbalanced | -der(5), + mar3 |
| 9 | CaPi circular | 2 | not done | 19 | der(8) | unbalanced | -der(10) |
| 10 | CaPi circular | 9 | not stable | 20 | 2 | unbalanced | complex (Z4, der(7), der(8), der(10)) |
| 11 | CaPi linear | 13 | not done | 20 | 1 | unbalanced | del der(7p) |
| 12 | CaPi linear | 5 | not done | 20 | Z8 | none | |
| 13 | CaPi linear | 4 | not stable | 20 | der(5) | none | |
| 14 | CaPi linear | 12 | not done | 20 | 8 | unbalanced | add(8p)/del(der(8p)) |
| 15 | CaPi linear | 4 | not done | 21 | der(7) | unbalanced | complex (4, Z4, M1) |
| 16 | CaPi linear | 1 | stable | 21, 31, 40, 41 | - | unbalanced | aneuploidy |

[a] MI: microinjection, CaPi: calcium-phosphate transfection with linear or circular DNA
[b] CHO DG44 chromosome number
[c] **t**: translocation, -:loss of chromosome, **del**:deletion, **add**:additional genetic material, **iso**: isochromosome.
**p**: short arm and **q**: long arm of the chromosome

*Figure 2.* Stability, cytogenetic and FISH analysis of the 16 CHO DG44 recombinant cell lines.

protein expression (cell line 5) were present in the collection of recombinant cell lines described here.

## 3. CONCLUSION

The results demonstrate that genetic alteration are frequently associated with the establishment of recombinant CHO DG44 cell lines. However, genetic instability did not directly correlate with the instability of recombinant protein expression.

## REFERENCE

Strutzenberger K, Borth N, Kunert R, Steinfellner W, Katinger H. 1999. Changes during subclone development and ageing of human antibody-producing recombinant CHO cells. Journal of Biotechnology 69:215-226.

# Efficient Cloning Method for Variable Region Genes of Antigen Specific Human Monoclonal Antibody

Shinei Matsumoto, Yoshinori Katakura, Makiko Yamashita, Yoshihiro Aiba, Kiichiro Teruya, Sanetaka Shirahata
*Graduate School of Systems Life Sciences, Kyushu University, Fukuoka, Japan.*

**Abstract:** We have developed an *in vitro* immunization protocol of human peripheral blood mononuclear cells (PBMC) for generating human antigen-specific antibodies. By using this protocol, B cells producing antigen specific antibody can be propagated within a week. In the present study, we tried to establish an efficient strategy to clone variable region genes of antigen specific human monoclonal antibody by applying *in vitro* immunized PBMC to the phage display method. To evaluate the efficiency of our strategy, heavy and light chain variable region genes were prepared by PCR from PBMC immunized *in vitro* with mite-extract (ME), and the combinatorial library (>$10^5$ members) cloned for display on the surface of a phage. Phage was subjected to a streptavidin magnetic bead panning procedure. After concentrating the ME-specific phage antibody by panning, the phage antibodies specific for ME were detected by ELISA. Results show that rapidly production of antigen-specific human monoclonal antibody from smallish (1.6 x $10^5$) library is possible by using *in vitro* immunized PBL.

**Key words:** *in vitro* immunization, human monoclonal antibody, phage display

## 1. INTRODUCTION

Human monoclonal antibodies (mAbs) have a great potential for diagnosis and treatments of cancer, allergy and other diseases. However, we cannot immunize human with antigen by ethical problems. Monoclonal antibodies from mouse origin are relatively easy to produce, however, their therapeutic availability is restricted by their antigenicity. At present, human monoclonal antibodies are mainly produced by humanizing mouse monoclonal antibodies. But it is difficult to completely remove antigenicity derived from mouse. Therefore, we have developed an *in vitro* immunization



*R. Smith (ed.), Cell Technology for Cell Products*, 447–449.

(IVI) protocol of human peripheral blood lymphocyte (PBL) for generating human antigen specific antibodies. By using this protocol, B cells producing antigen specific can be propagated within a week. In the present study, we tried to establish an efficient strategy to clone variable region genes of antigen specific human monoclonal antibody by applying *in vitro* immunized PBL to the phage display method.

## 2. METHODS

### 2.1 *In vitro* immunization

Human PBL were cultured for 8 days in ERDF medium containing 10% heat inactivated fetal bovine serum, CpG-ODN, IL-2, IL-4, 2-mercaptoethanol and Mite-Extract (ME). Antibody production in the supernatant of cultured PBL was measured by ELISA.

### 2.2 Construction of phage antibody library

The VH and VL genes of cultured PBL were prepared by RT-PCR. The VH and VL DNA fragments were joined with a linker DNA. The assembled ScFv DNA was digested with *Sfi* I and *Not* I for cloning into the pCANTAB5E phagemid vector. The ligated vector was introduced into competent *E. coli* TG1 cells. Phagemid-containing bacterial colonies were infected with M13KO7 helper phage to yield phage-displayed antibody ScFv fragments.

### 2.3 Detection and production of ME-specific antibody

The phage antibodies were selected by panning against ME. ME-specific phage antibodies bound to biotinylated ME on streptavidin magnetic beads and non-specific phage were washed off. *E. coli* TG1 cells were infected with the binding phage. After panning, 60 individual phage antibodies were picked. Using ME-coated 96 well plates, ME-reactive phage antibodies were detected by ELISA. The VH and VL genes were isolated from the positive clones. The resulting DNA were co-transfected to the CHO cells.

# 3. RESULTS AND DISCUSSION

15 positive clones were obtained from random picked 60 phage clones of *in vitro* immunized library (Figure 1). The VH and VL genes were isolated from the positive clone B11, D12, and E11. The resulting DNA were co-transfected to the CHO cells and secreted IgG in the supernatant were purified and analyzed by Western blotting. ME could be detected by the purified IgG.

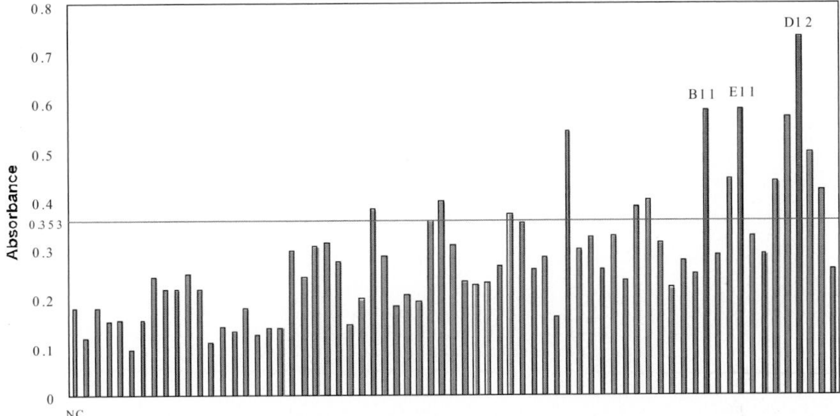


*Figure 1.* Detection of ME-specific scFv (ELISA).

Our results show that it is possible to rapidly produce antigen-specific human monoclonal antibody from smallish ($1.6 \times 10^5$) library by using *in vitro* immunized PBL. We expect that *in vitro* immunization and phage display method make a strong combination to produce human monoclonal antibody.

# REFERENCES

Ichikawa, A., Katakura, Y., Teruya, K., Hashizume, S. and Shirahata, S. (1999) In vitro immunization of human peripheral blood lymphocytes: establishment of B cell lines secreting IgM specific for cholera toxin B subunit from lymphocytes stimulated with IL-2 and IL-4. Cytotechnology 31, 131

Marks JD, Hoogenboom HR, Bonnert TP, McCafferty J, Griffiths AD, Winter G. (1991) By-passing immunization. Human antibodies from V-gene libraries displayed on phage. Journal of Molecular Biology. 1991 Dec 5;222(3):581-97.

# Design Approach Regarding Humanization and Functionality of an Anti-CD18 Monoclonal Antibody

Cristina A. Caldas[1,4]; Diorge P. Souza[2,]; Maria Teresa A. Rodrigues[3,4]; Andréa Q. Maranhão[2,4]; Ana M. Moro[3,4]; Marcelo M. Brigido[2,4]
[1]Laboratório de Imunologia, InCor, USP, Brasil; [2]Laboratório de Biologia Molecular, UnB, Brasi; [3]*Laboratório Biofármacos em Célula Animal, Instituto Butantan,Brasil*; [4]*Instituto de Investigação em Imunologia - iii, Institutos do Milênio, Brasil*

**Abstract:** In this work the closest human germline sequence was used as the framework on to which to graft murine CDRs. The proposed fully humanized version displays the same staining pattern to different cell subpopulations like CD4, CD8 and memory CD45RO lymphocytes, when compared with the original anti-CD18 MAb, reinforcing the germline approach as a successful strategy to humanize antibodies with maintenance of affinity and selectivity.

**Key words:** antibody, humanization, CD-18, adhesion, germline, framework, integrin, graft, HAMA, CDR, MAb, FACS, binding, scFv, hypermutation

## 1. INTRODUCTION

Due to HAMA (human anti-mouse antibody) response evoked by the use of murine MAbs, antibody humanization technology was developed for the production of a new molecule that maintains the capacity to recognize the antigen but resemble more like a human antibody. The purpose of this work was to humanize the 6.7 murine anti-human CD18 antibody (David *et al.*, 1991). CD18 is a β2 integrin family component and an antibody that blocks its adhesion property has potential application in a large panel of clinical manifestations through the migration's inhibition of neutrophils and other leucocytes. The huMAb antibody could be used as an anti-inflammatory agent, to decrease post ischemic effects.



*R. Smith (ed.), Cell Technology for Cell Products*, 451–454.

## 2. METHODOLOGY AND RESULTS

The strategy used was to choose the closest human germline sequence (Tomlinson *et al.*, 1992) onto which to graft the murine CDRs. The germline sequence does not present somatic hypermutation of rearranged V regions, which can be unique for that specific antibody and thus be seen as immunogenic to patients. The data bank search indicated HG3 as the closest human germline sequence (Rechavi *et al.*, 1983) to receive the anti-CD18 CDRs sequences. The framework residue Ala$^{71}$ was maintained as in the original murine antibody, for the assembly of a correct H2 loop in the humanized antibody (Tramontano *et al.*, 1990). Analysis of the crystallized Fab with sequence most similar to HG3 (1AD9) showed that residue 45 is located in a position of low flexibility near to H-CDR2 and L-CDR3, seen as critical for the maintenance of the H2-H3 loop structure. Based on this assumption, two versions of humanized VH were constructed: Proline version (HP) of murine origin and Leucine version (HL) of the closest germline human sequence (Caldas *et al.*, 2000). The FACS analyses of the variations showed higher binding for the HL version, confirmed by the blocking of the 6.7 original MAb.

<table>
<tr><th></th><th>Median IF</th><th>Blocking (%)</th><td rowspan="4">Hemi-humanized scFvs fragments block the binding of MAb 6.7 FITC to CD18+ cells. scFvs were added to the cells prior the addition of the 6.7 FITC.</td></tr>
<tr><td>6.7 FITC</td><td>40,32</td><td>0</td></tr>
<tr><td>Fv-VH (L)</td><td>9, 65</td><td>76,3</td></tr>
<tr><td>Fv-VH (P)</td><td>12,52</td><td>69,0</td></tr>
</table>

For the light chain humanization the same strategy was followed and the humanized VL gene fragment was expressed together with the selected version Leu$^{45}$ of the humanized VH (Caldas *et al.*, 2003). Two versions were analyzed in detail because of an expected loss of affinity. The version VQ contains glutamine$^{37}$ (present in 78% of both murine and human antibodies) while in the version VL the leucine residue of murine origin was maintained. The loss of affinity effect was shown by FACS analysis (Fig. 1) in comparison with the original 6.7 MAb. The capacity of the two versions' binding to the cell surface was further tested through its blocking effect on the binding of the original anti-CD18 MAb. We found that both the original anti-CD18 and the humanized LL version displayed the same pattern of staining to CD4+, CD8+ and to memory CD45RO+ lymphocytes (Fig. 2).

## 3. DISCUSSION AND CONCLUSION

For the HC, the FACS and further image analyses served as basis to choose the HL (leucine residue) for the position 45. In relation to the LC, a mutation far from the antibody's binding site was identified, resulting in a loss of affinity for the intact CD18 molecule on the cell surface. In the human Vκ germline sequence used in this work, a glutamine residue fills the position 37. This residue is found in many humanized antibodies, including our closest PDB crystal structure, 1AD9, in which the $Gln^{37}$ is mostly buried and it contacts the same set of residues as $Leu^{37}$ in the murine structure. The position 37 is related to the FR2 stabilization and is rarely variable. The only difference is a hypothetical H-bond between the amide group of $Gln^{37}$ and the hydroxyl group of $Tyr^{86}$. $Leu^{46}$ was maintained in the humanized version for its contact with LCDR2 and HCDR3. The construction was able to recognize PMNC harboring CD18, indicating successful humanization with transfer of the original binding capability. We have shown the complete humanization of the 6.7 mAb. The CDR-grafting using the closest germline, although very simple, is a reliable procedure for humanization of antibodies.

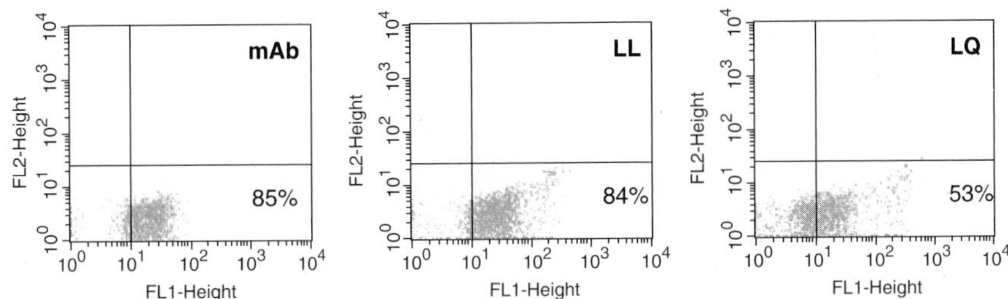


*Figure 1.* Anti-CD18 humanized versions binding to human lymphocyte cell surface; 85, 84 and 53% of the cells were positive when incubated with murine and humanized scFv versions.

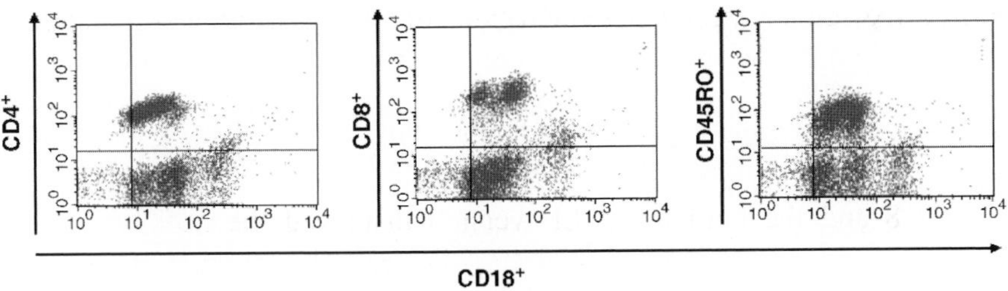


*Figure 2.* T cell subpopulations stained by the humanized LQ version.

# REFERENCES

Caldas C *et al.* (2000) Protein Engineering, 13: 353-360.
Caldas C *et al.* (2003) Mol. Immunol., 39: 341-952.
David D *et al.* (1991) Cel. Immunol., 136: 519-524.
Rechavi G *et al.* (1983). PNAS USA, 80:855-859.
Tomlinson IM *et al.* (1992) J. Mol. Biol., 227: 776-798.
Tramontano A *et al.* (1990). J. Mol. Biol., 215: 175-182.

# Development of a Transient Mammalian Expression Process for the Production of the Cancer Testis Antigen NY-ESO-1

David L. Hacker[1], Leonard Cohen[2], Natalie Muller[1], Huy-Phan Thanh[1], Elisabeth Derow[1], Gerd Ritter[2], and Florian M. Wurm[1]

[1]*Laboratory of Cellular Biotechnology, Ecole Polytechnique Federal de Lausanne (EPFL), 1015 Lausanne, Switzerland,* [2]*Ludwig Institute of Cancer Research, New York Branch at Memorial Sloan-Kettering Cancer Center, 1275 York Avenue, New York, NY 10021, USA*

**Abstract:** The cancer-testis antigen NY-ESO-1 is expressed in a wide range of tumor types while normal tissue expression is restricted to germ cells. It is highly immunogenic, frequently eliciting spontaneous humoral and cellular immune responses in patients with advanced tumors expressing the protein. A series of different NY-ESO-1 vaccines are currently being explored in early clinical trials worldwide. We have developed a reproducible process for the transient expression of full-length NY-ESO-1 in mammalian cells. Either Chinese hamster ovary (CHO) DG44 or HEK 293 cells were transfected with NY-ESO-1 DNA using polyethylenimine. NY-ESO-1 was expressed intracellularly as a histidine-tagged protein in agitated suspension cultures at volumes to 400 ml at levels up to 30 mg/L as estimated in crude cell lysates by Western blot. The protein was purified by immobilized metal affinity chromatography (IMAC) after detergent lysis and was shown to be reactive with sera from patients with tumors expressing NY-ESO-1.

**Key words:** Cancer-testis antigen, NY-ESO-1, vaccine, recombinant protein, Chinese hamster ovary cells, human embryonic kidney cells, polyethylenimine, transfection, transient gene expression, suspension cells, square bottles, histidine tag, affinity chromatography, antigenicity, ELISA.

## 1. INTRODUCTION

The cancer-testis (CT) antigen NY-ESO-1 is expressed in many tumor types but not in normal tissues except testis (Scanlan *et al.*, 2004). Both



*R. Smith (ed.), Cell Technology for Cell Products*, 455–458.

humoral and cellular immune responses can occur in patients with tumors that express this antigen (Scanlan *et al.*, 2004). In recent clinical trials in which patients with NY-ESO-1-positive tumors were vaccinated with recombinant NY-ESO-1 from *E. coli* in combination with an adjuvant both antibody and T cell responses were observed (Davis *et al.*, 2004). In this report we have demonstrated the feasibility of transient expression of intracellular histidine-tagged NY-ESO-1 (His-NY-ESO-1) in either CHO DG44 or HEK 293 cells in non-instrumented cultivation systems at volumes up to 400 ml. The partially purified His-NY-ESO-1 was shown to be reactive to sera from patients with NY-ESO-1-positive tumors. Transient expression in mammalian cells offers an alternative approach to the production of recombinant proteins for vaccines.

## 2. METHODS

For transfections in 1-L square bottles, suspension CHO DG44 or HEK 293 cells were seeded at 1 x $10^6$ cells/ml in 200 ml of RPMI 1640 medium (Derouazi *et al.*, 2004). The cells were transfected with pcH-NYESO1 in the presence of 25-kDa linear polyethylenimine (PEI) (Polysciences, Eppelheim, Germany) (Muller *et al.*, 2005). At 4 h post-transfection 200 ml of medium was added to each bottle. At 2 days post-transfection, 2 ml aliquots of the culture were collected. The cells were harvested by centrifugation, lysed in Mammalian Protein Extraction Reagent (M-PER) (Perbio Science, Lausanne, Switzerland) according to the manufacturer's recommendations, and the soluble fraction was analyzed by immunoblot using the mouse monoclonal antibody ES121 raised against NY-ESO-1. Antibody binding was detected with horseradish peroxidase-conjugated anti-mouse IgG (Sigma Chemical, St. Louis, MO).

## 3. RESULTS AND DISCUSSION

Histidine-tagged NY-ESO-1 (His-NY-ES0-1) was transiently expressed in both HEK 293E and CHO DG44 cells using PEI for DNA delivery. For both cell lines the highest steady state level of intracellular His-NY-ESO-1 was observed at day 2 post-transfection (data not shown). For production of the recombinant protein, cultures of 200 ml in agitated one-liter square bottles were transfected in serum-free RPMI 1640 (4). As shown in Fig. 1, the production process in 1-L square bottles was reproducible. The transfected cells were harvested at 2 days post-transfection and lysed with detergent. Partial purification of His-NY-ESO-1 was performed by IMAC.

Recovery of the protein from CHO DG44 was less efficient than from HEK 293 (data not shown). The antigenicity of His-NY-ESO-1 from HEK 293 cells was determined by ELISA using sera from cancer patients with tumors expressing NY-ESO-1. Importantly, HEK 293-derived His-NY-ESO-1 was equally reactive to *E. coli*-derived NY-ESO-1 (Table 1). Recombinant His-NY-ESO-1 transiently expressed in mammalian cells may become a valuable reagent for the development of NY-ESO-1 cancer vaccines. Potential applications include its use as a vaccine immunogen and as a reagent to monitor vaccine induced "in vitro" and "in vivo" immune responses to NY-ESO-1.

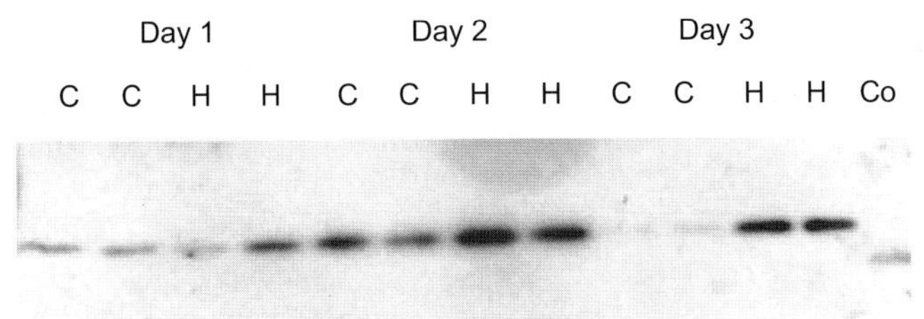


*Figure 1.* Transient His-NY-ESO-1 expression in HEK 293 (H) and CHO DG44 (C) cells. The cells were transfected with pH-NYESO1 and harvested at the times indicated. The lysates were analyzed by western blot using an antibody against NY-ESO-1. Recombinant NY-ESO-1 from E. coli (Co).

*Table 1.* ELISA reactivity of recombinantHis-NY-ESO-1 with sera from cancer patients containing NY-ESO-1 antibodies.

| Patient # | NY-ESO-1 (E. coli) | NY-ESO-1 (HEK) | Control |
|---|---|---|---|
| CA-001 | 1:5000 | 1:5000 | negative |
| CA-002 | negative | negative | negative |
| CA-003 | 1:16000 | 1:8000 | negative |
| CA-004 | negative | negative | negative |
| CA-005 | 1:25000 | 1:60000 | negative |
| CA-006 | negative | negative | negative |
| CA-007 | 1:60000 | 1:60000 | negative |
| CA-008 | negative | negative | negative |
| CA-009 | 1:60000 | 1:60000 | negative |
| Positive control | 1:1600000 | 1:1600000 | negative |
| Negative control | negative | negative | negative |

## REFERENCES

Scanlan, M.J., Simpson, A.J.G., and Old, L.J., 2004, The cancer/testis genes: review, standardization, and commentary. Cancer Immunity **4**:1-15.

Davis, I.D., *et al.*, 2004, Recombinant NY-ESO-1 protein with ISCOMATRIX adjuvant induces broad integrated antibody and CD4+ and CB8+ cell responses in humans. Proc. Natl. Acad. Sci. USA **101**:10697-10702.

Derouazi, M., Girard, P., Van Tilborgh, F., Iglesias, K., Muller, N., Bertschinger, M., and Wurm, F.M., 2004, Serum-free large-scale transient transfection of CHO cells. Biotechnol. Bioeng. **87**:537-545.

Muller, N., Girard, P., Hacker, D.L., Jordan, M., and Wurm, F.M., 2005, Orbital shaker technology for the cultivation of mammalian cells in suspension. Biotechnol. Bioeng. **89**:400-406.

# Effect of Different Materials used in Bioreactor Equipments on Cell Growth of Human Embryonic Kidney (HEK293) Cells Cultivated in a Protein-free Medium

Peter Aizawa[1], Göran Karlsson, Charlotte Benemar, Vera Segl, Kristina Martinelle, and Elisabeth Lindner
[1]*Biopharmaceuticals, Octapharma AB, SE-11275 Stockholm, Sweden*

**Abstract:** During process development in small-scale bioreactors, different materials used in the bioreactors were found to have a negative effect on cell growth and viability of HEK293 cells grown in protein-free medium. A variety of materials were tested including: stainless steel 316, Duran glass, ethylene propylene diene monomer (EPDM), silicone, C-flex, Viton, and low density polyethylene (LDPE). The most detrimental effect on cell growth and viability was seen with EPDM. Compared to the polystyrene control stainless steel and glass decreased cell growth while Viton, LDPE, and silicone showed the highest biocompatibility.

**Key words:** Human embryonic kidney 293 cell, biocompatibility, culture materials, protein-free medium

## 1. INTRODUCTION

Prior to inoculation (of cells) in a bioreactor, the cultivation medium is equilibrated with the desired physical conditions (e.g. pH, temperature, DO). During the equilibrium phase, the medium may come in contact with many different materials, such as glass, stainless steel, and polymers, which have a potentially detrimental effect on the medium and the cells. The aim of this study was to investigate the effect of different culture materials on cell growth and viability of HEK293 cells cultivated in protein-free medium.



*R. Smith (ed.), Cell Technology for Cell Products*, 459–462.

## 2. MATERIALS AND METHODS

The cell line used was a Human Embryonic Kidney (HEK293) cell line producing a recombinant protein. Cells were preadapted to grow in suspension in FreeStyle 293 expression medium (Invitrogen) or SF5 (a protein-free Octapharma AB proprietary medium). Samples of test materials were incubated in medium for 24 hours at 37°C in 75 $cm^2$ polystyrene (PS) T-flasks. Medium that was incubated in empty PS T-flasks was used as the control. Material-conditioned medium was transferred to new T-flasks and inoculated with cells. Cultivation was thereafter performed on shaker tables at 37°C in a 5%/95% $CO_2$/air incubator. Samples for viable cell density and cell viability were taken every day and analysed by the Trypan blue exclusion method with a CEDEX (Innovatis) cell counter. The ratio between the surface area of the material and the medium volume during incubation was measured and denoted as A/V-ratio [$cm^2/cm^3$ medium]. Analysis of leachables from EPDM was performed by RP-HPLC with a Waters Alliance 2695 HPLC system, including a YMC-Pack ODS-A column.

## 3. RESULTS

Growth was inhibited and viability decreased faster with medium conditioned in stainless steel 316 and glass compared to the polystyrene control (Figure 1). Similar results were obtained when the different culture materials were incubated in SF5 medium (data not shown). Among the tested polymers, C-flex and EPDM decreased cell growth and viability, while silicone, Viton, and LDPE performed similar to the polystyrene control (Figure 2). The A/V-ratio was kept at 0.3 [$cm^2/cm^3$ medium] for all polymers except for the LDPE bag, which, for practical reasons, required an A/V-ratio of 2.1 [$cm^2/cm^3$ medium]. The detrimental effect of EPDM on cell growth was clearly seen when the A/V-ratio was just 0.05 [$cm^2/cm^3$ medium] (Figure 3). With an A/V-ratio of 0.4 [$cm^2/cm^3$ medium] almost all cells were dead within 24 hours. Several leachables, possibly detrimental to cells, are seen in the chromatogram (Figure 4) from water conditioned in EPDM.

Effect of Different Materials used in Bioreactor Equipments on Cell Growth of Human Embryonic Kidney (HEK293) Cells Cultivated in a Protein-free Medium

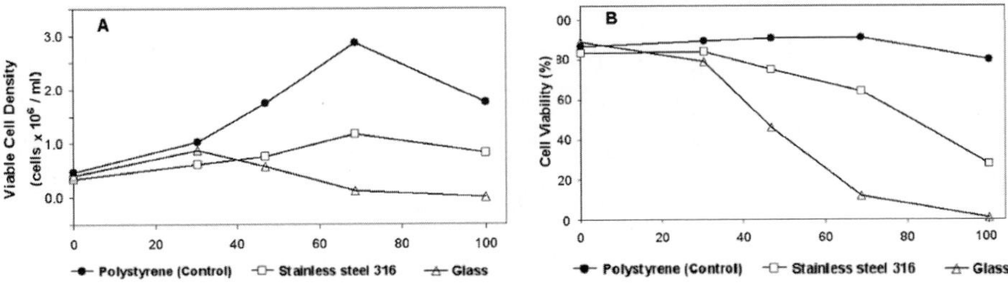


*Figure 1.* Cell growth (A) and cell viability (B) profiles. Stainless steel 316 and glass incubated in FreeStyle 293 expression medium.

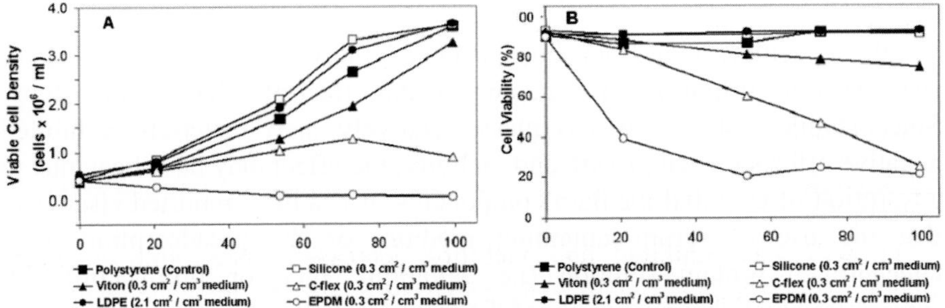


*Figure 2.* Cell growth (A) and cell viability (B) profiles. Different polymers incubated in FreeStyle 293 expression medium.

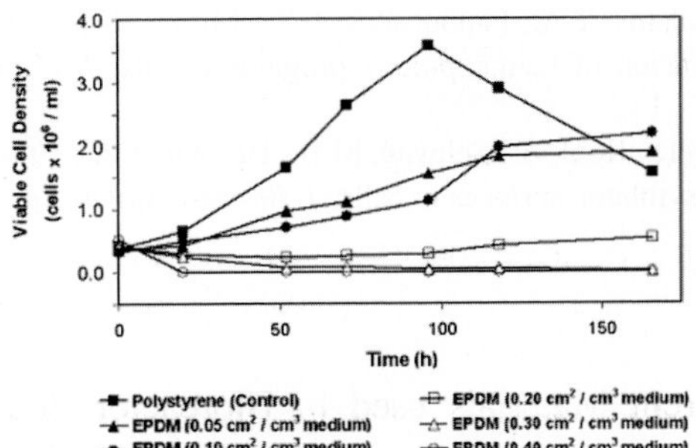


*Figure 3.* Cell growth profiles. EPDM incubated in FreeStyle 293 expression medium. Different ratios between the material surface area and the medium volume (A/V-ratio) were examined.

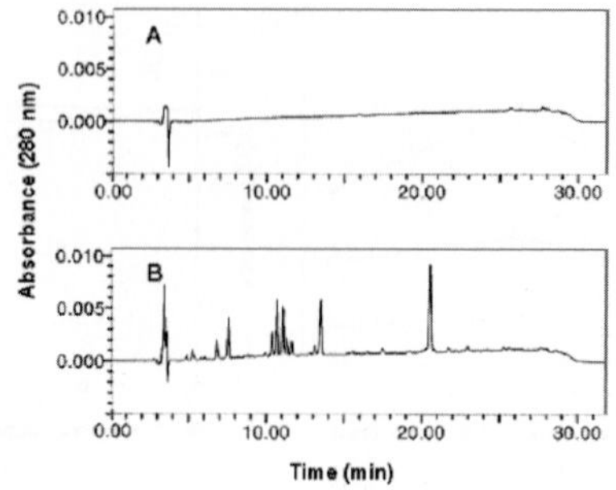


*Figure 4.* RP-HPLC analysis of leachables. Water control (A), an EPDM O-ring (B), incubated in water for 24 hours at 37°C.

# 4. DISCUSSION

We speculate that leachables, such as those seen in the RP-HPLC analysis, were responsible for the detrimental effect of EPDM in this study. Since stainless steel 316 and glass, relatively inert materials, exhibited negative effects on cell growth and viability, the effect may have been due to adsorption of essential medium components. It has been reported elsewhere that the use of serum-containing medium or the preadsorption of a concentrated protein solution (i.e., albumin, IgG, or fibronectin) can sometimes block the toxic effect of the materials.[1, 2]

# REFERENCES

1. LaIuppa, J.A., McAdams, T.A., Papoutsakis, E.T., Miller, W.M., 1997, Culture materials affect ex vivo expansion of hematopoietic progenitor cells, *J. Biomed. Mater. Res.,* **36**, 347-359
2. Ertel, S.I., Ratner, B.D., Kaul, A., Schway, M.B., Horbett, T.A., 1994, In vitro study of the intrinsic toxicity of synthetic surfaces to cells, *J. Biomed. Mater. Res.,* **28**, 667-675

# Antibody Production by GS-CHO Cell Lines Over Extended Culture Periods

A.J. Porter*, A.J. Dickson**, L.M. Barnes** and A.J. Racher*

** Lonza Biologics plc, Cell Culture Process Development, 228 Bath Road, Slough, Berkshire, SL1 4DX, United Kingdom. ** The Michael Smith Building, Faculty of Life Sciences, The University of Manchester, Oxford Road, Manchester, M13 9PT, United Kingdom.*

**Abstract:** When selecting a cell line for the production of a recombinant antibody it is important to assess the suitability of a cell line for the manufacturing process. Any assessment will include examining how the productivity of the cell line changes with increasing population doublings, as this is linked to process robustness and economics. The hypothesis that loss of productivity within antibody producing cell lines is due only to the outgrowth of a low-producer sub-population, with a higher specific growth rate than the high-producer sub-population, was tested.

**Key words:** GS, GS-CHO, stability, suitability, unstable, model, modelled behaviour, mathematical model, sub-population, low-producer, high-producer, non-producer, mixed population, specific growth rate, productivity, productivity kinetics, productivity characteristics, loss of productivity

## 1. INTRODUCTION

Observation of a large decrease in productivity with increasing population doublings (PD) may mean that a cell line is considered economically unsuitable for production of the recombinant antibody.

A possible explanation for loss of productivity with increasing PD is that the decrease in productivity is the result of the outgrowth of a non-producer (NP) or low-producer (LP) cell sub-population (1, 2). It has been suggested that such a sub-population will have a metabolic advantage and hence a growth advantage over a producer (P) or high-producer (HP) sub-population, allowing it to become the dominant sub-population (1, 2). As the proportion



*R. Smith (ed.), Cell Technology for Cell Products*, 463–466.

of NP/LP cells increases, their contribution to the productivity of the culture increases, resulting in a decrease in the apparent culture productivity.

A number of models describing the behaviour of cell lines containing a NP/LP cell sub-population and P/HP cell sub-population are reported in the literature (2, 3, 4, 5). However, the authors are aware of only one study (5) that tests a similar model in a commercially relevant system.

## 2. METHODS

Three mixed populations of LP and HP antibody-producing GS-CHO cells were established. Changes in parameters describing the growth and productivity kinetics with increasing PD were trended for the pure cell lines and the mixed populations. The behaviour of the mixed populations was compared against (i) the behaviour of the pure cell lines and (ii) a model developed to predict the behaviour of such a mixed population.

The growth of each population was modelled by the logistic growth equation (a) where t is time, X is the viable cell concentration, μ is the specific growth rate, $X_0$ the initial viable cell concentration and max $X_V$ the maximum viable cell concentration. Given the initial proportions of the two pure cell lines in a population, the equation models the cell concentration of the two sub-populations.

$$\text{(a)} \quad X = \mu\, t\, X_0 \left(1 - \frac{X}{\max X_v}\right)$$

## 3. RESULTS

All of the population mixes underwent a significant decrease in product concentration at harvest and specific production rate ($Q_P$). Compared to the pure cell lines, changes observed for the mixed populations were greater and faster (Figure 1).

The model predicts that a faster growing sub-population with a lower $Q_P$ will dominate the population (data not shown). If the model is a good representation of the behaviour in the mixed populations, there should be no substantial difference between the model and the observed data (i.e. linear regression analysis should return a slope of 1 and an intercept of 0), if the LP population has become the dominant population. Values for both the slope and the intercept of the observed data plotted against the model data (Figure 2) were compared with the values 1 and 0 respectively using the regression tool in Minitab® release 14 (Minitab Inc).

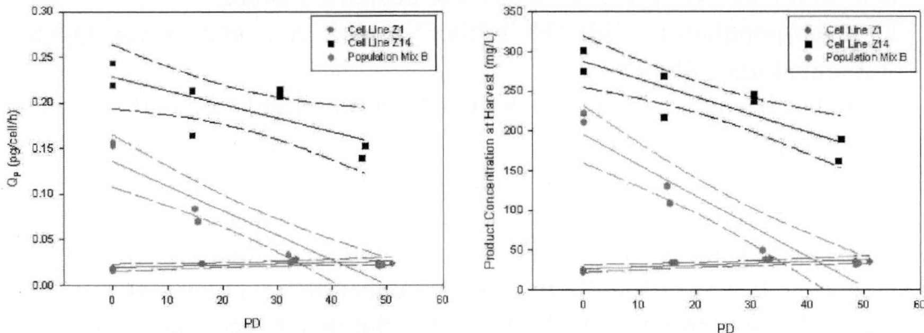


*Figure 1.* Example of the variation in kinetic parameters with increasing PD observed for two pure cell lines and the population mix established from them. Solid line is the regression line; dashed line represents 95% confidence intervals of the regression line.

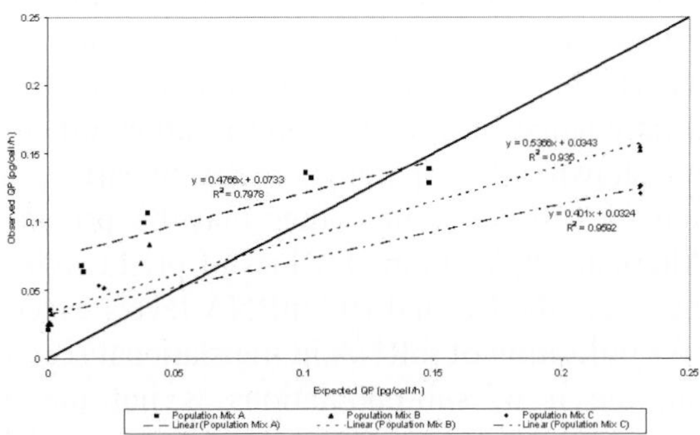


*Figure 2.* Plots comparing the specific production rate obtained from the batch culture assessments with predicted values from the model.

For all population mixes, a significance probability of 0.00 was calculated upon testing the hypotheses that the value of the intercept was zero and that the value of the slope was 1. Therefore, there was very strong evidence to reject both hypotheses. Consequently, the data do not support the initial hypothesis that loss of productivity is entirely due to the outgrowth of a faster growing LP sub-population.

## 4. DISCUSSION

The literature describing the behaviour of a cell line containing a NP/LP sub-population is contradictory. Several reports suggest that the NP/LP sub-population will become the dominant sub-population (1, 2), others report that they do not always become the dominant sub-population (6, 7).

A decrease in product concentration at harvest and $Q_P$ was seen for all population mixes. Possible reasons for the decrease include:

- The sub-population with the higher growth rate and lower $Q_P$ has dominated the culture.
- There has been a loss in productivity of the sub-population with the lower growth rate and higher $Q_P$.

Greater and faster changes were seen for mixed populations when compared to pure cell lines. For two population mixes, product concentration at harvest and $Q_P$ level out at a similar value to that of the pure LP cell line. This suggests that the decrease in product concentration at harvest and $Q_P$ is due to the LP sub-population becoming the dominant sub-population rather than a loss of productivity of the HP sub-population.

Differences between the observed data and the model indicate that the faster growing sub-population with the lower $Q_P$ did not dominate the population mix. Differences between the model and the observed data may be due to assumptions made during development of the model being incorrect. Minor fluctuations in $\mu$ and max Xv, parameters which the model assumed to be constant, were still observed for the pure cell lines – even when no significant changes in product concentration at harvest or $Q_P$ were observed (data not shown). Other parameters that impart a growth advantage to one population of cells over the other may be present, e.g., nutrient utilisation (8). Alternate explanations for loss of productivity may also be a contributing factor, e.g., the LC and HC mRNA levels exceeding a putative saturation point for utilisation of mRNA in translational/secretory events (9).

In conclusion, the $\mu$ of sub-populations is not the only parameter responsible for loss of productivity. A more complex model is required to describe the behaviour of mixed populations of LP and HP cells.

# REFERENCES

1. Frame, K.K. and Hu, W.S., 1990, Biotechnol. Bioeng. 35:469-476.
2. Kromenaker, S.J. and Srienc, F., 1994, Biotechnol. Prog. 10:299-307.
3. Frame, K.K. and Hu, W.S., 1991, Enzyme Microb. Technol. 13:690-696.
4. Lee, G.M., Varma, A. and Palsson, B.O., 1991, Biotechnol. Prog. 7:72-75.
5. Zeng, A.P., 1996, Biotechnol. Bioeng. 50:238-247.
6. Chuck, A.S. and Palsson, B.O., 1992, Biotechnol. Bioeng. 39:354-360.
7. Gardner, J.S., Chiu, A.L.H., Maki, N.E. and Harris, J.F., 1985, J. Immunol. Methods 85:335-346.
8. Kearns, B., Lindsay, D., Manahan, M., McDowall, J. and Rendeiro, D, 2003, BioProcess J. 2:52-57.
9. Barnes, L.M., Bentley, C.M. and Dickson, A.J., 2004, Biotechnol. Bioeng. 85:115-121.

# Growth Promoting Effect of a Hydrophilic Fraction of the Protein Hydrolysate Primatone on Hybridoma Cells

Michael Schomberg[1], Christian Hakemeyer[1], Heino Büntemeyer[1], Lars Stiens[2] and Jürgen Lehmann[1]

*1 Institute of Cell Culture Technology, University of Bielefeld, P.O. Box 100131, 33501 Bielefeld, Germany; 2 Roche Diagnostics GmbH, Nonnenwald 2, 82377 Penzberg, Germany*

**Abstract:** Animal derived protein hydrolysates are commonly used as a complex supplement for cell culture media because of their well known growth promoting properties. In an approach to find components being responsible for this effect we fractionated a protein hydrolysate by solid phase extraction (SPE) on a $C_{18}$-matrix. After lyophilization and further preparation the fractions were tested for their proliferation enhancing ability with a hybridoma cell line in a serum-free DMEM/F12 medium supplemented with human serum albumin and transferrin. We could demonstrate a growth promoting effect caused by a rather hydrophilic fraction. In further investigations preparative High Performance Liquid Chromatography (HPLC) was used to produce growth promoting fractions of less complexity for cell evaluation and analytical HPLC.

**Key words:** Primatone, protein hydrolysate, hybridoma, HPLC, solid phase extraction

## 1. INTRODUCTION

The use of protein hydrolysates is a popular and economical option to enhance cell growth and productivity, especially at large scale cultivation under serum and protein free conditions. Regardless of either plant or animal origin, the latter afflicted with heavy doubts about their biological safety, these hydrolysates are undefined mixtures consisting mainly of amino acids, peptides, other small organic substances and salts. Since we found a strong growth promoting effect of the protein hydrolysate Primatone HS/UF on MF20 hybridoma cells cultivated under insulin free conditions (unpublished



*R. Smith (ed.), Cell Technology for Cell Products*, 467–472.

data), an attempt to elucidate the responsible substances for this effect was made. Primatone HS/UF was fractionated by SPE or HPLC and the fractions were evaluated for their growth promoting property in cell culture under serum and insulin free conditions.

# 2. MATERIALS AND METHODS

## 2.1 Primatone HS/UF

For medium supplementation and SPE or HPLC fractionation a stock solution (20% w/v) of Primatone HS/UF (Quest Intl., Naarden, The Netherlands) was prepared.

## 2.2 Cell Culture

The mouse-mouse hybridoma cell line MF20 (Developmental Studies Hybridoma Bank, University of Iowa) was cultivated in serum-free DMEM/F12 (1:1) with enhanced concentrations of amino acids and pyruvate. The medium was supplemented with human serum albumin and transferrin. For pre-culture, the medium also contained insulin. Parallel cultivations were performed in 250 mL spinner flasks with 60 mL working volume stirred at 50 rpm in a $CO_2$-incubator (37°C, 5% $CO_2$). The cells were supplemented with different SPE or HPLC fractions of Primatone or with Primatone as positive control. Estimation of cell densities was carried out by automatically trypan blue staining with the CEDEX system (Innovatis AG, Bielefeld, Germany).

## 2.3 Fractionation by Solid Phase Extraction

10 mL Primatone were loaded on a $C_{18}$-SPE-cartridge and washed with 12,5 mL water six times followed by stepwise elution with 35 mL of acetonitrile / water mixtures of increasing acetonitrile concentration (10%, 30%, 50% and 95%, respectively). Each washing and elution step was collected as fraction, lyophilized, resuspended in 10 mL medium and sterile filtered for cell evaluation.

### 2.4 Fractionation by preparative HPLC

Preparative HPLC was performed on a Gilson-Abimed HPLC-system. 10 mL Primatone were loaded on a $C_{18}$-column and eluted by a water / acetonitrile gradient. Fractions were collected as indicated in the results, lyophilized, resuspended in 10 mL of medium and sterile filtered for cell evaluation.

### 2.5 Analytical HPLC of Primatone Fractions

Analytical HPLC was carried out using an automated Kontron RP-HPLC-system equipped with a $C_{12}$-column. Analytes were detected by UV at 214 nm.

## 3. RESULTS

### 3.1 Cell Culture Test of $C_{18}$-SPE Fractions

The SPE washing step was collected in six fractions. The impact on proliferation was evaluated for each fraction in a parallel cultivation experiment (Fig. 1). For positive and negative control cells were cultivated in medium supplemented with Primatone (0,8% w/v) and in pure medium, respectively. The first fraction collected in the washing step (designated "Washing Step 1") led to an enhanced proliferation, while other fractions did not influence the cell growth and performed like the negative control.

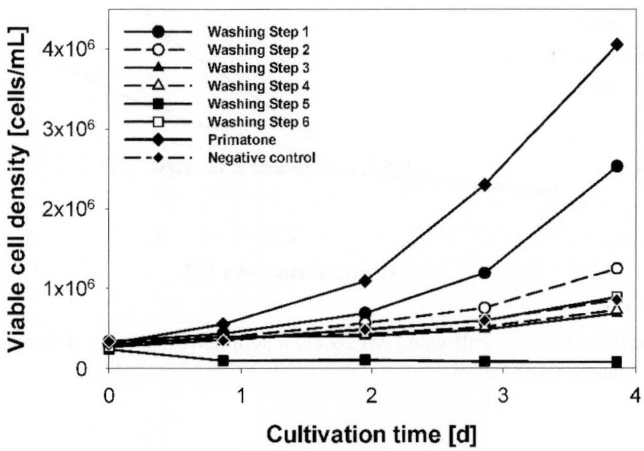


*Figure 1.* Growth of parallel cultivated MF20 cells supplemented with different C18-SPE fractions of Primatone (0,8% w/v).

The more hydrophobic elution fractions of the $C_{18}$-SPE were evaluated in a separate parallel cultivation. None of the fractions showed a better growth performance than the negative control (data not shown).

### 3.2 Cell Culture Test of $C_{18}$-HPLC Fractions

For the first preparative HPLC experiment a rough fractionation strategy was chosen, i.e. the elution volume from 0 to 800 mL was collected in four fractions of 200 mL, which were tested in cell culture (Fig. 2) against Primatone (0,5% w/v) and pure medium. Comparable to the results obtained with the SPE fractions, the fraction eluting in the first 200 mL showed the best growth promoting property, while more hydrophobic fractions collected later had no significant influence on cell growth.

In a subsequent experiment, the initial eluting 50 to 200 mL were collected in three subfractions and evaluated in cell culture. Here, the growth promoting property could be attributed to substances eluting between 100 and 150 mL (data not shown).The proliferation enhancing fraction from 100-150 mL was further subdivided into two smaller fractions (100-125 mL and 125-150 mL). The growth promoting effect could be assigned to the fraction eluting between 125 an 150 mL (Fig. 3).

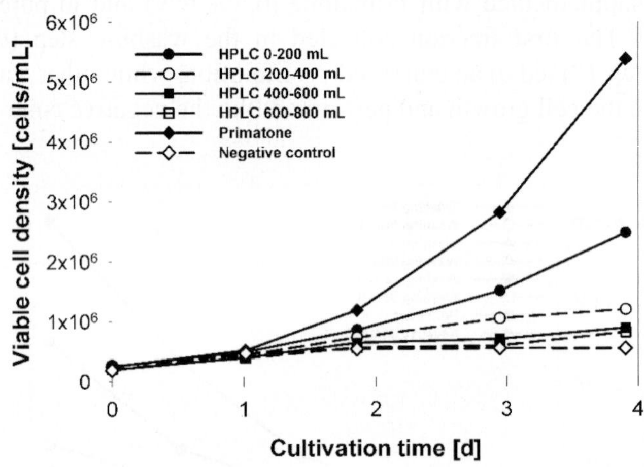


*Figure 2.* Growth of parallel cultivated MF20 cells supplemented with different C18-HPLC fractions of Primatone (0,5% w/v).

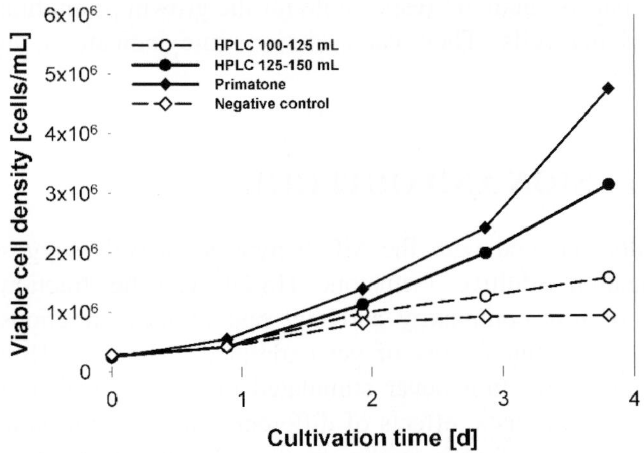


*Figure 3.* Growth of parallel cultivated MF20 cells supplemented with different C18-HPLC fractions of Primatone (1,0% w/v).

### 3.3 Analytical HPLC of Primatone Fractions

As demonstrated in Figures 1 and 4, respectively, the SPE fraction "Washing step 1" and the HPLC fraction "125-150 mL" both showed proliferation enhancing property. HPLC analysis demonstrated that all substances of the proliferation enhancing preparative HPLC fraction "125-150 mL" elute within the first 15 minutes (Fig. 4). While the whole

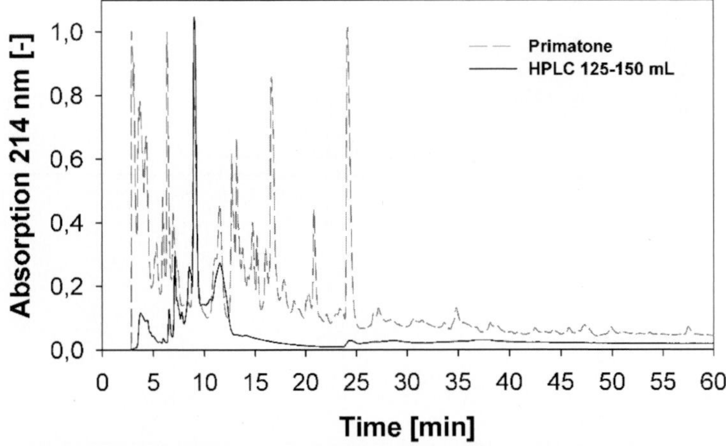


*Figure 4.* Chromatogram of the "125-150 mL" fraction prepared by HPLC (8% w/v). The chromatogram of Primatone (2% w/v) is plotted for reference.

hydrolysate contains a large mixture of different substances, we showed that only a small part of them are responsible for the growth promoting effect on MF20 hybridoma cells. Their early elution time indicates a hydrophilic character.

## 4. CONCLUSION AND OUTLOOK

The results obtained with the MF20 hybridoma cells suggest that the complex medium additive Primatone HS/UF can be fractionated into subfractions of less complexity and that one of these fractions contains proliferation enhancing factors of yet undefined character. However, the growth promoting fraction never stimulated growth as well as the whole hydrolysate, so synergistic effects of different beneficial substances can be assumed. Preparation of even smaller fractions by improving the separation methods and use of gas chromatography combined with mass spectrometry as analytical tool may allow to identify single substances being responsible for proliferation enhancing effects of protein hydrolysates.

# Effect of the Scaffold in Maintenance of Suitable Chondrcyte Morphology in Cartilage Regeneration

J. Rodríguez[1], X. Rubiralta[1], J. Farré[1], P.J. Fabregues[2], J.M. Aguilera[3] , J. Barrachina[3], F. Granell[3] , J. Nebot[2], J.J. Cairó[1], and F. Gòdia[1]
*[1]Departament d'Enginyeria Química. ETSE. Campus Universitat Autònoma de Barcelona, 08193, Bellaterra, Barcelona, Spain, [2]Unitat d'Anatomia i Embriologia. Facultat de Medicina. Campus Universitat Autònoma de Barcelona, 08193, Bellaterra, Barcelona, Spain, [3] Hospital Asepeyo Sant Cugat. Avda. Alcalde Barnils s/n. 08190, Sant Cugat del Vallés, Barcelona, Spain*

**Abstract:** Using chondrocytes in culture as starting cells, cartilage regeneration and its architectural profile achieved by means of induced extracellular matrix autoregeneration and the use of alginate, collagen and fibrinogen as polimerizable scaffolds is studied.

Alginate allowed to maintain a round cell shape, symilar to that of chondrocytes in cartilage, whereas the autoinduction of the extracellular matrix gave a fibroblastic morphology and the fibrinogen based scaffold yields a neuron-like morphology.

**Key words:** Articular cartilage; Tissue engineering; Biomaterials; Scaffolds

## 1. INTRODUCTION

The repair of defects in cartilage resulting from traumatic or degenerative processes can be envisaged by means of tissue engineering. As it has been previously described (Griffith, 2002), the tissue regeneration process uses specific combinations of cells, scaffolds and cell signalling factors, both chemical (such as growth factors and hormones) and physical (such as mechanic forces and electrical stimulation). There is a great dependence on the three-dimensional structure used as scaffold, and the functional and morphological characteristics of the resultant cell type.



*R. Smith (ed.), Cell Technology for Cell Products*, 473–477.

In the presented work, arthroscopically extracted chondrocytes were used as cell source. *In vitro* cultures were performed with the aim of studing the cartilage regeneration and the architectural profile achieved by means diferent strategies as induced extracellular matrix autoregeneration and also the use of alginate, fibrinogen and collagen as polymerizable scaffolds, in order to generate a healthy tissue for autologous transplant (Figure 1)

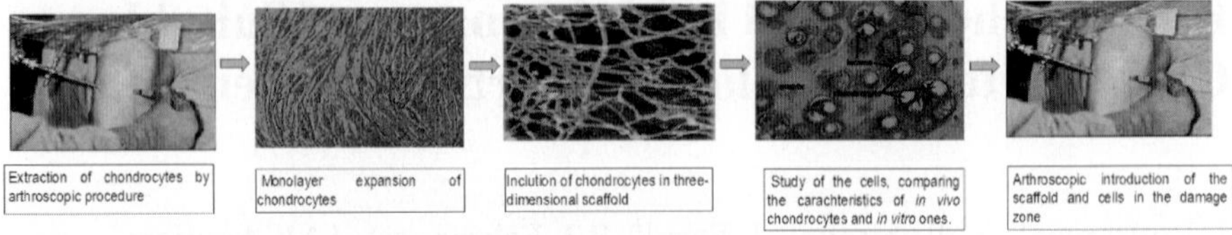
Extraction of chondrocytes by arthroscopic procedure
Monolayer expansion of chondrocytes
Inclution of chondrocytes in three-dimensional scaffold
Study of the cells, comparing the carachteristics of *in vivo* chondrocytes and *in vitro* ones.
Arthroscopic introduction of the scaffold and cells in the damage zone

*Figure 1.* Main goals of the study in order to generate a healthy tissue for autologous transplant.

## 2. MATERIALS AND METHODS

*Cell source and culture conditions.* Chondrocytes were derived from human healthy cartilage. Cartilage harvest was performed under aseptic conditions by arthroscopic procedure. Cells were isolated by digestion with type II collagenase (Roche) and pronase (Roche). They were cultured in Dulbecco's Modified Eagle Medium suplemented with 10% FCS (DMEM; Gibco) in T-75 culture flasks under an atmosphere of 5% $CO_2$ at 37ºC. They were passaged at least five times previous inclusion or culture onto supports.

*Scaffold preparation.* Auntoinduced laminar scaffold: Obtained by centrifugation of $1x10^5$ cell/ml at 500xg for 5 minutes in a 50 mL semicircular shaped tube. Fibrinogen scaffold: Equal volums of *ersta* and *once* (Tissucol duo; Baxter) were mixed with cell suspension to final concentration of $1x10^5$ cell/ml. Alginate scaffold: A solution of alginate (133mg/ml) was mixed with 0,1M $CaCl_2$ and cell suspension to final concentration of $1x10^5$ cell/ml. Collagen scaffold: A solution of rat tail collagen type I (4mg/ml Acetic acid 0,1%)(Sigma) was mixed with cell suspension to final concentration of $1x10^5$ cell/ml.

## 3. RESULTS AND DISCUSSION

The approach used for cartilage tissue regeneration was the inclusion of cultured cells into 3D scaffolds or the generation of an autoinduced matrix. In order to evaluate the performance of the proposed strategies for the obtention of funtional-like tissues, the morphological changes were studied. The obtained morphologies were compared to that observed in healthy cartilage tissues (Figure 3f) and approach is discused.

As starting point we studied cell growth, glucose consumption and lactate production of chondrocytes expanded in monolayer culture (Figure 2).

As we can see in the Figure 3a, chondrocytes expanded under this conditions, present a fibroblastic morphology. Mid-exponential phase monolayer cultures were posteriorly confined in 3D scaffolds or used for autoinduced matrix regeneration. From all studied cases, only alginate included chondrocytes (Figure 3c) pesented a morphological structure similar to the healthy hialine cartilage, with spheroid cellular shape. For the rest of the cases (Figures 3b, 3d, 3e), cultures in fibrinogen, collagen or autoinduced matrix, a fibroblastic morphology was observed, as in monolayer cultured chondrocytes.

Interestingly, in the fibrinogen scaffold it has observed how the cells evolved to a morphology resembling adipose cells (Figure 4), that rises the hypothesis of a probable dediferentiation of chondrocytes to adipose cells and vice versa

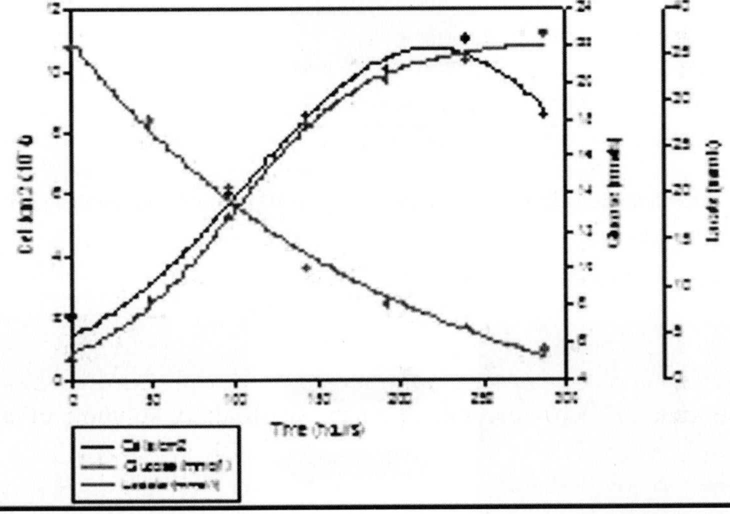

*Figure 2.* Cell growth, glucose consumption and lactate production of chondrocytes expanded in monolayer culture.

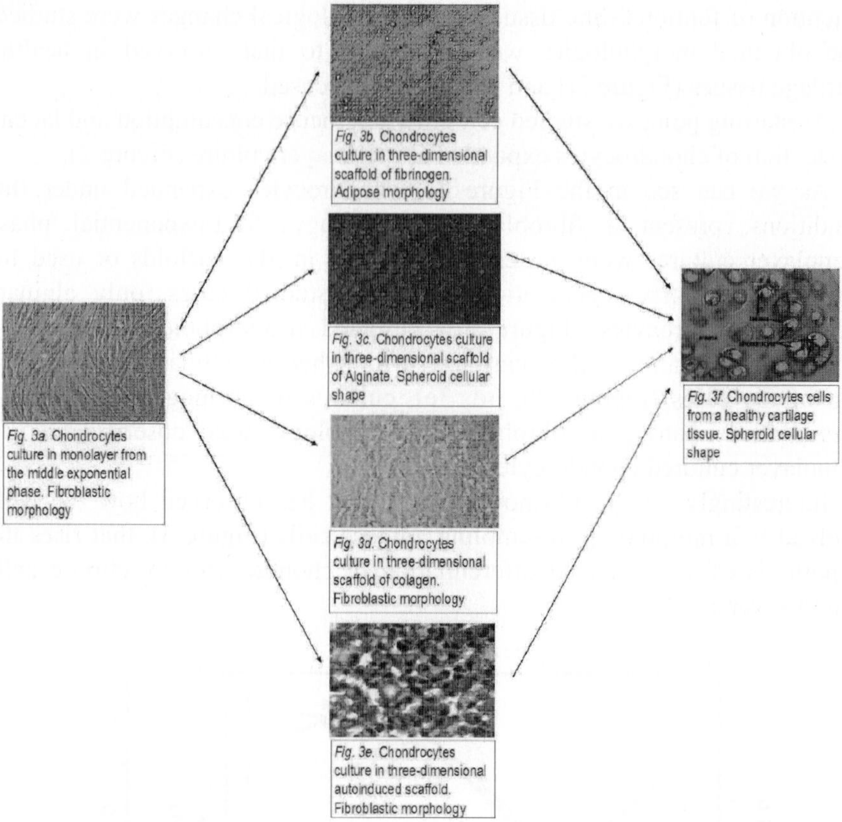

Fig. 3a. Chondrocytes culture in monolayer from the middle exponential phase. Fibroblastic morphology
Fig. 3b. Chondrocytes culture in three-dimensional scaffold of fibrinogen. Adipose morphology
Fig. 3c. Chondrocytes culture in three-dimensional scaffold of Alginate. Spheroid cellular shape
Fig. 3d. Chondrocytes culture in three-dimensional scaffold of colagen. Fibroblastic morphology
Fig. 3e. Chondrocytes culture in three-dimensional autoinduced scaffold. Fibroblastic morphology
Fig. 3f. Chondrocytes cells from a healthy cartilage tissue. Spheroid cellular shape

*Figure 3.* Morphological changes of chondrocytes into 3D scaffolds and autoinduced matrix.

Chondrocyte cells
Adipose cells

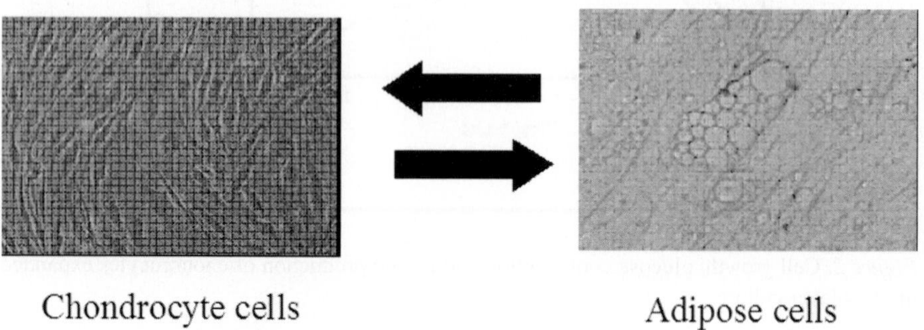

*Figure 4.* Probable deddiferentiation of chondrocytes to adipose cells and vice versa.

## 4. CONCLUTIONS

A great dependence has been observed between the chondrocyte morphology and the type of scaffold used to promote three-dimensional structure.

The case where the obtained three-dimensional structure is more similar to that found in vivo is for chondrocyte inclusion in alginate polymers.

Possible differentiation of chondrocytes to adipose cells will be further studied, since this could open new strategies for cartilage regeneration, with adipose tissue as cell source, enabling less invasive treatments.

## REFERENCES

Mitsuo Ochi, Nobuo Adachi, Hiroo Nobuto, Shinobu Yanada, Yohei Ito, and Muhammad Agung. 2004. Articular. Artificial Organs., **28**(1):28-32

X. Fei, B.-K. Tan, S-T. Lee, C.-L. Foo, D.-F. Sun, and S.-E. Aw. 2000. Effect. Transplantation Proceedings, **32**: 210-217.

Willem J.C.M. Marijnissen, Gerjo J.V.M. van Osch, Joachim Aigner, Henriette L. Verwoerd-Verhoef, Jan A.N. Verhaar. 2000. Tissue-engineered. Biomaterials, **21**: 571-580.

# Metabolic Changes Occuring During Adaptation to Serum Free Media

Sam Denby[1], Frank Baganz[1], David Mousdale[2]
*1The Advanced Centre for Biochemical Engineering, University College London, Torrington Place, London WC1E 7JE, UK 2Beocarta,Room 555, Dept. Of Bioscience,Royal College Building,204 George Street,Glasgow,G1 1XW*

**Abstract:** This work aims to investigate the changes occurring in metabolism of a cell population during adaptation of a cell line from serum containing to serum free media. In order to enable growth in a serum free media a supplement was added to DME:F12. Cells grown in 1.25% serum media were used to investigate the effect of the supplement, data for growth, antibody production and key metabolites are presented All data presented here are derived from cells banked at different serum concentrations during the adaptation and subsequently grown in 50mL culture in 250mL shaken ehrlenmyer flasks.

**Key words:** Hybridoma, Metabolism, Serum, Adaptation, Shake Flask

## 1. INTRODUCTION

Recent trends have moved mammalian cell culture away from traditional serum containing media towards the use of more defined media. In addition there are increasing cost pressures on the biochemical engineering insustry as cost of goods becomes a larger consideration. The adaptation to defined, serum and hydrolysate free medium has several benefits, in this instance it simplifies modeling of intracellular fluxes during the course of the culture. Metabolic modeling is being increasingly applied to media development and allows a rational approach to media design.

The cell line used for this work was VPM8 a murine murine hybridoma developed as part of a veterinary PhD project. The cell line in question did not adapt directly into a commercially available serum free medium, DMEF12 or supplemented DMEF12 medium. Over the course of multiple



*R. Smith (ed.), Cell Technology for Cell Products*, 479–482.

passages the cells became able to proliferate in DMEF12 medium supplemented with amino, acids, vitamins, lipids, insulin, transferring and other defined compounds. The aim of this ongoing work is to investigate what occurs during these passages and why they are necessary.

## 2. MATERIALS AND METHODS

All of the data presented here is gathered from 50mL working volume cultures grown in 250ml disposable ehrlenmyer flasks (corning). Cells were grown in DMEF12 media supplemented with serum or an in house chemically defined supplement. All components of the media, including the supplement, were supplied by Sigma. The data presented here is from cells banked at each stage of the adaptation process, revived and used to seed cultures which were grown simultaneously on a shaker (IKA) set to 120rpm in a $CO_2$ incubator (RS biosciences) maintained at 37 °C and 5% $CO_2$. Media was prepared immediately prior to the experiment from the same batch of non-supplemented media.

Viable cell counts were performed using a CASY TTC cell counter. Antibody quantitation was performed via an in house sandwich ELISA developed suring the project. Glutamine analysis was performed by revese phase HPLC along with other amino acid analysis (data not shown here). Lactate, and glucose analysis was carried out using a YSI select 2700 analyser, ammonia was quantified using the indophenol blue assay.

## 3. RESULTS/DISCUSSION

Figures 1 and 2 show that whilst there are significant differences in the viable cell concentrations achieved during the cultures, particularly between the serum containing and serum free conditions. In addition it shows that there is a real but small difference in the final antibody titre (mg/L) between cultures, ranging from 45 to 80 mg/L.

Figures 3,4 and 5 show the effect that the supplement added to DMEF12 is having on viable cell count, antibody concentration, ammonia, lactate, glutamine and glucose concentrations during the early phases of the culture. It can be seen that whilst the supplement leads to significantly higher viable cell counts it also leads to significantly lower antibody titres. There are no striking differences in glutamine or glucose consumption or lactate and ammonia accumulation that would account for this in the early stage of the fermentation and further investigation is being carried out.

Clear differences in observable events during the course of the cultures in different conditions have been demonstrated. Further analysis of the external metabolites from these cultures is being performed and will be used to determine intracellular fluxes. A greater understanding of the physiology behind them can only improve process control and optimisation

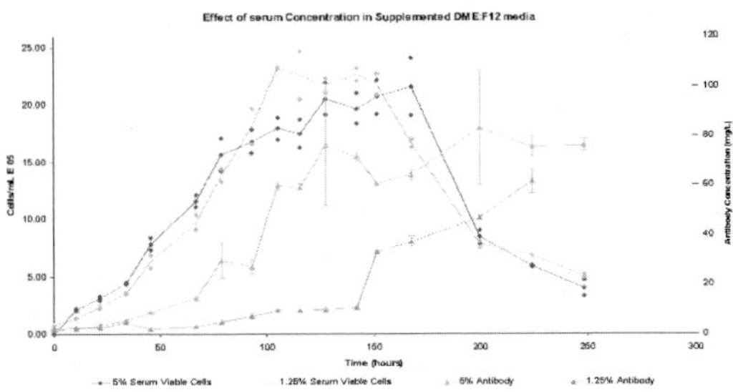


*Figure 1.* Comparison of 5% and 1.25% serum concentrations.

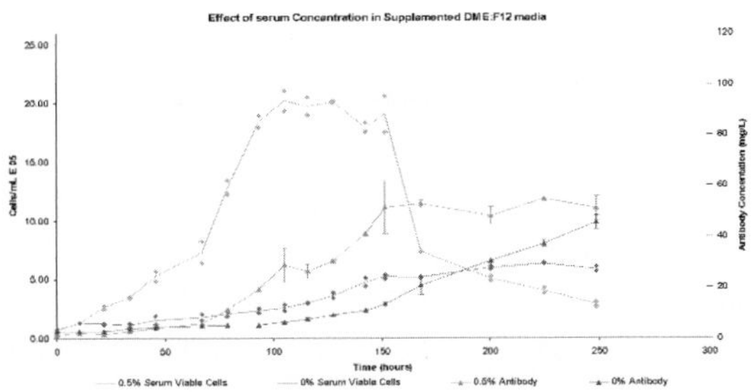


*Figure 2.* Comparison of 0.5% and 0% Serum concentrations.

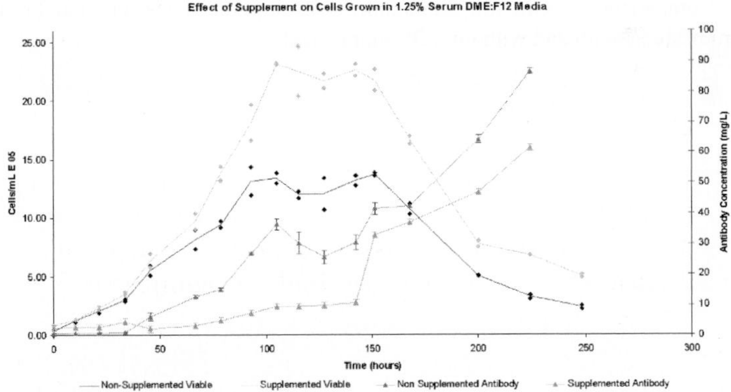


*Figure 3.* Comparison of Growth and Antibody Production in 1.25% Serum Media with and without CDS supplement.

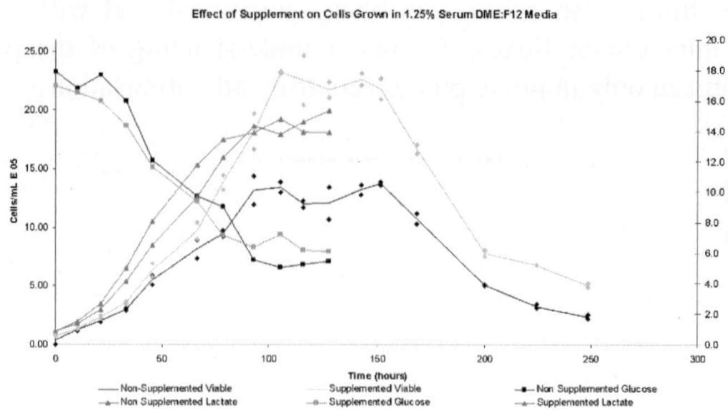


*Figure 4.* Comparison of Growth, Glucose Consumption and Lactate Production in 1.25% Serum Media with and without CDS supplement.

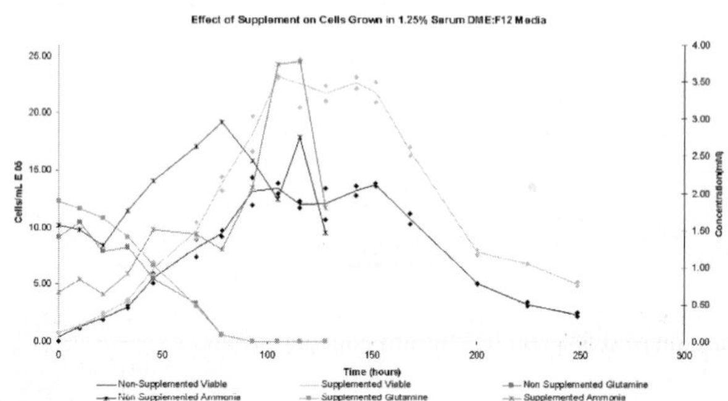


*Figure 5.* Comparison of Growth, Glutamine Consumption and Ammonia Production in 1.25% Serum Media with and without CDS supplement.

# The Expression of the *neo* Gene in GS-NS0 Cells Increases Their Proliferative Capacity

Farlan Veraitch[1,2] and Mohamed Al-Rubeai[1,3]
[1]*Department of Chemical Engineering, University of Birmingham, Birmingham B15 2TT, UK.* [2] *Present address: Biochemical Engineering, University College London, Torrington Place, London, UK.* [3]*Present address: Department of Chemical and Biochemical Engineering, University College Dublin, Belfield, Dublin 4, Ireland*

**Abstract:** Prior work has reported that cotransfecting a gene of interest with the selectable marker *neo* can seriously perturb a number of cellular processes. In this study the influence of the *neo* gene on the growth, death and metabolism of a murine myeloma NS0 cell line, expressing a chimeric antibody, was investigated. A pool of *neo* transfectants, 6A1-NEO, was selected with 500 μg/ml G418. Batch cultivation of 6A1-NEO showed that there was a 36% increase in maximum viable cell concentration, a 20% increase in the maximum apparent growth rate and a 134% increase in cumulative cell hours as compared with the parent, 6A1-(100)3. Batch cultivation of five randomly selected clones illustrated that 6A1-NEO's advantage over the parent was not due to clonal variation. Neither the use of G418 during the selection process nor the cultivation of cells in the presence of G418 were responsible. This implied that the *neo* gene product, APH(3')-II, was causing the changes in proliferative capacity. These results show that there was an increase in growth rate and proliferative capacity caused by the expression of recombinant APH(3')-II.

**Key words:** *neo*, aminoglycoside phosphotransferase, G418, NS0 myeloma cells, cell cycle, apoptosis, Bcl-2, metabolism



*R. Smith (ed.), Cell Technology for Cell Products*, 483–488.

# 1. INTRODUCTION

The *neo* gene has been commonly used as a selectable marker for the introduction of genes into mammalian cells. The aminoglycoside antibiotic G418 is used to select for cells which have successfully integrated *neo* into their chromosomal DNA (Colbere-Garapin *et al.*, 1981; Southern and Berg, 1982). However, this selection system assumes that the *neo* gene product is neutral and that G418 selection does not affect the cellular properties under investigation (Von Melchner and Housman, 1988). In eukaryotes, G418's primary function is to kill cells by binding to the 80S ribosome thereby blocking protein synthesis (Bar-Nun *et al.*, 1983). The *neo* gene codes for a bacterial aminoglycoside phosphotransferase (APH(3')-II) which inactivates G418 by phosphorylation (Davis and Smith, 1978; Shaw *et al.*, 1993).

Increasingly, immortalised mammalian cell lines such as CHO, NS0 and baby hamster kidney (BHK) are being used to produce recombinant proteins for clinical applications (Wurm, 2004). A number of attempts have been made to optimise cell growth and suppress cell death using genetic engineering techniques (reviewed in Fusseneger *et al.*, 1999; Arden and Betenbaugh, 2004). The majority of these studies have utilised the *neo* selection system for the over-expression of recombinant proteins such as Bcl-2, Bcl-$x_L$, XIAP and p21 (Itoh *et al.*, 1995; Mastrangelo *et al.*, 2000; Sauerwood *et al.*, 2002; Watanabe *et al.*, 2001). However, the authors are not aware of any studies which have compared the cell growth of a *neo* transfectant with the parent. The objective of this work was to investigate whether transfection with the *neo* gene altered key cellular processes such as proliferation, cell death, metabolism and cell cycle. The host was a NS0 cell line commonly used for the production of recombinant proteins. The effect of the *neo* transfection on the proliferative capacity of a pool of transfectants and five randomly selected clones was monitored in static, serum-supplemented batch cultures. We found that the pool of *neo* transfectants were capable of reaching a higher maximum viable cell concentration than the parent in batch cultures. In addition all five *neo* clones outgrew five randomly selected parent clones.

# 2. MATERIALS AND METHODS

## 2.1 Cell line

The parent cell line, 6A1-(100)3, was generated by transfecting a murine myeloma NS0 cell line with a GS expression vector (Lonza Biologics, Slough, UK) encoding the human–mouse chimeric antibody, cB72.3.

### 2.2 Transfection

The expression vector pEF-MC1*neo*pA encodes the *neo* gene under the control of a human thymidine kinase promoter (Cory, 1995). This vector does not contain any other the genes apart from *neo*. 6A1-(100)3 was transfected with pEF-MC1*neo*pA using lipfectamine (Gibco) to create the 6A1-NEO cell line

### 2.3 Cloning

5 random clones of 6A1-(100)3 and 6A1-NEO were derived using the limited dilution technique

## 3. RESULTS AND DISCUSSSION

### 3.1 Serum-supplemented batch cultures

6A1-(100)3 was transfected with pEF-MC1*neo*pA and selected with 500 μg/ml G418. The control cells, which had been treated with lipofectamine in the absence of any DNA, were less than 5% viable 4 days after the addition of the antibiotic. The pool of *neo* transfectants, named 6A1-NEO, was passaged until its growth was reproducible. The viable cell numbers of 6A1-(100)3 and 6A1-NEO were monitored in serum-supplemented batch cultures. 6A1-NEO grew faster than the parent and reached a higher maximum viable cell concentration (Figure 1(a)). After 144 hours there were 36% more viable 6A1-NEO cells than the parent. In addition 6A1-NEO's cumulative cell hours (CCH) at the end of the batch was $410 \pm 9 \times 10^9$ cell h/l, 134% higher than the parent's final CCH. This increase in proliferative capacity could have been caused by the expression of APH(3')-II, G418 acting as a growth enhancing supplement or the inadvertent selection of cells influenced by one or more of G418's secondary properties. 6A1-NEO was cultured for four weeks in the presence and absence of 500 μg/ml G418. Subsequent batch cultivation of 6A1-NEO with and without G418 revealed that the viable cell concentrations of both cultures were very similar (Figure 1(a)). This confirmed that G418 was not enhancing the growth of 6A1-NEO by acting as a growth additive. Furthermore, untransfected 6A1-(100)3 cells were allowed to recover from 4 days of treatment with 500 μg/ml G418. This population, 6A1-(100)3-GP1, was passaged in the absence of G418 until its viability had recovered. There were no significant differences between the growth of 6A1-(100)3-GP1 cells and the parent during batch cultivation (Figure 1(b)), indicating that the G418 selection process did not cause an increase in proliferative capacity. These results indicate that the expression of APH(3')-II was causing the increase in proliferative capacity exhibited by 6A1-NEO.

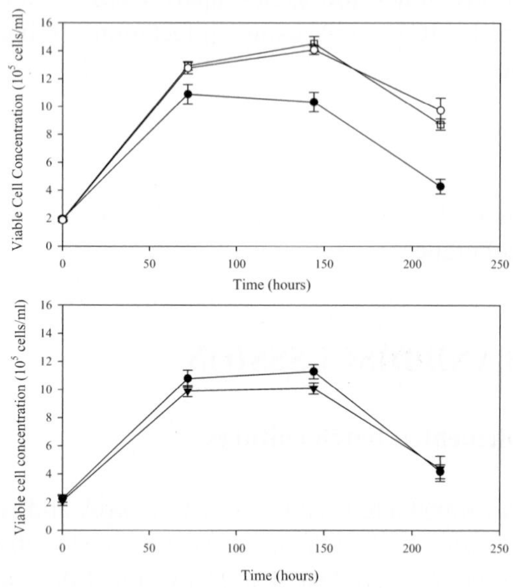


*Figure 1.* (a) shows serum-supplemented batch cultivation of 6A1-(100)3 (•), 6A1-NEO with G418 (○). 6A1-NEO was cultivated for four weeks without G418 before batch cultivation in the absence of the antibiotic (≤). (b) shows the batch cultivation of 6A1-(100)3 (•) and G418 treated parent cells, 6A1-(100)3-GP1 (▼). Viable cell concentration was measured using the trypan blue exclusion assay.

## 3.2 Clonal variation

6A1-(100)3 was transfected with the *neo* gene to create a mixed population of transfectants, 6A1-NEO. This population was cloned using the serial dilution technique and five clonal cell lines were selected at random. Control cell lines were generated by randomly selecting five 6A1-(100)3 clones. The viable cell concentration of all ten cell lines was monitored in serum-supplemented batch cultures. After 96 hours of cultivation all five 6A1-NEO clones had reached a higher viable cell concentration than any of the parent clones (Figure 2). The fastest growing clone, 6A1-NEO-4D5, reached a maximum viable cell concentration of 12.05 x $10^5$ cells/ml, whilst the fastest growing parent clone, 6A1-(100)3-1B6, only reached 8.25 x $10^5$ cells/ml. The parent clones progressed into the death phase whilst all five 6A1-NEO clones remained in the stationary phase. After 240 hours the average viable cell concentration of the *neo* clones was 8.15 x $10^5$ cells/ml, 5.6 fold higher than the average of the parent clones. Although there was some variability between the individual clones all five randomly selected clones exhibited the increase in proliferative capacity shown by the pool of transfectants, 6A1-NEO, in Figure 1(a). These results confirm that 6A1-NEO's advantage over the parent was not due to clonal variability.

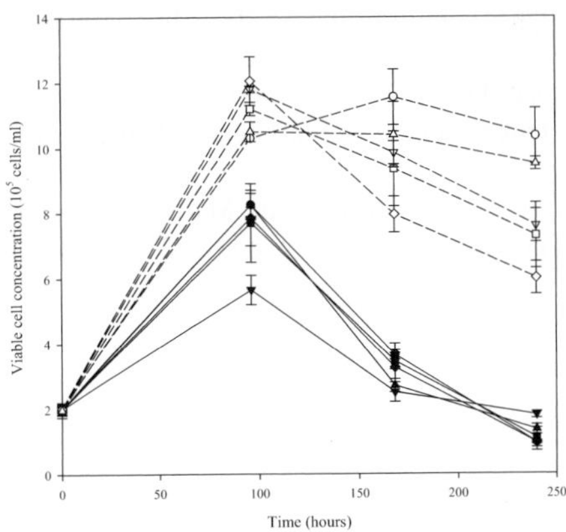


*Figure 2.* Batch cultures of five clones randomly selected from the pool of neo transfectants (dashed lines). The viable cell concentration of 6A1-NEO-1D6 (○), 6A1-NEO-2B10 (σ), 6A1-NEO-2F7 (≤), 6A1-NEO-4D5 (↓) and 6A1-NEO-4F4 (ρ) were measured using the trypan blue exclusion method. 6A1-(100)3 was cloned to create five control cell lines: 6A1-(100)3-1B6 (•), 6A1-(100)3-2C7 (▼), 6A1-(100)3-2F10 ('), 6A1-(100)3-3C7 (↲) and 6A1-(100)3-E10 (▲).

# REFERENCES


Arden N. and Betenbaugh, M.J., 2004, Life and death in mammalian cell culture: Strategies for apoptosis inhibition, Trends Biotechnol **22**:174-180

Bar-Nun, S., Shneyour, Y. and Beckmann, J.S., 1983, G-418, an elongation inhibitor of 80S ribosomes, Biochim Biophys Acta **741**:123-127

Colbere-Garapin, F., Horodniceanu, F., Kourilsky, P. and Garapin, A.C., 1981, A new dominant hybrid selective marker for higher eucaryotic cells, J Mol Biol **150**:1-14

Cory, S., 1995, Regulation of lymphocyte survival by the bcl-2 gene family, Annual Rev Immunol **13**:513-543

Davies, J. and Smith, D.I., 1978, Plasmid-Determined Resistance to Antimicrobial Agents, Ann Rev Microbiol **32**:469-508

Fussenegger, M., Bailey, J.E., Hauser, H. and Mueller, P.P., 1999, Genetic optimisation of recombinant glycoprotein production by mammalian cells, TIBTECH **17**: 35-42

Itoh, Y., Ueda, H. and Suzuki, E., 1995, Overexpression of *bcl-2*, apoptosis suppressing gene: prolonged viable culture period of hybridoma and enhanced antibody production, Biotechnol Bioeng **48**:118–122

Mastrangelo, A.J., Hardwick, J.M., Bex F. and Betenbaugh, M.J., 2000, Part I. Bcl-2 and Bcl-$x_L$ limit apoptosis upon infection with alphavirus vectors, Biotechnol Bioeng **67**:544-554

Sauerwald, T.M., Betenbaugh, M.J. and Oyler, G.A., 2002, Inhibiting apoptosis in mammalian cell culture using the caspase inhibitor XIAP and deletion mutants, Biotechnol Bioeng **77**:704-716

Southern, P.J. and Berg, P., 1982, Transformation of mammalian cells to antibiotic resistance with a bacterial gene under control of the SV40 early region promoter, J Mol Appl Genet **1**:327-341

Von Melchner, H., Housman, D.E., 1988, The expression of neomycin phosphotransferase in human promyelocytic leukemia cells (HL60) delays their differentiation, Oncogene **2**:137-140

Watanabe, S., Shuttleworth, J. and Al-Rubeai, M., 2002, Regulation of cell cycle and productivity in NS0 cells by the over-expression of $p21^{CIP1}$, Biotechnol Bioeng **7**:1–7

# 1000 Non-instrumented Bioreactors in a Week

## *Novel Disposable Technologies for Rapid Scale up of Suspension Cultures*

Matthieu Stettler, Maria De Jesus, Hajer Ouertatani-Sakouhi, Eva Maria Engelhardt, Natalie Muller, Sébastien Chenuet, Martin Bertschinger, Lucia Baldi, David Hacker, Martin Jordan and Florian M. Wurm
*École Polytechnique Fédérale de Lausanne, Faculty of Life Sciences, Integrative Bioscience Institute, Lausanne, Switzerland*

**Abstract:** Increasing interest for disposable non-instrumented bioreactors is due to the possibilities these systems offer to increase assay capacity for multi-parameter process development. A system based on disposable 50 ml shaken bioreactors is presented in this report. Up to 1000 bioreactor units were tested within a time period of one week. The experiment allowed the simultaneous analysis of 29 chemically defined medium and 20 protein hydrolysates on the growth and productivity of 3 distinct CHO cell lines. The feasibility of such a study opens clearly new perspectives in rapid and efficient cell line development.

**Key words:** small-scale, non-instrumented, process development, shaking technology, packed cell volume, relative fluorescence, disposable, CHO DG44, CHO YIgG3, CHO PPL11, peptone

## 1. INTRODUCTION

For adherent cell cultures many different types of disposable non-instrumented systems have been developed over decades to allow both high throughput (multiwell formats) and scale up (microlitres to hundreds of millilitres) in well-controlled incubators. Unfortunately, a similar strategy has not been attempted with suspension cultures. Recent developments prove that orbital shaking technology combined with a passive aeration mode can be successfully applied to animal cell cultures (Muller *et al.*, 2005 and



*R. Smith (ed.), Cell Technology for Cell Products*, 489–495.

Girard *et al.*, 2001). Using this approach, 50 ml bioreactors fitted with gas permeable caps proved to be highly efficient for cells grown in suspension. Their performance matched or exceeded that of culture systems widely used in industry (spinner flasks, instrumented bioreactors). Previous studies have partially defined the physico-chemical characteristics of these disposable 50 ml bioreactors (De Jesus *et al.*, 2004). Parameters such as optimal working volume, water evaporation, oxygen supply and release of carbon dioxide have been studied and optimized. Furthermore, the simplicity and flexibility of the system allows a large number of growth parameters to be tested simultaneously. An example of application that utilizes up to 1000 bioreactors operated by a small team is described here. The identification of optimal process parameters for cell line development thus facilitates the scale-up procedure. This is a consequence of the reliability and reproducibility of the data obtained with non-instrumented disposable cultivation systems at the scale which is presented in this study. At larger scales, such disposable bioprocess technologies are about to become increasingly important for the production of material for preclinical or even for clinical trials (Farid *et al.*, 2005).

## 2. MATERIALS AND METHODS

The experiment was set up to study four different parameters (variables): the cell line, the cell seeding density, the medium and the addition of different concentrations of protein hydrolysates (peptones). The variables are listed in Table 1. Each single bioreactor (filter tube 50 ml, TPP, Trasadingen, Switzerland) was seeded with 10 ml cell suspension and incubated for 6 days at 37°C in $CO_2$ and humidity controlled ISF-4-W shaking incubators (Adolf Kühner AG, Birsfelden, Switzerland). The shaking speed was set at 180 rpm with a 50 mm rotational diameter. The incubator with the largest capacity contained up to 468 single bioreactors on 4 individual shakers.

*Table 1.* Description of the experimental variables.

| Variable | Description |
|---|---|
| Cell line | CHO DG44, CHO YIgG3, CHO PPL11 |
| Cell seeding density | 2, 4, 5 and 8 $\times 10^5$ cells/ml |
| Medium | 29 commercial chemically defined medium from 5 suppliers |
| Peptone | 20 peptones from 4 suppliers |

At day 6, the resulting proliferation of the cells was assessed using the packed cell volume method (Mini PCV Tube, TPP, Trasadingen, Switzerland), a highly reliable and accurate biomass measurement method (Jordan, 2005). The recombinant cell line CHO YIgG3 (Hunt *et al.*, 2004) and PPL11 (Chenuet and Wurm, unpublished data) express a yellow and a green fluorescent protein (GFP), respectively. The fluorescence intensity was quantified at day 6 using a Cytofluor 4000 plate-reading fluorometer (PerSeptive Biosystems, Farmingham, MA) with an excitation wavelength of 485 nm and an emission wavelength of 530 nm. For the purpose of increasing the statistical relevance of the experiment, over 20% of the 1000 bioreactors were run with identical standard parameters. These control bioreactors were placed on different shakers. To avoid random errors, each single parameter was tested in triplicate for both packed cell volume and relative fluorescence intensity.

## 3. RESULTS AND DISCUSSION

The following results highlight only part of the data acquired during this experiment. Here, the objective is to show the potential benefit of non-instrumented bioreactors for optimizing growth conditions and for accelerated cell line development. The effect of the addition of peptones will be discussed as well.

### 3.1 The effect of media on cell growth and productivity

The effect on growth of 29 different media resulted in packed cell volume values between 0.4 and 1.4% for CHO PPL11 cells. Starting with a seeding density of $5 \times 10^5$ cells/ml, a maximal cell density of $5 \times 10^6$ cells/ml was obtained with the reference medium (code n° 1) at day 6 (Fig. 1). This medium has been developed to support high CHO cell density. In fact, only a few other media yielded similar results (code n° 5, 14 and 19). In contrast, the expression of GFP in suspension adapted CHO PPL11 cells had a different pattern (Fig. 1). At least 6 out of the 29 media provided increased fluorescence intensity compared to the reference medium (1.5-fold increase with media code n°5). The average standard deviation of the triplicates was low for both packed cell volume (average 6%) and fluorescence intensity (average 15%). Thus, in this context, the use of non-instrumented bioreactors allowed for a rapid comparison of the effect of a large number of different media on growth and recombinant protein expression.

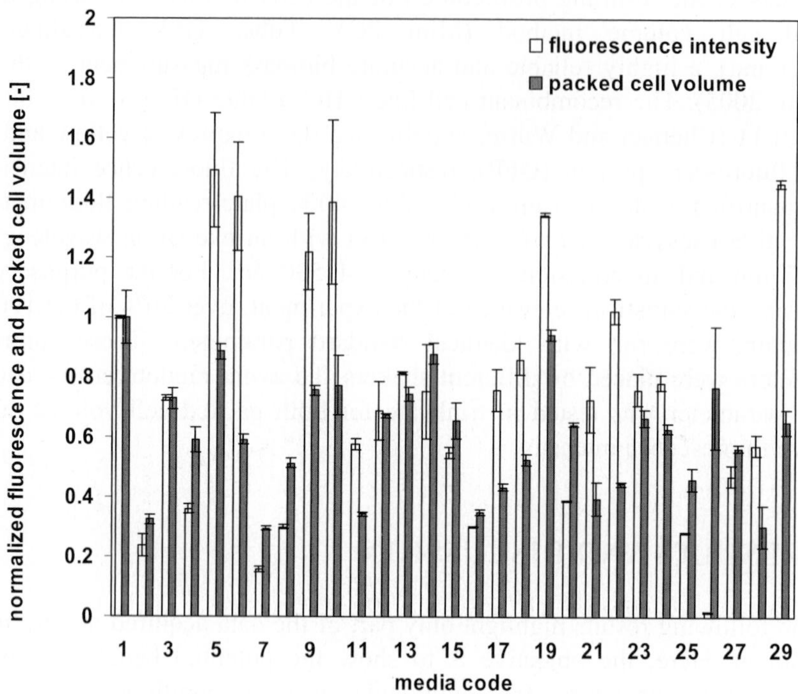


*Figure 1.* Normalized fluorescence intensity and packed cell volume on day 6 for CHO PPL11 cells grown in 29 different chemically defined media. Each data point was calculated from the average packed cell volume of 3 different bioreactors. The data were normalized to the values obtained with the reference medium (code n° 1).

## 3.2 A comparison between parental and stably transfected cell lines

The growth kinetic of a recombinant CHO cell line (CHO PPL11) was compared with the parental cell line (CHO DG44). Surprisingly, the response of both cell lines toward the different chemically defined media was frequently very different (Fig. 2). For example, medium n°15 supported excellent growth for DG44 cells but only moderate growth for the recombinant cell line. In contrast, some media supported better growth for the recombinant cells (code n° 5, 9, 13 and 24). In this experiment, the average standard deviation of the triplicates was less than 8% of the measured values for both cell lines. Again, this approach was reliable enough to detect even small differences. Furthermore, this experiment, due to the amount of data which was collected, helped to characterize a newly produced recombinant cell line and to further optimize its performance.

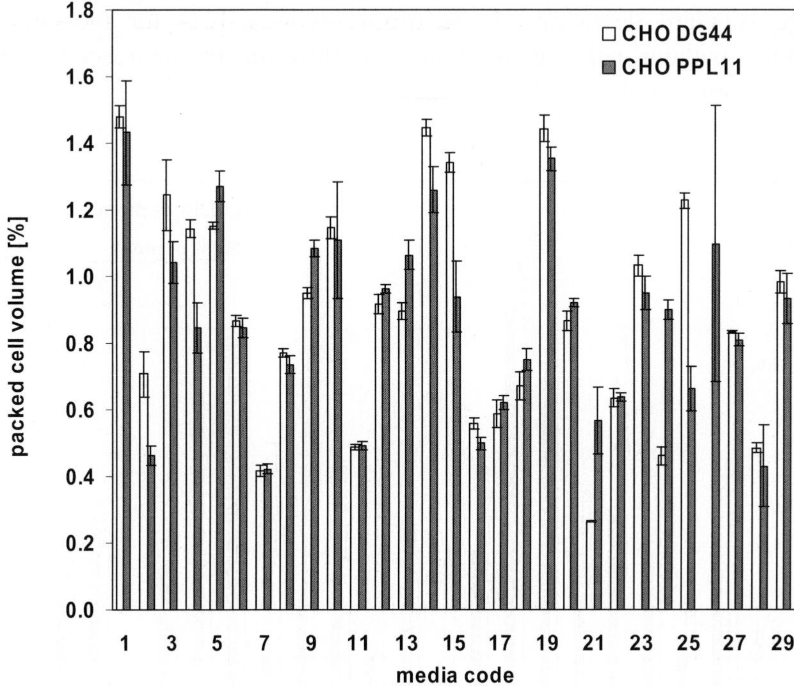


*Figure 2.* Packed cell volume on day 6 for CHO DG44 (parental cell line) and CHO PPL11 (DG44-derived recombinant cell line) grown in 29 different chemically defined media. Each data point represented the average packed cell volume of 3 different bioreactors.

## 3.3 The effect of protein hydrolysate addition on cell growth

Protein hydrolysates (peptones) were added to the growth medium at the time of seeding at concentrations of 0.5 and 1% (w/v). Hydrolysates from soy, wheat, rice, pea, casein and yeast provided by 4 different suppliers were tested. For this experiment, DG44 cells were cultivated in a chemically defined medium ensuring long-term growth, not high cell density, of CHO cells. This medium was selected because of the absence of protein hydrolysates and low total protein content. At day 6, the effect of the peptone addition on the growth of the cell population was assessed by the packed cell volume method (Fig. 3). More than a 2-fold increase of the packed cell volume was observed with 5 different peptones at a 0.5% (w/v) concentration (peptone code n° 4, 6, 7, 15 and 16). Peptones originating from soy hydrolysates outperformed the peptones from other origins. It was noticeable that some peptones had an inhibitory effect on cell division, especially at a concentration of 1% (w/v) (for example code n° 9 to 14). The average standard deviation of the triplicates was 12% for the control bioreactor without peptone addition and 16% for the bioreactors with peptone addition at 0.5 and 1% (w/w).

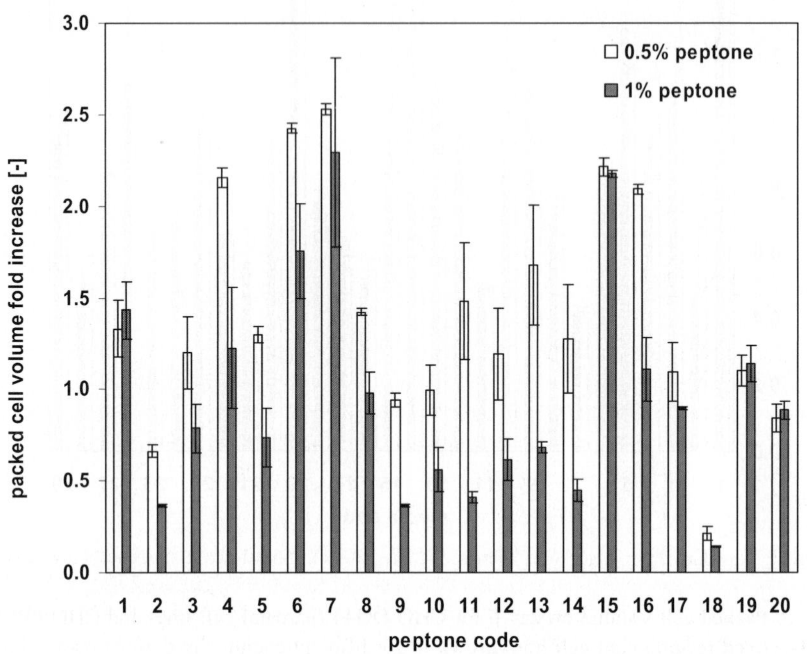


*Figure 3.* Fold increase on day 6 of the CHO DG44 packed cell volume due to the addition of peptones. The seeding density was 5 x $10^5$ cells/ml. At the time of seeding, 20 different peptones were added at concentrations of 0.5 and 1% (w/v). Each data point was calculated from the average packed cell volume of 3 different bioreactors.

# 4. CONCLUSION

The results demonstrated the powerful utilization of small-scale non-instrumented bioreactors for the simultaneous analysis of several cell growth parameters. In particular, the effect of 29 chemically defined media on the growth and productivity of a CHO recombinant cell line was rapidly tested. Also, the results showed that even very small metabolic variations between a parental and a transfected cell line can be revealed. This was due to the reliability of the system, which was assessed by statistical methods. The evaluation of the data provided good evidence that each single bioreactor reproduces the same physico-chemical growth conditions. Thus, by varying a single parameter, the effect on growth was assessed with a high degree of confidence. Applying this approach is expected to allow rapid identification of promising process conditions and to facilitate the scale-up procedure.

## AKNOWLEDGMENTS

The technical assistance of the following young collaborators (future students of a local engineering school) is gratefully acknowledged: Anne-Charlotte Bon, Sandrine Richard and Alexandre Super

## REFERENCES

De Jesus, M., Girard, P., Bourgeois, M., Baumgartner, G., Jacko, B., Amstutz, H., Wurm, F. M. (2004). TubeSpin satellites: a fast track approach for process development with animal cells using shaking technology. *Biochemical Engineering Journal* 17:217-223.

Farid, S. S., Washbrook, J., Titchener-Hooker, N. J. (2005). Decision-Support Tool for Assessing Biomanufacturing Strategies under Uncertainty: Stainless Steel versus Disposable Equipment for Clinical Trial Material Preparation. *Biotechnology Progress* 21:486-497.

Girard, P., Jordan, M., Tsao, M., Wurm, F. M. (2001). Small-scale bioreactor system for process development and optimization. *Biochemical Engineering Journal.* 7(2):117-119.

Hunt, L., Hacker, D. L., Grosjean, F., De Jesus, M., Uebersax, L., Jordan, M., Wurm, F. M. (2004). Low-Temperature Pausing of Cultivated Mammalian Cells. *Biotechnology and Bioengineering* 89(2):157-163.

Jordan, M. A novel disposable microtube for rapid assessment of biomass in cell cultures. (2005). in Gòdia, F. and Fussenegger, M., "*Animal Cell Technology Meets Genomics*". 5, 609-612.

Muller, N., Girard, P., Hacker, D. L., Jordan, M., Wurm, F. M. (2005). Orbital shaker technology for the cultivation of mammalian cells in suspension. *Biotechnology and Bioengineering* 89(4):400-6.

# Optimisation of Time-space-yield for Hybridoma fed-batch Cultures with an Adaptive Olfo-controller

Ralf Pörtner[1], Björn Frahm[2], Paul Lane[3], Axel Munack[3], Kathrin Kühn[4], Volker C. Hass[4]

[1]*Technische Universität Hamburg-Harburg, Bioprozess- und Bioverfahrenstechnik, Denickestr. 15, 21071 Hamburg, Germany,* [2]*Bayer Technology Services, Process Technology, 51368 Leverkusen, Germany,* [3]*Bundesforschungsanstalt für Landwirtschaft, Institut für Technologie und Biosystemtechnik, Bundesallee 50, 38116 Braunschweig, Germany,* [4]*Hochschule Bremen, Neustadtwall 30, 28199 Bremen, Germany*

**Abstract:** While fed-batch suspension culture of animal cells continues to be of industrial importance for the large scale production of pharmaceutical products, existing control concepts are still insufficient. The application of an adaptive Open-Loop-Feedback-Optimal (OLFO)-control provides a new approach for control of cell culture process which couples an efficient cultivation concept to a capable process control strategy. Applications of this strategy are discussed for optimization of hybridoma culture.

**Key words:** process control, fed-batch, hybridoma, monoclonal antibody, feeding strategy, adaptive control, OLFO, optimization, time-space-yield

## 1. INTRODUCTION

Batch or fed-batch suspension culture of mammalian cells is still the mostly applied mode of operation in industrial applications. Different fed-batch-techniques are in use, e.g. maintaining constant or predefined growth rates, maintaining a constant substrate concentration, substrate-limited fed-batch or $pO_2$-controlled feeding. All these strategies have limitations, as only few state variables like the dissolved oxygen concentration can be measured on-line. Other variables, like substrate concentrations or cell concentrations, can be measured off-line, whereby the results are available for control



*R. Smith (ed.), Cell Technology for Cell Products*, 497–501.

purposes only after some delay. The results of these control strategies are therefore often not optimal.

The Open-Loop-Feedback-Optimal (OLFO)-Controller is an adaptive model based controller consisting of an identification part which estimates the states and parameters of the process model using the available experimental data and an optimization part which calculates the feed trajectory based on the identified model status and a suitable optimization criterion.

An unstructured model has been used to describe the cultivation process. It consists of kinetic equations for growth and death kinetic as well as substrate uptake and metabolite production kinetics. The data could be fitted by identification of the model parameters using the simplex algorithm by Nelder and Mead.

In the presented work the hybridoma cell line IV F 19.23, producing monoclonal IgG antibody acting against penicillin-G-amidase, has been cultivated in a stirred bioreactor (2 L). For details see references.

## 2. OLFO-STRATEGY FOR FED-BATCH-CULTURES WITH CONTROLLED SUBSTRATE CONCENTRATION

The aim of this example for an OLFO-controlled fed-batch-experiment was to maintain low levels of the two main substrates, glucose and glutamine, during fed-batch mode at 1 and 2 mmol $L^{-1}$ by adjusting the feed rates of a concentrated glucose/amino acid feed and a glutamine feed (Figure 1). The experiment demonstrated the quality of the online prediction of important state variables like the concentrations of major substrates (glucose, glutamine) and the concentrations of toxic metabolites (lactate, ammonia). In the case of ammonia and glutamine, the measurement data were only available after the experiment has been completed. The produced ammonia inhibits cell growth at concentrations above 5 mmol $L^{-1}$, which finally led to the end of the cultivation. The final concentration of monoclonal antibodies was 41 mg $L^{-1}$.

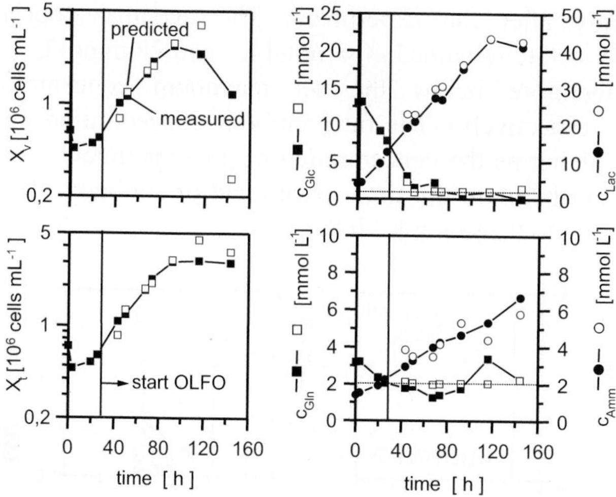


*Figure 1.* Fed-batch culture of a hybridoma cell line performed with an OLFO-strategy for control of glucose and glutamine concentration. Time course of viable cell concentration $X_v$, total cell concentration $X_t$, concentrations of glucose $c_{Glc}$, glutamine $c_{Gln}$, lactate $c_{Lac}$ and ammonia $c_{Amm}$. Measured (filled symbols) and predicted (open symbols) parameters.

# 3. OLFO-STRATEGY FOR FED-BATCH-CULTURES WITH OPTIMIZED TIME-SPACE-YIELD

Extention of the OLFO-concept for optimisation of a quality factor (e.g. time-space-yield) requires methods for calculation of an appropriate feed trajectories. This was accomplished by defining an optimal control problem with side conditions. As quality factor the time-space-yield of cells per volume and time was choosen. The optimisation algorithm has to calculate the feed rates of a glucose/nutrient feed and glutamine feed in a way that the time-space-yield reaches a maximum. The side conditions are required to consider minimal and maximal pump rates as well as the maximal culture volume. Furthermore they limit the range of the model by defining a tolerated range for the concentrations of glucose, glutamine, ammonia and lactate. As the model does not include terms for metabolite inhibition, the concentrations of ammonia and lactate were kept below inhibiting values. For optimization of the feed profiles the collocation method was used and implemented in the software DIRCOL.

An experiment performed with the described optimisation strategy is shown in Figure 2. The measured values of glucose and glutamine concentration coincided very well with the predetermined course. This is especially important for glutamine, as the samples were analysed for glutamine only after the experiment. The maximal concentrations of ammonia and lactate remained at 4 mmol $L^{-1}$ and 28 mmol $L^{-1}$, respectively and were therefore below the set maximum concentrations (4 and 40 mmol $L^{-1}$, respectively). The final antibody concentration was 57 mg $L^{-1}$, significantly higher as the concentration of the experiment shown in Fig. 2 (41 mg $L^{-1}$). Furthermore the time-space-yield of antibody was 70% higher (0.483 mg $L^{-1}$ $h^{-1}$ to 0.285 mg $L^{-1}$ $h^{-1}$).

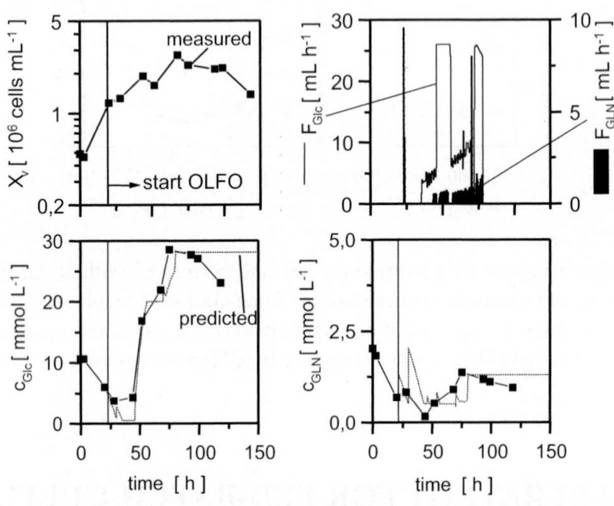


*Figure 2.* Fed-batch culture of a hybridoma cell line performed with an OLFO-strategy for optimization of time-space-yield. Time course of viable cell concentration Xv, concentration of glucose cGlc, glutamine cGln, feed rate for nutrient-concentrate-feed FGlc and glutamine FGln. Measured (filled symbols) and predicted parameters (dotted line).

## 4. EVALUATION OF CONTROLLER PERFORMANCE

The product yield (antibody concentration related to a corresponding batch experiment) was 2 - 3 times higher for the OLFO-experiment with controlled substrate concentrations and up to 4 times higher for the OLFO-experiment with optimised time-space-yield compared to other feeding strategies (fixed feed trajectories, control vial oxygen uptake rate, predetermined feed rates based on a kinetic model). The main reason for the higher productivity is thought to be due to the prolonged duration of the culture.

Our studies illustrate that the adaptive, model-based OLFO-controller is a valuable tool for the fed-batch control of hybridoma cell cultures. Due to its universal character, it can be transferred to different cell lines or adapted to different process strategies (e.g. enhanced fed-batch cultivation using dialysis. Sophisticated optimization criteria can be implemented.

## REFERENCES

Frahm, B., Lane, P., Atzert, H., Munack, A., Hoffmann, M., Hass, V.C., and Pörtner, R., 2002, Adaptive, model-based control by the Open-Loop-Feedback-Optimal (OLFO) Controller for the effective fed-batch cultivation of hybridoma cells, Biotechnol. Prog. **18**: 1095-1103

Frahm, B., Lane, Märkl, H., and Pörtner, R., 2003, Improvement of a mammalian cell culture process by adaptive, model-based dialysis fed-batch cultivation and suppression of apoptosis, Bioproc. Biosyst. Eng. **26**: 1-10

Pörtner, R., Schwabe, J.-O., and Frahm, B., 2004, Evaluation of selected control strategies for fed-batch cultures of a hybridoma cell line, Biotechnol. Appl. Biochem. **40**: 47-55

# Influenza Vaccines – Challenges in Mammalian Cell Culture Technology

Yvonne Genzel[1], Josef Schulze-Horsel[1], Lars Möhler[2], Yury Sidorenko[2], Udo Reichl[1,2]

[1] *Max Planck Institute for Dynamics of Complex Technical Systems, Magdeburg, Sandtor-str. 1, 39106 Magdeburg, Germany,* [2] *Lehrstuhl für Bioprozesstechnik, Otto-von-Guericke-Universität Magdeburg, Universitäts-platz 2, 39106 Magdeburg, Germany.*

**Abstract:** Experimental data as well as simulation results obtained by mathematical models for an influenza vaccine process with adherent MDCK cells clearly show that total cell number, specific virus replication rate and cell death due to virus infection (apoptosis) are the main factors to be taken into account for achieving high virus yields. In contrast, supply of cellular precursors and ATP for the synthesis of viral genome and virus specific proteins as well as multiplicity of infection seems not to limit virion formation.

**Key words:** vaccine, MDCK cells, microcarrier culture, stirred tank bioreactor, cell growth, virus replication, metabolites, mathematical models, bioprocess engineering, influenza virus, virus yield

## 1. INTRODUCTION

Viral vaccines play an important role in prevention, control and eradication of infectious diseases. Due to their enormous economical importance and difficulties in controlling existing but also up-coming risks, development of vaccines is still in the focus of today's research. Our main goal is to develop integrated concepts to optimize such processes. As an example we investigate influenza A virus replication in Madin Darby canine kidney (MDCK) cells in animal cell culture together with various downstream processing methods for the purification of inactivated viral harvests. Here, we present results with respect to the optimization of cell



*R. Smith (ed.), Cell Technology for Cell Products*, 503–508.

growth and virus titers in microcarrier systems. Experimental data are discussed on the background of simulation results obtained by mathematical models of virus replication dynamics.

## 2. MATERIALS AND METHODS

Madin-Darby canine kidney cells (MDCK) (ECACC No. 84121903) were cultivated in serum containing GMEM as described elsewhere [1]. Cells were infected with equine influenza A/Equi 2 (H3N8) Newmarket 1/93 (NIBSC) in GMEM medium without serum containing low levels of porcine trypsin (12.5 mg/L; Invitrogen). Virus seed was stored in aliquots of 1-10 mL (2.1-2.4 log HA units/100 μL) at -70 °C. For roller bottle experiments, virus was thawed and added with a multiplicity of infection (m.o.i.) of 0.0001-1.0 based on $TCID_{50}$/mL. For bioreactor cultivation a m.o.i. of 0.025 based on PFU/mL was used. Cells were grown in roller bottles (250 mL wv, Greiner) or in a 5 L stirred tank bioreactor (details see [1]) (B. Braun Biotech) on Cytodex 1 microcarriers (1.7-2 g/L) (GE Healthcare). After 4 days of cell growth, spent cell growth medium was removed and the remaining suspension was washed several times with PBS ($Ca^{2+}$/$Mg^{2+}$-free) before addition of virus medium. Culture conditions were controlled by a digital process control system (PCS7, Siemens) at 50 rpm, 37°C, pH 7.2-7.4 and addition of $O_2$ by pulsed aeration (minimum 40% $pO_2$).

Titration of influenza viruses by haemagglutination (HA) was based on the method described by Mahy and Kangro [2]. Serial two fold dilutions of test samples (100 μL) were made in duplicates in 96 well u-shaped microtitre plates containing 100 μL PBS. 100 μL of a fresh solution of chicken red blood cells (2.0 x $10^7$ cells/mL) in PBS were added and the last dilution showing complete haemagglutination was taken as the end point and recorded as dilution factor or log HA units/100 μL, respectively.

For simulation studies, parameter estimation and calculation of oxygen uptake rates MatLab/SimuLink (Mathworks©, Inc., USA) was used [3].

## 3. RESULTS AND DISCUSSION

After scale-up in roller bottles MDCK cells were inoculated at about 2 x $10^5$ cells/mL in a 5 L bioreactor. After cell attachment a maximum specific cell growth rate $\mu_{max}$ in the range 0.02 to 0.035 $h^{-1}$ was achieved before cell growth decreased due to contact inhibition after 4 days. Eventually, a maximum cell number of about 1.2 x $10^6$ cells/mL was achieved. During this time interval glucose as well as glutamine and glutamate were almost completely consumed. Neither lactate nor ammonia concentrations increased to inhibiting levels [1]. Also, none of the essential and non-essential amino acids reached critical levels. Average specific oxygen uptake rate at the end of the cell growth phase (75.4 to 94.0 h) was about 3.5 x $10^{-14}$ mol/(cell h) which is in the range typically reported for mammalian cells (data not shown).

About 16 h post infection (m.o.i. 0.025) HA increased strongly and achieved its maximum with 2.4 log HA/100 µL after 44 hours (Fig. 1). During the first hours post infection (100.5 to 106.4 h) specific oxygen uptake rate decreased from 7.7 x $10^{-14}$ to 2.5 x $10^{-14}$ mol/(cell h). In the mean 5.0 x $10^{-14}$ mol/(cell h) were consumed. The initial number of uninfected cells decreased rapidly within the first 4 h. Metabolism as well as oxygen consumption of the infected cells finally ceased and ended after 4 days (Fig. 2).

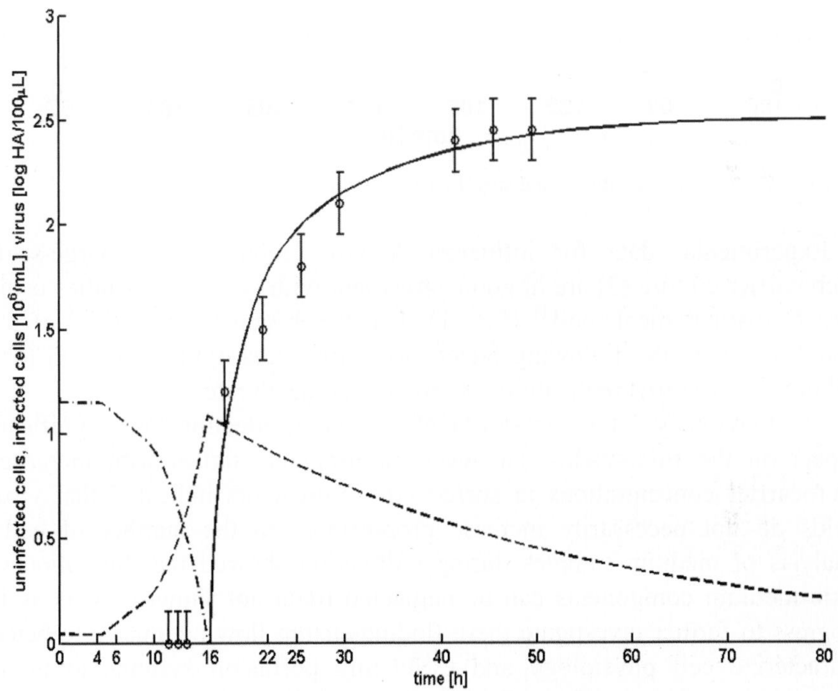


*Figure 1.* Experimental data and model simulations of cell infection and virus yield: (o) HA titer (mean ± standard deviation). Simulations: uninfected cells (-·-·-), infected cells (---) and HA (—).

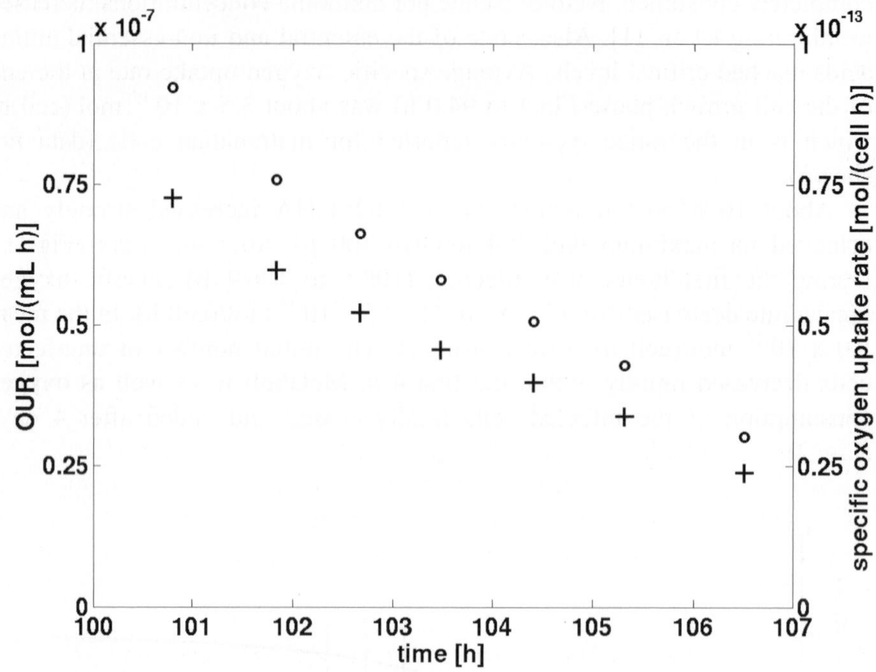


*Figure 2.* Oxygen uptake rate (o) and specific oxygen uptake rate (+) after infection.

Experimental data for influenza A virus replication in large-scale microcarrier culture [3] are in good agreement with simulation studies using a basic mathematical model (Fig. 1). Parameter sensitivity studies of the model indicated the following factors for virus yield optimization: a) total cell number, b) virus replication rate and c) cell death rate.

a) Obviously, total number of cells at time of infection has a significant impact on the final yields. However, preliminary studies with increased microcarrier concentrations in stirred tank bioreactors indicated that virus yields do not necessarily increase proportional to the number of cells. Analysis of medium samples during cultivation showed that limitation of basic medium components can be neglected (data not shown). Work is in progress to further investigate these findings using flow cytometry to better characterize cell physiology and evaluating perfusion systems to avoid accumulation of unspecific inhibitors of virus replication

b) Specific virus replication rates can be increased for example by the generation of reassorted viruses or by plasmid-driven systems to generate high yield virus strains. While this strategy might be useful for equine influenza vaccine production as substitution or addition of new strains is considered necessary only every 3 to 5 years, this strategy is usually not applicable for human vaccines to tight time schedules of annual campaigns.

c) Any increase in the average lifetime of cells should also increase virus yields. Mathematical models on a single cell level indicate that during the average lifetime of an infected cell neither the intracellular pools of precursors nor ATP production limit virus production [4].

An estimation of the cellular energy required for virion synthesis yields about 9.8 x $10^6$ ATP molecules required for virus specific compounds (viral genome, viral mRNAs and proteins). Most of this energy is required for viral protein synthesis (table 1). In addition, part of the apical cell membrane is also consumed for virion assembly, which accounts for another 1.0 x $10^6$ ATP molecules if replaced by the host cell. In total, virion synthesis requires about 2.0 x $10^7$ ATP molecules (viral carbohydrates neglected). Thus, if 10% of the produced ATP would be consumed for virion production during an average lifetime of an infected cell of 10 h (specific oxygen uptake rate of 5.0 x $10^{-14}$ mol/(cell h), see above; 2.5 $P_i$ per NAD(P)H) the amount of ATP synthesized per cell would allow to produce about 8355 virions (host cell membrane not replaced). If cell membrane were actively synthesized to replace losses due to budding, the total number of virions produced would be reduced to about 4053 virions.

*Table 1.* Energy demand for virion production of human influenza A virus A/PR/8/34.

| | ATP molecules [number] | virus specific % | cell derived % |
|---|---|---|---|
| **viral genome replication*¹** | 5.44 x $10^4$ | 0.55 | 0.27 |
| **vmRNA (0.8-1.0%)*¹** | 2.82 x $10^4$ | 0.29 | 0.14 |
| **viral protein (70%)*¹** | 9.73 x $10^6$ | 99.16 | 48.10 |
| **viral carbohydrates (5-8%)** | - | - | - |
| **membrane lipids *²**<br>**membrane proteins** | 1.04 x $10^7$ | - | 29.74<br>21.75 |
| **total** | 2.02 x $10^7$ | 100 | 100 |

*¹ assuming 2 $P_i$ and 4 $P_i$ per nucleotide and amino acid in synthesis of RNAs and proteins, respectively; *² assuming an average composition of the membrane of phosphatidylcholine (18%), phosphatidylserine (9%), phosphatidylethanolamine (11%), phosphatidylinositol (4%), sphingomyelin (14%), cholesterol (30%), others (14%) and membrane proteins (1 per 50 lipid molecules, 350 amino acids in average), virions 100 nm in diameter.

Obviously, extending the average lifetime of cells, i.e. the delay of influenza virus induced apoptosis, could increase virus yields. However, this will be a difficult task due to the enormous complexity of influenza virus induced influences on host cell signaling processes [5].

Additionally, our mathematical models of basic virus dynamics also indicate that m.o.i. does not influence overall virus yields [3]. This is supported by experimental results for the equine influenza subtype investigated here (Fig. 3). With decreasing m.o.i. increase in HA values starts later but achieves same maximum levels.

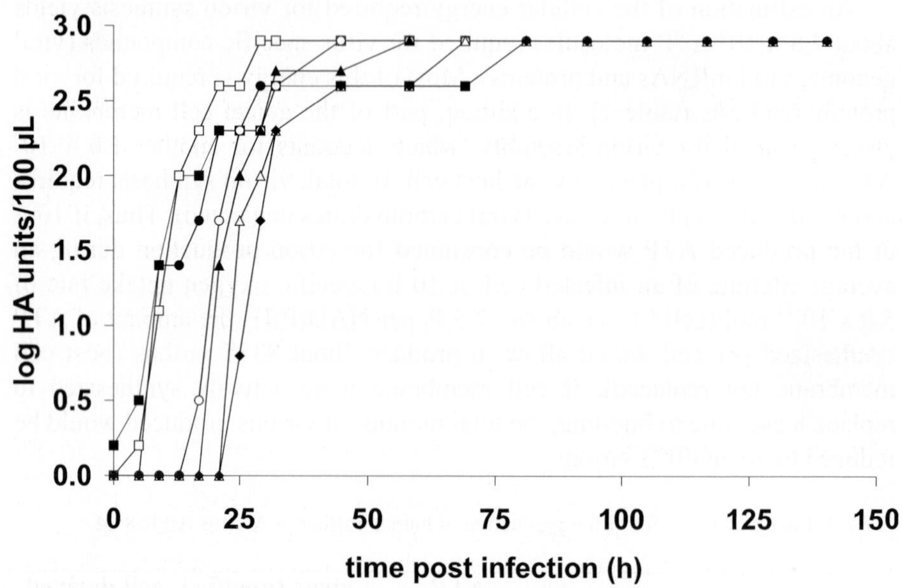


*Figure 3.* Virus titre (mean of duplicates) during equine influenza virus production (96-236 h) in MDCK cells in roller bottles. ■ moi=1; □ moi=0.5; ● moi=0.1; ○ moi=0.025; ▲ moi=0.01; △ moi=0.001; ◆ moi=0.0001.

# REFERENCES


[1] Genzel, Y., Behrendt, I., König, S., Sann, H., Reichl, U.; 2004, Vaccine, **22 (17-18)**: 2202-2208.

[2] Mahy, B.W.J., Kangro, H.O.; 1996. Virology methods manual. London:Academic press.

[3] Möhler, L., Flockerzi, D., Sann, H., Reichl, U.; 2005, Biotechnol. Bioeng., **90 (1)**, 46-58.

[4] Sidorenko, Y. and Reichl, U., 2004, Biotechnol. Bioeng., **88 (1)**, 1-14.

[5] Ludwig, S., Planz, O., Pleschka, S., Wolff, T.; 2003, Trends Mol. Med., **9(2)**, 46-52.

# Towards $pCO_2$-optimised Fermentations – Reliable Sampling, $pCO_2$ Control & Cellular Metabolism

Alexander Jockwer[1], Christian Klinger[2], Jochem Gätgens[1], Detlef Eisenkrätzer[3], Thomas Noll[1]

[1]*Research Centre Juelich GmbH, Institute of Biotechnology 2, Cell Culture Technology, Juelich, Germany;* [2]*University of Applied Sciences Weihenstephan, Weihenstephan, Germany;* [3]*Roche Diagnostics GmbH, Penzberg, Germany*

**Abstract:** Current industrial approaches using simple intermittent stripping of dissolved carbon dioxide ($pCO_2$) by air or nitrogen when exceeding certain levels may underestimate the influence of intracellular pH ($pH_i$) fluctuations in the cells' cytoplasms. In addition, commonly applied sampling procedures do not compensate for pressure and $pCO_2$ losses, respectively, resulting in erroneous reactor $pH_i$. To reliably investigate the influence of $pCO_2$ on industrial high cell density fermentations we developed a simple, yet effective and easily scalable pressure-controlled sampling device for representative display and maintenance of reactor conditions, e.g. whilst dye uptake of the cells for $pH_i$ measurement by FACS after sampling from pressurized fermenters.

**Key words:** dissolved carbon dioxide, $pCO_2$, $dCO_2$, intracellular pH, $pH_i$, pressure-controlled sampling device, glycosylation, large scale, FACS, recombinant protein, CHO

## 1. INTRODUCTION

In recent years dissolved carbon dioxide ($pCO_2$) has been identified as an important process parameter affecting cell growth, productivity and product quality (e.g. glycosylation) of recombinant proteins when exceeding critical levels (Garnier *et al.*, 1996; Gray *et al.*, 1996, Kimura *et al.*, 1996, Zhangi *et al.*, 1999), occurring especially in industrial large-scale cell culture processes due to the increased hydrostatic pressure. As $CO_2$ can easily pass the cellular membrane and thereby influence intracellular pH ($pH_i$)



*R. Smith (ed.), Cell Technology for Cell Products*, 509–518.

(Andersen and Goochee, 1994), important cellular processes (e.g. cell cycle regulation, enzymes of TCA cycle) are directly influenced by $pCO_2$ and dependent bicarbonate concentration. Consequently, process control strategies attend to keep $pCO_2$ within physiological range (Matanguihan *et al.*, 2001) Furthermore, the choice of $pCO_2$ control strategy directly influences cell culture performance, e.g. by $pH_i$ fluctuations. In addition, specially designed sampling devices are needed to reliably display reactor conditions as shown in this work.

## 2. MATERIALS AND METHODS

### 2.1 Cell Cultivation

The recombinant chinese hamster ovary cell line CHO-MUC1-IgG2a-PH3931 used in this work, produces a recombinant MUC1 protein, which is fused to an Fc part (IgG2a) of a murine IgG antibody and secreted into the media. Suspended cell cultivation was carried out in bubble-aerated bioreactors with 1.0 L (Applikon) and 10.0 L (Biostat ES, B.Braun) working volume at constant aeration rates of 0.083 vvm and 0.050 vvm, respectively. Bioreactor conditions were controlled to 37°C, pH 7.0, 40% DO, 200 rpm agitation rate and 75 kPa overpressure (for 10 L scale) in serum-free suspension media. Aeration gas ratios ($N_2$, $O_2$, $CO_2$) were mixed by mass flow controllers. *In-situ* measurement of the partial pressure of dissolved $CO_2$ was performed using an fiber optic chemical sensor (YSI 8500, YSI Inc.) (Pattison *et al.*, 2000).

### 2.2 Pressure-controlled sampling device

To investigate the influence of $pCO_2$ on industrial high cell density fermentations we developed a simple, yet effective and easily scalable pressure-controlled sampling device for representative display of reactor conditions. It allows adjustment of the sample's total pressure to the one of the reactor at the time of sampling and therewith it maintains $CO_2$ partial pressure whilst dye uptake for $pH_i$ measurement by FACS.

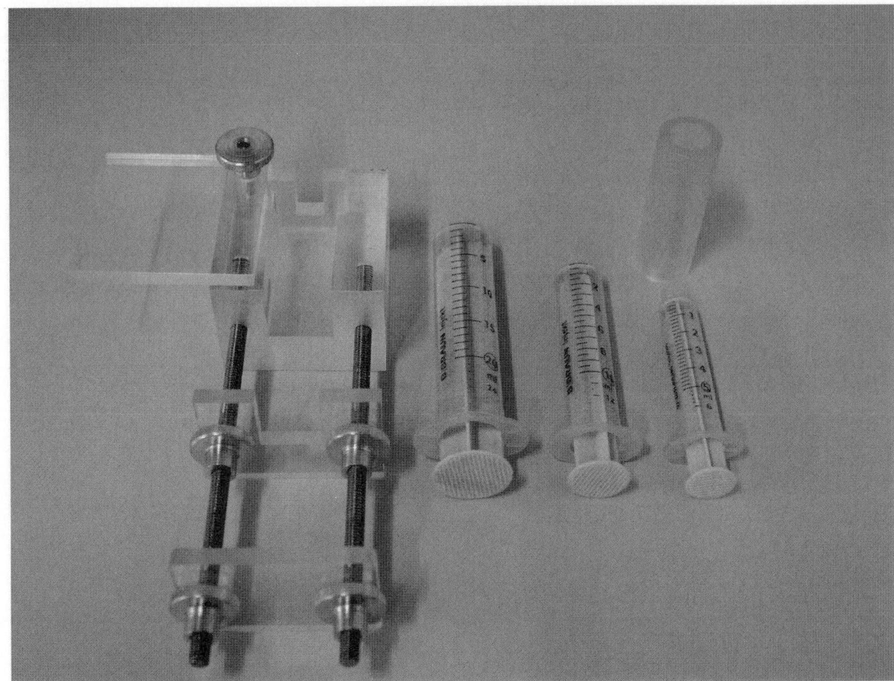

*Figure 1.* Construction of pressure-controlled sampling device: holding device (left) for use of different sized syringes with adaptor.

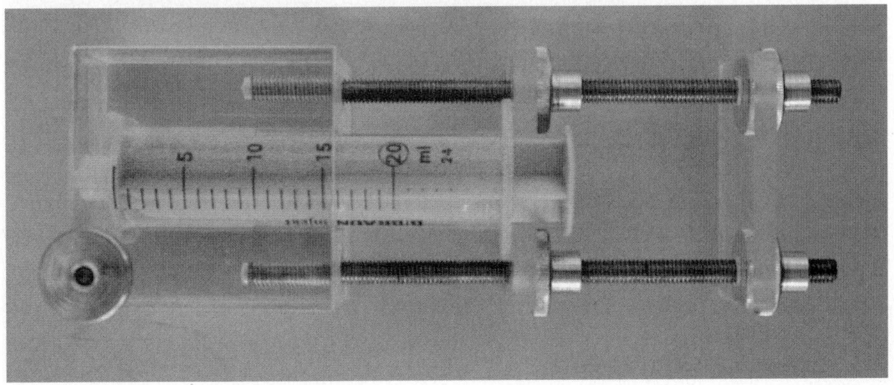


*Figure 2.* Syringe installed in holding device for administration of dye.

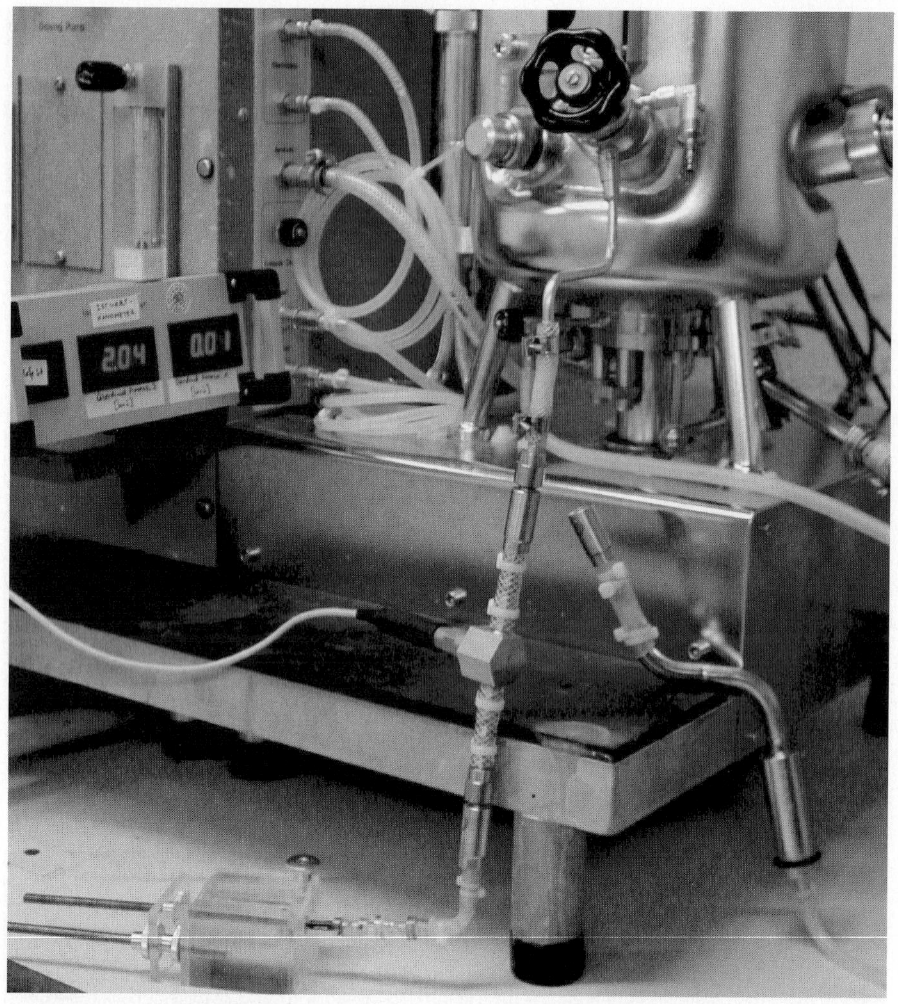


*Figure 3.* Sterile connection of the pressure-controlled sampling device pre-loaded with dye. Actual pressure of the sample is measured by a piezo-electric pressure transducer, digitally displayed and controlled after disconnection of the sampling unit.

## 2.3 Off-line Measurements

Cell concentration and viability were determined using a automatic cell counter (ViCell XR, BeckmanCoulter,). Off-line measurements of $pCO_2$, $pO_2$ and pH were performed by using a blood gas analyser (AVL Compact 3, Roche Diagnostics) immediately after pressure-controlled sampling and $pH_i$ determination by FACS analysis, respectively. For intracellular pH measurement by FACS, the pseudo-null point method using SNARF-1-AM

(Molecular Probes) as pH-sensitive dye was applied as previously described (Chow *et al.*, 1996; Bond and Varley, 2005) and modified for pressure-controlled sampling. Therefore, the developed sampling device was pre-loaded with an appropriate amount of SNARF-1-AM. After pressure-controlled-sampling the cells were incubated under bioreactor conditions and then analysed by FACS. For $pH_i$ calculations after FACS the actual pH values of the measured solutions were checked by using the blood gas analyser.

# 3. RESULTS AND DISCUSSION

Commonly applied sampling methods from pressurised bioreactors do not compensate for pressure losses of the sample and associated losses of dissolved gases like $CO_2$ (Figure 4).

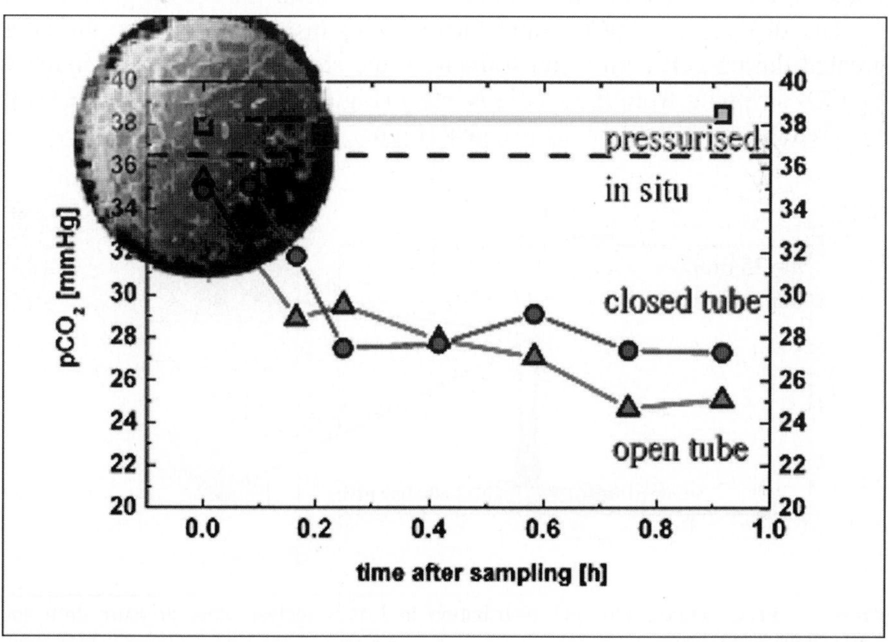


*Figure 4.* $PCO_2$ loss occuring at different sampling methods (triangles: open tube, circles: closed tube, squares: pressure-controlled sampling device, dashed line: *in-situ*).

This can lead to underestimation of $pCO_2$ concentrations in the cells' surrounding media. Even if there are no differences in $pCO_2$ concentration observable in the media, the intracellular $pCO_2$ content and dependent $pH_i$ may already be affected by the pressure drop created by common sampling procedures. The developed sampling device clearly prevents the media from degassing compared to commonly applied sampling methods (Figure 4). Although all sampling methods investigated lead to the same starting concentration of $pCO_2$ in the media when measured immediately after sampling (Figure 4), the cell populations show heterogenous $pH_i$ distribution in FACS analysis when incubated depressurised (Figure 5) whereas they show homogenous $pH_i$ distribution when incubated under bioreactor conditions in the pressure-controlled sampling device (Figure 6).

This is due to the intracellular $CO_2$ degassing whilst depressurisation and leads to time-dependent $pH_i$ shifts. Asychronous shift of the cell population's $pH_i$ at $CO_2$ degassing may be due to different physiological states of the cells.

Depressurisation causes even earlier $CO_2$ losses from inside the cells than from the surrounding media as it also occurs during $CO_2$ stripping from cell culture processes. The pressure-controlled sampling device is therefore needed for reliable measurement of important cell physiological parameters related directly to dissolved gas concentrations inside the cells, e.g. $pH_i$.

The difference of $pCO_2$ (and bicarbonate) inside and outside the cells created during cell cultivation influences the short-term $pH_i$ shifts induced by $CO_2$ stripping from the media *in vitro* (Figure 7). The more the $pCO_2$ is lowered the more alkaline the $pH_i$ gets (Figure 7) short-term.

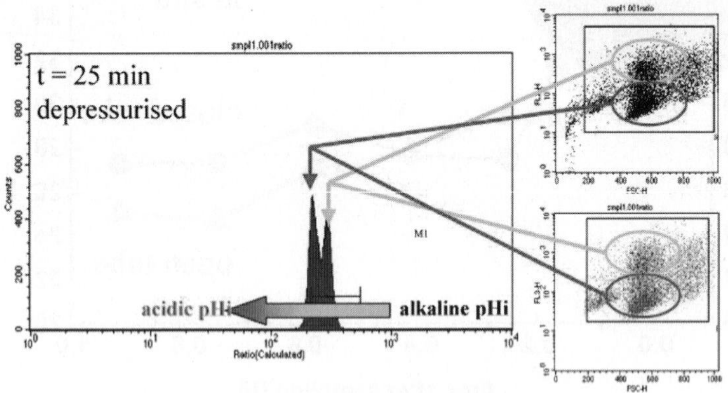


*Figure 5.* False heterogenous $pH_i$ distribution in FACS analysis after pressure drop and intracellular $pCO_2$ loss.

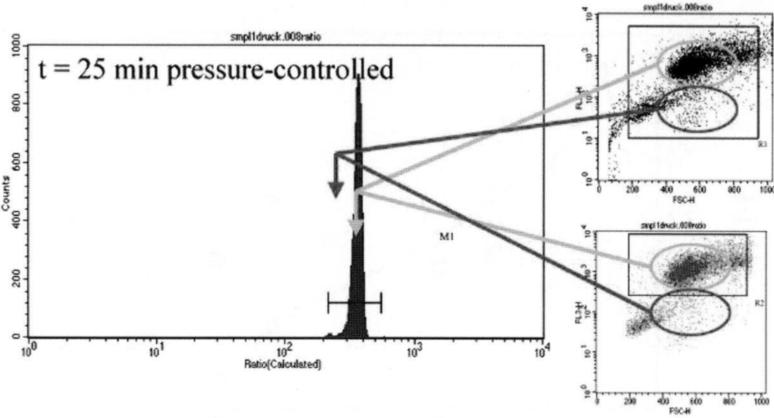


*Figure 6. Correct homogeneous* pHi distribution in FACS analysis after pressure-controlled sampling and incubation.

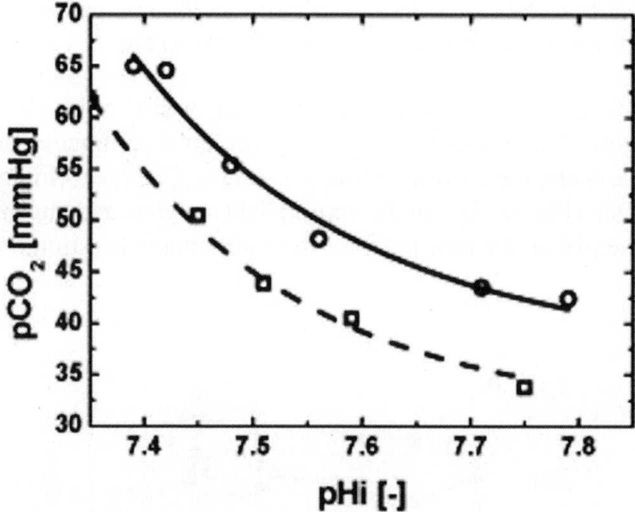


*Figure 7.* Chemical effect in vitro: withdrawal of the acidic compound $CO_2$ from media and cells, respectively, leads to short-term (t < 10 min) alkalinisation of cells' cytoplasm.

The following long-term metabolic effect of $CO_2$ stripping from cells *in vitro* (Figure 8) shows super-compensation of $pH_i$ shift after the loss of the acidic compound $CO_2$ (and bicarbonate) from the cells' cytoplasms.

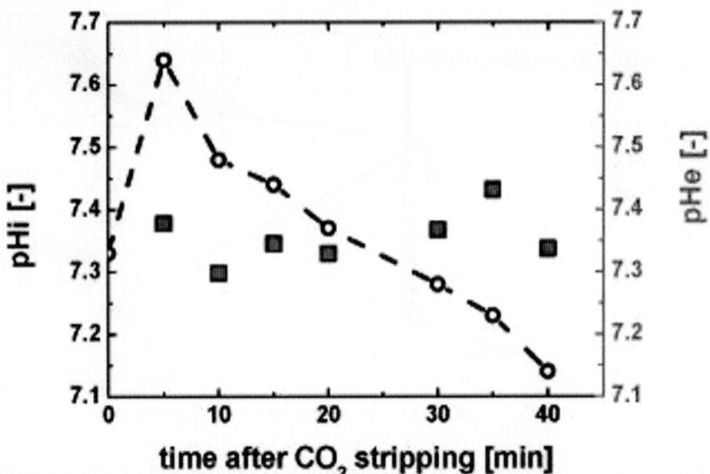


*Figure 8.* Metabolic effect in vitro: Cells super-compensate the short-term alkalinisation caused by the withdrawal of $CO_2$ from media and cells by stripping. $pH_i$ after stripping is more acidic than before (circles: intracellular pH, squares: external pH).

Inverse effects on $pH_i$ shifts can be seen when $pCO_2$ is stepwise increased. Figure 9 shows different $pCO_2$ levels in a continuous culture of the CHO cell line and the corresponding $pH_i$ values. Chemical (Figure 7) and metabolic effect (Figure 8) can be reproduced *in vivo* and the metabolic effect keeps the $pH_i$ on the new level for the following cultivation time.

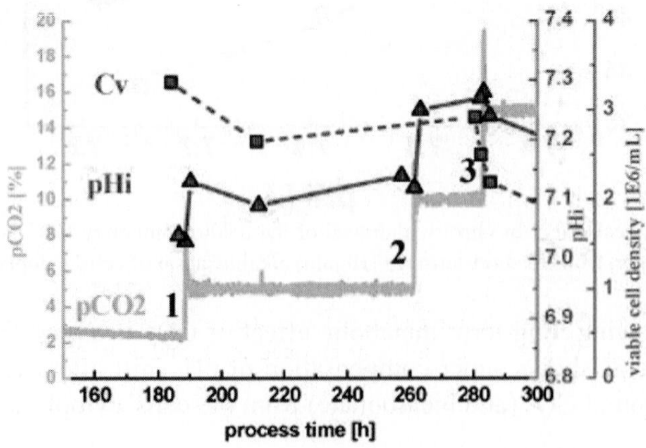


*Figure 9.* Intracellular pH shifts after $pCO_2$ steps in continuous CHO culture (1.0 L).

Corresponding bioengineering, metabolic and physiological details of this study will be published elsewhere.

## 4. CONCLUSIONS

For detailed investigation of cells' physiological parameters such as $pH_i$ which are directly influenced by dissolved $CO_2$ concentration, a special pressure-controlled sampling device is needed. Furthermore, when applying the novel sampling device to pressurised large-scale fermenters, we could show that stripping strategies to lower $pCO_2$ in the fermenter drastically influence the intracellular pH of the cells. Consequently, this may lead to sub-optimal glycosylation pattern of the recombinant proteins produced (Kimura and Miller, 1997; Zhangi *et al.*, 1999).

## ACKNOWLEDGEMENTS

Many thanks to ESACT for the bursary and to Roche Diagnostics GmbH (Penzberg, Germany) for funding.

## REFERENCES

Andersen, D. C., Goochee, C. F., 1994, The effect of cell culture conditions on the oligosaccharide structures of secreted glycoproteins, Curr. Opinion Biotechnol. **5**: 546-549.

Bond, J., Varley, J., 2005, Use of flow cytometry and SNARF to calibrate and measure intracellular pH in NS0 cells, Cytometry **64A**: 43-50.

Chow, S., Hedley, D., Tannock, I. 1996. Flow cytometric calibration of intracellular pH measurements in viable cells using mixtures of weak acids and bases, Cytometry **24** (4): 360-367.

Garnier, A., Voyer, R., Tom, R., Perret, S., Jardin, B., Kamen, A., 1996, Dissolved carbon dioxide accumulation in a large-scale and high-density production of TGF-Beta receptor with baculovirus infected Sf-9 cells, Cytotechnology **22**: 53-63.

Gray, D. R., Chen, S. Howarth, W., Inlow, D., Maiorella, B. L., 1996, $CO_2$ in large-scale and high-density CHO cell perfusion culture, Cytotechnology **22**: 65-78.

Kimura, R., Miller, W.M., 1996, Effects of elevated $pCO_2$ and/or osmolality on the growth and recombinant tPA production of CHO cells, Biotechnol. Bioeng.**52**, 152-160.

Kimura, R., Miller, W.M., 1997, Glycosylation of CHO-derived recombinant tPA produced under elevated $pCO_2$, Biotechnol. Prog. **13**: 311-317.

Matanguihan, R., Sajan, E., Zachariou, M., Olson, C., Michaels, J., Thrift, J., Konstantinov, K., 2001, Solution to the high dissolved $CO_2$ problem in high-density perfusion culture of mammalian cells, In: E. Lindner–Olsson *et al.* (eds.), Animal Cell Technology: From Target to Market, 399-402, Kluwer Academic Publishers, The Netherlands.

Pattison, R. N., Swamy, J., Mendenhall, B., Hwang, C., Frohlich, B. T., 2000, Measurement and control of dissolved carbon dioxide in mammalian cell culture processes using an in situ fiber optic chemical sensor, Biotechnol. Prog. **16**: 769-774.

Zhangi, J. A., Schmelzer, A. E.. Mendoza, T. P., Knop, R. H., Miller, W. M., 1999, Bicarbonate concentration and osmolality are key determinants in the inhibition of CHO cell polysialylation under elevated $pCO_2$ or pH, Biotechnol. Bioeng. **65**: 182-191.

# Cultivation of E-FL in Perfusion Mode using Wave® Bioreactor and Stirred-Tank Reactor for Parvovirus Vaccine Production

B. Hundt[1], S. Straube[1], N. Schlawin[1], H. Kaßner[1], U. Reichl[2,3]

[1] *Impfstoffwerk Dessau-Tornau GmbH, PF 400214, 06855 Roßlau, Germany,* [2] *Otto-von-Guericke-University, Chair of Bioprocess Engineering, Universitätsplatz 2, 39106 Magdeburg, Germany,* [3] *Max Planck Institute for Dynamics of Complex Technical Systems, Sandtorstraße 1, 39106 Magdeburg, Germany*

**Abstract:** Here we describe a 0.5 to 1 L wv perfusion process for cultivating adherent E-FL cells on microcarriers using a stirred-tank reactor and a Wave® Bioreactor for parvovirus vaccine production. In these systems cell yields were increased about 5-fold to maximum densities of 6.5 x $10^6$ cells/mL. For continuous parvovirus propagation and harvesting promising pre-results were achieved, but more optimization work needs to be done here for possibly establishing a continuous virus production process.

**Key words:** parvoviruses, mink enteritis virus, microcarriers, vaccine production, perfusion

## 1. INTRODUCTION

Adherent embryonic feline lung fibroblasts (E-FL) are used for the production of an inactivated mink enteritis vaccine at commercial scale. Mink enteritis virus (MEV) belongs to the group of independently replicating parvoviruses, the smallest known viruses with a diameter of 16-19 nm. It can only replicate in mitotic cells. Therefore, cells are usually infected at inoculation.

In previous work a production process was successfully transferred from roller bottles into microcarrier systems (> 10 L wv) with a repeated-batch strategy. This led to a significant increase in cell and virus yields and a simplification in process operation. Typical cell numbers were in the range



R. Smith (ed.), Cell Technology for Cell Products, 519–522.

of 1.0 x $10^6$ cells/mL with 2 g/L microcarriers compared with about 0.4 x $10^6$ cells/mL in static systems. Virus yields were about 5-fold higher than in static systems reaching maximum levels > $10^{7.5}$ $TCID_{50}$/mL.

However, we found limitations of these repeated-batch systems in respect to growth and average lifetime of cells. Obviously, the medium used was limited for some nutrients and the concentration of at least some metabolites reached growth-inhibiting levels. To further increase cell and virus yields the use of perfusion systems was investigated. As this strategy is close to the hitherto used repeated-batch process it may be easily implemented in larger scales and can be used both for an optimized pre-culture scheme in a production process and to establish a continuous MEV harvest.

## 2. MATERIALS AND METHODS

Cells were grown in a minimal essential medium (MEM) supplemented with 10% fetal bovine serum (FBS) and 4-6 mM L-glutamine. Cultivations were performed in a 0.5 L wv stirred-tank reactor (Infors, CH) and in a 1 L wv Wave® Bioreactor (Wave Biotech, CH). Typical seeding cell densities were in the range of 1.0-2.0 x $10^5$ cells/mL with a carrier concentration of 2 g/L Cytodex™ 1. For a high cell density experiment an inoculum of 5.0 x $10^5$ cells/mL and a carrier concentration of 5 g/L Cytodex™ 1 was chosen.

In the 0.5 L wv stirred-tank reactor (STR) cells were cultivated with 50 rpm at 37 °C, pH-value was corrected with 1 M NaOH to a level of 7.1 – 7.4, $pO_2$ was controlled at > 55% by pulses with pure oxygen. In the 1 L wv Wave® Bioreactor cultivations were performed with a tilt rate of 8 $min^{-1}$ and a tilt angle of 6°. Aeration was done with 0.1 L/min sterile air with 5-10% $CO_2$, pH was not controlled but always in physiological range of 6.8-7.6.

Perfusion was performed with self-designed retention devices which were adapted to the respective reactor. With those devices it was possible to retain the carriers with the immobilized cells and to exchange the medium with perfusion rates of 0.4–1.0 wv per day. Perfusion was started when the carriers were about 75% confluent (50-80 h). For MEV replication cells were infected with a MOI of 0.1 (based on $TCID_{50}$) at time of cell seeding.

Samples were taken in distinct intervals for offline analytics including cell count with trypan-blue method, automated enzymatic assays for quantification of glucose, lactate, glutamine and ammonia levels and quantification of MEV by a $TCID_{50}$ protocol.

## 3. RESULTS AND DISCUSSION

First experiments in perfusion mode were done in a 0.5 L wv STR with 2 g/L Cytodex™ 1, 2.0 x $10^5$ cells/mL inoculum and a perfusion rate of 0.75 wv per day. After 4 days a maximum cell number of 1.1 x $10^6$ cells/mL was obtained (Figure 1 a). Due to the constant nutrient supply and waste product removal by perfusion the cell number was kept constant on this level for a period of 7 days. Detachment of the cells as it was seen in previous repeated-batch experiments was not observed. The cells exhibited no morphological changes and were confluent on the microcarriers until the end of the cultivation.

After start of perfusion, concentrations of the substrates glucose and glutamine remained on a constant level of about 4 mM and 6 mM, respectively. Concentrations of the growth inhibiting products lactate and ammonia were nearly constant at 4 mM and 1 mM and thus far below cytotoxic levels (Figure 1 a). The results of this experiment showed that for our E-FL cell culture a carrier concentration of 2 g/L limits cell numbers to a maximum of 1.0-1.2 x $10^6$ cells/mL, regardless if batch, repeated-batch or perfusion mode were used. In contrast to batch and repeated-batch systems average lifetime of cells was significantly better in perfusion mode over the full period of 11 days. Results for STR and Wave® Bioreactor (data not shown here) were comparable in all aspects.

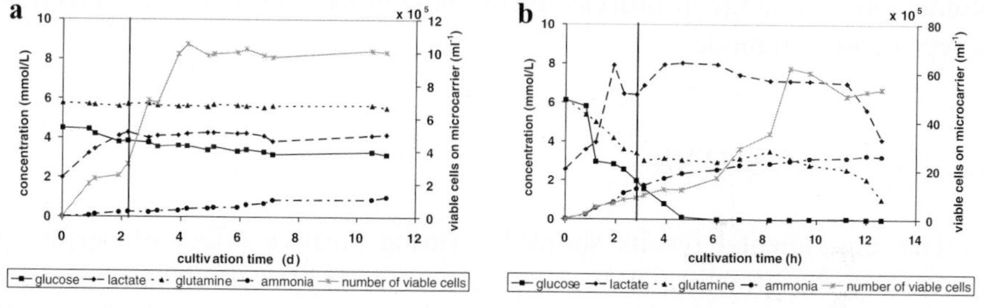


*Figure 1.* E-FL cell growth and metabolism in a) 0.5 L wv STR with 2 g/L Cytodex™ 1, 2.0 x $10^5$ cells/mL inoculum and 0.75 wv per day perfusion rate and b) in 0.5 L wv STR with 5 g/L Cytodex™ 1, 5.0 x $10^5$ cells/mL inoculum and 0.5 wv per day perfusion rate. Perfusion start after 52 h and 72 h (straight line).

In a second experiment, we tested the use of a higher carrier and inoculum concentration (5 g/L and 5.0 x $10^5$ cells/mL). Here, we obtained a maximum cell number of 6.5 x $10^6$ cells/mL after 9 days of cultivation, cell numbers were > 5.0 x $10^6$ cells/mL over a period of 5 days (Figure 1 b). Metabolic profiles reveal that nutrient supply, i.e. glucose concentration, was possibly a growth limiting factor (Figure 1 b). However, concentrations of lactate and ammonia did not reach inhibitory levels. In further experiments variations of perfusion rate and media composition should be tested to avoid nutrient limitations. Also the use of even high carrier concentrations in the range of 8-10 g/L should be evaluated to further increase cell concentrations. Although operational requirements for perfusion systems are significantly higher than those for batch processes, these very promising results should be taken into account for future optimization of pre-culture schemes for a vaccine production process in 100 L scale.

MEV replication in perfusion mode was investigated in a 0.5 L wv STR with 2 g/L Cytodex™ 1 and 2.0 x $10^5$ cells/mL inoculum. Cells were infected with MEV directly at inoculation with a MOI of 0.1. After 69 h perfusion was started with 1.0 wv per day, the harvest was pooled and stored at + 4°C in a closed system. Results show that average lifetime of cell is slightly higher in perfusion mode compared to experiments in repeated-batch mode. Although several samples had very high virus titres of $10^{7.8}$ $TCID_{50}$/mL, the average titre in the pooled harvest was $10^{5.2}$ $TCID_{50}$/mL. This means, possibly due to a suboptimal harvest scheme the total virus yield was lower in perfusion mode.

Modifications in perfusion rate and initial conditions (inoculum, start of perfusion, time of infection) should be tested to obtain virus yields on the same level or higher compared to repeated-batch systems. Nevertheless, for industrial application the closed system in perfusion mode represents a remarkable advantage in process operation compared to medium exchanges in repeated-batch mode.

## 4. CONCLUSIONS

The experimental results showed a strong positive effect of perfusion systems on the growth and average lifetime of E-FL cells in microcarrier systems, especially when higher carrier concentrations were used. These cell yields higher than 5.0 x $10^6$ cells/mL are well suitable for establishing an optimized pre-culture scheme. Results for MEV replication in perfusion systems were promising, but optimization is required for obtaining high virus yields in a closed system. This work should be done with support of flow cytometry data to gain more insight into virus-host cell interactions especially during the beginning of the process.

# Repeated Batch Fermentation Using a Viafuge Centrifuge

Kurt Russ, Peter Schlenke, Harald Huchler, Hans Hofer, Maya Kuchenbecker, Ralf Fehrenbach
*Rentschler Biotechnologie GmbH, Erwin-Rentschler-Straße 21, 88471 Laupheim, Germany*

**Abstract:** The described repeated batch mode combines the advantages of fed batch and perfusion processes. Harvesting is performed in preset intervals by a continuous-flow centrifuge (Viafuge from Carr, Pneumatic Scale INC:). The cells can be returned into the fermenter without loss of viability. The harvested medium is replaced by pre-heated fresh medium and the cell concentration for re-inoculation can be readjusted to the requested level

**Key words:** viafuge, repeated batch, fermentation, perfusion, process, Centrifugation

## 1. INTRODUCTION

A repeated batch fermentation method using a continuous-flow centrifuge was developed. The method combines the advantages of fed-batch and perfusion fermentations. By fermentation in the repeated batch mode the production period can be prolonged compared to standard fed-batch processes resulting in a significant increase of the final product yield. On the other hand, medium consumption is lower than in perfusion processes. Repeated batch fermentation shortened the time for process development considerably and has been applied for the GMP-production of several recombinant proteins using CHO cell lines. Cellular growth, viability, productivity as well as product quality in terms of glycosylation remained constant during the whole fermentation processes.



*R. Smith (ed.), Cell Technology for Cell Products*, 523–526.

## 2. FERMENTATION PROCESS

Processes in repeated batch fermentation modus using the Viafuge centrifuge (Carr, Pneumatic Scale INC, Fig. 1) have been applied for a wide range of different CHO cell lines and products at scales between 30 and 250L.

For harvesting, cells are separated from the supernatant by centrifugation, resuspended in fresh media and re-transferred into the fermenter. The cell density for re-inoculation can be adjusted, if required.

Depending on the process needs and on product stability, the harvesting is carried out in preset intervals (e.g. 1 – 4 days).

## 3. PROCESS IMPACT ON CELLS

The repeated batch process results in constant culture conditions over the whole fermentation period.

Figure 1 shows for example the growth characteristics of a culture over 33 days. After an initial growth phase of 7 days 26 daily harvests were performed

The viability of the culture remained high resulting in low amounts of residual HCPs and DNA in the harvests, what facilitates downstream processing.

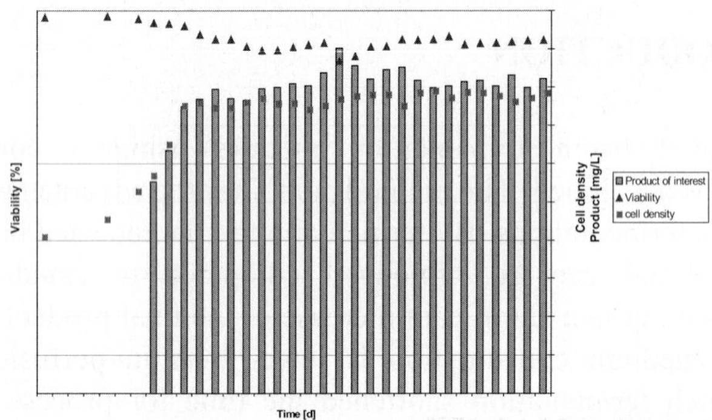


*Figure 1.* Cell density, viability and product-concentration over a fermentation period of 33 days (seven days growth phase, 26 harvest days).

Concentration of substrates and by-products was monitored throughout the fermentation run. Inhibition of cell growth or productivity or an impact on product quality due to the accumulation of toxic by-products like lactate or ammonia is avoided by the repeated batch mode (see Fig. 2 for example).

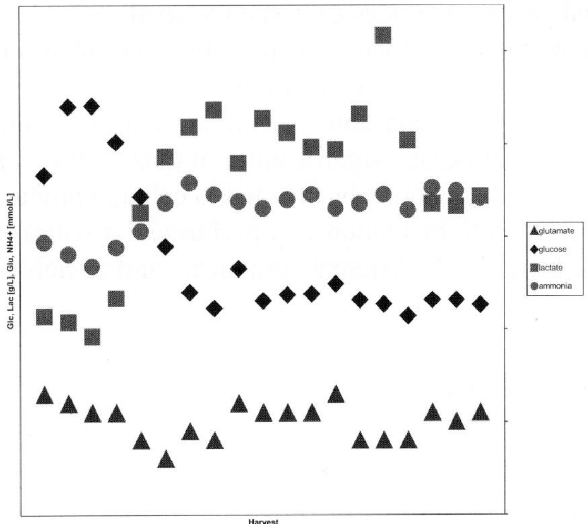


*Figure 2.* Concentration of glucose, lactate, glutamate and ammonia during the harvesting period.

Further genetic stability of the cell lines was unaffected. Figure 3 shows for example a Southern-Blot analysis using DNA isolated from cells at the beginning, during and at the end of the harvesting period (26 daily harvests).

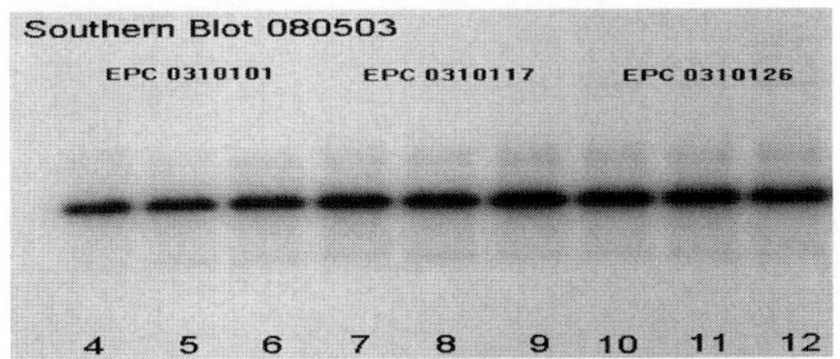


*Figure 3.* Analysis of gene copy number by Southern-Blotting over 26 harvests.

## 4. PRODUCT QUALITY

The N-glycans of different glycoproteins synthesised during repeated batch fermentations were analysed by HPAEC-PAD.

No differences in the N-glycan structures could be detected. Sialylation and antennarity remained constant over the whole production period (data not shown).

## 5. CONCLUSION

A repeated batch fermentation process combining the advantages of both fed batch and perfusion processes was developed.

This fermentation process results in controlled culture conditions without significant batch to batch variations.

Product quality was constant over the whole production process.

Further the process significantly reduced the time for process development and can be quickly adjusted to cell and product specific needs.

Repeated batch fermentation was performed at scales from 30L to 250L for the production of different products under non-GMP and GMP-conditions.

# Serum-free Transient Expression of Antibodies Using 40 Tray Cell Factories

Peter Schlenke, Katrin Langnickel, Thilo Grob, Frank Kohne, Hans Hofer, Ulrike Brändle, Kurt Russ and Ralf Fehrenbach

*Rentschler Biotechnologie GmbH, Erwin-Rentschler-Straße 21, 88471 Laupheim, Germany*

**Abstract:** A serum-free transient expression method for antibodies in HEK cells and a standardised purification protocol were developed. The methods described enable the fast and cost-effective production of hundreds of milligrams of different antibodies that can be used for test purposes. The N-glycan profile of the antibodies obtained is comparable to that of antibodies expressed in stable CHO cell lines, confirming the suitability of the system to produce material for target validation studies.

**Key words:** Transient expression, antibodies, HEK 293, N-glycans

## 1. INTRODUCTION

For target validation studies, significant amounts of antibodies are required. Because most of the biopharmaceuticals require specific posttranslational modifications for proper bioactivity, e.g. glycosylation, mammalian cells are mainly used as expression systems. The generation of stable cell lines is expensive and time consuming and therefore restricted to a small number of antibody candidates, consequently transient expression of the antibodies offers a suitable method.

We have developed a transient serum-free expression method in HEK293 cells using calcium-phosphate precipitation. The transfection of the cells is performed during cultivation in 40 Tray Cell Factories (CF), each with a culture area of about 25.000 $cm^2$. Due to the adherence of the cells,



*R. Smith (ed.), Cell Technology for Cell Products*, 527–530.

repeated harvesting is easily possible and results in a significant increase in the final product yield.

After a 3-step purification process a highly purified material is obtained. Analysis of the N-glycan structures released from the purified antibodies confirmed the suitability of the system to produce antibodies with characteristic glycosylation pattern.

## 2. RESULTS

### 2.1 Transient expression

Two IgGs were expressed in HEK293 cells cultivated in 4 CF40. In each case four harvests were performed repeatedly after 3 – 4 days of cultivation. Production volume was 4L/CF40, resulting in a total harvest volume of 64 L per IgG. The expression level reached up to 12 mg/L in serum-free culture conditions. For the two IgGs expressed a final product yield of about 530 resp. 430 mg was achieved. The results obtained are summarised in Table 1.

*Table 1.* Summary of the results obtained after large scale transient expression of two human IgGs

| | 1st IgG | | | 2nd IgG | | |
|---|---|---|---|---|---|---|
| Harvest | Titer [mg/L] | Volume [L] | Yield [mg] | Titer [mg/L] | Volume [L] | Yield [mg] |
| 1 | 12 | 16 | 192 | 10 | 16 | 160 |
| 2 | 9 | 16 | 144 | 9 | 16 | 144 |
| 3 | 8 | 16 | 128 | 6 | 16 | 96 |
| 4 | 4 | 16 | 64 | 2 | 16 | 32 |
| | | | Σ 528 | | | Σ 432 |

### 2.2 Purification

After transient large scale expression, the IgGs were purified from the 64 L cell culture supernatant using a standardised 3-step protocol after concentration by ultrafiltration. SDS PAGE analysis, isoelectric focusing and HPLC analysis (data not shown) confirmed, that the purification strategy resulted in a highly purified product (Fig. 1).

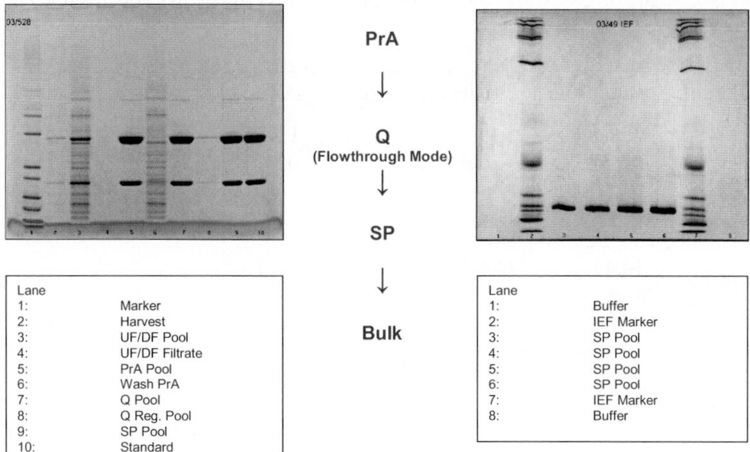


*Figure 1.* SDS PAGE analysis and isoelectric focusing. The IgG was analysed during and at the end of the purification process. After separation the IgG was visualized by silver staining.

## 2.3 Analysis of the N-glycan structures

The main N-glycans detected after desialylation were complex-type biantennary structures without terminal Gal (37%), with 1 terminal Gal (45%) and with 2 terminal Gal (15%). Only about 2% of the N-glycans were sialylated. The analysis confirmed the ability of the transiently transfected HEK293 cells to synthesise N-glycan structures similar to those obtained after stable expression in e.g. CHO cells (Fig. 2).

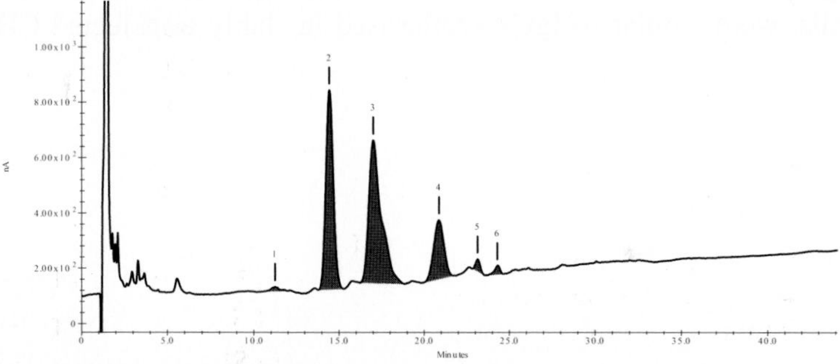


*Figure 2.* HPAEC PAD profile of the desialylated N-glycans released from IgGs after transient expression in HEK293 cells. The structures detected are summarised in Table 2.

*Table 2.* Desialylated N-glycan structures found on IgGs after transient expression in HEK293 cells.

| Peak | Retention [min] | Structure Code | Area [%] | Structure |
|---|---|---|---|---|
| 1 | 11.23 | CoreF1GN1G0S0 | 0,9 | |
| 2 | 14.43 | CoreF1GN2G0S0 | 36,6 | |
| 3 | 17.00 | CoreF1GN2G1S0 | 44,9 | |
| 4 | 20.87 | CoreF1GN2G2S0 | 15,1 | |
| 5 | 23.13 | unknown | 1.35 | |
| 6 | 24.30 | unknown | 1.18 | |

GlcNAc Man Gal Fuc

# 3. CONCLUSION

By transient expression of two human IgGs in 4 x 40 Tray Cell Factories, a final product titer of 530 respectively 430 mg was obtained.

A 3-step purification process using a standardised protocol resulted in a highly purified product.

The N-glycosylation pattern of the IgGs, after transient expression in HEK293 cells, were similar to IgGs synthesised in stably transfected CHO cell lines.

# Monitoring Cell Physiology in Influenza Vaccine Production by Flow Cytometry

Josef Schulze-Horsel [1], Yvonne Genzel [1], Udo Reichl [1,2]
[1] *Max Planck Institute for Dynamics of Complex Technical Systems, Magdeburg, Sandtorstr. 1, 39106 Magdeburg, Germany. E-mail: horsel@mpi-magdeburg.mpg.de*
[2] *Chair of Bioprocess Engineering, Otto-von-Guericke-University, Magdeburg, Germany*

**Abstract:** The cell cycle distributions of adherent Madin-Darby canine kidney cells grown for vaccine production were investigated. During cell growth phase cell cycle distributions for all cultivations were similar whereas during infection and mock-infection significant differences in S-phase contents were observed.

**Key words:** flow cytometry, cell cycle, cell physiology, influenza, DNA content, bioprocess engineering, vaccine production, virus, microcarrier culture, viral infection, Madin-Darby canine kidney, cell growth, cell attachment, Cytodex 1

## 1. INTRODUCTION

Equine influenza vaccine production was investigated as a model process [1]. Here, we describe differences in cell physiology during infection compared to a mock-infection (without virus seed).

For the examination of differences in cell cycle behaviour during cell growth and virus replication MDCK cells were grown on microcarriers in lab-scale bioreactors and harvested by trysinization for flow cytometric analysis.

## 2. MATERIALS AND METHODS

Adherent MDCK cells (ECACC, No. 84121903) were cultivated as described in [1]. After washing steps the cells were infected with equine influenza strain A/Equi 2 (H3N8) Newmarket 1/93 (NIBSC) at a multiplicity



*R. Smith (ed.), Cell Technology for Cell Products*, 531–533.

of infection (MOI) of 0.025 according to plaque forming units of the virus working seed.

In 0.5 L wv stirred-tank bioreactors (Sixfors, Infors), cells were seeded at an initial cell density of 1.9 to 2.8 x$10^5$ cells/mL to grow on 2 g/L Cytodex 1 microcarriers (G.E. Healthcare).

The cell cycle distribution of adherent cells and detached cells from the supernatant was measured by flow cytometry via cellular DNA content according to [2] and calculated from DNA histograms with Cylchred 1.0.2 software (Cardiff University) [3].

## 3. RESULTS AND DISCUSSION

To investigate the influence of virus on host cell physiology a mock-infected cultivation was used as negative control compared to an infected culture. Oxygen consumption and cell growth continued during the mock-infection phase (data not shown) whereas in the infected culture, the oxygen consumption decreased and the dissolved oxygen concentration constantly increased after 140 h due to virus-induced cell death (Fig. 1). Viral infection also caused increase of cells in supernatant and decrease of cells on microcarriers.

For the mock-infected culture we observed an increase of more than 30% in the S-phase fraction of MDCK cells within 24 h after seeding (Fig. 2). Although the cell cycle data of the inoculum for the infected culture is not available, the values were considered comparable because the pre-cultures were treated equally. After this maximum we observed a continuous decrease in the S-phase fraction to less than 8% at the end of the cell growth phase. Simultaneously, the fraction of cells in $G_{0/1}$-phase increased (data not shown).

After washing steps and addition of serum-free virus maintenance medium both mock-infected and influenza-infected culture showed an increase in the S-phase fraction. During mock-infection the increase in S-phase was parallel to an increase in cell numbers. Similar to the growth pattern of the cell growth phase after reaching a maximum of 16% the S-phase decreased again. However, for the infected culture the increase in S-phase continued up to 50% at 140 h and then decreased. At the time-point of maximum S-phase content all cells were considered to be infected. Whether this is due to the medium exchange or to the virus infection process has to be clarified.

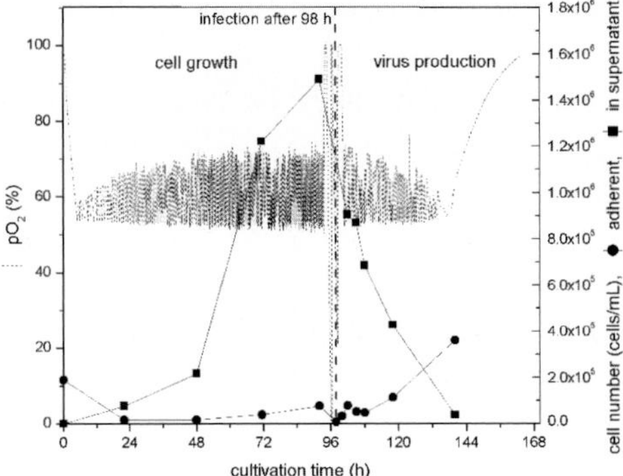


*Figure 1.* On-line pO2 value and cell numbers of cells on microcarriers and cells in supernatant of a virus-infected microcarrier cultivation.

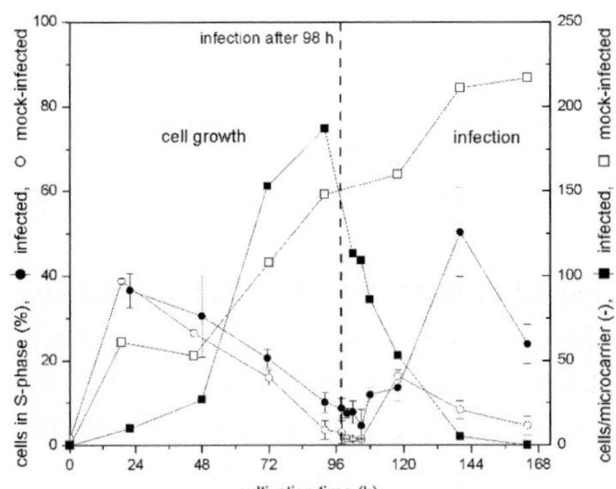


*Figure 2.* Distribution of S-phase and cell numbers on microcarriers for infected and mock-infected cultures.

# REFERENCES

[1] Genzel Y., Behrendt, I., Koenig, S., Sann, H., Reichl, U., 2004, Metabolism of MDCK cells during cell growth and influenza virus production in large-scale microcarrier culture, Vaccine 22 (17-18): 2002-2008.

[2] Darzynkiewicz, Z., Juan, G., 1997, DNA content measurement for DNA ploidy and cell cycle analysis, in: Robinson, J.P., Darzynkiewicz, Z., Dean, P.N., Hibbs, R.A., Orfao, A., Rabinovitch, P.S., Wheeless, L.L. (Eds.), Current protocols in cytometry, Wiley & Sons, Unit 7.5.

[3] Hoy T., 1999, http://cardiff.ac.uk/medicine/hematology/cytonetuk/documents/software.htm (2005/06/15).

# NucleoCounter – An Efficient Technique for the Determination of Cell Number and Viability in Animal Cell Culture Processes

Dimpalkumar Shah[1] Paul Clee[1], Sthen Boisen[2], Mohammed Al-Rubeai[1,3]
[1]*Department of Chemical Engineering, The University of Birmingham, Birmingham, U.K.,* [2]*ChemoMetec A/S, Gydevang 43, DK-3450 Allerod, Denmark.* [1,3]*Department of Chemical and Biochemical Engineering, University College Dublin, Dublin, Ireland*

**Abstract:** The NucleoCounter is a novel, portable device based on fluorescence microscopy principle for the determination of cell number and viability. The present work establishes its use with animal cells and checks its reliability, consistency and accuracy in comparison with other cytometric techniques. The work addresses and overcomes the problems of subjectivity, and some of the inherent sampling errors associated with using the traditional haemocytometer and Trypan Blue Exclusion Method. NucleoCounter offers reduced intra- and inter-observer variation as well as consistency in repetitive analysis that establishes it as an efficient and highly potential device for at-line monitoring of animal cell processes.

**Key words:** NucleoCounter, Elite EPICS Flow Cytometry, Trypan Blue Method

## 1. INTRODUCTION

Cell Count and Viability are important parameters in Animal Cell Culture Processes, industrial as well as research point of view, as productivity is directly related to the number of viable cells. There are number of techniques available in the market to monitor the status of Cell Culture. The conventional method of counting cells is manual counting of the cells in light microscope by Trypan Blue/Haemocytometer. This method has been used widely in the industry, as it is very cheap and easy to do. But the disadvantage of this method is large inter-user variability. The reasons for this might be counting too few squares, uneven distribution of cells in the



*R. Smith (ed.), Cell Technology for Cell Products*, 535–538.

sample or in the haemocytometer chambers (sampling errors), incorrect dilution, overfilling or variation of haemocytometer filling rate. [1] Flow Cytometry is a very important research tool for measuring the properties of the cells in culture, but until recently, it was very impractical for the routine use in the industry, as the instrument was used to be very large and required considerable operator training.

NucleoCounter [2, 3] (ChemoMetec A/S, Denmark) is the latest technique based on the fluorescence microscopy for determining the Cell Number and Viability in Animal Cell Processes. It is a portable machine with an external chamber to measure the cell number and viability without the internal movement of cell solution. 'NucleoCassette' is an external assembly to load the cell solution and by putting this 'NucleoCassette' in an external chamber of 'NucleoCounter', it gives a cell count directly on the digital screen. After measuring live and dead cell count separately, viability can be calculated mathematically.

With EPICS Elite Flow Cytometry [1], by the method and protocol according to the user manual, cell number and viability have been measured. Flow Cytometry can also be used to measure various parameters, like apoptosis, cell cycle, etc.

Techniques like, Trypan Blue Method, NucleoCounter and EPICS Elite Flow Cytometry were compared in terms of, Cell Count and Viability Curves with different cell lines, reproducibility, interpersonal variability, reliability/accuracy and time per reading and cost, by setting up various experiments with total of four cell lines, like TB/C3, NS0 myeloma, CHO and NS0 serum-free cell lines using appropriate media for all cell lines.

## 2. RESULTS AND DISCUSSION

### 2.1 Quantitative and Qualitative Comparison of different techniques for the determination of Cell Count and Viability in Animal Cell Culture Processes:

| | **NucleoCounter** | **EPICS Elite Flow Cytometry** | **Trypan Blue Method** |
|---|---|---|---|
| **Growth and Viability Curves** | For all four cell lines, growth and viability curves are smooth and conforming to the general growth curve format in terms of different phases of cell cycle. | Growth and viability curves are very curvy in nature and not smooth, not conforming to general growth curve nature of animal cells. | Slightly similar to the NucleoCounter, but still irregular shape of curve, implying variability in measurement due to different factors |

| Standard Deviations and Errors (Reproducibility) [4] | General Standard Deviations for similar experimental reading with NucleoCounter is lesser than other two techniques and also Standard Error (0.260 for Cell Count and 0.378 for viability measurement) is less than other two techniques for all the four experiments, in which these techniques were used for measurement. | General Standard Deviations are varying with different experiments and different readings and Standard Error (0.7 for Cell Count and 0.5 for viability measurement) is higher that the NucleoCounter for same experiments. | General Standard Deviations are highly varying from experiment to experiment and reading to reading and Standard Error (1.28 for Cell Count and 0.694 for viability measurement) for same experiment is highest than other two techniques. |
|---|---|---|---|
| Interpersonal Variability | Co-efficient of Variation is less with same operator as well as different operators. | - | C.V. is higher and highly varying with same as well as different operators. |
| Accu-racy | Accuracy is 98% with the beads solution of known concentration. | - | - |
| Time/reading | A.T. (Average Time) per reading is maximum 2 minutes (Cell Count and Viability). | A.T. per reading is 7-8 minutes after pre-running time of around 38 minutes (For alignment and cleaning). | Average minimum time per reading is 3.5 minutes for same user. |

## 3. CONCLUSION

The NucleoCounter can be applied as a rapid, precise and objective tool for routine measurement of cell number and viability in animal cell culture processes and it overcomes the problems of subjectivity and variability associated with using Trypan Blue Method. Thus the technique is highly suitable for at-line monitoring and could be automated for on-line monitoring.

# REFERENCES

M.Al-rubeai, K. Welzenbach, D.R. Lloyd, A.N.Emery; "A rapid method for evaluation of cell number and viability by flow cytometry";(1997); Cytotechnology; 24, 161-168

Nucleocounter[TM] – User's Guide; Chemometec Inc.

Official Site of Chemometec Inc., Denmark; www.chemometec.com

D.L. Streiner; ' Maintaining Standards: Differences between the Standard Deviation and Standard Error, and When to Use Each'; 1996; Can J Psychiatry; Vol 41, 498-502

# On-line Biomass Monitoring in Cell Culture with Scanning Radio-frequency Impedance Spectroscopy

J.P. Carvell and C. Cannizzaro
*Aber Instruments Ltd, Aberystwyth, UK; MIT, Center for SpaceResearch, Boston, USA*

## 1. INTRODUCTION

On-line radio-frequency (RF) impedance spectroscopy is now an accepted measurement system for the concentration of viable biomass in biotechnology. Using RF Impedance in either a fixed or dual frequency mode, the method provides useful information on the live cell concentration and is routinely used in both development and cGMP manufacturing systems in large scale cell culture. The on-line live cell concentration can be used to record or identify key changes in the process, or it can be used to control a constant level of live biomass in the reactor or the addition of feed reagents.

This paper describes how additional information can be provided by utilizing the full capacitance spectrum over a wide range of radio-frequencies (0.1-10MHz). Dielectric spectroscopy was used to monitor two CHO perfusion culture experiments. A plot of the capacitance at a low excitation frequency versus the value at a higher frequency proved to be an accurate indicator of the major transition points of the culture, i.e. maximum cell viability, end of lactate consumption, and point of zero viability.

Future developments of this method could provide an instantaneous reading of the metabolic state of the culture and an estimation of the mean cell diameter.



*R. Smith (ed.), Cell Technology for Cell Products*, 539–541.

# 2. THE USE OF VARIABLE FREQUENCY CAPACITANCE INSTRUMENTS TO DETECT CELL MORPHOLOGY CHANGES

## 2.1 CHO cell cultivation

Perfusion cultures of serum-free Chinese Hamster Ovary cells (CHO SSF3) were monitored with a Biomass Monitor (Aber Instruments Ltd, UK). For the full details for the methodology see Canizzaro *et al.*, 2003. Figure 1 shows an example of the glucose consumption and the lactate production/consumption profiles. Six distinct phases were present. The first phase (a) was batch growth on glucose with concomitant lactate production. The second phase (b) was the longest, and corresponded to the start of feeding until the consumption of the glucose. At the onset of feeding, glucose accumulated, but was then consumed and/or washed out as the cell density increased. There was an initial drop in the lactate concentration due to hydraulic dilution, but within 25 h it started to increase once more. The third phase (c) represents the time between the consumption of the glucose and the cessation of feeding. The fourth phase (d) corresponded to lactate consumption once the feed was stopped, while the fifth phase (e) began when accumulated lactate was consumed. The transition from the fifth to the sixth phase (e) was defined as the point of zero cellular viability. The viability started to decline during lactate consumption phase, and then rapidly fell to zero once lactate was consumed (Figure 2).

In Figure 2, a phase plot is shown, whereby the capacitance at 1.05MHz was plotted versus the capacitance at 0.35 MHz. Also noted are the defined phases a-f, and their transition times. The end of feeding (c-d transition), the end of lactate utilization (d-e transition), and the point of zero viability (e-f transition) are clearly elucidated. Graphical representation of the capacitance data in this way provides an online and intuitive representation of culture state, without the need for sampling. During the batch and perfusion phases (a-c), the capacitance at 0.35MHz was linearly correlated with that at 1.05 MHz. A linear calibration model for biomass established during this period would be relatively independent of chosen excitation frequency. However, it is quite clear from the phase plots that for the latter stages of cultivation, the capacitance is dependant upon excitation frequency.

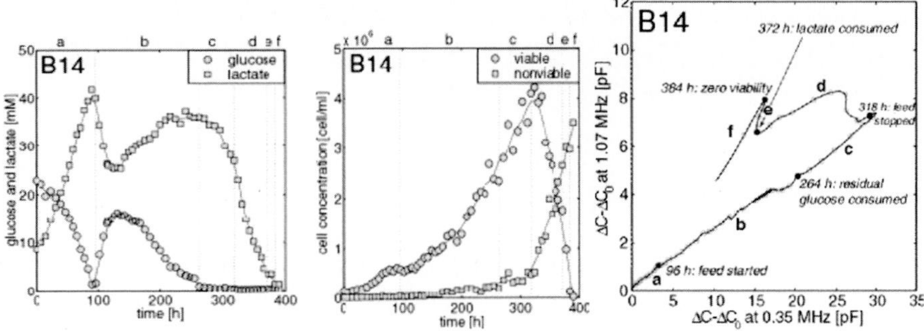


*Figure 1.* Glucose and lactate profiles for a CHO perfusion culture.

*Figure 2.* Viable and non-viable cell profiles for a CHO perfusion culture.

*Figure 3.* Capacitance at 1.05 MHz versus the capacitance at 0.35MHz for a CHO perfusion culture.

## 3. CONCLUSION

Dielectric spectroscopy has already been established as one of the best techniques for on-line biomass measurement. Here it is shown that the collection of the capacitance spectrum over a range of excitation frequencies greatly enhances its value. Significant variations were seen in the capacitance data collected at different frequencies. Changes in cell morphology due to budding and cell division can clearly be seen. Phase plots of the capacitance at high frequency versus the capacitance at low frequency provide information on the major transition points of a culture, and models can be designed to correlate biomass with cell number bioreactors, even where cell size changes are significant.

## REFERENCES

Canizzaro, C., Gugerli, R., Marison, I and von Stockar, U., 2003. On-line biomass monitoring of CHO perfusion culture with scanning dielectric spectroscopy. Biotechnology and Bioengineering 84. 597-610.

Ducommun, P., Ruffieux, P.A., Kadouri, A., Von Stockar, U. and Marison, I. 2001. Monitoring temperature effects on cell metabolism in a packed bed process, Biotechnology and Bioengineering 77, 838-842.

# The Use of a Mobile Robot for Complete Sample Management in a Cell Culture Pilot Plant

Martin Wojtczyk[1,2], Rüdiger Heidemann[1], Klaus Joeris[1], Chun Zhang[1], Mark Burnett[1], Alois Knoll[2] & Konstantin Konstantinov[1]

[1] *Bayer HealthCare, Biological Products, Process Sciences R&D,800 Dwight Way, Berkeley, CA 94710, USA,* [2] *Technische Universität München, Department of Informatics, Robotics & Embedded Systems, Boltzmannstr. 3, D-85748 Garching, Germany*
*email:rudiger.heidemann.b@bayer.com*

**Abstract:** A mobile robot system was developed that is capable of automating the complete sample management in a biotechnological laboratory. The robot consists of a wheeled platform and a mounted industrial robot arm with an attached gripper tool. The proper interaction with the biotechnological devices is accomplished by the use of a color camera for object recognition, a force/torque sensor to prevent damages and laser scanners for precise localization in the plant. Furthermore necessary changes to the environment of the robot are kept to a minimum. By providing a scripting language, the robot system can be easily adapted to new devices and further tasks. The use of this autonomous robot was successfully implemented and a fully automated sample management system established.

**Key words:** Automation, Cell Culture, Mobile Robot, Sample Management

## 1. INTRODUCTION

Sample management is an inevitable part during the development and subsequent production of biopharmaceuticals to keep track of the growth conditions; make general adjustments and optimize harvesting times. This finally will ensure the repeatability and consistency during the cell culture process. Monitoring and controlling cell culture processes often requires a 24 hours coverage schedule especially for continuous cultures.

To overcome this immense human involvement an autonomous mobile robot was previously developed at the University of Bielefeld which is



*R. Smith (ed.), Cell Technology for Cell Products*, 543–547.

capable of automating the entire sample management in a cell culture pilot plant (Lütkemeyer *et al.*, 2002, Scherer *et al.*, 2003, Knoll *et al.*, 2004, Scherer 2004 and Poggendorf 2004).

The robot's role in the sample management process is similar to that of a human operator. After drawing a sample from a bioreactor utilising a sampling device, the sample is brought to a pipetting station where the sample is aliquot and if necessary diluted and than handed to at a Cedex cell counter (Innovatis AG, Germany). The cells are separated from the supernatant using a centrifuge. The cell-free samples are then tagged with barcodes, registered at an appropriate scanner and finally stored in a freezer. Meanwhile the process control system evaluates the received data and adjusts the production parameters if necessary (see Figure 1 for the entire process routine).

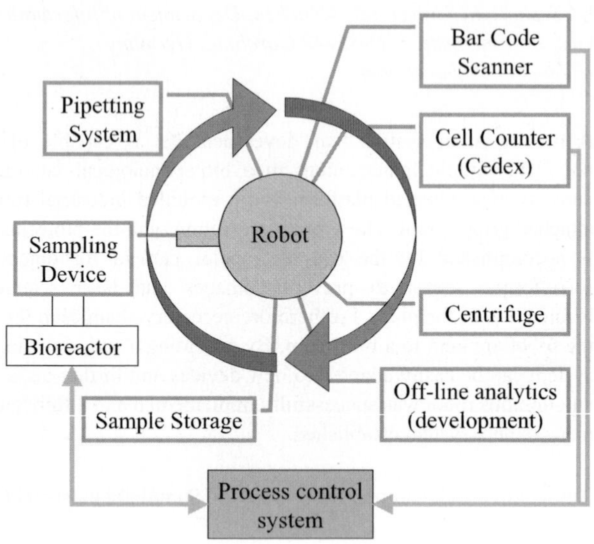


*Figure 1.* The automated sample management process.

## 2. ROBOT – BASIC REQUIREMENTS

A robot needs to accomplish certain basic requirements for the sample management in an environment which is usually made for human operators. It needs to reach the different devices, identify them in a reliable way and operates them without the need of costly modifications. Furthermore it is important to handle various types of sample tubes.

## 2.1 Hardware setup

A battery driven mobile platform MP-L655 from Neobotix (Germany) was chosen to provide the robot with the necessary mobility and autonomy. The platform is controlled by its own platform computer running a Linux operating system. Equipped with an industrial seven joints robot arm PA-10 from Mitsubishi Heavy Industries (Japan) and a two-fingered gripper tool attached to it the robot is able to reach and to operate the different devices and handle the desired tubes. Figure 2 shows the mobile robot and a close-up of the gripper tool.

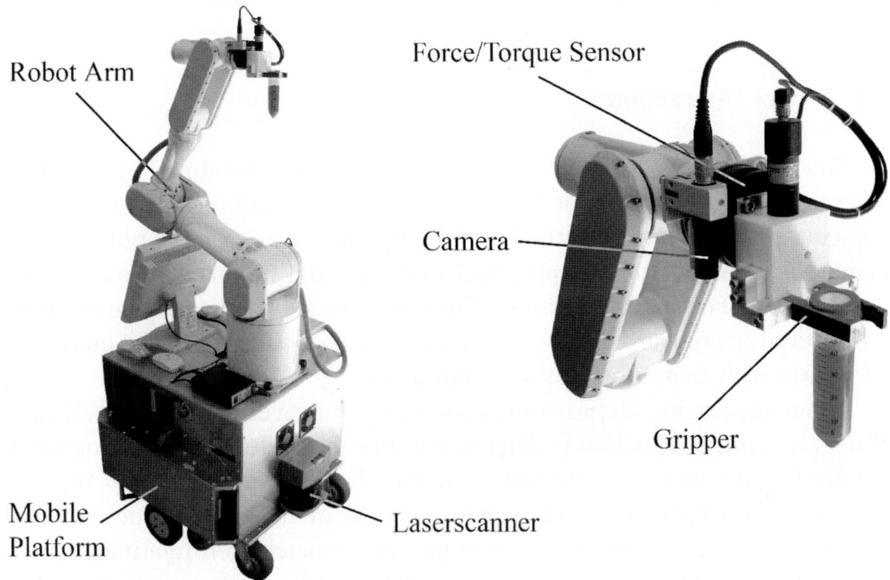


*Figure 2.* Mobile robot platform and close-up of the robot tool carrying a 50ml tube.

## 2.2 Localisation and Navigation

Two 180° laser scanners at the front and back side of the platform detect laser scan markers attached to the walls of a biotechnology laboratory whose positions are matched against the known entries in a map. This method enables the robot to estimate its position and orientation up to 38 times per second (see Figure 3 for the laboratory map). In conjunction with the given map the A* algorithm and tangent graph are applied to find the shortest non colliding way through the plant to the final positions. If dynamic obstacles - for example human personnel - are detected to be too close to the robot on its way to its target position, it stops immediately until the obstacle is out of scope again.

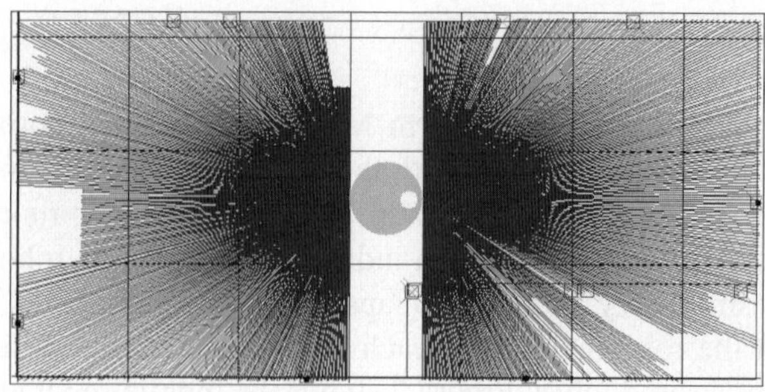

*Figure 3.* Localisation with the aid of laser scanners. The robot's environment seen from the top at the height of the laser scanners showing the acquired distance information.

### 2.3 Device Interaction

Since there is a need for high accuracy while operating the different laboratory devices, the robot tool is equipped with additional sensors. Since cameras provide a very natural and comprehensive sensor input, a CCD camera is mounted at the gripper tool to get a real-time image of the distinct device the robot is approaching. Thus it is possible to compensate little navigation errors caused by noise during the laser scans and to guide the robot arm with an accuracy of one millimeter.

Image processing often is time consuming, however it can be accelerated if unnecessary information is filtered out. For the current object recognition a color based region growing approach was chosen evaluating YUV images. The algorithm looks for compound segments of the colors blue, red and yellow and calculates their centers of gravity to match their relative position between each other against entries in the robot's database which makes object identification possible, where objects contain enough unique color information or at least are supplied with appropriate color markers (see Figure 4). The absolute position of the centers of gravity supplies the necessary information for fine positioning the robot arm.

Once the position and orientation of the current device or object is calculated, a scripted action is triggered which may be supplied with force constraints guarded by the force/torque sensor not to cause any damage to the equipment or even to human personnel. By the use of the general color based identification method and the scripting language, the robot can easily be adapted to new devices.

*Figure 4.* Image processing for object recognition. On the left the captured image of an optical marker and a 50ml tube, on the right the processed image showing the different segments for the search colors.

# 3. CONCLUSION

The goal of the project is to combine research results in both key technologies Robotics and Biotechnology. After the successful utilization in a cell culture pilot plant in Germany the robot was recently transferred to the Biological Products Division of Bayer HealthCare in Berkeley, CA, USA. The goal is to adjust the robot to the environment and Bayer-specific equipment like a new automated sampling valve and a LabView based supervisory system for the later use in development and ultimately GMP type production of biopharmaceuticals.

# REFERENCES

Lütkemeyer, D., Poggendorf, I., Scherer, T., Zhang, J., Knoll, A., and Lehmann, J., 2000, First steps in robot automation of sampling and sample management during cultivation of mammalian cells in pilot scale. Biotechnol. Prog., 16: 822-828

Scherer, T., Poggendorf, I., Schneider, A., Westhoff, D., Zhang, J., Lütkemeyer, D., Lehmann, J., and Knoll, A, 2003, A service robot for automating the sample management in biotechnological cell cultivations. Proceedings of the 30th Annual Conference of Computers in Cardiology, Lisbon, Portugal

Knoll, A., Scherer, T., Poggendorf, I., Lütkemeyer, D., and Lehmann, J., 2004, Flexible automation of cell culture and tissue engineering tasks. Biotechnol. Prog. 20: 1825-1835

Scherer, T., 2004, A Mobile Service Robot for Automisation of Sample Taking and Sample Management in a Biotechnological Pilot Laboratory. PhD thesis, University of Bielefeld

Poggendorf, I., 2004, Einsatz eines Serviceroboters zur Automatisierung der Probenentnahme und des Probenmanagements während Kultivierungen tierischer Zellen in einer Technikumsumgebung. [online] PhD thesis, University of Bielefeld. Available from: http://nbn-resolving.de/urn/resolver.pl?urn=urn:nbn:de:hbz:361-5651

# Closed Loop Control of Perfusion Systems in High-density Cell Culture

A. Bock[1] and U. Reichl[1, 2]
[1]*Max Planck Institute for Dynamics of Complex Technical Systems Magdeburg, Sandtorstr. 1, 39106 Magdeburg, Germany, e-mail: bock@mpi-magdeburg.mpg.de*
[2]*Otto-von-Guericke-University Magdeburg, Universitätsplatz 2, 39106 Magdeburg, Germany*

**Abstract:** The use of a perfusion system prevented unspecific inhibitions effects on cell growth for batch cultivations of adherent MDCK cells used for an equine influenza A vaccine production process. It also increased viable cell numbers. An enzyme sensor for monitoring and closed loop control of metabolites operated stable for at least 55 hours.

**Key words:** Bioprocess, vaccine, influenza, virus, microcarrier, mammalian cell culture, Madin-Darby canine kidney, cell growth, perfusion, closed loop, open loop, dilution rate, enzyme sensor, on-line monitoring, YSI 2700, bioreactor

## 1. INTRODUCTION

The aim of our group is the development and optimisation of integrated concepts for vaccine production with a process of influenza A virus as an example [1]. Clearly, cell number at time of infection is one of the most important factors for achieving high virus yields [4]. Here, we present results obtained for different control strategies of perfusion systems for growth of MDCK cells.

## 2. MATERIAL AND METHODS

Madin-Darby canine kidney (MDCK) cells were obtained from ECACC (No. 84121903). Cultivations were performed in stirred-tank bioreactors (5 L Biostat C, B.Braun; 0.5 L Sixfors bioreactor, Infors AG) using GMEM (Invitrogen) and Cytodex 1 microcarriers (GE Healthcare) under the



*R. Smith (ed.), Cell Technology for Cell Products*, 549–551.

following conditions: 37°C, 50 rpm; pH = 7.3; pulsed aeration of oxygen controlled at $pO_2$ = 40%.

## 3. RESULTS AND DISCUSSION

So far, batch cultivations with higher microcarrier concentrations of up to 5 g/L did not show an expected increase in cell numbers due to media limitations but also unspecific inhibition effects [1]. To overcome these limitations the use of a perfusion system was investigated.

As a starting point, we set up an open loop perfusion system with 4 g/L MC in a 5 L STR (Table 1: run 1) and observed an increase in cell number up to 95% of the theoretical maximum cell number (3.5 x $10^6$ 1/mL, data not shown).

Based on these results we performed two perfusion cultivations with 5 g/L MC and a closed loop control of lactate (run 2) or glucose (run 3) in 0.5 L bioreactors to reduce media consumption.

In run 2 we had chosen a set point for lactate concentration of 15 mM to avoid growth-inhibiting concentrations reported by Hauser [3]. For on-line monitoring a sample from the outlet of the bioreactor was analysed every 30 min in an enzyme sensor. We achieved up to 84% of theoretical maximum cell number (3.94 x $10^6$ 1/mL), but with a high medium consumption up to 14 reactor volumes. The enzyme sensor worked stable over a period of 55 hours.

In run 3 a set point of 10 mM glucose was chosen in respect to the quantitation limit of the enzyme sensor and results from Glacken [4] regarding changes in cellular metabolism at lower glucose concentration. We achieved cell numbers up to 99% of theoretical maximum cell number (4.67 x $10^6$ 1/mL, Figure 1). The on-line signal of glucose showed oscillations of about ±1 mM in respect to the set point due to the dead volume of tubing (≈ 5 mL) between bioreactor and enzyme sensor.

## 4. CONCLUSIONS

MDCK cells showed no growth limitations for lactate concentrations up to 40 mM. We observed a stable operation of the enzyme sensor during perfusion mode for about 55 hours. The cultivations in perfusion mode achieved clearly higher cell numbers in comparison to batch cultivations [1]. Work is in progress to improve monitoring of metabolites by the use of a dead-volume free, and automated sampling device [5].

*Table 1.* Configuration for perfusion cultivations of MDCK cells on microcarriers in different stirred-tank bioreactors.

| Run | vw (L) | D ($h^{-1}$) | Control of ... | Controller | Feed mode |
|---|---|---|---|---|---|
| 1 | 5 | 0.03 | None | None | Continuous |
| 2 | 0.5 | 0.25 (average) | Lactate | Integral | Continuous |
| 3 | 0.5 | 0.04 (average) | Glucose | Proportional | Pulsed |

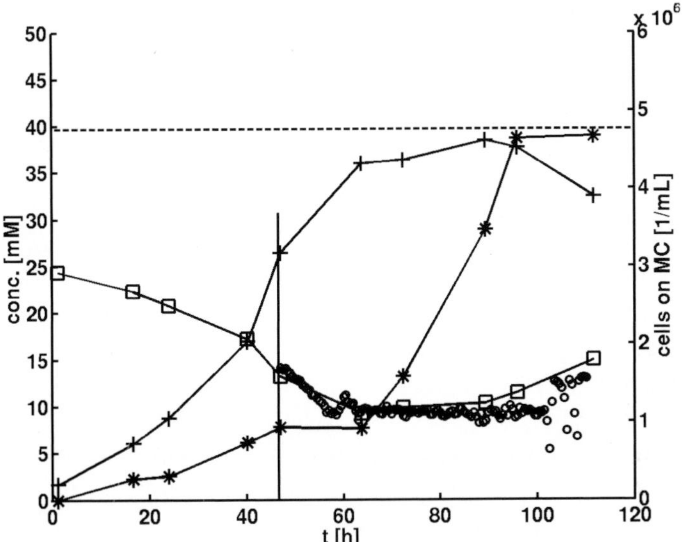


*Figure 1.* Cultivation of MDCK cells in perfusion mode with a closed loop control of glucose (5 g/L MC, run 3): Horizontal line: theoretical maximum cell number, vertical line (47 h): start of perfusion mode and on-line monitoring of glucose, cell number on MC (*) and metabolite profiles: glc off-line (□), lac off-line (+), glc on-line (o).

# REFERENCES

[1] Genzel, Y., Behrendt, I., König, S., Sann, H., Reichl, U., 2002, Metabolism of MDCK cells during growth and Influenza virus production in large-scale microcarrrier culture; Vaccine **22**(17-18): 2202-2208.

[2] Glacken, M. W., Fleischaker, R. J., Sinskey, A. J., 1985, Reduction of waste product excretion via nutrient control: possible strategies for maximizing product and cell yields on serum in cultures of mammalian cells, Biotech. Bioeng. **28**:1376-1389.

[3] Hauser, H., Wagner, R., 1997, Mammalian cell biotechnology in protein production, W. de Gruyther.

[4] Möhler, L., Flockerzi, D., Sann H., Reichl, U., 2005, A mathematical model of Influenza A virus production in large-scale microcarrier culture; Biotech. Bioeng. **90** (1): 46-58.

[5] Sann, H., Bock, A., Reichl, U., 2002, Device and method for extracting liquid samples, WO 2004/ 033077 A2.

# Optimization of Culture Medium with the Use of Protein Hydrolysates

Paul Ganglberger[1], Bernd Obermüller[1], Manuela Kainer[2], Peter Hinterleitner[2], Otto Doblhoff[2], Karlheinz Landauer[2]

[1]*University of Applied Sciences, Course Bio- and Environmental Technology, Wels, Austria*

[2]*igeneon AG, Vienna, Austria*

**Abstract:** In this study the effect of protein hydrolysates in order to increase cell viability, cumulative cell cell-days number, volumetric and specific productivity with respect to product quality was examined. 10 different complete non-animal protein hydrolysates (wheat gluten, rice, soy, pea, cotton, corn) were tested in standard grade roller bottles. The concentration levels ranged from $1 g \cdot l^{-1}$ to $5 g \cdot l^{-1}$. The experiments revealed a 30% – 50% increase of volumetric productivity compared to standard cultures.

The effect based mainly on the extended culture duration of the bioprocess. For characterizing product quality the techniques of anti-idiotypic ELISA with target antigen, complement dependent cytotoxicity, antibody dependant cellular cytotoxicity and isoelectric focusing were employed. It could be shown that the use of peptones did not change the product quality, potency or specificy. As conclusion, addition of peptones might be a successful strategy to enhance the performance of biological production processes.

**Key words:** peptones, protein hydrolysates, media optimization, hybridoma, cell growth, productivity, cell culture, serum-free, monoclonal antibodies, Animals, growth substances, cost-benefit, cancer, immuno therapeutic, economic

# 1. INTRODUCTION

The increasing awareness of the risk of contamination with mycoplasma, viruses and TSE caused by animal-derived components has led to serum-free media formulations. With the elimination of serum, e.g. Fetal Calf Serum or



*R. Smith (ed.), Cell Technology for Cell Products*, 553–557.

other animal derived additives for biological production processes, cell culture performance often decreases. Lots of different agents such as amino acids, trace elements, vitamins or complete vegetarian protein hydrolysates are used to overcome this problem. Especially small peptides like protein hydrolysates, which are also called peptones, protein fission products, peptides and hydrolyzed proteins, seem to be a viable option for the replacement of serum. In addition, peptones provide a stable source of glutamine and other amino acids.

The aim of this study was to show the effect of 10 different protein hydrolysates, which are using soy, rice, corn, wheat gluten, cotton or pea as raw material, added to a chemical defined medium on a recombinant SP2/0 hybridoma cell line in order to increase cell viability, cumulative cell number, volumetric and specific productivity with respect to product quality.

## 2. MATERIAL AND METHODS

The used cell line was a recombinant SP2/0 producing a humanized anti-Lewis Y antibody. As standard medium for all experiments the commercial available Ex-Cell™-Sp2/0 (JRH, UK) supplemented with 8mM L-glutamine 200mM (100x) (Gibco, UK) was used. Ex-Cell™-Sp2/0 is an animal-component-free, protein-free, and chemically defined liquid medium. The content of protein hydrolysate ranged between $1 g \cdot l^{-1}$ and $5 g \cdot l^{-1}$.

Cell culture was performed at 37°C in standard grade polystyrene cell culture flasks (Nunc, Denmark) in a humidified incubator with 7% $CO_2$. Tests were performed in 850 $cm^2$ standard roller bottles (Corning, Germany) with 200 to 450 ml medium in duplicates. The inoculation densities varied from $1 \cdot 10^5$ to $2.5 \cdot 10^5$ $cells \cdot ml^{-1}$. Sampling was performed on a one to three day basis. The cell number of the supernatant was determined with the trypan blue dye exclusion method and the concentration of produced antibodies were measured with size exclusion chromatography (SEC). The cultures were terminated, if viability dropped below 40%.

The quality of the produced antibody IGN311 was examined with the following assays: Lewis Y ELISA, IEF-Page, CDC and ADCC. The LewisY ELISA was used to evaluate the binding activity of antibodies to the carbohydrate antigen LewisY. Specifically bound immunoglobulins were detected with an anti-human immunoglobulin-enzyme conjugate which was inducing a photometrically measurable color change. For the IEF Page an Ampholine PAGplate™ was used with pH values between 3.5 and 9.5. ADCC against SKBR 3 cell line was analyzed in a $^{51}Cr$ release assay. CDC against SKBR 3 cell line was analyzed in a caspase enzyme assay.

# 3. RESULTS

## 3.1 Growth Characteristics

As it can be seen in Figure 1, HyPep4601 had an positive effect on the cumulative cell-days number Hence, HyPep4601 increased cumulative cell-days number by 25%. HyPep5603 had no significant influence on cumulative cell-days number (Figure 1) and viable cell number.

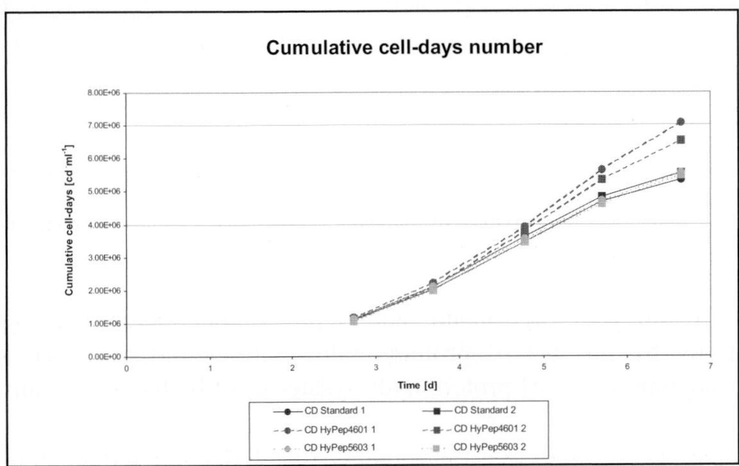


*Figure 1.* Cumulative cell-days number of standard cultures, cultures supplemented with HyPep4601 and cultures supplemented with HyPep5603.

6 out of 10 protein hydrolysates (HyPep1510, HyPep4601, PCE80B, CNE50M, WGE80M-UF and SE50MAF-UF) had an positive effect on the growth characteristics, which was characterized by the parameters maximum viable cell number, cumulative cell-days number, growth rate and cell culture duration, in comparison to the standard culture.

## 3.2 Productivity

Figure 2 depicts an increase of 56% in final volumetric productivity by the addition of HyPep4601 in comparison to the standard culture. The final mean endtiter of cultures supplemented with HyPep4601 was 323 $mg \cdot l^{-1}$. The standard cultures only had a final mean endtiter of 206 $mg \cdot l^{-1}$. HyPep5603 showed no influence on the cell culture in terms of volumetric productivity (211 $mg \cdot l^{-1}$).

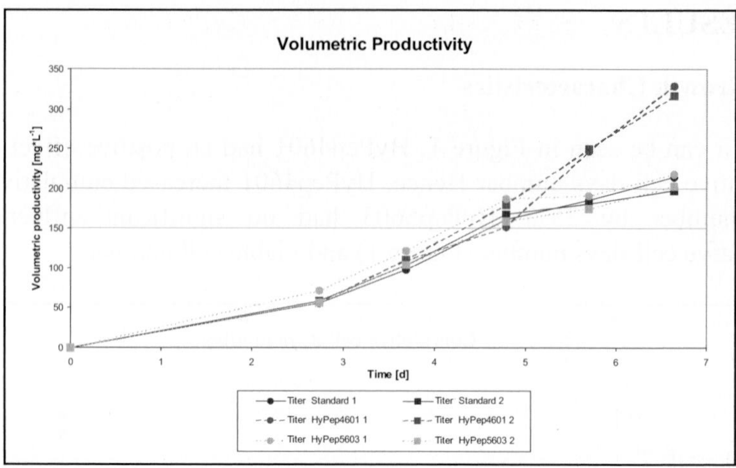


*Figure 2.* Volumetric Productivities of standard cultures and cultures supplemented with HyPep4601 and HyPep5603.

In our study a significant and reproducible increase in specific productivity by the use of protein hydrolysates could not been shown. Hence, the main effect of protein hydrolysates must be based on nutritional benefit.

### 3.3 Product Quality

The IGN311 produced under serum free conditions with the use of peptones had the same antigen binding activity and a slight increase in effector functions as measured by ADCC- and CDC-assays in comparison to the reference standard. The Lewis Y ELISA showed no significant differences between IGN311 ref. and the tested AB's.

The IEF Page of the antibody produced under serum containing conditions had one band less than the antibodies produced under serum free conditions and the use of protein hydrolysates. However, the cell line is robust with regards to productivity and product quality under the tested conditions (medium change, up-scaling).

## 4. CONCLUSION

7 out of 10 protein hydrolysates improved volumetric productivity (see Table 1). The effect based mainly on the extended culture duration of the bioprocess. The experiments revealed an increase up to 56% in terms of

*Table 1.* Summary selected parameters of all batch cultures.

| Name | Growth rate $[d^{-1}]$ | Cumulative cell-days $[cd \cdot ml^{-1}]$ | Duration [d] | Specific productivity $[pg \cdot cd^{-1}]$ | Volumetric productivity $[mg \cdot l^{-1}]$ |
|---|---|---|---|---|---|
| HyPep1510 | 0.41 / R=0.98 | $4.7 \cdot 10^6$ | 7 | 51 / R=0.99 | 264 |
| HyPep4601 | 0.36 / R=0.97 | $6.8 \cdot 10^6$ | 7 | 46 / R=0.98 | 323 |
| HyPep4602 | 0.36 / R= 0.99 | $4.3 \cdot 10^6$ | 7 | 32 / R= 0.97 | 211 |
| HyPep5603 | 0.26 / R= 0.93 | $5.5 \cdot 10^6$ | 7 | 32 / R= 0.93 | 162 |
| HyPep7401 | n.a. | n.a. | n.a. | n.a. | n.a. |
| HyPep7504 | 0.28 / R=0.97 | $4.0 \cdot 10^6$ | 7 | 53 / R= 0.99 | 288 |
| SE50MAF-UF | 0.44 / R=0.96 | $8.0 \cdot 10^6$ | 8 | 27 / R= 0.95 | 241 |
| WGE80M-UF | 0.40 / R=0.98 | $8.2 \cdot 10^6$ | 8 | 28 / R=0.93 | 263 |
| CNE50M | 0.51 / R= 0.95 | $9.3 \cdot 10^6$ | 9 | 33 / R= 0.97 | 319 |
| PCE80B | 0.43 / R=0.96 | $8.7 \cdot 10^6$ | 10 | 43 / R= 0.96 | 321 |
| Standard | 0.40 | $5.2 \cdot 10^6$ | 7 | 40 | 209 |

volumetric productivity compared to the standard culture in a chemically defined medium. The best results could be obtained with HyPep4601. This hydrolysate, based on wheat gluten, increased volumetric productivity by 56%, cumulative cell-days by 25% and specific productivity by 36%.

The quality of the monoclonal antibody produced in peptone-supplemented cell cultures, analyzed by ELISA and IEF was not affected. In CDC-tests a slightly positive influence on the potency of the antibody produced in protein-hydrolysate containing medium could be revealed.

As conclusion, addition of peptones might be a successful strategy to replace animal-derived components and to enhance the performance of biological production processes.

# Optimization of a Fed-batch Strategy for Production of Mab Ign 311 in Small Scale Experiments

P. Hinterleitner, P. Ganglberger, B. Obermüller, B. Antes, O. Doblhoff, K. Landauer
*Igeneon AG, Brunner Strasse 69, Obj. 3, 1230 Vienna, Austria*

**Abstract:** We demonstrate here an optimization of a feeding strategy in standard grade roller bottles in a 400 mL scale. Knowledge of critcal parameters, e.g. nutrients shortage, was incorporated in our feeding strategy and thereby final product concentration could be increased by 75% compared to non-feeded standard cultures. Combined with optimized media increases in product titers of about 125% could be achieved. These optimization steps had no negative impact on product characteristics. At the end, a strategy ready for further experiments in bench-top bioreactors could be presented.

**Key words:** Feeding strategy, fed batch, small scale, roller bottles, bench-top bioreactor, consumption rates, SP2/0, IGN311, hydrolysate, Hypep 4601, product quality

## 1. INTRODUCTION

Igeneon is currently testing a humanized monoclonal antibody (IGN311) specific for Lewis Y, a carbohydrate antigen, in clinical trials of passive cancer immunotherapy. A Phase I trial was just finished, and a Phase II study has to be supported with sufficient material. To meet this increasing demand the production process has to be adjusted.

As not all experiments can be performed in large scale, we tested a variety of process parameters, e.g. the feeding strategy, in roller bottles.



*R. Smith (ed.), Cell Technology for Cell Products*, 559–562.

## 2. MATERIAL AND METHODS OF ROLLER BOTTLE EXPERIMENTS

Production cell line for this antibody is a murine Sp2/0. Tests were performed in 850 cm² roller bottles (Corning, Germany) with a starting volume of 400 mL. Inoculation density was 1.5 x $10^6$ cells $mL^{-1}$. Experiments were performed at 37 °C in an incubator with 7% CO2.

Antibody concentration of samples was analyzed by SEC-HPLC. Standard was IGN311 Batch 213601 (Igeneon, Austria). This material was produced in a hollow fiber process using a medium which was supplemented with 10% FCS. Feeding media was a concentrate of our cultivation media, glutamine was supplemented separately.

## 3. DATA OF ROLLER BOTTLE EXPERIMENTS

First experiments were aimed to describe the effect of adding L-glutamine at a later stage in cultivation. As expected feeding prolonged cultivation time and cell densities (2.0 x $10^6$ cells $mL^{-1}$ on average in fed batches versus 1.6 x $10^6$ cells $mL^{-1}$ for the batch). Product titers were increased by later addition of L-glutamine by 20%, and by the use of Hypep (HP) 4601 in the feeding media by 30%. These results were very promising for the further design of our fed batch strategy, and in our next experiments we combined these two effects.

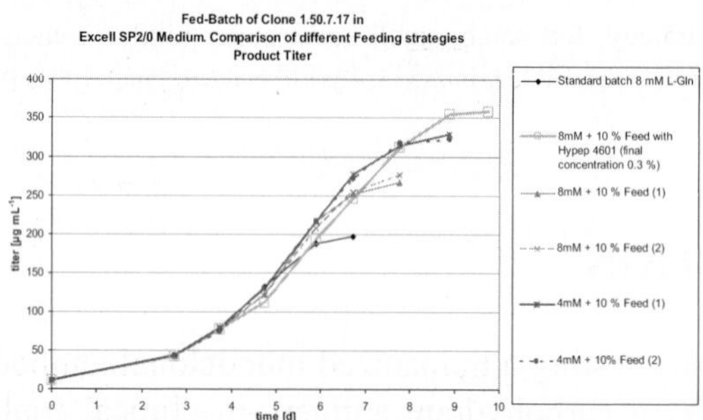


*Figure 1.* Cultures containing 4mM L-Glutamine were fed on day 4 and 4mM L-Gln was added on day 6. Cultures containing 8 mM L-Gln were fed on day 6. Longer cultivation time reflects in higher product titers as can be seen in this graph.

In the next experiment positive effects could be verified, titers could be increased about 15% just by supplementation of HP 4601 to the feeding media, and up to 35% by using the hydrolysate as component of the cultivation media. As a consequence it was decided that HP 4601 will be taken as standard supplement for further fed batch experiments.

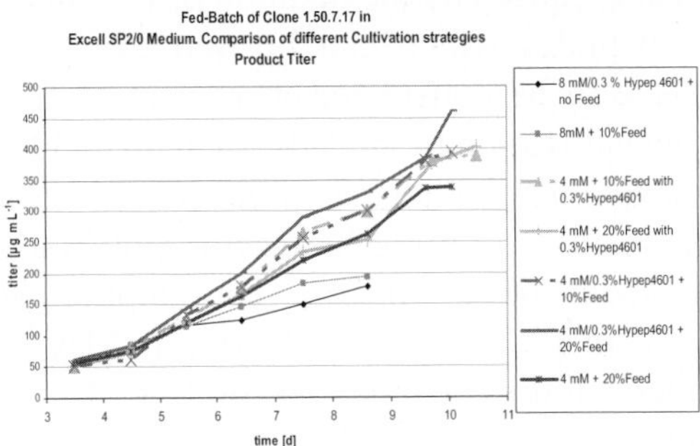


*Figure 2.* Combination of addition of HP 4601 and delayed supplementation of L-Glutamine (was generally added to 4mM -cultures on day 6). Longer cultivation time and higher maximum cell number accumulates in final titers.

## 4. QUALITY TESTING OF IGN311

Analyses of the binding affinity by anti-idiotypic ELISA and anti-Lewis Y ELISA revealed no changes in affinity between IGN311 standard and IGN311 produced with up to 0.5% Hypep 4601. When compared to IGN311 standard, material derived from cultures containing Hypep 4601 (batch and fed-batch) displayed ADCC activities generally 2 to 4 times higher, CDC activity was elevated between 10% and 30%. The IEF pattern was not changed by whatever kind of feeding strategy or supplementation. Compared to IGN 311 standard, all samples own two additional band in the alkaline region (pI 8.4 – 8.1). This is mainly caused by the switch from FCS containing production media (for standard) to serum free media.

## 5. CONSUMPTION AND PRODUCTIVITY RATES AND BIOREACTOR CULTURE

Specific productivities stayed constant throughout cultivation in all roller bottle experiments and were in the usual range of variation of standard cultures (40 ± 4 pg $cell^{-1}$ $d^{-1}$).

We calculated average consumption rates based on cumulative cell-days of cultures in basal media with 8 mM L-Gln. Afterwards we compared them with consumption rates in the early stage of a discontinuous fed-batch containing 0.5% HP 4601 in its media. This fed batch was performed in a small scale bioreactor (Minifors 2.5L, Infors, Switzerland).

In this bioreactor culture consumption rates of almost all amino acids were reduced. This effect is caused by the addition of HP 4601, as it presents a resource of various peptides which can be utilized by our host cells. Critical amino acids have to be considered in respect of feeding, as their content decreased during cultivation in the fed- batch up to 50%.

Glutamine is the mayor energy resource of our IGN311 clone. We found that its consumption is increased during bioreactor cultivation in HP 4601 containing media by about 60%. Glucose consumption stayed almost unchanged in the bioreactor (2.2 ± 0.1 $\mu$mol*$10^{-6}$cells*day$^{-1}$) compared to average values for roller cultures without HP 4601 (2.4 ± 0.1 $\mu$mol*$10^{-6}$ cells*day$^{-1}$]. Although feeding rate in this bioreactor fed-batch was not fully optimized, maximum viable cell number of 5 x $10^6$ cells mL$^{-1}$, and a final product titer of about 500 $\mu$g mL$^{-1}$ could be achieved.

## 6. CONCLUSIONS

- Roller bottle experiments were able to support assessment of feeding strategies and verification of selection of suitable media additive (Hypep 4601).
- In roller bottles increases in product titers of almost 125% were achieved, maximum cell number could be increased up to 25%. Cultivation time could be prolonged by 2 to3 days.
- Material derived either from batch or fed-batch cultures showed slightly enhanced lytic capacity compared to standard produced in serum-containing media.
- Critical amino acids (Asn, Leu, Lys) could be identified and an increase in Gln-consumption could be observed in a small-scale bioreactor cultivation with discontinuous feed.

# Serum-free Influenza Vaccine Production with MDCK Cells in Wave-bioreactor and 5L-stirred Tank Bioreactor

Yvonne Genzel[1], Marlies Fischer[1,2], Ruth Maria Olmer[1,3], Bastian Schäfer[1], Claudia Best[4], Susanne König[1], Boris Hundt[4], Udo Reichl[1,4]

*[1]Max Planck Institute for Dynamics of Complex Technical Systems, Magdeburg, Sandtorstr. 1, 39106 Magdeburg, Germany, [2]Lehrstuhl für Bioverfahrenstechnik, Universität Stuttgart, Stuttgart, Germany, [3]Lehrstuhl für Zellkulturtechnik, Universität Bielefeld, Bielefeld, Germany, [4]Lehrstuhl für Bioprozesstechnik, Otto-von-Guericke-Universität Magdeburg, Magdeburg, Germany.*

**Abstract:** A serum-free process for influenza virus vaccine production (equine and human) in roller bottles and microcarrier systems in 5L-stirred tank and 2L-wave bioreactor (Cytodex 1) is described. MDCK cells were adapted from growth in serum containing GMEM medium to serum-free Ex-Cell MDCK medium. Virus titers of 2.0-2.9 log HA units/ 100 μL were obtained. Omission of the medium exchange before infection has clearly simplified the process.

**Key words:** serum-free media, influenza vaccine, MDCK cells, wave bioreactor, microcarrier, stirred tank bioreactor, cell attachment, metabolites, bioprocess engineering, virus, roller bottles

## 1. INTRODUCTION

The switch from serum containing media to serum-free media in mammalian cell culture has become a major issue in the last years [1]. Especially for virus vaccine production processes such as influenza, where the infection phase has to be serum-free, even when cultivating the cells in serum containing media, a complete serum-free process has many advantages.

Here, we present the successful adaptation of an influenza vaccine production process from serum containing GMEM medium [2] to serum-free



*R. Smith (ed.), Cell Technology for Cell Products*, 563–565.

Ex-Cell MDCK medium for microcarrier systems (Cytodex 1). Cultivations in roller bottles, 2L-Wave and 5L-stirred tank bioreactor are compared.

## 2. MATERIALS AND METHODS

Madin-Darby canine kidney cells (MDCK) (ECACC No. 84121903) were cultivated in serum containing medium (SC) as described elsewhere [2]. For serum-free cultivation (SF) MDCK cells had been adapted to Ex-Cell MDCK medium (JRH Bioscience) supplemented with 2 mM glutamine (Sigma).

Cells were infected with either equine influenza A/Equi 2 (H3N8) Newmarket 1/93 or human influenza A/PR/38 (H1N1) (NIBSC) in GMEM medium without serum or in Ex-Cell MDCK medium both containing low levels of porcine trypsin (12.5 mg/L; Invitrogen) at 37°C. Virus seed was stored in aliquots of 1-10 mL (2.1-2.4 log HA units/100 μL) at -70 °C, thawed and added with moi of 0.1-0.025 based on plaque forming units of the virus seed. Cells were grown in roller bottles (250 mL wv) (Greiner) (start 1 x $10^5$ cells/mL), 2L-Wave bioreactor (1 L wv) (Wave Biotech) or 5L-stirred tank bioreactor (STR) (start 2 x $10^5$ cells/mL) (details see [2]) (B. Braun Biotech) on Cytodex 1 solid microcarriers (1.7-2 g/L) (GE Healthcare).

In all cultivation vessels the cells were grown to confluency after 4 days of cultivation. For serum containing cell growth the medium was removed and the remaining suspension was washed several times with PBS (without $Ca^{2+}/Mg^{2+}$) before virus medium addition and infection. In serum-free medium infection was directly without washing and medium exchange.

## 3. RESULTS AND DISCUSSION

After adaptation of MDCK cells to serum-free Ex-Cell medium (SF) roller bottle experiments were carried out to compare metabolite profiles and virus titers with results in serum containing medium (SC).

Further scale-up into a microcarrier process in a 5L-stirred tank bioreactor and a wave bioreactor was then tried. Here, the adhesion to the chosen Cytodex 1 carriers [2] was most problematic. Data on corresponding typical runs are summarized in Table 1.

Although SF medium contained higher starting glucose concentrations less glucose was consumed than in SC medium. Thus, no limiting concentrations even without washing steps and medium exchange before infection were obtained. Correspondingly, higher lactate concentrations were reached in SC medium. In SC medium gln uptake after virus infection was about 0.7-1.0 mM compared to only 0.1 mM in SF medium. For the cell growth phase ammonia release correlated with the glutamine uptake. However, during virus infection the increase in ammonia did not correspond to the glutamine uptake. The glutamate starting concentration was clearly higher in SF than in SC medium. The glutamate profile for cell growth as seen in SC medium with a release phase and almost complete uptake after 50 h [2, 3] could not be observed in SF medium. Instead a steady increase of glutamate was found. After infection a stronger release of glutamate than during growth was observed directly after infection. Finally, virus titers and profiles were similar for all cultivations with 2.4-2.9 log HA units/100 μL. Only for the SF wave bioreactor low HA-values of 2.0 were found. This was probably due to difficulties in pH control, resulting in a final pH of 6.25. In SF medium for the overall process (growth + infection) lower glucose uptake and ammonia release was found in the wave bioreactor compared to the stirred tank bioreactor. This could result from additional viable cells (suspension) that were present in the supernatant of the stirred tank bioreactor (data not shown).

*Table 1.* Serum containing and serum-free influenza production in different cultivation vessels.

| vessel | med.[a] | end[b] | moi | gln[c] | $NH_3$[c] | gluc[c] | lac[c] | glu[c] | HA[d] |
|---|---|---|---|---|---|---|---|---|---|
| RB | SC | n.a | 0.1 | -1.2/-0.7 | +0.7/+1.1 | -10.5/-6.4 | +13.6/+12.5 | -0.2/+0.5 | 2.6 |
| RB | SF+ | n.a | 0.1 | -1.1/-0.4 | +1.0/+1.1 | -5.3/-5.3 | +10.6/+10.1 | +0.2/+1.1 | 2.4 |
| RB | SF | n.a | 0.1 | -1.1/-0.1 | +1.1/+1.1 | -3.5/-5.6 | +10.3/+8.5 | +0.2/+0.5 | 2.4 |
| wave[e] | SC | 2.8 | 0.05 | -1.9/-1.0 | +1.5/+1.7 | -18.4/-13.3 | +33.8/+24.8 | -0.4/+0.8 | 2.6 |
| wave[e] | SF | 0.9 | 0.05 | -1.5/+0.1 | +1.7/+1.3 | -13.0/-4.7 | +20.3/+1.6 | +0.5/+0.6 | 2.0 |
| STR | SC | 1.2 | 0.03 | -1.6/-0.8 | +1.6/+0.8 | -23.2/-12.4 | +38.3/+22.3 | -0.3/+0.4 | 2.4 |
| STR[f] | SF | 1.3 | 0.05 | -1.2/-0.1 | +0.9/+0.8 | -13.1/-11.8 | +27.3/+20.8 | +0.4/+0.3 | 2.9 |

amedium used; bend cell number on microcarriers (x 106 cells/mL); coverall consumed (-) or released (+) gln, NH3, gluc, lac & glu (mM) for cell growth phase/virus replication; dvirus titer in log HA units per 100 μL (4 days p.i.); ewave angle 7°, frequency 15 min-1, 2-5% CO2 mixed with air at 0.1 mL/min; fhere human influenza instead of equine; n.d.: not applicable; SF +: Ex-Cell cultivation with medium exchange.

## REFERENCES


[1] Merten, O.W.; 2002, Dev Biol Stand, **111**: 233-257,

[2] Genzel, Y., Behrendt, I., König, S., Sann, H., Reichl, U.; 2004, Vaccine, **22 (17-18)**: 2202-2208,

[3] Genzel, Y., Ritter, J.B., König, S., Alt, R., Reichl, U.; 2005, Biotechnol. Progr., **21 (1)**: 58-69.

# Rotary Bioreactor for Recombinant Protein Production

Stephen Navran
*Synthecon, Inc. 8042 El Rio, Houston, TX 77054*

**Abstract:** The Rotary Cell Culture System (RCCS) developed at NASA has been used primarily for tissue engineering applications. Because it creates a low shear, high mass transport environment simulating microgravity, we tested the ability of the RCCS to enhance the production of recombinant proteins compared to a stirred bioreactor. A human cell line was transfected with LacZ or glycodelin and cultured in the RCCS and a spinner flask. The RCCS culture produced a 7-fold greater yield of beta galactosidase and 3-fold greater yield of glycodelin than the spinner flask suggesting that the low shear conditions in the RCCS promoted increased protein production.

**Key words:** cell culture, microgravity, recombinant protein, post-translational modify-cation,

## 1. INTRODUCTION

The original purpose of the Rotary Cell Culture System (RCCS) was to carry cell into space to study the effects of microgravity. The design incorporates a horizontally rotating, cylindrical culture vessel with a coaxial tubular oxygenator (Figure 1). The cells experience a microgravity-like environment with extremely low fluid shear stress (1). In other culture systems, cells are suspended by agitation or sparging which creates high shear and results in cellular damage. We hypothesized that transfected cells would produce higher levels of recombinant protein in the low shear conditions of the RCCS. To test this hypothesis, we transfected LacZ and glycodelin, a human glycoprotein gene, into a human cell line and compared protein production in the RCCS and a spinner flask.



*R. Smith (ed.), Cell Technology for Cell Products*, 567–569.

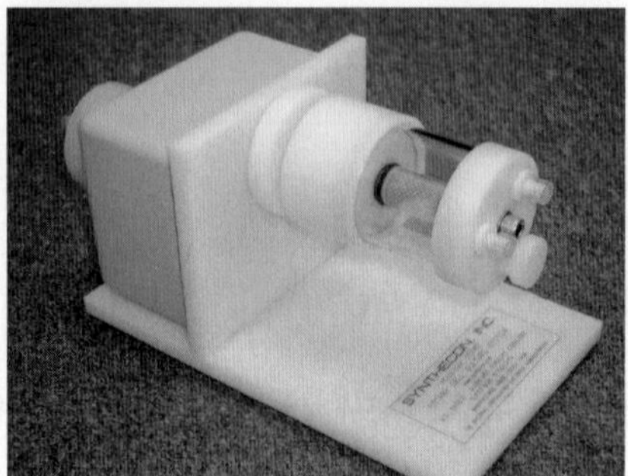

*Figure 1.* The Rotary Cell Culture System.

## 2. METHODS

cDNA's for LacZ and glycodelin were cloned into an expression vector, pcDNA/Myc-HIS (Invitrogen) and transfected into K562 cells. After selection with blasticidin, clones were screened for protein production. Transfected cells from a single clone were grown in flasks and then transferred to a spinner flask and an RCCS and cultured in RPMI 1640 media with 10% FCS. The media was changed every 48 hours for 7 days. At the end of the culture cells were harvested and analyzed for beta galactosidase and the media was analyzed for glycodelin. The biological activity of glycodelin was measured by the ability to inhibit IL-2 secretion from activated Jurkat cells.

## 3. RESULTS

Beta-galactosidase extracted from transfected K562 cells cultured in the RCCS was increased 7-fold compared to the spinner flask (Figure 2).

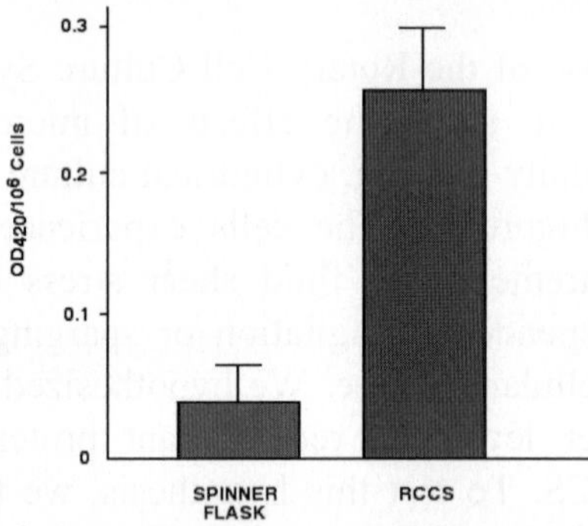


*Figure 2.* Comparison of beta-galactosidase production in a Rotary bioreactor (RCCS) and a spinner flask. K562 cells transfected with LacZ were initially seeded in the RCCS and a spinner flask. Media was changed at 48 hour intervals. After 7 days the cells were harvested and assayed for beta-galactosidase activity. N=3 Glycodelin secreted into the media was increased 4.2-fold in the RCCS compared to the spinner flask (Figure 3).

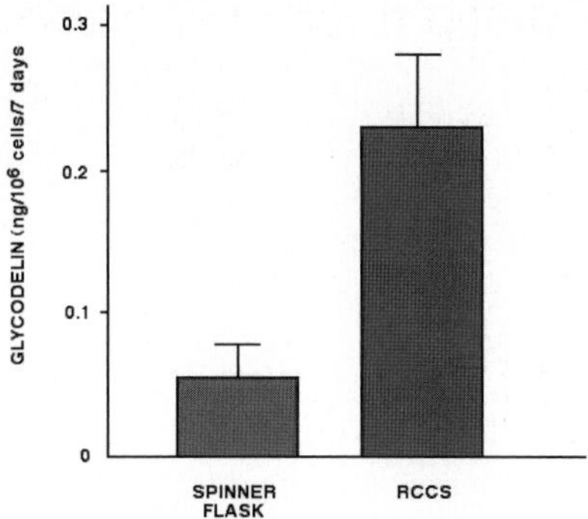


*Figure 3.* [Production of recombinant glycodelin in K562 cells cultured in a spinner flask and a Rotary Cell Culture System (RCCS). K562 cells transfected with glycodelin cDNA were seeded in an RCCS and a spinner flask. Media was changed at 48 hour intervals. After 7 days, the media was harvested and glycodelin isolated and quantitated. N=3].

# 4. CONCLUSIONS

The low shear stress environment of the RCCS facilitates higher levels of recombinant protein production compared to a spinner flask. The biological activity of the recombinant protein appears to be unaffected by the type of bioreactor in which it is produced.

Since the RCCS is capable of culturing virtually any cell type at high density, the choice of cell lines for recombinant protein production is greatly expanded. In cases where a recombinant protein might require species-specific or even cell-specific post-translational modifications, the RCCS offers the opportunity to use a cell line which may not be adaptable to high shear stress bioreactors.

# REFERENCES

Hammond TG, Hammond JM. Optimized suspension culture: the rotating-wall vessel. Am. J. Physiol. Renal Physiol. 281: F12-F25, 2001

# Flow Cytometry as a Tool for Process Monitoring of Virus-cell Systems

K.S. Sandhu[1] and M. Al-Rubeai[1,2]
[1]*Department of Chemical Engineering, The University of Birmingham, Edgbaston, B15 2TT, UK.* [2]*Department of Chemical and Biochemical Engineering and Centre for Synthesis and Chemical Biology, University College Dublin, Belfield, Dublin, Ireland. E mail: m.al-rubeai@ucd.ie*

**Abstract:** Physiological and morphological changes upon infection by viruses have been observed in many cell lines used for the production of proteins and viruses. Here we demonstrate the utility of single-cell analysis to monitor various cellular parameters in insect cells-baculovirus and HEK293-adenovirus systems, including DNA content, cell size and cell granularity and their correlation with the propagation of virus and production of recombinant protein. Analysis of infected cells at the single-cell level by flow cytometry was shown to be an effective method to monitor the changes in virus titre and intracellular protein. The results show that flow cytometry is an effective tool for process monitoring of virus-cell systems that allows the design of optimum harvest strategies.

**Key words:** Flow Cytometry, Adenovirus, Baculovirus, HEK293, Sf9, GFP, Cell size, Granularity, DNA.

## 1. INTRODUCTION

Physiological and morphological changes upon infection by viruses have been observed in many cell lines used for the production of proteins and viruses. Measurements of morphological and physiological parameters, primarily based on measuring fluorescence can be simply achieved by flow cytometry. The technique has the facility for detection and quantification of specific changes in each cell within population. This may provide powerful evidence in the assessment of effects of viral infection on cellular activity



*R. Smith (ed.), Cell Technology for Cell Products*, 571–573.

and enable the identification and characterisation of heterogeneous cell populations. Here we utilise single-cell analysis to monitor DNA content, cell size and cell granularity in HEK293-adenovirus and insect cells-baculovirus systems and correlate these parameters with the propagation of virus and production of recombinant protein.

## 2. RESULTS AND DISCUSSION

### 2.1 Animal cell (HEK293)-Adenovirus(Ad5GFP) results

The relationships between Ad5GFP titres & relative DNA content, relative cell size & relative granularity in the cells after infection with MOI's 1, 10, and 50 are shown in Table 1. The correlation coefficients ($R^2$) indicate that best correlation is obtained with DNA content.

*Table 1.* The correlation parameters between cellular attributes and virus titre.

| Parameter correlated with Ad5gfp titre | Equation | $R^2$ |
|---|---|---|
| DNA content | Y=3E-08x +486.13 | 0.910 |
| Cell Size | Y=3E-08x +307.08 | 0.840 |
| Granularity | Y=3E-08x +246.07 | 0.748 |

### 2.2 Insect cell (Sf9)-baculovirus(AcNPV-lacZ) results

Table 2 shows a decrease in viable cell number with increase in MOI. However, there is an increase in %DNA content of cells proportional to the increase in MOI. Similar increase was found in cell size and granularity as well as in β-galactosidase production, all of which are proportional to the increase in MOI. The results suggest that flow cytometric monitoring of cellular parameters can give a good indication of productivity. FC could detect the increasing DNA content as the virus multiplies with time relative to the MOI and in a manner that would predict infectivity and productivity level.

*Table 2.* Changes in cellular attributes at various MOI's with time as a % of MOI 50 value.

| Parameter measured at 27 hr pi | MOI 0.1 | MOI 1 | MOI 10 | MOI 50 |
|---|---|---|---|---|
| DNA content | 66 | 74 | 90 | 100 |
| Cell Size | 79 | 82 | 94 | 100 |
| Granularity | 58 | 71 | 91 | 100 |
| Production of B-galactosidase | 25 | 37 | 62 | 100 |
| Viable Cell Number X10E5/ml | 10.9 | 9.3 | 6.7 | 5.3 |

## 3. CONCLUSION

Analysis of infected cells at the single-cell level by flow cytometry was shown to be an effective method to monitor the changes in virus titre in HEK293 cells and intracellular recombinant protein in insect cells.

The results show that flow cytometry can be an effective tool for process monitoring of virus-cell systems that may allow the design of optimum harvest strategies. Understanding and prediction of the virus infection and multiplication process are possible with the aid of DNA, cell size and cell granularity analysis which can also provide the data necessary to address the problem of modelling cell population-virus infection dynamics. For example by measuring the DNA content, the replication of adenoviruses and baculoviruses in cells can be predicted and specific control actions recommended.

We also found that the naturally fluorescing reporter gene product, GFP showed a good correlation with viability, virus titre and the proportion of GFP positive cells in infected HEK293 heterogeneous populations (data not shown) which supports the evidence that FC is clearly important technique for cell culture process identification and a sophisticated early process sensor.

# The Use of Multi-parameter Flow Cytometry for Characterisation of Insect Cell-baculovirus Cultures

B. Isailovic[1], R. Hicks[2], I.W. Taylor[2], A. Nienow[1], C.J. Hewitt[1]*
[1]*Department of Chemical Engineering, The University of Birmingham, Edgbaston, B15 2TT, United Kingdom,* [2]*AstraZeneca, Mereside, Alderley Park, Macclesfield, Cheshire, SK10 4TG, United Kingdom*

**Abstract:** Bacteria and mammalian cells have been traditionally used as hosts for commercial recombinant protein production. In recent years, the insect cell-baculovirus system has emerged as a potentially attractive recombinant protein expression vehicle. This route is attractive because baculovirus-infected insect cells are able to perform post-translational modification while accommodating very abundant expression of recombinant protein. Although flow cytometry has been used widely for analysis of mammalian and microbial cell physiology and morphology, there is very little information on applications of this powerful and highly efficient technique in insect cell culture.

Here we have compared cell ratiometric counts and viability of Sf-21 cell cultures using a flow cytometer to those determined by more traditional methods using a haemocytometer and the trypan-blue exclusion dye. There was good agreement between the two counting methods but the former technique proved to be a more reliable and statistically robust viability indicator (stains used: propidium iodide and calcein AM).

Flow cytometry has also been used to monitor various parameters during fermentations of Sf21 cultures infected with the recombinant *Autographa californica* Nuclear Polyhedrosis Virus (AcNPV). This recombinant baculovirus contains the inserted nucleic acid sequence amFP486 coding for AM-Cyan coral protein, which emits natural green fluorescence. A good correlation has been obtained between parameters such as mean green fluorescence and AmCyan positively stained cells linked to cell viability. Additionally, DNA content, cell size and granularity have proven to be good indicators of baculovirus infection.

We have also simplified and optimised methods of cell treatment prior to flow cytometric analysis.

**Key words:** Baculovirus, Flow cytometry, Sf-21, insect cell, AmCyan.



*R. Smith (ed.), Cell Technology for Cell Products*, 575–577.

## 1. INTRODUCTION

In recent years, the insect cell-baculovirus system has emerged as a potentially attractive protein-production option. This route is attractive because insect cells are able to perform post-translational modification while expressing reasonable quantities of recombinant protein. Although insect cell-baculovirus systems have been discussed in numerous research papers there is very little information on the use of flow cytometry for characterisation of baculovirus fermentations.

## 2. RESULTS AND DISCUSSION

Table 1 compares the Sf-21 growth and viability curves obtained by a more traditional method using haemocytometer (in conjunction with trypan blue exclusion dye) with the ones obtained by flow cytometry (FC). Both total cell density curves and viability are in a good agreement. However, after 100 hours, the superiority of viability analysis and cell count by FC to the other method is emphasised. Table 2 shows infectivity and AmCyan protein yield for Sf-21 cultures infected at MOI=5 pfu/cell and MOI=1 pfu/cell respectivelly. The infection dynamics is similar at both MOIs, which suggests that culture is saturated with viruses at MOIs above 1 pfu/cell. However, the AmCyan maximum yield is reached 12 hours sooner at MOI=5 due to higher virus titer, which, in turn, causes more frequent secondary infections and at an earlier stage.

## 3. CONCLUSIONS

Flow cytometry (FC) proved to be less laborious, less operator dependant and statistically more accurate for cell count and viability analysis than the haemocytometer and trypan blue exclusion method. Additionally, the former proved to be a very powerful tool in monitoring infection and recombinant protein production. Reliable FC protocols were developed.

*Table 1.* Comparison of flow cytometry (FC) count and viability analysis with haemocytometer and trypan blue exclusion method.

| Time (h) | Viable cell density (cell/ml) (Haemocytometer) | Viability (%) (Haemocytometer) | Total cell density (cell/ml) (Haemocytometer) | Total cell density (cell/ml) (FC) | Viability (%) (FC) | Viable cell density (cell/ml) (FC) |
|---|---|---|---|---|---|---|
| 0 | 2.00E+05 | 97.05% | 2.06E+05 | 2.48E+05 | 98.94% | 2.45E+05 |
| 27 | 4.30E+05 | 96.63% | 4.45E+05 | 4.42E+05 | 96.72% | 4.27E+05 |
| 52 | 1.34E+06 | 98.53% | 1.36E+06 | 1.15E+06 | 96.88% | 1.12E+06 |
| 65 | 2.15E+06 | 96.85% | 2.22E+06 | 2.25E+06 | 96.06% | 2.16E+06 |
| 76 | 3.54E+06 | 96.59% | 3.67E+06 | 3.85E+06 | 94.47% | 3.64E+06 |
| 101 | 7.82E+06 | 96.31% | 8.12E+06 | 8.50E+06 | 91.86% | 7.81E+06 |
| 113 | 9.00E+06 | 95.54% | 9.42E+06 | 8.85E+06 | 89.22% | 7.89E+06 |
| 125 | 8.44E+06 | 91.34% | 9.24E+06 | 8.54E+06 | 83.27% | 7.11E+06 |
| 137 | 8.40E+06 | 89.36% | 9.40E+06 | 7.95E+06 | 66.68% | 5.30E+06 |
| 148.5 | 7.86E+06 | 84.15% | 9.34E+06 | 7.91E+06 | 40.35% | 3.19E+06 |

*Table 2.* Comparison of infectivity and AmCyan recombinant green fluorescent protein at an MOI of 1 and 5 (analyses performed by flow cytometry).

| Time (hour) | % Infectivity MOI=1 | Mean green fluorescence (GFU) MOI=1 | AmCyan Yield (GFU/ml) MOI=1 | Time (hour) | % Infectivity MOI=5 | Mean green fluorescence (GFU) MOI=5 | AmCyan Yield (GFU/ml) MOI=5 |
|---|---|---|---|---|---|---|---|
| 0 | 0.00% | 0.4 | 7.99E+05 | 0 | 0.00% | 0.3 | 7.32E+05 |
| 11 | 0.00% | 0.4 | 1.47E+06 | 12 | 0.00% | 0.3 | 9.58E+05 |
| 24 | 1.15% | 0.4 | 1.83E+06 | 24 | 6.21% | 0.6 | 2.94E+06 |
| 35 | 34.15% | 2.9 | 1.32E+07 | 30 | 9.59% | 0.9 | 4.04E+06 |
| 48 | 99.09% | 14.2 | 5.31E+07 | 36 | 33.06% | 2.8 | 1.19E+07 |
| 59 | 99.55% | 28.6 | 9.35E+07 | 41 | 70.25% | 4.8 | 1.99E+07 |
| 72 | 99.77% | 48.4 | 1.49E+08 | 48 | 99.65% | 19 | 7.06E+07 |
| 83 | 99.96% | 51.9 | 1.52E+08 | 60 | 99.97% | 62.5 | 1.96E+08 |
| 97 | 99.92% | 39.4 | 1.10E+08 | 72 | 99.98% | 62.1 | 1.78E+08 |

# Understanding Metabolic Needs of EB14 Cells in Culture

J. Hartshorn,[1] F. Guehenneux,[2] K. Kolell,[1] Z.-H. Geng,[1] E. Atarod,[1] T. Mohabbat,[1] C. McCormick,[1] P. Sutrave,[1] C. Hernandez,[1] M. Avery,[1] S. Luo,[1] M. Mehtali,[2] K. Etchberger[1]

[1] *JRH Biosciences, Inc., Lenexa, Kansas, USA* ; [2] *Vivalis, Nantes, France*

**Abstract:** $EB_{14}$ is a genetically stable diploid cell line derived from chicken embryonic stem cells. $EB_{14}$ cells demonstrate great potential in virus and recombinant protein production for prophylactic and therapeutic purposes. The present study aims at identifying metabolic needs of the $EB_{14}$ cells so that specific nutrients can be provided adequately and in a timely manner. $EB_{14}$ cells exhibit unique metabolic patterns in various serum-containing and serum-free media, and are highly sensitive to nutritional supply and waste accumulation in culture. Typically, $EB_{14}$ cells require frequent replenishment of culture media to stay viable. The ratio of aerobic to anaerobic metabolism was calculated based on carbon and nitrogen energy source consumption rates in $EB_{14}$ cells. As a result, optimal media formulations and culture processes were developed to support high viable cell growth supporting recombinant and attenuated viral vaccine production. Chemical and cellular analysis of $EB_{14}$ cell cultures with specific viable cell density (VCD), culture longevity and viral productivity will be discussed.

**Key words:** Chicken Embryonic Stem Cells, $EB_{14}$, Media Development, Metabolic Analysis, Influenza Vaccine.

## 1. INTRODUCTION

The current flu vaccine manufacturing process issues have caught the world's attention and revealed concerns about the ability to provide an adequate supply of vaccines. Currently the flu vaccine is produced primarily



*R. Smith (ed.), Cell Technology for Cell Products*, 579–581.

in chicken embryonic fibroblasts (CEFs). Processes based on CEFs are highly dependent on the tedious process of isolating and culturing CEFs from chicken eggs. $EB_{14}$ cells provide a brand new cell culture platform to the vaccine industry with immediate benefits. This avian stem cell line was first successfully isolated in long-term culture by Vivalis (Nantes, France). JRH Biosciences and Vivalis scientists are the first team to develop a serum-free media and process to support high cell density and viral production in $EB_{14}$ cells. More interestingly, JRH and Vivalis scientists investigated $EB_{14}$ cell's energy expenditure and metabolic pathway control through media nutrient optimization.

## 2. MATERIALS AND METHODS

### 2.1 Cells and Culture Media

Stock cultures of $EB_{14}$ (Nantes, France) were maintained in a 37°C, 7.5% $CO_2$ humidified incubator in EX-CELL™ 60947, 65318, 65319, 65320, and 65126 (JRH Biosciences, Inc., Lenexa, KS) and competitor Medium X. The Influenza A/New Caledonia/20/90 (H1N1) was obtained from the Centers for Disease Control and Prevention (CDC). Madin Darby Canine Kidney (MDCK) cell line is from American Type Culture Collection (ATCC CCL-34).

### 2.2 Process Development

Bioreactors (Applikon® Biotechnology, Sciedam, Holland) were seeded at 0.2e6 cells/mL in EX-CELL™ 60947 and Medium X. Dissolved oxygen (DO) and pH were monitored online and controlled using air, $O_2$, $N_2$, $CO_2$, and 7.5% $NaHCO_3$, respectively.

### 2.3 Viral Infection

Twenty four hours prior to infection, $EB_{14}$ cells were seeded at 0.4e6 cells/mL in 125mL Erlenmeyer flasks. Cells were inoculated with a multiplicity of infection (MOI) of 0.01 of Influenza A/New Caledonia and allowed to adsorb for 1 hour at 34°C. After 1 hour, additional media was added and the cultures were incubated at 34°C for 5 to 6 days. Trypsin was added at 3μg/mL each day while sampling cultures.

### 2.4 Western Blot Analysis

Ten microliters of supernatant from infected cells were used for a western blot using WesternBreeze Chromogenic Immunodetection Kit

(Invitrogen). Hemagglutinin (HA) Monoclonal Antibody (QED Bioscience Catalog# 20301).

### 2.5 Analytical Methods

Cell density and viability was determined by Cedex (Innovatis, Bielefeld, Germany) and by trypan blue exclusion method with hemacytometer. Metabolites dynamics was monitored offline with the BioProfile® 100 (Nova Biomedical Corporation, Waltham, MA).

## 3. RESULTS AND DISCUSSIONS

$EB_{14}$ cells grew significantly higher in EX-CELL™ 60947 to greater longevity as compared to in Medium X (data not shown). EB14 cells in EX-CELL™ 60947 reached a maximum cell density of approximately 3e6 cells/ml on Day 2 and maintained this density through the Day 4 whereas rapidly declined in Medium X on day two. A similar pattern was reproduced in bioreactor runs (data not shown).

Metabolite analysis indicated the rate of lactate and ammonia production was much higher for the competitor Medium X than all EX-CELL™ 60947 formulations. However, the rate of glucose and glutamine consumption remained comparable among all formulations tested.

Western blot analysis indicates that EX-CELL™ 60947 supports the production of Influenza A/New Caledonia, as demonstrated by the presence of HA0 and HA2 (data not shown).

## 4. CONCLUSIONS

We conclude, thus, 1) EX-CELL™ 60947 is a serum-free, animal component free medium that supports higher cell densities, improved viabilities and greater culture longevity in comparison with Medium X; 2) Optimized EX-CELL™ 60947 supports Influenza Virus production.

# Virus Production in Vero Cells Using a Serum-free Medium

K. Kolell, J. Padilla-Zamudio, B. Schuchhardt, S. Gilliland, S. McNorton, B. Dalton, S. Luo, K. Etchberger
*JRH Biosciences, Inc., Lenexa, Kansas, USA*

**Abstract:** The manufacture of viral vaccines has historically been accomplished using animal products such as chicken eggs, or cell cultures using fetal bovine serum. To reduce regulatory concerns in vaccine production, serum-free cell culture processes are being embraced by the vaccine industry. The Vero cell line initiated from the African green monkey is an excellent cell line for the production of animal and human prophylactic viral vaccines. We have developed a serum-free (SF) and animal-component free (ACF) medium for the production of viral vaccines using the Vero cell line. This medium supports growth of Vero cells on microcarriers in a controlled bioreactor environment and virus production equivalent to serum-containing cultures. These characteristics make this an ideal medium for vaccine production using the Vero cell line.

**Key words:** Serum-Free Medium, Vero, Virus, Microcarrier, Bioreactor

## 1. INTRODUCTION

The Vero cell line, isolated from the kidney of a normal adult African Green monkey (*Cercopithecus aethiops*), has been well characterized and is instrumental in the biotechnology sector for virus replication studies, viral plaque assays, $TCID_{50}$ determinations and production of viral vaccines.

JRH Biosciences has developed a new serum-free medium specifically for the Vero cell line. EX-CELL™ Vero is serum-free and free of animal-derived components. The medium contains a plant-derived hydrolysate and low levels of recombinant proteins, but does not contain phenol red or Pluronic® F-68. In these studies, we show that EX-CELL™ Vero supports



*R. Smith (ed.), Cell Technology for Cell Products*, 583–585.

high-density cell growth in both stationary flasks and on microcarriers in bioreactor culture.

## 2. RESULTS

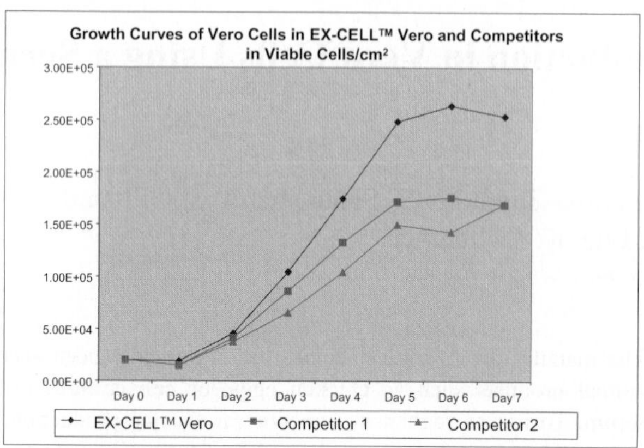


*Figure 1.* Vero cells were seeded in triplicate 25cm$^2$ T-flasks at 2x10$^4$ cells/cm$^2$. Cells were harvested from flasks and viable cells were counted daily for 7 days. Culture viabilities remained above 95% in all media during the study.

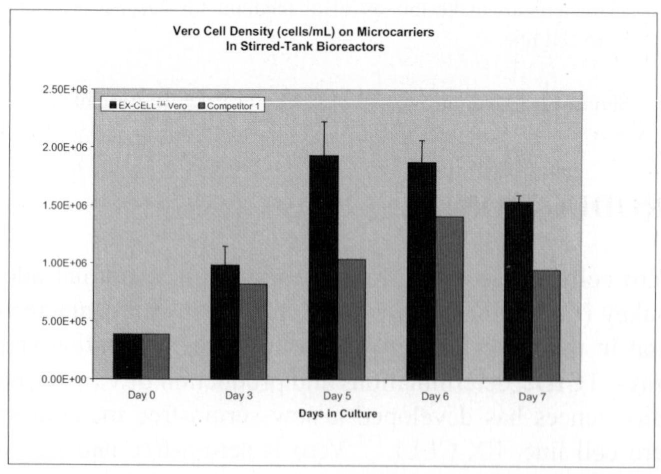


*Figure 2.* Vero cells were inoculated in Applikon stirred tank bioreactors at a 1L working volume. Reactor temperature was set to 37° C, the agitation speed was 70-85 rpm, dissolved $O_2$ was maintained at 50%, and pH was maintained at 7.0-7.3 with $CO_2$. Reactors were monitored and samples obtained daily for seven days; cell growth was determined by counting released nuclei using a crystal violet staining procedure. Cells grew to confluence

achieving maximum densities of approximately 1.9x10$^6$ cells/mL (~1.5x10$^5$ cells/cm$^2$) in EX-CELL™ Vero and 1.4x10$^6$ cells/mL (~1.1x10$^5$ cells/cm$^2$) in competitor's medium.

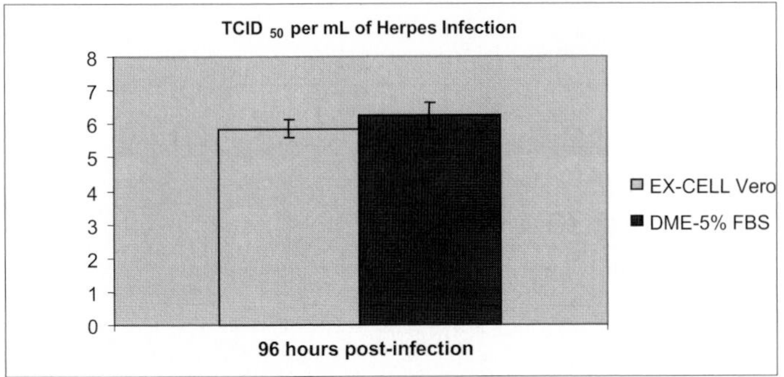


*Figure 3.* Vero cells were infected with Herpes Simplex Virus II (HSV-II) 96 hours post-seeding at a MOI of 0.01. Flasks displayed greater than 75% cytopathic effect 96 hours post-infection. Flasks were harvested and results represent $TCID_{50}$/ mL at harvest point.

# 3. CONCLUSION

EX-CELL™ Vero is a new serum-free medium designed and optimized for high-density Vero cell growth in adherent-stationary and adherent-suspension conditions. EX-CELL™ Vero is a regulatory-compliant medium, free from all animal-derived components, and contains only recombinant proteins. These studies indicate that adaptation to EX-CELL™ Vero from basal medium with serum can easily be accomplished and EX-CELL™ Vero supports high-density Vero cell growth, superior to competitor formulations.

EX-CELL™ Vero supported HSV-II production in Vero cells. The cells exhibited classic CPE in culture and produced HSV-II titers in the range of 106.0 $TCID_{50}$/mL, comparable to serum-supplemented cultures.

For further information regarding EX-CELL™ Vero, please contact Technical Services.

# Computational Fluid Dynamics Applied to Laboratory Scale Cell Culture Reactors for the Prediction of Hydrodynamic Stress

Eric Olmos[1], Nicolas Fischbach[1], Annie Marc[1]
[1] *Laboratoire des Sciences du Génie Chimique UPR CNRS 6811-INPL; 2, avenue de la forêt de Haye BP172, 54505 Vandoeuvre les Nancy CEDEX France*

**Abstract:** Using Computational Fluid Dynamics, a method for the design of animal cell culture reactors is proposed. This approach is based on the concept of Fluid Stress Distribution and on the calculation of animal cell paths.

**Key words:** Computational Fluid Dynamics, hydrodynamics, culture reactor, Fluid Stress Distribution

## 1. INTRODUCTION

### 1.1 Context of the study

Large scale animal cells culture is still a challenge regarding engineering considerations. A global process optimization must take into account the reactor design and the operating parameters. To do this, local phenomena have to be well understood and, if possible, precisely controlled. Among phenomena which influence animal cell culture kinetics, hydrodynamic stress due to mechanical agitation and bubble rupture have been the subject of numerous studies. Nevertheless, in these studies, these hydrodynamic stresses are generally fairly estimated and the effective stress quantity in terms of intensity and duration is not well known.

### 1.2 Aim of the work

Computational Fluid Dynamics (CFD) is used to simulate the velocity fields in two laboratory reactors: a roller bottle and a Taylor-Couette reactor (TCR, dedicated to long term cultures with low-level hydrodynamic stress).



*R. Smith (ed.), Cell Technology for Cell Products*, 587–590.

Then, Fluid Stress Distributions (FSD) of these reactors and their variability within the operating conditions and within the reactor are predicted. These distributions provide global but relevant information on the local constraints. Animal cell trajectories in the reactor (and consequently local cell concentration) are calculated in order to detect a possible poor mixing and/or local cell accumulation.

## 2. COMPUTATIONAL APPROACH

Computational Fluid Dynamics consists in solving the Navier-Stokes equations. In this study, the finite-volume method is used. Steady-state flow and incompressibility are supposed. When turbulence occurs, the Reynolds Stress turbulence model is used. The transport of animal cells in the calculated flow is performed by integrating the force balance on the cell in a Lagrangian frame. The force balance equates the particle inertia with the forces acting on the particle. Both buoyancy and drag force are taken into account (Morsi and Alexander 1972). The equations are solved with the commercial software Fluent 6.2. The Fluid Stress Distributions are obtained by post-processing numerical simulation results with the Scilab freeware.

## 3. RESULTS

Concerning the roller bottle, the calculation of animal cell trajectories puts into evidence possible cell accumulations which may lead to reactor malfunction due to a high concentration of toxics or a lack of nutriments (Figure 1).

The numerical simulations of the TCR allow the prediction of the transition from the laminar to the Taylor-Couette regime with a unique model, which is an original contribution. Moreover, the velocities calculated obtained are well validated by those measured experimentally (Haut *et al.*, 2003). Another original result concerns Fluid Stress Distributions (FSD). Starting from sophisticated numerical simulations, an easy tool is proposed to test *a priori* the operation of the reactor (Figure 2).

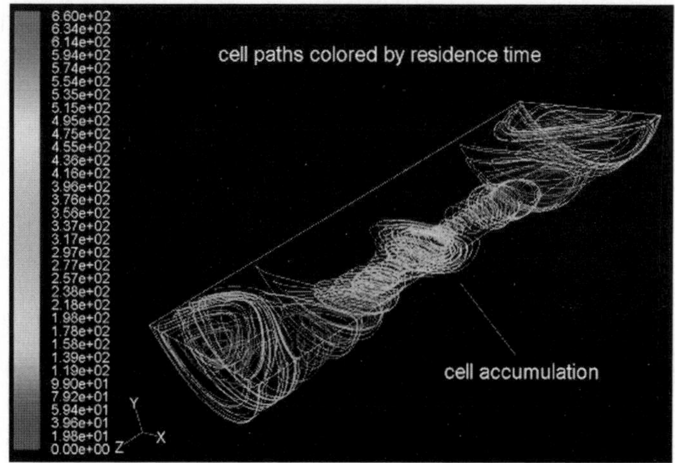


*Figure 1.* Animal cell hydrodynamic paths in the roller bottle.

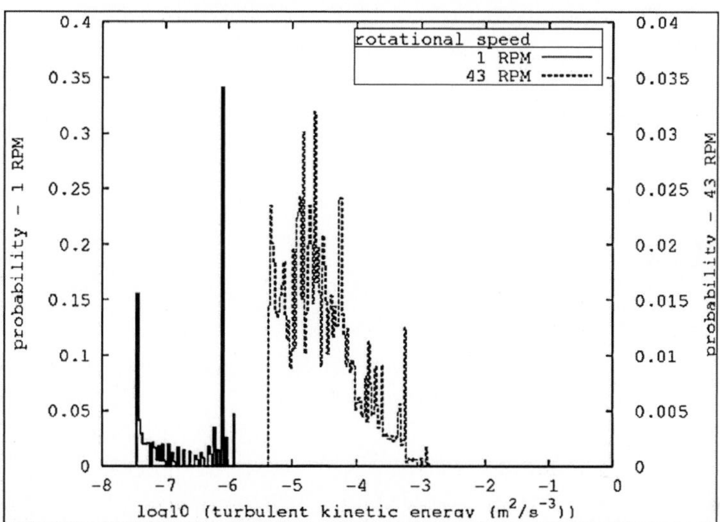


*Figure 2.* Fluid Stress Distributions in the Taylor-Couette reactor.

# 4. CONCLUSION AND PERSPECTIVES

Using CFD, we put into evidence possible culture reactor malfunction (poor mixing, cell accumulation) and we proposed an innovating method to numerically test the efficiency of a given reactor, in terms of hydrodynamic stress distributions. This work has now to be completed by: a house made experimental validation by Particle Image Velocimetry (PIV) and Laser Doppler Anemometry (LDA), the calculation of the amount of stress received by a cell during a classic culture, the coupling of hydrodynamics with culture kinetics and the generalization of the method to industrial stirred reactors for an optimized design.

## REFERENCES

Haut , B., Ben Amor, H., Coulon, A., Jacquet, A., Halloin, V. 2003. Hydrodynamics and mass transfer in a Couette-Taylor bioreactor for the culture of animal cells, Chem. Eng. Sci. 58 777-784.

Morsi, S.A. and Alexander, A.J. 1972, An investigation of particle trajectories in two-phase flow systems, J. Fluid Mech. **55** 193-208.

# Using Genomic Tools for the Identification of Important Signaling Pathways in Order to Facilitate Cell Culture Medium Development

Daniel W. Allison, Kathryn A. Aboytes, Terrell K. Johnson, Danny K. Fong, Stacy L. Leugers, Melissa L. Hoepner, and Laurel M. Donahue
*Clinical Cell Culture R&D, Sigma-Aldrich, St. Louis, MO, United States.*

**Abstract:** Given the diversity of cells grown in culture and the fact that each of these cultures is poised to respond to a variety of stimuli, we set out to establish a method to identify these important pathways to accelerate the development of cell culture media. Many of the pathways required for cell functions (i.e. cell proliferation, protein production, cell adhesion, etc.) have been elucidated in detail, but there are undoubtedly unknown pathways, which may also be involved. Identification of these resources allows us to better understand these processes and potentially manipulate them to our advantage.

In order to validate this concept, we identified the mRNA expression profile for a variety of proteins within cells in culture using cDNA microarrays. In one example, we identified a specific growth factor receptor not previously known to be expressed *in vitro*. When the ligand for this receptor is added to the culture, the cells are poised to respond and proliferate at an increased rate. This method has been used on a variety of culture systems to either reduce (or eliminate) FBS requirements or to improve the performance of already serum-free formulations. This more targeted approach to medium development allows us to perform less of the random screening approaches of the past, thereby decreasing the investment of time and resources into this endeavor.

**Key words:** Cell culture, medium, microarray, cell signaling, genomic tools, growth factors, receptors, CHO.

## 1. INTRODUCTION

One of the biggest concerns for the development of new cell culture media is the investment of time required to optimize these products for the



*R. Smith (ed.), Cell Technology for Cell Products*, 591–593.

intended culture. It is not atypical for the development of a new cell culture product to take more than one year. We have tried to provide a more targeted/focused approach to the design of cell culture products by using rational methods of identifying candidates for inclusion in a medium.

The influx of genomic and proteomic research into the scientific community has led to an increased number of tools that can be applied to the development of cell culture products. These tools enable the rational design of more robust media formulations by providing insights into the stimuli the cells are poised to respond to *in vitro*. These methods could include genomic tools such as microarrays and quantitative PCR or proteomic tools such as antibody-based arrays. These approaches allow us to look either at mRNA or protein levels within cells in culture and predict based on the expression patterns what might elicit a response. The expression patterns of receptors (for growth factors, cytokines, etc.), adhesion molecules, or cell signaling components indicate important pathways for us to explore. These pathways could have beneficial effects on a wide range of functions, such as regulation of proliferation, apoptosis, differentiation, adhesion, or production.

## 2. RESULTS AND DISCUSSION

In order to test whether or not we could use information about the expression profile of a cell culture to determine factors that the cells would respond to, we examined the profile of many growth factor/cytokine receptors in several different cell lines and/or culture conditions. We were able to identify lists of receptors that were expressed within the different cultures (see Table 1 below for a list of receptors from a CHO cell line expressing alkaline phosphatase (CHO-AP)). Upon addition of the ligands for these receptors (or alternatively endogenous intermediates for the pathways), we were able to manipulate cellular functions such as proliferation, productivity and adhesion. Figure 1 demonstrates this effect with the addition of interleukin 1 to CHO-AP, leading to both increased proliferation and productivity of the recombinant protein. This methodology not only allows us to improve on existing serum-free cell culture medium formulations, but also allows us to reduce or eliminate FBS from FBS-dependent cultures.

The use of genomic or proteomic tools appears to be a powerful tool for use in medium development, allowing us to speed development time and correspondingly reduce costs. Based on the accumulation of data from various cell lines, we can apply a variety of development tools (microarray, macroarray, qPCR, antibody arrays, etc.) to cells in culture in order to enhance our ability to quickly design cell culture media. These tools allow us to provide a more targeted approach that is both reproducible, with higher throughput and will have a significant impact on the development time.

*Table 1.* CHO-AP: List of positive receptor:ligand pairs identified from microarray.

| Receptor | Ligand | Receptor | Ligand |
|---|---|---|---|
| Interleukin 12 receptor, β 2 | IL-12 | TGF, β receptor II | TGFβ |
| CSF 1 receptor | CSF1 | Activin A receptor, type II | Activin A |
| Burkitt lymphoma receptor 1 | BLC | Macrophage stimulating 1 receptor | MSP |
| Chemokine receptor 9 | TECK | Interferon γ receptor 2 | IFNγ |
| Interleukin 11 receptor, α | IL-11 | BMP receptor, type II | BMP2 |
| BMP receptor, type IA | BMP2 | G protein-coupled receptor 9 | MIG |
| FGF receptor 4 | aFGF | Activin A receptor, type I | Activin A |
| Interleukin 1 receptor-like 1 | IL-1 | FGF receptor 1 | bFGF |
| Chemokine receptor 4 | MDC | IGF 2 receptor | IGF2 |
| Chemokine receptor 1 | MCP3 | Autocrine motility factor receptor | PGI |
| PDGF receptor, β polypeptide | PDGF AB | GDNF family receptor α 3 | Artemin |

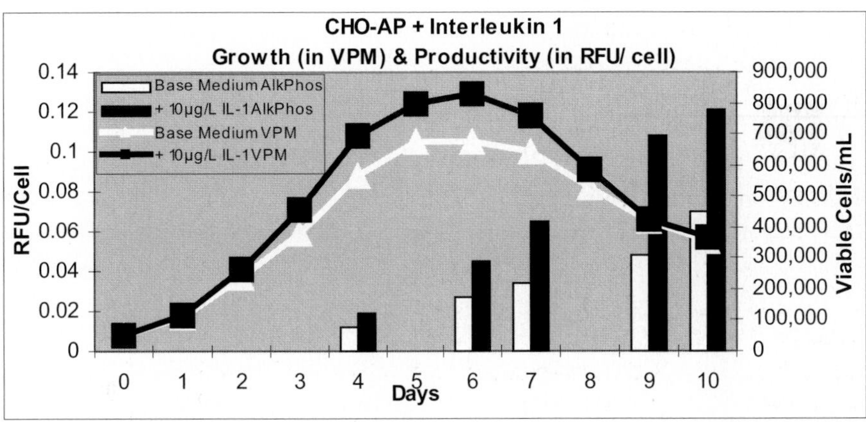


*Figure 1.* Growth and productivity of CHO-AP cells in spinner culture is enhanced with the addition of interleukin 1.

# Development and Application of an Animal-component Free Single-cell Cloning Medium for Chinese Hamster Ovary Cell Lines

N. Lin, J. L. Beckmann, J. Cresswell, J. S. Ross, B. Delong, Z. Deeds and M. V. Caple
*Cell Culture R&D, Sigma-Aldrich, St. Louis, MO, United States.*

**Abstract:** Single-cell cloning is a critical process in the generation of recombinant protein-producing mammalian cell lines. This process traditionally requires 10 – 20% fetal bovine serum (FBS) or other sera. Due to the presence of sera, single-cell cloning is a potential source of contamination from animal viruses and other adventitious agents. In order to address the regulatory needs of the biopharmaceutical industry, we have developed an animal-component free (AF) medium designed for single-cell cloning of Chinese Hamster Ovary (CHO) cells. We utilized Design of Experiment (DOE) methodology to optimize the levels of six groups of nutrients (amino acids, trace metals, plant-derived hydrolysates, lipids, vitamins and selenium). The optimized formulation was tested with three recombinant CHO cell lines and was shown to generate comparable results, in terms of clonal survival and growth (80% of positive control in average), to the 10% FBS control. The clones generated using the AF cloning medium from three recombinant CHO cell lines were successfully scaled up to spinner or shaker flask cultures in animal-component free culture media. The AF cloning process demonstrated improvement of growth and/or productivity. Transfection of a parental CHO K1 cell line and the subsequent selection and cloning processes were also evaluated in this AF cloning medium.

**Key words:** Cell culture, medium, single-cell cloning, selenium, CHO, clonal growth.



*R. Smith (ed.), Cell Technology for Cell Products*, 595–597.

# 1. MATERIALS AND METHODS

The stock cultures of the test cells lines (parental CHO K1, CHO AP expressing Secreted Alkaline Phosphatase, Recombinant CHO Line 2 expressing recombinant IgG) were maintained in suspension culture in animal-component free media (Sigma-Aldrich C8862 or proprietary formulation). The Basal Medium supplemented with 10% fetal bovine serum (FBS) was used as positive control. Single-cell cloning was performed using limiting-dilution method. The plating density was 0.5 – 0.9 cell/well in 200 μL media. The clonal survival and growth results were reported as Wells with Growth (% of Positive Control) from duplicate 96-well plates. Results of the two Factorial Matrix Experiments were analyzed using Design Expert® software (Stat Ease). Normal probability plots were used to illustrate the significant factors (plots not shown here). The parental CHO K1 cells were transfected with a proprietary Green Fluorescence Protein (GFP) expression vector using Escort II Transfection Reagent (Sigma-Aldrich L6037) in the AF Cloning Medium (C6366). The transfectants were cultured in 1:1 mixture of C6366 and CHO DHFR Medium (Sigma-Aldrich, C8862) containing 200 μg/ml G418, and single-cell cloned in C6366. Selected clones were expanded to T75 flasks in C6366 and C8862 (1:1).

# 2. RESULTS AND DISCUSSION

In brief, in Matrix Experiment 1, hydrolysates, amino acids and iron were statistically significant based on Day 6 growth (ANOVA $p<0.05$) therefore beneficial to clonal survival. The interaction between amino acids and iron was statistically significant based on Day 14 growth (ANOVA $p<0.05$). In Matrix Experiment 2, lipids, vitamins, selenium and the interaction between lipids and vitamins were statistically significant based on Day 7 growth (ANOVA $p<0.01$). Lipids, selenium and the interaction between lipids and selenium were statistically significant based on Day 14 growth (ANOVA $p<0.01$). Selenium was subsequently tested at higher concentrations (data not shown). The final formulation contains, in Percentage of Medium A (proprietary), 12.5% Hydrolysates, 5% amino acids, 7.5% iron, 50 nM selenium and optimized chelator concentrations. The final formulation was tested with three recombinant CHO cell lines. The clonal survival and growth performance was 80% in average of the positive control (Figure 1).

In order to demonstrate that the clones arisen from this Cloning Medium are expandable, and the cloning process using this AF Cloning Medium is able to improve growth and/or productivity comparing to the bulk cultures prior to cloning, selected clones generated from Recombinant CHO Line 2 cells were expanded in a 1:1 mixture of C6366 and C8862. Twenty-one recombinant CHO Line 2 clones were evaluated for growth and productivity in two separate spinner experiments in presence of selection pressure (200 mg/ml G418). Fourteen out of these 21 clones reached higher peak viable cell density than the bulk control (Data not shown here). Nine out of 21 clones demonstrated higher accumulative IgG productivity (42.8% of the selected clones). The CHO K1 cells stably transfected with the GFP expression vector was single-cell cloned using C6366. Out of the 40 clones that were expanded to T75 cultures, 36 were visually evaluated under the fluorescence microscope in order to eliminate mixed population cultures and the clones with very low fluorescence. Twenty-seven clones homogeneously expressed GFP. These clones (in T75 static cultures) were then evaluated for GFP expression levels by measuring mean fluorescence intensity (in Relative Fluorescence Units) on a fluorescence plate reader (Molecular Devices). The RFU values from each clone were normalized by per $10^6$ viable cells. Twenty-six out of 27 clones evaluated demonstrated 1.8 to 5.5-fold higher normalized mean fluorescence than the heterogeneous culture.

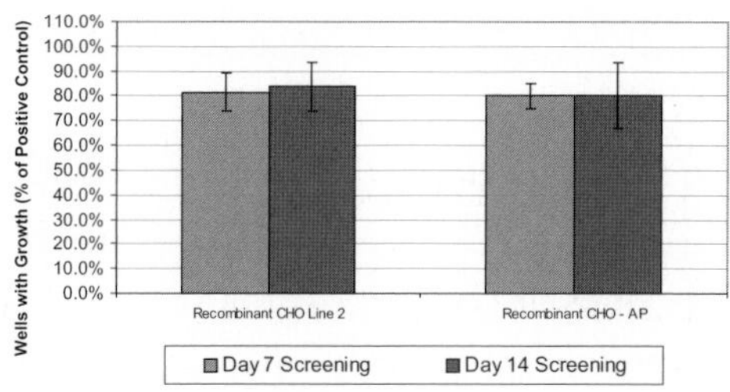


*Figure 1.* Cloning Performance of the Final Formulation in Two rProtein Producing Cell Lines.

# 3. CONCLUSIONS

The animal-component free cloning medium supports CHO cell clonal survival and growth comparable to 10% FBS supplemented Basal Medium. Selected clones from two recombinant CHO cell lines arisen from the AF cloning medium were successfully expanded and demonstrated to have improved growth or productivity over the heterogeneous culture. The AF Cloning medium can be used for other cell line generation procedures such as transfection and selection. The DOE approach is applicable for custom animal-component free cloning media optimization.

# Plant Peptones: Nutritional Properties Sustaining Recombinant Protein Secretion by CHO Cells

Johann Mols, Nicolas Cavez, Spiros Agathos and Yves-Jacques Schneider
*Cellular bioengineering group, Institut des Sciences de la Vie and Université catholique de Louvain; 5, place Croix du Sud; B-1348 Louvain-la-Neuve, Belgium. yjs@uclouvain.be.*

**Abstract:** Plant protein hydrolysates or plant peptones have been widely utilized as culture media additives to support cell growth and recombinant protein production. Unfortunately, their molecular targets remain largely unknown and it remains controversial whether plant peptones act as simple nutritive supplements providing various amino-acids to the cells or as inducers of some signal transduction pathways.

**Key words:** CHO cells, plant peptones, peptide transporters, protein-free, suspension.

# 1. INTRODUCTION

Chinese hamster ovary (CHO) cells are considered as one of the most versatile cell line for the production of recombinant proteins and as such are widely utilized. When the protein production is directed towards therapeutic purposes, biosafety concerns have to be taken into account. As such, serum and animal-derived proteins are progressively banned from the composition of culture media, while plant proteins and plant protein hydrolysates appear as very attractive substitutes.

In this study, an interferon-gamma (IFN-γ) secreting CHO cell line (CHO-320 cells) was utilized as a model to study the effect of plant peptones on the physiology of the cells during cultivation in suspension and in protein-free media.



*R. Smith (ed.), Cell Technology for Cell Products*, 599–601.

## 2. RESULTS AND DISCUSSION

CHO-320 cells were inoculated in a homemade protein-free medium designated as BDM for *Basal-defined medium* with or without rice peptones ($2g\ .l^{-1}$) at $3.\ 10^{5}\ ml^{-1}$ and cultivated in suspension at 100 rpm. Living cell density was determined by Trypan blue assisted manual counting and IFN-γ secretion by ELISA (R&D Systems) (Figure 1). Mainly, rice peptones induce a doubling of IFN-γ concentration in the culture medium after ~120 hours of culture whereas the cell density remains weakly affected by the addition of these peptones to the culture medium.

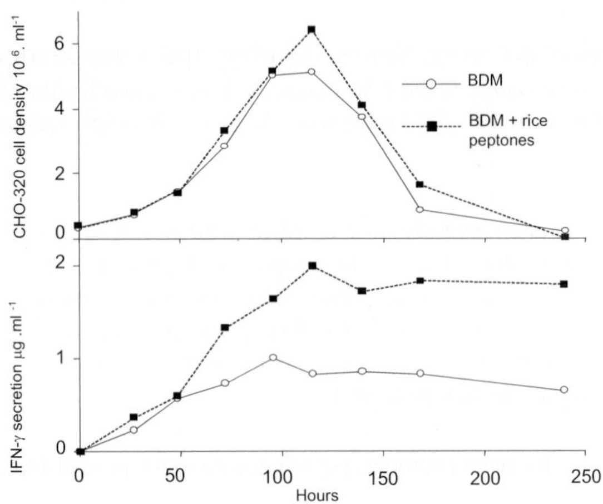


*Figure 1.* Plant peptones support IFN-γ secretion.

To investigate whether some peptides present in the rice peptones could be directly internalized as di- or tripeptides or if some larger peptides could be partially digested and afterwards taken up by the cells, some substrate competition assays were undertaken (Figure 2).

The activities of two membrane exopeptidases (dipeptidyl peptidase IV, DPPIV and tripeptidyl peptidase II, TPPII) were measured using para-nitroanilide (*p*NA)-conjugated peptides. It appears that only the TPPII activity was reduced by an increase of rice peptones concentration in the extracellular medium, suggesting that these peptones could be substrates of TPPII. As such, this suggests also that TPPII could release some tripeptides in the medium. Therefore, the peptide transport activity (Figure 2B) was measured. It appears that the cultivation of CHO-320 cells in the presence of the rice peptones does not dramatically modify the transport activity as peptones were removed for the assay by washing the cells, suggesting that the expression of the transporters were not modified. In addition

(Figure 2D), the presence of rice peptones during the assay strongly affect the peptide transport activity suggesting that some peptide present in the peptones could act as substrates of the peptide transport proteins. Finally, Figure 2C shows that a typical peptide resulting from the activity of TPPII (Ala-Ala-Phe) has a similar inhibitory activity towards the peptide transport.

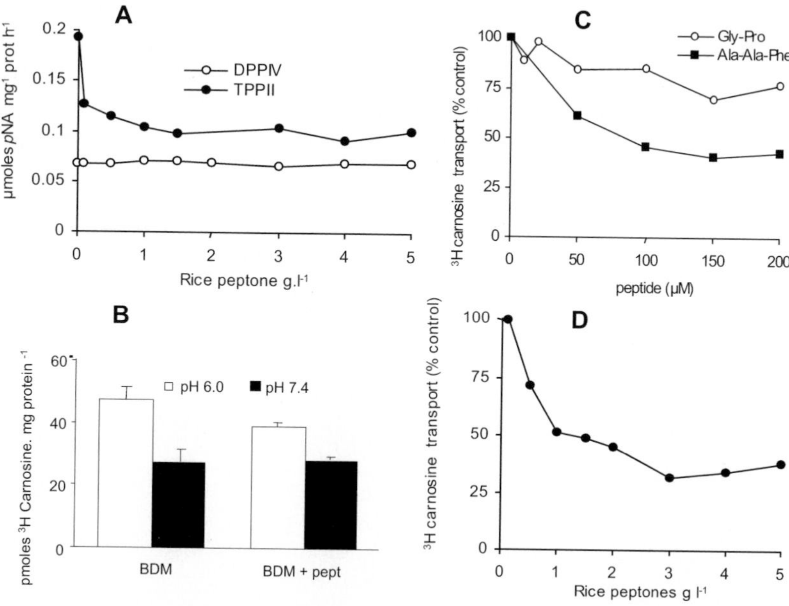


*Figure 2.* Rice peptones interfere with the activity of TPPII as well as with the peptide transport activity through the plasma membrane.

To conclude, these results strongly suggest that some peptides present in the rice peptones could be directly taken up by some peptide transporters or indirectly after partial digestion into tripeptides by TPPII. Consequently, these results support the idea that at least rice peptones have some nutritive properties.

Rice peptones were kindly provided by Kerry-Quest international (Naarden, the Netherlands). This work was supported a grant from the Wallon region (DGTRE) and J. Mols was a fellow from the Belgian FRIA.

# Monitoring of Extracellular TCA Cycle Intermediates in Mammalian Cell Culture

Joachim B. Ritter[1], Yvonne Genzel[1], Udo Reichl[1,2]
[1] *Max Planck Institute for Dynamics of Complex Technical Systems, Magdeburg, Sandtorstr. 1, 39106 Magdeburg, Germany.* [2] *Lehrstuhl für Bioprozesstechnik, Otto-von-Guericke-Universität Magdeburg, Magdeburg,Germany*

**Abstract:** Some intracellular intermediate metabolites can also be found in the medium supernatant at micromolar concentrations. In this work, we are investigating extracellular concentrations of five organic acids (succinic, malic, fumaric, citric and isocitric acid) during cell growth and after viral infection of MDCK cells.

**Key words:** influenza, vaccine, mammalian, MDCK cells, adherent, TCA cycle, organic acids, metabolism, extracellular, intracellular, batch cultivation, monitoring, T-flask, spinner, bioreactor, chromatography, anion-exchange

## 1. INTRODUCTION

One research focus of our group is the optimization of an influenza vaccine production process with the adherent cell line MDCK (Madin-Darby canine kidney) on microcarriers [1]. In addition to monitoring basic metabolites the measurement of extracellular amino acid levels was already established [1, 2]. Further information about metabolism could be achieved by analyzing intracellular intermediates. But some of these metabolites can also be found in the medium supernatant in cell culture [3]. Since no active transport for organic acids is reported for our cell line, they might be released by disruption of dead cells, leakage by protonated diffusion through membranes or transporters for e.g. monocarboxylates [4, 5] or amino acids [6]. Our goal is to gain additional information about metabolism by the investigation of the accumulation of intermediates from central carbon metabolism in the culture broth.



*R. Smith (ed.), Cell Technology for Cell Products*, 603–605.

## 2. MATERIALS AND METHODS

Cell culture, virus infection and standard analytics was done according to Genzel *et al.* [1]. Cells (start concentration 0.4-1.3 x $10^5$ cells/mL) were grown in T-flasks (175 mL, 125 mL wv) (Greiner), spinner bottles (200 mL wv) or 5L-stirred tank bioreactor (B. Braun Biotech) on Cytodex 1 microcarriers (2-3 g/L, GE Healthcare).

Cultivation supernatant was mixed with methanol/formic acid solution in a ratio 3:8, followed by drying. Samples were dissolved in Milli-Q $H_2O$ prior to analysis. Chromatography for malate, succinate, fumarate, citrate, isocitrate was performed with a method adapted from Koswig *et al.* [7] on a DX-320 system (Dionex) using an AS-11 (2 x 250 mm) column with conductivity and UV-detection. A KOH gradient was generated automatically by an eluent generation system, no $NaHCO_3$ was added. Injection volume was 20 µL; flow rate was set to 0.25 mL $min^{-1}$.

## 3. RESULTS AND DISCUSSION

Peaks were identified by standard addition, UV-signal and retention times. For quantification, standards with seven different concentrations were randomly inserted between samples for each sequence. Clear differences could be seen in typical chromatograms of samples at the beginning and the end of cultivation. Identified peaks were quantified (see Fig. 1), but also several not identified peaks showed major changes.

For the three different cultivation modes, cell as well as glucose, lactate, glutamine, glutamate and ammonia concentrations followed typical profiles of batch cultivations. For the bioreactor experiment, metabolite concentrations were changed to starting concentrations by medium exchange before virus infection. After virus infection, a clear decline of viable cells could be seen accompanied by an increase of dead cells. Figure 1 shows the concentrations of the investigated organic acids during the cultivations. At t = 0 h, all samples already contained organic acids coming from serum and peptone. The level of all organic acids was constant during the first 48 h of cultivation, followed by an increase for approximately the next 48 h. Other experiments from our group have shown clear changes in amino acid metabolism [1] and cell cycle distribution in this time interval. Both the experiment in the T-flask and in the spinner-flask showed for malate, fumarate and citrate a continuous accumulation until the end of cultivation, whereas succinate and isocitrate seemed to reach a constant level between 96 and 120 h. There is no indication that the concentration changes of the organic acids were correlated to dead cells, because succinate and isocitrate stopped increasing during the end of cultivation, when dead cells accumulated. Additionally, calculations showed that intracellular concentrations of metabolite pools (values from literature) do not suffice to result in the change of concentration after lysis of dead cells. Furthermore, in chromatograms from samples after virus infection additional peaks appeared, which were probably substances released by dead cells. With regard to these results, we assume that depending on the intracellular concentrations of organic acids, a corresponding unspecific leakage occurred. The exact mechanism however cannot be explained so far.

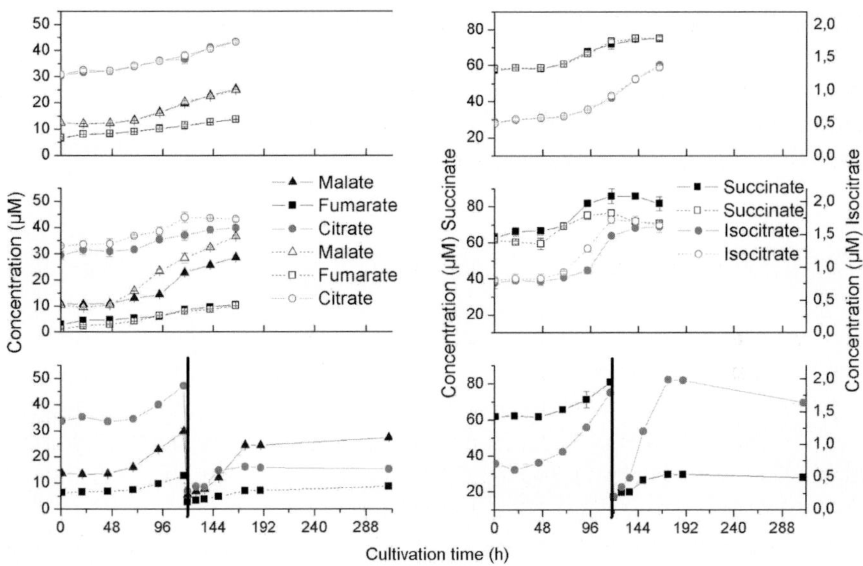


*Figure 1.* Concentrations of organic acids during cultivation: first row: T-flask, second row: spinner flask, third row: bioreactor (virus infection at t = 120 h, vertical line).

## REFERENCES


1. Genzel, Y. *et al.*, 2004, Vaccine, **22(17-18)**: 2202-2208.
2. Genzel, Y. *et al.*, 2005, Biotechnol Progr., **21 (1)**: 58-69.
3. Weigang, F. *et al.*, 1989, J Chromatogr., **497**: 59-68.
4. Putnam, W. S. *et al.*, 2002, J Pharm Sci., **91 (12)**: 2622-2635
5. Rosenberg, S.O. *et al.*, 1993, Biochem J., **289**: 263-268.
6. Roy, G., Banderali, U., 1994, J Exp Zool., **268**: 121-126.
7. Koswig, S. *et al.*, 9-21 in Kettrup, Antonius: Spurenanalytische Bestimmung von Ionen, 1997, Landsberg: ecomed

# A Comparison of the SixFors Fermenter System and Lab Scale Fermenters For Clone Selection

Giles Wilson, Annette Jørgensen
*Novo Nordisk A/S*

**Abstract:** A basic method of selecting clones uses a 96 well plate format whereby cells are assessed and selected based on growth rate and specific production rates. However, such data does not allow us to select a clone based on criterion which may make it more amenable to production in a fermenter environment. Because of this we evaluate a number of clones in fermenters at 5 litre scale before making a final selection on a production candidate. Clones are assessed at 5 litre scale using a modified draw-fill fermentation. In this process, cells are grown in a fed-batch format for 5 days. This is designated as cycle 1. After this time 80% of the volume is removed and replaced with fresh medium. This first harvest is used for purification and characterisation of the product. In the second cycle of the fermentation the process is run in fed-batch mode to destruction and this data is used to characterise the cell performance and process kinetics. Fermenter time is a limited resource and we have evaluated an alternative strategy.

In the alternative strategy we use wave bioreactors in a 5 day batch mode to replace cycle 1 of the modified draw fill. We have also used a SixFors fermenter system to replace the second cycle of our modified draw-fill. The wave bag provides material for purification and characterisation and the Sixfors provides data to characterise the cell line and process kinetics. Data will be presented to show the comparability of these two techniques to data generated at 5 litre scale.

**Key words:** CHO, small scale fermentation, animal component free medium, clone selection, fermenter, SixFors.



*R. Smith (ed.), Cell Technology for Cell Products*, 607–607.

# Comparison of Cultivations of CHO Cell Line in Chemostat and Acceleratostat Experiments

Rasmus Bjerre Nielsen, Giles Wilson
*Novo Nordisk A/S*

**Abstract:** Models of culture behavior are strong tools for optimization of cultivations of both microbial and animal cells. Continuous steady state cultivations are the most widely employed method for experimental estimation of model parameters due to the ease at which reliable results can be obtained from such cultures. The standard mode of operation for laboratory scale continuous cultures is the chemostat in which the dilution rate is varied stepwise in order to investigate behavior at different specific growth rates. Depending on the cell line, long periods of time may be needed before a chemostat culture reaches a new steady state and in the intervening time, measurements must be considered dynamic. This means that very few steady state values are obtained during a complete chemostat run. In accelerostats, the dilution rate of the culture is gradually varied at an acceleration or deceleration sufficiently low for the cells to remain at steady state. In this study, we investigate the use of the accelerostat method as a tool for parameter estimation and show that the obtained results correspond completely with those obtained from a chemostat.

**Key words:** CHO, chemostat, acceleratostat, fermentation, serum free, animal component free, continuous culture, mathematical modelling.



*R. Smith (ed.), Cell Technology for Cell Products*, 609–609.

# Evaluation of a Novel Micro-bioreactor System for Cell Culture Optimisation

Giles Wilson
*Novo Nordisk A/S*

**Abstract:** BioProcessors SimCell™ high throughput cell culture process development system is a robotically controlled micro-bioreactor system capable of conducting over 1000 parallel bioreactor experiments. The animal cell cultures are grown in approximately 600μl volumes in a format capable of simulating environmental control and stresses that are present in conventional fermenters. Each SimCell Micro Bioreactor Array contains 6 x 600 μl mini-fermenters and allows the operator to rapidly screen many different parameters using a robotic control system. In this case we operated the dual chamber SimCell micro-bioreactors in a perfusion mode with a CHO cell line in serum free conditions for over 10 days and examined the effects of temperature and pH in 50 different combinations. With 9 replicate data points for statistical analysis this study, in the normal course of events, would have utilised ten 5 litre tanks for 1 year. In comparison to 5 litre perfusion fermentations (both suspension cell perfusion and microcarrier) the microbioreactors gave equivalent levels of growth and production, demonstrating that the micro-scale system can mimic a larger fermenter. An optimisation of pH and temperature showed that culture pH has a very strong effect on productivity and that small changes in pH can have a large effect upon culture performance. The scalability of these observations made in the SimCell MicroBioreactor Arrays were confirmed in our validated 5L bench top fermenter model.

**Key words:** CHO, animal component free medium, pH, temperature, SimCell, fermenter, screening, optimisation.



*R. Smith (ed.), Cell Technology for Cell Products*, 611–611.

# Comparison of Different Design Configurations of a 1-L Multi-bioreactor Station for Cell Culture Process Development

Heiko Schütte, Ulrike Kolrep, Liane Franck, Roland Wagner
*Miltenyi Biotec GmbH, Bioprocess Science, Robert-Koch-Straße 1, D-17166 Teterow, Germany*

**Abstract:** Different design modifications based on the multi-bioreactor station Biostat Q® using HEK293-EBNA suspension cells were compared. The various system configurations were combined with different agitation methods using magnetic stir bars of different radius and shape as well as the two standard 4-blade Rushton impellers, commonly used in microbial fermentations. The design was additionally modified by removing most of the baffles resulting in a minimum shear stress to the cells. Aeration, either performed by sparging with bubbles, microbubbles or only by headspace aeration alone showed substantial differences in growth rate, viability and metabolic rates. Highest growth rate, viability and productivity were achieved with a baffle-free configuration using bubble aeration at low rates and egg-shaped stir bars.

**Key words:** Multi-bioreactor system, HEK293-EBNA, serum.-free medium, process development

## 1. INTRODUCTION

Multi-bioreactor systems consisting of 4 to 16 bioreactor vessels have the advantage of saving considerable amounts of space for parallel investigations by maintaining the configuration of a single lab-scale bioreactor station. ***A significant number of experiments can be performed simultaneously using low volumes of culture per vessel.*** However, the sizes



*R. Smith (ed.), Cell Technology for Cell Products*, 613–617.

of the bioreactor vessels are comparably low, mostly limited to 0.5 to 1 L, and the stirrer, the electrodes and the temperature sensor displace a substantial part of the working volume thus generating a comparably high density of different inevitable baffles affecting the fluid dynamics of the culture broth. The system-associated characteristics can reduce the spectrum of applications of multi-bioreactor systems for process development studies to pilot and production scale. For evaluating the efficiency of the different design modifications of mini-bioreactors, the multi-bioreactor station Biostat Q® and HEK293-EBNA suspension cells were used . These variations of the system configuration were combined with different agitation methods using magnetic stir bars of different radius and shape as well as the two standard 4-blade Rushton impellers, commonly used in microbial fermentations. The design was additionally modified by removing most of the baffles resulting in a minimum shear stress to the cells.

## 2. MATERIALS AND METHODS

Cell line and medium. A recombinant mouse interferon-α-producing HEK293-EBNA cell line, kindly provided by Dr. Werner Müller (GBF, Braunschweig, Germany), was cultivated in suspension using Ex-Cell293 serum-free medium (JRH Biosciences, Lenexa, KS, USA) supplemented with 8 mmol $L^{-1}$ ultraglutamine (Cambrex, East Rutherford, NJ, USA) and 0.1% Pluronic F68 (Sigma-Aldrich, St. Louis, MO, USA).

Cultivation. Precultures were carried out in spinner flasks (200 mL working volume) stirred at 30 rpm (Techne, Cambridge, UK). Cultures were incubated at 37°C under 90% air/ 7.5% $CO_2$ atmosphere and 99% humidity (Revco Ultima, Revco Lindberg, Ashville, NC, USA). Fermentation was performed in 1.0 L-stirred tank vessels (0.9 L maximum working volume) of the multi-bioreactor station Biostat Q® (B. Braun Biotech International, Melsungen, Germany) equipped with a DCU 3 control unit and a gas mixing station modified to alternatively mix air/$N_2$/$CO_2$, $O_2$/$N_2$/$CO_2$ or air/$O_2$/$CO_2$ for guaranteeing a sufficient oxygen supply under constant low aeration rates of 1 L $h^{-1}$ (Fig. 1). Cell numbers were counted in a hemocytometer (Neubauer, improved) and cell viability was estimated using the trypan blue exclusion method (Biochrom, Berlin, Germany). Glucose, lactate, glutamine, glutamate and ammonium were quantified using a multibiosensor analysis system BioProfile 200 (Nova Biomedical, Waltham, MA, USA).

## 3. RESULTS AND CONCLUSION

The multi-bioreactor station Biostat Q® consisting of four 1-L glass vessels was used (Fig. 1).

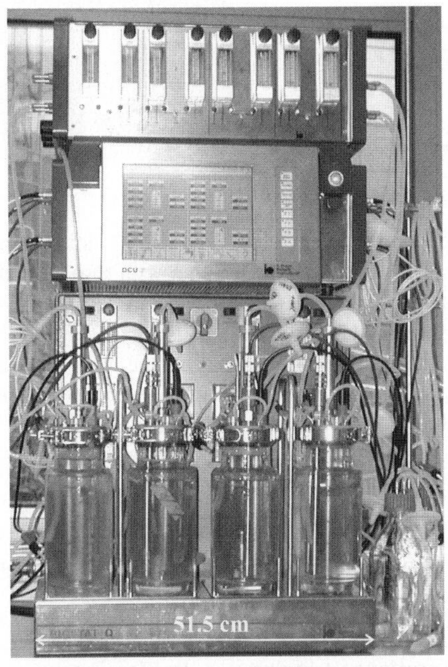


*Figure 1.* Multi-bioreactor system Biostat Q®. The system need a minimum area of 0.35 $m^2$.

Different agitation methods were applied using magnetic stir bars of different radius and shape (Fig. 2 left) as well as the two standard 4-blade Rushton impellers, commonly used in microbial fermentations. The design was additionally modified by removing most of the baffles to guarantee a minimum shear stress to the cells (Fig. 2 right). Thus, a substantial reduction of the inner equipment could be reached. For comparative experiments cells were inoculated at a concentration $1\times10^5$ $mL^{-1}$ and repeated batch cultivations were performed. When glucose was limited (< 1 g $L^{-1}$) 400 to 700 mL culture broth were harvested and replaced by fresh medium according to cell concentration and growth rate obtained under the respective culture conditions. As expected from the small surface of 40 $cm^2$ (H/D = 2.5), cells did not grow under exclusive surface aeration even when pure oxygen was supplied (Table 1). The opposite extreme, aeration by microbubbles, resulted in highest mitotic rate ($\nu_{max}$ = 0.048 $h^{-1}$) and lowest population doubling time (g = 20.8 h) for the first phase of culture, but the cells started to continuously reduce their viability after one week of

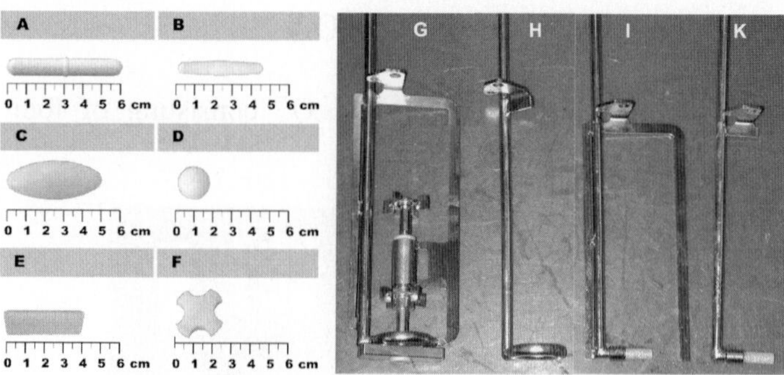


*Figure 2.* Left side: Structure of different TEFLON® stirring bars to be used as mixing devices in culture vessels. A = polygonal spin bar with ring around center, B = polygonal stir bar with extended center, C = egg-shaped stir bar, D = stir ball, E = triangular wedge-shaped spinwedge stir bar, F = stirring head stir bar.
Right side: Different design configurations of the gas mixing devices. G = originals configuaration with Rushton impeller and baffles, H = sparge ring, Rushton impeller and baffles were removed, I = original microsparger with baffles, K = microparger without baffles.

*Table 1.* Bioreactor configurations at a glance. A sparge ring supported by a polygonal stir bar gave highest cell concentrations and viability.

| Bioreactor Configuration | Aeration | Stirrer | Baffles | $c_{max}$ [$mL^{-1}$] | $v_{max}$ [$h^{-1}$] | g [h] | Av. Viab. [%] |
|---|---|---|---|---|---|---|---|
| **Spinner-like** | Surface | Egg-shaped stir bar | No | No growth | - | - | - |
| **Microbubbles** | Microsparger | Egg-shaped stir bar | No | $1.5\times10^6$ | 0.048 | 20.8 | 91 |
| **Microbial-like** | Sparge ring | Rushton impeller | Yes | $3.5\times10^6$ | 0.027 | 37.2 | 83 |
| **Sparging** | Sparge ring | Polygonal stir bar | No | $2.5\times10^6$ | 0.032 | 30.8 | 93 |

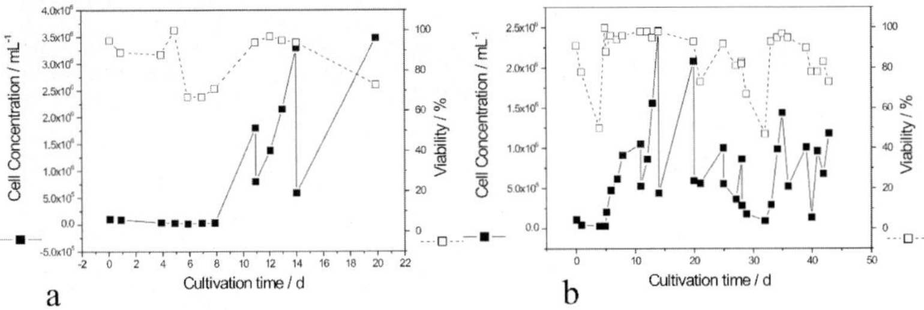


*Figure 3.* a. Cultivation of HEL293-EBNA cells at microbial-like bioreactor configuration (see Table 1). Cells needed a 6-days lag phase for adapting to the shear conditions. b. Cultivation of HEK293-EBNA cells at sparging bioreactor conditions (see Table 1) Cells showed a growth comparable to T-flasks and spinner flasks.

cultivation due to the high shear stress introduced by the microbubbles and the resulting foam formation.

A bioreactor equipped with a sparge ring for aeration and a Rushton impeller showed a maximum mitotic rate of $\nu_{max} = 0.027\ h^{-1}$ and a population doubling time of g = 37.2 h. A continuous high mitotic rate ($\nu_{max} = 0.027\ h^{-1}$), a low population doubling time (g = 30.8 h) and highest viability (93%) were obtained using a sparge ring and a polygonal or egg-shaped stir bar. All the other stir bars as indicated in Figure 2, especially the stirring head stir bar led to a reduction in cell viability (data not shown). A lag-phase of 6 days was observed when the original bioreactor configuration was chosen which was derived from microbial fermentation (Fig. 3).

# The Effect of Genetically Modified Insulin Growth Factor on Cell Proliferation and Adenovirus Productivity in Serum Free Medium

Angela Buckler[1] and Mohamed Al-Rubeai[1,2]
*[1]Department of Chemical Engineering, University of Birmingham, Birmingham B15 2TT, UK,*
*[2]Department of Chemical and Biochemical Engineering and Centre for Synthesis and Chemical Biology, University College Dublin, Belfield, Dublin 4, Ireland*

**Abstract:** Data clearly show that Long$^{TM}$ $R^3$ IGF-1 was able to increase HEK293 cell proliferation significantly more than other growth factors when added individually or in combination to serum free media. However, serum free media with or without growth factors produced similar amount of adenovirus indicating that a simple infection medium containing no growth factors could be adequate for virus production phase. This would provide a cost effective approach, since media containing no growth factors would be less expensive.

**Key words:** HEK293, Insulin, LONG$^{TM}$ $R^3$ IGF-1, Serum Free Media, Adenovirus.

## 1. INTRODUCTION

Insulin is known to be essential for cell proliferation in serum free culture. Long$^{TM}$ $R^3$ IGF-1 (insulin-like growth factor 1) is used to provide an inexpensive yet high quality potent IGF-1 analog in serum free or reduced serum culture media. Insulin and IGF-1 have very similar amino acid sequences and tertiary structures (1). Long$^{TM}$ $R^3$ IGF-1 was created for use in cell culture by substituting the third amino acid of IGF-1 with an arginine, it also has a 13 amino acid N-terminal extension. Previous work shows that Long$^{TM}$ $R^3$ IGF-1 is significantly more potent than insulin *in vitro* (1).

The aim of this study was to investigate the effect of Long$^{TM}$ $R^3$ IGF-1 compared to insulin and IGF-1 on proliferation of HEK 293 cells. Infection and specific productivity of Adenovirus was also studied.



*R. Smith (ed.), Cell Technology for Cell Products*, 619–622.

## 2. RESULTS AND DISCUSSION

### 2.1 The Effect of Insulin and Insulin Like Growth Factors on Propagation

Figure 1a shows the cell growth of HEK 293 cells in EX-CELL™ 293 with growth factors on day 3 and day 7 of a 10 day batch. These data clearly show that Long™ $R^3$ IGF-1 was able to increase cell proliferation significantly more than other growth factors when used individually or in combination. The viability for all the cultures drops consistently throughout the growth curve (data not shown). Fig. 1b shows growth in DMEM + 5% FCS with growth factors on day 2 and day 5 of a 10 day batch. When IGF-1 and Long™ $R^3$ IGF-1 are added with insulin, the cell growth is higher than other cultures. The viability for all the cultures decreases at a similar rate throughout the 10-day period (data not shown).

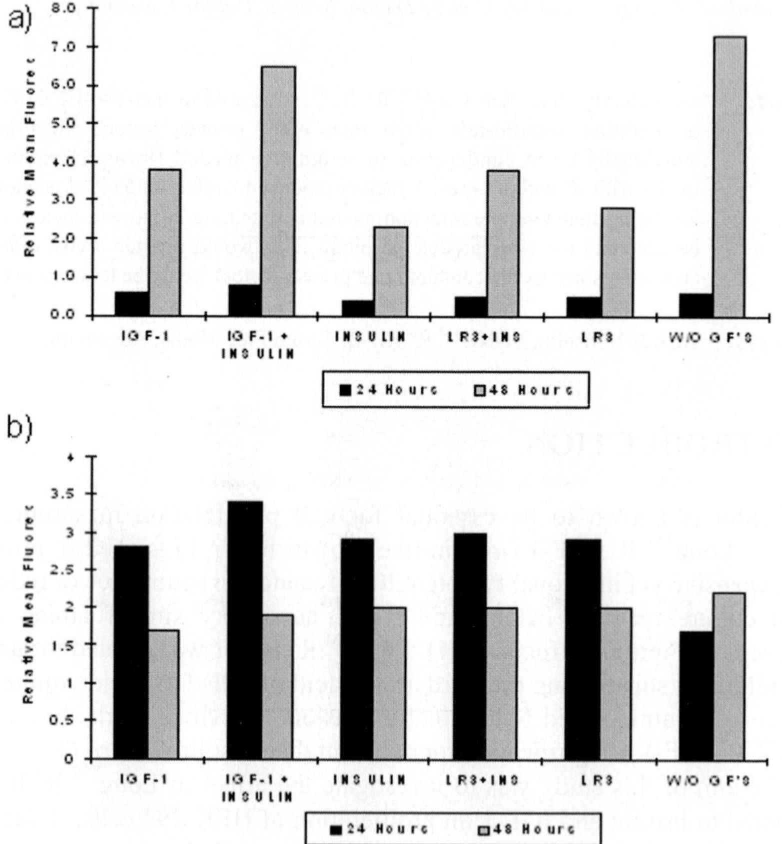


*Figure 1.* a) Cell Growth of HEK 293 Cells in EX-CELL™ after Growth Factor Addition b) Cell Growth of HEK 293 Cells in DMEM+FCS after Growth Factor Addition.

## 2.2 The Effect of Insulin and Insulin Like Growth Factors on Virus Infectivity

Figure 2a shows the mean intensity of GFP, which equates to more virus within the cell population. This data is consistent with the percentage of infected cells, ie the higher the mean fluorescence value the higher the percentage of infected cells (data not shown). More virus is produced in cultures containing both IGF-1 and Insulin or when there are no growth factors present. It also Fig. 2b shows the mean GFP intensity in DMEM + FCS.

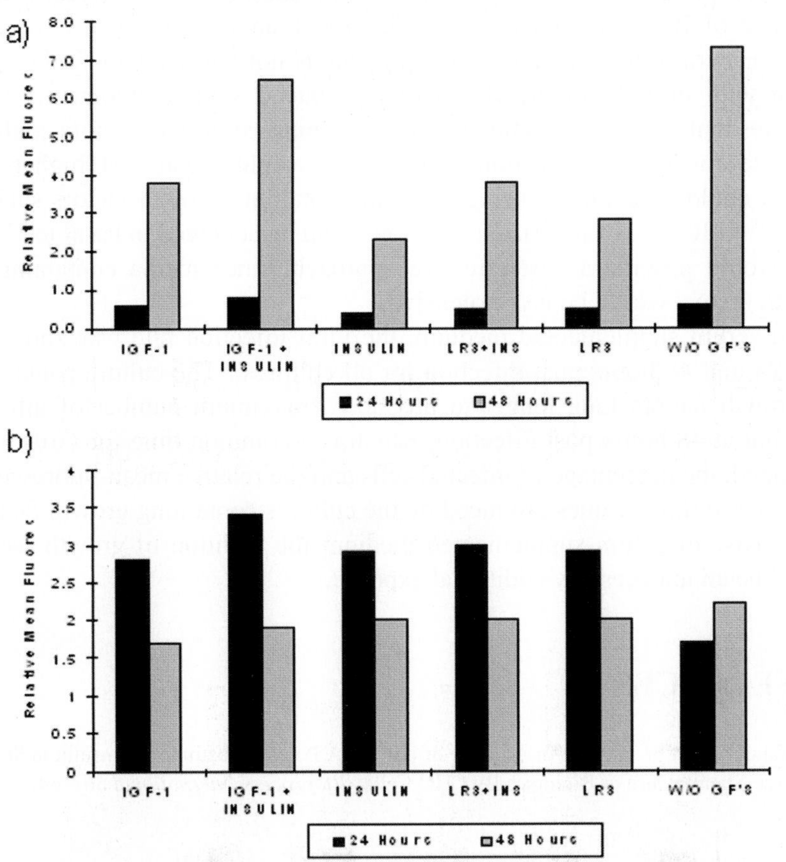


*Figure 2.* a) Mean GFP Intensity in EX-CELL™ after Growth Factor Addition. b) Mean GFP Intensity in DMEM+FCS after Growth Factor Addition.

# 3. CONCLUSION

By monitoring cell growth and viability of HEK293, we have shown that Long$^{TM}$ $R^3$ IGF-1 promotes cell proliferation more effectively than IGF-1 and/or insulin in serum free conditions.

In serum-supplemented medium, the addition of growth factors resulted in increased cell proliferation, although Insulin took longer to have this effect. When insulin is added in combination with either IGF-1 or Long$^{TM}$ $R^3$ IGF-1, a synergistic effect was observed resulting in optimum cell proliferation throughout the culture.

Under serum free conditions, the viral infection rate was increased in the presence of IGF-1 whereas Long$^{TM}$ $R^3$ IGF-1 appears to have a negative effect on viral infection. The reason for this is not known. Our finding that media with or without growth factors produced similar amount of virus indicates that a simple infection media containing no growth factors could be adequate for virus production phase. This would mean that higher cell growth could be achieved using medium containing growth factors such as Long$^{TM}$ $R^3$ IGF-1 while virus production could be achieved in basal medium. This would provide a cost effective approach, since media containing no growth factors would be less expensive.

In serum supplemented medium, the viral infection rate was similar at both 24 and 48 hours post infection for all cultures. The culture containing no growth factors took longer to produce a maximum number of infected cells but at 48 hours post infection, which is a common time for virus to be harvested, the percentage of infected cells and the relative mean fluorescence is similar to those values produced by the cultures containing growth factors. In the case of serum supplemented medium the addition of growth factors would be an unnecessary additional expense.

# REFERENCES

(1) Yandell, C *et al.* (2004) An Anolgue of IGF-1. A Potent Substitute for Insulin in Serum-Free Manufacture of Biologics By CHO Cells. *Bioprocess International* 56 – 64.

# Evaluation of a Scalable Disposable Bioreactor System for Manufacturing of Mammalian Cell Based Biopharmaceuticals

Silke Langhammer, Marco Riedel, Hikmat Bushnaq-Josting, René Brecht, Ralf Pörtner*, Uwe Marx
*ProBioGen AG, Goethestraße 54, 13086 Berlin, Germany, * Hamburg University of Technology, Denickestr. 15, 21071 Hamburg; Germany*

**Abstract:** The use of disposable equipment in biopharmaceutical up- and downstream processes has increased in the last few years. This results in higher process flexibility and safety at reduced CIP and SIP efforts. The lack of robust scaleable disposable bioreactors initiated our own engagement in this area.

A new generation of membrane-based bioreactor was designed. This family of reactors starts with a small scale device, where eight bioreactors operate simultaneously for focused multiple process parameter optimisation. The up scaling concept includes a pilot scale bioreactor for the production of clinical trial materials in flexible pilot manufacturing plants; and a commercial scale unit for market supply. To demonstrate early feasibility at moderate costs, a lab scale prototype was specially designed for this purpose.

This new generation of bioreactors works with an efficient oxygen supply to support high cell densities in a continuous perfusion process. The cell retention is integrated in the bioreactor, thereby ensuring continuous cell-free harvesting over several months.

The development of a mathematical model describing the background of efficient nutrient supply is in progress, as well as a financial model for the operation of these new disposable bioreactors in manufacturing plants. Results of the biocompatibility tests of the disposable system components, as well as long term performance studies on the membrane behaviour are summarized. Test runs were carried out by the cultivation of a CHO suspension cell line expressing a recombinant protein and the feasibility data will be discussed.

**Key words:** disposable, scalable, bioreactor, membrane based, perfusion, biopharmaceuticals



*R. Smith (ed.), Cell Technology for Cell Products*, 623–628.

## 1. INTRODUCTION

The continuing specification of indications as well as an individualization of medicine request more flexibility in manufacturing of biopharmaceuticals. Further more the increasing efficiency of biopharmaceuticals causes smaller batch amounts. To meet these requirements with future manufacturing of biopharmaceuticals, a new kind of bioreactor is necessary which allows a high flexibility in a multi product facility with moderate COG's. The solution would be a robust, mobile, disposable and scalable bioreactor applicable to multi product facilities.

## 2. WORKING PRINCIPLE OF THE BIOREACTOR

The bioreactor consists of two parts, one disposable unit including all cell touching parts and one stationary unit with pumps, measurement technique, control devices and so on. The heart of the disposable part is a disposable module which contains a number of identical membrane surrounded micro compartments. The sum of all inner volumes of the micro compartments forms the cell culture volume. The cylindrical module (vessel) itself is half filled with media and will continuously be flown through by media and gas.

Caused by the rotation of the module every cell containing micro compartment is periodically exposed either to media or to gas. So nutrient feed as well as oxygen supply is ensured.

The membrane of the micro compartments is permeable for nutrients and oxygen but also for the produced protein. By this way the protein can be harvested continuously and cell-free. A ultra filtration unit will be installed in the outflow line, to concentrate the protein solution directly.

## 3. MODELING OF CELL YIELD DEPENDENT ON MEMBRANE DESIGN

By equation of the Fick's law for the oxygen flow through the membrane and the oxygen uptake of the cells the relation between cell concentration in the culture volume and membrane thickness can be calculated.

$$n = D_{50} \cdot A \cdot \frac{\Delta c}{\Delta x} \qquad n = q_{O2} \cdot X \cdot V_{CV} \qquad \Delta x = \frac{4 \cdot D_{50} \cdot \Delta c}{q_{O2} \cdot X \cdot d_i}$$

n-mass transfer, $D_{50}$-diffusion coefficient, A-inner membrane surface, $\Delta$c-oxygen concentration gradient over the membrane, $\Delta$x-thickness of the membrane wall, $q_{O2}$-specific oxygen uptake rate, X-cell concentration, $V_{CV}$-culture volume, $d_i$-inner diameter of the cylindrical compartments

Assumptions:
- oxygen transport through the membrane exclusively by diffusion
- diffusion coefficient $D_{50}$ over the membrane is 0,05cm$^2$/h (50% of pure water)
- specific oxygen uptake rate $q_{O2}$: low 1,6*10$^{-9}$mg/c*h; high 6,4*10$^{-9}$mg/c*h

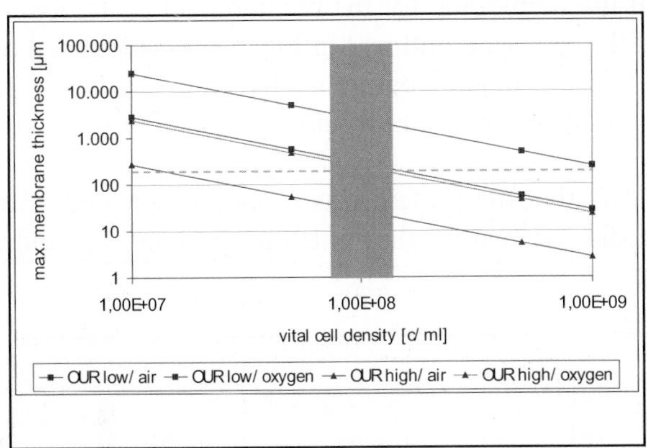


*Figure 1.* Allowed membrane thickness dependent on the cell concentration for unlimited oxygen supply.

The gray area in Fig. 1 marks the management target for vital cell concentration. The dotted line marks the thickness of the used membrane. With the 200µm membrane the target could be reached with cells of a low oxygen uptake rate. A gasing with pure oxygen would increase the vital cell concentration tenfold.

## 4. METHODS AND RESULTS

For feasibility testing two kinds of prototypes were designed. The Test Scale with ~4 ml Culture Volume and 100 ml vessel volume and the lab scale with 110 ml culture volume and 15 L vessel volume. The vessel volume of 15 L could supply the tenfold culture volume but the amount of cells should stay limited during the test phase.

For the tests a CHO line producing a recombinant protein was used. The daily protein concentration in outflow was measured by a specific ELISA, the productivity was calculated based on the amount of purified protein.

The test scale was supplied with regular media exchange and gasing, there was no continuous perfusion. In the test scale the cultivation of different cell lines was tested. Therefore the always used CHO line as well as a second CHO clone and a Hybridom line, all producing the same recombinant protein were cultivated. The experiment lasted 5 days. The processes were not optimized in any way. The goal was only to hold the cells vital. This was reached in all cases. The Hybridom culture even increased its vitality (Fig. 2).

The lab scale was continuously flown through by media and gas and run 10 to 23 days. In experiments with single micro compartments the good permeability of the membrane for the recombinant protein was proved. The productivity of the lab scale outlined in Fig. 3 was calculated by the amount of purified protein out of the continuous harvest. The horizontal bar marks the management target for feasibility proof. This goal was surpassed in two processes.

Figure 4 shows that the protein was produced constantly. In spite of the continuous media perfusion the outflow protein concentration stayed constant.

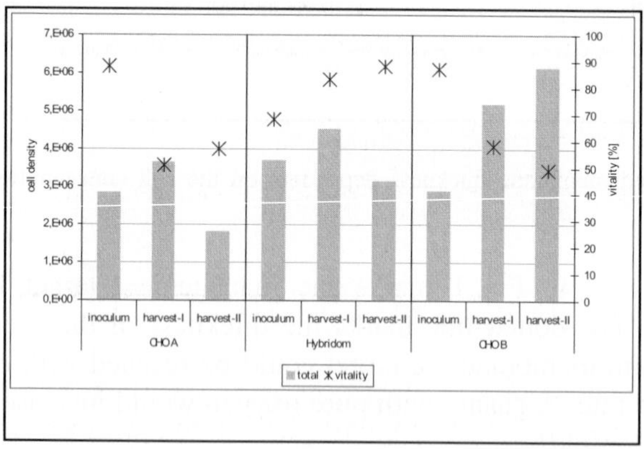


*Figure 2.* total cell density and vitality of different cell lines cultivated over five days in test scale.

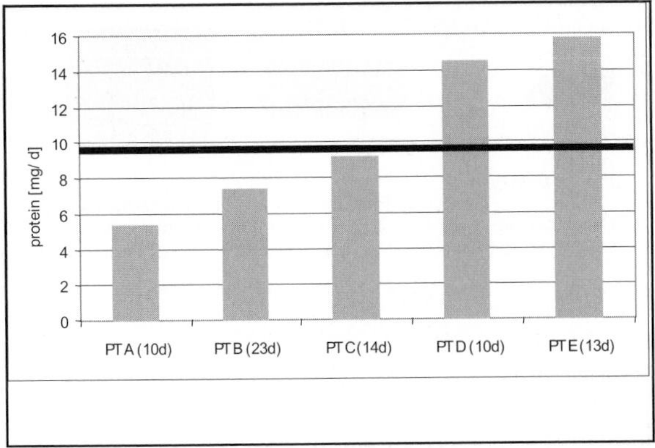


*Figure 3.* Produced protein per day in different lab scale runs.

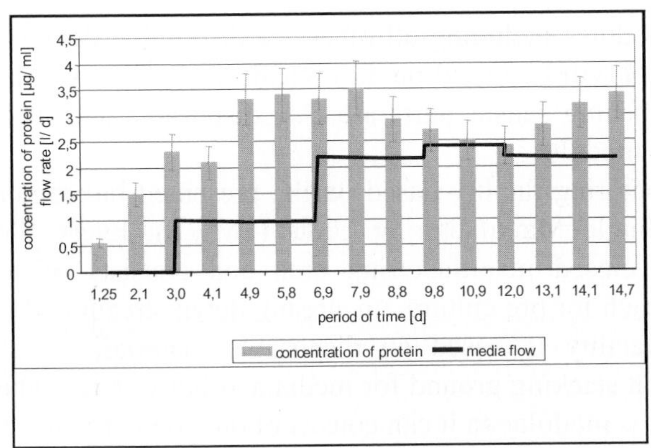


*Figure 4.* protein concentration in the outflow media of one lab scale run.

# 5. ISSUES OF SCALABILITY AND FACILITY CONCEPT

An important request on future protein manufacturing systems is the scalability from units for process optimization to reactors for market supply. Fig. 5 shows the design of the bioreactor family. With the Process Development Device it is possible to run eight processes in parallel with different parameters. The pilot scale can be used for manufacturing of material for toxicity studies and pre clinical phase to phase II. The commercial scale will be used for market supply and will work with the same stationary hardware unit as the pilot scale. This increases the flexibility of a facility very much.

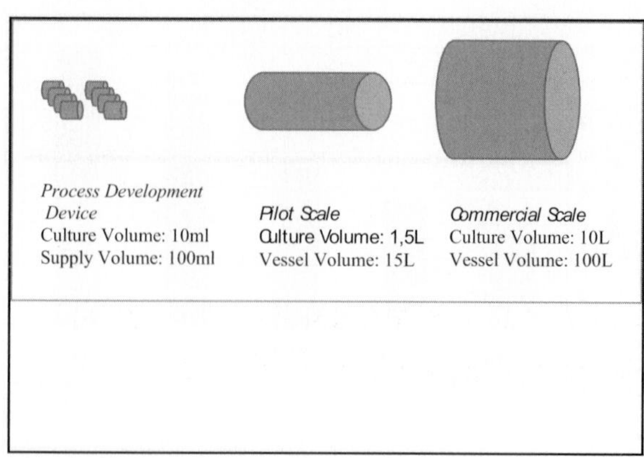


*Figure 5.* Design of the reactor family.

All the modules including all tubes are disposable devices. The pH and oxygen control will be based on an optical measurement technique. At this the media touching sensor parts are also disposable and included into the pre-sterilized module.

A manufacturing facility based on the presented bioreactor technique is divided in Single Manufacturing Units (SMU's). Every SMU forms an independent, closed manufacturing area and consists of four dedicated rooms. One each for pre culture, up stream, down stream and final filtration. The whole facility is based on disposable material, there is no steam installation but stacking ground for media and buffer bags. The construction of the facility is modular so it can consist of one, two or many SMU's.

## 6. SUMMARY

The developed bioreactor is a continuous system with integrated membrane based cell retention system. This causes short residence times for the produced protein in the system. Furthermore the bioreactor technique is scalable and disposable.

Pilot and commercial scale bioreactors support the construction of flexible multi product facilities with very low capital expenditures and moderate COG's.

# Purification of Rabies Virus Produced on Vero Cells Using Chromatography Techniques

Mohamed Chtioui, Khaled Trabelsi and Héla Kallel

*Viral Vaccines Research & Development Unit. Institut Pasteur de Tunis. 13, place Pasteur. BP74. 1002 Tunis. Tunisia*

**Abstract:** Rabies encephalomyelitis is a fatal disease. Fifty to seventy thousands death and 10 millions post-exposure treatment are reported yearly among humans. This infection remains a major public health concern, particularly in developing countries. Pre and post exposure vaccination remains the most efficient way to protect humans.

Safe and potent vaccines produced by cell culture techniques, are available in developed countries. However due to economical and technical hurdles, these vaccines are not used in developing countries where tissue brain infected vaccines are used instead.

The aim of this work is to develop a novel process to produce a human rabies vaccine using Vero cells grown on microcarriers. In previous work, we have optimized cell growth as well as virus production phases (Trabelsi *et al.*, 2005). In the current work, we studied the purification of rabies virus harvests obtained through the culture of Vero cells on 6 g/l Cytodex1, in a 7-l bioreactor, using chromatography techniques.

To optimize virus recovery yield, we tested different chromatography media (Sephacryl S200, Sephacryl S300 and Sepharose 4FF). Furthermore, we investigated the effects of pH, ionic strength and virus charge of the sample. The purity of virus suspension was estimated by SDS-PAGE analysis whereas the quantity of virus was evaluated using the RFFIT technique. Data obtained showed that the range of pH (6-8) and ionic strength (0.2-0.5 M NaCl) tested were not critical. The highest overall yield was equal to 40% and was obtained under the following conditions : chromatography medium : Sepharose 4 FF, purification flow rate : 8 ml/min, sample volume : 8 ml, maximal virus charge : $0.6x10^6$ FFU/ml support .

Purification of rabies virus by expanded bed chromatography was also investigated. Two supports : streamline SPXL (cation) and streamline QXL (anion) were evaluated. Using the anion exchange chromatography support, we obtained a very low recovery rate (2%). In the case of the cation exchange



*R. Smith (ed.), Cell Technology for Cell Products*, 629–634.

chromatography support, we reached a recovery rate equal to 17% after the incubation of virus suspension at a pH of 6.1 for 24 hours at 4°C.

**Key words**: rabies virus, purification, Vero cells, chromatography techniques

# 1. MATERIALS AND METHODS

## 1.1 Virus suspension production

LP 2061 rabies virus strain was used to infect Vero cells cultivated in a 7-l bioreactor as described by Trabelsi *et al.* (2005)

## 1.2 Clarification & concentration of rabies virus harvests

Rabies virus harvests were first clarified by filtration through 8 µm filter then concentrated with a 0.1 µm hollow fiber cartridge (Model : H1MP01-43, ref. 1853, Millipore).

## 1.3 Gel filtration chromatography

XK 16/20 columns from Amersham Biosciences were packed with the medium to be tested according to the manufacturer instructions. Unless indicated elsewhere, phosphate buffer pH 7, 0.2 M NaCl was used during the purification. Fractions of 1.5 ml, were collected throughout. Purification was monitored by the measurement of the optical density at 280 nm (OD 280 nm), SDS-PAGE 10% analysis and by virus titration.

## 1.4 Streamline chromatography

### 1.4.1 Method scouting

Streamline SP XL and Streamline Q XL from Amersham Biosciences, were used through out this study. To determine optimal conditions for virus binding and elution for each streamline adsorbent, a wide pH range was tested : 6.6, 6.8, 7, 7.2, 7.4, 7.6, 7.8, 8, 8.2 and 8.4. The experiments were carried out in 5 ml tubes containing 1 ml of the support to be tested and 0.2 ml of virus suspension that has a virus titer equal to $7.7 \times 10^6$ FFU/ml. Phosphate buffer pH 7 1 M NaCl was used for elution. Absorbance at 280 nm and SDS-PAGE 10% analysis of the fractions collected before and after elution, were performed.

### 1.4.2 Breakthrough studies

XK 16/70 column (Amersham Biosciences) packed with either Streamline SP XL or Streamline Q XL was loaded with increasing volume of desalted virus suspension at 4 ml/min. Elution was carried out at 4 ml/min with phosphate buffer pH 7.6 ( for Streamline Q XL) or pH 7 (for Streamline SP XL) 1 M NaCl. The content of the collected fractions was analyzed by SDS-PAGE 10%. Absorbance of the collected fractions at 280 nm was also carried out.

### 1.5 Analytical methods

Rabies virus titration : Virus titer was determined according to a modified RFFIT method as described by Trabelsi *et al.* (2005) and expressed in Fluorescent Focus Units per ml (FFU/ml).

## 2. RESULTS

We studied rabies virus purification using either a multi-steps or a single-step chromatography process.

### 2.1 Multi-steps Chromatography process

To purify concentrated rabies virus suspensions by gel filtration, different chromatography media : Sephacryl S200, Sephacryl S300 and Sepharose 4FF, were evaluated. Virus suspensions were first clarified then concentrated as indicated in Materials & Methods. Data obtained showed that Sepharose 4FF is the most suitable support for rabies virus purification. Using this support, we obtained the highest virus yield with the better separation between virus particles and the other contaminants.

We have also investigated the effect of virus charge and protein concentration on the efficiency of rabies virus purification. Different amount of rabies virus : $15.2 \times 10^6$ FFU, $7.6 \times 10^6$ FFU and $3 \times 10^6$ FFU, were loaded into 25 ml column packed with Sepharose 4FF. Virus recovery yield remains almost constant under the different conditions investigated. We obtained 39, 40 and 35% during the purification of 8, 4 and 1.6 ml of virus suspension respectively.

To investigate the effect of flow rate during purification, we purified 8 ml of rabies virus at a virus titer of $1.9 \times 10^6$ FFU/ml, at different flow rate : 8, 4 and 2 ml/min. Data obtained indicate that no significant difference was noticed under the different operating conditions. Therefore 8 ml/min was selected for further experiments.

We have also studied the effect of different pH (7; 6.1; 8) and ionic strength (0.2M NaCl; 0.5M NaCl). OD 280 nm profiles obtained under the different operating conditions were similar. Phosphate buffer at a pH ranging from 6 to 8 and NaCl concentration varying from 0.2 to 0.5 M can be used for the purification of rabies virus on Sepharose 4 FF without any decrease of the performance.

In conclusion, Sepharose 4 FF appears to be the optimal chromatography support to separate rabies virus from the contaminants present in the virus harvest obtained during Vero cells infection. Capacity study shows that the amount of virus loaded into the column should not be too high, $0.64 \times 10^6$ FFU/ml Sepharose 4 FF, seems to be optimal. Increasing the purification flow rate from 2 to 8 ml/min did not affect the virus recovery yield. Under optimal conditions, the virus yield was equal to 40%. However compared to purification by sucrose zonal centrifugation, the recovery yield is low.

### 2.2 Single step chromatography process

The aim of this work is to study the purification of rabies virus harvests without any preliminary step such as clarification, concentration, etc. For this purpose, we evaluated an anion and a cation exchange streamline chromatography supports : QXL and SPXL.

#### 2.2.1 Method scouting

Test tube studies were carried out using streamline QXL and SP XL as indicated in Materials & Methods. OD 280 nm of the supernatant before and after elution were estimated. Data obtained showed that no interaction occurs when Streamline SP XL support was used. Streamline QXL seems to be more suitable for rabies virus purification. We selected this support to carry out break through studies. Data are shown in Table 1.

#### 2.2.2 Streamline QXL support

Table 1 indicates that the recovery yield is very low. This is due to the competition that exists between virus particles and albumin. Indeed albumin pHi is equal to 4.8 ; under these conditions the protein is negatively charged, therefore it will bind to the support.

*Table 1.* Capacity studies of Streamline QXL support. Experiments were carried out in packed bed using 90 ml of support and a flow rate of 4 ml/min. Phosphate buffer pH 7.6 1 M NaCl was used for elution.

| **Volume of virus suspension (ml)** | 40 | 80 | 120 | 160 |
|---|---|---|---|---|
| **Recovered virus (FFU)** | 2.58 $10^6$ | 9.1 $10^6$ | 1.84 $10^7$ | 2.5 $10^7$ |
| **Loaded virus (FFU)** | 3.08 $10^8$ | 6.16 $10^8$ | 9.24 $10^8$ | 1.23 $10^9$ |
| **Yield ( % )** | 0.84 | 1.47 | 1.99 | 2 |
| **Support capacity (FFU/ml Streamline Q XL)** | 2.86 $10^4$ | 10.1 $10^4$ | 2.4 $10^5$ | 2.68 $10^5$ |

Table 1 indicates that the recovery yield is very low. This is due to the competition that exists between virus particles and albumin. Indeed albumin pHi is equal to 4.8 ; under these conditions the protein is negatively charged, therefore it will bind to the support.

### 2.2.3 Streamline SP XL

Gaudin (2000) and Gaudin *et al.* (1993) have shown that rabies virus incubation at pH 6.1, results in a conformational change of the glycoprotein which becomes positively charged. Such modification is reversible by re incubating the virus particle at neutral pH (7.15). We used this property to study the purification of rabies virus on streamline SPXL support. We first compared OD280 nm level of non incubated and incubated virus suspension at pH 6.1, 4°C for 24 hours. Then we evaluated the capacity of this support.

Data obtained showed that incubation of virus suspension at pH 6.1, 4°C for 24 hours results in an increase of absorbance at 280 nm. We expect an increase of the amount of binded virus to the support.

*Table 2.* Capacity studies of Streamline XLSP support. Experiments were carried out in packed bed using 33 ml of support and a flow rate of 4 ml/min. Phosphate buffer pH 7. 1 M NaCl was used for elution.

| **Virus suspension Volume (ml)** | 40 | 80 |
|---|---|---|
| **Recovered virus (FFU)** | 5.43 $10^7$ | 6.78 $10^7$ |
| **Loaded virus (FFU)** | 3.08 $10^8$ | 6.16 $10^8$ |
| **Recovery yield (%)** | 17.6 | 11 |
| **Support capacity (FFU/ml Streamline XL SP)** | 1.65 $10^6$ | 2.05 $10^6$ |

Table 2 shows that the use Streamline XL SP support combined with rabies virus incubation at pH 6.1, 4°C for 24 hours, increased the recovery yield. Compared to the anion exchange support, the yield was 8.5 fold higher. Nevertheless, the overall yield remains low.

## REFERENCES

Gaudin Y (2000) Rabies virus-induced membrane fusion pathway. J Cell Biol, 150, 601-612

Gaudin Y, Ruigrok RW, Knossow M and Flamand A (1993) Low-pH conformational changes of rabies virus glycoprotein and their role in membrane fusion. J. Virol, 67, 1365-1372

Trabelsi K, Rourou S, Loukil H, Majoul S and Kallel H (2005) Comparison of various culture modes for the production of rabies virus by Vero cells grown on microcarriers in a 2-l bioreactor. Enzyme & Microbial Technology, 36, 514-519

# Human Growth Hormone (HGH) Purification From CHO-Cell Culture Supernatant Utilizing Macroporous Chromatographic Media

Johanna-Gabriela Walter[1], Alexander Tappe[1], Cornelia Kasper[1], Robert Zeidler[2], Oscar-Werner Reif[3], Thomas Scheper[1]

[1]*Institut für Technische Chemie, Callinstr. 3, 30167 Hannover, Germany,* [2]*Vivascience AG, Feodor-Lynen-Str. 21, 30625 Hannover, Germany,* [3]*Sartorius AG, Weender Landstr. 94-108, 37075 Goettingen, Germany*

**Abstract:** The human growth hormone is responsible for the growth process of human beings. In addition it plays an important role in metabolism, protein synthesis and cell proliferation. Since microsomia is one of the main consequences of hGH deficiency and occurs in one of 5000 children, the recombinant protein is the third major drug produced by biotechnology. The purification of biological active compounds such as human growth hormone from complex cell culture samples is a crucial step in proteomic research and down-stream processing of biotechnological products. Traditional chromatographic methods are based on packed columns and often require several steps resulting in time consuming and costly procedures. In contrast, other chromatographic media offer some advantages: Due to their macroporous structure, monoliths and membrane adsorber based chromatographic media allow high flow rates without causing high back pressure. Thus, shear stress to fragile structures is avoided. In the media there are no long diffusion paths, mass transfer takes place through convection rather than diffusion. As a result, monolith and membrane adsorber based protein purification enables a high throughput and time effective performance. In our work we show the expression of hGH in Chinese hamster ovary (CHO) cells and its isolation from cell culture supernatant via monoliths and membrane adsorbers utilizing ion exchange and affinity chromatography.

**Key words:** human growth hormone, protein purification, membrane adsorber, Vivawell 8-strip, membrane chromatography, convective interaction media (CIM) disks, high performance monolithic disk chromatography (HPMDC), ion exchange chromatography, affinity chromatography



*R. Smith (ed.), Cell Technology for Cell Products*, 635–641.

# 1. INTRODUCTION

Human growth hormone (hGH, somatotropin) is produced by the pituitary gland and is responsible for the growth process of human beings. HGH also plays an important role in metabolism, protein synthesis and cell proliferation (Kostyo and Isaksson 1977). It consists of a single polypeptide chain with a molecular weight of 21.5 kDa. Microsomia is one of the main consequences of hGH deficiency, thus the protein is of pharmaceutical interest. Therefore, hGH is one of the top selling drugs produced by biotechnology. The hormone is available through recombinant DNA technology, either expressed in mammalian or bacterial cells. The main challenge in purification of recombinant proteins consists in the elimination of DNA, pyrogenic substances and proteins originating either from the host cell or the culture medium (Ribela *et al.,* 2003). In addition to the stringent purity requirements, the purification process should be fast and cost-effective.

The most common purification methods for hGH are based on ion exchange chromatography (Olson *et al.,* 1981) and affinity chromatography (Gray *et al.,* 1985; Jonsdottir *et al.,* 1986). For the preparation of clinical grade hGH, a set of chromatographic steps is required (De Oliveira *et al.,* 1999, Flodt 1986).

Traditional chromatographic techniques are based on packed columns and often require lengthy procedures, which can lead to the degradation of sensitive proteins. Another disadvantage of packed columns is the high pressure drop across the column and the slow diffusion of solutes within the matrix (Ghosh 2002). In contrast, chromatography based on macroporous media like membrane adsorbers and monolithic disks offers several advantages: There are no long diffusion paths in the media, mass transfer takes place through convection rather than through diffusion. Due to this fact, membrane adsorbers and monolithic disks enable a time-effective performance with high flow rates without high back pressure (Zou *et al.,* 2001).

The membrane adsorber units utilized in this work are designed as 8-strips, to allow the simultaneous screening of several loading and elution conditions on a 96-well plate in parallel. Membranes with the strong cation exchanger group sulfonic acid (S) and the strong anion exchanger group quarternary ammonium (Q) are used for ion exchange chromatography.

The different types of specific membranes for IEX applications were investigated for their ability to bind hGH and the achieved purification effect was examined. Additional an epoxy activated CIM disk (Tennikova, 2000) was used to create an affinity matrix for hGH purification.

The applicability of both devices for the isolation of hGH and for the screening for optimal purification conditions was investigated.

# 2. MATERIALS AND METHODS

## 2.1 Materials

The cell line was donated by CCS (Hamburg, Germany), medium was purchased from Cambrex (Verviers, Belgium), the cross flow filtration unit from Vivascience (Hannover, Germany). The Vivawell 8-strips were provided by Vivascience AG (Hannover, Germany). The epoxy activated CIM disks were provided by BIA Separations (Ljubljana, Slovenia). The molecular mass marker was purchased from MBI-Fermentas (ST Leon-Rot, Germany). HGH concentrations were determined with the hGH-ELISA (Roche, Penzberg, Germany). SDS-PAGE was performed using 15% precast gels purchased from BIO-RAD (Muenchen, Germany). Mouse anti-hGH was purchased from ANOGEN (Mississauga, Canada), hGH standard was from Santa Cruz Biotechnology (Santa Cruz, California, USA). All buffer salts and other reagents were purchased from Fluka (Buchs, Switzerland) and Sigma (Taufkirchen, Germany) and were of p.a. quality. Ultrapure water was produced using the arium 611 VF water purification system purchased by Sartorius (Goettingen, Germany).

## 2.2 Cell culture and preparation of crude extract

The CHO cells were cultivated under serum free conditions in ProCHO4-CDM at 37°C, 5% $CO_2$ in spinner flasks. The medium was supplemented with 4 mM L-glutamine.

Cells were separated from the supernatant by centrifugation at 400 g. The supernatant was concentrated via cross flow filtration (Vivaflow 50, 10 kDa MWCO).

## 2.3 HGH purification utilizing membrane adsorber based 8-strips

The isolation of hGH was performed utilizing membrane adsorber based devices with cation (S) and anion (Q) exchanger functionalities. All steps like loading, washing and eluting of the bound proteins were performed in a centrifuge by spinning the 96-well plate with the attached 8-strips at 1000 g, per centrifuge run a volume of 300 µl was processed. In 8 individual experiments the screening for the optimal binding and elution conditions was performed on one single 96-well plate, using the cation and the anion exchanger membranes at 4 different pH values (25 mM sodium acetate, pH 4; 25 mM sodium phosphate, pH 6; 25 mM Tris/HCl, pH 7.5; 25 mM sodium-bicarbonate, pH 9). To avoid high salt concentrations and to adjust the sample to the appropriate pH value, the cell culture supernatant was diluted 1:5 in the respective binding buffer. After washing with binding buffer, two elution steps with increasing salt concentration (300 and 600 mM NaCl in each binding buffer) were performed. For the identification and quantification of hGH an hGH-ELISA was carried out.

### 2.4 HGH purification utilizing monolithic disks

The purification was performed utilizing epoxy activated CIM disks. Prior to use, the disk was washed in a methanol-water mixture (50:50 vol. %) and afterwards immersed into 50 mM $Na_2CO_3$, pH 9.3 for 2 h. The disk was then incubated in 1 ml of an anti-hGH solution (5 mg/ml in 50 mM $Na_2CO_3$, pH 9.3) for 16 h at 30°C. Unbound antibody was washed away with the same buffer. The residual epoxy groups were quenched with 1 M ethanolamine (1 h). Finally, the disk was washed with 10 mM PBS, 150 mM NaCl, pH 7.

The purification of hGH was performed via HPLC (flow rate: 2 ml/min). The concentrated hGH-containing cell culture supernatant was diluted 5 times in 10 mM PBS, 150 mM NaCl, pH 7, 1 ml was injected. The mobile phase for adsorption and washing steps was 10 mM PBS, 150 mM NaCl, pH 7, for desorption of unspecifically bound proteins 2 M NaCl was used. Elution of hGH was performed with HCl, pH 2.0. The eluates were neutralized with 50 mM $Na_2CO_3$, pH 9.3. For the quantification of hGH in the different fractions an hGH-ELISA was carried out.

In order to demonstrate the applicability of the disk for screening purposes, different mobile phases for adsorption and washing steps were tested (100 mM PBS, 150 mM NaCl; 200 mM PBS, 150 mM NaCl; 10 mM PBS, 150 mM NaCl).

The long time stability of the affinity disk was investigated by applying an hGH dilution series onto the disk in the course of 12 weeks. Between the measurements, the disk was stored at 4°C.

## 3. RESULTS AND DISCUSSION

### 3.1 Purification of hGH utilizing membrane adsorber based 8-strips

As a result of the screening procedure, the best purification effect was achieved using a cation exchanger (S) membrane at pH 4.5 (Figure 1, lane 1).

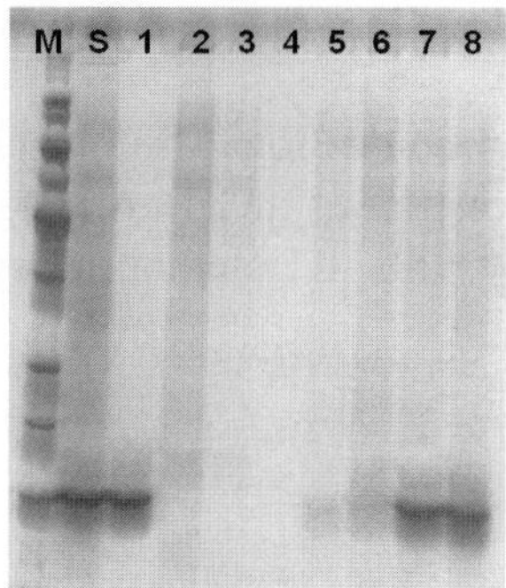


*Figure 1.* 15% SDS-Gel, silver stained. Screening for optimal purification conditions via Vivawell S 8-strip (cation exchanger) and Vivawell Q 8-strip (anion exchanger). The following samples were applied to the gel under reducing conditions: (M) molecular weight marker, (S) diluted cell culture supernatant, (1) – (8) elution fractions using 300 mM NaCl in corresponding buffer, (1) S-membrane, pH 4.5, (2) S-membrane, pH 6, (3) S-membrane, pH 7.5, (4) S-membrane, pH 9, (5) Q-membrane, pH 4.5, (6) Q-membrane, pH 6, (7) Q-membrane, pH 7.5, (8) Q-membrane, pH 9.

Since the isoelectric point of hGH is close to 5.0, it exhibits a positively net charge at pH 4.5, thus binding to the cation exchanger. According to Figure 2, hGH has bound to the membrane under these conditions, the flow through (lane 1) and in the wash fractions (lane 2-4) no hGH was found. In the first elution fraction, which was achieved with 25 mM sodium acetate buffer, pH 4.5 containing 300 mM NaCl (lane 5), hGH was obtained in considerable purity. The protein concentration in this fraction was 35.8 μg $ml^{-1}$, resulting in a total amount of 7.74 μg protein. The total protein recovery was 93%.

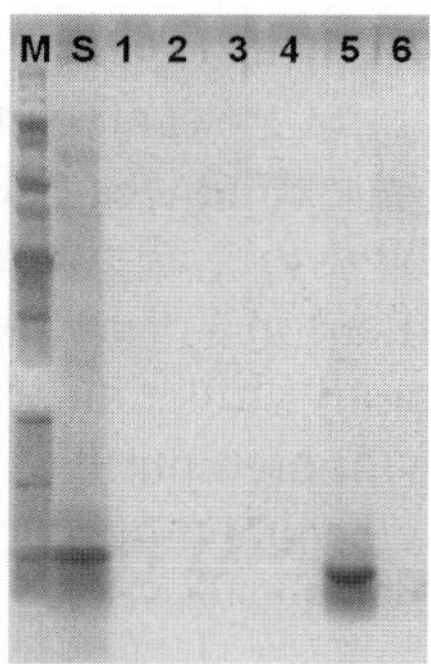


*Figure 2.* 15% SDS-gel, silver stained. Purification of hGH utilizing Vivawell S 8-strip at pH 4.5. The following samples were applied: (M) molecular weight marker, (S) diluted cell culture supernatant, (1) flow through, (2) – (4) wash fractions, (5) elution (300 mM NaCl), (6) elution (600 mM NaCl).

### 3.2 Purification of hGH utilizing monolithic disks

As a result of the immobilization of anti hGH, 0.9 mg antibody was bound on the disk. The isolation of hGH utilizing the custom made affinity disk results in a high purity with all tested buffer systems (Figure 3).

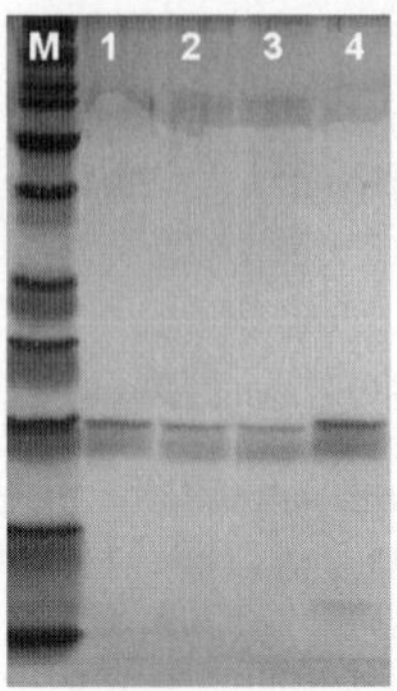


*Figure 3.* 15% SDS-gel, silver stained. Examination of different loading buffers via CIM disk. The following elution fractions were applied: (1) 100 mM PBS, 150 mM NaCl, (2) 200 mM PBS, 150 mM NaCl, (3) 10 mM PBS, 150 mM NaCl, (4) hGH-standard.

As a result of the investigation of the long time stability of the affinity disk, the signal intensity after the elution of hGH shows no significant loss in the first 4 weeks. Afterwards it decreases, after 12 weeks only 38% of the initial bound hGH can be recovered in the eluate.

## 4. CONCLUSION

The purification of hGH from CHO-cell culture supernatant via Vivawell 8-strips with cation exchanger membranes results in a considerable purity within one step. More over, the device allows a highly parallel screening for the optimal purification conditions in a decent amount of time.

The affinity chromatography utilizing CIM disks succeeded in a purification within one step with high purity. Due to the fast procedure, the device is also applicable for screening procedures. Another advantage of this approach is the possibility to reutilize the disks over a long period of time. In summary, both devices have shown to be suitable tools for the purification of hGH.

## ACKNOWLEDGEMENTS

We thank CCS Cell Culture Services GmbH (Hamburg, Germany) for kindly donating the CHO cell line and Vivascience AG (Hannover, Germany) for the support and for providing the membrane adsorber based devices.

## REFERENCES

De Oliveira, J.E., Soares, C.R.J., Peroni, C.N., Gimbo, E., Carmago, I.M.C., Morganti, L., Bellini, M.H., Affonso, R., Arkaten, R.R., Bartolini, P., Ribela, M.T.P.C., 1999, High-yield purification of biosynthetic human growth hormone secreted in *Escherichia coli* periplasmic space. J. Chromatogr. A **852**: 441-450.

Flodt, 1986, Human growth hormone produced with recombinant-DNA technology – development and production. Acta. Paedriatr. Scand. Suppl. **325**: U1-U1.

Gosh, R., 2002, Protein separation using membrane chromatography: opportunities and challenges. J. Chromatogr. **A 952**: 13-27.

Gray, G.L., Baldrige, J.S., Mc Keown, K.S., Heyneker, H.L., Chang, C.N., 1985, Peryplasmic production of correctly processed human growth hormone in Escherichia coli: natural and bacterial signal sequences are interchangeable. Gene **39**: 247-254.

Jonsdottir, I., Skoog, B., Ekre, H.P.T., Pavlu, B., Perlmann, P., 1986, Purification of pituitary and biosynthetic human growth hormone using monoclonal antibody immunoadsorbent. Mol. Cell. Endocrinol. **46(2)**: 131-135.

Kostyo, J.L., Isaksson, O., 1977, Growth hormone and regulation of somatic growth. Int. Rev. Physiol. **13**: 288-274.

Olson, K.C., Fenno, J., Lin, N., Harkins, R.N., Snider, C., Kohr, W.H., Ross, M.J., Fodge, D., Prender, G., Stebbing, N., 1981, Purified human growth hormone from Escherichia-coli is biologically-active. Nature **293**: 408-411.

Ribela, M.T.C.P., Gout, P.W., Bartolini, P., 2003, Synthesis and chromatographic purification of recombinant human pituitary hormones. J. Chromatogr. B **790**: 285-316.

Tennikova, T.B., 2000, An introduction to monolithic disks as stationary phases for high performance biochromatography, J. High Resol. Chromatogr. **23(1)**: 27-38.

Zou, H., Luo, Q., Zhou, D., 2001, Affinity membrane chromatography for the analysis and purification of proteins. J. Biochem. Biophys. Methods **49**: 199-240.

# Process Development for a Veterinary Vaccine Against Heartwater Using Stirred Tanks

Isabel Marcelino[1], Marcos F.Q. Sousa[1], Ana Amaral[1], Cristina Peixoto[1], Célia Verissimo[1], Antonio Cunha[1], Manuel J.T.Carrondo[1,2], Paula M.Alves[1]

*[1]ITQB/IBET, Apt.12, 2781 Oeiras, Portuga,l [2]FCT/UNL, 2825 Monte da Caparica, Portugal*

**Abstract:** The work presented herein reports the improvements achieved on endothelial cell culture conditions and the production of Ehrlichia ruminantium (ER) under stirring conditions towards the development of a cost effective process for the widespread application of the inactivated vaccine against Heartwater. The effect of cell origin, inoculum size and microcarrier type upon maximum cell concentration was evaluated. Afterwards, using the optimised parameters for cell growth, ER production in stirred tanks was validated for two bacterial strains (Gardel and Welgevonden). Critical bioprocess parameters related with the infection strategy such as serum concentration at time of infection, multiplicity of infection, and medium refeeding strategy were analysed. The results obtained indicate that it is possible to produce ER in stirred tank bioreactors and that the production yields can be increased by a factor of 6.5 using a serum-free medium during and after the infection process. To expand the scale-up process and improve ER production yields, ER growth kinetics was characterised showing that this stirring culture system is capable of efficient bacterial amplification with maximum titers going up to 2.2 log10ER. The suitability of this process was validated up to a 2L scale. A downstream processing for the purification of ER is also described, taking into account ER recovery yields and the number of steps. Overall, these results open "new avenues" for the production of other ehrlichial species, with emerging impact in human and animal health, in a fully controlled and scaleable environment.

**Key words:** *Ehrlichia ruminantium*, heartwater vaccine, serum-free medium, stirred tank culture strategies, growth kinetics, purification, cost analysis.



*R. Smith (ed.), Cell Technology for Cell Products*, 643–648.

# 1. INTRODUCTION

Heartwater is a tick-borne disease of domestic and wild ruminants caused by the obligate intracellular bacterium *Ehrlichia ruminantium* (ER), ranking as one of the most economically important diseases in sub-Saharan Africa and in the West Indies [1,2]. Currently, a vaccine based on the chemically inactivated ER elementary bodies produced in finite culture of ruminant endothelial cells is the best candidate for protection against Heartwater [1,2]. To overcome the socio-economical impact of Heartwater, there is a need to establish an industrial cost-effective production process that will permit the widespread application of the vaccine.

# 2. MATERIALS AND METHODS

Cell Culture: Caprine Jugular (CJE) and Bovine Aortic Endothelial cells (BAE), supplied by Dr. D. Martinez (CIRAD/EMVT, France), were cultivated in static conditions as described previously [2]. Experiments in stirred tanks were performed with 125 ml spinner vessels (Wheaton, USA) using two non-porous (Cytodex 3 from Amersham Pharmacia, Sweden, and 2D MicroHex from Nunc, Denmark) and a porous microcarrier (Cultispher-S from Percell Biolytica AB, Sweden).

*Ehrlichia ruminantium* culture: ER Gardel (ERG) was supplied by Dr. D. Martinez and ER Welgevonden (ERW) by Dr. E. Zweygarth (OVI, South Africa). ER was cultivated in static conditions as described elsewhere [2,3]. For stirred tank ER production, two culture media were tested: a Glasgow based medium supplemented with 10% FBS and a special serum-free formulation based on the commercial DMEM/HAM'S F12 [3]. Optimised parameters for ER culture were up-scaled to 2 L bioreactors (BBraun International, Germany), under fully controlled conditions (37°C, pH 7.2 and 30% of dissolved oxygen) [4].

*Ehrlichia ruminantium* purification: ER were purified using a multistep centrifugation strategy, as described elsewhere [2]. Experiments with filters were performed with depth filters (MaxiCap from Sartorius, Germany), cassettes (Sartocon from Sartorius) and hollow fibers (Amersham Pharmacia). To determine the recovery yields of intact ER, bulk ER suspensions were treated with DNAse I to reduce the background of DNA from lysed ER [6].

# 3. RESULTS AND DISCUSSION

## 3.1 Mass Production of Ruminant Endothelial Cells

Microcarrier cell culture has been previously used by our group to produce large quantities of ruminant endothelial cells [7]. The results presented in Table 1 clearly show that maximum cell density ($1.4\times10^6$ cells/ml) was only achieved for BAE cells, using an inoculum of $0.25\times10^6$ cells/ml seeded on non-porous Cytodex 3 (at 6 g/1), without any addition of expensive medium supplements such as fibroblast or epidermal growth factors, normally used for the culture of most endothelial cells.

*Table 1.* Optimised bioreaction parameters for CJE and BAE cells under stirring conditions.

| | CJE cells | BAE cells | | | |
|---|---|---|---|---|---|
| Microcarrier designation | Cytodex 3™ | Cytodex 3™ | | CultiSpher S™ | 2D MicroHex ™ |
| Microcarrier weight | 6 g/l | 6 g/l | | 3 g/l | 20 g/l |
| Inoculum size (x10$^6$ cells/ml) | 0.10 | 0.10 | 0.25 | 0.25 | 0.25 |
| Maximum Cell Concentration (x10$^6$ cells/ml) | 0.68 | 0.96 | 1.38 | 1.30 | 0.80 |
| Days for max. cell .conc. | 8* | 5 | 4 | 5 | 5 |

(* Two complete media refeed were required to stimulate cell growth

## 3.2 *Ehrlichia ruminantium* Life Cycle in Stirred Tank Culture Conditions

In static culture conditions, ER has a biphasic life cycle which consists of an extracellular infectious form designated by elementary body (EB) and a non-infectious, metabolically active, intracellular form designated by reticular body (RB) [8]. To ensure that ER was able to colonise and proliferate inside the host cells, its growth was monitored using both phase contrast microscopy and real time PCR (Fig. 1). After a 36h period for ER attachment/internalisation (Fig. 1E), the host cells exhibited a substantial swelling when compared to non-infected cells (Fig. 1.A); this is an indicator of successful infection and ER development inside host cells. EBs maturation into RBs, forming inclusion bodies called morula, can be visualized in Fig. 1 C and the exponential growth is observed until 120 hpi, with a net increase of up to 2.2 orders of magnitude (Fig. 1E). After successive rounds of binary fission, non-infectious RBs reorganize to EBs that are released from the host cell after complete cell lysis (Fig. 1 D). This is the first evidence of successful ER growth under stirring conditions, supporting the development of a large-scale process for ER production.

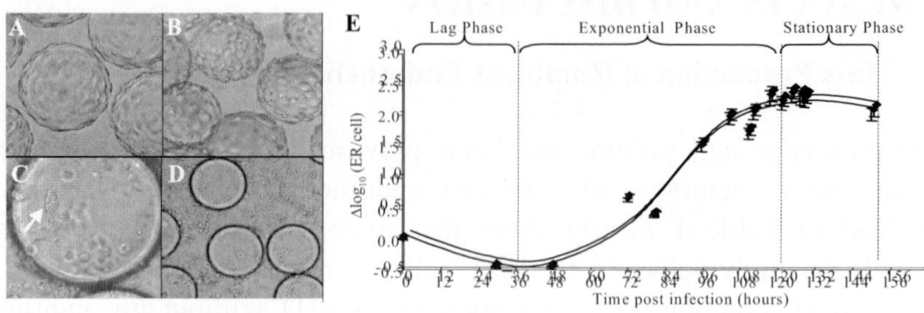


*Figure 1. E. ruminantium* Gardel life cycle in microcarrier cells culture. A - non infected BAE cells; B – infected cells; C - reticulate bodies (morula) ; D – elementary bodies release; E - ER growth kinetics.

Considering the quality constraint of the final product (ER vaccine), complementary studies showed that the ER suspension should be harvested at 113 hpi, when maximum amount of intact ER can be obtained.

### 3.3 Mass Production of *Ehrlichia ruminantium*

Two chemically distinct culture media were tested in their ability to support two ER isolates (ERW and ERG): a serum-containing medium based on Glasgow MEM and a serum-free medium [4]. Results showed that not only higher ER production yields were obtained when serum-free medium was used during and after infection (Fig. 2A), but it was also possible to achieve a 6-fold reduction in ER inoculum size (from 400 to 66 ERs cell$^{-1}$) (Fig. 2B). The elimination of serum oligosaccharides or glycoproteins, that normally inhibit host cell-bacterium interaction, could have contributed for an increase in the susceptibility of endothelial cells to ER infection therefore augmenting the final amount of produced bacteria. Considering the industrial demands for a biotechnological process, the use of this medium will reduce the production costs and minimise the interference of serum proteins on vaccine dosage (still based on total protein content).

Using these optimised parameters, a three-step medium reefed strategy was substituted for only one medium exchange at infection time (data not shown), reducing the risks of contamination and also the costs of the process.

The optimised culture conditions described above were validated up to a 2L scale and approx. 3800 vaccine doses per litre were produced. The first doses produced at this mid-scale were used to vaccinate cattle and sheep in Kenya and Burkina Faso and the preliminary results obtained were quite satisfactory, demonstrating the effectiveness of the vaccine in inducing protection against homologous challenge (data not shown).

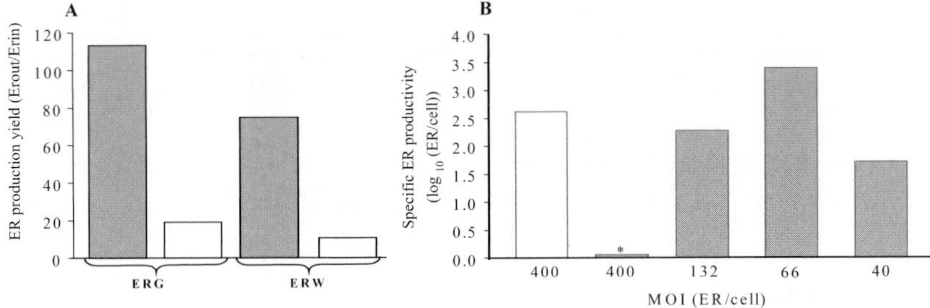


*Figure 2.* Effect of serum-free medium on: (A) ER Gardel and Welgevonden production yields and (B) on the multiplicity of infection (MOI).

An overall estimation of cost *per* vaccine dose (Table 2) was performed assuming the lowest dose that proved effective in field trials.

The values presented herein include not only the costs involved with the consumables but also those related to labour and equipment required for the production of the host cells and the bacteria. Since the use of roller bottle is a traditional cell-culture approach for vaccine production, this device was also included in this cost analysis; the productivity was assumed to be identical to the one obtained for T-flasks.

*Table 2.* Cost estimation of one heartwater vaccine dose produced in different types of culture.

| Culture method | Price/vaccine dose (€) |
|---|---|
| T-flasks | 0.12 |
| Spinner | 0.18 |
| Bioreactor | 0.11 |
| Roller bottle | 0.14 |

## 3.4 ER purification

Until recently, a multi-step centrifugation was the best purification strategy [2]. Since this method is time consuming and can have detrimental effects on ER, a downstream process based on depth and tangential filtration was envisaged taking into account ER recovery yields and the number of steps. The filtration devices were chosen according to the differences in particles size. Preliminary results have shown that the use of filter leads to an efficient removal of microcarriers, as it also reduces the time of operation and the endothelial protein contamination, permitting to achieve higher yields of pure bacteria compared to the previous centrifugation strategy (Table 3).

*Table 3.* Large scale purification of E.ruminantium: comparison of different strategies.

| Purification Process | Time (min) | ER recovery (%) | Total Protein (mg) |
|---|---|---|---|
| Multi-step centrifugation | 115 | 11.2 | 2.5 |
| Depth filter +Cassette + high speed centrifugation | 39 | 14.2 | 1.5 |
| 2 Depth filters + Cassette + high speed centrifugation | 37.3 | 10.1 | 0.3 |
| Depth filters + Hollow fibers | 6.5 | Approx. 0% | - |

(*cell death due to cytopathic effect of ER inoculum).

## REFERENCES


[1] Totté P *et al.*, 1999.Parasitol Today 15: 286-90.

[2] Martinez D. *et al.*, 1994. Vet Immunol Immunop. 41, 153-163.

[3] Zweygarth E. & A.I. Josemans. 2001.Onderstepoort J Vet Res 68: 37-40.

[4] Marcelino I. *et al.*, 2005. Vaccine (in press).

[5] Peixoto C.C. *et al.*, 2005. Vet Microbiol 107: 273-278.

[6] Marcelino I. *et al.*, 2005. Vet Microbiol (in press).

[7] Moreira J.L. *et al.*, 2003. ACS Symposium Series, pp124-141.

[8] Jongejan F. *et al.*, 1991. Onderstepoort J Vet Res 58:227-237.

# Working Towards a Chemically-defined Replacement for Hydrolysates

Zachary W. Deeds, C. Steven Updike, Benjamin J. Cutak, Matthew V. Caple
*SAFC-JRH Biosciences, 3050 Spruce Street, Saint Louis, MO 63103 USA*

**Abstract:** Protein hydrolysates are commonly used in cell culture processes either as a component of a complete medium formulation or as part of a feeding supplement for a fed-batch bioreactor process. Due to the undefined nature of hydrolysates, there is a push to develop a chemically-defined substitute. It is well known that hydrolysates serve a nutritive function as a source of free amino acids, small peptides, carbohydrates, vitamins and many other potential contaminants. They also have been reported to contain bioactive peptides[1] that could potentially activate specific pathways within the cells. The data presented here are the initial results of studies performed to elucidate the essential components of commercially available wheat gluten and meat hydrolysates. Reverse-phase HPLC was used to initially separate the complex mixture into fractions (66 each) containing either classes of components or individual components so they could be subsequently tested in a cell culture system utilizing Chinese Hamster Ovary cells.

**Key words:** Hydrolysates, peptides, CHO, chemically-defined, fed-batch

## 1. INTRODUCTION

As an integral part of many cell culture media formulations and bioreactor feeding supplements protein hydrolysates have multiple advantageous functions. First and foremost, hydrolysates serve as a source of nutrition in the form of free amino acids and small peptides. Contaminants such as carbohydrates and vitamins are also often present and undoubtedly have a positive impact. However, there are other benefits that have been reported that cannot be attributed solely to nutritional enhancements.



*R. Smith (ed.), Cell Technology for Cell Products*, 649–651.

Schlaeger *et al.*, report that a meat hydrolysate, Primatone RL, has an anti-apoptotic effect[2], while Franek *et al.*, suggest that peptides within the hydrolysates can mimic growth and survival factors[1].

The data presented here represent the initial steps that have been taken to identify the active components in two commercially available hydrolysates.

## 2. RESULTS

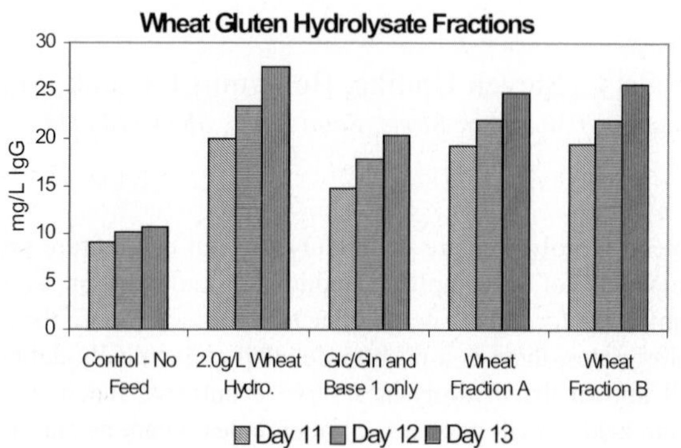


*Figure 1.* Wheat Gluten Hydrolysate Fraction Productivity.

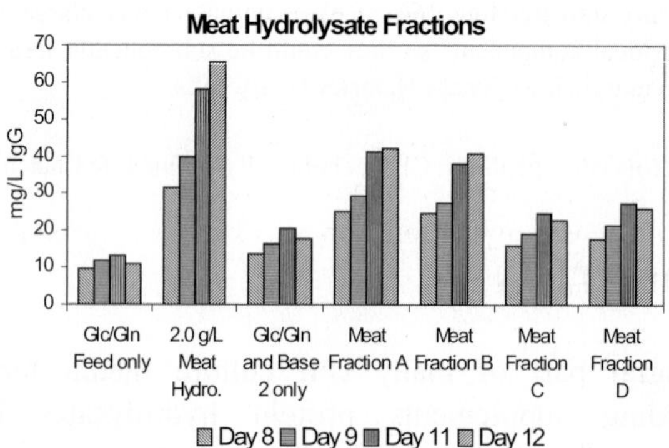


*Figure 2.* Meat Hydrolysate Fraction Productivity.

## 3. DISCUSSION

In order to define the critical components in protein hydrolysates, the many functions that they serve must be addressed. The first of which is a nutrient source and was accounted for by creating a "Hydrolysate Base".

The Hydrolysate Base was originally developed to mimic the levels of amino acids, vitamins and metals present in a commercially available hydrolysate. Further optimization with matrix-style experiments utilizing Design-Expert® software led to the development of Hydrolysate Base 2.

The next step is to identify the other active components within the hydrolysates. Reverse-phase HPLC fractionation was used to aid with the separation of these compounds.

The complete fractionation of a commercially available wheat gluten hydrolysate was subsequently analyzed in experiments designed to demonstrate a correlation between cell culture productivity and the presence of bioactive peptides. These experiments showed that different fractions enhanced antibody production with a recombinant Chinese Hamster Ovary (CHO) cell line (as seen in Figure 1). While fractions A and B did not improve cell growth like the hydrolysate control (data not shown), they did have a positive impact on productivity.

A commercially available meat hydrolysate was also fractionated and analyzed in experiments as performed above. Four fractions were found to enhance cell proliferation and/or antibody production. Only two of the fractions, A and B, yielded enhanced cell growth when compared to the control (data not shown). Additionally, fractions A and B had a significant positive impact on production (see Figure 2). Fractions C and D had a minor impact on productivity.

With the isolation of six hydrolysate fractions displaying positive biological effects further analytical experiments can now be performed to identify the active component(s) contained in the fractions.

*Design-Expert® is a registered trademark of Stat-Ease, Inc.

## REFERENCES

Franek F. *et al.*, 2000, Plant protein hydrolysates: preparation of defined peptide fractions promoting growth and production in animal cells cultures, Biotechnol. Prog. **16**:688-692.

Schlaeger, E.J. *et al.*, 1996, The protein hydrolysate, Primatone RL, is a cost-effective multiple growth promoter of mammalian cell culture in serum-containing and serum-free media and displays anti-apoptotic properties, J. Immunol. Methods **194**:191-199.

# Effect of Vitamins or Lipids Addition on Adenovirus Production at High Cell Densities

Tiago B. Ferreira[1], Manuel J. T. Carrondo[1,2] Paula M. Alves[1]
*[1]IBET/ITQB, Ap. 12, 2781-901 Oeiras, Portugal [2]FCT/UNL 2825-114 Monte da Caparica, Portugal*

**Abstract:** Methodologies for production of concentrated adenovirus vectors (AdV) with warranted purity and efficacy at low cost are needed as the market for AdV is expanding fast. However, production of AdV that maintain a high specific yield is limited to cell densities in the range of $1\times10^6$ cell/ml; the nature of the factors for this remains unknown. The aim of this work is to study the effect of addition of vitamins or lipids at the time of infection at high cell densities in order to see if these nutrients are limiting. No improvement on AdV productivity was observed. Furthermore, a significant decrease on cell specific productivity after vitamin addition was observed.

**Key words:** Adenovirus, high cell densities, vitamins, lipids.

## 1. INTRODUCTION

AdV became one of the vectors of choice for delivery and expression of foreign proteins for gene therapy and vaccination purposes (Ferreira *et al.*, 2005a). Nevertheless, the production of AdV is currently limited by the so-called "cell density effect" (Nadeau and Kamen, 2003). Several approaches have been made in order to overcome this drop in cell specific productivity concomitant with increased cell concentration at infection (CCI): Garnier *et al.* (1994), demonstrated that medium replacement at infection and the addition of glucose at 24 hours post infection (hpi) together with periodical pH adjustments, allowed a sustained maximum specific productivity at $1.6\times10^6$ cell/ml; whereas Nadeau *et al.* (1996), further improved this strategy by also adding essential amino acids at 24 hpi thereby stabilising volumetric productivity at cell densities above $2\times10^6$ and below $3\times10^6$ cell/ml. Such results hint at the existence of substrate limitation and/or byproduct



*R. Smith (ed.), Cell Technology for Cell Products*, 653–656.

inhibition at high cell densities. In our latest study, by adapting the cells to non-ammoniagenic medium, a 1.8 fold increase on AdV volumetric productivity at CCI $3\times10^6$ cells/ml was obtained (Ferreira *et al.*, 2005b), although the "ideal" situation would have been an increase of 3 times at this CCI; thus, ammonia is an important parameter to be considered for infection at high cell densities, but not the only one.

Future improvements to the culture process may come from analysis of other nutrients than the usual glucose, lactate, ammonia and amino acids. A cocktail of different nutrients are present in the culture media that may have an important impact on AdV production as vitamins, lipids, hormones and growth factors. Thus, in this work, addition of vitamins and lipids at the time of infection at high cell densities was evaluated in order to see if these nutrients are limiting the AdV production at high cell densities.

## 2. MATERIALS AND METHODS

Suspension adapted 293 cells (ATCC-CRL-1573) were grown and infected in non-ammoniagenic medium in a 2L bioreactor (Braun) as described in Ferreira *et al.* (2005b). For the fed-batch operation mode, Vitamins and Lipids cocktail solutions (all from Gibco) were used; AdV titration was performed by the end-point dilution method ($TCID_{50}$) using 96 well plates and 293 cells. The titer was calculated according to the method of Spearman and Kraber, as described in Darling *et al.* (1998).

## 3. RESULTS AND DISCUSSION

Vitamins and lipids are important nutrients for cell growth. Moreover, lack of such nutrients may affect cell productivities. Thus, two different fed-batch operation modes were tested with addition of: (1) vitamins or (2) lipids cocktail at the time of infection at CCI $3\times10^6$ cell/ml. As can be observed in Figure 1, neither the addition of vitamins nor of lipids, at the time of infection, improved the cell specific productivity at this CCI. On the other hand, a significant cell specific productivity decrease was observed after the addition of vitamins. Following the cell metabolism after infection it was possible to observe that none of the measured parameters was shown to be responsible for this significant decrease (Table 1), suggesting that vitamins may have an effect on AdV stability.

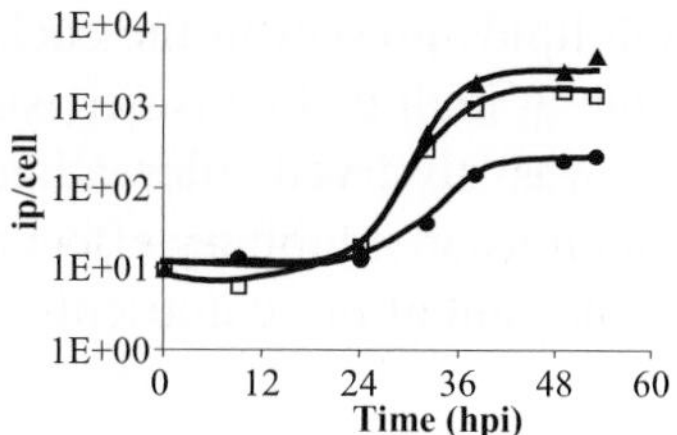


*Figure 1.* Cell specific productivity after addition of vitamins (•) or lipids (□) at the time of infection, at CCI 3×106 cell/ml. The control (▲) was performed without any nutrient addition.

*Table 1.* Measured parameters during infection at CCI 3×106 cell/ml concomitant with addition of vitamins.

| ***Parameters*** | ***Result*** |
|---|---|
| Glucose | Not limiting |
| Lactate | Below inhibitory values |
| $Y_{Lac}/_{Glc}$ | Unchanged |
| Amino acids | Not limiting |
| Ammonia | Below inhibitory values |
| Osmolality | Below inhibitory values |
| Cell growth | No effect |

In order to evaluate the effect of vitamins on AdV stability, AdV were incubated with the vitamins cocktail at different concentrations during 48h at 37ºC and pH 7.2 (Figure 2). Although a small decrease on AdV titer was observed this was not statistically different from the ones in the control, meaning that vitamins have no influence upon AdV stability.

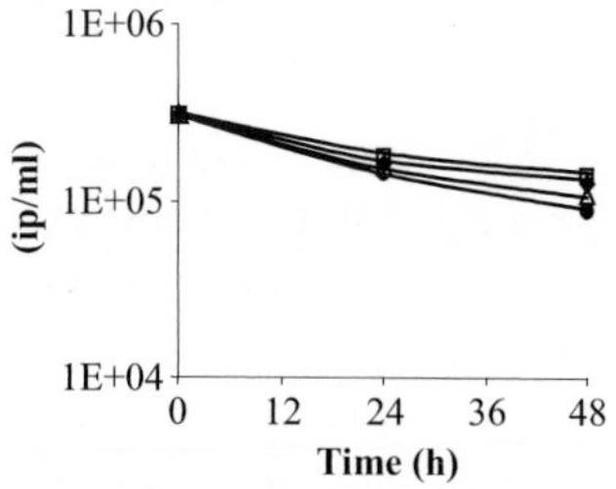


*Figure 2.* AdV stability during incubation with medium supplemented whit 0× (•) (control), 1× (□), 1.5× (Δ) and 2× (♦) of vitamins.

Several studies reported some vitamins as virus inactivators, e.g. riboflavin, present in our vitamins cocktail, was shown to selectively induce damage only to nucleic acids (Dardare and Platz, 2002) and probably this may explain the significant decrease on cell specific productivity, since AdV DNA damage might be occurring during AdV replication.

In conclusion, from the lipids present in the cocktail no one seems to be limiting on AdV production at high cell densities, since no improvement on cell specific productivity was observed after addition of these nutrients. Concerning addition of vitamins an inhibitory effect appears to be present on AdV production after the addition of these nutrients.

## REFERENCES

Dardare N., Platz M.S., 2002, Binding affinities of commonly employed sensitizers of viral inactivation, Photochem Photobiol **75**:561-4.

Darling A.J., Boose J.A., Spaltro J., 1998, Virus assay methods: accuracy and validation, Biologicals **26**:105-10.

Ferreira T.B., Alves P.M., Aunins J.G., Carrondo M.J.T., 2005[a], Use of adenoviral vectors as veterinary vaccines, Gene Ther (*in press*).

Ferreira T.B., Ferreira A.L., Carrondo M.J.T., Alves P.M., 2005b, Effect of refeed strategies and non-ammoniagenic medium on adenovirus production at high cell densities, J Biotechnol (*in press*).

Garnier A., Cote J., Nadeau I., Kamen A., Massie B., 1994, Scale-up of the adenovirus expression system for the production of recombinant protein in human 293S cells, Cytotechnology **15**:145-155.

Nadeau I., Garnier A., Côté J., Massie B., Chavarie C., Kamen A., 1996, Improvement of recombinant protein production with human adenovirus/293S expression system using fed-batch Strategies, Biotechnol Bioeng **51**:613-623.

Nadeau I., Kamen A., 2003, Production of adenovirus vector for gene therapy, Biotechnol Adv **20**:475-489.

# Observation and Analysis of Lab Scaled Microcarrier Cultivation by In-situ Microscopy with Image Processing Tools

G. Rudolph, A. Gierse, P. Lindner, C. Kasper, B. Hitzmann, T. Scheper
*Institut für Technische Chemie, Callinstraße 3, 30167 Hannover, Germany*

**Abstract:** The online monitoring of biomass for concentration determination or morphological studies is of main interest in modern bioprocess analysis. Different methods are available to monitor biomass concentration but no system is able to analyze concentration and morphology simultaneously online. In-situ microscopy offers the possibility for both. This work presents the possibility to analyze lab scaled microcarrier cultivation with a new in-situ microscope generation. Therefore, the model cell line NIH-3T3 was cultivated on Cytodex 1 microcarriers.

**Key words:** online monitoring, biomass concentration, morphology, bioprocess analysis, in-situ microscopy, microcarrier cultivation

## 1. INTRODUCTION

The online monitoring of biomass for concentration determination or morphological studies is of main interest in modern bioprocess analysis. Different methods are available to monitor biomass concentration (Ulber *et al.*, 2003) but no system is able to analyse concentration and morphology simultaneously online. In-situ microscopy based on image analysis offers the possibility for both. Different in-situ microscopy applications for yeast-, BHK- and CHO-cell cultivations have already been described (Bittner *et al.*, 1998; Joeris *et al.*, 2002; Camissard *et al.*, 2002; Brueckerhoff *et al.*, 2003; Guez *et al.* 2004). This work focuses on the observation and analysis of lab scaled microcarrier cell cultivation.

Microcarriers are widely used in cell culture and are well suited to achieve high-density cell cultures. Adherent cells grow on the particle's surface, so that the cells can be cultivated in a volume instead of a two-



*R. Smith (ed.), Cell Technology for Cell Products*, 657–661.

dimensional surface as monolayers. The accessible cell densities in culture volume are up to 20 times higher compared to a monolayer bottle.

Since cell concentration and plating efficiency cannot be measured directly in microcarrier culture in-situ microscopy was utilized to monitor the level of cell colonization on the microcarriers in real-time. Cytodex 1 microcarriers were selected throughout the cultivations because its transparent properties make them ideal for in-situ microscopy. Cytodex 1 Microcarriers attached with adherent cells can easily be distinguished from non-overgrown microcarriers.

## 2. MATERIAL AND METHODS

### 2.1 In-situ microscope hardware

The hardware consists of the in-situ microscope (ISM) fitted out with a CCD-camera, a microcontroller and a PC containing software to operate the ISM and image processing tools for the analysis of the grabbed images. The ISM is based on transmitted light microscopy (bright field) and designed to fit with its front-end sampling zone into a bioreactor's standard 25 mm side port (Figure 1 below). The sampling zone is immerged directly into the culture broth to acquire image sequences of the cells and microcarriers passing it. The sampling zone is defined by a fixed and an adjustable sapphire window working as slide and cover slip. Sterilization of the microscope is possible by either autoclaving or steam-in-place (SIP) cleaning. For long-term fermentation processes the ISM system is arranged with a retract mechanism similar to the industry standard InTrac that allows manual cleaning of the sampling zone without influence on the sterility of the process.

The system is equipped with finite corrected optics and can be used with two different objectives depending from the object's size: a 4-fold and a 10-fold magnifying achromatic objective (NDA4X; NDA10X, Olympus) with a numerical aperture of 0,10/0,25. The CCD-camera (Sony XCD SX-910 1/2" monochrome; digital interface IEEE 1394) has a maximum resolution of 1280 x 960 pixels. Using the 10-fold objective magnification (overall magnification 400-fold) each pixel represents an image area of 0,137 $\mu m^2$.

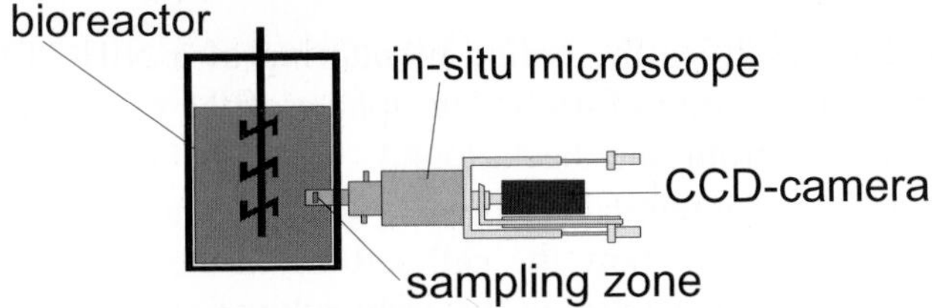


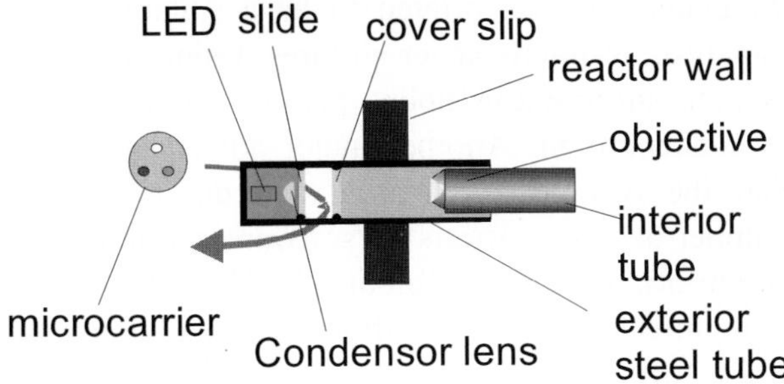


*Figure 1.* ISM side port implementation into a bioreactor and sampling zone principle.

## 2.2 In-situ microscope software

The in-situ microscope's software is divided into two parts. *In-situ Control* controls the microscope's basic functions: illumination intensity, sampling zone volume flow, object focussing and the CCD-camera's exposure time. The image series acquired by *In-situ Control* can simultaneously be analyzed by *In-situ Analysis*. The current version of *In-situ Analysis* contains algorithms for the identification and cell counting of mammalian cells while it can detect and eliminate air bubbles passing the sampling zone.

## 2.3 Microcarrier cultivation

Cultivations were performed in a self-built 2 L steel tank reactor with a 25 mm port for the ISM. The model cell line NIH-3T3 (*mouse musculcus*, fibroblasts) was used throughout the cultivations. The chosen medium consisted of 50% DMEM and 50% RPMI with 3,25 g/L glucose and 10% newborn calf serum due to the adhesion behaviour of the NIH-3T3 cells. The Cytodex 1 (Amersham, Sweden) microcarrier concentration was 2 g/L with an inoculation cell density of $1\cdot10^5$ cells per mL. During the first 3 h of cultivation an intermittent stirring was chosen with 3 min stirring at 75 rpm followed by 30 min rest interval to allow the cells attaching to the microcarrier's surface. Afterwards the cultivation was switched to continuous stirring at 75 rpm.

## 3. RESULTS AND DISCUSSION

Figure 2 exemplifies the online monitoring of a NIH-3T3 microcarrier cultivation over a period of 48 h. The sphere of the microcarrier can easily be distinguished from the background. The first image shows a plain microcarrier for an intermediate mode value prior to cell inoculation. The following image series show the cell settling to the microcarrier's surface and their morphology changes in the early cultivation state. After 48 h some microcarriers are already completely overgrown.

The new ISM generation has proven to be a suitable system for the observation and analysis of microcarrier cultures. Using in-situ microscopy it is possible to gain information about plating efficiency, morphology changes and cell concentration. Another major advantage arises from the facility to calculate the average colonization state due to the possibility to analyze a high number of microcarriers by software within a short time. In vaccine production information about colonization state is essential since the cells are going to be infected by viruses afterwards and an optimal infection time is needed.

The next step will be the implementation of an algorithm able to detect and calculate the microcarrier's colonization level. One successful strategy to accomplish this is to compare the mode value of a plain microcarrier to overgrown microcarriers from later states of the cultivation. One effect that also could be useful for this calculation is that the transparency of a microcarrier decreases while the grey value distribution rises.

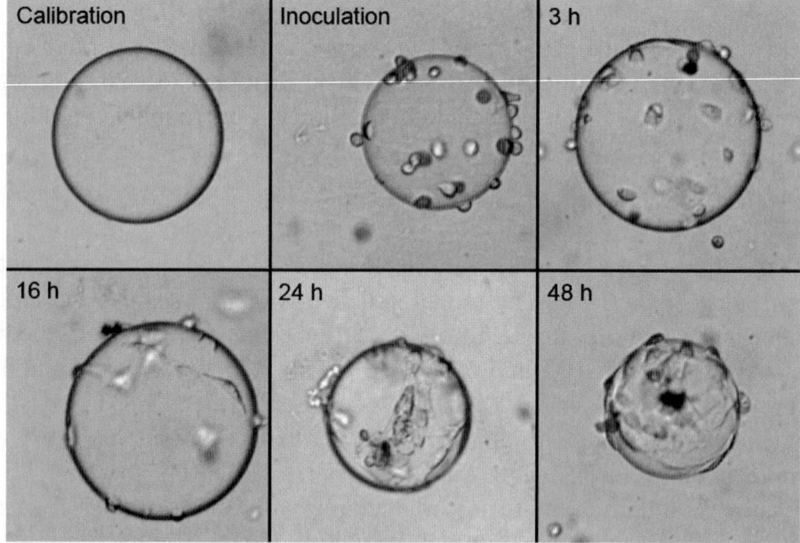


*Figure 2.* Online monitoring of NIH-3T3 growing on Cytodex 1 microcarriers (in-situ microscope, 4-fold magnification).

# REFERENCES

Bittner, C., Wehnert, G., Scheper, T., 1998, In situ microscopy for on-line determination of biomass, Biotechnol. Bioeng. **60** (1), 24-35.

Brueckerhoff, T., Frerichs, J.-G., Joeris, K., Konstantinov, K., Scheper, T., 2003, Image analysis based realtime-control of glucose concentration; Proceedings of 18th ESACT meeting Granada, May 11-14 2003, 589-592.

Camisard, V., Brienne, J.-P., Baussart, H., Hammann, J., Suhr, H., 2002, Inline characterization and cell volume in agitated bioreactors using in situ microscopy: application to volume variation induced by osmotic stress, Biotechnol. Bioeng. **78**, 73-80.

Guez, J.S., Cassar, J.Ph., Wartelle, F., Dhulster, P., Suhr, H., 2004, Real time microscopy for animal cell-concentration monitoring during high density culture in bioreactor, J. Biotechnol. **111 (3)**, 335-343.

Joeris, K., Frerichs, J.-G., Konstantinov K., Scheper, T., 2002, In-situ microscopy: online process monitoring of mammalian cell culture, Cytotechnology **38**, 129-134.

Ulber, R., Frerichs, J.-G., Beutel, S., 2003, Optical sensor systems for bioprocess monitoring, Anal Bioanal Chem **376**, 342-348.

# Rapeseed Peptidic Fractions as Medium Additives in Animal Cell Bioreactors

B. Farges[1], B. Tessier[1], I. Chevalot[1], C. Harscoat[1], S. Chenu[2], I. Marc[1], J.L. Goergen[1], A. Marc[1]
[1] *Laboratoire des Sciences du Génie Chimique, UPR CNRS 6811, ENSAIA-INPL - 2, avenue de la Forêt de Haye, 54 505 Vandoeuvre-lès-Nancy, France.* [2] *Centre d'Immunologie Pierre Fabre, 5 av. Napoléon III, BP 497, F-74164 St Julien-en-Genevois*

**Abstract:** To limit contamination risks by animal-derived components, we designed a new serum-free medium supplemented with rapeseed peptidic fractions. Compared to a reference serum-free medium this new medium allows a 3-fold increase in cell density, a 2-times extended longevity, a 5 times higher final product titer, Modifications of glucose metabolism were also observed. Addition of free amino-acids did not lead to the same positive effect, suggesting that peptides act not only as carbon, nitrogen and energy source but also as growth or survival factors. This rapeseed fraction also allows a very simple formulation of a protein-free medium.

**Key words:** rapeseed peptides, peptidic fractions, protein-free medium, bioreactor kinetics

## 1. OBJECTIVES

The aims of the present work are: i) to compare various peptide fractions of a rapeseed protein hydrolysate as supplements in a simple serum-free medium; ii) to study kinetics of cell growth and protein production in presence of one selected rapeseed fraction; ii) to evaluate the role of the peptide mixture on cell metabolism; iii) to set up a very simple protein-free medium suitable for cell freezing, cell propagation and suspension cell culture.



*R. Smith (ed.), Cell Technology for Cell Products*, 663–665.

## 2. MATERIAL AND METHODS

CHO-C5 suspension cells producing gamma-interferon [1]. Media: reference (RPMI basic + TIB mixture), rapeseed (reference + 4 g/L rapeseed fraction), amino-acid (reference + free amino-acid), protein-free (basic + 4 g/L rapeseed fraction). Rapeseed fractions: enzymatic hydrolysis [3] of proteins from cattle-cake [2], fractionation by membrane processes [4].

## 3. RESULTS

### 3.1 Screening of the peptide fractions

Individual rapeseed fractions vary significantly in their growth promoting effects. Thus, fractionation of peptidic hydrolysate is of great interest to screen some cell growth stimulating effect. The presence of a balanced mixture of peptide sizes (<5000 Da) seems necessary to yield the best results.

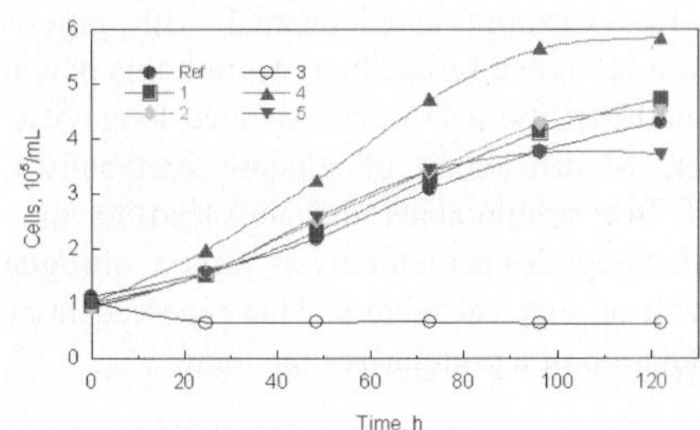


*Figure 1.* Cultures with the reference medium supplemented with different fractions.

### 3.2 Cell growth and IFN production in bioreactors

Supplementation of reference medium with the rapeseed peptidic fraction n°4 leads to a 2.5-fold increase of the maximal cell concentration, a 3.5-times extended culture longevity, an important enhancement of interferon production (x 5), a reduction of the metabolism of carbohydrates.

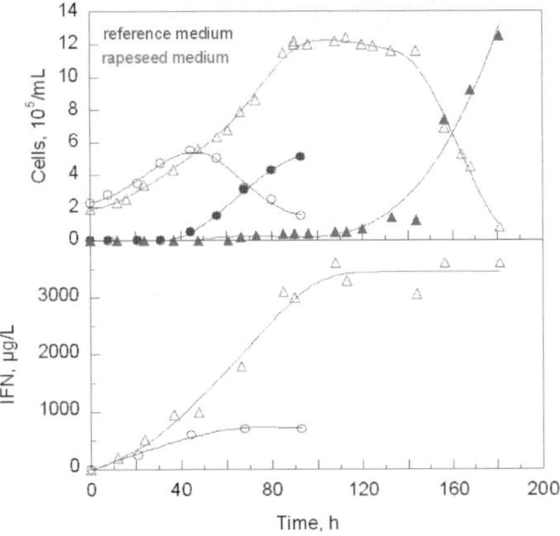


*Figure 2.* Kinetics of bioreactor cultures.

## 3.3 Effect of rapeseed peptides versus free amino-acids

A higher stimulation of cell growth and IFN production is observed in rapeseed medium, than in the amino-acid medium (same content than in totally hydrolysed rapeseed peptides), suggesting growth or survival-factor activities provided by larger peptides.

## 3.4 Validation of a simple protein-free medium

The media supplemented, either with TIB proteins (reference) or with only rapeseed peptides (protein-free) show the same maximal cell concentration but a lower cell death rate in presence of rapeseed peptides. The rapeseed fraction allows a very simple formulation of a protein-free medium. This protein-free medium is also suitable for suspension cell growth, cell propagation and cell freezing.

# REFERENCES


[1] L. Monaco, A. Marc, A. Eon-Duval, G. Acerbis, G. Distefano, M. Soria, D. Lamotte, J.M. Engasser, N. Jenkins. Cytotechnology, 22, 197-203 (1996).

[2] V. Deparis, A. Jestin, A. Marc, J.L. Goergen. Biotechnol. Prog., 19, 624-630 (2003).

[3] B. Tessier, M. Schweizer, F. Fournier, X. Framboisier, I. Chevalot, R. Vanderesse, C. Harscoat, I. Marc. Food Res. Intern., 38, 577-584 (2005).

[4] B. Tessier. PhD thesis, INPL, Nancy (2004).

# Influence of Rapeseed Protein Hydrolysis Conditions on Animal Cell Growth in Serum-free Media Supplemented with Hydrolysates

G. Chabanon[1], I. Chevalot[1], B. Farges[1], C. Harscoat[1], S. Chenu[2], J.-L. Goergen[1], A. Marc[1], I. Marc[1]

[1] *Laboratoire des Sciences du Génie Chimique, UPR CNRS 6811, ENSAIA-INPL - 2, avenue de la Forêt de Haye, 54 505 Vandoeuvre-lès-Nancy, France.* [2] *Centre d'Immunologie Pierre Fabre, 5 av. Napoléon III, BP 497, F-74164 St Julien-en-Genevois*

**Abstract:** Different protein hydrolysates have been prepared from enzymatic hydrolyses of a rapeseed isolate (> 90% protein content) using commercial enzymes of non-animal origin. The extent of hydrolysis has been controlled to prepare various hydrolysates corresponding to different hydrolysis degree (DH) from 5 to 30. The results showed CHO promoting-growth effects of several hydrolysates especially those obtained at high hydrolysis degree. However, different hydrolysates exhibiting same peptide size patterns and similar amino acids compositions did not improve the cell growth in the same way.

**Key words:** rapeseed, hydrolysates, culture medium, CHO, supplements, extensive hydrolysis, protease specificity

## 1. INTRODUCTION

One of the key steps for the production of any recombinant proteins of clinical or diagnostic value is the optimization of culture medium. Total elimination of any animal or human derived substances becomes the main objective for the development of serum-free media. Peptones of soy, rice, wheat gluten are yet commercially available. Rapeseed is a potential source of protein containing well-balanced amino acids and until now, a single study reported the use of rapeseed protein hydrolysates as cell culture medium supplements (Deparis *et al.*, 2003). The aim of this present work is



*R. Smith (ed.), Cell Technology for Cell Products*, 667–669.

to study the influence of various rapeseed hydrolysates obtained by enzymatic hydrolyses of a rapeseed isolate on CHO cell growth.

## 2. RESULTS

A protein isolate is obtained from rapeseed cakes and is composed of proteins (91.4%), lipids (6.6%), fibers and sugars (1.0%), ash (0.8%), polyphenols (0.2%) and phytic acid (0.02%). Apparently, this protein isolate is almost free of non protein components which could play a role in the cell growth.

Hydrolyses of this isolate were carried out with different enzymes and the hydrolysis extent was evaluated by the degree of hydrolysis (DH). Chineese Hamster Ovary cells (CHO-C5) cultures were performed at 37°C in 96 wells plates. Media were composed of a reference one (RPMI 1640, albumin, transferrin, insulin, glutamine and minerals) added of either rapeseed protein hydrolysate or water (reference). Cell growth was followed by measuring total cell density using the Cellscreen system (Innovatis) by image analyses (Table 1).

*Table 1.* Molecular size distribution of hydrolysate peptides (SE-HPLC) and percent of increase or decrease of cell density in media supplemented with hydrolysates at 2 g/L compared to cell density in the reference at 70 h of culture, measured by image analyses in microplates.

| Protease | DH (%) | > 10kDa (%) | 1-10 kDa (%) | <1 kDa (%) | Cell density increase (%) |
|---|---|---|---|---|---|
| Alcalase 2.4L | 5.4 | 26 | 56 | 18 | -15 |
| | 8.5 | 15 | 59 | 26 | 3 |
| | 12.5 | 9 | 49 | 42 | 8 |
| | 20.4 | 4 | 39 | 57 | 17 |
| | 31.6 | 1 | 22 | 77 | 33 |
| Esperase 7.5L | 5.9 | 18 | 61 | 21 | -15 |
| | 9.6 | 12 | 61 | 27 | -8 |
| | 15.1 | 6 | 51 | 43 | -15 |
| | 27.9 | 2 | 29 | 69 | 25 |
| Purified Pronase | 11.4 | 8 | 40 | 52 | 0 |
| | 18.1 | 4 | 28 | 68 | 31 |
| | 23.4 | 1 | 18 | 81 | 56 |
| Neutrase 0.8L | 10 | 13 | 48 | 39 | -5 |
| | 14.7 | 9 | 41 | 50 | 11 |
| | 23.2 | 4 | 25 | 71 | 33 |
| Orientase 90N | 6.4 | 16 | 54 | 30 | 0 |
| | 10.8 | 13 | 50 | 37 | 12 |
| | 15.6 | 8 | 42 | 50 | 4 |
| | 34.2 | 1 | 20 | 79 | 20 |

After 70 h of culture, hydrolysates characterized by a high degree of hydrolysis (DH > 20%) significantly increase the cell density. Cell growth kinetics were then performed with hydrolysates composed of 70-80% < 1 kDa peptides (Figure 1).

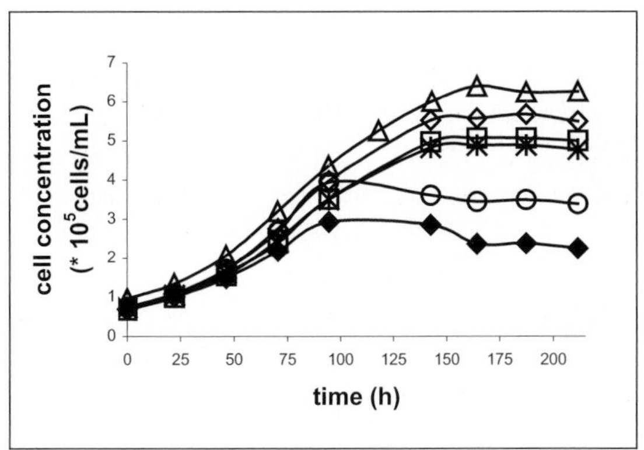


*Figure 1.* Effect of the supplementation at 4 g/L of a serum-free medium with different rapeseed protein hydrolysates obtained with different proteases: (Δ) Pronase and DH = 23.4%. (◊) Alcalase 2.4L and DH = 31.6%. (□) Esperase 7.5L and DH = 27.9%. (*) Orientase 90N and DH = 34.2%. (○) Neutrase 0.8L and DH = 23.2%. (♦): without hydrolysate.

The supplementation with hydrolysates from extensive hydrolysis process allowed a dramatic increase of the maximal cell density. Some peptides seemed to play the role of growth or survival factors.

# 3. CONCLUSION

The results show that rapeseed protein hydrolysates, especially those characterized by an extensive degree of hydrolysis, support the growth of CHO cells cultivated in a simple and defined serum-free medium. The increase of the cell density and survival duration depend on the enzyme specificity and consequently on the composition of produced peptides.

# REFERENCES

Deparis V., Durrieu C., Schweizer M., Marc I., Goergen J.L., Chevalot I., and Marc A., 2003, Promoting effect of rapeseed proteins and peptides on Sf9 insect cell growth, Cytotechnology, **42** : 75-85

# Rapid Media and Process Optimization Using DOE: A Case Study

Stacy Holdread,* Ji Hee Kim,* Toyin Oshunwusi,* Jagdish Kuchibhatla,** Cindy Hunt,* James W. Brooks *
**AMDS, BD Biopharm & Industry, Baltimore MD, USA - **AMDS, BD Biopharm & Industry at Singapore*

**Abstract:**

**Key words:** media, CHO, component optimization, process optimization, peptones, hydrolysates, protein production, IgG, antibody, DOE, Statistical Designs, supplements, animal-free, chemically-defined, HTS

## 1. INTRODUCTION

As production goals continue to increase, it has become increasingly necessary to optimize key production factors that directly affect cell performance. Since media selection has a major impact on culture performance, a medium formulation optimized for a particular cell line can significantly increase protein production. The performance of the new medium can be further enhanced when coupled with an effective feed strategy applied under ideal process conditions. Optimized media and processes can be rapidly identified when comprehensive component screens are conducted using relevant statistical design.

In this study, the performance of two CHO lines was significantly improved through the application of efficient optimization strategies.



*R. Smith (ed.), Cell Technology for Cell Products*, 671–682.

## 2. MATERIALS

Cell Lines

CHO Line 1 and CHO Line 2, both producing human IgG

Culture Media and Supplements

- BD Proprietary Chemically Defined Basal Media (Library Media) and BD AutoNutrient™ Medium CD1000
- Commercially Available CHO Control Media
  BD Animal-Free Peptones – TC Yeastolate UF, Yeast Extract UF, Phytone™ Peptone UF, Difco Springer DS100 Soy Peptone UF (DS100), Select Soytone

Equipment and Reagents

- Falcon Microtiter Plates
  Proliferation – Alamar Blue™ (BioSource)
  Production – BD Hu IgG ELISA with
  BD Biocoat™ plates and Rockland
  Immunological Reagents
- 125 ml Shaker Flasks, 5L Stir Tank Bioreactor
  Proliferation – Trypan Blue
  Production – Protein A HPLC

## 3. METHODS AND RESULTS

CHO line 1 was initially screened in 96 well microtiter plates against BD's library of 45 chemically defined basal media formulations. Cell proliferation and IgG production were evaluated and, based on these results, five media formulations were selected for performance verification in shaker flasks. These formulations, along with a commercially available chemically defined CHO control medium, were scaled into 125 ml shaker flasks and seeded with CHO line 1 without any prior media adaptation. Data for both cell proliferation (Figure 1) and IgG production (Figure 2) showed that Medium 2 performed equivalently to the control CHO medium. Therefore, Medium 2 was selected as the basal medium to use for further optimization with a panel of animal-free peptones (listed above in the Materials section).

Five animal-free peptones were titrated from 1-17 g/L into Medium 2 and screened, in 96 well microtiter plates, against CHO line 1 to determine the concentration for each peptone that would exhibit optimal cell performance (data not shown). The peptone titration data was then used to produce a statistical design in which the five peptones were blended in various combinations in Medium 2. These peptone blends in Medium 2 were screened, in 96 well microtiter plates, against CHO line 1 and cell proliferation and IgG production were monitored (data not shown). The best performing peptones or combinations in Medium 2 from both screens were identified and scaled into 125 ml shaker flasks. The cell proliferation data (Figure 3) shows that, in general, all of the peptone containing media achieved lower numbers of viable cells compared to the chemically defined CHO control media. However, the presence of peptones, either individually or as a blend, in Medium 2 (Figure 4) resulted in higher IgG production levels (bars) and specific productivities (line) compared to the control medium. This data illustrates and supports the fact that cell growth does not always directly correlate with protein production. Therefore, it is critical to evaluate both parameters at every step of the optimization process. After additional experimentation, Medium 2 supplemented with TC Yeastolate UF was carried forward in the optimization process. Prior to beginning feed study experiments, several rounds of optimization were performed on the basal Medium 2 formulation (data not shown). The resulting chemically defined formulation was renamed BD AutoNutrient™ Medium CD1000 (CD1000).

With CD1000 supplemented with TC Yeastolate UF (CD1000/TCY) established as the optimal medium, Feed Study #1 was designed to evaluate eleven different feed strategies. In this study, the first three flasks were given CD1000/TCY as a feed, the next six flasks were given chemically defined feeds based upon spent media analysis, and the final two flasks were given TC Yeastolate UF alone as a feed on different days (Flask Key in Figure 5). The IgG production levels (bars) and specific productivity (line) data shown in Figure 6 indicated that a feed with TC Yeastolate UF (Flasks 10 and 11) resulted in 1.2 g/L IgG, which was the highest performance for this cell line and base medium combination.

Spent medium analysis from Feed Study #1 indicated that CHO line 1 was rapidly consuming the nucleosides provided by the TC Yeastolate UF (data not shown). Feed Study #2 was then designed to evaluate the effect of nucleoside supplementation, as well as finalize the starting glucose and peptone feed concentrations. The details of each condition are outlined in the Flask Key (Figure 7). Flask 9, in which the glucose level remained at a middle concentration and 3 g/L TC Yeasolate UF was fed on day 5, exhibited the highest level of IgG production at almost 1.4 g/L (Figure 8).

With the optimal feed established, CHO line 1 was scaled into a 2 liter culture in a 5 liter stirred tank bioreactor to confirm the improved performance observed in shaker flasks with the CD1000/TCY and a TC Yeastolate feed. On day 5, TC Yeastolate UF was added as a bolus and key growth parameters were maintained throughout the course of the culture.

The culture remained viable with acceptable cell densities over the entire 21 day run (Figure 9) and the level of IgG production was improved to 1.7 g/L (Figure 10). To evaluate IgG product quality, bioreactor supernatant samples were run on an SDS-PAGE gel under reducing conditions and found to be equivalent to the IgG control (Figure 11). In summary, by applying a systematic optimization strategy, the level of IgG production from CHO line 1 was increased almost 3-fold, from a starting level of 600 mg/L to 1.7 g/L, in less than 9 months.

CHO line 2 was initially screened in 96 well microtiter plates against BD's library of 45 chemically defined basal media formulations. Cell proliferation (data not shown) and IgG production (Figure 12) were evaluated and, based on these results, five media formulations were selected for performance verification in shaker flasks. These formulations, along with a commercially available chemically defined CHO control medium, were scaled into 125 ml shaker flasks, seeded with CHO cell line 2, and cell proliferation and IgG production were measured throughout the culture. Based on the IgG production levels, three media candidates out-performed the commercially available CHO control medium (Figure 13). Upon further experimentation, it was determined that Medium C would be carried into the next phase of optimization. Once Medium C was selected, the cells were adapted and passaged in the new formulation.

A Base Media Optimization DOE was created to evaluate the concentrations of several key media components. CHO line 2 was screened, in 96 well microtiter plates, against the formulations generated from this design (data not shown). Several media candidates were selected for scale up into 125 ml shaker flasks. Four of the Base Media formulations out-produced the control media (CHO control and Medium C) with Medium 5 formulation achieving a production level of 619 mg/L (Figure 14). Additionally, all of the new media achieved higher specific productivities than the commercially available CHO control medium. The production levels of the CHO control medium and Medium C were elevated in this shaker study compared to the results obtained in the initial evaluation (Figure 12). This increased production could be attributable to the cells being adapted to Medium C. By identifying a new base medium and performing the appropriate DOE, the production level of CHO line 2 was increased from 250 mg/L to 619 mg/L in 3 months. The performance of this cell line will be further optimized through peptone supplementation and feed strategy development.

# 4. RESULTS

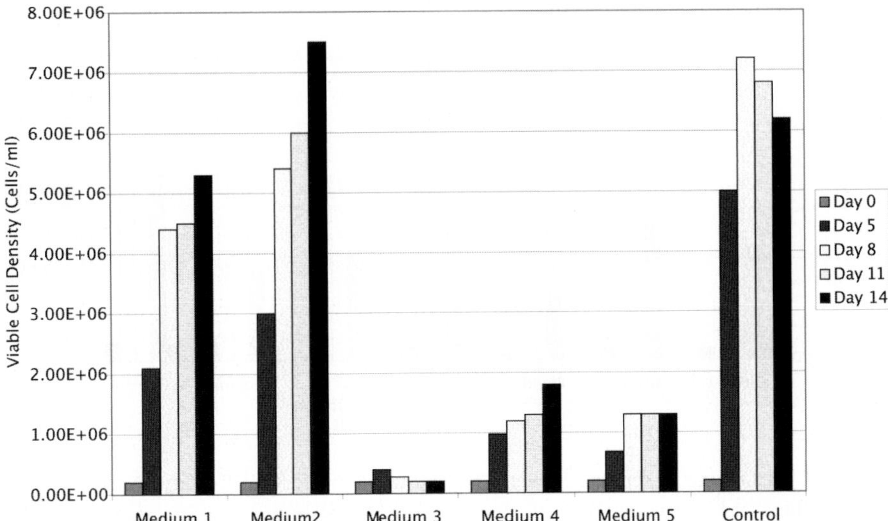


*Figure 1.* CHO Line 1 Media Library Shaker Study Viable Cell Density Determined by Trypan Blue.

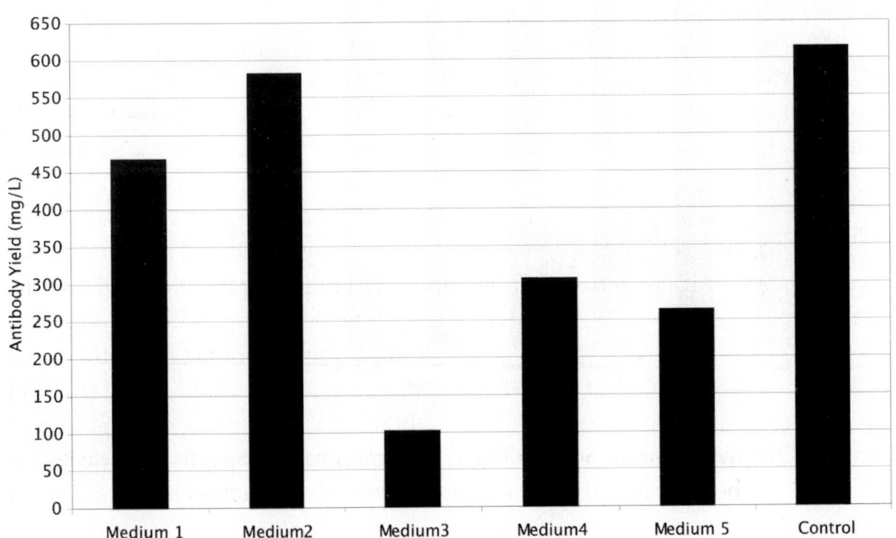


*Figure 2.* CHO Line 1 Media Library Shaker Study Day 14 Production Determined by Protein A HPLC.

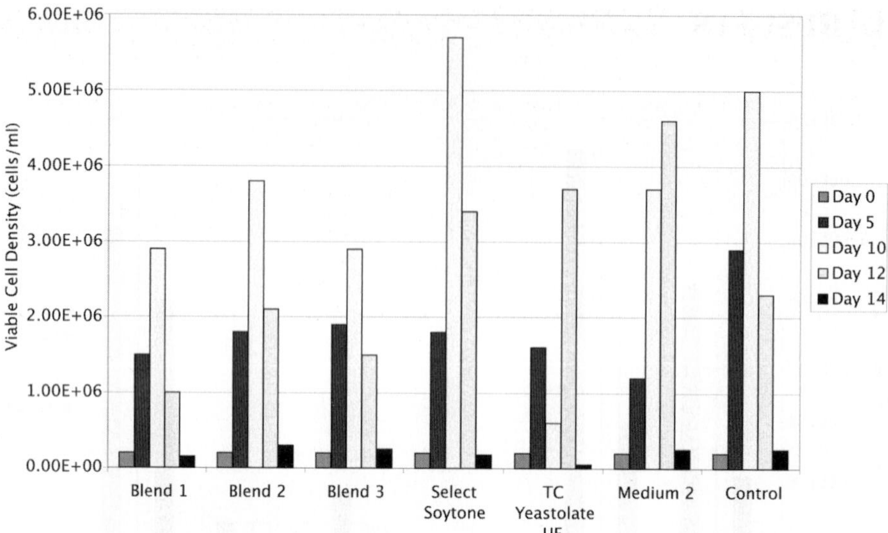


*Figure 3.* CHO Line 1 Peptone Screen Shaker Study Viable Cell Density Determined by Trypan Blue.

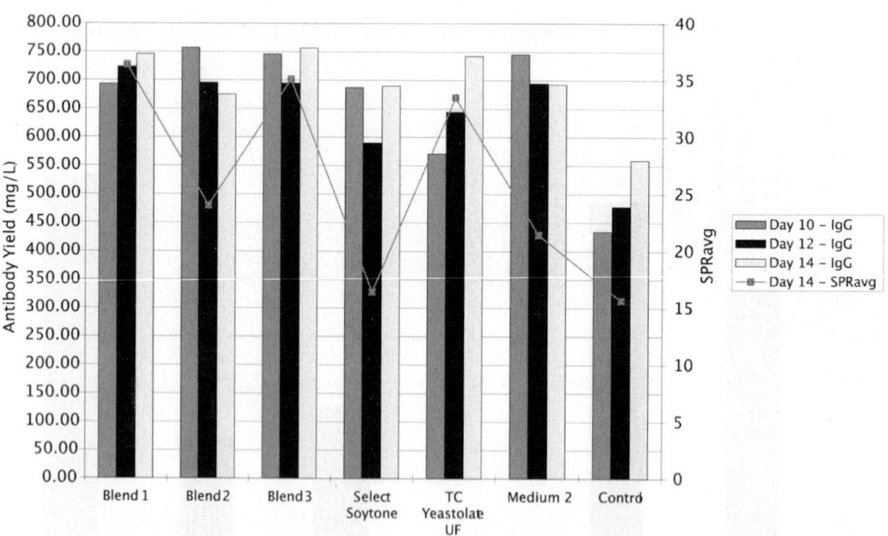


*Figure 4.* CHO Line 1 Peptone Screen Shaker Study Production and Specific Productivity.

| Flask | Starting Medium | Starting Culture Volume | Method of Feed Supplement | Feed Addition |
|---|---|---|---|---|
| 1 | CD1000 with TC Yeastolate UF | 50 ml | 5 ml CD1000 with TC Yeastolate UF | Added on Days 1-10 |
| 2 | CD1000 with TC Yeastolate UF | 50 ml | 5 ml CD1000 | Added on Days 1-10 |
| 3 | CD1000 | 50 ml | 5 ml CD1000 with TC Yeastolate UF | Added on Days 1-10 |
| 4 | CD1000 with TC Yeastolate UF | 100 ml | CD Feed 1 | Day 5 Bolus |
| 5 | CD1000 with TC Yeastolate UF | 100 ml | CD Feed 1 | Days 3, 5, 7, 9, and 11 |
| 6 | CD1000 with TC Yeastolate UF | 100 ml | CD Feed 2 | Day 5 Bolus |
| 7 | CD1000 with TC Yeastolate UF | 100 ml | CD Feed 2 | Days 3, 5, 7, 9, and 11 |
| 8 | CD1000 with TC Yeastolate UF | 100 ml | CD Feed 3 | Day 5 Bolus |
| 9 | CD1000 with TC Yeastolate UF | 100 ml | CD Feed 3 | Days 3, 5, 7, 9, and 11 |
| 10 | CD1000 with TC Yeastolate UF | 100 ml | TC Yeastolate UF | Day 5 Bolus |
| 11 | CD1000 with TC Yeastolate UF | 100 ml | TC Yeastolate UF | Days 3, 5, 7, 9, and 11 |
| 12 | CD1000 with TC Yeastolate UF | 100 ml | None | NA |

*Figure 5.* Flask Key for Figure 6.

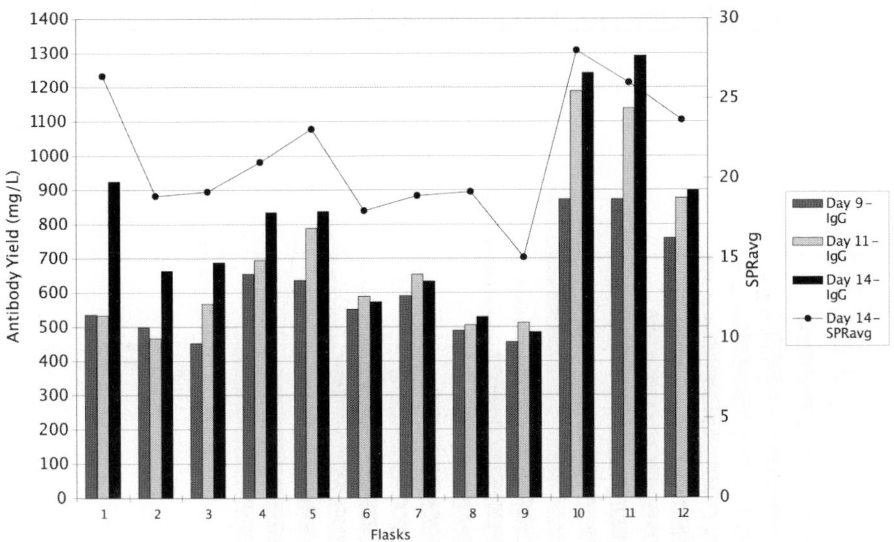


*Figure 6.* CHO Line 1 Feed Study #1 Shaker Study Production and Specific Productivity.

| Flask | Starting Medium | Glucose Starting Volume | Feed Supplement | Method of Feed Addition |
|---|---|---|---|---|
| 1 | CD1000 with TC Yeastolate UF | Middle | 0.2X Nucleosides | Bolus |
| 2 | CD1000 with TC Yeastolate UF | Middle | 0.2X Nucleosides | Multiple Days |
| 3 | CD1000 with TC Yeastolate UF | Middle | 1.0X Nucleosides | Bolus |
| 4 | CD1000 with TC Yeastolate UF | Middle | 1.0X Nucleosides | Multiple Days |
| 5 | CD1000 with TC Yeastolate UF | Middle | 5.0X Nucleosides | Bolus |
| 6 | CD1000 with TC Yeastolate UF | Middle | 1 g/L TC Yeastolate UF | Day 3 |
| 7 | CD1000 with TC Yeastolate UF | Middle | 1 g/L TC Yeastolate UF | Day 5 |
| 8 | CD1000 with TC Yeastolate UF | Middle | 3 g/L TC Yeastolate UF | Day 3 |
| 9 | CD1000 with TC Yeastolate UF | Middle | 3 g/L TC Yeastolate UF | Day 5 |
| 10 | CD1000 with TC Yeastolate UF | Middle | 7 g/L TC Yeastolate UF | Day 3 |
| 11 | CD1000 with TC Yeastolate UF | Middle | 7 g/L TC Yeastolate UF | Day 5 |
| 12 | CD1000 with TC Yeastolate UF | Low | None | NA |
| 13 | CD1000 with TC Yeastolate UF | Low | 3 g/L TC Yeastolate UF | Day 5 |
| 14 | CD1000 with TC Yeastolate UF | Low | 5.0X Nucleosides | Day 5 |
| 15 | CD1000 with TC Yeastolate UF | High | None | NA |
| 16 | CD1000 with TC Yeastolate UF | High | 3 g/L TC Yeastolate UF | Day 3 |
| 17 | CD1000 with TC Yeastolate UF | High | 5.0X Nucleosides | Day 3 |
| 18 | CD1000 with TC Yeastolate UF | Middle | None | NA |
| 19 | CD1000 with TC Yeastolate UF | Middle | None and No Glucose Supplementation | NA |

*Figure 7.* Flask Key for Figure 8.

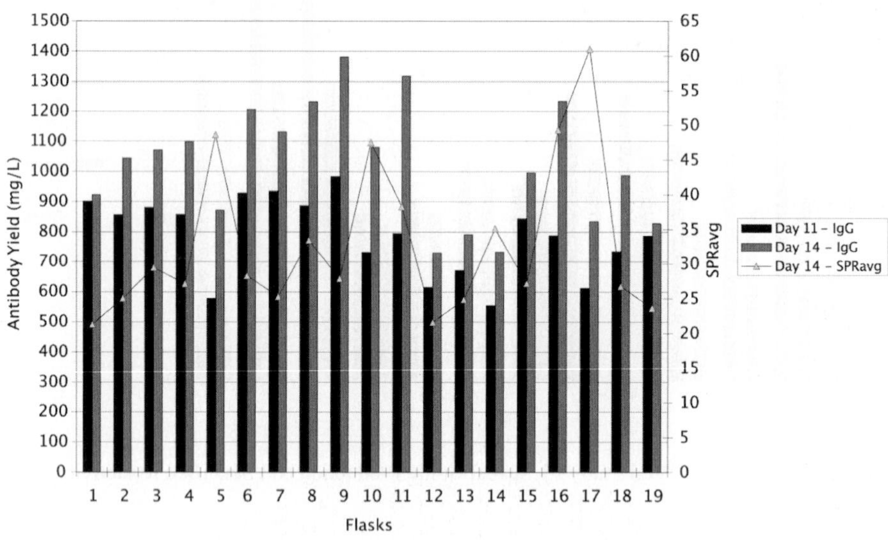


*Figure 8.* CHO Line 1 Feed Study #2 Shaker Study Production and Specific Productivity.

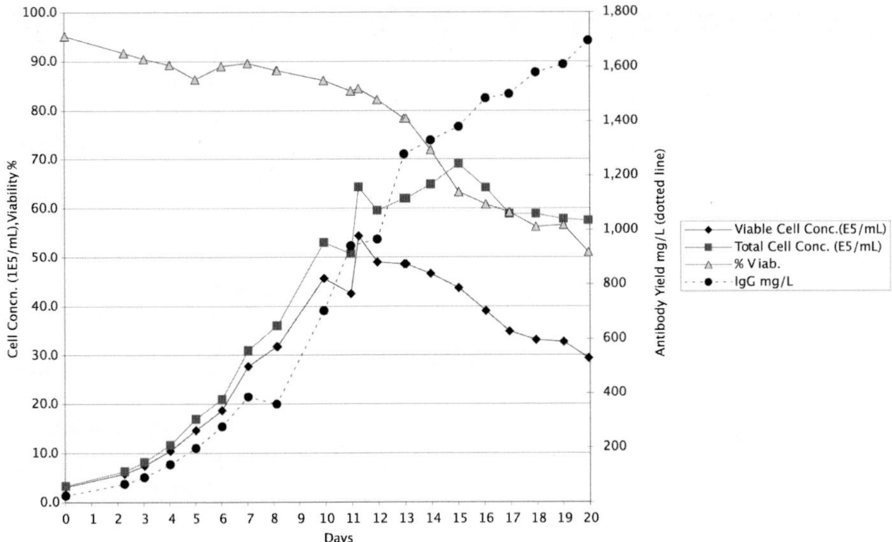


*Figure 9.* CHO Line 1 Bioreactor Proliferation and Production.

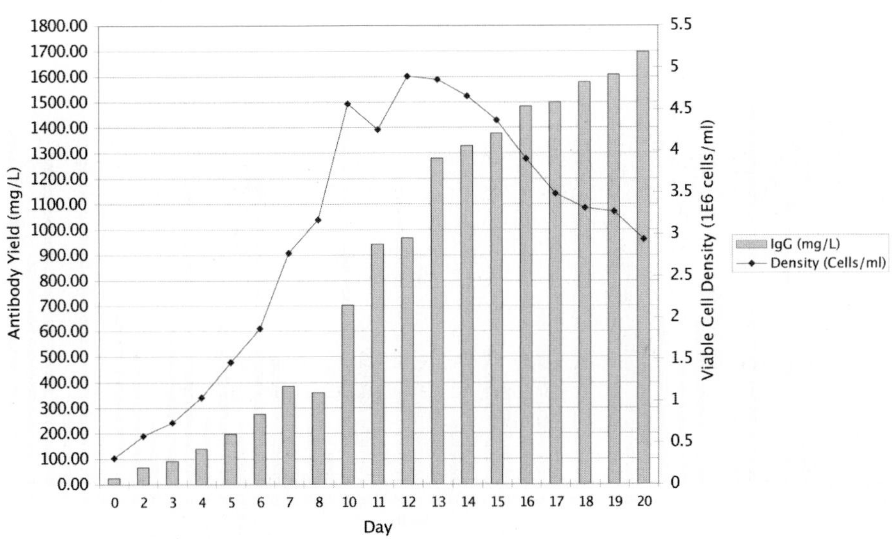


*Figure 10.* CHO Line 1 Bioreactor Production and Proliferation.

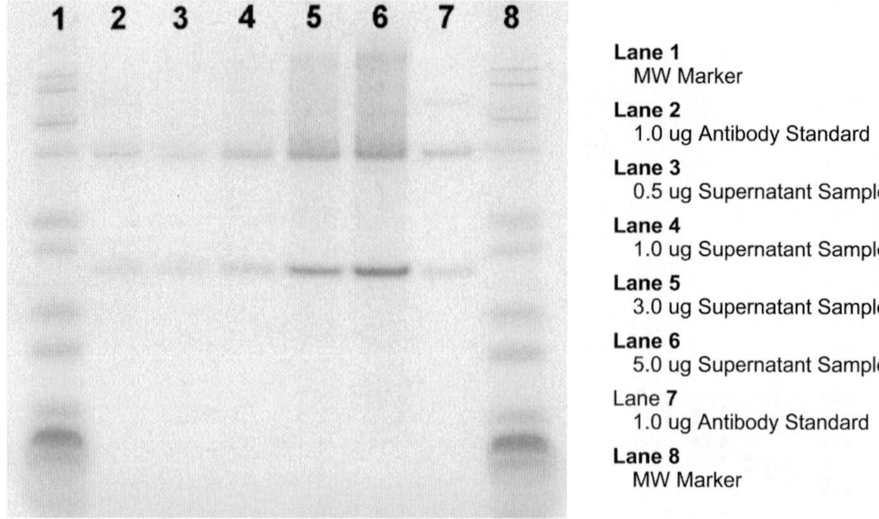


*Figure 11.* SDS PAGE of Bioreactor Samples.

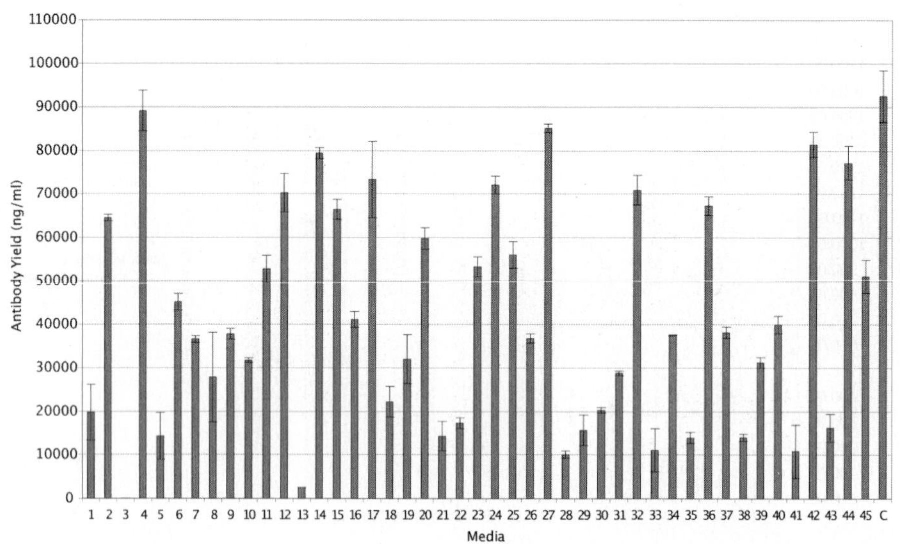


*Figure 12.* CHO Line 2 Media Library Screen in Microtiter Plates Day 7 Production Determined by ELISA.

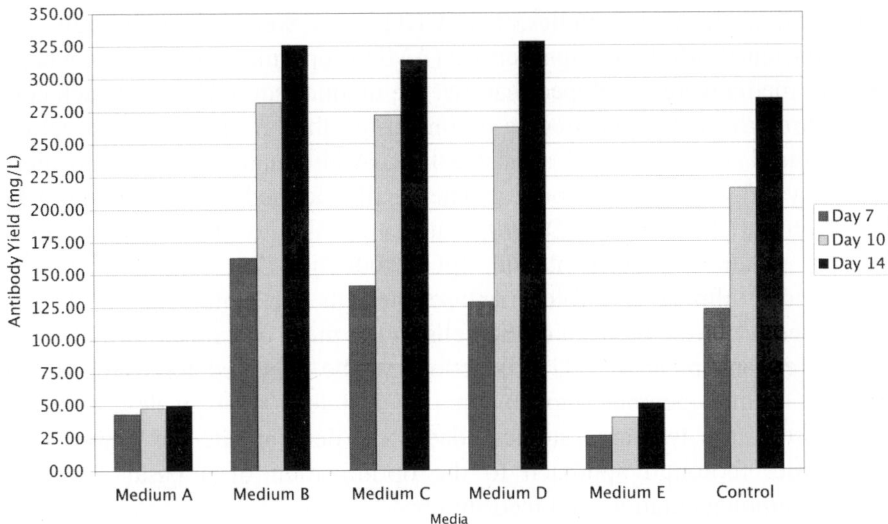


*Figure 13.* CHO Line 2 Media Library Shaker Study Production Determined by Protein A HPLC.

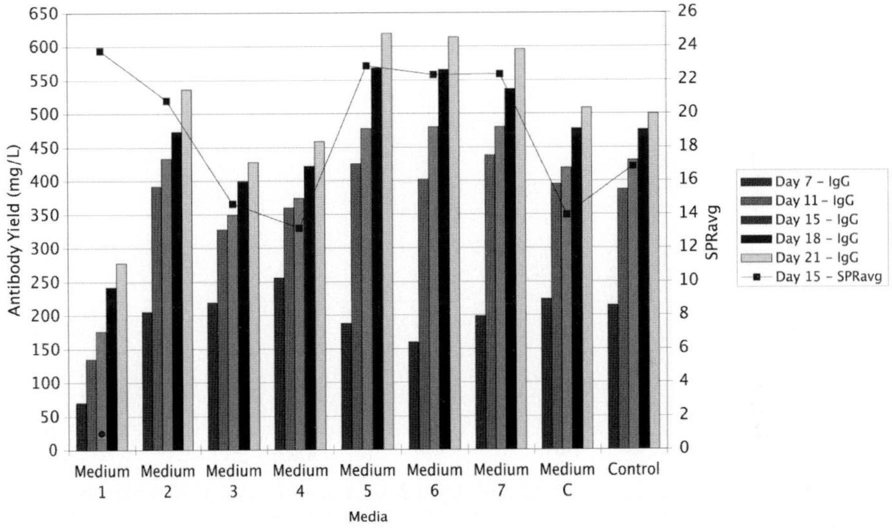


*Figure 14.* CHO Line 2 Base Media Optimization Shaker StudyProduction and Specific Productivity.

## 5. CONCLUSIONS

Dramatic increases in therapeutic protein production can be rapidly achieved through the use of a methodical approach to cell culture media optimization. In these studies, two CHO lines were taken through the AutoNutrient™ Media Design Service (AMDS) optimization strategy, where optimal media were developed that met the unique requirements of each cell line. Through media and process optimization, the IgG production level of CHO line 1 was increased almost 3-fold. Additionally, this optimization project yielded a high performing chemically defined basal medium, BD AutoNutrient™ Medium CD1000, that is amenable to further optimization through peptone supplementation. IgG production for CHO line 2 also experienced almost a 3-fold increase through the optimization of key components and the passage of the cells in the more optimized formulation. Further enhancement of IgG production can be expected for CHO line 2 through optimal peptone supplementation and feed strategy development. Therefore, these two case studies adequately demonstrate the power of a systematic, thorough approach to the optimization of a basal medium, peptone supplementation, and feed strategy.

# Developing Chemically Defined Media Through DOE: Complete Optimization with Increased Protein Production in Less than 8 Months

Kimesha Hammett*, Jagdish Kuchibhatla†, Cindy Hunt*, Stacy Holdread*, James W. Brooks*
*AMDS, BD Biopharm and Industry* **BD at Baltimore – †BD at Singapore*

**Abstract:**

**Key words:** media, CHO, component optimization, process optimization, peptones, hydrolysates, protein production, IgG, antibody, DOE, Statistical Designs, supplements, animal-free, chemically-defined, HTS

## 1. INTRODUCTION

Cell culture media are one of the most vital components in the production of biopharmaceutical proteins. A high performance, chemically defined cell culture medium is the ultimate goal of many in this area. Historically, the development of a chemically defined medium has been achieved through significant investments of time and money and, usually, with the outcome of reduced protein production. Although many chemically defined media are commercially available, they may not adequately meet the specific nutrient requirements of individual cell lines. Therefore, the most effective approach to the development of a chemically defined media, is to design the media formulation based on the specific nutritional needs of a cell line and to perform this process in both an economical and efficient manner.

Toward this end, we have developed and implemented a rapid and effective component optimization method for chemically defined media, which significantly decreases the typical development time while



*R. Smith (ed.), Cell Technology for Cell Products*, 683–691.

maintaining or improving protein production. Here, we present data on the utilization of this approach to develop a high performance medium for a CHO cell line producing a hu mAb. Through the use of proprietary DOE and our AutoNutrient Media Design Service (AMDS), we were able to identify a chemically defined medium formulation that increased cell proliferation and protein production within 8 months. Peptone supplementation and process optimization is being investigated to further increase cell line performance.

## 2. MATERIALS

Proprietary Chemically Defined Basal Media (Library Media), Chemically Defined Control Media (Commercially available medium, Media 15 and revisions), Alamar Blue™ (BioSource), microtiter plates (BD Falcon™), 125 mL shaker flasks (Corning), ELISA Capture plates (BD BioCoat™), pNPP Substrate (Sigma), PBS-Tween 20 (Sigma), anti-hu Ig (G,A,M) (Rockland Immunologics), Protein A HPLC (Waters 2695), ViCell™ Cell Viability Analyzer (Beckman Coulter), NOVA Bioprofile 400 (NOVA Biomedical)

## 3. METHODS AND RESULTS

In order to identify an appropriate starting base medium, 33 proprietary chemically defined library media were screened with the CHO cell line in 96 well microtiter plates. During the 6 day culture period, cell proliferation was assayed by Alamar blue on days 0, 3, 4, 5 and 6 (Figure 1), and day 6 IgG production was determined by ELISA (Figure 2). Based on proliferation and production levels, as compared to control medium, multiple media candidates were chosen for further analysis. Cell performance was evaluated in shaker flasks for each of the candidate media over an 11 day batch culture. Cell counts were performed throughout the culture and day 11 IgG levels were determined by Protein A HPLC (Figure 3). Based on the specific productivity (SPRavg), Medium 15 (M15) was identified as the best base medium candidate for component optimization.

A proprietary mixture design used to approximate a response surface design of 5 to 11 media components was used to optimize M15 through several rounds of component optimization. During each round, 5-11 media components were chosen and the resulting 78 media design points were screened with the CHO cells in 96 well microtiter plates for a 7 day culture period. The best media candidates, based on proliferation (Alamar blue) and IgG production (ELISA), along with multiple statistical predictions were scaled to shaker flasks for 14 day batch culture performance confirmations. Proliferation was assayed by cell counts and IgG levels were determined by Protein A HPLC.

The proliferation data (Figure 4) and day 7 IgG production data (Figure 5) from the 96 well microtiter plate screen for Round One was imported into our statistical package for analysis. Based on proliferation only, day 7 production only, or the combination of proliferation and day 7 production, three media formulations were predicted to improve performance (DOE-1 Growth, DOE-1 IgG, and DOE-1 IgG+Growth media, respectively). Figure 6 is an example of the prediction profiler results from analysis of the day 7 production data. The contour map of the three media chosen by the prediction profiler is shown in Figure 7.

Along with the 3 statistically predicted media, 2 media from the screen (Medium 4 and Medium 7), and a negative control (Medium 34) were chosen to be scaled to shaker flasks for confirmation of performance as compared to Medium 15. Based on the proliferation and IgG production data (Figure 8), the medium predicted to maximize growth (DOE-1 Growth Medium) gave the highest level of proliferation as well as the highest IgG production. Therefore, this medium was chosen as the new base for the second round of component optimization.

Three subsequent rounds of component optimization were conducted, in which 11 components were evaluated per round in the process described above. Again, based on cell proliferation and IgG production, the optimal medium formulation was identified and used as the control for the following round. In Round Two (Figure 9), DOE-2 IgG Media was identified as the best performer in shaker flasks. In Round Four (Figure 10), DOE-4 IgG Media was identified as the best performer in shaker flasks. In Round Five (Figure 11), Media 62 was identified as the best performer in shaker flasks. Note: Round Three data was eliminated due to precipitation problems with the media preparations.

Following completion of the multiple rounds of component optimization, the final step in determining the optimal chemically defined medium was to conduct a verification shaker study, which included the best performing medium from each round of optimization. Prior to shaker flask seeding, the CHO cells were adapted through 3 passages in each medium formulation. During the 21day shaker flask study, glucose levels were monitored and adjusted to maintain concentrations above 2 g/L. This verification study indicated that, following adaptation in each medium formulation, the DOE-2 IgG Medium had one of the greatest effects on cell proliferation and the greatest enhancement of IgG production (Figure 12).

Spent media analysis was conducted on samples from multiple time points to determine if there were any limiting amino acids potentially affecting media performance. Amino acid analysis of DOE-2 IgG Medium (Figure 13) clearly shows that serine and asparagine levels will need to be addressed either in the base medium or through a feed strategy.

# 4. RESULTS

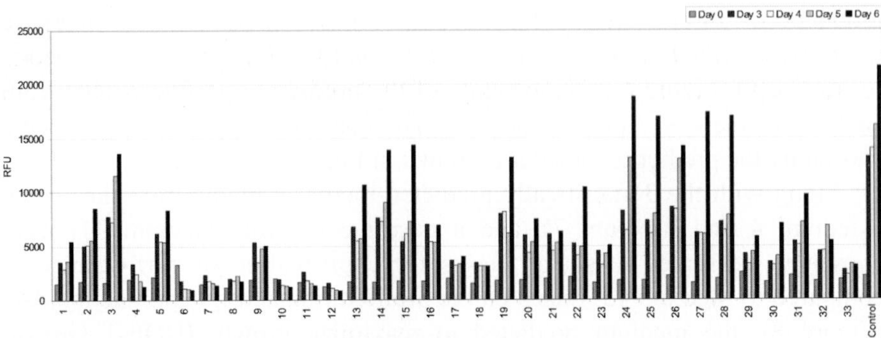


*Figure 1.* Library Screen for CHO Cell Line Alamar Blue Proliferation Data.

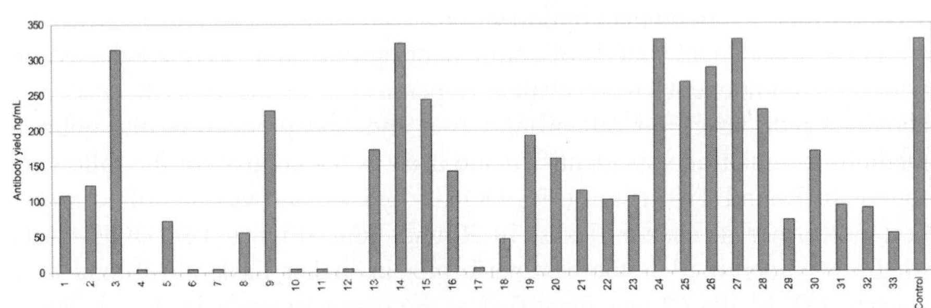


*Figure 2.* Library Screen for CHO Cell Line Day 6 ELISA Data.

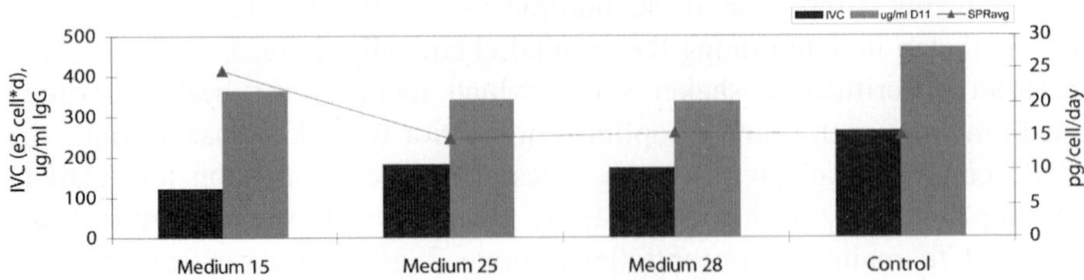


*Figure 3.* Library Screen For CHO Cell Line Shaker Flask Summary.

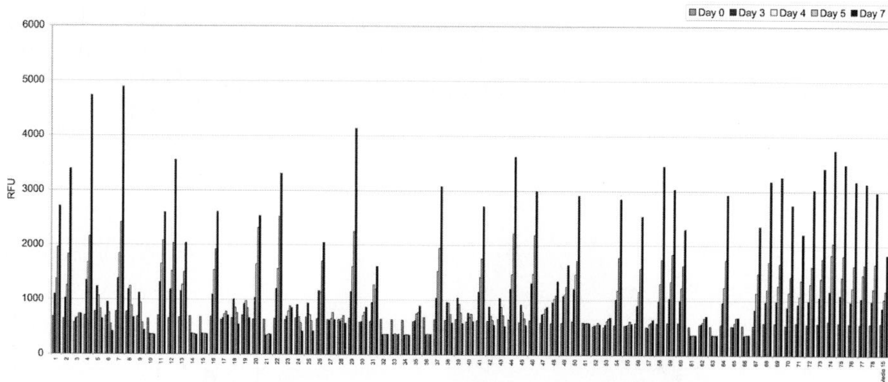


*Figure 4.* Round One Base Media Optimization 96 Well Plate Proliferation Data.

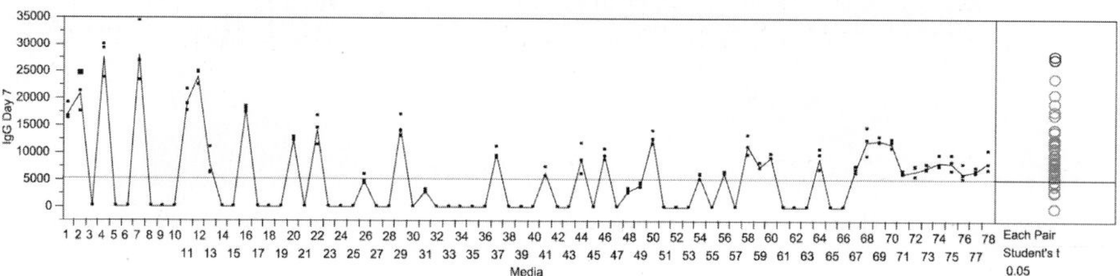


*Figure 5.* Round One Base Media Optimization 96 Well Plate Day 7 Antibody Production (ng/mL).

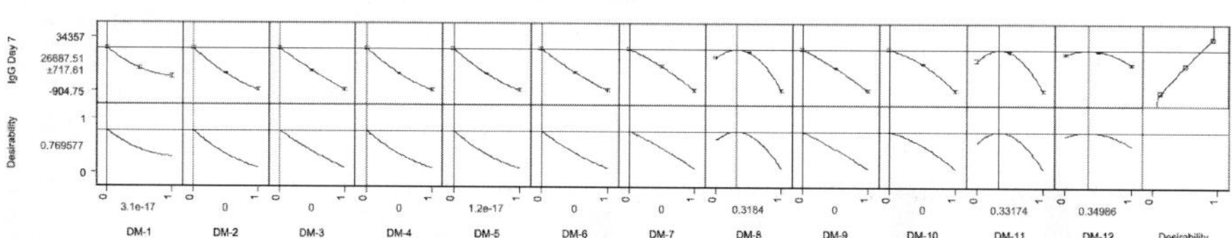


*Figure 6.* Prediction Profiler for Round One Base Media Optimization 96 Well Plate Day 7 Antibody Production Data (ng/mL).

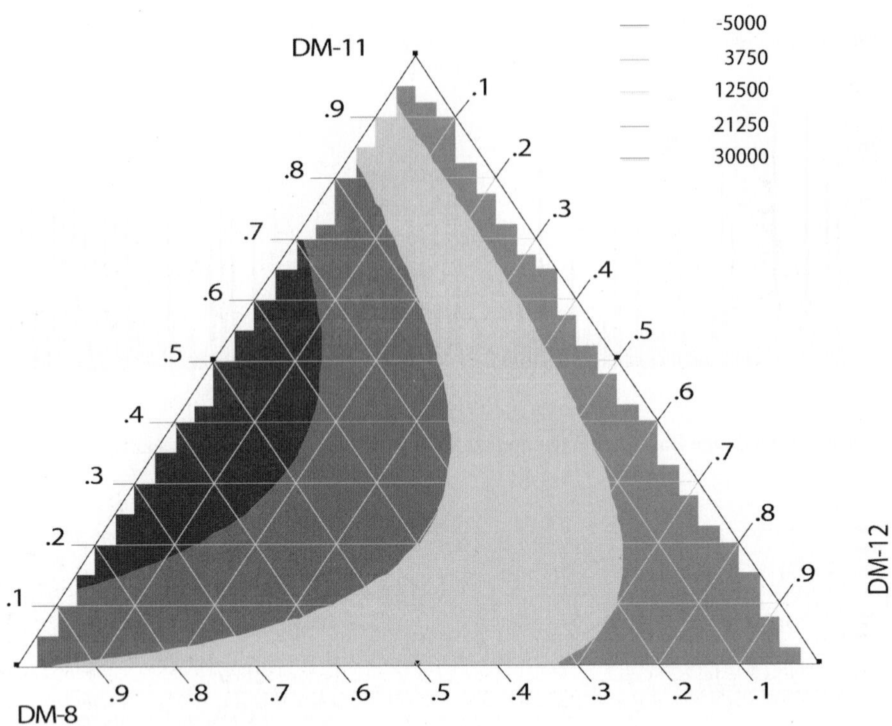


*Figure 7.* Round One Base Media Optimization Contour Plot of DOE Media 8, 11, and 12.

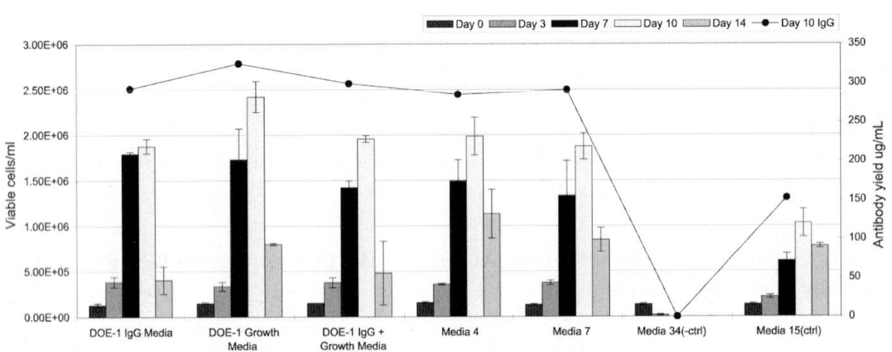


*Figure 8.* Round One Base Media Optimization Shaker Study for CHO Cell Line.

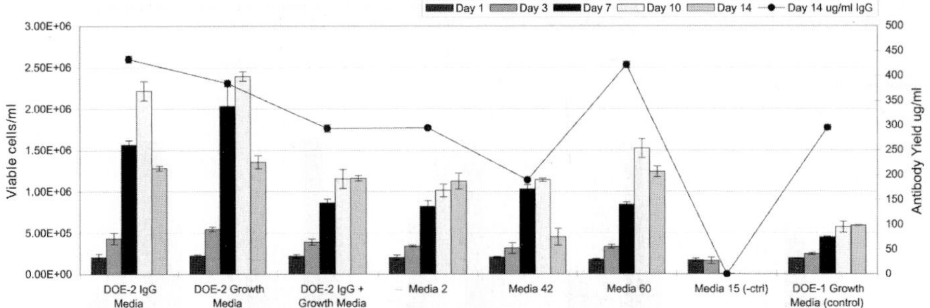


*Figure 9.* Round Two Base Media Optimization Shaker Study for CHO Cell Line.

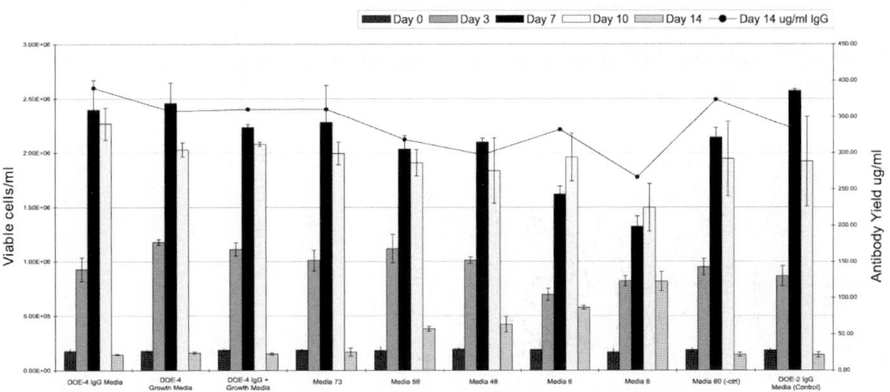


*Figure 10.* Round Four Base Media Optimization Shaker Study for CHO Cell Line.

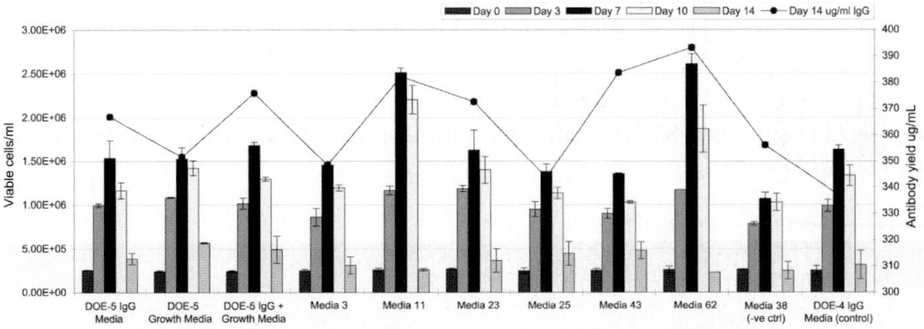


*Figure 11.* Round Five Base Media Optimization Shaker Study for CHO Cell Line.

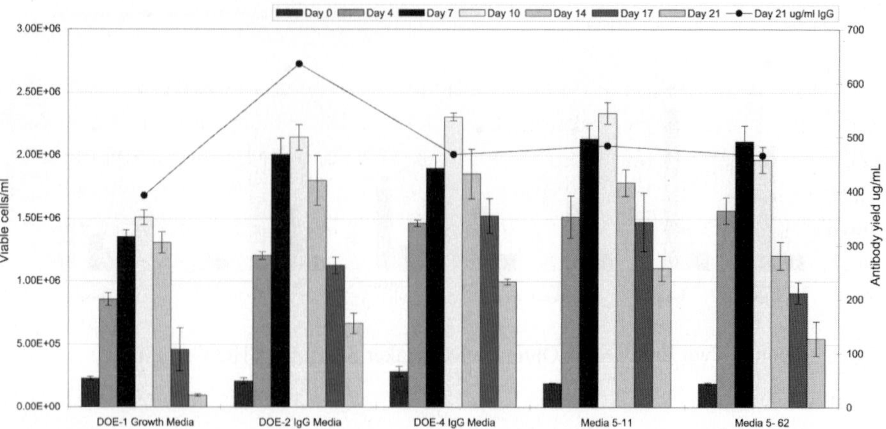


*Figure 12.* Verification Shaker Study after Cell Adaptation Best Media Candidates from each Optimization Round.

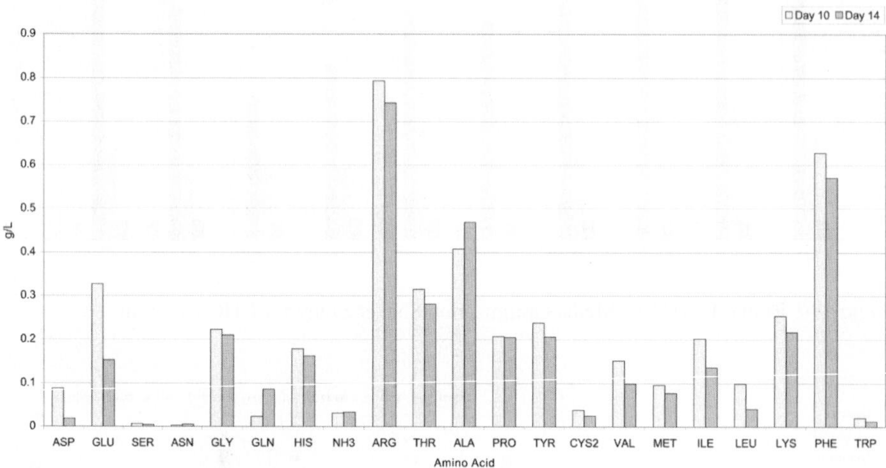


*Figure 13.* Verification Shaker Study after Cell Adaptation Free Amino Acid Analysis of DOE-2 IgG Medium.

| Media ID | Days in culture | Antibody yield ug/mL |
|---|---|---|
| Base Media 15 | 14 | 343 ug/mL |
| Final Media DOE-2 IgG | 14 | 579 ug/mL |
| | 21 | 636 ug/mL |

*Figure 14.* Shaker Flask Performance Comparison of DOE-2 IgG Medium to Medium 15.

## 5. CONCLUSIONS

Each round of component optimization required 4 weeks of assays and cultures. In 2 rounds of optimization (8 weeks), we were able to dramatically increase the IgG production level of the CHO cell line (Figure 14) in shaker flask cultures. The last 2 rounds did not result in an increase in production, but did provide data indicating that there were no critical components having negative effects on cell performance. Scale up into bioreactors is in progress, with the assumption that the IgG production level in reactors will be higher than those obtained in an uncontrolled shaker flask environment. Additional protein production enhancements may also be possible through supplementation with animal free hydrolysates either at initiation of the culture or as a feed supplement.

# Bioreactor Process Strategies for rNS0 Cells

Delia Fernandez, Jennifer Walowitz, William Paul, Lia Tescione and Stephen Gorfien
*GIBCO Invitrogen Corporation 3175 Staley Road Grand Island New York 14072 USA*

**Abstract:** rNS0 cell line has been grown in protein-free, chemically-defined medium using stirred-tank bioreactors in batch and fed-batch mode. Various operating parameters, including pH and temperature, were investigated to assess their effect on growth and IgG production in a chemically defined, protein-free culture system. Spent medium analysis was used to determine the effect of bioreactor operating parameters on nutrient metabolism and IgG production. To increase the total yield, optimum bioreactor operating conditions found in batch culture were then applied to a fed-batch process. The effect of bioreactor conditions on nutrient utilization rates and feeding strategy will be discussed.

**Key words:** pH, temperature, NS0, IgG, protein production, stirred tank reactor, STR, batch mode, fed-batch mode, amino acids, glucose, nutrient utilization, nutrient metabolism

## 1. INTRODUCTION

The Biopharmaceutical industry continues investing in process development research in an effort to increase yield and hence decrease the cost of the process. Cell line, media, reactor mode and culture conditions are some of the factors that have an influence on the product titers achieved. In the present work, the interaction between reactor parameters, in batch and fed-batch mode, and metabolism with a recombinant NS0 (rNS0) cell line was studied.



*R. Smith (ed.), Cell Technology for Cell Products*, 693–695.

## 2. MATERIAL AND METHODS

Batch and fed-batch experiments were performed with a proprietary non-GS rNS0 producing IgG. CD Hybridoma basal medium was prepared from concentrated liquid stock solutions, and a variant with low glucose (2g/L) and low glutamine (2mM) used in the fed-batch mode experiments, was prepared as described by Walowitz *et al.*, 2003. Glucose, glutamine (2mM), cholesterol lipid concentrates (CLC) and a 6 amino acid (AA) (arg, cys, leu, met, tyr and val) solution were used as feeding supplements. Glucose concentration was measured daily using a YSI analyzer (Yellow Spring Instruments, OH), and was adjusted up to 1g/L. Immunoglobulin content was quantitated using protein G HPLC (Waters) as described by Walowitz *et al.,* 2003.

Batch and fed-batch studies were performed in 7.5L CelliGen Plus® (New Brunswick Scientific Co., Inc., NJ) stirred tank reactors with 4L working volume at 30% dissolved oxygen and an agitation speed of 60 rpm. Temperature (37, 35, 33, 32°C) and pH (6.8, 7.0, 7.1, 7.3) were studied in batch mode and the best combination was then applied to the previously optimized fed-batch mode.

## 3. RESULTS AND DISCUSSION

By modifying pH and/or temperature in batch mode, comparable peak cell densities were achieved and the culture was extended by up to 4 days relative to the control (Figure 1A). The extended death phase mirrored observations reported by Yoon *et al.*, 2005 and Osman *et al.*, 2001, and suggest involvement of cell cycle and apoptotic pathways. pH 7.1 or pH shifted from 7.3 to 7.0 at day 2 and temperature shifted from 37 to 33°C at day 2 were found to be the minimum optimal pH and temperature for this rNS0 cell line. The combination of the optimal pH (7.1) and temperature shift (37 to 33°C), as observed when single parameters were studied, resulted in comparable peak cell densities to the control but the culture was extended by 3-4 days.

The increased culture longevity resulted in higher total IgG yields. By modifying pH, a 1.5 to 2-fold higher volumetric IgG production was achieved. Temperature had a smaller effect by itself, but when combined with pH, the effects were additive and resulted in a 2.3-fold increase in IgG production (Figure 1A).

The effect of pH and/or temperature on amino acid consumption was evaluated in batch mode and the optimal bioreactor parameters were applied in a previously optimized fed-batch mode (Walowitz *et al.*, 2003). Amino acid utilization rates were lower at pH 7.1 than at pH 7.3 (data not shown), indicating that the time for feeding needed to be delayed. A temperature shift on day 2 from 37 to 33°C did not affect amino acid consumption.

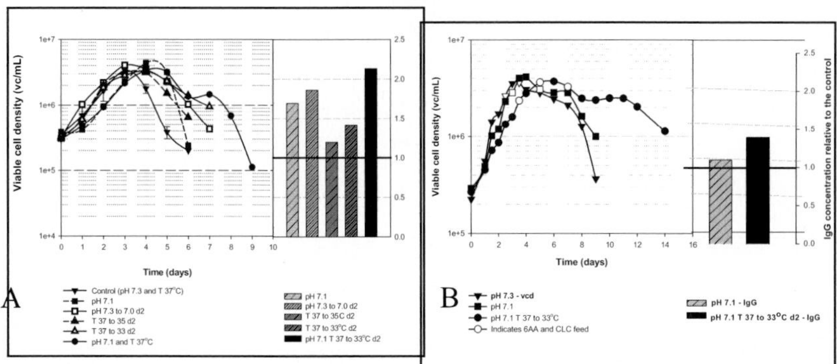


*Figure 1.* Cell growth and relative peak IgG production at different pH and temperature in batch mode (A) and in fed-batch mode (B).

Cell growth in fed-batch cultures at pH 7.1 was comparable to pH 7.3 (Figure 1B). By combining pH and temperature changes in fed-batch mode, the culture was extended by 6-7 days and 1.4-fold and 3.1-fold higher IgG yields were achieved (when compared to pH 7.3 and temperature of 37°C) in fed-batch and batch mode, respectively.

## 4. SUMMARY/CONCLUSIONS

Optimization of pH and temperature extends the culture by slowing down the death phase. The effects on IgG production in fed-batch mode followed similar trends to batch mode, but were less remarkable, suggesting that recombinant protein expression might be limited by other factors such as toxic by-product accumulation and/or depletion of other nutrients.

## REFERENCES

Walowitz J, Tescione L, Paul W, Jayme D and Gorfien S. Optimized Feeding Strategy for NS0 cells. In: Animal Cell Technology Meets Genomics, Proceedings of the 18th ESACT Meeting, Granada, Spain 2003,ed: Godia and Fussenegger, pp 711-714, Springer (Dordrecht, Netherlands), 2005.

Osman J.J., Birch J. and Varley J. 2001 The Response of GS-NS0 Myeloma Cells to pH Shifts and pH Perturbations. Biotechnol Bioeng, Vol. 75, 1: 63-73

Yoon S.K., Choi S.L., Song J.Y. and Lee G.M. 2005 Effect of Culture pH on Erythropoietin Production by Chinese Hamster Ovary Cells Growth in Suspension at 32.5 and 37°C. Biotechnol Bioeng. Vol. 82, 3: 289-297

# The Efficient Induction of Human Antibody Production in *in Vitro* Immunization by CPG Oilgodeoxynucleotides

Yoshihiro Aiba[1], Makiko Yamashita[1], Yoshinori Katakura[1,2], Sin-ei Matsumoto[1], Kiichiro Teruya[1,2], Sanetaka Shirahata[1,2]
[1]*Graduate school of systems life siences, Kyushu University Fukuoka, Japan;* [2]*Department of Genetics Resources Technology, Faculty of Agriculture, Kyushu University,Fukuoka, Japan*

**Abstract:** We developed an *in vitro* immunization (IVI) protocol for generating antigen-specific human monoclonal antibodies from human peripheral blood lymphocytes (PBLs), cytokines, soluble antigen, and muramyl dipeptide (MDP) as adjuvant. In this study, we examined CpG oligodeoxynucleotides (CpG ODNs) as a new adjuvant to improve IVI. "K"-type CpG ODNs induced significant polyclonal antibody production and enhanced the titer of antigen-specific antibody production. On the other hand, "D"-type CpG ODNs efficiently induced antigen-specific antibody production. Furthermore, combined use of "D"-type and "K"-type CpG ODN augmented efficient antigen-specific production. These results would offer a novel method to efficiently generate antigen-specific human monoclonal antibodies in IVI.

**Key words:** antigen-specific production, CpG oligodeoxynucleotides, *in vitro* immunization

## 1. INTRODUCTION

We developed an IVI protocol for generating antigen-specific human monoclonal antibodies from human PBLs, cytokines, soluble antigen, and MDP as adjuvant. However, immune response and antigen-specific production in IVI was very low. Recently, it has been shown that synthetics ODNs with unmethylated cytosine-phosphate-guanine motif (CpG motif) activated mammalian immune response via Toll-like receptor 9 (1). In



*R. Smith (ed.), Cell Technology for Cell Products*, 697–701.

human, it has been reported two distinct types of CpG ODNs (2). K-type CpG ODNs trigger the activation of plasmacytoid dendritic cells (pDCs) (3) and the proliferation and activation of B cells (2). D-type CpG ODNs directly induce the secretion of IFN-α from pDCs, which indirectly supports to differentiate monocytes into DCs (4). In this study, we examined the effects of "K"-type and "D"-type CpG ODNs as new adjuvant in IVI.

## 2. MATERIALS AND METHODS

### 2.1 Adjuvants

MDP was purchased from BACHEM (Bubendorf, Switzerland). All ODNs were purchased from Sigma-Genosys (Hokkaido, Japan). The sequences used were: "K"-type CpG ODNs, 5'-tcgaggttctcC-3'; "K"-type ctrl CpG ODNs, 5'-tgcaggcttctcC-3'; "D"-type CpG ODNs, 5'-ggTGCATCGATGCAGGGGggG-3'; "D"-type ctrl CpG ODNs, 5'-gGTGCATCTATGCAGGGGggG-3'as previously reported (2). Capital and lowercase letters indicate bases with phosphorodiester and phosphorothioate-modified backbones, respectively. All ODNs were diluted in PBS or sterilized water and was used a concentration at 1μM.

### 2.2 Cell preparation

PBMCs were isolated from healthy volunteers by centrifugation using lymphocyte separation medium (LSM: Organ Teknika Durham, NC). PBMCs were depleted of NK cells and monocytes by incubation with 0.25 mM LLME (BACHHEM, Torrance, CA) in ERDF (KYOKUTO Pharmaceutical, Tokyo, Japan). After washing, the remaining cells, $5\times10^{6}$cells, were cultured in ERDF supplemented with 10□ fetal bovine serum (FBS: Trace Scientific, Melbourne, AUS) and IL-2 (Genzyme, Cambrige, MA), IL-4 (Pepro Tech, London, U. K.), 2-merucaptoethanol (Life Technologies, Grand Island, NY), mite extract (ME: LSL, Tokyo, Japan) and adjuvants.

### 2.3 ELISA and ELISPOT

**ELISA:** For determination of concentrations of total antibody production in culture supernatants, 96-well microtiter plates (Nunc, Naperville, IL) were coated with anti-human IgM or IgG antibodies (TAGO, Burlingame, CA) 2hrs at 37°C. The plates were washed with phosphate buffered saline (PBS) containing 0.05% Tween-20 (Wako, Osaka, Japan)

(PBS-T) and blocked with PBS containing 1% ovoalubumin (Wako) 2 hrs at 37 °C. After washing with PBS-T, various dilutions of culture supernatants and standard were added to the plates and incubated overnight at 4 °C. Next, the plates were washed with PBS-T thoroughly and then incubated with appropriately diluted horseradish peroxidase-conjugated goat antibodies human IgM or IgG (TAGO) 2hrs at 37°C. Subsequently, the plates were washed with PBS-T and developed with 0.1 M citrate buffer (pH 4.0) containing 0.003% $H_2O_2$ and 0.3 mg/ml 2.2'-azinodi diammonium salt (Wako) substrate solution.

**ELISPOT**: ME-specific IgM secreting cells were detected by enzyme-linked immunospot assay (ELISPOT) on day 6 of culture. Briefly, 96-well Millititer plates (Millipore, Bedford, MA) were coated with 10 µg/ml ME overnight at 4 °C. Plates were washed with PBS and blocked with PBS containing 5% FBS 2 hrs at 37 °C. After washing with PBS, appropriate number of cells was added to the plates and incubated 24hrs at 37 °C. The plates were washed with PBS-T and incubated with appropriately diluted horseradish peroxidase-conjugated goat antibodies human IgM 2hrs at 37 °C. The plates were washed with PBS-T and PBS thoroughly and then developed with 3.3'-diamiinobenzidine tetrahydrochloride. Developed plates were washed with water and counted using a light microscope.

# 3. RESULTS AND DISUCUSSION

## 3.1 K-type CpG ODNs induce significant total antibody in IVI

At first, we examined the effects of "K"-type and "D"-type CpG ODNs on total antibody production in IVI. "K"-type CpG ODNs induced significant total antibody production compared to MDP. In addition, K ctrl CpG ODNs induced little increase of total antibody production. In contrast to these results, "D"-type CpG ODNs could not induce increase of total antibody production (Table .1). Above results were consistent with published reports (2).

*Table 1.* Effect of adjuvants on total antibody production in IVI.

| Treatment | IgM (µg/ml) | IgG (µg/ml) |
|---|---|---|
| Medium | 0.28 | 1.16 |
| MDP | 0.80 | 1.76 |
| K ctrl | 1.45 | 2.40 |
| K | 19.83 | 13.60 |
| D ctrl | 0.32 | 1.07 |
| D | 0.59 | 1.71 |

### 3.2 D-type CpG ODNs induce efficient antigen-specific IgM in IVI

It has been reported that "K"-type CpG ODN directly activates pDCs, but not myeloid DCs (3). "D"-type CpG ODNs were directly differentiated monocytes into mature DCs and indirectly supports maturation of DCs through IFN-α production from pDCs (4). Therefore, we investigated the effects of "K"-type and "D"-type CpG ODNs on antigen-specific antibody production in IVI. ME-specific antibody production of "D"-type CpG ODNs was almost equal to those of "K"-type CpG ODNs. However, considering total antibody production, "D"-type CpG ODNs enhanced efficient ME-specific antibody production compared to K-type CpG ODNs (Table .2).

### 3.3 Combined use of K-type and D-type CpG ODNs antigen-specific antibody production in IVI

Above results demonstrate that "K"-type CpG ODNs induced significant polyclonal antibody production and enhanced the titer of antigen-specific antibody production and "D"-type CpG ODNs enhanced antigen-specific immune responses in IVI. Therefore, we hypothesized to trigger more efficient antigen-specific antibody production by combined use of "K"-type and "D"-type CpG ODNs in IVI. "K"-type CpG ODNs treatment after "D"-type CpG ODNs pretreatment (D p + K) efficiently augmented ME-specific antibody production (Table. 2). These results would offer a novel method to efficiently generate antigen-specific human monoclonal antibodies in IVI.

*Table 2.* Effect of adjuvants on antigen-specific antibody production in IVI.

| Treatment | ME-specific IgM spots |
|---|---|
| Medium | - |
| MDP | + |
| K ctrl | - |
| K | ++ |
| D ctrl | - |
| D | ++ |
| D p + K | +++ |

## REFERENCES


1. Bauer, S., Kirschning, C. J., Hacker, H, Redecke, V., Hausman, S., Akira, S., Wagner, H., Lipford, G, B. 2001, Human TLR9 confers responsiveness to bacterial DNA via species-specific CpG motif recognition. Proc. Nalt. Acad. Sci. USA. **98**: 9237-42.
2. Verthelyi, D., Ishii, K. J., Gursel, M., Takeshita, F., Klinman, D. M. 2001, Human peripheral blood cells differentially recognize and respond to two distinct CPG motifs. J. Immunol. **166**: 2372-77.

3. Krug, A., Towarowski, A., Britsch, S., Rothenfusser, S., Hornung, V., Bals, R., Giese, T., Engelmann, H., Endres, S., Krieg, A. M., Hartmann, G. 2001, Toll-like receptor expression reveals CpG DNA as a unique microbial stimulus for plasmacytoid dendritic cells which synergizes with CD40 ligand to induce high amounts of IL-12. Eur. J. Immunol. **31**: 3026-37.
4. Gursel, M., Verthelyi, D., Klinman, D. M. 2002, CpG oligodeoxynu- cleotides induce human monocytes to mature into functional dendritic cells. Eur. J. Immunol. **32**: 2612-17.

# Rapid Culture Media Optimization to Gain Multiple Fold Increase in Recombinant Protein Production in Two CHO Cell Lines

Scott D Storms, Aaron Chen, Jenny Yu, and Tom Fletcher
*Irvine Scientific*

**Abstract:** Culture media optimization can lead to significant increases in product titers from recombinant CHO cell lines. When faced with limited development time for optimization, rigorous selection of methods and scales must be applied. A media optimization strategy has been developed to meet this need. Overlapping phases allow flexibility in optimization of process conditions, basal media, supplements, and feed strategies. A case study shows culture media optimization for two CHO cell lines using this process. Growth and productivity of both cell lines was significantly increased when compared with starting media. These results show that productivity can be rapidly increased using this methodology that hinges on the coordination of appropriate methodologies and scales during each phase of the optimization process.

**Key words:** CHO, culture media, media optimization

## 1. INTRODUCTION

Culture media optimization is one means that can lead to increased productivity of a recombinant cell line. Rational Culture Media Design™ is a strategy developed to rapidly and efficiently meet this need. It is modular allowing conduct of multiple phases concurrently. Experimental approaches (methods and scales) are chosen at each phase to meet the project criteria. In this case study, data from representative experiments are shown from three phases (Basal Medium Discovery, Medium Optimization, and Hydrolysate Optimization) of a media optimization project for two different recombinant CHO cell lines.



*R. Smith (ed.), Cell Technology for Cell Products*, 703–705.

## 2. BASAL MEDIUM DISCOVERY

Goal: Discover a well-performing basal medium for each of the given cell lines starting with off-the-shelf media.

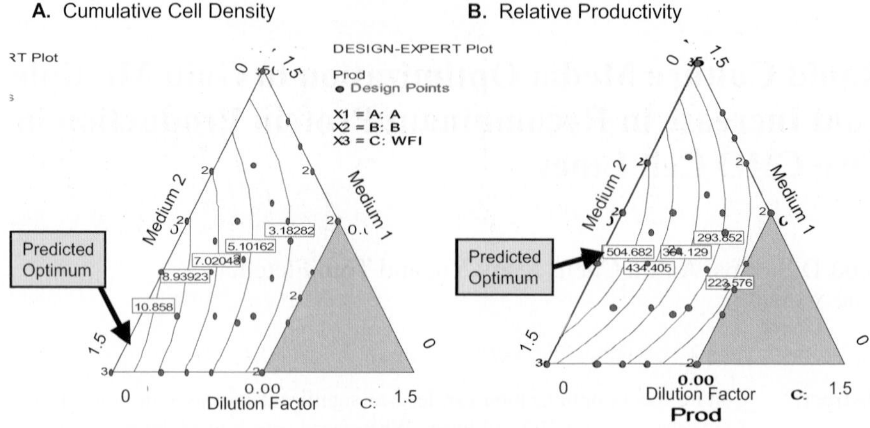


*Figure 1.* Combined Blending and Concentration Experiment. Cumulative Cell Density (Panel A) and Relative Productivity (Panel B) of recombinant CHO cell line 2 in shake flask culture. The experiment was a statistically designed mixtures blend of two media and a dilution factor.

## 3. BASAL MEDIUM OPTIMIZATION

Goal: Optimize well-performing basal medium for the given cell lines.

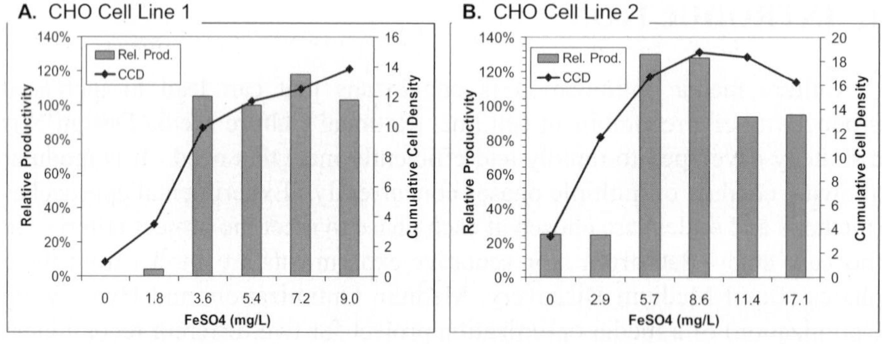


*Figure 2.* Nutrient Optimization Experiment – Iron. Effects of iron on cumulative cell density (CCD) and relative productivity (Rel. Prod.) of two CHO cell lines in shake flask culture (Panels A & B).

## 4. HYDROLYSATE OPTIMIZATION

Goal: Find well-performing hydrolysate blends for the given cell lines.

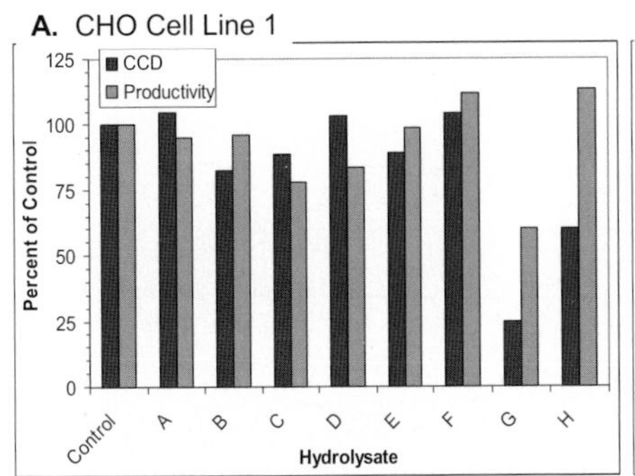


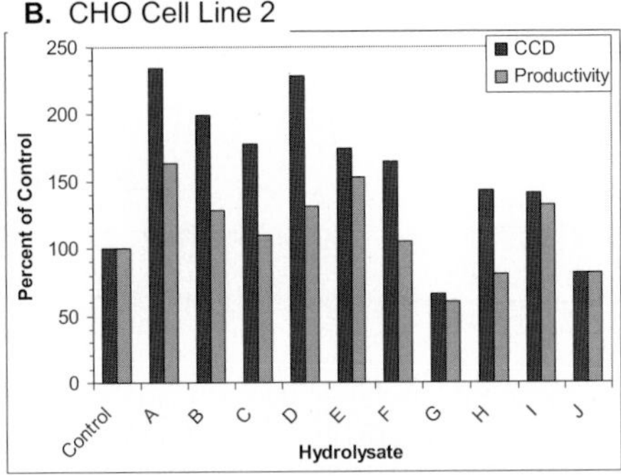


*Figure 3.* Hydrolysate Survey Experiment. Effects of different hydrolysates on cumulative cell density (CCD) and relative productivity of two CHO cell lines in shake flask culture (Panels A & B).

## 5. SUMMARY

1. Media optimization was achieved for both recombinant CHO cell lines using Rational Culture Media Design™ methodologies (Table 1).
2. CHO Cell Line 1 productivity was increased 3.6x and CHO Cell Line 2 productivity was increased 1.8x.
3. Each cell line responded differently to off-the-shelf media and further optimization underlining the different nutritional needs of each recombinant cell line.

*Table 1.* Summary of media optimization for two recombinant CHO lines. Each optimization phase builds on the previous phase.

| **Optimization Phase** | **Hydrolysate ?** | **CHO Cell Line 1** | | **CHO Cell Line 2** | |
|---|---|---|---|---|---|
| | | **CCD** | **Product** | **CCD** | **Product** |
| Control Medium | Yes | 17 | 100% | 25 | 100% |
| Basal Medium Discovery | No | 14 | 97% | 30 | 89% |
| Medium Optimization | No | 19 | 130% | 35 | 111% |
| **Hydrolysate Optimization** | **Yes** | **36** | **363%** | **27** | **183%** |

# Strategies for Large Scale Production of *Parapoxvirus Ovis NZ-2*

M. Pohlscheidt[1], U. Langer[1], B. Bödeker[1], D. Paulsen[2],
H. Rübsamen-Waigmann[2], U. Reichl[4], H.J. Henzler[3], H. Apeler[1]
*[1]Bayer HealthCare AG, Pharma-Operations, Biotechnology, Wuppertal, Germany; [2]Bayer HealthCare AG, Antinfectiva Research, Wuppertal, Germany; [3]Bayer Technology Services, Wuppertal, Germany; [4]Max Planck Institute for Dynamics of Complex Technical Systems, Magdeburg, Germany*

**Key words:** *Bovine Kidney*, *Parapoxvirus Ovis NZ-2*, Adaptation, Batch, Fed-Batch, Dialysis, Scale-Up

## 1. INTRODUCTION

The production of recombinant proteins and vaccines in animal cells for therapeutic use is of increasing importance. Due to the characteristics of mammalian cells the cultivation requires complex media composition (chemically defined favored), special bioreactor design and a sophisticated operation mode.

For the production of a chemically inactivated *Parapoxvirus ovis NZ-2* (*PPVO NZ-2*) an adherent *Bovine Kidney* cell line is cultivated on Cytodex® -3 microcarriers in suspension culture. The inactivated and purified virus particles have shown antiviral activity in several animal models. *PPVO NZ-2* is produced by a biphasic batch process at the 3.5 and 10 L scale. Oxygenation is realised by shear less membrane oxygenation via a tube stator with a central 2-blade anchor impeller. In order to achieve higher efficiency, process robustness and safety, of the established process was optimised.



*R. Smith (ed.), Cell Technology for Cell Products*, 707–710.

# 2. RESULTS

## 2.1 Adaptation of Cell Line

The used *Bovine Kidney KL 3A* cell line was adapted from human protein fraction (0.1 g/L) and insulin (570 IU/L) containing in-house medium (C7PMF 3.2292C) to animal, herbal and human compound free media. Adaptation was performed by subsequent decrease of original medium (~25%). Two commercial available media (competitor I and II) were compared to an in-house formulation without human protein fraction (C7PMF 3.2295A V). For all media containing recombinant insulin the adaptation was successful regarding growth rate, viability and morphology. Significant differences in virus production of competitor I and II compared C7PMF 3.2292C were found in T-flasks (~1 $log_{10}$ level in $TCID_{50}$). Cells adapted to C7PMF 3.2295A V medium depicted comparable virus production.

Conducted test runs in biphasic batch process (V =10 L) showed no differences concerning attachment of the cells on microcarriers, growth on microcarriers and virus production. The new cell line adapted to low cost in house media C7PMF 3.2295A V was named *Bovine Kidney KL 3A V*.

## 2.2 Analysis of Biphasic Batch Process and Description of Cell Line

The biphasic batch process was analyzed and depicted possible substrate limitations in virus propagation phase. Glucose and glutamine were depleted and evaluated lactate was consumed. To prepare necessary changes in operation mode specific rates of the cell line were determined. For process management the direct expansion of the cell line from covered carriers to freshly added carriers was of major interest and found to be possible in a ratio of ~1:5 without trypsination or pH shift. The results, supported by a mathematical model in MatLab 6.5, led to changes in operation mode.

Among other operation modes fed-batch and dialysis fermentations were performed. In summary, both operation modes increased virus production significant (Fig. 1), but depicted disadvantageous (high deviations in fed-batch, complex peripheral equipment in dialysis). Based on these results a new, safe, robust and high effective process was developed: "Volume-Expanded-Fed" batch avoids substrate limitations and possible inhibitions by subsequent dilution of infected cells in next (final) process scale (Fig. 2). The productivity related to biphasic batch process was significantly increased (Fig. 1) without loss of biologic activity and significant decrease of residual Host Cell Protein and DNA-content of formulated bulk.

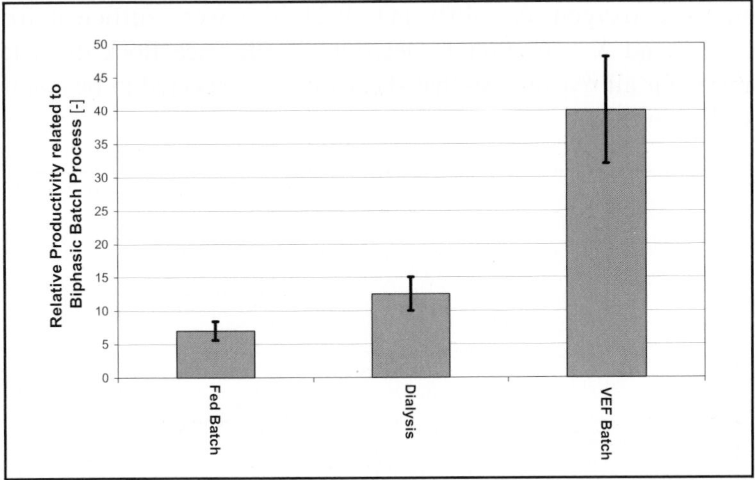


*Figure 1.* Comparison of relative productivity related to biphasic batch process.

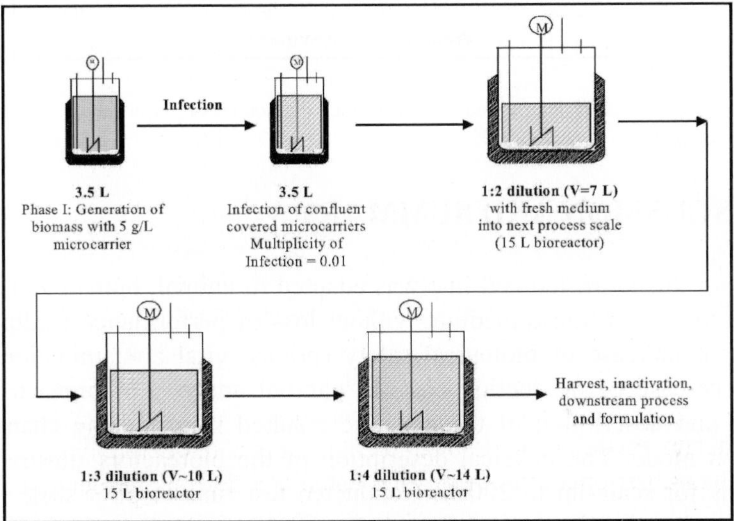


*Figure 2.* Scheme of "Volume-Expanded-Fed" Batch (Exemplary from 3.5 L into 15 L).

## 2.3 Scale-Up

In order to realise sophisticated scale-up to possible production scale (200 L) a physical description of several bioreactors (3.5 L up to 200 L) in different geometries (H/D= 1–2.3, d/D=0.45–0.79) regarding homogeneous distribution (determination of homogenization number $c_H$), shear force (model particle system) and oxygen supply was performed. In summary the results depicted no problems for scale-up to 200 L. Homogenization and aeration (pure oxygen, $c_L$= 40% $pO_2$=0.21 bar) were sufficient for 5g/L microcarrier and X = 6 e06/ml. Determined reference flock diameter was comparable for all reactors, so that shear force is expected to be comparable in all scales (Fig. 3).

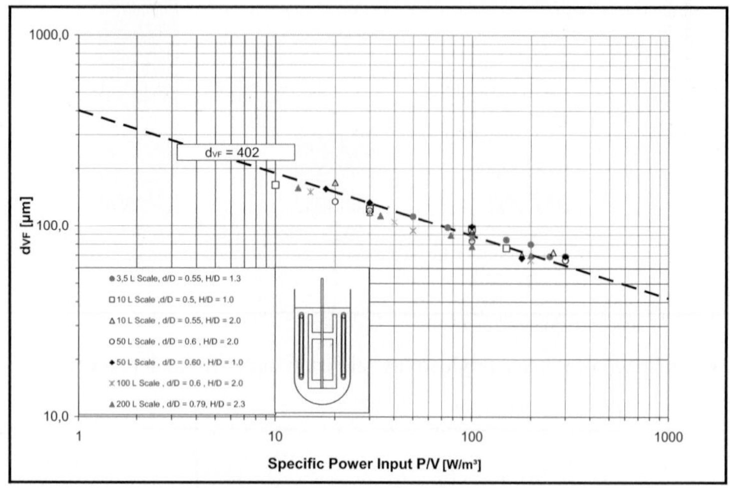


*Figure 3.* Comparison of determined reference flock diameter dVF in different scales.

## 3. DISCUSSION AND SUMMARY

Within this work the cell line was adapted to animal, human and herbal compound free in-house medium without loss of performance leading to a significant increase of biological safety (prions, viral contamination, etc.) and a decrease of production cost. In parallel analysis of biphasic batch process and description of the cell line resulted in promising changes of operation mode. The physical description of the bioreactors illustrated no problems for scale-up to 200 L. Conducted test runs in pilot scale (50 L) including direct expansion of the cell line (3.5 L to 10 L to 50 L) confirm results.

This integrated approach led to a new safe, robust and high productive large scale production process, called "Volume-Expanded-Fed" batch. Based on the results a possible large scale production process up to 800 L was designed in theory.

# Proteomic Investigation of Recombinant CHO Cells in High Density Culture

J.R. Cresswell, J.M. Zobrist, J. Wildsmith, N. Lin, R. Valdes-Camin, and M. Caple
*Sigma-Aldrich Biotechnology, 2909 Laclede Ave., St. Louis, MO 63103. U.S.A.*

**Abstract:** Bioreactor optimization is a complex task, involving both media and process development. This task is further complicated by variations in recombinant CHO cell lines. The intracellular effects of different media or reactor conditions are largely undetermined. Study of intracellular responses could provide information to improve recombinant cell performance in two ways. First, accumulated data could help predict requirements of specific cell lines, increasing the efficiency of media and process optimization. Also, such investigation may identify targets of cell engineering to increase performance of recombinant lines. A proteomics approach was used to study the protein expression changes in a recombinant IgG-producing CHO cell line throughout a fed-batch culture. Growth, productivity and metabolism were monitored, and protein expression profiles of growth and production phases were analyzed by 2-dimensional gel electrophoresis. Differentially expressed proteins were selected using Phoretix™ 2D Expression software and identified using in-gel tryptic digest and MALDI-MS. This study provides insight into the intracellular processes affecting growth and productivity of recombinant CHO cell lines.

**Key words:** 2-D electrophoresis, proteomics, CHO, bioreactor

## 1. MATERIALS AND METHODS

Stock culture of a recombinant IgG producing cell line was maintained in animal component free suspension culture under G418 selection. A 5L bioreactor was seeded at 7.5e4 cells/mL and maintained as fed-batch culture until viability dropped below 70%. The culture was sampled daily and



*R. Smith (ed.), Cell Technology for Cell Products*, 711–713.

monitored for growth, viability, IgG productivity, and metabolic profile. Culture was supplemented with glucose (to 3g/L), glutamine (to 4mM), and a concentrated CHO feed (C1415, every other day).

Cell pellets (5 x $10^7$ total cells) were extracted in 1 mL of Cellular and Organelle Membrane Solubilizing Reagent (C0356) to which 10 mM spermine and 5 mM TCEP had been added. The extractions were incubated for 1 hour with mixing at room temperature, then centrifuged for 30 minutes at 16,000 rcf. Samples were reduced with tributylphosphine and alkylated with iodoacetamide. Total protein concentration was determined by Bradford assay. Aliquoted samples were TCA precipitated then re-suspended in the extraction solution. IPG strips (11cm, pH 3-10) were rehydrated with 250 µg samples and were focused at 8,000 Volts for 85,000 Volt hours. After equilibration for 20 minutes with IPG equilibration buffer (I7281), SDS-PAGE was performed using 4-20% Tris-HCl precast gels. Protein spots were visualized in the gels using EZBlue gel stain (G1041) and images were analyzed using Phoretix™ 2D Expression Software (Nonlinear Dynamics). In-gel digestion of protein spots was performed using Trypsin Profile IGD Kit (PP0100). The tryptic digests were dried and the samples resuspended in a solution of 10 mg/mL α-cyano-4-hydroxycinnamic acid in 70% acetonitrile, 0.03% trifluoroacetic acid. Matrix assisted laser desorption ionization mass spectrometry (MALDI-MS) data was acquired in positive ion reflectron mode using an Axima-CFR+ mass spectrometer (Shimadzu Biotech).

## 2. RESULTS

The bioreactor culture reached peak cell density of 8.0 x $10^6$ cells/mL on day 10. Percent of viable cells remained above 90% through day 14. Culture was ended on day 18 when viability dropped below 70%. Recombinant IgG production increased rapidly from day 6 through day 12. Once productivity reached approximately 200mg/L, the rate of production decreased sharply, even though the culture maintained stationary phase for several more days. Days 7, 9, 12 and 14 of culture were chosen as time points for 2-D analysis as these represented beginning, middle and end of recombinant protein production for this culture.

Stained gels were analyzed using the Phoretix™ 2D Expression Software. Protein spots with normalized volume change of 1.8 fold or greater were identified as up or down regulated. Approximately forty spots were chosen as potential candidates for identification and further analysis. The expression profiles for three protein spots are shown in Figure 1.

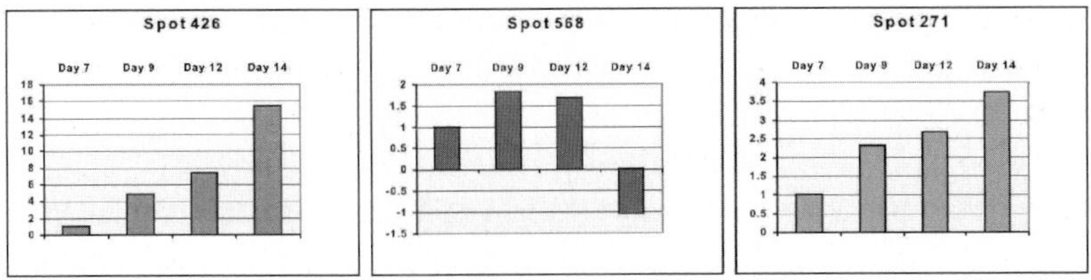


*Figure 1.* Expression profiles for three protein spots normalized to day 7 to demonstrate the relative change over the course of the culture.

Gel spots of interest were excised from the 2-DE gels and digested with trypsin. The digested samples were analyzed by MALDI-MS to generate peptide mass fingerprints. Protein identifications for three of these samples are shown in Table 1.

*Table 1.* Peptide mass fingerprints from MALDI-MS analysis were submitted to MASCOT database for protein identification ($p<0.05$).

| Spot Number | Protein Identification | Mascot Score | Sequence Coverage |
|---|---|---|---|
| 426 | Cofilin | 71 | 77% |
| 568 | ERP57 | 79 | 49% |
| 271 | Aldose Reductase | 103 | 43% |

# 3. CONCLUSIONS

Proteomic analysis was successfully utilized to study changes in protein expression during a recombinant CHO bioreactor culture. Three proteins with altered expression levels were identified. Cofilin, an ADF family protein that depolymerizes actin filaments, was upregulated during the culture. Another upregulated protein, aldose reductase, converts glucose to sorbitol in the polyol pathway. An endoplasmic reticulum chaperone ERP57 (GRP58) was upregulated during log phase and early stationary growth phases and then downregulated late in culture. The decreased expression of this protein coincides with decreased IgG production rates and decreased viability. MALDI-MS analysis is pending on additional protein spots.

Study of protein expression profiles in this cell line has identified proteins that may be important to support production of recombinant protein.

# Comparative Proteomic Study of Mechanisms Involved in the Expression of Recombinant Monoclonal Antibody in Non Amplified NS0 Myeloma Cell Line

Luis E. Rojas[1], Kathya R. de la Luz Hernández[1], Svieta Victores[1], Lila Castellanos[2], Simon Gaskell[3], Adolfo Castillo[1] and Rolando Pérez[1]

*[1]Research and Development Division. Center of Molecular Immunology. Havana, Cuba, [2]Physic-Chemistry Division. Center of Genetic Engineering and Biotechnology. Havana. Cuba [3]Michael Barber Center for Mass Spectrometry. UMIST. Manchester. United Kingdom.*

**Abstract:** In the present work we have carried out a comparative proteomic study of protein expression patterns by two-dimensional gel electrophoresis (2DE) in the pH range of 3-10 between the host cell line NS0 and cell lines that produce different recombinant monoclonal antibodies (rMabs). We have studied the adapted and non adapted to protein free medium (PFM) versions of cell lines producing two different rMabs that were also humanized using different methods and vector constructions. We have observed a pattern with a preferential over expression of spots in the host cell lines (adapted and non adapted to PFM) compared to producer cell lines. Protein spots detected with a statistical different expression levels were identified using MALDI TOF mass spectrometry (MS). Most of the spot with statistically differential expression levels were down regulated in the group of producer cell lines. The higher amount of upregulated protein that have been characterized in NS0 host cell line are mainly related with cell proliferation, protein synthesis and carbohydrate metabolism.

**Key words:** Two-dimensional electrophoresis; mass spectrometry; NS0 myeloma cell line.

## 1. INTRODUCTION

The identification of factors affect the expression levels in recombinant cell lines is a topic of major interest for production process development based on mammalian cell cultures (Barnes. L, 2001; Kim y cols. 1998B).



*R. Smith (ed.), Cell Technology for Cell Products*, 715–721.

Possible changes in the protein expression of the immunoglobulin expression from those associated with the expression of selection genes and phenotypic variability induced by transfection and cloning procedures maybe could be implicate in the level of expression of interesting protein. The two-dimensional electrophoresis technique is the powerful procedure that allows studied the differential present in the proteome of cell in different condition (Huber L. A, 2003; Graves P, Haystead, A.J.T, 2002), and facilitated the posterior analysis in the mass spectrometric technique.

## 2. MATERIAL AND METHODS

### 2.1

Cell line: NS0 host. NS0rh: Producer rMabs with selection marker Hyg in $C_L$ and gpt in $C_H$. NS0rt: Producer rMabs with selection marker in gpt $C_L$ and histidinol in $C_H$ (Mateo C, 1997; Coloma MJ, 1992). All of them are performed in adapted and non adapted at serum free medium respectively.

### 2.2

Two dimensional electrophoresis: Two-dimensional polyacrilamide (15%) gel carried out in electrophoresis chamber (Biorad, USA). 17 cm strip in the pH range 3-10 (Biorad, USA) were run in IEF Cell (Biorad, USA). Gels were analysed using Melanie 5 software. Spots were characterized using Q-TOF mass spectrometer.

## 3. RESULTS

With the aim to analyze the proteome related with immunoglobulin expression, different groups of cell lines were created considering firstly if produce rMab and the type of rMab produced in the medium that they grow. Each group was compared from the detection of spot with differential percent of volume by Student test. Three groups are considered: Producers vs Non producers, specific NS0rh producer vs Non producers and specific NS0rt producer vs Non producers (see Figure 1). Three gels by each condition for each cell line were evaluated. The Figure 2 shows typical two-dimensional gel.

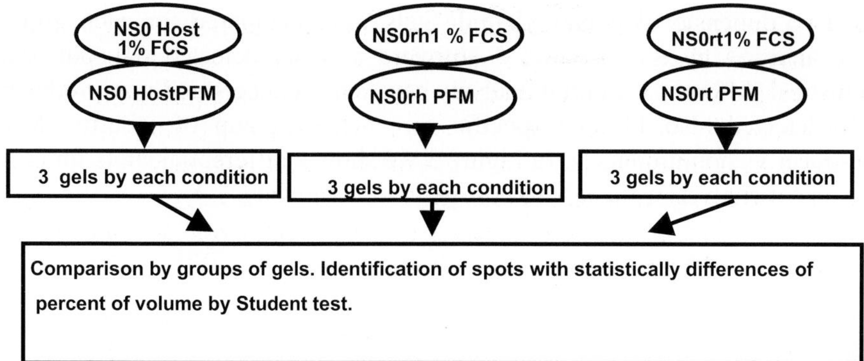


*Figure 1.* Protocol for comparison between NS0 host and recombinant cell lines, that produce two different humanised Mabs in 1%FCS and PFM medium.

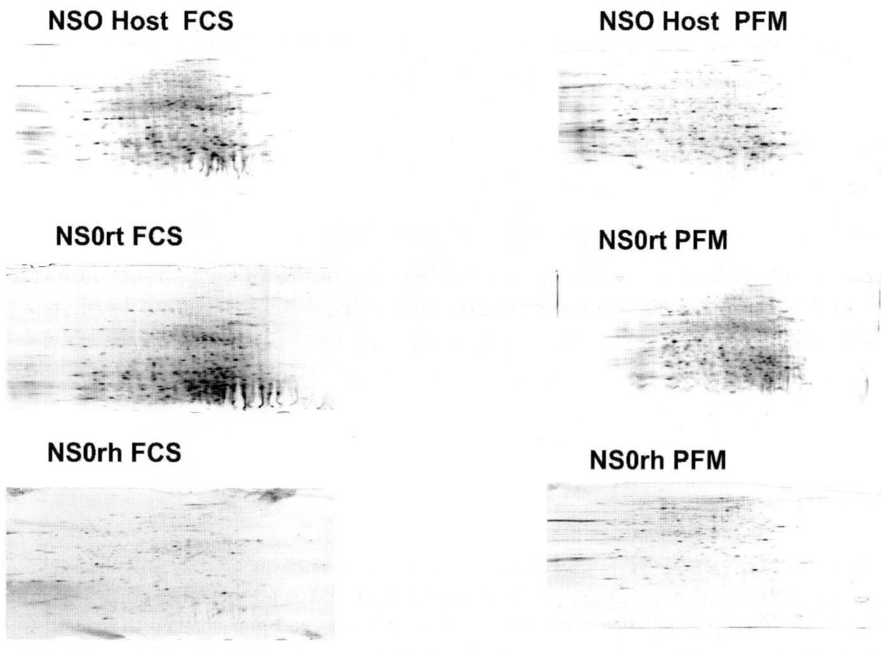


*Figure 2.* Two-dimensional polyacrylamide (15%) gels of producer cell lines (NS0rh and NS0rt adapted and non-adapted to protein free medium) and non producer cell lines (NS0 adapted and non adapted). 17 cm strips in the pH range 3-10 (BioRad, USA) were run in IEF Cell (Biorad, USA).

Two-dimensional polyacrylamide gels were compared for each group. The analysis by the Melanie 5 software allowed detected 61 spot with statistical differential in total analysis between producers and non producer. We detected also 11 spot specific for each of group of specific rMab producer vs non producers. In Figure 3 we show a differential spots for each groups of analysis using a NS0 host gel with references.

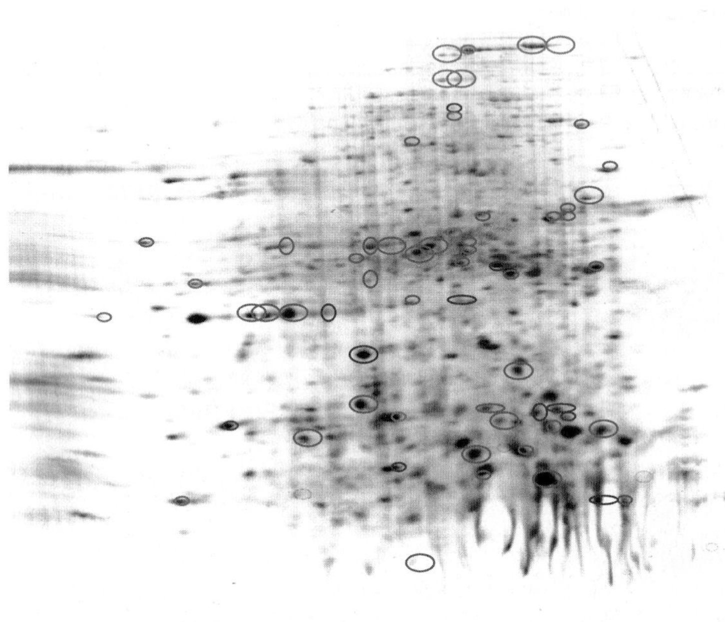

*Figure 3* One 2DE gel from non adapted NS0 host cell line was used as reference gel and spots with differential expression levels. Gels were analyzed using Melanie 5 software. RED CIRCLES: Producer cell lines (NS0rh and NS0rt adapted and non adapted to protein free medium) Vs. non producer cell lines (NS0 adapted and non adapted), BLUE CIRCLES: Specific spots in NSOrh cell line (adapted and non adapted) Vs. non producer cell lines (NS0 adapted and non adapted), PINK CIRCLES: Specific spots in NSOrt cell line (adapted and non adapted) Vs. non producer cell lines (NS0 adapted and non adapted).

The Figure 4 shows example of spot over expressed in non producer cell lines were compared producers and non producers groups.

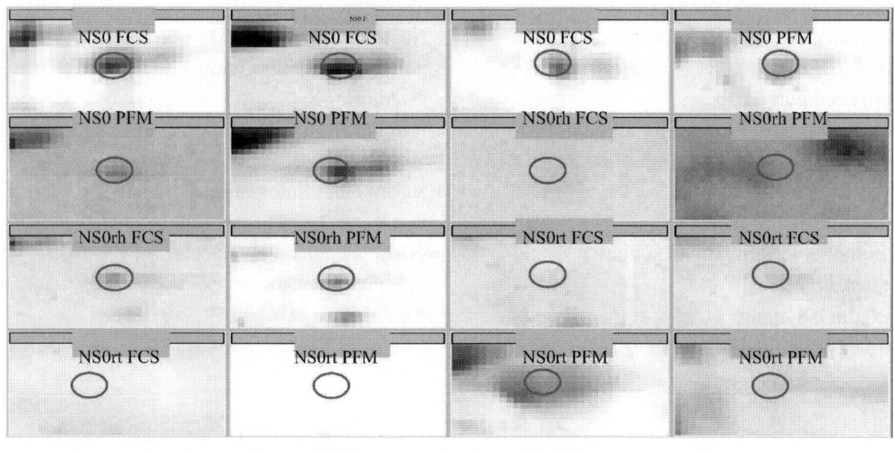

*Figure 4* Example of spot overexpressed in non producer cell lines (both adapted and non adapted to PFM) to respect to producer NS0rh and NS0ht (adapted and non adapted to PFM).

Up to now, we have identified by MALDI-TOF and Q-TOF mass spectrometry analysis a percent of statistical differential spot for each group, 36% of total between producers and non producers, 18% of specific NS0rh producers vs non producer and 55% of specific NS0rt producers vs non producers (See Figure 5). Table 1 shows the relation of identified protein. Figure 6 shows an example of analysis of spot by Q-TOF mass spectrometry.

*Table 1.* identified protein by MS.

| | |
|---|---|
| 78 KDa glucose-regulate protein precursor | D-3 phosphoglycerate dehydrogenase |
| Actin cytoplasmic 1 | Elongation factor 1 -1α |
| Aldolasa reductasa | Elongation factor 1- 2α |
| ATP synthase α chain mitochondrial precursor | ATP synthase β chain mitochondrial precursor |
| Dynein light chain 2 cytoplasmic | Eucaryotic initiacion factor 4A-I |
| Gamma enolasa | α enolasa |
| Glyceraldehyde-3-phosphate dehydrogenase | Heat shock protein 75KDa mitochondrial precursor |
| Heat shock cognate 71KDa protein | Heat shock protein 90α |
| Heterogeneous nuclear ribonucleoprotein L | Proliferation associated nuclear element protein 1 |
| Heat shock protein HSP 90β | Histone H2A |
| IgE-binding protein | Histone H2B F |

| Inhibitor of growth protein 4 | Muskelin |
|---|---|
| Nuclear pore complex protein Nup 153 | Nucleolin(protein C23) |
| Peroxiredoxin 1 | Poly(rC) binding protein 1 |
| Peroxiredoxin 4 | Prohibitin |
| Potassium chanel subfamily K member 15 | Protein-disulphide isomerase A3 precursor |
| Pyruvato kinasa | Serum albumin precursor |
| Stomatin-like protein 2 | Syndecan-3 precursor |
| T complex Protein 1(zeta subunit) | Triosephosphato isomerasa |
| Transketolasa | Tubulin α 2 chain |
| Tubulin β 1 chain | α actin |
| Tumor protein p53-inducible nuclear protein 2 | Ubiquitin-like protein SMT3A |

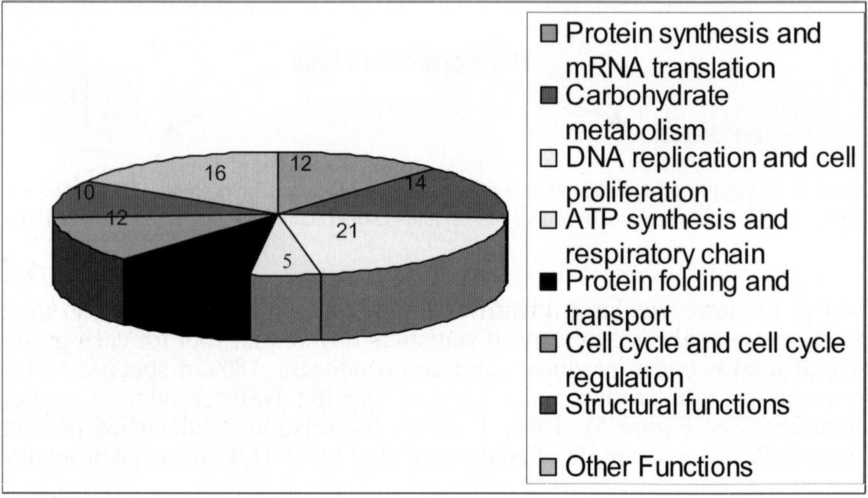


*Figure 5.* Percent of identified protein by specific function.

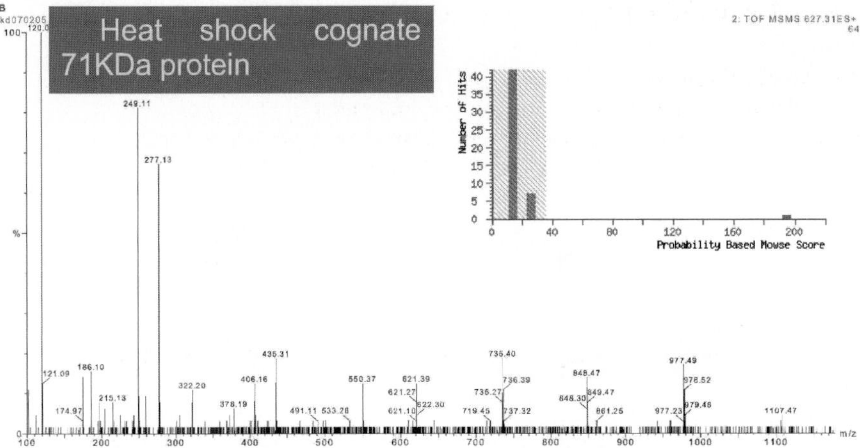


*Figure 6.* Excised spots have been characterized using Q-TOF mass spectrometer.

# 4 CONCLUSIONS

Most of the spots with statistically differential expression levels were down regulated in the group of producer cell lines (59 spots) when compared with upregulated (2) spots. The groups of spots observed specific for NS0rh or for NS0rt respectively could be related with the selection system employed in both cell lines. The higher amount of upregulated proteins in NS0 host cell line that have been characterized up to now are mainly related with cell proliferation, protein synthesis and carbohydrate metabolism.

# REFERENCES

Barnes L. (2001). Characterization of the stability of recombinant protein production in the GS-NS0 expression system. Biotechnology Bioeng. May 20; 73(4):261-70.

Coloma MJ, Hasting A, Wims LA, Morrison SL. Novel vector for the expression of antibody molecules using variable regions generate by polymerize chain reaction. J Immunol Method 152: 89-104, 1992.

Graves P., Haystead A.J.T, 2002 Molecular Biologist Guide to Proteomics. Microbiology and Molecular Biology Reviews, Mar. 39–63.

Huber L.A, 2003. Is Proteomics Heading In The Wrong Direction? Nature Reviews Molecular Cell Biology 4:74-80

Kim S. (1998.B) Clonal variability whitin dihidrofolate reductase-mediated gene amplified Chinese Hamster Ovary Cell: Stability in the absence of selective pressure. Biotechnol. Bioeng. 90: 679-688.

Mateo C, Moreno E, Amour K, Lombardero J, Harris W, Pérez R. Humanization of mouse monoclonal antibody that block the EGF-R: recovery of antagonistic activity. Immunotechnology 3(1): 71-81, 1997.

# Monitoring Nutrient Limitations by Online Capacitance Measurements in Batch & Fed-batch CHO Fermentations

Sven Ansorge[1] Geoffrey Esteban[2] Charles Ghommidh[3] Georg Schmid[1]
*[1]F. Hoffmann-La Roche AG, Pharmaceuticals Division, Pharma Research Basel, Protein Sciences, Basel, Switzerland [2]FOGALE nanotech, Nîmes, France [3]University of Montpellier, Montpellier, France*

**Abstract:** Capacitance measurements can give valuable information on the physiological state of a culture. Sequential nutrient limitations for batch and fed-batch CHO fermentations were monitored online.

**Key words:** on-line biomass determination, capacitance, nutrient limitations

## 1. INTRODUCTION

The viable, i.e metabolically active, cell density is one of the most important parameters to monitor in any cell culture process. In development and production it can be useful e.g. for the timing of cell transfers in the inoculum train or from inoculum to production reactors and for the timing of feed additions in fed-batch processes. The online measurement of culture turbidity, fluorescence, or capacitance properties have been used to estimate the biomass concentration (1-5). Currently available and often employed optical density probes have limitations when cell viabilities decrease as they measure culture turbidity rather than viable cell density (2).

On the other hand, capacitance measurements have been described as to belong to the best methods available although the statement that "honestly, there is no ideal in situ on-line biomass sensor available today" still seems to be valid (5,6).

In previous experiments we performed fermentations with a CHO host cell line using the Fogale Biomass System® to monitor the capacitance or



*R. Smith (ed.), Cell Technology for Cell Products*, 723–726.

permittivity online. We obtained a linear relationship of viable cell count and capacitance for the complete time course of the CHO host cell fermentation (similar to Figure 1) whereas the turbidity signal also measured dead cells and debris.

In recent experiments we extended our evaluation of the usefulness of the permittivity signal for online monitoring. It was found to be influenced by the physiological state, e.g. nutrient limitations of a culture. Several batch and fed-batch fermentations were performed at different scales in fully instrumented bioreactors using the Fogale Biomass System® for online permittivity measurements.

## 2. MATERIALS AND METHODS

Batch and fed-batch fermentations were carried out in 2 L bioreactor (Biostat MCD, Braun) equipped with a 12 mm Fogale Biomass System® (Fogale nanotech, Nimes, France). Cultivation conditions for the CHO dhfr$^-$ host cell line were 100 rpm, 37°C, pH 7.2 using a proprietory serum-free medium.

## 3. RESULTS AND DISCUSSION

Figure 1 demonstrates how the on-line permittivity signal can be used to monitor viable cell density during a batch culture of CHO host cells. The permittivity signal is providing a linear correlation with the viable cell count over the complete timecourse of the culture. It is obvious from Figure 2 that changes in cell metabolism can be monitored by means of the permittivity

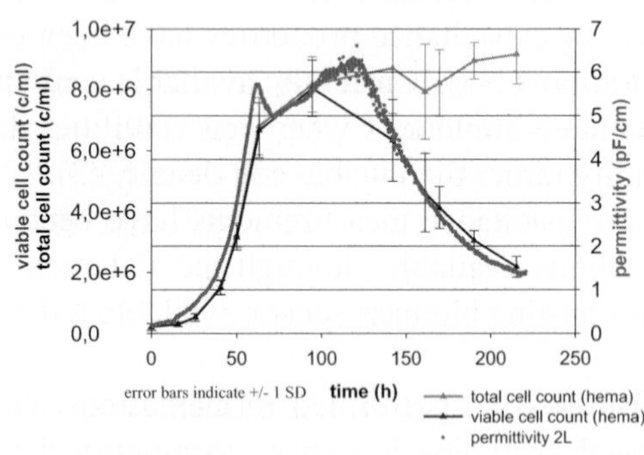


*Figure 1.* Permittivity, viable and total cell count for a batch fermentation of CHO host cells.

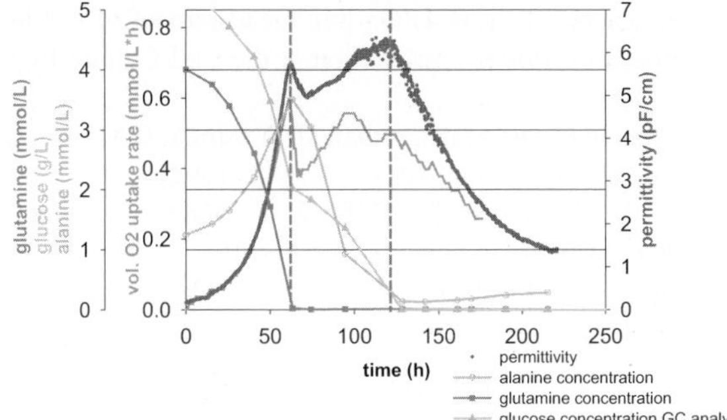


*Figure 2.* Permittivity, vol. O2 uptake rate and main nutrient/byproduct concentrations for a batch fermentation of CHO host cells.

measurement. The first drop in permittivity corresponds to a metabolic shift when glutamine becomes a limiting nutrient; the oxygen uptake rate also decreases significantly. Alanine is accumulating until glutamine is depleted and is subsequently consumed. The second drop in permittivity coincides with the exhaustion of alanine and glucose that marks the beginning of the death phase. Fed-batch fermentations gave similar results (data not shown).

## 4. CONCLUSIONS

Viable cell counts and the permittivity signal correlate well during the growth and death phases of a batch culture for CHO host cells. Nutrient limitations for batch and fed-batch CHO fermentations can be monitored online by capacitance measurements.

## ACKNOWLEDGEMENTS

During his Diploma thesis Sven Ansorge was supported by a fellowship from Hoffmann-La Roche. We thank M. Foggetta, J.-M. Vonach, P. Stohler, H. Kurt, B. Wipf and D. Zacher for their various contributions to this work.

## REFERENCES

1. P. Wu, S.S. Ozturk, J.D. Blackie, J.C. Thrift, C. Figueroa, and D. Naveh, Biotechnology and Bioengineering, 45, 495-502 (1995).
2. R. Beri, J. Wayte, J. Swift, S. Abraham, and M. Brown, ACS Meeting, Boston, Sept. 1998.

3. Schmid, G., Ansorge, S., Zacher, D. (2004) On-line measurement of viable cell density in animal cell culture processes. Poster presentation at the Cell Culture Engineering IX, Cancun, Mexico
4. K.B. Konstantinov, R. Pambayun, R. Matanguihan, T. Yoshida, C.M. Perusich, and W.-S. Hu, Biotechnology and Bioengineering, 40, 1337-1342 (1992).
5. I. Cerckel, A. Garcia, V. Degouys, D. Dubois, L. Fabry, and A.O.A. Miller, Cytotechnology, 13, 185-193 (1993).
6. Sonnleitner, B., Locher, G., Fiechter, A. Biomass determination. Journal of Biotechnology 25, 5-22 (1992).

# Characterisation of Shake Flasks for Cultivation of Animal Cell Cultures

G. Jänicke, C. Sauter, R. Bux, J. Haas
*Boehringer Ingelheim Pharma GmbH & CO. KG, Process Science / Upstream Development, Birkendorfer Str. 65, D 88397 Biberach/Riss*

**Abstract:** This study investigated the oxygen transfer processes and general correlations between culture performance and the operating conditions for shake flask fermentations of animal cell cultures. This involved both the online measurement of the oxygen transfer rate and the continuous recording of measurements for the dissolved oxygen concentration in the shake flask.

**Key words:** shake flask, mass transfer, gas-liquid, oxygen, power input, oxygen transfer rate, sulphite system, cell culture, CHO cells, RAMOS, PreSens, dissolved oxygen, mammalian cells, bioreactor, mass transfer coefficient

## 1. INTRODUCTION

Shake flasks have increasingly been used in recent years in cell culture technology. This system is easy to manage and has proved increasingly effective in terms of the use of different raw materials, tests for limitations in respect of differing substances and the effect of differing carbon dioxide contents on cell culture performance. The growing importance of this system in cell culture technology can also be attributed to the possibility of running parallel test batches in a confined space. To obtain a better understanding of the oxygen transfer processes in cell culture fermentations in shake flasks, we investigated the dependence of the oxygen transfer rate (OTR) on the shake conditions (shaker speed, shake diameter, flask size and filling volume) by the sulphite method using the RAMOS (Respiration Activity MOnitoring System) in an aqueous system. The results obtained were applied to the cell culture processes with the aim of enabling fermentation



*R. Smith (ed.), Cell Technology for Cell Products*, 727–731.

processes in shake flasks to be interpreted in respect of oxygen supply and of determining a correlation factor for the OTR between both systems.

## 2. METHODS AND MATERIALS

The RAMOS (Respiration Activity Monitoring System), jointly developed by Anderlei and Büchs [1], is used to measure the respiratory activity of a cell culture directly and online under sterile conditions, including the oxygen transfer rate OTR, carbon dioxide transfer rate CTR and the resulting respiration quotient RQ.

The correlation between the oxygen transfer rate and the operating parameters was investigated using a chemical oxygen consumer (1 M sulphite system) in 250 ml, 500 ml and 1000 ml flasks. A recombinant CHO cell line was used to perform the fermentations. The dissolved oxygen concentration was measured on the basis of luminescence quenching using the multichannel oxygen meter OXY-4 from PreSens.

## 3. RESULTS

The investigations of the oxygen transfers process in the shake flask by the sulphite method showed distinct correlations for the operating parameters (Fig. 1). Thus, an increase in the filling volumes of the flasks and larger flask diameters had a negative effect on the oxygen transfer rate, while an increase in the shake frequency and shake diameter had a positive effect on the OTR. Using the developed correlation formula, the OTR and thus the volumetric mass transfer coefficient $k_La$, could be calculated prospectively for known operating parameters.

Since it was not possible to measure power input with the existing measuring device, this was calculated theoretically using a correlation equation published in the literature and applicable to this operating range. This showed that a power input can be achieved in the shake flask (10 – 350 $W/m^3$) that easily exceeds, in some cases, the operating range for stirred bioreactors operating with animal cell cultures (approx. 10 - 40 $W/m^3$) (data not illustrated).

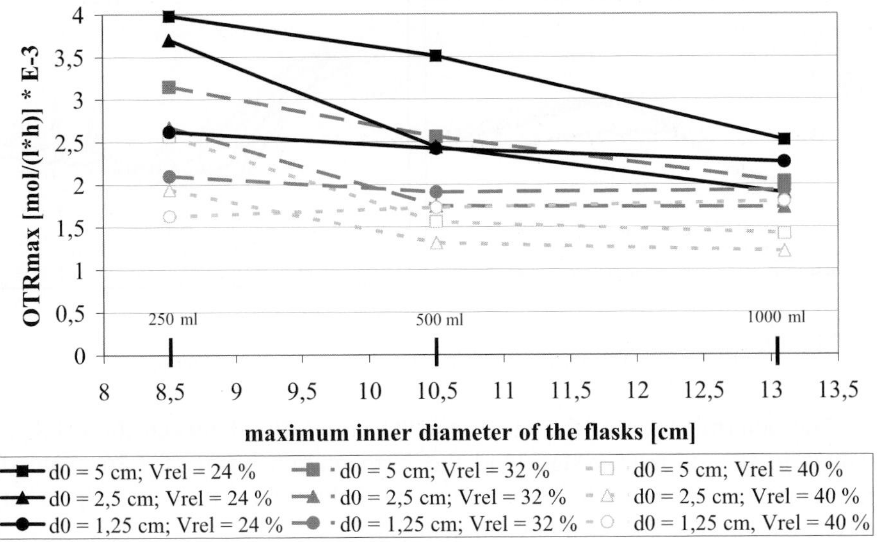


*Figure 1.* [Dependence of the OTR].

In addition to the measurement of the oxygen transfer rate by RAMOS during the cell culture fermentations (Fig. 2), the dissolved oxygen concentration DO was also measured in order to obtain more detailed information about the respiratory activity of the cells and enable the maximum oxygen transfer rate, and thus the $k_La$, to be calculated. With a shake diameter of 2.5 cm and a speed of 120 1/min, a maximum oxygen transfer rate of 1.95 * $10^{-3}$ mol/(l*h) was calculated in the 250 ml flask with a 40% filling volume. Measurement of the dissolved oxygen concentration showed that, even on the second day of fermentation, a sampling time of 10 – 15 min during which the flasks were not gassed was sufficient for producing oxygen limitation in the culture. In the flask with 60% filling volume, the culture was in a state of oxygen limitation from the third to the sixth day.

Compared to non-oxygen-limited fermentations, this resulted in differences - marked differences in some cases - in culture performancein respect of cell concentration, viability and product concentration (Figs. 4, 5, 6).

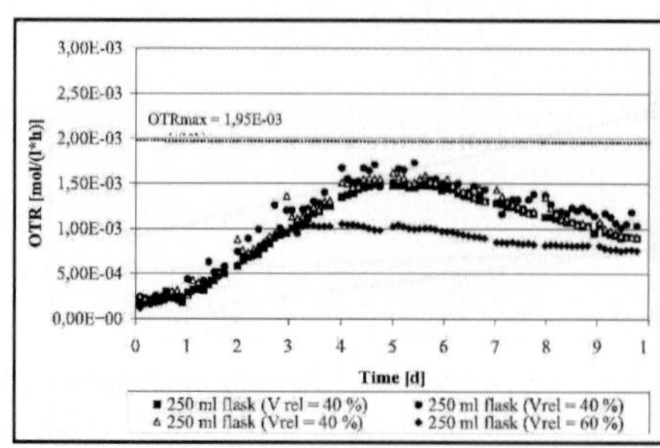


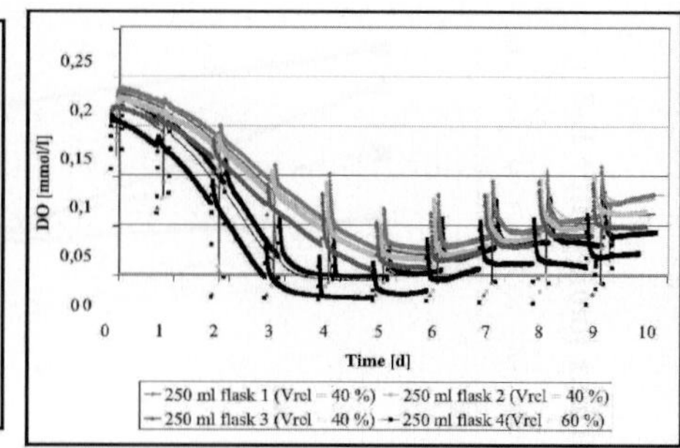


*Figure 2.* [OTR max and DO in cell culture during fermentation].

The determination of the factor for the correlation between the $OTR_{max}$ values in the sulphite system and the fermentation broth of 1.36 enabled fermentation processes to be interpreted in the shake flask in respect of oxygen supply for a known oxygen demand.

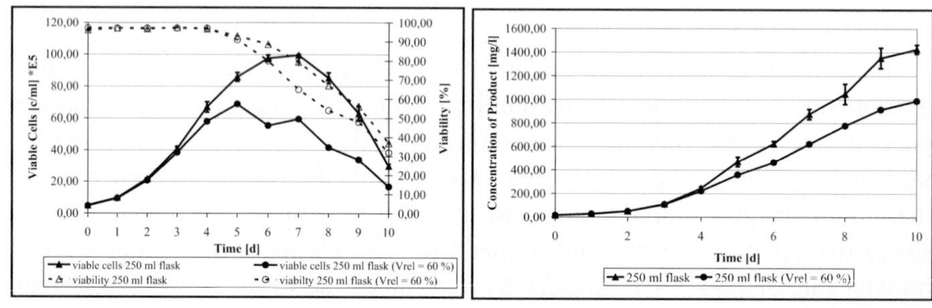


*Figure 3.* [Cell culture performance].

# 4. SUMMARY

The gas-liquid mass transfer was analysed by the sulphite method. As expected, increasing the shake frequency and the shake diameter produced higher OTR's and thus better oxygen transfer. Increasing the flask diameter and filling volume had a negative effect on the gas-liquid transfer. Furthermore, a factor of 1.36 was determined for converting the $OTR_{max}$ sulphite system to the $OTR_{max}$ cell culture.

The effect of the oxygen availability in the culture medium on the cells and their production of the target protein could be demonstrated simply by combining the OTR measurement and the DO measurement.

Key results:

- power input in the shake flask comparable with, or higher than, that in the stirred bioreactor at Boehringer Ingelheim
- mass transfer in the shake flask comparable with stirred bioreactor (depending on the size and operating parameters)
- the operating parameters for the shake flask fermentations have a crucial influence on the oxygenation of the medium and thus on culture performance
- the operating conditions in shake flasks should be chosen carefully in order to rule out possible artefacts in cell culture behaviour resulting from oxygen limitation.

## REFERENCE

1. Anderlei, T., Büchs, J. (2001), Device for sterile online measurement of the oxygen transfer rate in shaking flasks. Biochemical Engineering Journal 7, 157-162.

# Manufacturing of Lots of Human T-Lymphocytes for Cell Therapy

*A New Process Based on Disposable Bioreactors*

Nadia De Bernardi, Silvia Trasciatti[1], Antonio Orlandi, Fausto Gaspari, Ilaria Giuntini, Tiziana Bacchi, Sonia Castiglioni, Elena Muru, Marta Galagano, Claudia De Mattei, Luigi Cavenaghi, Maria Luisa Nolli

*Areta International S.r.l., Via R. Lepetit 34, 21040 Gerenzano (VA), Italy, [1]Abiogen Pharma Research Center, Via del Paradiso 6, Migliarino Pisano (PI), Italy*

**Abstract:** About 10 years ago nobody would have believed that cell therapy would become one of the future promises of therapy, and human cells one of the hopes for the cure of emerging diseases not treatable with conventional drugs. The advent and potentiality of stem cells and other types of cells as drugs is revolutionizing the production and regulatory fields. Producer and regulators alike have to be prepared to face with this new concept of therapeutics in terms of cell expansion related to the lot definition, bioreactor used, safety and QC tests for lot release. Here we describe a new process for the GMP manufacturing of human T lymphocytes not HLA restricted, for Phase I and Phase II clinical trials of different types of tumors. The process is based on the use of the modular and disposable bioreactor Cell Factory and was conceived to maximize the potentiality of the bioreactor related to the physiology of the cells. All the steps of the process from MCB thawing to the bags filling are discussed.

**Key words:** T-lymphocytes, GMP, disposable bioreactors, cell therapy, stem cells, cell expansion, cell factory, suspension cell line

## 1. INTRODUCTION

The TALL-104 cells are a human cell line, IL-2 dependent, of T cytotoxic lymphocytes not restricted by the Major Histocompatibility Complex (MHC) with a potent tumoricidal effect. The cells originally isolated from the peripheral blood of a child, affected from a Acute



*R. Smith (ed.), Cell Technology for Cell Products*, 733–737.

Lymphoblastoid Leukemia, can be expanded in vitro and maintained in culture in the presence of IL 2 (1). The TALL-104 were demonstrated, by the group of D. Santoli to be able to control the tumoral growth and to increase the survival in different animal models with spontaneous or induced malignant tumors (2).

Their MHC non restricted killer activity can overcome some limitations associated with the transfer of a patient's own effector cells and can not require the concomitant administration of exogenous toxic cytokines, such as high doses of rhIL-2. The potential broad range of use of these cells needs the set up and implementation of industrial method to cultivate in large scale the TALL to be able to produce therapeutic doses of the cells in GMP conditions. We describe here new process for the GMP manufacturing of human TALL cells, for Phase I and Phase II clinical trials of different types of tumors. The process is based on the use of the modular and disposable bioreactor Cell Factory and was conceived to maximize the potentiality of the bioreactor related to the physiology of the cells. All the steps of the process from MCB thawing to the bags filling are discussed.

TALL cells after one month in culture are shown in Figure 1, they show the NK phenotype, necessary for their cytotoxic activity.

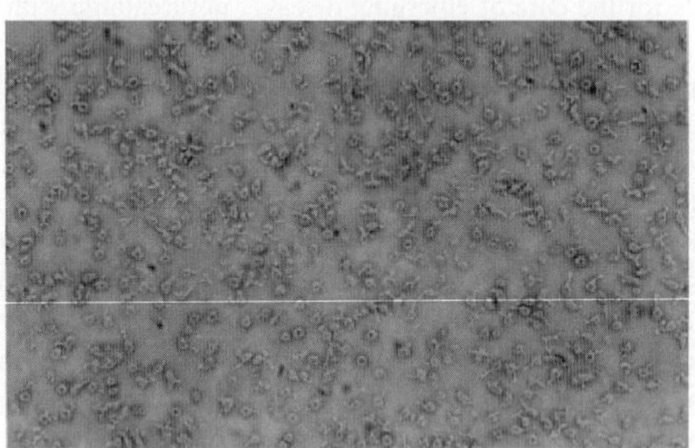

*Figure 1.* TALL cells after one month in culture.

## 2. PROCESS DEVELOPMENT AND MANUFACTURING OF LOTS OF TALL CELLS

All the operations, including the process development, were carried out in the high containment cell culture plant of Areta International that operates in respect to ISO 9001 and GMP certification. The plant was authorized in 2004 by the Italian Ministry of Health to produce lots of human cells for cell therapy for phase I/II clinical trials.

### 2.1 Set Up of Cell Culture Process

Cells after thawing were initially cultivated in flasks in FCS medium supplemented by IL-2. After 15 days the FCS was substituted with AB human serum.

Since the cells are very sensitive to cell density a kinetics of growth in flask was performed in order to study the glucose consumption and lactate production during cell growth.

Then we tried to cultivate the cells in different types of disposable bioreactors to find the more appropriate for the large scale up: the miniPERM, the Spinner Flask, and the Cell Factory. The parameters monitored were: viability, expansion, glucose consumption. Both the miniPERM and the spinner flask resulted not appropriate for the scale up of Tall cells, while in the Cell Factory, originally designed for anchorage dependent cells, TALL had a good expansion and were able to maintain the viability of 1 x $10^6$/ ml of culture.

### 2.2 Bioreactor and Mode of Culture

Cell Factory (CF, Figure 2) is a high surface volume ratio bioreactor, sterile, mono use, easy to use, modular, designed for anchorage dependent cells. We have demonstrated that the CF can be successfully used for scale up production of the human TALL suspension cell line (3).

The modularity of this bioreactor is ideal for the GMP production of cells for cell therapy. Its use avoids cross contamination, cleaning validation. Either the 10 trays, the 40 trays or their multiples may be used for tailored made supply of lots for Phase I/II clinical trials.

*Figure 2.* Cell Factories.

### 2.3 Gmp Production of Lots of Tall Cells

The industrial production of TALL cells means that you must have a system able to guarantee you expansions of cultures of at least $10^9$ cells in an homogeneous system constituted of 1 module of 10 trays-CF. The recovery from each 10 trays-Cell Factory after 8 days from the inoculum is from 2.5 to 5 x $10^9$ TALL cells. 1.5 x $10^{11}$ cells were produced in 72 CF10.

Switching to the 40 trays CF there is the possibility to recover 2-4 x $10^{10}$ TALL cells from one unit after 8 days. According to this method, because all the CF have the same distribution and aeration pattern, the number of cells needed can be obtained through an appropriate number of CF. This is of great advantage both in terms of less manipulation and quality of the final product.

After thawing of a single vial of the MCB, TALL cells were expanded initially in flasks and when the NK phenotype started to appear (approximately after one month in culture) they were moved to the CF (10 trays module) for the production of the final quantity requested.

The parameters monitored during the production were the phenotype, viability and sterility.

Once recovered from the bioreactor, the TALL cells were irradiated using a non radioactive beta-particle accelerator (Betatrone).

After irradiation TALL cells were prepared for filling. One sample was taken for the Quality Control and Safety tests.

### 2.4 Filling and Media Fill

TALL cells were resuspended in the appropriate freezing medium and the bags filled with respectively 1 x $10^8$, 5 x $10^8$, 2.5 x $10^9$ cells under the laminar flow hood in class B sterile room. The cycle Cell Factory-Irradiation-Bags is repeated until reaching the number of cells needed.

The filling in aseptic conditions was validated by the filling of bacteria culture medium in the appropriate number of bags.

### 2.5 Analytical Tests

The set up of a quantitative test to follow the production and the quality of the final product is a fundamental prerequisite to generate reliable data. Any delay in producing dedicated quantitative assays at this level can produce terrific delays in the project.

Regarding the Quality Control tests, both bulk and final product (after irradiation) were tested for the Viability, Biological Activity, Endotoxin, Phenotype, Proliferation, Sterility test (EurPh). Moreover, microbiological monitoring is routinely carried out during production on media, air and personnel. The results (Table 1) show that the TALL lot is compliant with the specification required.

*Table 1.* Analytical Results.

| TEST | METHOD | SPECIFICATION |
|---|---|---|
| Viability | Nucleocounter | > 80% |
| Biological Activity | Cytotoxic Test | > 60% Lysis |
| Endotoxin | L.A.L. | < 0.5 Eu/Ml |
| Phenotype | Immunofluorescence | > 90% $CD3^+$, $CD8^+$, $CD56^+$ |
| Proliferation | H3-Tdr | No Incorporation |
| Sterility | 14 Days Test | No Growth |

## 3. CONCLUSION

The procedure described for TALL cells, based on the single use, modular and sterile CF bioreactor is a new method for industrial GMP production of human cells for Phase I and Phase II clinical trials of cell therapy that guarantees a high quality and reliable product.

## REFERENCES

1. O'Connors R., Cesano A., Lange B., Finan J., Nowell P.C, Clark S.C., Raimondi S.C., Rovera G. and Santoli D., 1991, Growth factor requirements of childhood acute T-lymphoblastic leukaemia: correlation between presence of chromosomal abnormalities and ability to grow permanently in vivo. Blood **77**: 1534-1546
2. Cesano A., Visonneau S., Jeglum K.A., Owen J., Wilkinson K., Carner K., Reese L., and Santoli D., 1996, Phase I clinical trial with a human major histocompatibility complex nonrestricted cytotoxic t-cell line (Tall-104) in dogs. **56,** 3021-3029, 1996
3. PCT/ EP 03/11024 – Procedure for the scale-up of T-Lymphocytes in an homogeneous system. Trasciatti S, Nolli M L, Cavenaghi L, De Bernardi Nadia

# Oxygen Control in Static Cell Cultures

## *Comparison Between Different Cell Lines in Two Cell Factory Systems*

Nadia De Bernardi, Edwin Schwander[1], Antonio Orlandi, Ilaria Tano, Sonia Castiglioni, Elena Muru, Fausto Gaspari, Marta Galgano, Claudia De Mattei, Luigi Cavenaghi, Maria Luisa Nolli
*Areta International S.r.l., Via R. Lepetit 34, 21040 Gerenzano (VA), Italy*
*[1]NUNC A/S, Roskilde, Denmark*

**Abstract:** Oxygen control in static cell cultures is one of the most critical parameters of a process, influencing the metabolism of the cells and ultimately the final yield of the product. With the aim to improve the cultures of different types of cells, both anchorage dependent and suspension , we have carried out several experiments evaluating the behaviour of some types of cells, (CHO, CaCo2, hybridomas) in a new version of Cell Factory bioreactors (with gas and with gas + motion). The type of culture was the batch mode and the parameters checked were: 1. Time to attach to the surface, 2. Time to arrive to confluence, 3. Cells number at confluence, 4.Viability, 5. Glucose consumption, 6. Lactate production, 7. Product yield after purification. The results demonstrate that the gas blown into the bioreactor is beneficial for the cells in terms of time to reach confluence, cell growth and productivity, while the motion does not increase significantly the production and cell growth compared with static conditions. Other experiments are in progress to confirm these encouraging data by evaluation of gas mixing both in 10 and in 40 trays modules.

**Key words:** Gas Cell Factory, Oxygen control, Mab production, bioreactor, productivity, cell growth, hybridomas, suspension cells, anchorage dependent cells, CHO.

## 1. INTRODUCTION

Oxygen control in static cell cultures is one of the most critical parameters of a process, influencing the metabolism of the cells and ultimately the final yield of the product. This problem becomes of particular importance when culture scale-up is needed for industrial production of a



*R. Smith (ed.), Cell Technology for Cell Products*, 739–744.

biological. On the market there are a lot of systems which have been designed for the oxygenation of animal cells cultures trying to solve the problem of potential damaging effects of air-liquid interface which the bubbles make with the culture fluid, the delivery of oxygen to high concentration cell masses and the transfer of all the necessary oxygen required by the breathing cells (1). With the aim to improve the cultures of different types of cells, both anchorage dependent and suspension, we have carried out several experiments evaluating the behaviour of some types of cells, (CHO, CaCo2, hybridomas) in a new version of Cell Factory (CF) bioreactors (with gas and with gas + motion – Figure 1) (2). The results demonstrate that the gas blown into the bioreactor is beneficial for the cells in terms of time to reach confluence, cell growth and productivity, while the motion does not increase significantly the production and cell growth compared with static conditions. Other experiments are in progress to confirm these encouraging data by evaluation of gas mixing both in 10 and in 40 trays modules.

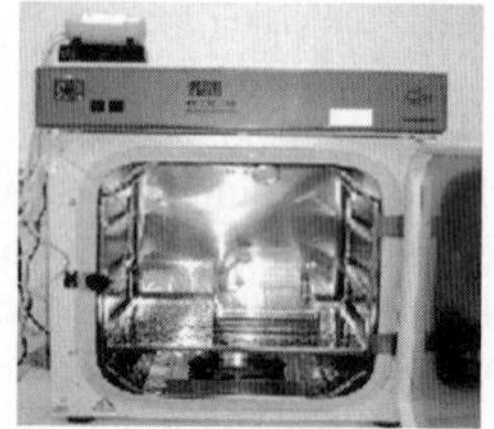

*Figure 1.* CF1, CF2, CF3.

All the operations, including the process development, were carried out in the high containment cell culture plant of Areta International that operates in respect to ISO 9001:2000 and GMP certification.

## 2. BIOREACTORS AND MODE OF CULTURE

CF is a high surface volume ratio bioreactor, sterile, mono use, easy to use, modular, designed for anchorage dependent cells. We have recently demonstrated that the CF can be successfully used for scale up production of suspension cells (3).

# 3. EXPERIMENTAL DESIGN

In order to evaluate two new prototypes of the CF bioreactor (with gas and with gas and motion) on the metabolism of different types of cells in culture we have carried out different experiments in batch mode of culture of two anchorage dependent cell lines (recombinant CHO and CaCo2) and of a suspension cell line (hybridoma). The following parameters were checked: 1. time to attach to the surface, 2. time to confluence, 3. cells number at confluence, 4. viability, 5. glucose consumption, 6. lactate production. 7. production

The first experiments aimed to evaluate the behaviour of an anchorage dependent cell line (recombinant CHO cells) in two new prototypes of CF bioreactor (with gas and with gas + motion), they then continued using other cell lines both in suspension (hybridomas) and with strong anchorage dependence (CaCo2).

Three prototypes were used in the experiments: the conventional CF with no gas (CF1) was compared with the CF with gas (CF2) and with the CF with gas and motion (CF3). The modules used were the 4 trays.

The gas pump connected with CF2 and CF3 was set on two litres of gas blown into the CF for one minutes every four minutes.

## 3.1 CHO cells

In each Cell Factory $1x10^4$ cells/cm$^2$ in 400 ml of medium (Alpha-MEM + 10% FBS+ 1% Glutamine + 2% pen/strep + 1% NEAA) were seeded.

The movement on CF3 started the day after the seed to allow the cells to attach to the surface.

A sample of supernatant was taken daily from each Cell Factory and every day cells were observed at microscope. After three days supernatants were withdrawn and cells recovered after trypsinization to evaluate the culture parameters in the different prototypes.

## 3.2 CaCo2

The density of cells inoculated in each CF was $1x10^4$ cells/cm$^2$ in 400 ml of medium (DMEM + 10% FBS+ 1% Glutamine + 2% pen/strep + 1% NEAA). Both pumps for CF2 and CF3 have been set on two litres of gas blown into the CF for one minute every four minutes.

The movement on CF3 started the day after the seed to allow the cells to attach to the surface.

During culture, the medium was sampled daily from each CF to test glucose consumption and the lactate production.

Every day cells were observed at microscope.

### 3.3 Hybridoma

In this experiment we tested an hybridoma cell line, producing a monoclonal antibody and growing in suspension.

The antibody-producing hybridoma was inoculated in each CF at the concentration of $1x10^4$ cells/$cm^2$ in 400 ml of medium (SFM + 1% Glutamine + 2% pen/strep + 1% Hybridmax).

During culture, the medium was sampled daily from each CF to test both glucose consumption and the lactate production.

Every day cells were observed at microscope.

## 4. RESULTS

The results on CHO cells show that the gas blown into the CF is beneficial for the cells as shown in the Figure 2. Furthermore in both the CF with gas there was a small number of cells in suspension that in any case were viable ($1x10^7$ cells were found in suspension in CF2 and $1.5x10^7$ cells in CF3). Concerning the metabolite parameters, glucose consumption is similar in all three CF prototypes, arriving to about 100% after 3 days in culture. The same is true for the lactate production.

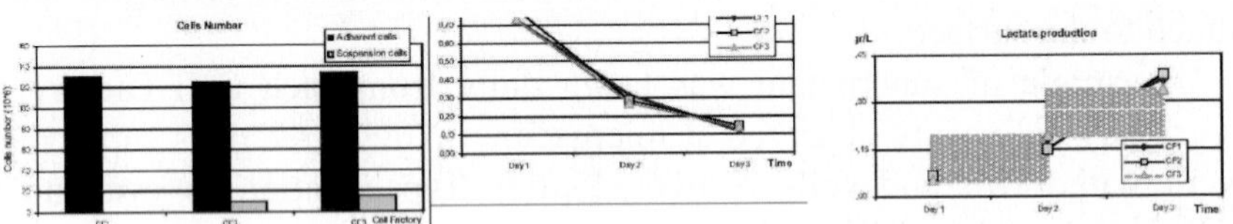


*Figure 2.* CHO Experiments.

Regarding the CaCo2 (Figure 3), the first culture that reached the confluence was the CF2. The movement in CF3 supports the cells growth in suspension.

A negligible amount of cells in suspension has been found in CF1 and CF2. In CF3 cells in suspension were 18% of total cells.

Concerning the metabolic parameters, glucose consumption is similar for all three CF, reaching about 100% after three days in culture, while in CF1 lactate production is less than in other CF, due to a lower number of cells in the bioreactor.

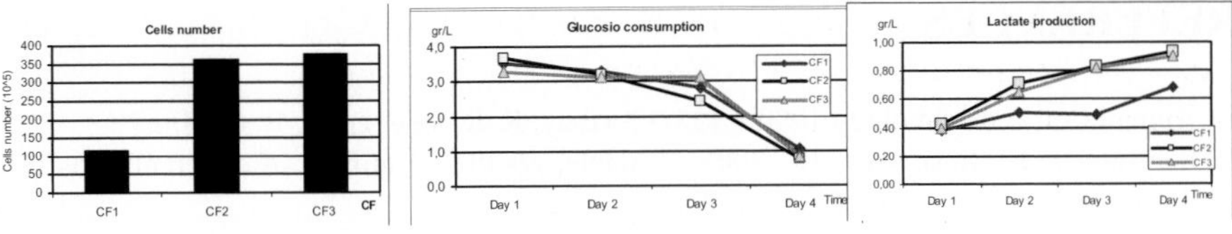


*Figure 3.* CaCo2 Experiments.

The results obtained on the hybridoma secreting an IgG1 Mab (Figure 4) are the following: cells in CF2 and CF3 grew up more rapidly and appeared healthier than those in CF1. Concerning the metabolic parameters, CF2 and CF3 have higher glucose consumption and lactate production than CF1. As in the third experiment, in CF1 glucose consumption followed cells growth: at the end of this experiment there was a higher glucose level than in other CF. Lactate production in CF1 is lower than in other CF.

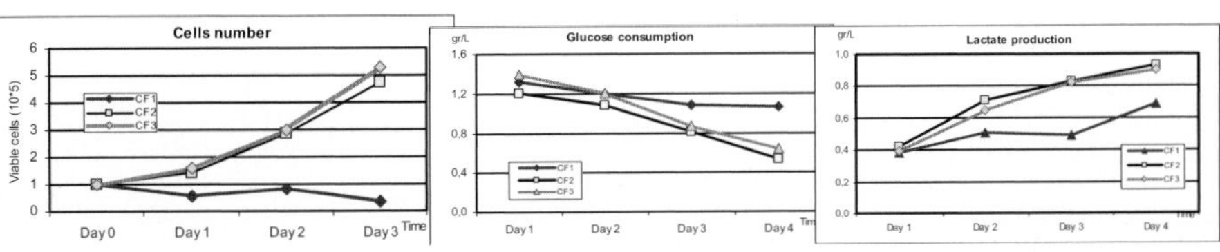


*Figure 4.* Hybridoma Experiments.

The supernatant of all cultures were harvested and the antibody purified. Here is the antibody recovery (mg/l) from each CF after purification: CF1 1,9 mg/l - CF2 3 mg/l - CF3 3 mg/l. Growing the cells in SFM without serum, the antibody recovery after purification is only due to the monoclonal antibody secreted by the hybridoma. As the antibody production is the same for CF2 and CF3, although suspension cell cultures with gas improve the time to reach cells confluence, the movement did not show a significant effect on the productivity.

## 5. CONCLUSION

The results demonstrate that the gas blown into the CF bioreactor is beneficial for the cells in terms of time to reach confluence, cell growth and productivity, while to evaluate the beneficial effect of motion some more experiments are required.

Such results underline the importance of the quantity of gas for the development of large-scale processes for both suspension and adherent cells.

# REFERENCES

1. Animal Cell Biotechnology 1990 4 123-32 ed by: R. E. Spier and J. B. Griffith) .
2. Schwander, E., Rasmusen, H., 2005, Scalable, controlled growth of adherent cells in a disposable, multilayer format. Gen. Eng. News **25**, 29.
3. PCT/ EP 03/11024 – Procedure for the scale-up of T-Lymphocytes in an homogeneous system. Trasciatti S, Nolli M L, Cavenaghi L, De Bernardi Nadia.

# The Effects of Cell Separation with Hydrocyclones on the Viability of CHO Cells

Rodrigo C. V. Pinto[1], Ricardo A. Medronho[2], Leda R. Castilho[1]
*Federal University of Rio de Janeiro (UFRJ), [1]COPPE, Chemical Engineering Program, Cx.P. 68502, 21941-972 Rio de Janeiro/RJ, Brazil [2]Dept. Chem. Eng., School of Chemistry, CT, Bloco E, 21949-900 Rio de Janeiro/RJ, Brazil*

**Abstract:** Five different hydrocyclone geometries were tested for the separation of CHO cells and presented separation efficiencies over 97%. Cell passage through the hydrocyclone did not induce apoptosis. The high separation efficiencies coupled to the low viability losses obtained in the present work confirm that hydrocyclones are suitable for animal cell retention in perfusion processes.

**Key words:** Hydrocyclone, cell retention device, perfusion, CHO.K1 cells, separation efficiency, cell viability, apoptosis.

## 1. INTRODUCTION

Perfusion cultures of animal cells have several advantages over batch or fed-batch cultures (Castilho & Medronho, 2002). Hydrocyclones are very simple devices that can be employed in such systems (Jockwer *et al.*, 2001; Elsayed *et al.*, 2003; Medronho *et al.*, 2005). In the present work, different techniques were employed to investigate if cell separation in hydrocyclones presents any effects on CHO cell viability and on apoptosis induction.

## 2. MATERIALS AND METHODS

CHO.K1 cells were grown in spinners using DMEM + Ham's F12 (1:1) medium with 1% FCS. The separation efficiency (fraction of cells recovered in the underflow) in hydrocyclones (HC) specially designed for cell retention was determined. HC 3020 (with 0.3 cm underflow and 0.2 cm overflow



*R. Smith (ed.), Cell Technology for Cell Products*, 745–748.

diameters) was further used at 1 bar to assess eventual apoptosis induction. Cell viability and concentration were measured by trypan blue exclusion and by crystal violet staining of cell nuclei. LDH activity was determined at 340 nm. Apoptosis was monitored by fluorescence microscopy, using acridine orange and ethidium bromide, and a set of filters suitable for fluorescein.

## 3. RESULTS AND DISCUSSION

CHO separation efficiencies obtained in the present work using the different hydrocyclone geometries ranged from 97.1% for HC 2015 to 99.9% for HC 3015 (Figure 1). These separation levels were higher than those reported in literature. Wen *et al.* (2000) obtained a separation efficiency of 88% using two sequential settlers. Iding *et al.* (2000), studying perfusion cultures of CHO cells using a spin-filter, obtained efficiencies in the range of 75-95%. Apparent viability drops (trypan blue exclusion) of 2.9%, 5.7%, and 5.8% were observed for HC 3020, 3015 and 2010, respectively, while greater losses were observed for HC 2015 and 2020 (17.8% and 14.4%, respectively). These results show that viability loss is a function of the hydrocyclone geometry. The LDH levels measured in the supernatant were compatible to trypan blue viability values (Figure 2). These viability drops are in the same range (6.8% to 12.3%) as those obtained by Jockwer *et al.* (2001), also working with CHO cells.

Table 1 shows the results for CHO.K1 separation using HC 3020 and the cultivation after HC operation of cells separated in the underflow. Cell separation efficiency was 98.7% and cell viability remained above 93% in the first 6h after HC separation, increasing to 96% and 99% after 24h and 48h. Necrotic cell concentration measured by fluorescence microscopy increased after HC separation, but started to decrease already after 3h, reaching within 48h a level of 7.5%, which is significantly lower than that in the culture before HC separation (18.8%). This necrotic cell increase observed immediately after HC operation, but followed by a rapid recovery of the culture, may be related to the disagreggation of cell clumps inside the HC, releasing necrotic cells that were in the culture inside cell agreggates. LDH values increased slowly within 24h after HC operation, and this is probably due to the fact that the release of LDH by necrotic cells occurs slowly and gradually. Quantification of apoptotic cells showed that a single passage through the HC does not induce apoptosis in CHO.K1 cells, since VA and NVA cell concentrations before and after HC operation remained initially at the same range and then gradually decreased, reaching quite low values after 48 h (Table 1). These results show that shear stress levels inside a hydrocyclone, coupled to the short residence times inside the device (0.03-0.1 s, according to Castilho and Medronho, 2002), do not induce apoptosis.

A comparison of cell viability in the HC overflow (82%) and underflow (97%) streams (Table 1) indicates a selective retention of viable cells and a removal of dead cells, what is a desirable characteristic for a cell retention device to be used in animal cell perfusion cultures.

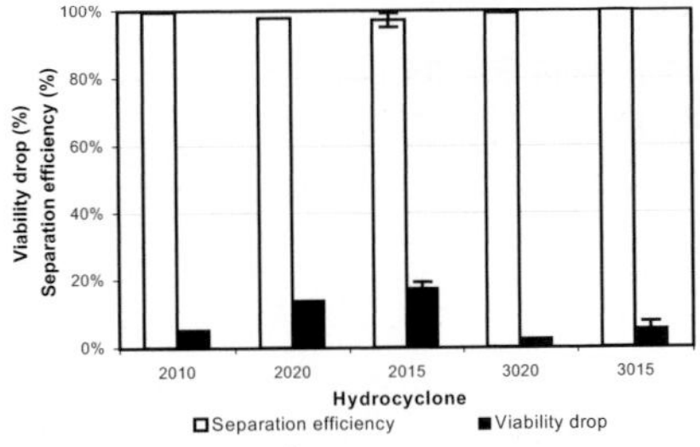


*Figure 1.* Underflow viability drops and separation efficiencies in different HCs.

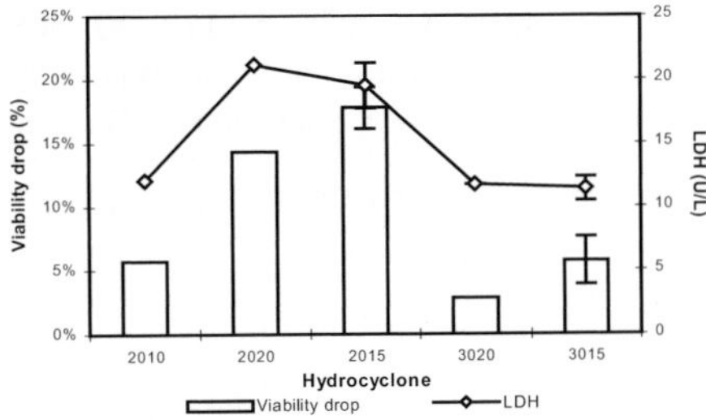


*Figure 2.* Underflow viability drops and LDH activity.

*Table 1.* Data of cell samples taken before (culture) and after (0, 3, 6, 24 and 48 h) a single passage through HC 3020. VNA, viable non-apoptotic cells; VA, viable apoptotic cells; NVA, non-viable apoptotic cells; NEC, necrotic cells and CF, cromatin-free cells.

| ***Sample*** | ***Cell concentration ($10^5$ cells/mL)*** | ***Viability (%)*** | ***LDH (U/L)*** | ***VNA (%)*** | ***VA (%)*** | ***NVA (%)*** | ***NEC (%)*** | ***CF (%)*** |
|---|---|---|---|---|---|---|---|---|
| Culture | 8.0 | 99 | 71.5 ± 6.9 | 74.9 | 1.6 | 3.8 | 18.8 | 0.9 |
| 0 h | 7.7 ± 0.1 | 97 | 95.9 ± 6.1 | 64.4 | 0.6 | 5.1 | 29.9 | 0.0 |
| 3 h | 8.7 ± 0.0 | 93 | 110.4 ± 11.5 | 70.1 | 0.3 | 1.5 | 28.1 | 0.0 |
| 6 h | 9.2 ± 0.8 | 93 | 116.1 ± 18.4 | 71.2 | 1.7 | 2.5 | 24.6 | 0.0 |
| 24 h | 8.1 ± 0.5 | 96 | 132.0 ± 12.4 | 78.5 | 0.8 | 2.5 | 17.7 | 0.5 |
| 48 h | 11.7 ± 0.4 | 99 | 126 ± 17.9 | 91.2 | 1.0 | 0.0 | 7.5 | 0.3 |
| Overflow | 0.8 ± 0.0 | 82 | 83.4 ± 3.3 | - | - | - | - | - |

# 4. CONCLUSIONS

High separation efficiencies (over 97%) were obtained using hydrocyclones for CHO.K1 retention. Cell viability loss after a single passage through the hydrocyclone was dependent on HC geometry, and small viability losses, as low as 2%, were obtained. Fluorescence microscopy showed that the use of hydrocyclones does not induce apoptosis in the culture. Finally, hydrocyclones were able to selectively retain viable cells.

## REFERENCES

Castilho LR, Medronho RA, 2002, Adv. Biochem. Eng. Biotechnol. **74**:129-169.

Elsayed A, Piehl GW, Nothnagel J, Medronho RA, Deckwer WD, Wagner R, 2003, In: Gòdia F (ed.), Animal Cell Technology Meets Genomics, Kluwer Academic Pub., Dordrecht, 383-386.

Iding K, Lütkemeyer D, Fraune E, Gerlach K, Lehmann J, 2000, Cytotechnology **34**:141-150.

Jockwer A, Medronho RA, Wagner R, Anspach FB, Deckwer WD, 2001, In: Olsson EL, Chatzissavidou N, Lüllau E (eds.), Animal Cell Technology: From Target to Market. Dordrecht, Kluwer Academic Pub., 301-305.

Medronho RA, Schuetze J, Deckwer WD, 2005, Latin Am. Appl. Res. **35**:1-8.

Wen Z-Y, Teng X-W, Chen F, 2000, J. Biotechnol. **79**:1-11.

# Erythropoietin Purification Employing Affinity Membrane Adsorbers

Maria Candida M. Mellado[1], David Curbelo[2], Ronaldo Nobrega[1], Leda R. Castilho[1]

*[1]Federal University of Rio de Janeiro (UFRJ), COPPE, Chemical Engineering Program, P.O. Box 68502, CEP 21941-972, Rio de Janeiro/RJ, Brazil; [2]Center of Molecular Immunology (CIM), Calle 16 y 15, P.O. Box 16040, Havana 11600, Cuba.*

**Abstract:** In the present work, rhEPO was purified from crude CHO cell culture supernatant either with Sartobind-Cibacron Blue (CB) or Sartobind-IDA-$Cu^{+2}$ affinity membranes. Purity degrees were of 55 and 75%, respectively.

**Key words:** Erythropoietin, affinity chromatography, membrane adsorbers, purification.

## 1. INTRODUCTION

The major limitation of packed-bed chromatography is the dependence on intra-particle diffusion for the transport of molecules to their binding sites, increasing process time and requiring large eluent volumes (Ghosh, 2002). An approach to overcome these limitations is to use microporous membranes as chromatographic supports (Castilho *et al.*, 2002). Apart from lower manufacturing costs, membrane chromatography is a convection-controlled process, allowing shorter process times and employing lower eluent volumes. Affinity membrane adsorbers rely on the selectivity and reversibility of ligand-ligate interactions. Recombinant human erythropoietin (rhEPO) has already been purified with affinity chromatographic packed columns, but this is the first report on the use of membrane adsorbers to purify rhEPO.



*R. Smith (ed.), Cell Technology for Cell Products*, 749–752.

## 2. MATERIALS AND METHODS

Sartobind-Cibacron Blue and Sartobind-IDA (75 $cm^2$) were kindly provided by Sartorius (Göttingen, Germany). Sartobind-IDA was washed with 0.1 M $CuSO_4$ in order to immobilize $Cu^{+2}$ ions. After membrane equilibration, crude cell culture supernatant containing rhEPO (Center of Molecular Immunology, Cuba) was applied. After membrane washing with the respective equilibration buffer, elution was carried out with 0.4 M (E1) and 2.5 M (E2) NaCl (Sartobind-CB) or at pH 4.1 (Sartobind-IDA-$Cu^{+2}$). The eluate samples were analyzed for total protein concentration (Bradford assay), SDS-PAGE and reverse phase-HPLC (RP-HPLC).

## 3. RESULTS

The chromatograms obtained for rhEPO purification with the affinity membranes investigated in this work are shown in Figure 1.

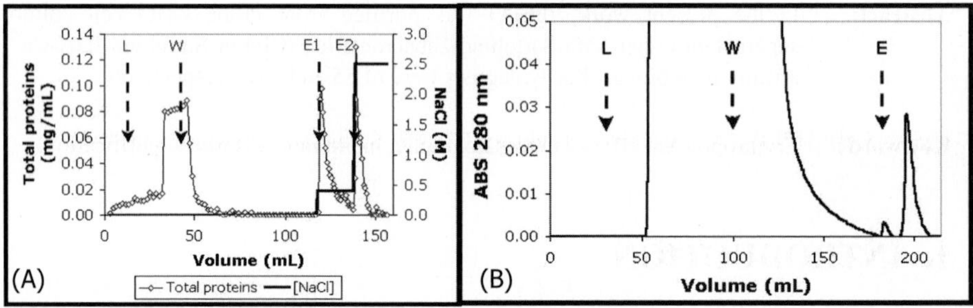


*Figure1.* Chromatogram of rhEPO purification employing Sartobind-CB (A) and Sartobind-IDA-Cu+2 (B) membranes. Arrows indicate sample loading (L), washing (W) and elution (E).

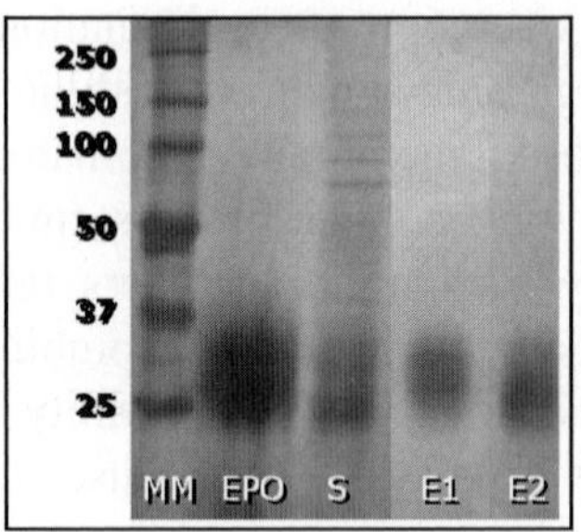


*Figure 2.* Silver-stained SDS-PAGE (MM: molar mass marker; EPO: standard sample; S:supernatant; E1 and E2: eluates from the Sartobind-CB chromatographic run).

Eluates (E1 and E2) from Sartobind-CB were analysed by silver-stained SDS-PAGE (Figure 2), which confirmed the presence of rhEPO in both eluates and the absence of other contaminating proteins from the cell culture supernatant. The eluates from both membranes were further analysed by RP-HPLC (Figure 3), and levels of purity of 55% (Sartobind-CB) and 75% (Sartobind-IDA-$Cu^{+2}$) were obtained.

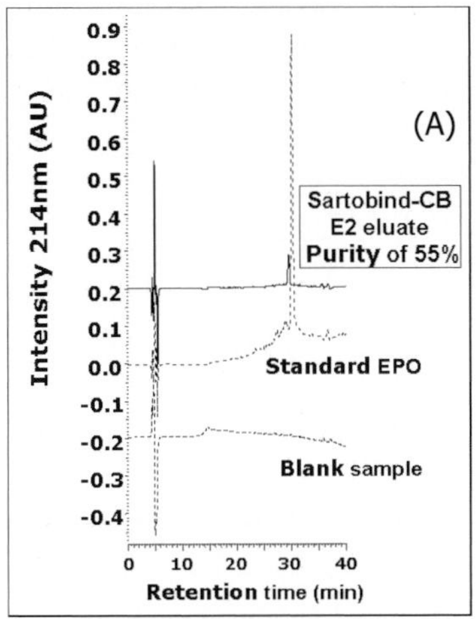


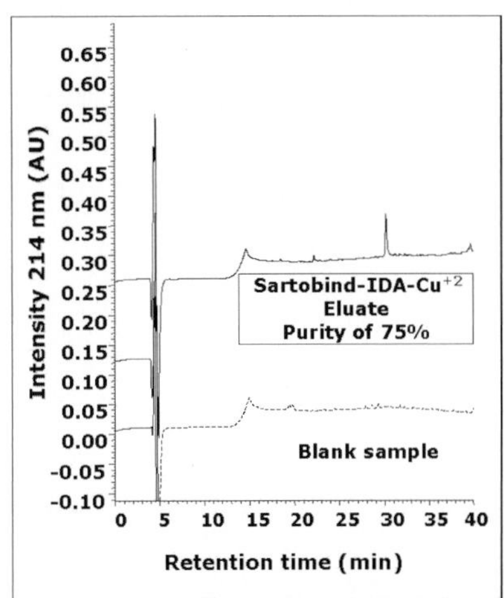


*Figure 3.* RP-HPLC: (A) Sartobind-CB eluate (E2); (B) Sartobind-IDA-Cu+2 eluate.

## 4. CONCLUSION

In the present work, rhEPO was partially purified from crude CHO cell culture supernatant using affinity membrane adsorbers. A one-step chromatographic purification using Sartobind-Cibacron Blue membranes gave a purity degree of 55%, whereas the use of a metal chelate membrane adsorber (Sartobind-IDA-$Cu^{+2}$) resulted in a purity degree of 75%.

## ACKNOWLEDGEMENTS

The financial support from the Brazilian research agencies CNPq and FAPERJ is gratefully acknowledged.

## REFERENCES

Castilho, L.R., Anspach, F.B., Deckwer, W.-D., 2002, Comparison of affinity membranes for the purification of immunoglobulins, J. Membrane Sci. 207:253-264.

Ghosh, R., 2002, Protein separation using membrane chromatography: opportunities and challenges, J. Chromatogr. A **952**:13-27.

# Cultivation of Vero Cells on Microporous and Macroporous Microcarriers

Marta Cristina O. Souza[1], Marcos S. Freire[2], Leda R. Castilho[1]
[1]*Federal University of Rio de Janeiro - UFRJ/COPPE - Chemical Engineering Program, P.O. Box 68502, 21941-972, Rio de Janeiro, Brazil* [2]*Oswaldo Cruz Foundation - FIOCRUZ - Bio-Manguinhos, Rio de Janeiro, Brazil*

**Abstract:** The manufacture of immunobiologicals, such as viral vaccines, is largely based on animal cell technology. Large-scale production of viral antigens depends on efficient cell proliferation and virus replication, both in terms of quality and quantity. This can be achieved by choosing adequate culture conditions and cultivation systems. Microcarriers are a tool to provide a large adhesion area in homogeneous bioreactor systems, allowing high densities of anchorage-dependent cells to be obtained. The aim of the present work was to study the use of different microcarriers and culture conditions to obtain high-density Vero cell cultures for the production of viral antigens.

**Key words:** Vero cells, microcarriers, culture conditions, immunobiologicals, vaccine, Cytodex 1, Cytodex 3, Cytoline, Cytopore, experimental design.

## 1. INTRODUCTION

The cell line Vero is a continuous, adherent cell line, which has been recommended by the WHO for the production of human vaccines. Vero cells are presently used in the commercial manufacture of human rabies and polio vaccines (Butler *et al.*, 2000). Cultivation systems are usually based on microcarriers, which are microspheres that serve as support for the cells. In the microporous carriers, cells grow on the external surface of the beads, whereas in the macroporous carriers the inner pore surface is also available for cell adhesion, increasing the total available area for cell proliferation. The use of microcarriers in stirred bioreactors allows high cell densities and, consequently, high virus titres to be achieved (Yokomizo *et al.*, 2004). Thus,



*R. Smith (ed.), Cell Technology for Cell Products*, 753–755.

the aim of the present work was to study the influence of severals variables - serum concentration in the medium, agitation rate and cell-to-carrier ratio - on the growth of Vero cells on different microporous (Cytodex 1 and Cytodex 3) and macroporous (Cytopore 2 and Cytoline 1) carriers.

## 2. MATERIALS AND METHODS

- Cell line: Vero CCL 81 obtained from ATCC;
- Medium: DMEM high-glucose, supplemented with FBS;
- Cultivation: in 100 mL spinner flasks containing 3 mg/mL of microcarriers, at 37° C and 5% $CO_2$;
- Statistical experimental design (Cytodex 1 and 3): use of a $2^{3-1}$ fractional factorial design to investigate the effects of agitation rate, FBS concentration and cell-to-carrier ratio (at inoculation) on cell growth.

## 3. RESULTS

The experimental conditions tested, as well as the maximum cell densities obtained in the different runs, are shown in Table 1 for the microporous (Cytodex 1 and Cytodex 3) microcarriers.

*Table 1.* Maximum cell density obtained in the 23-1 experimental design using Cytodex 1 and Cytodex 3 microcarriers.

| *Experi-ment* | *Agitation rate (rpm)* | *Cell-to-carrier ratio (cells/μg)* | *FBS concentration (% v/v)* | *Maximum cell density ($10^6$ cells/mL)* | |
|---|---|---|---|---|---|
| | | | | *Cytodex 1* | *Cytodex 3* |
| 1 | 40 | 10 | 1 | 1.33 | 0.90 |
| 2 | 70 | 70 | 1 | 1.91 | 1.60 |
| 3 | 70 | 10 | 5 | 1.47 | 1.54 |
| 4 | 40 | 70 | 5 | 3.53 | 2.73 |
| 5 | 55 | 40 | 3 | 1.75 | 1.26 |
| 6 | 55 | 40 | 3 | 1.70 | 1.39 |
| 7 | 55 | 40 | 3 | 1.80 | 1.47 |

The cell densities were highest (3.53 and 2.73 x $10^6$ cells/mL, for Cytodex 1 and 3, respectively) in experiment 4, when agitation remained low (40 rpm), but serum concentration and cell-to-carrier ratio were at their highest levels (5% and 70 cells/μg carrier, respectively). Under these conditions, the propagation of Vero cells was compared with two macroporous (Cytopore 2 and Cytoline 1) carriers, as shown in the Figure 1.

Although the bead mass was the same in all experiments, Cytopore has a surface area of 11000 $cm^2/g$, while Cytodex 1, Cytodex 3 and Cytoline have 4400, 2700 and >3000 $cm^2/g$, respectively. In spite of that, maximum cell densities were obtained with Cytodex 1 and Cytodex 3 microcarriers. Thus, a comparative study of the propagation of Vero cells on Cytodex 1 and Cytodex 3 carriers was carried out using MC concentrations that resulted in an equal total surface area of 1320 $cm^2$ for both carriers (Figure 2). Also in this case the highest cell concentration was obtained with Cytodex 1.

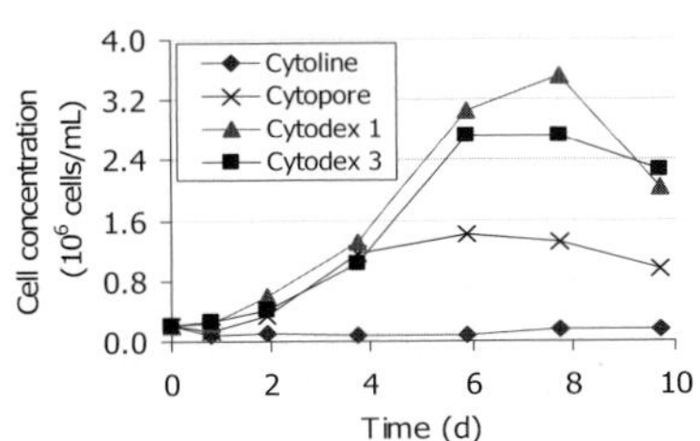


*Figure 1.* Cell concentration under the conditions of experiment 4, using a MC concentration of 3 mg/mL.

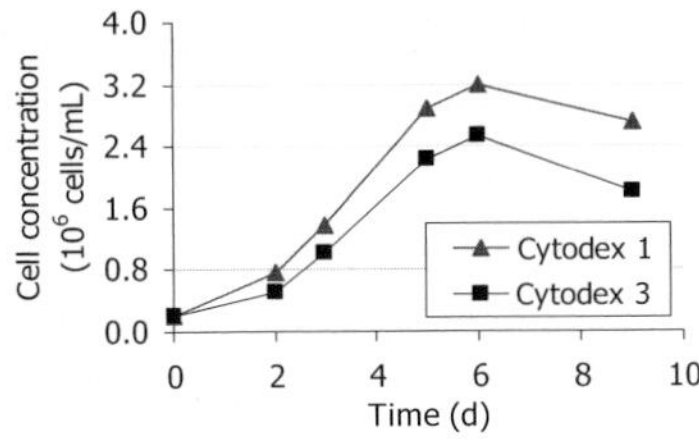


*Figure 2.* Vero cell growth under the conditions of experiment 4, using a total MC surface area of 1320 $cm^2$ for both MCs.

## 4. CONCLUSIONS

Within the studied range, a high cell-to-microcarrier ratio at inoculation, a high FBS concentration and a low agitation speed provided the highest Vero cell densities. Under these conditions, Cytodex 1 was the best microcarrier, giving cell concentrations approximately 30% and 52% higher than those obtained with Cytodex 3 and Cytopore 2, respectively. The microcarrier Cytoline 1 showed to be inadequate for Vero cell propagation under the studied conditions.

## REFERENCES

Yokomizo, A.Y., Antoniazzi, M.M., Galdino, P.L., Jorge, S.A.C., and Pereira, A.A, 2004, Rabies virus production in high Vero cell density cultures on macroporous microcarriers, Biotechnol. Bioeng. **85**:506-515.

Butler, M., Burgener, A., Patrick, M., Berry, M., Huzel, N., Barnabé, N., and Coombs, K., 2000, Application of a serum-free medium for the growth of Vero cells and the production of reovirus, Biotechnol. Prog. **16**: 854-858.

# CFD-Aided Design of Hollow Fibre Modules for Integrated Mammalian Cell Retention and Product Purification

Romi L. Machado[1], Alvio Figueredo[2,3], Danilo G. P. Carneiro[2], Leda R. Castilho[1] and Ricardo A. Medronho[2]

*[1]Federal University of Rio de Janeiro (UFRJ), COPPE, Chem. Eng. Program, Cx.P. 68502, 21941-972 Rio de Janeiro/RJ, Brazil; [2]Federal University of Rio de Janeiro (UFRJ), School of Chemistry, Chem. Eng. Dept., CT, Bloco E, 21949-900 Rio de Janeiro/RJ, Brazil; [3]Center of Molecular Immunology (CIM), P.O.Box 16040, Habana 11600 Cuba.*

**Abstract:** Membrane-based affinity chromatography is a powerful technique for protein purification. The use of membrane adsorbers allows combining cell separation and primary product purification in one step. However, a challenge is to find module geometries that can avoid both membrane fouling and premature product breakthrough. In the present work, computational fluid dynamics was employed as a tool to evaluate the flow pattern and pressure profiles in hollow fibre membranes as a function of feed velocity, with the final aim of designing a membrane device that is adequate for carrying out integrated perfusion/ purification processes.

**Key words:** computational fluid dynamics, CFD, hollow fibre membranes, perfusion processes, cell retention, product purification

## 1. INTRODUCTION

Mammalian cell cultivation in perfusion systems provides higher productivities than batch or fed batch processes (Castilho and Medronho, 2002). A way of increasing productivity even more is through integrated perfusion/purification processes. This can be done using, for instance, membrane affinity chromatography (Castilho *et al.*, 2002). Since in tangential-flow modules premature product breakthrough may occur as a consequence of non-homogeneous flow through the membrane, leading to local saturation of adsorption sites and low ligand utilisation efficiency, the



*R. Smith (ed.), Cell Technology for Cell Products*, 757–760.

use of hollow fibres for this task would only be possible if modules with adequate hydrodynamics are developed. In this work, computational fluid dynamics (CFD) was employed to evaluate the flow pattern and pressure profiles along the membrane fibres for different feed velocities.

## 2. METHODOLOGY

A hollow fibre membrane with internal and external diameters of 350 and 600 μm, respectively, permeability of $6.5 \times 10^{-16}$ $m^2$ and length of 1 cm was simulated using the CFD package CFX 5.7, from Ansys®. The partial differential equations for mass conservation and transport of momentum (Bird *et al.*, 2001) were solved using the finite-volume method. All simulations were performed in a time independent manner (steady state assumption). Density and viscosity of the medium were assumed to be constant, which corresponds to an isothermal approach. The simulated feed fluid velocities were 1 cm/s and 1 m/s, which result in laminar flows. The three-dimensional grid representing the hollow fibre and its surroundings were built with 45318 nodes generating 226066 tetrahedral cells.

## 3. RESULTS

Figure 1 shows the pressure and velocity profiles in the hollow fibre, considering that the culture medium enters into the hollow fibre at the bottom of the figure, leaving the fibre at the top (the parallel white lines represent the fibre walls). The simulated profiles are shown for different feed velocities (1 m/s and 1 cm/s), which are in the range of velocities used in perfusion and in chromatography processes carried out with hollow fibre modules. Pressure and velocity levels decrease along the fibre length, this being related to fluid permeation through the fibre wall.

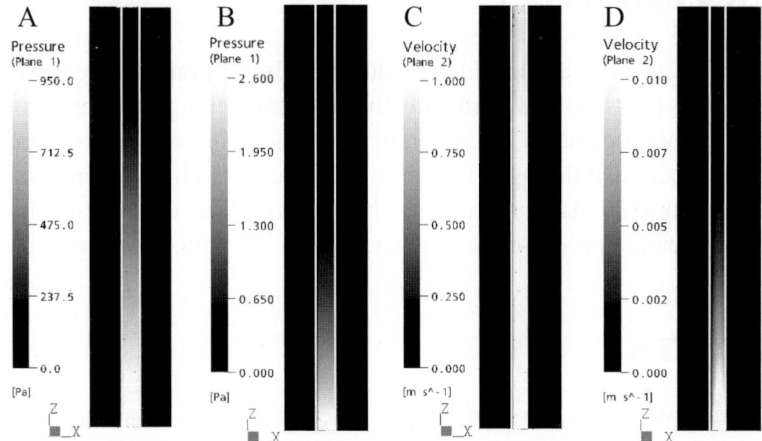


*Figure 1.* Pressure profiles in the hollow fibre for fluid velocities of 1 m/s (A) and 1 cm/s (B), and velocity profiles for 1 m/s (C) and 1 cm/s (D). Flow direction inside the fibre is upwards.

Figures 2A and 2B show the transmembrane pressure and the permeate flux along fibre length, respectively. The decrease in both variables along fibre length follows the well-known Darcy's law, since permeate flux is directly proportional to pressure drop accross the porous membrane. This non-homogeneous permeate flow through the membrane along fibre length is the reason for the low ligand utilization efficiency and premature product breakthrough observed for tangential-flow membrane modules. Saturation of adsorption sites occurs rapidly in the surroundings of the fibre inlet, while ligands located near the outlet of the fibre remain unoccupied. In order to use hollow fibres in perfusion/purification processes for both cell retention and membrane chromatography, it is necessary to design a hollow fibre module able to give a rather constant transmembrane pressure along the fibre length. To achieve this task, more simulations are being carried out with different module geometries and operational conditions.

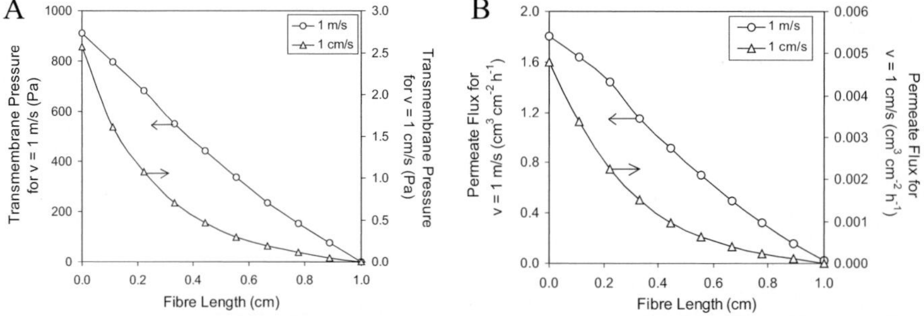


*Figure 2.* Transmembrane pressures (A) and permeate fluxes (B) at 1 cm/s (△) and 1 m/s (o).

## 4. CONCLUSIONS

CFD is able to give a good insight into the flow through hollow fibres. Transmembrane pressure and permeate flux decrease along the fibre length, promoting non-homogeneous saturation of adsorption sites and, consequently, premature product breakthrough. Thus, further simulations of different module geometries and operational conditions are being carried out to allow integrated perfusion/purification processes to be carried out using hollow fibre modules.

## REFERENCES

Bird, R.B., Stewart, W.E., and Lightfoot, E.N., 2001, Transport Phenomena, John Wiley and Sons, New York.

Castilho, L.R., Anspach, F.B. and Deckwer, W.-D., 2002, An integrated process for mammalian cell perfusion cultivation and product purification using a dynamic filter. Biotechnol. Prog. **18**:776-781.

# Comparison of Manufacture Technologies for RAV12 Monoclonal Antibody in Production Medium Containing No Animal Derived Proteins

Irene Shackel, Amy Bass, Arno Brewer, Peter Brown*, Mary C. Tsao, Lucille W.S. Chang

*Raven Biotechnologies, inc., 1140 Veterans Blvd., South San Francisco, CA 94080*

**Biotechnology Solutions, 20 Woodcrest Dr., Orinda, CA, 94563*

**Abstract**: Cells engineered to express proteins using rDNA technology have enabled the commercialization of therapeutic proteins. A variety of production technologies have been shown to increase the total viable biomass for the engineered cells, enabling production of sufficient materials for clinical trials. RAV12 is a high-affinity internalizing chimeric monoclonal antibody being developed as an anti-cancer therapeutic. We developed a defined production medium containing no animal-derived proteins for our RAV12 production cell line. Using this defined medium, we evaluated four modes of production for RAV12: two fed-batch and two perfusion systems. The four production modes were WAVE, 2L-2500L bioreactors, settler cell retention device, and hollow fiber. We monitored the growth, yield, and metabolites for the four production modes. The antibodies produced were purified and characterized to examine differences in their physicochemical properties, including glycosylation, and biological activity.

## 1. INTRODUCTION

The process development group at Raven was asked to meet several needs of the company for RAV12 production: 1) develop a closed, disposable system for the in-house GMP production facility currently being established and qualified at Raven, 2) generate enough RAV12 material for purification and assay development as well as PK studies, and 3) develop a fed-batch process for RAV12 with a yield greater than 100mg/L.

These requirements led Raven to evaluate and develop several different modes of production for RAV12, including: 2-15L fed-batch bioreactors, 5L



*R. Smith (ed.), Cell Technology for Cell Products*, 761–763.

fed-batch Wave bioreactor, cell settler perfusion system, and BIOVEST's Maximizer hollow fiber perfusion system.

## 2. PROCESS DEVELOPMENT

An optimal medium formulation for RAV12 was developed in-house by balancing nutrients, salts, additives, and growth factors to obtain an animal component free growth and production medium.

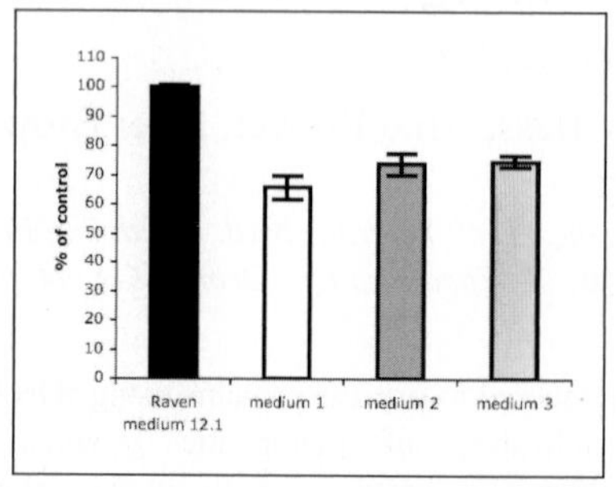


*Figure 1.* Comparison of growth on day 5 of the RAV12 producing CHO cell line in Raven Basal Medium 12.1 (RBM12.1) and 3 commercially available serum-free CHO media.

After development of RBM12.1, the following parameters were evaluated to improve production: temperature shift (33°-37°), osmolarity shift (290 mOsm-425 mOsm), chemical induction with sodium butyrate, glucose/medium/nutrient feed formulations for fed-batch process, feed schedule for fed-batch process, duration of fed-batch process, pH shift (6.8-7.3), and inoculum density ($2x10^5$ VC/mL-$1x10^6$ VC/mL).

## 3. RESULTS

Decreasing temperature, increasing osmolarity, increasing inoculum density, pH shifts, and varied feed schedule showed no increase of final titer (data not shown). Addition of a nutrient feed (AVP), as well as chemical induction (sodium butyrate), both showed an increase of final titer.

*Table 1.* shows fold increase on day 8 for each condition over the control condition, which is no nutrient feed and no induction by sodium butyrate.

| Condition | Viable Biomass | RAV12 Titer | Specific Productivity | RAV12 Titer (mg/L) |
|---|---|---|---|---|
| Control | 1 | 1 | 1 | 126 |
| +AVP | 2.7 | 2.4 | 1 | 164 |
| +Sodium Butyrate | 1 | 2.3 | 2.6 | 187 |
| +AVP+Sodium Butyrate | 1.8 | 3.4 | 3.1 | 299 |

Glycosylation analysis of RAV12 material from the Wave, 2L bioreactor, 2500L bioreactor, settler perfusion and hollow fiber perfusion showed no major difference in glycosylation. The method used has a standard deviation of 13%, based on triplicates.

## 4. CONCLUSION

By using the cell settler perfusion system (Biotechnology Solutions) coupled with an Applikon 2L stirred tank, Raven was able to generate enough RAV12 material for purification development and PK studies in 30 days. A fed-batch process was developed including a nutrient feed and sodium butyrate induction which yielded 3.4-fold higher titer from the same cell population, and with a final titer of 350 mg/L RAV12 on day 10, exceeded the company goal of 100 mg/L.

*Table 2.*

| Mode of production | Method of production | Specific productivity (ug/1e6VC/day) |
|---|---|---|
| Fed-batch | 2L Bioreactor | 10.51±1.2 |
| Fed-batch | 2500L Bioreactor | 9.6±4.4 |
| Fed-batch | Wave | 9.3±3.2 |
| **Mode of production** | **Method of production** | **Volumetric Productivity (mg/L)** |
| Perfusion | Hollow Fiber Maximizer* | 313.0 |
| Perfusion | 2L Bioreactor with Settler | 42.0 |

*High packed cell density in the EC region of the hollow fiber produces more concentrated antibody per unit volume: harvest rate is only 480mL/day.

Purified RAV12 from the 2500L fed-batch process and the settler perfusion system was run in Raven's in-house GLP potency assay, and was found to be equally active for both modes of production.

# Highly Efficient Serum-free Production of Biopharmaceuticals in CHO-Cells

S. Zahn, K. Abst, N. Palmen, S. Schindler, A. Herrmann
*Celonic GmbH, Karl-Heinz-Beckurts-Str. 13, 52428 Jülich, Gerrmany*

**Abstract:** This study describes the development of a highly efficient method for the fast production of biopharmaceuticals using new CHO-K1 derived suspension host cells adapted to various protein-, serum-free and chemically defined media in combination with optimized nucleofection protocols and high density cell cultivation in fluidized bed bioreactors. Finally, these new CHO-K1 derived host cells growing in chemically defined media are suitable for large scale processes with respect to a short population doubling times, high cell densities of approx. 7 x $10^6$ cells/ml as shown in spinner cultures, and high productivities of 6 pg/c/d without performing any subcloning steps as shown with a highly glycosylated P-Selektin/IgG-fusion protein.

**Key words:** mammalian cell, CHO-K1, host-cell, biopharmaceutical protein, glycosylation, nucleofection, production cell line, protein-free media, suspension

## 1. INTRODUCTION

The CHO-K1 cell line is one of the most widely used host cell for the production of biopharmaceuticals. Since methods for the efficient transfection and limited dilution of CHO-cells growing in suspension in serum- and protein-free medium are hardly available, the initial cell-line development is mostly performed using anchorage dependent cells. After the establishment of a single cell derived cell-line by limited dilution, this cell-line is usually adapted to serum-free medium in a long-lasting and cost-intensive procedure which is very often associated with a loss of cell-specific productivity and a significant change of product quality with respect to posttranslational modifications.

Therefore, in a first step CHO-K1 host cells were adapted to various regulatory compliant serum-, protein-free, or chemically-defined media by



*R. Smith (ed.), Cell Technology for Cell Products*, 765–767.

different adaptation strategies and further subcloned in order to obtain CHO-K1 suspension host cells suitable for large scale production of biopharmaceutical proteins. Secondly, a highly efficient nucleofection method for these suspension cells in protein-free medium has been established leading to transfection efficiencies of up to 95%.

The finally selected host cell-lines are able to grow at very high cell densities (up to $7x10^6$ cells/ml) even in spinner cultures and could be successfully transfected for the production of as much as 0.5g/l of a highly glycosylated fusion protein using a fluidized bed bioreactor without performing any subcloning step after transfection.

## 2. METHODS AND RESULTS

### 2.1 Adaptation of cells to protein-free media

CHO cell-lines from different sources (ATCC, DSMZ) were adapted to various serum-, protein-free and chemically-defined media using direct and stepwise adaptation strategies (CHO-PFM and CHO-DHFR, Sigma; EX-Cell-325 and EX-Cell CD-CHO, JRH; BD-CHO, Becton Dickinson; CD-CHO, Invitrogen; HyQ CDM4CHO, Perbio). Those cell-lines successfully adapted to suspension, were used for further cloning steps in order to select for clones suitable for large scale productions, i.e. growing in suspension with high cell densities, high viabilities and short population doubling times. One of the finally selected CHO-K1 derived suspension clone growing in serum-free BD-CHO medium grow at high cell-densities of $7x10^6$ cells /ml with short population doubling times of about 20-24 hours as shown in spinner cultures.

### 2.2 Optimized transfection method in suspension

Subsequently, methods for the highly efficient transfection of these cells in suspension using the Nucleofector™ technology (Amaxa) have been developed by optimizing each of the following parameter: cell number, pulse, solution and DNA concentration. The Nucleofection™ was performed with different reporter genes (secreted alkaline phosphatase, SEAP; enhanced green fluorescent protein, eGFP) as well as using a highly glycosylated fusion protein (P-Selektin/IgG1) for proof of concept investigations. All reporter genes were expressed under the control of an enhanced CMV-promoter. The optimal pulse for the transfection of these subclones in suspension was U24.

A head to head comparison was using done using optimized protocols for Nucleofection® and Lipofection (Lipofectamin-2000), and SEAP and

P-Selektin/IgG-fusion protein as reporters in combination with selected CHO-subclones growing in protein-free medium. This study demonstrates that Nucleofection® of those suspension cells leads to much more higher transfection and production rates as Lipofection. As much as 1g/l SEAP could be produced in 5 days without any clonal selection.

### 2.3 Fluidized bed bioreactor versus stirred vessel

Furthermore, two production runs were performed for proof of concept in $pO_2$-, pH-, and temperature controlled spinner flasks and in a Fluidized Bed Bioreactor E500 using a cell-pool selected for 20 days with G418 (600 µg/ml). Levels of 6 pg/c/d could be achieved in the resulting clone pool without performing any subcloning step. The volumetric productivities are 30 mg/d in spinner flasks and 550 mg/d in the Fluidized Bed Bioreactor (1 L scale each)

## 3. SUMMARY AND CONCLUSIONS

- CHO-K1, DG44 and HEK-293 (data not shown) host cells have been adapted to and further subcloned in different protein-free media with respect to their suitability for large scale production processes (high cell densities, short PDT) and their use for the generation of regulatory compliant high yield production cell-lines via limited dilution.
- Methods for the highly efficient Nucleofection™ (>98% positive cells) of these CHO-derived (CHO-K1 and DG44) subcloned host cell-lines have been developed (best pulse U24, optimal DNA-quantity: 5µg).
- Nucleofected host-cells are able to produce high levels of recombinant proteins (up to 6 pg/c/d) without any further subcloning step.
- Proof of concept material could be produced in gram-scale by FBR-technology within 8 weeks after getting the plasmid. This significantly shortens the time and costs for the proof of concept studies in animals.
- Using these serum-free cell-lines in combination with the optimized protocols for highly efficient transfection shortens the time for the establishment of regulatory compliant, high-yield production cells by about 6 months and reduces the risk of loss of productivity and change of product quality inherent in classical adaptation strategies.

# A New Host Cell-line for the Fast, Reproducible and Efficient Production of Biopharmaceuticals

B. Greulich, H.F. Abts, S. Zahn, A. Herrmann
*Celonic GmbH, Karl-Heinz-Beckurts-Str. 13, 52428 Jülich*

**Abstract:** We describe here the proof of concept of a new, innovative "wildcard"-strategy (termed TaCEx-system, **ta**g, **c**haracterize and **ex**change) for the generation of CHO-based cell-lines for the production of biopharmaceuticals using site-specific recombinase-mediated cassette exchange. Core element of the TaCEx-system is a pre-formed host cell-line. It contains a tagged reporter cassette by which the expression capacity of the integration site has been characterized and which allows the exchange of the reporter against a gene of interest (GOI). Such a new host cell facilitates the cloning of a GOI into predetermined genomic expression hot spots within days in order to get reproducible, highly efficient production cell-lines.

**Key words:** recombinant protein expression, cell line generation, recombinase, biopharmaceuticals, production, FLP, resolvase

## 1. INTRODUCTION

The CHO cell-line is one of the most widely used host cell for production of biopharmaceuticals. Classically, the host cell is transfected with an expression construct containing the GOI coding for the recombinant protein and a selection marker such as neomycine or methotrexate. Afterwards, stably transfected cells are selected under selection pressure of the antibiotic compound. In order to get a regulatory compliant single cell derived cell-line, a subcloning step by limited dilution is performed subsequently. Although several hundreds of single cell derived clones are analyzed, the expression level was often still weak, which requires the optimization of transfection procedures or the amplification of initially integrated copies of the expression construct, e.g. by increasing concentration of methotrexate or



*R. Smith (ed.), Cell Technology for Cell Products*, 769–771.

another amplifying system. The whole procedure takes at least 6 months, sometimes up to a year.

In order to decrease the time for the generation of a high producer cell-line and to maximize the rate of success, a system for the targeted insertion of any GOI into a hot spot of expression of the genome would be desirable. Such a targeted recombination could be done by site-specific recombinases such as Cre-, FLP- or the γδ-resolvase, once a target site has been tagged by an appropriated vector containing a reporter gene and the required recognition sites for the used recombinase. While the reporter gene allows the characterization of the expression capacity of the randomly tagged integration site, the recognition sites allow the exchange of the reporter against the GOI. We demonstrate here the feasibility of such a strategy termed TaCEx-system.

# 2. METHODS AND RESULTS

## 2.1 Site-specific recombinases

In this proof of concept for the TaCEx-system the , FLP-recombinase of *Saccharomyces cerevisiae,* was used. Site-specific cassette exchange was facilitated by the use of heterotypic FLP-recognition target (FRT) sites flanking the cassette intended to be exchanged.

In principal the realisation of the TaCEx-strategy is also feasible with other site-specific recombinases or site-specific integration strategies. Currently this system is modified for use with specific mutated γδ-resolvase variants (Celonic proprietary; IP covered).

## 2.2 Vector systems

Target-vector (TV) contains as selectable marker a hygromycin$^R$ gene and a reporter gene (human secreted enhanced alkaline phosphatase, hSEAP or other) flanked by heterotypic FRT sites.

Exchange-vector (EV) contains a promoter-less neomycin$^R$ gene and the GOI (here the green fluorescent protein (eGFP)) driven by a CMV-promoter. The neoR-gene is only expressed after successful exchange, since this brings the gene under control of the SV40 promoter in the tagged locus.

## 2.3 Transfection, clonal selection

Transfection of suspension CHO cells with TV was performed in serum-free medium. Cells with stably integrated target-vector were identified by and expanded under hygromycin selection and further subcloned by limited dilution. Expression capacity of the randomly tagged integration site was monitored by activity of the reporter gene. Reporter expressing clones were selected as "wildcard" cells for the recombinase-mediated exchange of the reporter gene against the gene of interest.

### 2.4 Site-specific exchange reaction

For the exchange reaction the "wildcard" cell was cotransfected with the EV and a FLP-expression vector. Successful cassette exchange by FLP results in $neo^R$ and GFP positive cells, in which the original reporter and selection marker is deleted. The system has been tested with adherent and suspension cultures of the "wildcard" cell-line. By selection with G418 stable, high expressing clones for eGFP could be obtained. In the absence of FLP recombinase no G418 resistant cells could be obtained.

## 3. SUMMARY AND CONCLUSIONS

This study demonstrates that the TaCEx-system allows the rapid generation of a homogenous, genetically uniform production cell line. Initially a universal "wildcard" cell-line with a tagged genomic locus, suitable for high-expression of a foreign CDS is generated. The system allows than the reproducible and efficient integration of a GOI into this predetermined genomic locus, thus, generates reliable a new production cell-line within short time.

The strategy offers several advantages for the generation of high-producer cells necessary for the efficient and economically production of biopharmaceuticals:

- Time saving: a stable production cell-line could be obtained within a few weeks starting with a pre-made "wildcard" cell-line.
- Due to integration into pre-identified expression hot spots, the expression level of the production cell-line can be predicted more reliable and could be reached reproducible for different GOIs.
- High efficiency since only cells expressing the GOI will be obtained after the exchange.
- No additional integration of dispensable vector sequences during site-specific integration.
- Production of material for proof of concept experiments could be done within a few months. There will be no change in product quality at later development phases because a homogenous and genetically identical cell population will be available from the very beginning.

# Proteomic Characterization of Recombinant NS0 Cell Clones Obtained by an Integrated Selection Strategy

Adolfo Castillo[1], Kathya r. De La Luz[1], Svieta Victores[1], Luis Rojas[1], Yamilet Rabasa[1], Simon Gaskell[2] and Rolando Perez[1].
[1]*Research and Development Division, Center of Molecular Immunology, Havana, Cuba,*
[2]*Michael Barber Center for Mass Spectrometry, Manchester University, United Kingdom.*

**Abstract:** In this work several clones of a recombinant myeloma NS0 cell that were isolated in protein-free medium and selected by an integrated methodology that takes in account specific production and growth rates, stability of expression and resistance to apoptosis induced by nutrient limitation were characterized by two-dimensional gel electrophoresis in the pH range of 3-10. We have determined a group of 5 spots that correlated well with the values of selection indexes calculated for different clones.

**Key words:** NS0, proteomics, cell line selection, production rate, stability, apoptosis.

## 1. INTRODUCTION

One of the main goals during production process development of mammalian cell based products is the selection of the 'right' cell line before to start with the cell banking procedures. The selected cell line should have not only good growth properties, but also high and stable expression levels of the product of interest with the desired physico-chemical and biological characteristics. The lack of high throughput clone selection procedure that could render with high efficiency clones with desired characteristics for further scale-up process (i.e. high specific production rates) have been pointed out. However the detailed molecular basis of these desired characteristics have not been well established yet. For these aims new postgenomic technologies, as proteomics could be of great impact to identify



*R. Smith (ed.), Cell Technology for Cell Products*, 773–776.

and study the mechanisms involved in the generation of particular cell phenotype.

Previously several clones of a recombinant myeloma NS0 cell line producing a humanized monoclonal antibody were isolated in protein-free medium and an integrated methodology was developed for the final cell line selection. In order to characterise the molecular basis of obtained phenotypes we have carried out a comparative study of protein expression patterns by two-dimensional gel electrophoresis (2DE) in the pH range of 3-10 between clones with different values of Ks. We have determined 5 spots that correlated with the Ks values for different clones. In summary these results offer useful information about the complex mechanisms and protein expression patterns that are related with improved phenotypic characteristics needed for industrial cell lines.

# 2. RESULTS

## 2.1 Characterization of phenotypic differences

Several clones of a recombinant myeloma NS0 cell line producing a humanised Mab were obtained in protein-free medium (PFM) and those with IgG concentration values higher than 3 mg/mL in 96-well culture plates were selected. Antibody concentration and cell proliferation were measured in 24-well culture plates by ELISA and colorimetric assay respectively. Samples of cells population at the end of exponential phase in spinner flasks batch cultures were taken and apoptosis measured using the Anexin V Assay kit. The integrated selection index (K*s*) was calculated as reported previously (1).

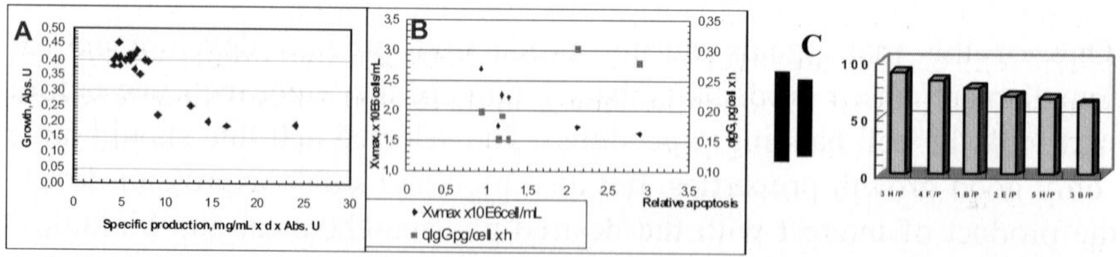


*Figure 1.* Correlation between: (A) specific production rates qab and growth values measured in 24-well plates and (B) maximum viable cell density (Xvmax) and qab measured in spinner flasks versus percent of apoptotic population relative to non apoptotic cell population (positive control), C) Selection indexes calculated for different clones in protein free medium.

## 2.2 Proteomic characterization of different phenotypic behaviours expressed in calculated selection index

Bidimensional gels for each clone was carried out by triplicate employing 17 cm, pH 3-10 linear IPG gel strips for IEF first dimension employing a total of 75 000 V*h. Second dimension SDS-PAGE was performed using 12,5% acrylamide gels and afterward gels were fixed and silver stained.

Obtained gels were analysed using the Melanie IV software: 1550 spots were detected, but only 878 were matched across all replicates for all clones.

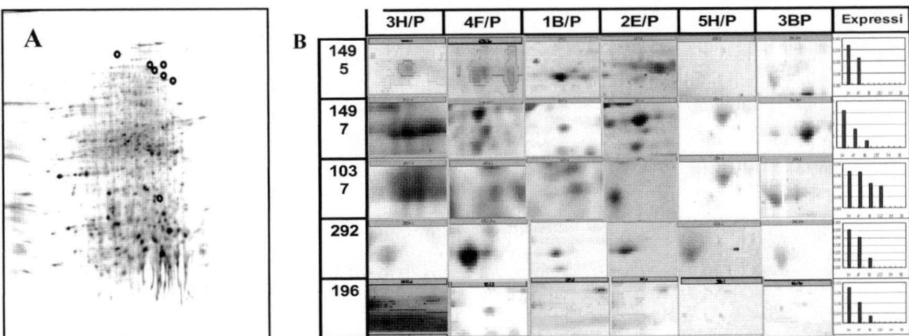


*Figure 4.* A) Spots with lineal correlation coefficient higher than 0.8 between selection index and values of percent of total volume intensity. B) Expression profile for five of selected spots.

# 3. CONCLUSIONS

During cell line selection process we observed the generation of clones with phenotypic differences. We have observed an inverse correlation between proliferation and production rates measured in 96-wells plates.

There exists also an inverse correlation between maximum viable cell number at the end of exponential phase and percent of cell population in apoptosis at this stage. The clones with the highest production rates showed also the highest values of apoptosis induction, that suggests that this phenotype is more sensible to nutrient limitation.We have considered the influence of these factorss in order to select cell lines for process scale-up.

We have found five spots that showed a high correlation coefficient between their expression intensities and calculated selection index, that could be related with the given phenotype for each analysed clone. Interesting to note that in all cases the expression of these spots decrease for clones with lower selection indexes.

# REFERENCE

Castillo A. J., Víctores S., Faife E., Rabasa Y., de la Luz K., Development of an integrated strategy for recombinant cell line selection. Animal Cell Technology

# Analysis and Modeling of the Metabolism of Lepidopteran Cell Lines

Jean-Christophe Drugmand[1], Yves-Jaques Schneider[2], Spiros Agathos[1]
*Université Catholique de Louvain, Institut des Sciences de la Vie[1], Unit of Bioengineering, Place Croix du Sud, B-1348 Belgium[2], Unit of Cellular Biochemistry, Place Louis Pasteur[1], B-1348 Belgium*

**Abstract:** The network metabolic model of High-FiveTM insect cells based on the consumption / production of metabolites and the protein, DNA/RNA, lipid and carbohydrates content of the cells has made it possible to study the internal fluxes of these cells' metabolism.

**Key words:** Insect Cells, Lepidopteran Cells, Metabolism, Model, High-FiveTM

## 1. INTRODUCTION

In the last two decades, the Insect Cell - Baculovirus Expression Vector System (BEVS) has become popular for producing most recombinant (glyco-)proteins. The simplicity of insect cell cultivation in serum-free media and the easy construction of recombinant baculovirus vectors have made the BEVS quite an effective transient expression system. Among insect cell lines, High-Five™ and Sf-9 cells have great potential as technology platforms.

Metabolic flux models are based on the knowledge of the cells' metabolism. In our study, the analysis of pathways was deducted from the literature on insect cells. All concentrations of intracellular intermediates were assumed as constant. We measured the macroscopic extracellular consumption / production rate of each metabolite by enzymatic test or HPLC for amino acids. The cellular content was determined by standard tests



*R. Smith (ed.), Cell Technology for Cell Products*, 777–780.

(Bradford test for protein, Hoechst test for DNA, RNA fluorometric test, Soxhlet test for total lipids and the anthrone test for total carbohydrates).

## 2. ANALYSIS OF THE METABOLISM

The metabolism of in vitro cultivated insect cells is known to be different from that of mammalian cells. Typically insect cells consume more glucose, oxygen, glutamine (Gln) and asparagines (Asn) than mammalian cells. Moreover, the production of lactate is lower than commonly seen in mammalian cells. The consumption of amino acids and lipids, and the production of ammonia are more intense.

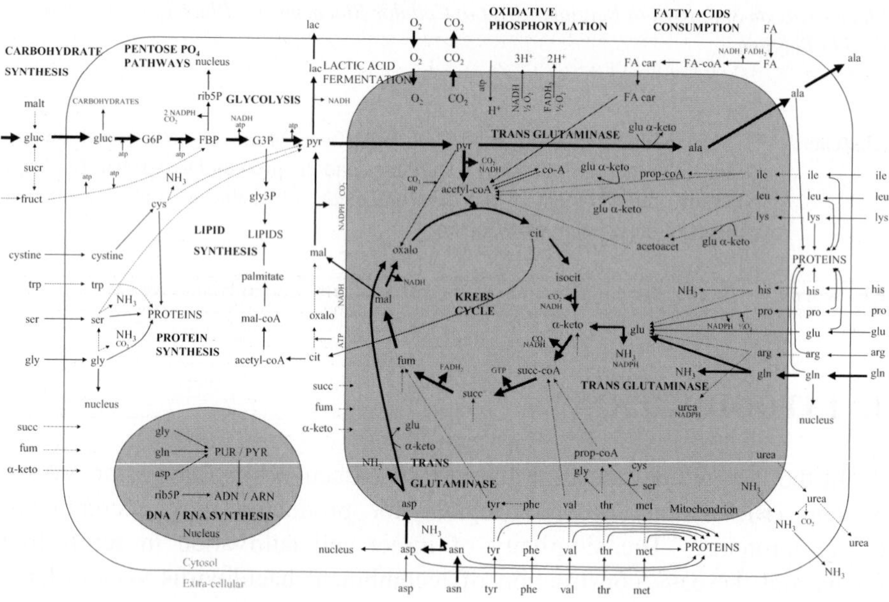


*Figure 1.* Catabolism and anabolism of High-FiveTM insect cells. The thickness of the lines represents the magnitude of the flux.

In exponential phase, we found that High-Five cells consume more glucose and Asn than Sf-9 cells. The latter consume more Gln and produce lower lactate than High-Five cells (data not shown).

The consumption of Asn and Gln bring high energy to the cells. Moreover, we found that the feeding of the Krebs cycle by Asn and Gln caused an efflux of malate out of the mitochondrion (15% of the malate flux), which was converted to pyruvate in the cytosol. In case of cultivation with high Asn (25 mM), this spillover increased by 2.5 times and produced a

3-fold increase of lactate, while keeping the same alanine (Ala) production (Table 1 and Figure 1).

We found that cells consume more nutrients (glucose, Gln, Asn) when we added them in higher concentrations. This may be due to an increase of these compounds' influx when the concentration gradient (between cytosol and media)rises. With high Asn and Gln concentrations, cells consume more glucose, and produce more lactate than in lower glucose concentration. The majority of nitrogen ($NH_3$ and Ala) produced comes from Asn except in presence of high concentration of Gln. During the lag phase, cells consume high concentrations of amino acid via an anaplerotic pathway (Gln for Sf-9 and Asn for High-Five) and thus obtain high energy levels.In a standard batch culture, we found that the cells used 73% of the ATP produced for their growth and 27% for their maintenance.

*Table 1.* Effect of different media compositions on High-FiveTM cells' consumption of glucose, Gln, Asn, the production of lactate and the source of ATP and Nitrogen produced.

| | | High Gluc | Standard Gluc | Low Gluc | High Asn | High Gln | With exogenous Lac |
|---|---|---|---|---|---|---|---|
| Media composition at the inoculation (mM) | glucose | 55 | 34 | 12 | 55 | 55 | 55 |
| | lactate | 0 | 0 | 0 | 0 | 0 | 5 |
| | asparagine | 12 | 12 | 12 | 25 | 12 | 12 |
| | glutamine | 12 | 12 | 12 | 12 | 25 | 12 |
| Specific consumption rate of (% of the standard culture) | glucose | **1.15** | **1.00** | **0.88** | **1.37** | **1.45** | **1.39** |
| | lactate | **0.54** | **1.00** | **1.64** | **1.71** | **3.91** | **1.57** |
| | asparagine | **1.60** | **1.00** | **0.21** | **3.84** | **1.70** | **1.50** |
| | glutamine | **0.94** | **1.00** | **0.40** | **2.24** | **15.79** | **2.80** |
| Ernegy : ATP production (% of the total Atp production) | Krebs Cycle | **66.90** | **60.60** | **59.70** | **66.05** | **56.74** | **71.67** |
| | Glycolysis | **17.80** | **16.20** | **16.00** | **17.58** | **15.13** | **19.11** |
| | Lactic fermentation | **0.10** | **0.20** | **0.40** | **2.98** | **0.58** | **0.28** |
| | Asn consumption | **6.65** | **3.70** | **0.50** | **7.26** | **15.39** | **4.00** |
| | Gln consumption | **3.80** | **3.40** | **1.40** | **1.86** | **8.73** | **0.80** |
| | Lipid consumption | 3.96 | 4.10 | 4.80 | 3.40 | 2.99 | 3.65 |
| | others pathway | 0.80 | 11.80 | 17.20 | 0.47 | 0.45 | 0.48 |
| Nitrogen production (% of the total Atp production) | Asn consumption | **44.50** | **33.00** | **5.10** | **67.00** | **32.00** | **56.20** |
| | Gln consumption | **1.90** | **0.50** | **6.50** | **6.20** | **60.00** | **1.80** |
| | others pathway | 53.60 | 66.50 | 88.40 | 26.80 | 8.00 | 42.00 |
| Ratio NH3 + urea / ala | ratio NH3 + urea / ala | 1.05 | 1.64 | 1.93 | 2.71 | 2.02 | 1.23 |

# 3. CONCLUSION

The analysis of this metabolic flux network with High-Five and Sf-9 cells has shown different rates in the metabolic pathways in batch cultivation and in infection phase (data not shown) under different operational conditions (e.g. exponential growth phase, lag phase, limitation of glucose, Gln and Asn, etc.).

Based on this flux model, we have constructed a general metabolic pathway network of these insect cells and we have developed a predictive model (data not shown) that can simulate the batch growth, the metabolism (glucose, lactate, Gln, Asn, amino acids, and oxygen) of the cells, plus the time course of the pH and osmolarity during cultivation.

# REFERENCE

Ikonomou, L., G. Bastin, Y.-J. Schneider, and S.N. Agathos. 2001. Design of an efficient medium for insect cell culture and recombinant protein production. *In Vitro Cell. and Develop. Biol.-Animal,* **37**:549-559.

# Growth of Mammalian and Lepidopteran Cells on BioNOC® II Disks, a Novel Macroporous Microcarrier

Jean-Christophe Drugmand[1], Jean-François Michiels J.-F.[2],
Spiros Agathos[1], Yves-Jaques Schneider[2]
*Université Catholique de Louvain, Institut des Sciences de la Vie[1] Unit of Bioengineering, Place Croix du Sud, B-1348 Belgium[2], Unit of Cellular Biochemistry, Place Louis Pasteur[1], B-1348 Belgium*

**Abstract:** The use of BioNOC II microcarriers in a fixed bed bioreactor setup allows to produce high protein levels in a CHO-320 cell line expressing constitutively interferon-γ and also with insect cells used with a baculovirus transient expression vector.

**Key words:** Fixed bed, Microcarriers, BioNOC, CHO cells, Insect cells

## 1. INTRODUCTION

Microcarrier technology has the advantages of easy sampling, relatively large surface area and is useful in perfusion culture. Insect cells are well-established for the production of recombinant proteins by the baculovirus expression vector system (BEVS). CHO cells are widely used for the production of therapeutic proteins. Plant peptones are the safest supplement for protein-free media. They exert a nutritional effect, and they contain bioactive peptides, responsible for increases in growth and production.

BioNOC II™ (Cesco Bioengineering Co, Taichung, Taiwan) cell culture disks are new 100% pure PET nonwoven macroporous carriers for the growth of animal, mammalian and insect cells. BelloCell-500 is a disposable bioreactor bottle. It consists of a chamber holding 6.5 g of BioNOC II



*R. Smith (ed.), Cell Technology for Cell Products*, 781–784.

carriers and a lower compressible chamber for mixing and surface aeration by an up/down reciprocal motion.

CHO-320 cells producing recombinant interferon-γ were cultivated in BDM serum-free media (Burteau *et al.*, 2003) supplemented with 0.5-4 g/l cotton peptone from Quest (Naarden, The Netherlands). High Five™ insect cells were cultivated in YPR serum-free media and infected with a AcMNPV baculovirus r-βgal (MOI 2) (Ikonomou *et al.*, 2001). Celligen-Plus 2 L and BelloCell-500 bioreactors were used. The cell biomass was estimated by a protein assay.

## 2. CHO CELLCULTIVATION IN BelloCell-500

Two BelloCell-500 batch cultivations were conducted simultaneously with and without peptone (1 g/L). Although the growth was slower (Figure 1), interferon-γ production was slightly higher when peptone was added.

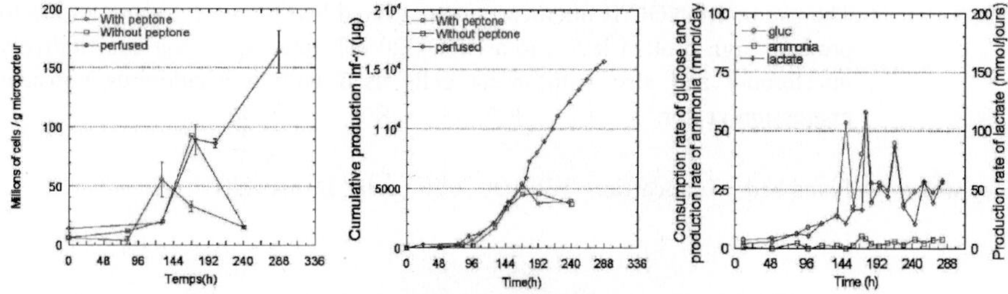


*Figure 1.* CHO cell cultivation in BelloCell-500. The bioreactor parameters were culture up/down speed = 1/1 mm/s and up/down hold time = 0/3 min.

## 3.

In a BelloCell-500 perfusion cultivation medium was supplemented with 1 g/L cotton peptone. The medium was changed for the first time after 72 hours of culture, when the glucose concentration was low. After 108 hours, medium was changed twice a day. The time course of glucose consumption and of the lactate and ammonia and ammonia production shows that most of the glucose was converted to lactate which decreased the pH, while ammonia production was low. Interferon-γ production was 3-fold higher than in batch.INSECT CELL cultivation in fixed bed.

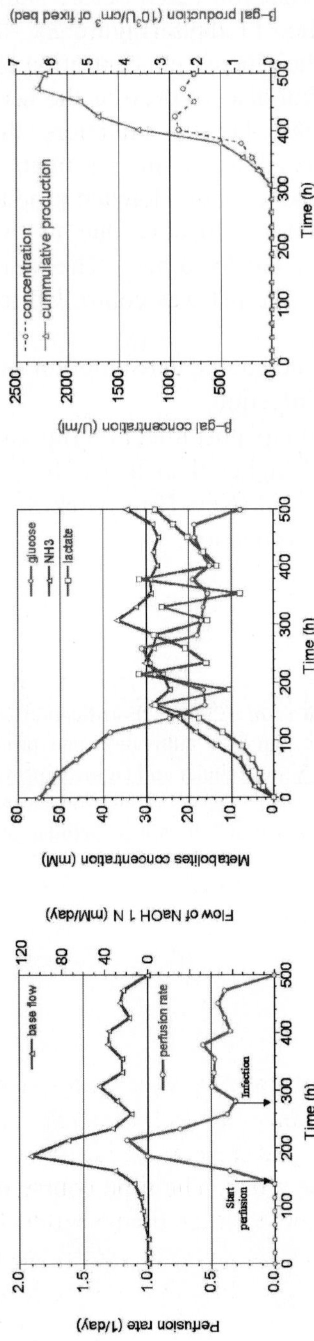


*Figure 2*. Insect cell perfusion cultivation in Celligen-Plus 2 L bioreactor in fixed bed basket configuration (1.35 L working volume) with 20 g of BioNOC II (bed volume 500 cm3, 80 rpm, 28°C, 60% DO). The pH was controlled with the addition of 1 N NaOH.

A fixed bed reactor in perfusion mode was used with BioNOC II. The cells were cultivated for 6 days in batch before starting the perfusion, and the cells were infected after 11 days (Figure 2). The perfusion rate was adapted to keep not only the glucose concentration below 20 mM (a third of the initial concentration) but also to prevent the accumulation of ammonia and lactate above 25 mM. During infection, this perfusion rate was decreased to prevent the dilution of r-protein and keep by-product concentrations in an acceptable range. Lactate production was high (about 1 mol lactate / 1 mol glucose) probably due to the lack of oxygen and accumulation of $CO_2$ inside the fixed bed. The ratio oxygen / glucose of 3 supports this observation. The pH was controlled to prevent its dropdue to lactate production.

Cell density (based on metabolite consumption) was estimated at $400.10^6$ cells/g of BionocII before infection.

We obtained after 200 hours post infection (hpi) a high production ($3x10^6$ U β-gal /run) i.e. 2.5 fold higher than in batch. In experiments using the same setup but with Fibra-Cell (New Brunswick Scientific, Edison, NJ), we obtained the same level of production.

## REFERENCES

Burteau, C. C., F. R. Verhoeye *et al.* (2003). "Fortification of a protein-free cell culture medium with plant peptones improves cultivation and productivity of an interferon-γ-producing CHO cell line." In Vitro Cellular and Developmental Biology **39**: 291-296.

Ikonomou, L., G. Bastin *et al.* (2001). "Design of an efficient medium for insect cell culture and recombinant protein production." In Vitro Cellular and Developmental Biology-Animal **37**: 549-559.

# Screening Strategies for Iron Chelators in Serum Free Media

Joanne Keenan[1] Dermot Pearson[2] Martin Clynes[1]
[1]*National Institute for Cellular Biotechnology, Dublin City University, Ireland*
[2]*Delta BiotechnologyLtd., Nottingham, UK*

**Abstract:** A variety of iron chelators, including recombinant human transferrin, were evaluated as replacements to holo human transferrin on a number of industrially relevant cell lines (MDCK, BHK-21 and Vero). The results showed that short term screening over one passage is not necessarily adequate to select a suitable transferrin replacement.

**Key words:** transferrin, iron chelators, recombinant human transferrin, SFM, mammalian cell culture

## 1. INTRODUCTION

The development and routine use of serum-free media (SFM) in large scale production plays an important role in achieving optimal production, reducing complexity of down-stream processing and has an impact on biosafety especially with the removal of animal-derived components from serum-free media (Salis *et al*., 2002). The use of high-throughput systems to analyse new components and how they interact with serum-free components has greatly improved the efficiency of selection.

Transferrin, one of the animal-derived proteins that is regularly used in serum-free formulations can be replaced by alternative iron chelators (Darfler, 1990; Keenan and Clynes, 1996). Many of these have been chosen by screening methods because not single chemical entity will work for all cell lines. In this report, we compare the short term screening strategies to longer subculture methods for selecting suitable transferrin replacements.



*R. Smith (ed.), Cell Technology for Cell Products*, 785–788.

# 2. MATERIALS AND METHODS

## 2.1 Cell lines and materials

MDCK cells were obtained from the American Type Culture collection. BHK-21-PPI-C16 (Gammell *et al.*, 2003) and Vero-PPI (O'Driscoll *et al.*, 2002) were previously engineered to express human preproinsulin. Media was supplied by Gibco. All components for SFM were supplied by Sigma (Poole, UK) except DeltaFerrin$^{TM}$ recombinant human transferrin which was supplied by Delta Biotechnology Ltd (Nottingham, UK). For MDCK, the SFM designed by Taub (1979) was used. For BHK-21-PPI-C16, the SFM designed by Bradshaw (1994) was supplemented with 25ng/ml prostaglandin E1 and 50nM hydrocortisone. For Vero-PPI the Bradshaw SFM was supplemented with 5pM Tri-iodothreonine and 5ug/ml fibronectin. The iron chelators tested included holo human transferrin, DeltaFerrin$^{TM}$, ferrous ammonium sulpate (FAS) and ferrous sulphate ($Fe_2SO_4$)

## 2.2 Iron chelator screen

Miniaturised 96-well plate assays were used to screen iron chelators. Cells were washed and incubated for 24 hours in basal medium to remove excess transferrin. After incubation at 37°C and 5% $CO_2$ for 5 – 7 days growth when confluency was being approached, cell growth was estimated using the acid phosphatase assay (Martin and Clynes, 1991). All assays were carried out in triplicate with 8 replicates per plate.

## 2.3 Subcultures

Subcultures were carried out by growing the cells in the respective SFM for 4 to 5 days and performing cell counts using a haemocytometer. The cells were then used to reset up a fresh flask. All assays were carried out in triplicate with two replicates per iron chelator.

# 3. RESULTS

Initial screening assays show the three cell lines to have varying dependence on transferrin. MDCK was most dependent with at least a 5-fold increase in growth on inclusion of transferrin in serum-free media. BHK-21-PPI-C16 showed a 2-fold increase on transferrin inclusion and Vero-PPI showed littlereliance with only a maximum of 20% increase on inclusion of transferrin.

For MDCK, the growth promoting activity of holo human transferrin and DeltaFerrin$^{TM}$ are similar in both screening and subculture assays. For the iron chelators, FAS and $Fe_2SO_4$, the screening assays do not reflect the subculture results. $Fe_2SO_4$ supported less than 40% of the growth achieved by transferrin in the screen but was marginally less effective than transferrin in the subcultures. In contrast, FAS appeared similar to transferrin in the screening assay but was only 20% as effective by the fifth subculture.

For BHK-21-PPI-C16, screening showed all the factors to be as good as holo human transferrin, however only DeltaFerrin™ and $Fe_2SO_4$ were comparable to the human transferrin after subculturing. As with the MDCK, FAS could not support continued growth.

Vero-PPI was shown in screening assays to be least dependent on transferrin. For this reason, Vero-PPI were subcultured in SFM with no iron chelator, to see if the cells could continue growing long term without any additional iron chelator. After 5 passages, Vero-PPI without an iron chelator only showed 20% growth achieved with transferrin. Of the iron chelators tested on Vero-PPI, the screening and the initial subculture assays correlated well (transferrins and FAS). $Fe2SO_4$ was not included as it was shown to be inhibitory in comparison to the blank.

## 4. DISCUSSION AND CONCLUSION

The results shown here indicate that short term screening assays for selecting growth promoting factors in SFM, may not always reflect the long term requirements of cells. This is especially true in the case of circulating factors like transferrin. Transferrin transports iron to growing cells via transferrin receptors. Once internalized, the transferrin is then depleted of iron and exported out of the cell where it can circulate to bind any free iron (Richardson and Ponka, 1997). Results in short term assays may be compromised by circulating residual transferrins. In these short term assays and prior to subculturing, circulating bovine transferrin should not have presented a significant problem as particular attention was paid to ensure that the cells were depleted as much as possible of bovine transferrin (several washings and incubation overnight in transferrin-free media).

The short term screens correlated well with long term results from the subcultures for the transferrins for the three cell lines. For FAS, subculturing revealed this chelator unable to support growth similar to transferrin for MDCK and BHK-21-PPI-C16 despite being as effective in screenings. FAS was a suitable chelator for Vero-PPI (both in short term screens and subcultures) perhaps due to the low dependence of Vero-PPI on transferrin.

Results for screening and subculturing for $Fe_2SO_4$ were consistent for Vero-PPI and BHK-21-PPI-C16 but were not consistent for MDCK. Based solely on the screening, FAS may have been selected as a suitable replacement but for MDCK and BHK-21-PPI-C16, $Fe_2SO_4$ would have been a better choice based on the subculture results.

Overall, from these results it can be seen that screening of transferrin replacements may be useful for eliminating unsuitable chelators but should not be used as the sole basis on which to select a transferrin replacement.

The level of transferrin-dependence of the cell line may also be important in the choice of assays. For Vero-PPI, with low level dependence on transferrin, screening reflected the results obtained in long-term subcultures. For MDCK and BHK-21-PPI-C16, with greater dependence on transferrin, the screening did not reflect subculture results. From the work presented here, the most appropriate format for selecting a transferrin replacement is a combination of screening and subculturing for a number of passages, depending on the reliance of the cells on transferrin.

## REFERENCES

Bradshaw GL, Sato GH, McClure DB and Dubes GR (1994). The growth requirements of BHK-21 in serum-free culture. J Cell Physiol. Feb 114(2); 215-21.

Darfler FJ (1990). A protein-free medium for the growth of hybridomas and other cells of the immune system. In Vitro 26, 769-778.

Gammell P., O.Driscoll L. and Clynes M. (2003). Characterisation of BHK-21-PPI-C16-21 cells engineered to secrete human insulin. Cytotechnol. 41: 11-21

Keenan J and Clynes M (1996). Replacement of transferring by simple iron compounds for MDCK cells grown and subcultured in serum-free medium. In Vitro Sep 32(8) 451-3.

Martin A and Clynes M (1993). Comparison of 5 microplate colorimetric assays for in vitro cytotoxicity testing and cell proliferation assays. Cytotechnology 11:49-58.

O'Driscoll L., Gammell P. and Clynes M. (2002). Engineering Vero-PPI cells to secrete human Insulin. In Vitro 38: 146-153

Richardson DR and Ponka P (1997). The molecular mechanism of the metabolism and transport of iron in normal and neoplastic cells. BBA 1331, 1-40.

Taub M, Chuman L, Saier, MH, Sato G (1979). Growth of MDCK in hormonally defined serum-free medium. PNAS 76 (7): 3338-3342.

# Production of Human Growth Hormone in a Mammalian Cell High Density Perfusion Process

T. Nottorf, W. Hoera, H. Buentemeyer, S. Siwiora-Brenke, A. Loa*, J. Lehmann
*Institute of Cell Technology, University Bielefeld, Germany; *Cell Culture Service GmbH, Hamburg, Germany*

**Abstract:** A perfusion culture with an inclined cell settler for cell retention was used to produce human growth hormone. Cell densities of $9.0 \cdot 10^6$ cells/mL could be maintained in steady state over a designated period of time to increase the process productivity. The cell extraction and perfusion rates had to be finely adjusted to maintain steady state conditions. Analyses of growth hormone concentrations were performed by ELISA and RP-HPLC. The downstream procedure comprises several techniques including centrifugation, ultra-diafiltration and hydrophobic interaction chromatography.

**Key words:** perfusion, human growth hormone, inclined settler

## 1. INTRODUCTION

In vivo the predominant variant of human growth hormone (hGH) is a non glycosylated, 22 kDa single chain protein consisting of 191 amino acids. HGH, as a therapeutic agent, is mostly produced in prokaryotes. The main advantage of the classical microbial processes is a high product concentration yield. Unfortunately, hGH is accumulated as inclusion bodies or secreted in the periplasm. Mammalian cells, like the $CHO^{SFS}$ cell line, which was used, secrete correctly folded hGH in the medium. The $CHO^{SFS}$ cell line was growing under serum free conditions to facilitate purification.



*R. Smith (ed.), Cell Technology for Cell Products*, 789–793.

In this work a perfusion process with an inclined cell settler for cell retention was performed. Cell densities of $9.0 \cdot 10^6$ Cells/mL could be maintained in steady state over a designated period of time to increase the process productivity. Therefore cell extraction and perfusion rates had to be periodically adjusted to maintain steady state conditions. Analyses of product concentrations were performed by ELISA and RP-HPLC. The downstream procedure comprises several techniques including centrifugation, ultra-diafiltration and hydrophobic interaction chromatography.

## 2. MATERIALS AND METHODS

A 2 L lab-scale bioreactor from Sartorius BBI Systems was used for each cultivation. Bubble free aeration and mixing were ensured using a tumbling membrane stirrer. The CS20 laminar settler (Biotechnology Solutions, Ca, USA) is made of medical grade steal, containing two chambers, each with four plates. Before entering the settler, the cell suspension was cooled to ambient temperature. The hGH-concentrations were determined by RP-HPLC and UV detection at 214 nm. Furthermore, a hGH-specific ELISA (Roche Diagnostics) was used.

## 3. RESULTS

### 3.1 Perfusion process

A perfusion process was performed with the purpose of analysing long term stability steady states in perfusion mode. In batch mode cells grew exponentially until day 6.8, when cell extraction was commenced (Figure 1). The cell density decreased, although $\mu_{max}$ (0.41/d) was higher than the adjusted cell extraction rate. This could be attributed to a limitation of the proliferation stimulating factor in the medium. The limitation was eliminated by adding of the limiting component at day 14. As a result, the cells resumed growth at cell extraction rate of 0.39/d. On day 17, the cell extraction rate was adjusted to 0.42/d. Thereupon the cell density slowly decreased.

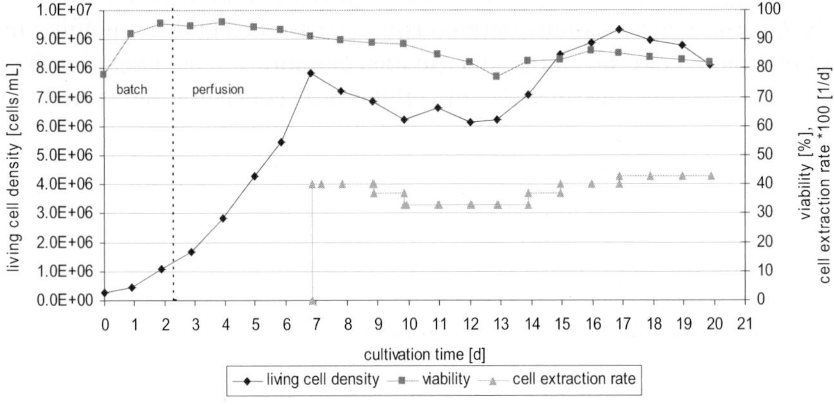


*Figure 1.* Living cell density, viability and cell extraction rate during perfusion process.

Altogether it can be concluded that the steady-state could not be maintained at a constant cell extraction rate. However, it would be possible to confine cell density within an upper and a lower limit. Upon exceeding or falling below this limit, the cell density must be adapted by raising or declining the cell extraction rate. Furthermore, it was observed that after overfeeding the cells the viability did not decrease as much as in a substrate limited cultivation.

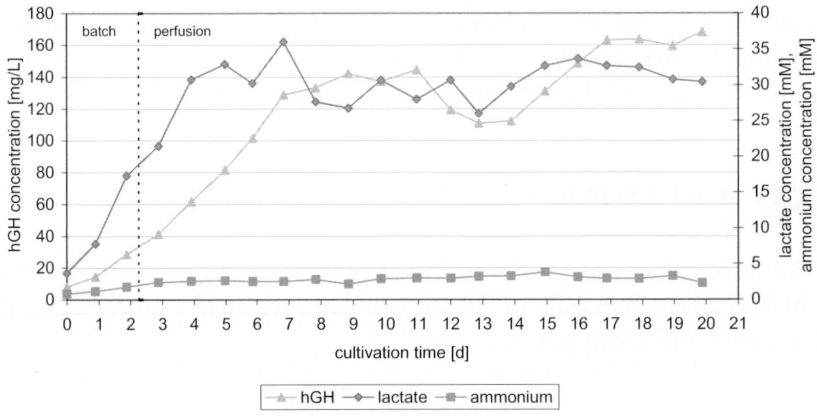


*Figure 2.* hGH-, lactate- and ammonium concentrations during perfusion process.

Figure 2 shows that hGH concentrations could be maintained between 111 and 168 mg/L during perfusion mode. The lactate concentration varied between 21 and 36 mM, while ammonium concentration stayed below 5 mM.

Thus, both metabolites did not reach growth inhibiting concentrations. The glucose concentration continuously decreased from the beginning of the cultivation. At day seven the glucose concentrations ranged between 0.3 and 1.3 mM. During the process the glutamine concentration stayed above 400 μM.

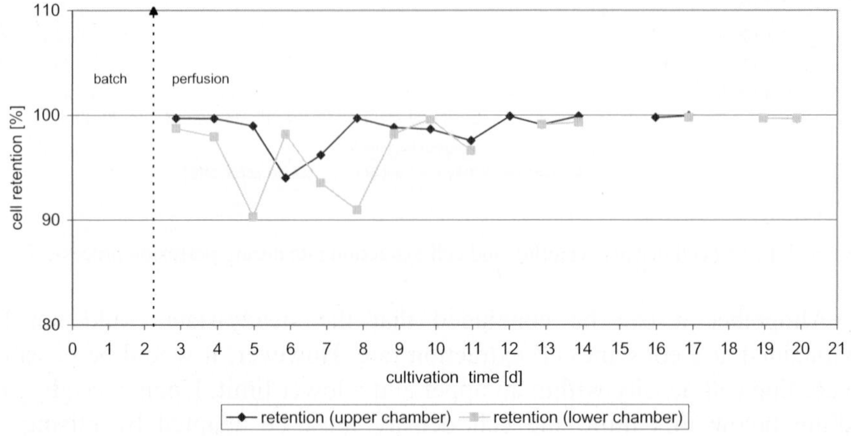


*Figure 3.* Cell retention during perfusion process. During this run the plates in the upper chamber were mirror polished and the plates in the lower chamber were Teflon coated.

As can be seen in Fig. 3, cell retention in both chambers was high, with cell retention in the upper chamber being slightly higher. As experiments with interchanged plates (data not shown) indicate, this was not due to coating of plates with Teflon, but rather a geometric artefact.

## 4. CONCLUSIONS

We were able to establish a perfusion process, with a productivity of 450 mg hGH/d, at a cell density of $9.0 \cdot 10^6$ cells/mL, which is 9 times higher than in a conventional batch process.

## REFERENCES

Büntemeyer, H. (2000)
Off-line analysis in animal cell culture, Methods
Encyclopedia of Cell Technology, Ed. R.E. Spier, Wiley, New York, 945-959
Stiens, L.R., Büntemeyer, H., Lütkemeyer, D., Lehmann, J., Bergmann, A., Weglöhner, W. (2000)

Development of serum-free bioreactor production of recombinant human thyroid stimulating hormone receptor.
Biotechnol. Prog. 16, No. 5, 703-709
Büntemeyer, H., Siwiora, S., Lehmann, J. (1997)
Inhibitors of cell growth: Accumulation and concentration.
Animal Cell Technology, From vaccines to genetic medicine. Eds. M.J.T. Carrondo, B. Griffiths, J.L.P. Moreira. Kluwer Acad. Publ. 651-655

# Development of an Integrated Qualitative and Quantitative Approach to Study Expression-Secretion Bottlenecks of Mammalian cDNAs in HEK293 EBNA Cells

Loïc Glez, Christine A. Power, Thierry Battle, Georg Feger
*Serono Pharmaceutical Research Institute 14, Chemin des Aulx, 1228 Plan Les Ouates Geneva, Switzerland. e-mail:* loic.glez@serono.com

**Key words:** Animal cells, HEK293-EBNA, Transient, PEI, Transfection, Recombinant Protein, qRT-PCR, mRNA Expression level, Secretion Pathway, Rate Limiting Steps.

## 1. INTRODUCTION

The goal was to establish an analytical tool kit allowing quantitative and qualitative analysis of transient gene expression process on HEK293-EBNA cells.

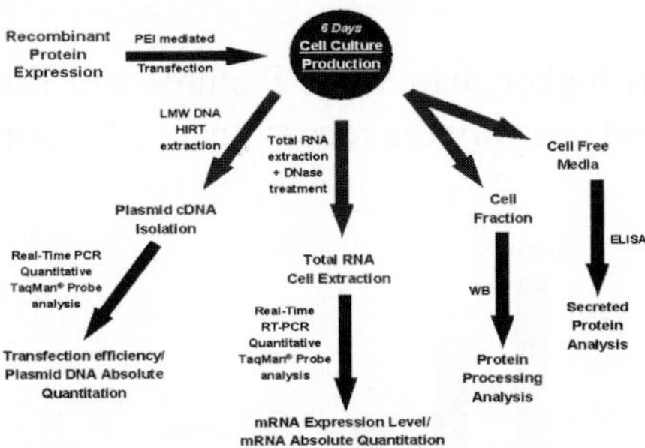


*Figure 1.* Integrated qualitative and quantitative investigations.



*R. Smith (ed.), Cell Technology for Cell Products*, 795–898.

By this strategy we aimed at identifying recombinant secreted protein expression rate limiting steps.

## 2. MATERIAL AND METHODS

Ref. Protein: AS900739, affinity tagged human rec. protein of 24 kDa.

Cell line and Culture: HEK293-EBNA cells were transiently transfected using the JetPEI™ (PolyPlus Transfection) transfection reagent (N/P = 5).

Vector: Modified pEAK 12, carries a fluorescent reporter gene.

QRT-PCR: TaqMan reagent on the 3'UTR allows absolute quantification of both incorporated plasmid cDNA and transgene mRNA.

Analyticals: Tagged products were visualised by WB and specific ELISA measured specific productivity.

## 3. RESULTS AND DISCUSSION

Cell population reach a maximum of $1.5*10^6$ cells/ml after 6 days

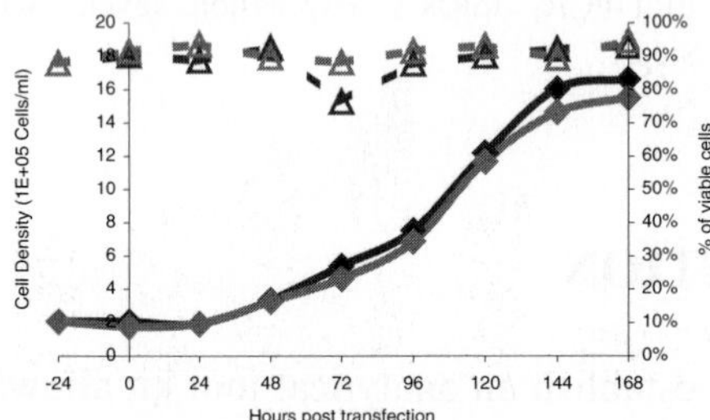


*Figure 2.* Cell concentration and viability curves.

Viability was still higher than 90%. Pictures and fluorescence read out allowed to monitor and quantify the recombinant cells population

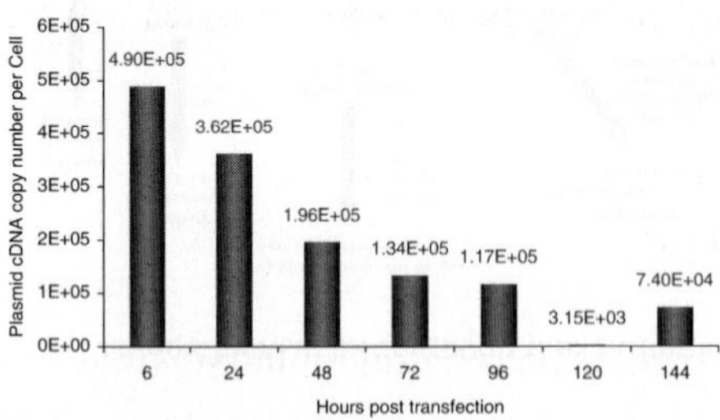


*Figure 3.* Plasmid cDNA copy number per cell.

More than 490,000 plasmid copies (>73%) on the 675,000 initially exposed per cell were « incorporated » in the first 24 hours. Following the cell division the plasmid copy number per cell progressively decreased to a minimum of 74,000 copies.

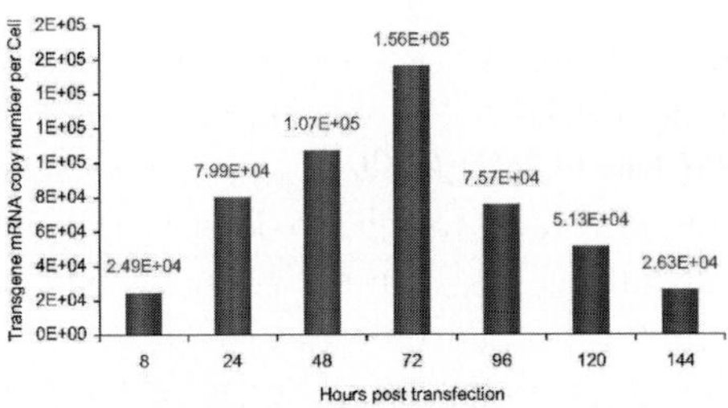


*Figure 4.* Transgene mRNA expression level per cell.

Transgene mRNA expression level reaches a maximum of 156,000 copies per cell at day 3. Following the reduction of plasmid cDNA copy number per cell, the copy number of transgene mRNA falled down to 26,300 copies per cell at day 6.

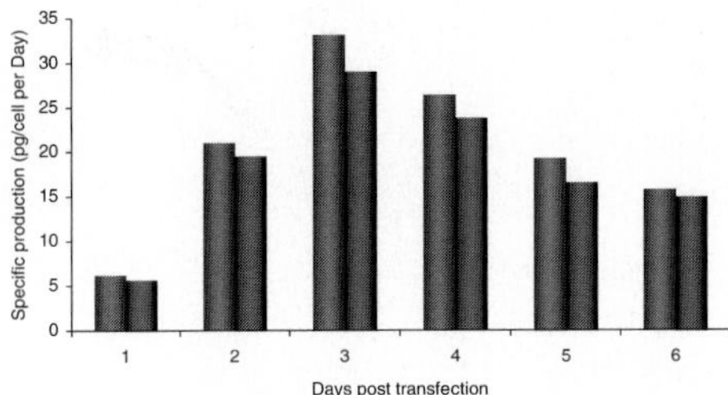


*Figure 5.* Specific Productivity.

Specific productivity was higher than 30 pg/cell per day during day 2 and 3. Four days after transfection, the specific productivity dropped down to 5 pg/cell per day. The final concentration reaches a maximum of 22 mg/l in the cell free media.

# 4. CONCLUSION

This integrated qualitative and quantitative strategy was applied to a set of 60 (predicted) secreted proteins, our data will address the different rate limiting steps of recombinant protein expression/secretion.

## ACKNOWLEDGMENTS

The authors would like to thank Patricia De-Lys (immunology, SPRI), for setting and running the ELISA.

## REFERENCE

Hirt B. J Mol Biol. 1967 June 14;26(2):365-9.

# Rotating Bed Bioreactors: a Novel Way to Natural Proteins by Tissue-like Cultivation of Mammalian Cells

Hans P.A. Hoffmeister,
Ziegeleistr. 7, 16727 Eichstädt, Germany

**Abstract:** A new type of bioreactor with a rotating bed of porous ceramic material is presented. The rotating bed technology allows cultivation of anchored cells without loss of vitality and productivity for 3 to 6 month. This BIOSTAT® Bplus-RBS reactor system works in perfusion mode and can be scaled up from 5 ml to 3000 ml bed volume. Cells grow imbedded in their own ECM to tissue-like densities. Supply of adhered cells with oxygen and nutrients is outstanding, cell stress is minimized. Cells need less supplements to produce the same amount of glycoprotein, cell broth is easier to handle, compared to suspension culture. Excellent glycosylation patterns were observed when a human EPO was produced in a BIOSTAT® RBS reactor using a $CHO^{dhfr-}$ cell clone for expression. The protein backbone and overall glycosylation structures of this glycoprotein were analysed by IEF, oligosaccharid mapping and MALDI/TOF-MS techniques after immunoaffinity isolation from culture harvest and compared to the international EPO reference standard (BRP batch II).

**Key words:** Rotating bed bioreactor; porous ceramic bed; anchored cell culture; adherent mammalian cells, high density cell growth; CHO-DHFR-cells, extra cellular matrix; post processing; glycosylation pattern; human erythropoietin

## 1. INTRODUCTION

Recombinant human proteins for pharmaceutical applications with complex glycosylation patterns need to be expressed in mammalian cell lines for appropriate post-processing. Glycoproteins produced in suspension culture often show deviation in their glycosylation structure compared to



*R. Smith (ed.), Cell Technology for Cell Products*, 799–801.

human wild type substances. One cause may be that most of the interesting glycoproteins derive from cells naturally organized as tissues. That means cells being imbedded in their specific ECM. The rotating bed reactors (BIOSTAT® Bplus-RBS reactors, see Fig. 1) offer an excellent way to expand and culture anchored mammalian cells tissue-like. It was presumed that cultivation as realized in this reactor system may be advantageous with respect to their glycosylation patterns of expressed glycoproteins, compared to production in suspended cell culture. In case of human EPO the hypothesis was proven and partly verified.

## 2. CHARACTERISTICS OF THE ROTATING BED SYSTEM AND CERAMIC SPONCERAM® DISCS

Figure 1 shows a BIOSATAT® Bplus-RBS 500 reactor used for the cultivation of the $CHO^{dhfr-}$ cell clone. The solid bed consists of thin Sponceram® discs rotating in the reactor glass vessel which is half filled with medium. Sponceram® is a special ceramic material possessing large surfaces of open macro pores covered with soaking micro pores. Anchored cells attach fast at the surface without any preparing. Characteristics of this technique are:

- Inoculation on the porous skeleton of Sponceram® discs leads to fast adherence and equal distribution of cells on the surface (no hot spots)
- Alternating contact of cells with medium and gas atmosphere guarantees excellent supply with oxygen and nutrients
- Gentle moving of cells on discs through medium minimize cell stress
- Cells grow up to high density (up to $2 \times 10^{10}$ per fully equipped rotating bed)
- Cells form their typical ECM and show tissue-like organisation
- Perfusion mode minimises presence time in detrimental environment
- Cells remain viable and secrete recombinant glycoproteins in constant high yields and with excellent glycosylation pattern
- Cell broth with target glycoprotein contains nearly no suspended cells and debris

## 3. LONG-TERM CULTIVATION OF A CHO DHFR-CLONE

Cells were cultivated for 80 days in serum free medium. Sponceram® discs became fully covered during this time with cells imbedded in ECM. Glucose consumption and specific EPO production can be seen in Fig 2.

EPO was isolated from cell broth by immunoaffinity capture. It showed excellent glycosylation pattern compared to human EPO BRP standard (native oligosaccharid mapping, antennarity etc).

*Figure 1.* [BIOSTAT Bplus-RBS bioreactor with magnetic drive, operated in a GMP Breeder (Zellwerk), controlled by a BIOSTAT® Bplus unit (SBBI)].

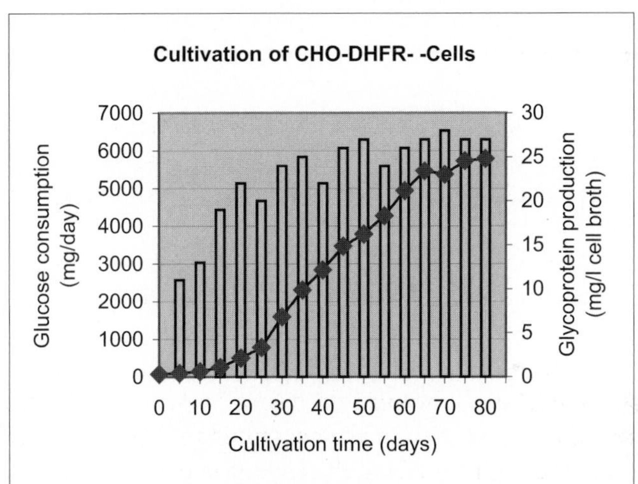


*Figure 2.* [Glucose consumption (left scale, dots) and specific EPO production rate (right scale, columns) during long term cultivation of a CHO clone in a BIOSTAT® Bplus RBS bioreactor].

# Minireactors for Animal Cell Culture: Monitoring of Cell Concentration and Cellular Activity

A. Soley[1], J. Gálvez[1], E. Sarró[1], M. Lecina[1], A. Fontova[2], R. Bragós[2], J.J. Cairó[1], F. Gòdia[1]
*[1]Departament d'Enginyeria Química, ETSE, Universitat Autònoma de Barcelona (UAB), Bellaterra, Barcelona, Spain, [2]Departament d'Enginyeria Electrònica, Universitat Politècnica de Catalunya (UPC), Barcelona, Spain.*

**Abstract:**

**Key words:** Bioreactor, Minireactor, High throughput screening, Hybridoma, Vero

## 1. INTRODUCTION

The application of the technology based on the use of Genomics, Proteomics and also combinatorial chemistry is favouring the development of improved organisms, new substances and processes in the areas of biomedicine, biotechnology, food and environmental industry. This trend will increase in the future since the potentiality of these technologies. Moreover, in the area of tissue engineering, there is a need to screen cell capabilities and the proper architecture, physiology and functionality of the repaired tissue.

In order to evaluate the mentioned properties of the new products it has been developed a versatile, robust, low cost, high performance (simultaneous sample analysis, different environmental conditions, etc) and highly automated minireactor. The optical and electrical sensors allow the possibility to follow both concentration and activity of cells or enzymes in culture. The performance of the minireactors using suspension and adherent animal cell cultures will be presented and discussed in this abstract.



*R. Smith (ed.), Cell Technology for Cell Products*, 803–806.

## 2. MATERIALS AND METHODS

Description of the equipment (**PCT/ ES 2003/000607**). The system consists of a battery of sterile and disposable minireactors made of plastic material in order to carry out multiple simultaneous tests.

The equipment is composed of two parts:

1. Single-use sterile plastic plate with 6 separated minireactors (Figure 1A). Each 10-15 ml working volume minireactor (Figure 1B) is fully equipped with non invasive probes: optical probe (measure of pH and turbidimetry) and electrical probe (measure of pO2). It also includes miniaturized probe ports, stirring bar, aeration outlet and inlet, filters and inoculation septum.
2. Workstation that provides common stirring, temperature regulation and individual measurement and control of pH, pO2, OUR, optical density and visible absorption spectrum. A user-friendly, interactive and intuitive graphic interface allows data acquisition and analysis of growth and metabolism of different cellular models.

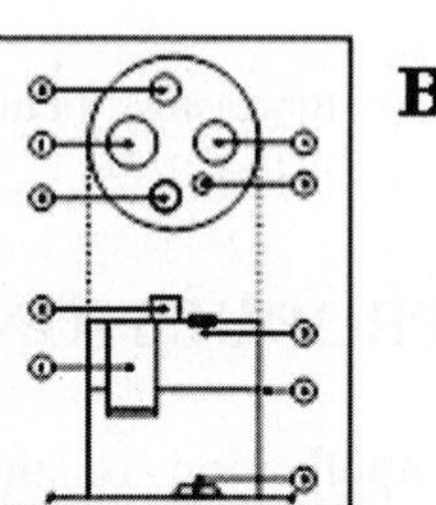


*Figure 1.* Equipment 3D model (A). Description of a minireactor (B): 1) Optical port. 2) Headspace aeration. 3) Septum. 4) Polarographic port. 5) Stirring bar. 6)Liquid level.

Cell line, medium and culture conditions. The suspension cell line used was KB 26.5 hybridoma; the adherent cell line used was a Vero cell. Both were cultured as described (Sanfeliu *et al.*, 1997). The basal media used were DMEM (Sigma) + 4%FCS (Biological industries).

Analytical methods: *off-line* variables. Density and viability were assessed by the trypan blue exclusion method using a hemocytomer (Neubauer improved, Brand). Adherent cells were trypsinized before counting. Glucose and lactate were measured using an automatic analyser (YSI, 2700 Select).

# 3. RESULTS AND DISCUSSION

The equipment described above has been applied to culture different animal cell lines in order to determine the most accurate probes to follow both the concentration and cellular activity:

## 3.1 Suspension Cell Cultures

Both optical and electrical probes can provide information about main variables of the culture. Slow dynamic in oxygen consuming rate allows the measure of OUR using the polarographic probe. On the other hand, the optical probe shows a significant variation in the pH profile and an increasing in the absorbance rate (Figure 2A).

These results can be explained with the off-line profiles (Figure 2B). Glucose depletion and biomass growth increase absorbance rate while pH is decreased by lactic acid production.

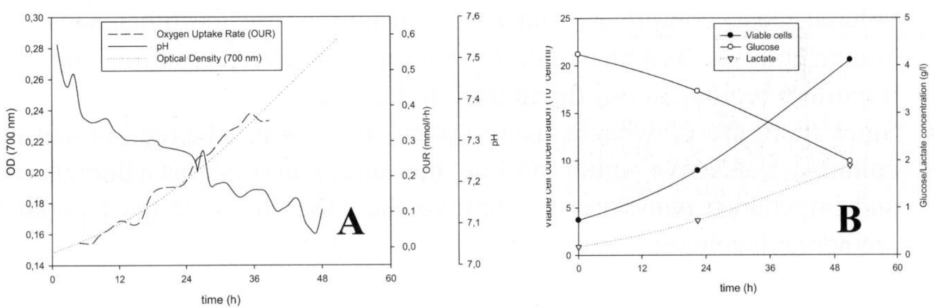


*Figure 2.* Suspension cell culture (KB-26.5). On-line parameters (A) and off-line parameters (B) of the culture.

## 3.2 Adherent Cell Cultures

The cultivation of adherent cells in monolayer denies the possibility of performing absorbance measurements, since the cell growth happens to be on the bottom of the minireactor and interferes with the light signal. So, only the polarographic sensor provides information about the culture, and the measure of OUR indicates biomass growth (Figure 3A).

In this case, the off-line profiles can also explain the results (Figure 3B).

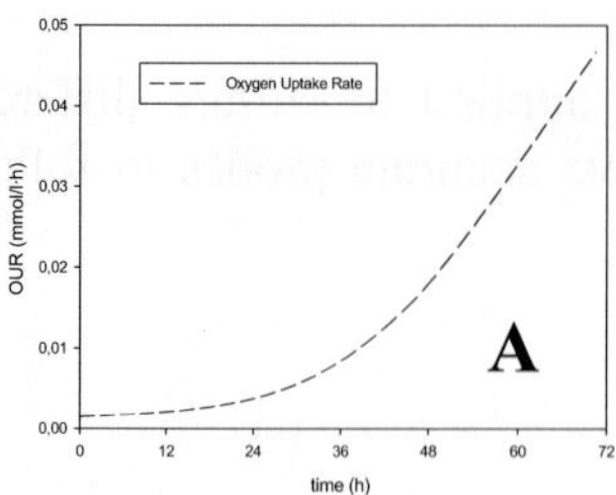


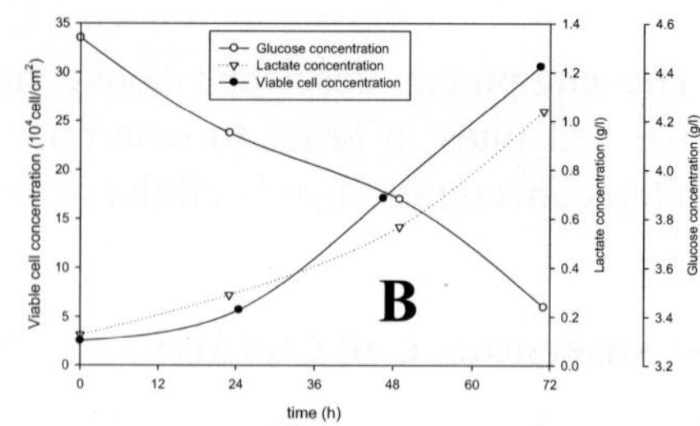


*Figure 3.* Adherent cell culture (Vero cell). On-line parameters (A) and off-line parameters (B) of the culture.

# 4. CONCLUSIONS

- The obtained results open a brand new way to carry out optimization experiments in the system in order to replace bioreactors of a higher volume, cost, complexity and time consuming. On the other hand, this system allows keeping culture parameters under control and gets important biological information from them.
- Since there are drawbacks using optical sensors in adherent animal cell cultures, nowadays, other on-line measurement systems (fluorescence and impedance detectors) to improve the capabilities of the equipment are being developed.

# REFERENCES

Sanfeliu A., Paredes C.,Cairó J.J., Gòdia F.. Indentification of key patterns in the metabolism of hybridome cells in culture. Enz. Micr. Tech. 21 (1997), 427-428.

Kostov Y., Harms P., Randers-eichhorn L., y Rao, G.. Low-cost microbioreactor for high-throughput bioprocessing. Biotechnology and Bioengineering 72, 3: 346-352 (2001).

De León A., Mayani H., Ramírez y O.T.. Design, characterisation and application of a minibioreactor for the culture of human hematopoietic cells under controlled conditions. Cytotechnology 28:127-138 (1998).

# Air-lift Bioreactor Improvement of Scale-up Linearity Between Shake Flasks and 25-L Scale in the Insect Cells–baculovirus Expression System

Frédéric Bollin, Christophe Losberger, Laurent Chevalet, Thierry Battle
*Serono Pharmaceutical Research Institute 14, Chemin des Aulx, 1228 Plan Les Ouates Geneva, Switzerland.*

**Abstract:** The insect cells-baculovirus system can accommodate to various production formats, ranging from shake flasks up to large-scale reactors. In this study, careful monitoring of the dissolved oxygen profiles in shake flasks allowed us to devise better settings for our 25-L air-lift reactor.

## 1. INTRODUCTION

Scaling up of insect cell-based expression from shake flasks to cell culture reactors sometimes results in a decrease of production yields. In our 25-L airlift reactor (ALR), we suspected that aeration conditions affected cell growth and, consequently, protein expression.

A special probe-equipped flask was designed to monitor dissolved oxygen (DO) in the small-scale cultivation system, which was compared to the regulated DO profile of the ALR.

## 2. MATERIAL AND METHODS

**High Five cells** (Invitrogen corp.,USA) are cultivated at **27°C** in the **Excell 405 medium** (JRH, UK).

The equipped shake flask is stirred at **170 rpm**.



*R. Smith (ed.), Cell Technology for Cell Products*, 807–809.

The ALR is supplied with conventional aeration conditions *(cf. Table 1).*

In the second part of the study, the ALR is supplied with modified aeration conditions *(cf. Table 1).*

**Cultivation of infected cells:** Cell density reaches **1.3-1.5 E6 cells/ml.**

Cells are then infected with a recombinant baculovirus expressing a soluble kinase, at **MOI=5.0**.

The culture is sampled at **48 hours post-infection** and spun at 500xg. The cell pellet is separated from the supernatant and frozen before being treated for densitometric quantification (using Coomassie gel electrophoresis).

## 3. RESULTS AND DISCUSSION

A significant difference of cell growth rate and maximal cell density reflects the discrepancy between the shake flask and the 25-L bioreactor.

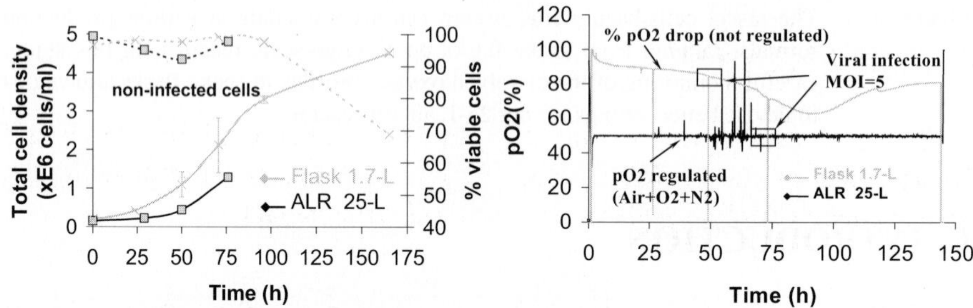


*Figure 1.* Equipped shake flask vs. 25-L bioreactor.

DO is naturally dropping in the shake flask from 100% air saturation. On the other hand, DO is regulated at 50% in the reactor during the whole cultivation process. This led us to propose new aeration settings for the large-scale bioreactor, in order to be closer to the natural aeration profile of the shake flask:

*Table 1.* Modification of the aeration conditions in the 25-L bioreactor.

| | Old | New |
|---|---|---|
| **Starting DO (%)** | 50 | 100 |
| **Starting Airflow (NL/min)** | 1.5-2.0 | 1 |
| **Max. O2 flow rate (NL/min)** | 3 | 1.5 |
| **DO Set point (%)** | 50 | 30 |

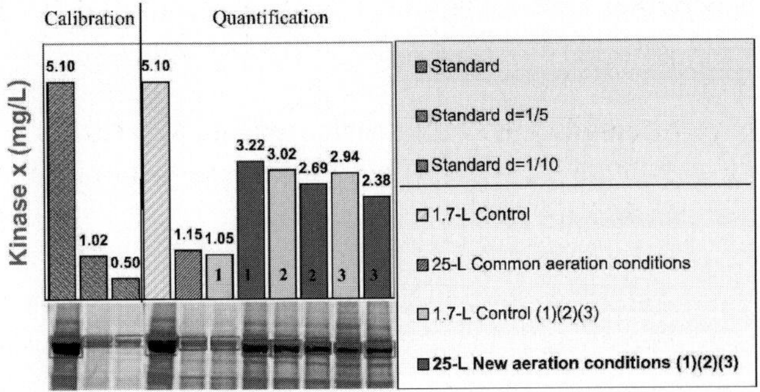


*Figure 2.* Batch to batch comparative expression levels at various conditions and scales.

After fine-tuning of cultivation conditions, cell density and viability were improved by 40% and 2.2% respectively. The protein production yield in the 25-L reactor was raised to a level comparable to that observed in shake flasks.

## 4. CONCLUSION

The new starting DO and the delayed DO regulation minimized air and $O_2$ sparging in the air-lift reactor. This was shown to contribute to the improvement of cell growth and protein expression levels.

## ACKNOWLEDGEMENTS

The authors would like to thank H. HEINE (Upstream processing) for technical advice, K. ROULIN and J. SHAW for purification and fruitful discussion.

# Author Index

# Subject Index